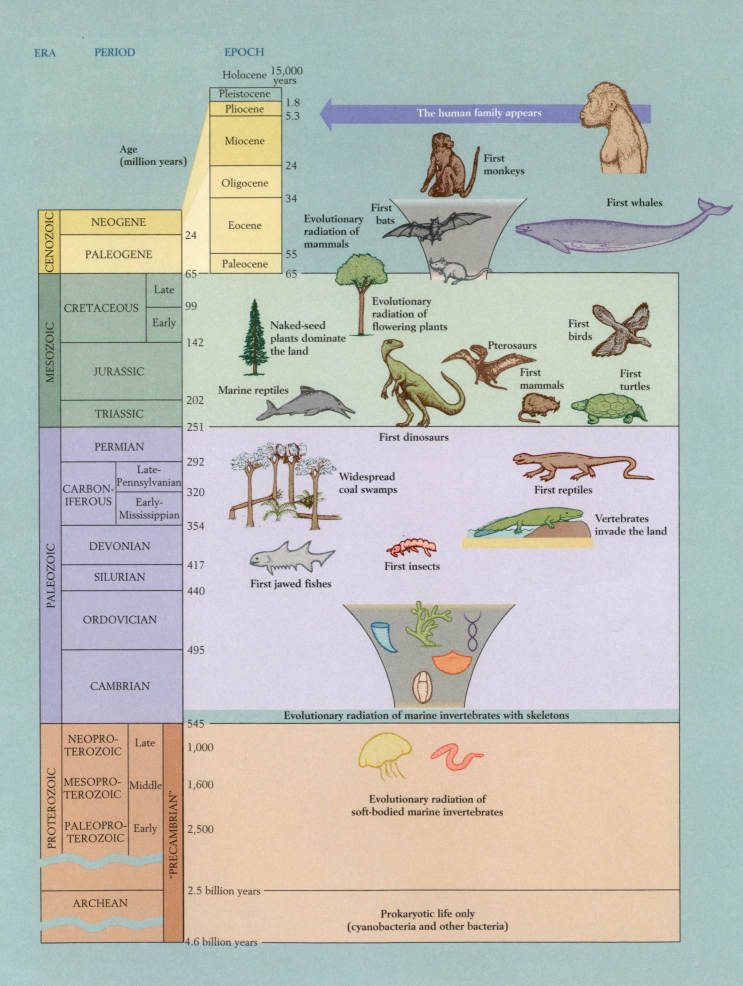

ERA	PERIOD		EPOCH	Age (million years)

The human family appears

First monkeys

First whales

First bats

Evolutionary radiation of mammals

Evolutionary radiation of flowering plants

Naked-seed plants dominate the land

First birds

Pterosaurs

First mammals

First turtles

Marine reptiles

First dinosaurs

Widespread coal swamps

First reptiles

Vertebrates invade the land

First jawed fishes

First insects

Evolutionary radiation of marine invertebrates with skeletons

Evolutionary radiation of soft-bodied marine invertebrates

Prokaryotic life only (cyanobacteria and other bacteria)

MIDDLE SILURIAN

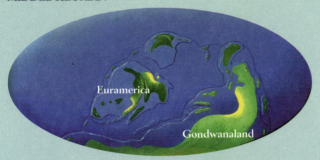

Euramerica

Gondwanaland

Laurentia and Baltica are
sutured to form Euramerica

MIDDLE MIOCENE

Continents are dispersing

MIDDLE ORDOVICIAN

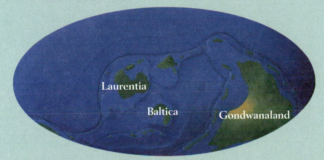

Laurentia

Baltica

Gondwanaland

Microcontinents and island
arcs are sutured to Laurentia

LATE CRETACEOUS

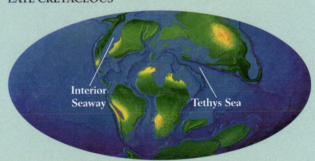

Interior
Seaway

Tethys Sea

Pangaea is fragmenting

LATE CAMBRIAN

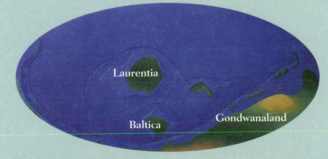

Laurentia

Baltica

Gondwanaland

The Proterozoic supercontinent
has fragmented

LATE PERMIAN

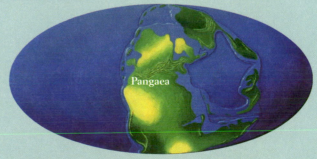

Pangaea

Gondwanaland is sutured to
Euramerica to form Pangaea

Earth System History

Earth System History

SECOND EDITION

Steven M. Stanley

Johns Hopkins University

W. H. Freeman and Company · New York

For Sveta, who has done a great job.

Acquisitions Editor: Valerie Raymond
Development Editor: Jeff Ciprioni
Publisher: Susan Finnemore Brennan
Assistant Editor: Sharon Merritt
New Media/Supplements Editors: Victoria Anderson and Amy Shaffer
Marketing Director: John Britch
Project Editor: Georgia Lee Hadler
Copy Editor: Norma Roche
Cover and Text Designer: Diana Blume
Illustration Coordinator: Bill Page
Illustrations: Precision Graphics, Fine Line Studios, Roberto Osti
Photo Editor: Patricia Marx
Photo Researcher: Dena Diglio Betz
Production Coordinator: Susan Wein
Composition: Matrix Publishing Services, Marsha Cohen
Printing and Binding: RR Donnelley & Sons Company

Library of Congress Cataloging-in-Publication Data
    Stanley, Steven M.
    Earth system history/Steven M. Stanley-2d ed.
        p.   cm.
      Includes index.
      ISBN 0-7167-3907-0
    1. Historical geology. 2. Physical geology. I. Title.
      QE28.3.S735 2004
      551.7—dc22   2004053253
EAN: 9780716739074

Printed in the United States of America.

First Printing

W. H. Freeman and Company
41 Madison Avenue
New York, NY 10010
Houndmills, Basingstoke RG21 6XS, England
www.whfreeman.com

Contents in Brief

Contents

Contents

Preface

A Modern Approach

I have not only an intellectual enthusiasm for my work but also an emotional and aesthetic excitement about it. The scale of Earth's history evokes these emotions—both its spatial scale, which encompasses our entire planet, and its temporal scale, which entails what John McPhee has termed "deep time." Deep time is what separates geology from physics, chemistry, and biology, endowing our field with a special role in science. Remarkable events have profoundly altered Earth as it has progressed through deep time to the present. Understanding Earth's long history is humbling, and yet it is also enlightening. It reveals our roots and our place in nature. After all, we humans have trod our planet for only a minute fraction of its history.

Adhering to the tradition of the first edition of *Earth System History* and my textbooks that were ancestral to it, this new edition is based on the conviction that the physical and biological history of Earth are inextricably intertwined. Not only have changes in the nonbiological aspects of Earth affected life profoundly, but the reverse is also true: changes in life have transformed the oceans, the crust, and the atmosphere. Such interrelations provide the rationale for viewing our planet holistically as one great dynamic system. The rock cycle is part of that system, but so are the water cycle and the carbon cycle. Any change in one component of the system can exert a powerful effect on other components that, on the face of things, might seem far removed from it. Chapter 10 of this book, without counterpart in other books on Earth History, epitomizes this connectivity in explaining the nature of large-scale chemical cycles.

Students can gain a comprehensive view of the Earth system only by studying past states and perturbations of this system that have never been mimicked in all of recorded history, let alone in their lifetimes. For example, most meteorologists are nonplussed when they hear that alligators flourished inside the Arctic Circle 45 million years ago, or that continental glaciers spread widely near the equator slightly more that half a billion years ago. To understand how Earth works, we must understand such phenomena: we must understand Earth system history. This book is designed to make the subject accessible even to students with no previous background in geology.

How This Book Is Organized

Part I of this book shows how the upper part of the solid Earth and its oceans, atmosphere, and biosphere function interactively, and it explains how we bring the history of the Earth system to light by studying rocks and fossils.

Part II of the book takes the student on a voyage through Earth's history—a voyage guided by the facts, principles, and methodologies introduced in Part I. The present world emerges from this history. As students come to understand the origins of the world in which they live, they will develop a sense of their own position in relation to deep time and vast space. Through this experience, they will come to appreciate the ephemeral nature of environments and ecosystems and the fundamental fragility of their habitat.

PART I

The facts, concepts, and scientific methods that students must grasp to comprehend Earth's history.

Each of the nine chapters following the first is designed to encapsulate those elements of a field of science that students must comprehend in order to address the topics of Part II, which is a narrative of Earth system history. Some of the subject matter in Part I will represent a review for students who have studied physical geology, but most of the material—even much that is contained in the chapters on sedimentary environments, plate tectonics, and mountain building—will be new to those students.

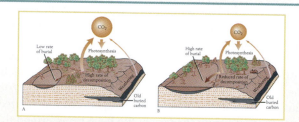

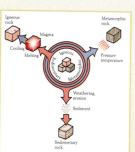

- Chapter 1 introduces the kinds of materials and concepts we employ to reconstruct the physical and biological history of Earth.

- Chapter 2 describes the minerals and rocks that are central to the study of Earth's history.

- Chapter 3 presents an overview of life, introducing important groups of organisms with which many students will be unfamiliar.

- **Unique!** Chapter 4 summarizes basic principles of ecology and reviews the kinds of ecosystems that exist on Earth today and also those that existed in the past.

- Chapter 5 explains how geologists reconstruct ancient environments by studying sedimentary rocks.

- Chapter 6 addresses how geologists position rocks within the framework of geologic time.

- Chapter 7 reviews the nature of biological evolution and the ways in which the fossil record contributes to our understanding of this ubiquitous process.

- Chapter 8 covers plate tectonics, emphasizing how geologists developed this unifying concept and how it explains many phenomena that seemed problematic before its advent.

- Chapter 9 shows how plate tectonics explains mountain building and other processes that alter continents.

- **Unique!** Chapter 10 outlines major chemical cycles on Earth, highlighting the intimate connection between the physical and biological components of our planet.

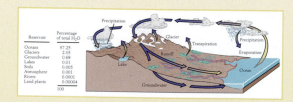

PART II

The story of our planet, beginning with its origin and ending with the events that ushered in the modern world.

Many of the most entrancing events in Earth's history, such as the initial diversification of animals and the catastrophic demise of the dinosaurs, occurred before the Cenozoic Era. Nonetheless, the final three chapters of this book convey extremely important messages because they explain how the world we now inhabit came into being. They also illustrate especially well the key role of geologic processes in global change (for example, Cenozoic glaciation in the Southern Hemisphere, and perhaps also in the Northern Hemisphere, resulted from plate tectonic movements). In addition, these chapters portray remarkably rapid climatic changes in the recent past—changes that must give us pause as we contemplate the possible future effects of human activities on the global climate.

- Chapter 11, covering the earliest interval of Earth's history, begins with our planet's birth and then shows how Earth differed from its present state during its early history and how life may have originated in hot, watery settings along mid-ocean ridges.

- Chapter 12 describes how Earth was, in effect, modernized during the Proterozoic Era, with continents expanding, oxygen building up in the atmosphere, and multicellular life appearing. It also addresses the concept of a "snowball Earth."

- Chapter 13 necessarily focuses on the explosion of life early in the Paleozoic Era, but it also explains mountain building in the Appalachian region and interprets the terminal Ordovician mass extinction, which was associated with abrupt climatic change and glaciation.

- Chapter 14, covering Middle Paleozoic time, emphasizes the effects of the diversification of land plants on terrestrial habitats and animal life—as well as on atmospheric carbon dioxide (perhaps causing glaciation and the Late Devonian mass extinction). It also reviews episodes of mountain building.

- Chapter 15, on the late Paleozoic, features the expansion of coal swamps, their effects on Earth's climate, and their disappearance as climates became more arid. This chapter gives special attention to origin of the great ice age, centered in the Southern Hemisphere, that was associated

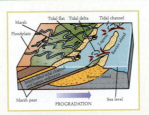

with a temporary transformation of the global ecosystem. Another central topic is the formation of Pangaea, which included the final episode of Appalachian mountain building. This chapter also provides an up-to-date view of the greatest mass extinction of all time, which brought the Paleozoic Era to a close.

- Chapter 16 portrays the great expansion of new forms of life, including the dinosaurs, early in the Mesozoic Era. It shows how the mass extinction at the end of the Triassic Period, perhaps related to massive volcanism associated with the breakup of Pangaea, allowed the dinosaurs to rise to dominance. It also begins coverage of extensive mountain building in the American West.

- Chapter 17, on the Cretaceous Period, continues the coverage of mountain building in the American West. It shows how the Cretaceous world was transitional between an earlier primordial state and the present. This chapter also offers a thorough account of the mass extinction, resulting from an asteroid impact, that swept away the dinosaurs and many other forms of life to end the Mesozoic Era.

- Chapter 18 shows how Earth took a large step toward its present state early in the Cenozoic Era—first, with the expansion of modern forms of life following the terminal Cretaceous mass extinction, and second, with the climatic deterioration that resulted from plate tectonic movements that isolated Antarctica over the south pole.

- Chapter 19 describes how Earth moved even further toward its present state during Neogene time with the onset of the modern Ice Age of the Northern Hemisphere. It also presents evidence of startlingly rapid climatic change in the recent past, and it offers an up-to-date picture of human evolution.

- Chapter 20, **unique in its emphasis** on the Holocene, transports the student into the modern world, explaining how the last glacial maximum came haltingly to an end and how numerous large mammal species died out at that time, leaving the modern world biotically impoverished. This chapter shows how the world that we know came into being and how that world is still changing—and may change even more dramatically in the near future as a result of our own activities.

Earth System Shift 14-1 | Plants Alter Landscapes and Open the Way for Vertebrates to Conquer the Land

Plants began to move onto land from the sea before Devonian time, but their effect on terrestrial environments was minimal (p. 000). During the Devonian Period, however, plants' more extensive invasion of the land had profound consequences.

One result of the global spread of terrestrial vegetation early in Devonian time was that, for the first time in Earth's history, plants carpeted the soil and gripped it with their roots, thereby stabilizing it against erosion. Precambrian and early Paleozoic rocks are characterized by braided-stream deposits, which reflect rapid erosion (p. 000). Only after vegetation stabilized the land and confined rivers to discrete channels could rivers meander and deposit sediment in orderly cycles (see Figure 5-16). Thus it is only in rocks of Early Devonian age or younger that we find meandering river deposits in the geologic record. Preserved root traces attest to the spread of vegetation over the floodplains of rivers during the Devonian Period.

The next step, the origin of plants that grew to tree proportions, took place in Late Devonian time. Probably the adaptive breakthrough responsible for the increase in the size of land plants was the evolution of broad leaves, which captured sunlight effectively and allowed for a higher overall rate of photosynthesis. In Late Devonian time, *Archaeopteris* formed Earth's first forests. These forests were restricted to marshy environments, however, because *Archaeopteris* was a spore plant that required moisture to reproduce (see p. 000). The origin of seed plants near the end of the Devonian Period was another evolutionary breakthrough that had enormous ecological consequences: it permitted forests to expand into well-drained habitats far from lakes and rivers. For the first time in Earth's history, forests spread widely over the landscape.

It is probably no accident that amphibian animals did not arise until shortly before the end of Devonian time. The earliest amphibians must have had

Figure 1 Braided stream deposits characterize pre-Devonian nonmarine rocks. This block of the Silurian Shawangunk Formation in eastern Pennsylvania includes cross-bedded sand and gravel deposited by braided streams that flowed from uplands formed by the Taconic Orogeny. The vertical dimension of this block is about 30 centimeters (one foot). (Courtesy of Steven M. Stanley.)

Figure 2 Traces of roots of early land plants. These traces were found in Devonian rocks along Gaspé Bay, Quebec, Canada. (Courtesy of Jennifer M. Elick; J. M. Elick, S. G. Driese, and C. I. Mora, *Geology* 26:143–146, 1998.)

NEW! Earth System Shift Boxes

The most important new features of this edition of *Earth System History* are the Earth System Shift boxes, at least one of which appears in every chapter of Part II. As I approached this revision, it seemed evident that many pivotal events of Earth's history that were causally related should be treated in one place to emphasize their interrelations: hence the Earth System Shift boxes. In fact, these new features epitomize the central theme of this book, usually focusing on global changes that entail both physical and biological components of the Earth system. Elements of the Earth System Shifts portrayed in the boxes are noted in appropriate sections of the general text, but the boxes contain more detail and emphasize interrelationships between events. Perhaps the Earth System Shift that will prove to be most appealing to students will be the one that depicts the rise of the dinosaurs and reviews the evidence as to who these storied creatures actually were. Several other Earth System Shift boxes provide up-to-date accounts of particular mass extinctions and the likely causes of these revolutionary events.

New Methods and New Knowledge

I have made every effort to incorporate the latest research into this revision—both new scientific approaches and new results. Happily, we have learned that most instructors view elementary applications of stable isotopes as being essential to their teaching of Earth's history. Accordingly, I have explained the most important isotopic applications in terms that students should be able to comprehend—and I have shown what these applications have recently revealed about Earth's history. Also new in this edition are elementary discussions of molecular clocks and molecular phylogenies (the latter being essential to our understanding of the diversification of Bacteria and Archaebacteria in Archean time); the importance of methane hydrates as sources of greenhouse gas; and the use of stomate densities in fossil leaves to reveal past concentrations of carbon dioxide in the atmosphere.

Among the new topics of Earth history are the proposal that a "snowball Earth" existed in Neoproterozoic time; the observation that a new state of the marine ecosystem arose during the late Paleozoic ice age; and the recognition that the terminal Triassic mass extinction—and subsequent ascendancy of the dinosaurs—may have resulted from extensive igneous activity associated with the fragmentation of Pangaea. The discussion of human evolution features the most recent fossil discoveries and includes evidence that modern humans and Neanderthals represent distinct lineages that diverged from a common ancestor at least half a million years ago.

New Modes of Presentation

The integration of related topics into Earth System Shift boxes is only one of several new features in this edition that are designed to aid students:

A More Intuitive Art Program

• Retained from the last edition, the popular Visual Overviews provide students with vivid illustrations of the key themes and concepts of Chapters 2 through 20. In this edition, the Visual Overviews contain numerous captions that explain the illustrations more fully to further enhance student understanding.

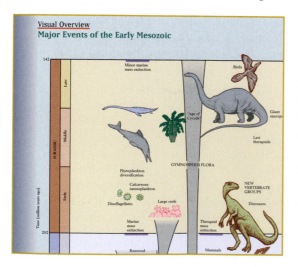

**Visual Overview
Major Events of the Early Mesozoic**

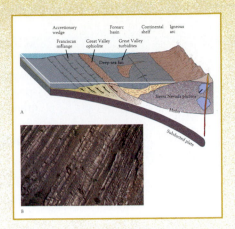

- To facilitate their recognition, particular features of illustrations, such as continental crust, volcanic rocks, and the asthenosphere, have been portrayed with uniform patterns throughout the book.
- Timelines at the beginnings of Chapters 11-20 have been standardized to help students maintain a temporal perspective.

More Student Study Support

- The chapter summaries now take the form of a series of questions and answers, providing students with a more powerful way to review the material.
- Second-level headings now take the form of complete sentences to provide students with a succinct summary of the contents of substantial sections of the text.

Supplements

Instructor Resources

Instructor's Resource CD-ROM

NEW! To help instructors create lecture presentations, Web sites, and other resources, this CD-ROM allows instructors to search and export all the following resources by key term or by chapter:

- All text images in JPEG, GIF, and PowerPoint format
- Animations, interactive exercises, and PowerPoint files
- Instructor's Resources material
- Test bank questions

PowerPoint Lecture Presentation

NEW! A set of online lecture presentations created in PowerPoint allows instructors to tailor their lectures to suit their own needs using images and notes from the text.

Overhead Transparency Set

NEW and expanded full-color transparency program, based on key text illustrations from the new art program, are available to qualified adopters.

Instructor's Resource Web Site

The Instructor's Resource material, available at www.whfreeman.com/esh2e consists of lecture outlines and teaching tips for each chapter.

Test Bank

This resource contains 25 **NEW** multiple-choice questions per chapter, many of which are supported by geologic diagrams found in the text.

Computerized Test Banks

The Diploma test bank allows instructors to easily create tests, write and edit questions, and create study sessions for students. Instructors can add an unlimited number of questions, scramble questions, and include pictures, equations, and multimedia links. Tests can be printed in a wide range of formats or administered to students with Brownstone's network or Internet testing software. The software's unique synthesis of flexible word processing and database features creates a program that is extremely intuitive and capable.

Online Courses

As a service for adopters, W. H. Freeman and Company will provide content files in Blackboard and WebCT course formats, including the instructor and student resources for this text. The files can be used as is, or they can be customized to fit specific needs. Course outlines, pre-built quizzes, links, activities, and an array of other materials are included.

Student Resources

Earth System History, Second Edition, Web site
http://www.whfreeman.com/esh2e

- Provides enhanced and updated interactive exercises, key terms, and review questions
- All **NEW** online quizzes feature additional questions not found in the text to help students further their study, review, and preparation for exams
- Provides numerous links to other Web sites related to geology
- An Online Lecture Notebook, based on the instructor's PowerPoint files, is available for download to aid students in taking notes

Acknowledgments

By way of thanks, I must begin with my acquisitions editor, Valerie Raymond, who was remarkably astute and creative in helping me plan this new edition. She was always there, from start to finish, keeping things on track. Jeff Ciprioni was a pleasure to work with as a developmental editor. With Susan Brennan in the background as publisher and Sharon Merritt in the foreground as assistant editor, I knew there would be solid support throughout the effort. Norma Roche did the best job of copyediting I have ever experienced. Dena Diglio Betz was highly efficient and effective in her photo research, and Patricia Marx was very helpful as photo editor. Precision/Fine Line produced attractive and accurate line art. Bill Page, the illustration coordinator, kept the myriad figures under control, and Susan Wein, as production coordinator, did the same for the entire project. Diana Blume designed a beautiful book, and I am indebted to Ron Redfern for permission to adorn it with his splendid photo of the White Cliffs of Dover. Finally, I thank Georgia Lee Hadler, who as project editor, kept me on schedule going down the stretch, as she has done so reliably many times before.

In thanking reviewers of the manuscript, I must first single out Annie Holmes, who was brought to the project because of her sharp eye and general knowledge of the subject. She carefully read and re-read page proofs, occasionally noting small errors that the rest of us had missed.

On a subject as vast as Earth history, it is essential to obtain critiques of chapters by experts, in order to root out mistakes and identify conspicuous omissions. I express gratitude to all of the following for sharing their expertise and taking time to vet my manuscript:

Lloyd Burckle, Columbia University

Beth A. Christensen, Georgia State University

C. J. Collom, Mount Royal College

William J. Frazier, Columbus State University

A. Kem Fronabarger, College of Charleston

David H. Griffing, University of North Carolina, Charlotte

Stephen T. Hasiotis, University of Kansas

Alisa K. Hylton, Central Piedmont Community College

Markes E. Johnson, Williams College

David T. King, Jr., Auburn University

James Lamb, Wake Technical Community College

Larry T. Spencer, Plymouth State University

Ron Stieglitz, University of Wisconsin, Green Bay

Nicholas W. Taylor, Minneapolis Community and Technical College

Joseph Pachut, Indiana University/Purdue University, Indianapolis

Michael C. Pope, Washington State University

Mark A. Wilson, College of Wooster

Margaret M. Yaccobucci, Bowling Green State University

I am also grateful to colleagues who reviewed material for previous editions: Barbara Brande, University of Montevallo; Beth Nichols Boyd, Yavapai College; Fred Clark, University of Alberta; William C. Cornell, University of Texas at El Paso; John W. Creasy, Bates College; R. A. Davis, College of Mount St. Joseph; Louis Dellwig, University of Kansas; Steven R. Dent, Northern Kentucky University; Stanley C. Finney, California State University, Long Beach; Jay M. Gregg, University of Missouri, Rolla; John P. Grotzinger, Massachusetts Institute of Technology; Bryce M. Hand, Syracuse University; Leo J. Hickey, Yale University; Calvin James; Markes E. Johnson, Williams College; James O. Jones, University of Texas, San Antonio; Andrew H. Knoll, Harvard University; Karl J. Koenig, Texas A&M University; David Liddell, Utah State University; Michael T. May, Western Kentucky University; Robert Merrill, California State University, Fresno; Cathryn R. Newton, Syracuse University; Anne Noland, University of Louisville; Geoffrey Norris, University of Toronto; Mark e. Patzkowsky, Pennsylvania State University; Lisa Pratt, University of Indiana, Bloomington; Gregory J. Retallack, University of Oregon; Charles A. Ross, Western Washington University; Frederic J. Schwab, Washington and Lee University; Peter Sheehan, Milwaukee Public Museum; Randall Spencer, Old Dominion University; John F. Taylor, Indiana University of Pennsylvania; David K. Watkins, University of Nebraska, Lincoln; Neil A. Wells, Kent State University.

PART I

Materials, Processes, and Principles

Earth system history is a story of epic proportions about changes in the physical and biological features of our planet over billions of years. It is the work of scientists to use the geologic evidence of these changes, combined with Earth processes known today, to bring Earth's history to light. The first ten chapters of this book introduce the components of the Earth system and the tools that scientists use in their historical quest.

A Hawaiian volcano from which fiery lava flows and black cinders of volcanic rock shoot into the air. (Al Merrill/Breck P. Kent.)

CHAPTER 1

Earth as a System

Few people recognize, as they travel down a highway or hike along a mountain trail, that the rocks they see around them have rich and varied histories. Unless they are geologists, they probably have not been trained to identify a particular cliff as rock formed on a tidal flat that once fringed a primordial sea, to read in a hillside's ancient rocks the history of a primitive forest buried by a fiery volcanic eruption, or to decipher clues in lowland rocks telling of a lofty mountain chain that once stood where the land is now flat. Geologists can do these things because they have at their service a wide variety of information gathered during the two centuries when the modern science of geology has existed. The goal of this book is to introduce enough of these geologic facts and principles to give you an understanding of the general history of our planet and its life. The chapters that follow describe how the physical world assumed its present form and where the inhabitants of the modern world came from. They also reveal the procedures through which geologists have assembled this information. Students of Earth's history inevitably discover that the perspective this knowledge provides changes their perception of themselves and of the land and life around them.

Knowledge of Earth's history can also be of great practical value. Geologists have learned to locate subterranean reservoirs of petroleum and water, for example, by ascertaining where the porous rocks of these reservoirs tend to form in relation to other bodies of rock. Geologists have also helped to discover deposits of coal and metallic ores and other natural resources buried within Earth.

Exploring the Earth System

The rocks of Earth's outer regions amount to a vast archive that we can read and interpret in order to unravel the planet's long history. By studying Earth's history, we learn how our planet functions as a complex system. This is not a purely academic exercise. An understanding of the system will help us to address problems caused by changes that are now taking place in the world, or that will soon be occurring.

The components of the system are interrelated

The Earth system has both physicochemical and biological components. We can reconstruct many aspects of the planet's physical history, including the growth and destruction of mountains, the breakup and collision of continents, the flooding and reemergence of land areas, and the warming and cooling of climates. We can also trace the evolution of life from an early world inhabited largely by bacteria through the origins of plants and animals in ancient seas to the invasion of the land, the rise and fall of dinosaurs, and ultimately the ascendancy of humans. We cannot understand either the physical history of Earth or the history of life in isolation, however, because the two have been tightly intertwined: the physical environment has influenced life, and life, in turn, has influenced the physical environment. For example, as we shall see in Chapter 4, climatic patterns control distributions of plants on land. At the same time, plant life affects climates. Forests warm regional climates by trapping heat, for instance, and plants also affect global climates by altering the chemistry of the atmosphere. The geologic record reveals that the histories of land plants and climates have shifted in concert for hundreds of millions of years. Many other factors, including continental movements and the rising and falling of seas, have influenced climates as well. The present state of Earth is a momentary condition that is the product of a long and complex history.

Armed with knowledge of Earth system history, we can more effectively address problems caused by changes that are now taking place in the world. Consider the shifting of coastlines as sea level rises or falls. The geologic record of the past few thousand years documents a general rise in sea level as huge glaciers have melted and released water into the ocean. The geologic record near the edge of the sea reveals how coastal marshes have shifted as sea level has changed. These marshes are very important to humankind; they cleanse marginal marine waters and sustain forms of animal life that are valuable to us. Study of the geologic history of coastal marshes will help us to predict their fate as sea level continues to change in the decades and centuries to come.

The system is fragile

The geologic record of the history of life also provides a unique perspective on the numerous extinctions of animals and plants that are now resulting from human activities. Humans are causing extinctions by destroying forests and other habitats, but our collective behavior also affects life profoundly in less direct ways. Very soon human activities will cause average temperatures at Earth's surface to rise in many areas of the world. The geologic record of ancient life reveals how climatic change has affected life in the past—how some species have survived by migrating to favorable environments, for example, and how others that failed to migrate successfully have died out. To the surprise of many biologists, geologic evidence has revealed that many of the natural assemblages of species that populate the world today are not ancient associations of interdependent species. Instead, they are associations that have developed very recently (on a geologic scale of time) as cli-

matic changes have caused many species to shift about independently of one another.

As we come to understand the speed and power of natural environmental change and the temporary nature of assemblages of species, we begin to appreciate the fragility of the world we live in. More generally, having studied the past, we can make more intelligent choices as we contemplate the future of our changing planet.

Before we launch into our detailed examination of the history of Earth and its life, however, an introduction to some of the basic facts and unifying concepts of geology is in order. The first ten chapters lay this groundwork, and the chapters that follow trace out Earth system history.

The Principle of Uniformitarianism

Fundamental to the modern science of geology is the principle of **uniformitarianism**—the understanding that there are inviolable laws of nature that have not changed in the course of time. Of course, uniformitarianism applies not only to geology but to all scientific disciplines—physicists, for example, invoke the principle of uniformitarianism when they assume that the results of an experiment conducted on a given day will be applicable to events that take place a day, a year, or a century later. Geologists hold this principle in particularly great esteem, however, because, as we shall see, it was the widespread adoption of uniformitarianism during the first half of the nineteenth century that signaled the beginning of the modern science of geology.

Actualism employs the present as the key to the past

The principle of uniformitarianism informs geologists' interpretations of even the most ancient rocks on Earth. It is in the present, however, that many geologic processes are discovered and analyzed. The application of these analyses to the study of ancient rocks, in accordance with the principle of uniformitarianism, is sometimes called **actualism**. When we see ripples on the surface of an ancient rock composed of hardened sand (sandstone), for example, we assume that they formed in the same way that similar ripples develop today—under the influence of certain kinds of water movement or wind (Figure 1-1). Similarly, when we encounter ancient rocks that closely resemble those forming today from volcanic eruptions of molten rock in Hawaii, we assume that the ancient rocks are also of volcanic origin. We cannot observe rocks twisting into contorted configurations like those seen in mountains,

A

B

Figure 1-1 Ripples in sediments and sedimentary rocks. *A.* Sand along a beach, exposed to the air at low tide. *B.* A large block of sandstone showing similar ripples. (*A*, Raymond Siever; *B*, Reg Morrison/Auscape International.)

but we can witness the breaking, bending, and uplift of rocks during earthquakes, and we can easily imagine that the immense forces that produce these effects can strongly contort rocks deep within Earth and elevate them into mountains.

Actualism is commonly expressed by the phrase "The present is the key to the past." This statement is only partly true, however. Although it is universally agreed that natural laws have not varied in the course of geologic time, not all kinds of events that occurred in the geologic past have been duplicated within the time span of human history. Many researchers believe, for example, that the impact of very large asteroids (meteorites or comets) may explain certain past events, such as the extinction of the dinosaurs 65 million years ago.

In Chapter 17 we will review evidence that the dinosaurs' reign on Earth ended when a massive asteroid—one perhaps 10 kilometers (6 miles) in diameter—plunged through the atmosphere and ocean and penetrated the seafloor along the coast of Mexico. Such an impact should have produced a huge wave that would have crashed over coastlines thousands of kilometers from the impact site. Nonetheless, because we have never observed the arrival of such a large extraterrestrial object, we do not know exactly what else would have happened. Some scientists have suggested that dust injected into the upper atmosphere might have blocked the sun's rays from Earth's surface for many days. Others have asserted that when this dust later settled, the total kinetic energy (energy of motion) of the dust par-

ticles would have been converted to heat, warming Earth's surface dramatically. It is easy to imagine that these and other consequences of a huge meteorite impact would have wiped out many species around the world. Even so, because humans have never observed such an event, we must rely on theoretical considerations to surmise what actually happened. Because the details remain uncertain in this case, actualism does not apply.

Similarly, geologists have found that certain types of rocks cannot be observed in the process of forming today. In such cases, geologists usually make one of the following three assumptions:

1. The rocks in question formed under conditions that no longer exist.

2. The conditions responsible for the formation of the rocks still exist, but at such great depths beneath Earth's surface that we cannot observe them.

3. The conditions exist today, but produce the rocks only over a long interval of geologic time.

Many iron ore deposits more than 2 billion years old, for example, are of types that cannot be found in the process of forming today. It is believed that when the iron ore formed, chemical conditions on Earth differed from those of the present world and, furthermore, that the rocks underwent slow alteration after they were formed. The existence of these iron ore deposits does not negate the principle of uniformitarianism inasmuch as there is no evidence that natural laws were broken. It does, however, present geologists with a problem they cannot solve by applying the principle of actualism, because a human lifetime is only a small fraction of the time needed to study the rocks' development.

In an attempt to address some of these problems, geologists have learned to form certain kinds of rocks in the laboratory by simulating the conditions that prevail at great depths within Earth. They expose simple chemical components to temperatures and pressures many times greater than those at Earth's surface. Such experiments indicate the range of conditions under which a particular type of rock could have formed in nature. In conducting these experiments, geologists are, in a sense, expanding the domain of actualism by using as a model not only what is happening in nature today, but also what happens under artificial conditions and may have happened under natural conditions long ago.

Nineteenth-century geologists rejected catastrophism

Until the early nineteenth century, many natural scientists subscribed to the concept of **catastrophism**, which proposed that floods caused by supernatural forces formed most of the rocks visible at Earth's surface. Late in the eighteenth century, Abraham Gottlob Werner, an influential German professor of mineralogy, claimed that most rocks had been formed as a result of the precipitation of minerals from a vast sea that periodically flooded and retreated from the surface of Earth. These ideas were largely speculative.

Near the end of the eighteenth century, not long after Werner published his ideas, James Hutton, a Scottish gentleman farmer, established the foundations of uniformitarianism in his writings on the origins of rocks in Scotland. Hutton came to the conclusion that the rocks had formed as a result of the same processes that were currently operating at or near the surface of Earth—processes such as volcanic activity and the accumulation of grains of sand and clay under the influence of gravity.

Central to Hutton's view of Earth's history was vast geologic time. For the processes that were constantly shaping and reshaping the planet, he envisioned "no vestige of a beginning, no prospect of an end." Everyday processes, he proposed, had created and destroyed large bodies of rock, elevated and leveled mountains, and left remnants of their workings in an immense geologic record.

After extensive debate, Hutton's principle of uniformitarianism came to dominate the science of geology after Charles Lyell, an Englishman, popularized it in the 1830s in a three-volume book titled *Principles of Geology*. Lyell was a more effective writer than Hut-

ton, and the world was more receptive to the concept of uniformitarianism when he wrote than in Hutton's day. Like James Hutton, Lyell understood that volcanoes, floods, and earthquakes transform Earth. He argued that these events transform Earth in piecemeal fashion, and that they operate on local or regional scales, as do more subtle agents of change, such as the wearing away of old rocks and the accumulation of sand and mud to form new ones. In the eyes of Hutton and Lyell, Earth resembled an enormous machine that was always churning but retained its basic features.

Uniformitarianism, with its implication that Earth is very old, remains the central principle of modern geology. Nonetheless, as we shall see, Lyell's rejection of global catastrophes was unjustified, and he and Hutton also were mistaken in believing that the Earth system had remained unchanged in all of its basic features. Before addressing these points, we must learn more about the materials and processes of geology.

The Nature and Origin of Rocks

Rocks consist of interlocking or bonded grains of matter, which are typically composed of single minerals. A **mineral** is a naturally occurring inorganic solid element or compound with a particular chemical composition or range of compositions and a characteristic internal structure. Quartz, which forms most grains of sand, is probably the most familiar and widely recognized mineral; the materials we call limestone, clay, and asbestos consist of other minerals. Most rocks are formed of two or more minerals.

Rocky surfaces that stand exposed and are readily accessible for study are generally designated as **outcrops** or **exposures**. Scientists also have access to rocks that are not visible in outcrops. Well drilling and mining, for example, allow geologists to sample rocks that lie buried beneath Earth's surface.

Igneous, sedimentary, and metamorphic rocks can form from one another

On the basis of modes of origin, many of which can be seen operating today, early uniformitarian geologists, led by Hutton and Lyell, came to recognize three basic types of rocks: igneous, sedimentary, and metamorphic.

Igneous rocks, which form by the cooling of molten material to the point at which it hardens, or freezes (much as ice forms when water freezes), are composed of bonded grains, each consisting of a particular mineral (Figure 1-2). The igneous rock most familiar to nongeologists is granite. Molten material, or **magma**, that turns into igneous rock comes from great

Figure 1-2 Interlocking grains in granite. The pink and white grains are two kinds of feldspar, the gray grains are quartz, and the black grains are mafic minerals. (Breck P. Kent.)

Figure 1-3 Intrusive igneous rock. The dark bodies are pieces of surrounding rock that magma incorporated before it solidified into igneous rock. The light-colored diagonal bands on the left are veins that formed when a second body of magma intruded the main body of igneous rock. (Martin G. Miller/Earth Lens.)

depths within Earth, where temperatures are very high. This material may reach Earth's surface through cracks and fissures in the crust and then cool to form **extrusive**, or **volcanic**, igneous rock, or it may cool and harden within Earth to form **intrusive** igneous rock (Figure 1-3).

Even intrusive rocks that form deep within Earth can eventually be exposed at the surface if they are uplifted by earth movements and overlying rocks are stripped away. **Weathering** is a collective term for the chemical and physical processes that break down rocks of any kind at Earth's surface (Figure 1-4). Water carries some products of weathering away in solution. Solid products are removed by **erosion**, the process that loosens pieces of rock and moves them downhill. After erosion sets these pieces of rock in motion, moving water, ice, or wind transports them to a site where they accumulate as sediment.

Sediment is material deposited on Earth's surface by water, ice, or air. Grains of sediment accumulate in a variety of settings, ranging from surfaces of desert dunes to river channels, lake bottoms, sandy beaches, and the floor of the deep sea. After grains have accumulated as loose sediment, they can become bonded together to form solid **sedimentary rock** by either of two processes: the grains may become mutually attached by compression of the sediment after burial, or they may be glued together by precipitation of mineral cement from watery solutions that flow through the sediment. The processes that turn loose sediment into solid rock are collectively termed **lithification**. There are three principal kinds of rock-forming sediments:

Figure 1-4 Granite weathering in a mountainous region of Morocco. (Anne E. Hubbard/Photo Researchers.)

Figure 1-5 Horizontal bedding in the Kaibab Formation, bordering the Grand Canyon. The top of this formation forms the Kaibab Plateau, which marks the horizon. (Tom Bean.)

1. Most sedimentary rocks are formed of the kind of sediment described above—debris generated by weathering of preexisting rocks. The most common grains produced in this way are particles of sand and clay. Sand particles are grains that weathering releases from preexisting rocks, generally without chemical alteration. Sand grains are globular, and they do not stick together well when compacted. Loose sand therefore becomes solid **sandstone** only when cement precipitates between adjacent grains, locking them together. Tiny clay particles form by the chemical breakdown of certain minerals: they are chemical products of weathering. Clay is a flaky material that compacts to form the soft rock known as **shale**.

2. Other sedimentary rocks consist of fragments of skeletons of once-living organisms. Many **limestones** are formed of such material, including bits of broken seashells. Cementation turns accumulations of this limey debris into solid rock.

3. Still other grains that form rocks are precipitated chemically from water. The salt deposits that we mine for a variety of purposes form in this way when bodies of water evaporate in dry climates.

Sediments usually accumulate in discrete episodes, each of which forms a tabular layer known as a **stratum** (plural, **strata**) or **bed**. A breaking wave can create a stratum, for example, and so can the spreading waters of a flooding river. Even after lithification, a stratum tends to remain distinct from the one above it and the

one below it because the grains of adjacent strata usually differ in size or composition. Because of such differences, the contacting surfaces of the strata usually adhere to each other only weakly, and sedimentary rocks often break along these surfaces. As a result, sedimentary rocks exposed at Earth's surface often can be seen to have a steplike configuration when viewed from the side (Figure 1-5). **Stratification** and **bedding** are the synonymous words used to describe the arrangement of sedimentary rocks in discrete layers.

Metamorphic rocks form by the alteration, or **metamorphism**, of rocks within Earth under conditions of high temperature and pressure. By definition, metamorphism alters rocks without turning them to liquid. If the temperature becomes high enough to melt a rock and the molten rock later cools to form a new solid rock, this new rock is by definition igneous rather than metamorphic. Metamorphism produces minerals and textures that differ from those of the original rock and that are characteristically arrayed in parallel wavy layers (Figure 1-6). The two groups of rocks that form at high temperatures—igneous and metamorphic rocks—are commonly referred to as **crystalline rocks.**

Figure 1-7 summarizes the various possible relationships among igneous rocks, metamorphic rocks, and sedimentary rocks that are composed of debris from other rocks. Any body of rock can be transformed into another body of rock belonging to the same group (metamorphic, igneous, or sedimentary) or to either of the other two groups. In other words, any kind of rock can be metamorphosed, or it can be melted to produce magma or weathered to produce sediment.

Figure 1-6 Metamorphic rock. The rock shown here is a coarse-grained kind known as gneiss. (Dan Guravich/Photo Researchers.)

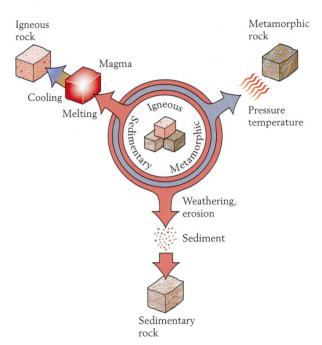

Figure 1-7 Relationships in the rock family. Any of the three basic kinds of rock—igneous, sedimentary, or metamorphic—can be transformed into another rock of the same kind or either of the other two kinds.

Bodies of rock are classified into formal units

Geologists also classify rocks into units called **formations**. Each formation consists of a body of rocks of a particular type that formed in a particular way—for example, a body of granite, of sandstone, or of alternating layers of sandstone and shale. A formation is formally named, usually for a geographic feature such as a town or river where it is well exposed.

The Kaibab Limestone is a typical formation. It forms the rim of a large portion of the Grand Canyon, and its upper surface forms much of the surface of the Kaibab Plateau, which borders the canyon and gives the formation its name (see Figure 1-5). The Kaibab Limestone is composed of fragments of shells and other skeletal debris. These and other distinctive features of the formation, including its color and the characteristic thickness of the beds within it, permit geologists to recognize the Kaibab wherever it occurs. Other limestones that occur below the Kaibab in the Grand Canyon region display different features.

Smaller rock units called **members** are recognized within some formations. Similarly, some formations are united to form larger units termed **groups**, and some groups, in turn, are combined into **supergroups**.

Steno's three laws concern sedimentary rocks

Forming on Earth's surface, sedimentary rocks provide most of our information about the history of life and en-

vironments on Earth. It is therefore important that we understand their distribution and their age relationships. The study of stratified rocks and their relationships in time and space is known as **stratigraphy**.

In the seventeenth century, Nicolaus Steno, a Danish physician who lived in Florence, Italy, formulated three sensible axioms for interpreting stratified rocks. Steno's first principle, the principle of **superposition**, states that in an undisturbed sequence of strata, the oldest strata lie at the bottom and successively higher strata are progressively younger (Figure 1-8A). In other words, in an uninterrupted sequence of strata, each bed is younger than the one below and older than the one above. This is a simple consequence of the law of gravity, of course, as is Steno's second principle, the principle of original horizontality.

The principle of **original horizontality** states that all strata are horizontal when they form. As it turns out, this principle requires some modification. We now recognize that some sediments, such as those of a sand dune, accumulate on sloping surfaces, forming strata that lie parallel to the surface on which they were deposited. Sediments seldom accumulate at an angle greater than 45° to the horizontal, however, because they slide down slopes that are steeper than that. Therefore, a reasonable restatement of Steno's second principle would be that almost all strata are initially more nearly

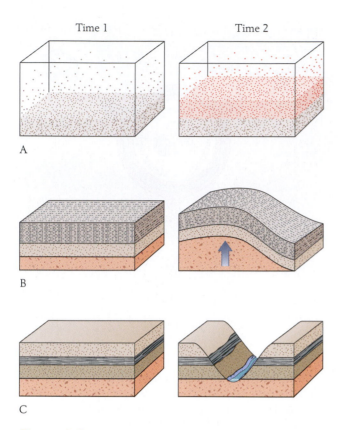

A

B

C

Figure 1-8 Steno's three principles. *A.* The principle of superposition: at time 2, sediment builds up on top of other sediment that was deposited earlier, at time 1. *B.* The principle of original horizontality: strata that were horizontal at time 1, shortly after being deposited, have been uplifted and tilted by time 2. *C.* The principle of original continuity: by time 2, strata that were continuous at time 1 have been divided into two bodies of strata by a river that has cut through them.

horizontal than vertical. Thus we can conclude that any strongly sloping stratum was tilted by external forces after it formed (Figure 1-8*B*).

Steno invoked his third principle, the principle of **original lateral continuity**, to explain the occurrence on opposite sides of a valley (or some other intervening feature of the landscape) of similar rocks that seem once to have been connected. Steno was, in effect, pointing out that strata originally are unbroken flat expanses, thinning laterally to a thickness of zero or abutting the walls of the natural basin in which they formed. The original continuity of a stratum can be broken by erosion, as when a river cuts downward to form a valley (Figure 1-8*C*).

The rock cycle relates all kinds of rocks to one another

After rocks form, they are subject to many kinds of change. Central to the uniformitarian view of Earth is the **rock cycle**—the endless pathway along which rocks of various kinds change into rocks of other kinds. Two simple principles are useful for recognizing steps of the rock cycle. The principle of **intrusive relationships** states that intrusive igneous rock is always younger than the rock that it invades. The principle of **components** states that when fragments of one body of rock are found within a second body of rock, the second body is always younger than the first. The second body may be a body of sedimentary rock in which the fragments have come from another body of rock, or it may be a body of igneous rock that contains distinctive pieces of older rock that magma engulfed before it cooled (see Figure 1-3).

The rock cycle is actually a complex of many kinds of cycles in which components of any body of rock—

Figure 1-9 The rock cycle. On the left is an igneous intrusion formed by magma that melted sedimentary rock and incorporated it. Some of the magma containing the melted sedimentary rock was extruded from volcanoes. Those volcanoes are now inactive and are eroding, along with exposures of the intrusive igneous rock and the metamorphic rock formed during the intrusive activity. The resulting sediment is accumulating in water nearby. Thus sediments have become igneous and metamorphic rocks, and those rocks have yielded younger sediment, completing the cycle. The volcano on the right is still active.

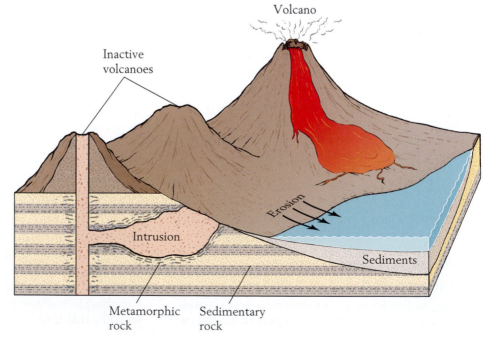

whether igneous, sedimentary, or metamorphic—can become part of another body of rock of the same kind or either of the other two kinds. In other words, as partly illustrated by Figure 1-7, any rock may be (1) melted to form magma that later cools to form igneous rock, (2) incorporated in magma without melting, (3) weathered to form debris that becomes part of sedimentary rock, or (4) turned into metamorphic rock by exposure to high temperatures and pressures. Figure 1-9 illustrates the rock cycle with a hypothetical example that includes igneous, sedimentary, and metamorphic rocks.

Movements of Earth play a key role in the rock cycle. When mountains rise up, for example, weathering and erosion wear them down to expose rocks that formed deep within the planet. Over vast stretches of time, these destructive processes level mountains, and streams and rivers carry the resulting sediments to faraway depositional settings.

Global Dating of the Rock Record

Thus far, we have discussed only age relations between bodies of rock that are in close proximity to one another. Geologists use additional techniques to piece together the history of Earth on a broader scale, showing, for example, that a body of rock located on one continent is older or younger than a body of rock located on another continent.

Fossils and physical markers indicate relative ages of rocks

Remnants of ancient life, called **fossils**, are useful for comparing the ages of bodies of sedimentary rock throughout the world. The term *fossil* is usually restricted to tangible remains or signs of ancient organisms that died thousands or millions of years ago. Because few fossils can survive at the high temperatures at which igneous and metamorphic rocks form, almost all fossils are found in sediments and sedimentary rocks. Fossils range in size from cells of tiny bacteria to massive dinosaur bones. They include such things as shells of invertebrate animals and teeth and bones of vertebrate animals—and also leaves of plants and impressions of soft-bodied animals.

Fossils provide one valuable means of establishing the relative ages of rocks that lie far apart. William "Strata" Smith, a British surveyor, noted late in the eighteenth century that fossils are not randomly distributed in rocks. When Smith studied large areas of England and Wales, he found that fossils in sedimentary rocks occurred in a particular vertical order ("vertical" in terms of the succession of one layer above another). To the surprise of less experienced observers, Smith could

Figure 1-10 Characteristic bivalve mollusks from the Paleozoic (bottom), Mesozoic (middle), and Cenozoic (top) eras. (Courtesy Smithsonian Institution, photo by Chip Clark.)

predict the vertical ordering of fossils in areas he had never visited. We now recognize that this ordering, known as **fossil succession**, reflects the sequence of organic evolution—the natural appearance and disappearance of species through time. Figure 1-10 illustrates the fossil succession of bivalve mollusks, a group of animals that includes clams, scallops, and mussels. Each of the geologic eras during which bivalve mollusks lived was populated by its own unique kinds of these animals. Although Smith, like nearly all of his contemporaries, had no knowledge of evolution, he was able to use his knowledge of fossil succession to determine where isolated outcrops of sedimentary rocks fitted into the general sequence of strata in England and Wales.

Since the time of William Smith, geologists have extended the use of fossils to establish the relative ages of rocks on a global scale. They have also discovered other kinds of markers showing that strata in many parts of the world were deposited simultaneously. One such marker is a high concentration of the element iridium at a stratigraphic level just above the uppermost dinosaur fossils. Iridium is much more common in meteorites than in nearly all rocks at Earth's surface. The unusual iridium-rich layer occurs throughout the world. As Chapter 17 will explain, this occurrence is what first suggested that an asteroid struck Earth and killed off the dinosaurs. The impact sent up a huge cloud of dust, spiked with telltale iridium, that spread around the globe. In more practical terms, the layer that is rich in iridium, an element very rare on Earth, allows geologists to locate, in sedimentary deposits throughout the world, the precise stratigraphic level that marks the end of the Age of Dinosaurs.

Radioactive decay provides actual ages of rocks

Although the principles outlined in the preceding section allow us to establish the relative ages of many bodies of rock, they do not permit us to determine the actual ages of rocks measured in thousands or millions of years. As we shall see in Chapter 6, some sedimentary beds are produced annually, like the rings in a tree trunk. Unless the latest of a continuous sequence of annual beds is currently forming, however, it is impossible to count backward to determine precisely how many years ago an older bed formed. In other words, if a sequence of this type formed long ago, we cannot tell the actual ages of its beds. Fortunately, "geologic clocks," in the form of chemical components that undergo radioactive decay, provide us with good estimates of the actual ages of ancient rocks. Naturally occurring radioactive materials decay into other materials at known rates. By measuring the amount of radioactivity in a radioactive material that has been decaying since it became part of a rock, we can estimate the age of the rock.

The geologic time scale divides Earth's history into formal units

During the nineteenth century, long before the discovery of radioactivity, it became apparent that very old sedimentary rocks contain no identifiable fossils. Beginning with these rocks and examining progressively younger rocks in any region, early geologists discovered that fossils became abundant at a certain level. This level became the boundary at which all of geologic time was divided into two major intervals (Figure 1-11). The oldest rocks with conspicuous fossils were designated as Cambrian in age, and still older rocks became known as Precambrian rocks. Today the Precambrian designation is still used informally, but the Precambrian interval is formally divided into the Archean Eon and the Proterozoic Eon, with the boundary between these two eons placed at 2.5 billion years ago. Subsequent geologic time, from Cambrian on, constitutes the Phanerozoic Eon, meaning the "interval of well-displayed life." An **eon** is the largest formal unit of geologic time.

Figure 1-11 The geologic time scale. The numbers on the right represent the ages of the boundaries between periods and epochs in millions of years. The Holocene Epoch (the past 15,000 years or so) is also known as the Recent.

Phanerozoic time is divided into three primary intervals, or **eras**, which the history of life on Earth serves to define. The earliest is the "interval of old life," or the Paleozoic Era. This era is followed by the "interval of middle life," or the Mesozoic Era, which is commonly

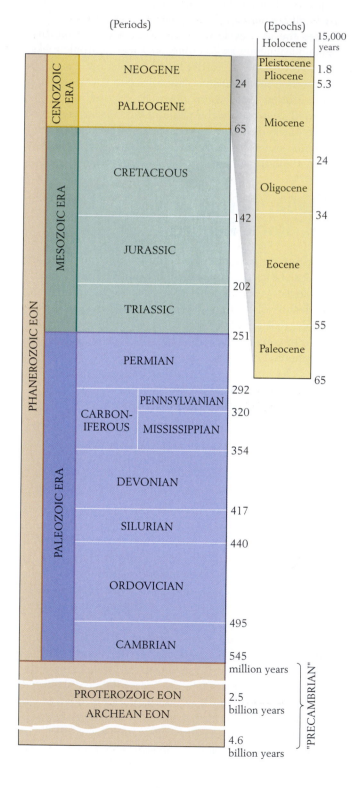

called the Age of Dinosaurs, and by the "interval of modern life," or the Cenozoic Era, which is informally designated as the Age of Mammals. Figure 1-11 depicts these eras and the intervals within them, known as the geologic **periods**. Periods are divided into **epochs**. Figure 1-11 lists epochs for the Cenozoic Era.

Figure 1-11 also indicates when each period began and ended, as determined by radioactive materials in rocks whose ages approximate period boundaries. Note that the Phanerozoic interval began about 545 million years ago. A human lifetime is so short in comparison that geologic time seems too vast for us to comprehend; experience does not permit us to extrapolate from the time scale familiar to us, measured in seconds, minutes, hours, days, and years, to a scale suitable for geologic time. Geologists therefore use a separate scale when they think about geologic time—one in which the units are millions of years. If the Phanerozoic interval of time were compressed into a year, we would find animals with backbones crawling up onto the land for the first time in mid-April, dinosaurs inheriting Earth in early July but then suddenly dying out in late October, and humans appearing on Earth about 2 hours before midnight on New Year's Eve.

Dating of rocks by means of radioactive materials reveals that some rocks on Earth are more than 3.8 billion years old. Many major geologic events span millions of years, but on the scale of geologic time they are only brief episodes. We now know, for example, that the Himalaya, the tallest mountain range on Earth, formed within the past 15 million years or so, but this period of time represents less than one-third of 1 percent of Earth's history. Destructive processes have also yielded enormous changes within a tiny fraction of Earth's lifetime. Mountains that were the precursors of the Rockies in western North America were leveled just a few million years after they formed, and much of the Grand Canyon of Arizona was cut by erosion within just the past 2 million or 3 million years. We will examine these events in greater detail in later chapters.

Intervals of the geologic time scale are distinctive

In the nineteenth century, when the geologic periods were first distinguished as unique intervals of time, geologists did not know even approximately how long ago each period had begun or ended. Each period was defined simply as the undetermined interval of time represented by a body of rock called a **geologic system**. The Cambrian Period, for example, was simply an interval of time that corresponded to those rocks that were designated as the Cambrian System. A geologic system is not to be confused with the Earth system, which encompasses all aspects of our dynamic planet.

Although the order in which the geologic systems were designated was haphazard, the total body of rock assigned to each system was not chosen arbitrarily. Two criteria were most important in these decisions. One was the occurrence of unique groups of fossils. Most systems contain many fossils that differ considerably from the fossils found below and above them. Major extinctions have caused the most striking contrasts between systems. A system representing an interval that followed a great extinction lacks many fossil groups that are well represented in the preceding system, and the younger system contains many new groups that evolved to replace those that died out.

Another feature that led early geologists to recognize some bodies of rock as systems was the nature of the rocks themselves. Most of the distinctive lithological features of geologic systems have some relation to the history of life. The Cretaceous System, for example, was designated to include the thickest deposits of chalk in the world. Chalk is soft, fine-grained limestone. The abundance of chalk in the Cretaceous System reflects the fact that during the Cretaceous Period there was a great proliferation of the kinds of organisms whose skeletons produce the particles of calcium carbonate that form chalk: small, single-celled organisms whose descendants float in the sea today, but in reduced abundance. No early scientist had the means to study the entire sequence of rocks on Earth, from the most ancient to the most modern; a single person could study only those promising rock sequences that were accessible. Thus the Cretaceous System was formally designated in 1822, whereas the much older Cambrian and Silurian systems did not gain formal recognition until 1835. System after system was added up and down the sequence until, finally, all the Phanerozoic rocks of Europe were included.

It seems remarkable today that all of the geologic systems of the Phanerozoic Eon were first designated during a brief interval of the nineteenth century in one small region of the world: Great Britain and nearby areas of western Europe. It was a lucky circumstance that modern geology came into being in this particular region. Few other geographic areas of comparable size display such large volumes of sediment and rock representing all of the Phanerozoic systems.

Imaging Earth Below

Most of our knowledge about the structure of Earth's deep interior derives from the study of *seismic waves*, large vibrations that travel through Earth as a consequence of natural earthquakes or artificial disturbances, such as those produced by nuclear explosions.

An earthquake always begins at a *focus*, a place within Earth where rocks move against other rocks

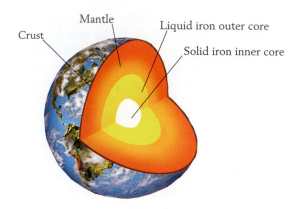

Figure 1-12 Zonation of Earth's interior. The crust, which includes Earth's continents, rests on the mantle. The mantle, in turn, rests on the core. The outer core is liquid, but the inner core is solid.

along a fault and produce seismic waves. A **fault** is a surface along which rocks have broken and moved. Earthquake foci lie within Earth's outer layers, far from its center—but the waves emitted from foci often pass great distances through Earth to emerge at the surface, where they can be detected with instruments called *seismographs*. Geophysicists can then evaluate movements deep within Earth by recording the times at which the waves of an earthquake arrive at many locations.

Earth's density increases with depth

The denser the material, the more rapidly seismic waves travel through it. Study of the rates at which waves travel in various directions from a focus reveals that the materials that form the central part of Earth are much more dense than those near the surface. The density gradient from the surface to the center of Earth is not gradual, however; instead, the planet is divided into several discrete concentric layers (Figure 1-12). At Earth's center is the **core**, whose solid, spherical inner portion and liquid outer portion consist primarily of iron. Form-

ing a thick envelope around the outer core is the **mantle**, a complex body of less dense rocky material. Finally, capping the mantle is the **crust**, which consists of still less dense rocky material. As we shall see in Chapter 11, the density gradient from the core to the crust developed early in Earth's history, when molten materials of low density rose to float on materials of higher density. The passage of seismic waves from the rocks of the crust to the denser rocks of the mantle is signaled by an abrupt increase in velocity known as the **Mohorovičić discontinuity**, or **Moho** for short (Figure 1-13). Because continental crust is much thicker than the crust beneath the oceans, the Moho dips downward beneath the continents.

Rocks that form oceanic crust are the type known as **mafic**—a label whose first three letters indicate that these dark rocks are rich in magnesium (Mg) and iron (Fe). Mafic rocks are much less common in continental crust than are the lighter-colored, less dense rocks known as **felsic**—an adjective derived from the first three letters of *feldspar*, the name of the most common mineral of continental crust. In comparison with mafic rocks, felsic rocks are rich in silicon and aluminum and poor in the heavier element iron. Rocks of the mantle are even richer in iron than the oceanic crust—hence their great density—and they are known as **ultramafic** rocks.

Continental crust not only stands above oceanic crust but also extends farther down into the mantle (see Figure 1-13). The continental crust extends even farther down beneath a mountain range than it does elsewhere. Isostatic adjustment, the upward or downward movement that keeps the crust in gravitational equilibrium as it floats on the mantle, is responsible for this phenomenon (Figure 1-14). In effect, the root beneath a mountain acts to balance the mountain; this balance is known as **isostasy**. An iceberg illustrates the same principle: because it is only slightly less dense than the water in which it floats, only a small fraction of its mass stands above the surface.

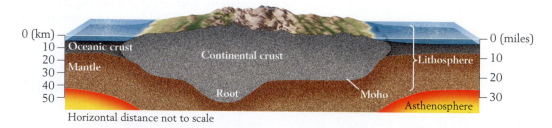

Figure 1-13 **The structure of the upper part of Earth.** Continental crust is much thicker than oceanic crust, and it is especially thick beneath mountains, where deep roots extend downward. The Moho separates the crust from the mantle. The crust and the upper mantle together form the rigid lithosphere.

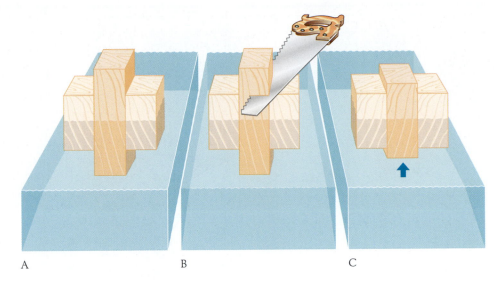

Figure 1-14 The principle of isostasy. *A.* Three blocks of wood float next to one another. Because the wood is half as dense as water, half of each block lies above the surface of the water and half lies below. The weight of a block of wood is equivalent to the weight of the volume of water that it displaces. *B.* When the top of the center block is cut off, the weight of this block no longer balances the weight of the displaced water. *C.* As a result, the block bobs upward until it is balanced once again, lying half above and half below the water. The center block is like the part of a continent where a large mountain is present—the low-density elevated crust must be balanced by a root.

Solid plates of lithosphere move over the slushlike asthenosphere

Although the crust and the upper mantle differ in density, they are firmly attached to each other, forming a rigid layer known as the **lithosphere** (see Figure 1-13). Below the lithosphere is a layer called the **asthenosphere**, which is also known as the "low-velocity zone" of the mantle because seismic waves slow down as they pass through it. This property tells us that the asthenosphere is composed of partially molten rock—slushlike material consisting of solid particles with liquid occupying the spaces in between. Although the asthenosphere represents no more than 6 percent of the thickness of the mantle, the mobility of this layer allows the lithosphere to move over it.

Plate Tectonics

The lithosphere does not move as a unit; instead, it is divided into sectors called **plates**, which move in relation to one another. Some plates carry continents with them as they move, while others carry only oceanic crust. The movement of lithospheric plates is known as **plate tectonics**. Throughout the first half of the twentieth century, most geologists believed that Earth's continents occupied fixed positions above the mantle. During the 1960s, however, geologists established beyond a reasonable doubt that plates—some carrying continents—migrate around the globe. This discovery ushered in one of the greatest scientific revolutions of the twentieth century. Plate tectonics accounts for many kinds of geologic phenomena that previously defied explanation, including the elevation of great mountain ranges.

Plates spread apart where they form, slide past one another, and eventually sink

Figure 1-15 illustrates the distribution of lithospheric plates in the modern world. An average plate moves about as rapidly as your fingernails grow: about 5 centimeters (2 inches) per year. When we recognize that plates have moved tens of thousands of kilometers at this pace, we can appreciate the immensity of geologic time. One of the most important of the forces that drive plate movements is convection, which operates on a large scale within the asthenosphere. **Convection** is the process by which material that is heated deep within the asthenosphere rises to displace cooler, less dense material nearer the surface. The same process causes rotational motion in a pot of water when it is heated (Figure 1-16).

Plate boundaries are dynamic regions of the lithosphere. In fact, most earthquakes originate when rocks move along faults at or close to these junctures. Three types of relative movement can occur between two

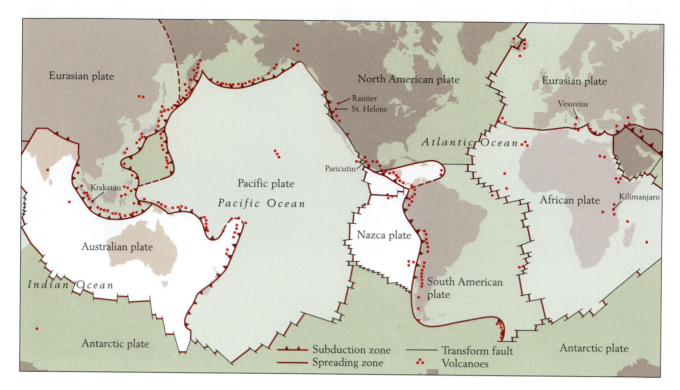

Figure 1-15 The distribution of lithospheric plates over Earth's surface. There are eight large plates and several small ones on Earth today. The three kinds of plate boundaries— subduction zones, spreading zones, and transform faults—are shown here (they will be discussed later in the chapter). Most volcanoes lie near subduction zones or spreading zones.

lithospheric plates that are in contact: the plates can be sliding past each other, or they can be moving toward or away from each other (see Figure 1-15).

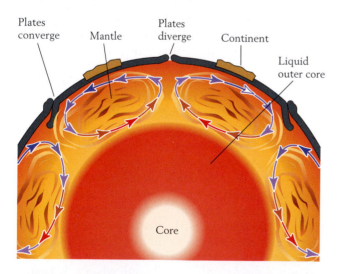

Figure 1-16 Convection. When a pot of water is heated from beneath, the water at the bottom warms and expands. Because of its reduced density, the warmed water rises, and cooler, denser water from above sinks to replace it. The result is the circular motion known as convection. Earth's mantle also experiences convection because it is heated from below, while the upper mantle loses heat to the crust.

Plates move apart along **spreading zones**, where new lithosphere forms as mafic material of relatively low density rises up from the ultramafic asthenosphere and cools. Heat that rises from the asthenosphere along a spreading zone swells the newly formed lithosphere to create a **mid-ocean ridge**, such as the Mid-Atlantic Ridge (Figure 1-17). The lithosphere on each side of the ridge slides laterally away from the ridge axis. Thus the two plates that are growing along a ridge are also moving apart.

Lithospheric plates move back down into the asthenosphere along deep-sea trenches, which floor the deepest zones of the oceans. The regions where plates descend are termed **subduction zones**. Because Earth is neither shrinking nor growing, the total rate at which plates are descending into subduction zones must balance the rate at which they are growing at spreading zones.

Plates slide past each other along surfaces called **transform faults**. As Figure 1-15 shows, these faults offset mid-ocean ridges throughout their length.

Volcanoes along plate boundaries Most volcanoes occur along those boundaries where plates are moving apart or together. Figure 1-17 shows the reason for this pattern. Large fractures in the lithosphere that parallel the axis of a spreading zone, such as the Mid-Atlantic Ridge, act as conduits through which magma rises from

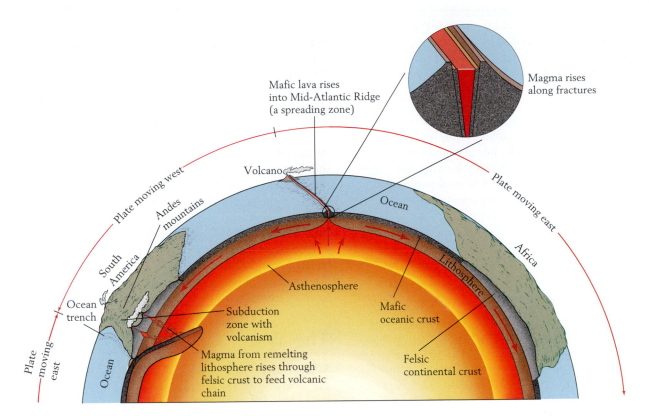

Mafic lava rises
into Mid-Atlantic Ridge
(a spreading zone)

Magma rises
along fractures

Plate moving west

Plate moving east

Volcano

Ocean

Andes
mountains

Africa

South
America

Lithosphere

Ocean
trench

Asthenosphere

Mafic
oceanic crust

Subduction
zone with
volcanism

Plate
moving
east

Magma from remelting
lithosphere rises through
felsic crust to feed volcanic
chain

Felsic
continental crust

Ocean

Figure 1-17 **Cross section of the lithosphere and the asthenosphere in the vicinity of the South Atlantic Ocean.** Note that the lithospheric plate that includes South America is moving westward from the Mid-Atlantic Ridge. At the same time, the lithospheric plate beneath the eastern Pacific Ocean is moving eastward; when it meets South America along a deep-sea trench, this oceanic plate moves downward into the asthenosphere, and the collision produces the Andes.

the asthenosphere and is added to the lithosphere. These fractures result from divergence at the spreading zone. A spreading zone stands above the seafloor as a ridge because it swells as heat flows upward along it as magma ascends. Some of the material cools to form shallow intrusive rock, and some makes its way through the lithosphere to form a zone of new seafloor along the spreading zone. In most places, the lava emerges along cracks as broad flows, but in some places it emerges through pipelike conduits to build tall, cone-shaped volcanoes. As newly formed lithosphere moves away from a ridge, it cools and subsides—and its surface becomes part of the broad, nearly planar expanse that floors most of the deep sea.

Volcanoes are also abundant along subduction zones. There they result from partial melting of the lithosphere that is moving down into the asthenosphere. Temperature increases with depth in the planet, and a segment of the descending lithosphere partly melts when it reaches a critical temperature. The resulting magma is less dense than the surrounding asthenosphere, and rises. Much of this magma reaches the seafloor and erupts to form volcanoes, some of which eventually rise above the sea surface to form islands.

Mountain building along subduction zones When a plate is subducted beneath continental lithosphere, which is thicker than oceanic lithosphere, much of the magma that rises from the subducted plate margin cools within the overlying lithosphere to form intrusive rock. Some of it, however, makes its way to the surface; thus volcanoes form many of the peaks along a mountain chain such as the Andes (see Figure 1-17). It is not only igneous activity that produces a mountain chain, however, when a continent encounters a subduction zone. Under great pressure, the continental margin crumples into folds that elevate the land. As we shall see in Chapter 9, the continental crust also breaks under the pressure here, and large slices of it slide along faults. These slices pile up on top or each other, thickening the crust.

Heat from radioactive decay fires the engine of plate tectonics

Heat from deep within the asthenosphere drives many plate tectonic processes. It causes convection within the asthenosphere (see Figure 1-16) and creates the magma that rises along spreading zones. For centuries, miners have known that Earth's interior is hotter than its surface, because the temperature increases as one descends through a mine shaft. Of course, hot springs and geysers, such as those of Yellowstone National Park, also give evidence of great heat deep within the planet. Late in the nineteenth century, most scientists believed that this heat was primordial heat—heat that remained after the planet cooled from a molten state early in its history. Soon after the Frenchman Henri Becquerel discovered radioactivity in 1896, however, geologists came to a new understanding of Earth's heat budget. They learned that radioactive material is so abundant within Earth that its decay supplies an enormous amount of the planet's heat. Thus, while Earth has been gradually losing its original heat, it has been generating a large amount of new heat.

Plate tectonics plays a role in the rock cycle

Plate tectonic processes exemplify the rock cycle on a grand scale (Figure 1-18). As Figure 1-18A illustrates, most oceanic lithosphere that forms where magma rises from the asthenosphere to form new seafloor ultimately descends into the asthenosphere again. This happens when that seafloor, having moved away from the spreading zone, encounters a subduction zone. In time, the area of lithosphere that is subducted, termed a **slab**, breaks away from the plate to which it originally belonged. The isolated slab then sinks farther and melts to become part of the asthenosphere. It may seem surprising that the slab sinks, because it consists largely of

mafic material, whereas the asthenosphere is ultramafic, but the slab is denser than the ultramafic material because it is cold. If, perhaps hundreds of millions of years later, convection moves the melted material from the slab upward along a spreading zone, it will again serve as a source of new lithosphere and thus complete a large cycle of movement.

Figure 1-18B shows how plate tectonics contributes to the rock cycle at shallower depths. Lithosphere that has formed along a spreading zone is eventually subducted, along with sediment. Some of the subducted material melts and returns to the surface through volcanism. Erosion of the resulting volcanoes yields new sediment, some of which is subducted and melts to produce a new generation of magma that rises to form new volcanic rock. This rock, in turn, erodes and sheds sediment, some of which is subducted. In this manner, the cycling continues.

The Water Cycle

Water moves over and through Earth in a complex cycle of its own. The **water cycle** is simpler than the rock cycle, however, in the sense that the cycled material is a single chemical compound. Water is so abundant on Earth that it is too easily taken for granted. Actually, its abundance is quite remarkable. Our planet is large, so its gravitational attraction holds water vapor in its **atmosphere**, the envelope of gases that surrounds Earth. At the same time, Earth's range of

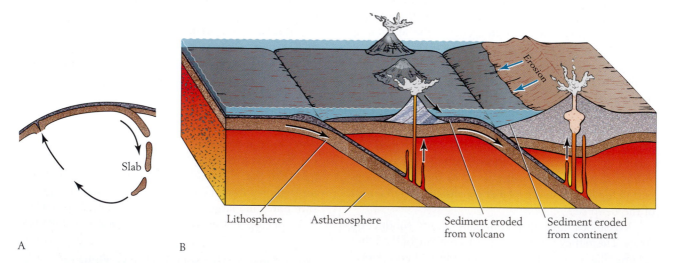

A   B

Lithosphere   Asthenosphere   Sediment eroded   Sediment eroded
                                      from volcano   from continent

Figure 1-18 **Plate tectonics and the rock cycle.** *A.* A subducted slab of lithosphere sinks, melts, and is incorporated into the asthenosphere. Convection carries the combined material upward, and some of the subducted material emerges along a spreading zone as new lithosphere, completing the cycle. *B.* A shallower cycle operates where sediments are subducted and melted to contribute to magma that is extruded

through volcanoes of an island arc (left). Sediment eroded from these volcanoes accumulates on the seafloor and is subducted and melted, completing a cycle. The cycling continues as melted material emerges through volcanoes along a mountain chain that borders a continent (right); erosion of the volcanic rock produces another generation of sediment.

surface temperatures makes it the only planet of the solar system to have abundant liquid H_2O on its surface. Here this special liquid plays many crucial roles. Water is essential for the existence of life as we know it because it is a major component of the cells that are the basic units of organisms. It also serves as an external medium for all those forms of life that we refer to as "aquatic." A high heat capacity is another important feature of water; in other words, water holds a large amount of heat at a given temperature. For this reason, water currents in the oceans spread heat much more effectively over Earth's surface than atmospheric winds can do.

Water moves between reservoirs

The water cycle entails the endless flow of H_2O between natural reservoirs at or near Earth's surface. Most of these reservoirs, including oceans, lakes, and rivers, contain liquid water. In addition, however, the atmosphere serves as a reservoir for water vapor, and **glaciers**—large, slowly flowing masses of ice—amount to solid reservoirs of H_2O.

Figure 1-19 illustrates the major features of the water cycle. Whereas heat within Earth drives plate movements and igneous processes, it is primarily heat from the sun that drives the water cycle. Solar heat evaporates water, sending water vapor into the atmosphere. It also causes the oceans and atmosphere to circulate, transporting H_2O over vast distances.

The oceans cover more than two-thirds of Earth's surface and contain more than 97 percent of the H_2O within the water cycle. It may seem surprising that glaciers contain most of the remaining H_2O—more than 2 percent. Glaciers are confined to high latitudes in lowland areas, but they occur throughout the world in mountainous regions, where the air is cold. Most of the water locked up in glaciers today occupies two ice caps, the one that covers most of Greenland and the one that

covers most of Antarctica. If all of the glaciers now present on Earth were to melt, sea level would rise about 100 meters throughout the world.

Lakes, though more abundant than glaciers, contain only about one-tenth of 1 percent of the H_2O of the water cycle. The atmosphere contains even less, only about one-hundredth of 1 percent, and rivers and streams account for only one-thousandth of 1 percent.

Much more water occupies pores and cracks within the solid Earth than runs over its surface. Soil contains about 50 times as much water as do rivers and streams, and the upper 4 kilometers of the lithosphere, below the soil, contain more than 100 times as much water as the soil itself. Wells supply humans with what is known as *groundwater* from this vast lithospheric reservoir. Land plants also take up groundwater through their roots, and although they use some of this water to produce sugars for their own use, they release most of it from their leaves. In this process, termed **transpiration**, plants convey moisture from subsurface reservoirs to the atmosphere, augmenting evaporation from bodies of water and moist soil. Although land plants contain only a minute fraction of the water in the water cycle, they transpire about as much water to the atmosphere during any period of time as is discharged by all the world's rivers.

The sizes and locations of the various reservoirs of the water cycle are constantly changing, and during many intervals of geologic time these reservoirs have differed markedly from their present configuration. At times, glaciers have expanded where none existed before, moist land has turned into desert, or the ocean has spread over dry land—and such changes have later been reversed or followed by equally profound changes of some other kind. The water cycle is constantly reshaping Earth in concert with the rock cycle and plate tectonics. The study of Earth's history focuses on these changes.

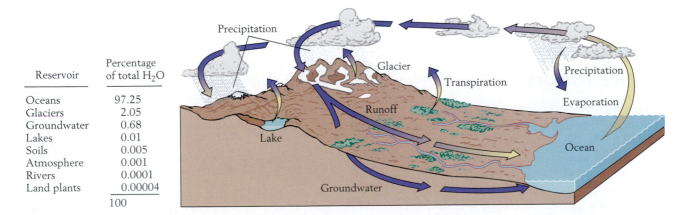

Reservoir	Percentage of total H_2O
Oceans	97.25
Glaciers	2.05
Groundwater	0.68
Lakes	0.01
Soils	0.005
Atmosphere	0.001
Rivers	0.0001
Land plants	0.00004
	100

Figure 1-19 **The water cycle.** The relative sizes of reservoirs are indicated next to the diagram.

The water cycle and the rock cycle are inseparable

Water is actually a component of certain minerals. For this and other reasons, the water cycle and the rock cycle are interconnected. Earth's surface is in a dynamic state not only because of plate movements and igneous activity, but also because of weathering, erosion, and deposition. Watery solutions weather rocks chemically and transport the dissolved products. Water and ice also erode rocks and sediments, and they transport the solid material released by weathering and erosion. Water contained within oceanic lithosphere and deep-sea sediments is even subducted into the asthenosphere—and then returned to the lithosphere and atmosphere by rising magma.

Water as pure as the distilled water produced in a laboratory does not exist in nature. All natural waters contain chemical constituents other than H_2O, which find their way into the water cycle from such processes as weathering at Earth's surface and chemical leaching of subsurface rocks and sediments by groundwater. When natural waters evaporate, chemical components dissolved in them form what are termed *salts*. Salts that accumulate in layers constitute sedimentary deposits known as **evaporites** (see p. 41). Sodium chloride, which we use as table salt, is the most abundant kind of evaporite. When mined from evaporite deposits, it is called rock salt. We refer to natural water as **fresh water** if it contains less than 0.5 percent salt by weight. Most lakes, like most rivers and streams, contain fresh water. In a dry climate, however, a lake from which no water flows may become quite salty over a long period of time as streams continually replace the water that evaporates from its surface. Consider the Great Salt Lake of Utah. The salt concentration within it varies, but averages about five times that of the ocean. The ocean itself might be viewed as a giant salty lake that is billions of years old.

In Chapter 10 we will examine how chemical compounds necessary for life cycle through the lithosphere, the oceans, the atmosphere, and living things. We will also learn how changes in the abundance of these compounds profoundly affect environments and life.

Directional Change in Earth's History

Although Charles Lyell made an enormous contribution to science when he argued for uniformitarianism, his view of Earth's history was flawed. Lyell viewed Earth simplistically, as a huge machine that underwent no net directional changes in the course of geologic time. In his view, the planet's changes were entirely cyclical: they were embodied in the rock cycle and the water cycle. Geologists today recognize that, although these cycles are always operating, Earth has also experienced many long-term directional changes. Lyell also believed, erroneously, that life had not changed significantly in the course of geologic time. He understood from the fossil record that species had disappeared and were somehow replaced by others, but he did not envision a progression from simple, primitive forms of life early in Earth's history to the present array of species, which includes many complex forms.

We will first examine what was wrong with Lyell's view of the history of life. Then we will review some of the physical aspects of our planet that have changed markedly in the course of geologic time.

Evolution reshapes life drastically and irreversibly

Environmental changes in the course of Earth's history have had enormous impacts on life. In fact, the word *environment* means "that which surrounds some form of life." Early in the nineteenth century, the prevailing view of literate people in Europe and North America was that, although many species of animals and plants had become extinct in the past, no species had ever given rise to another species before disappearing. According to this view, although environments on Earth had undergone significant changes—at least cyclical changes—populations of plants and animals had not. Charles Darwin overturned this belief in 1859 by providing powerful evidence for the concept of organic evolution. In fact, he showed that life changes even in the absence of changes in the physical environment. Particular kinds of plants and animals, Darwin concluded, gave rise to other kinds by a process that other scientists later termed **evolution**. Darwin reasoned that, although many forms of life have become extinct, many of them have first given rise to other forms that have outlived them. The preservation of these sequences of ancestral and descendant species in the stratigraphic record has produced the pattern of fossil succession that geologists use to date rocks.

Darwin's conception of organic evolution resulted from observations that he made as an unpaid naturalist aboard the *Beagle*, a ship that sailed around the world on an exploratory voyage, making many landings along the way. Darwin embarked on the voyage in 1831, at the age of 23. He had been well tutored in geology by Adam Sedgwick, an expert on the early Paleozoic rocks and fossils of England.

The principle of uniformitarianism played a key role in Darwin's thinking. Just before he set sail on the *Beagle*, one of his teachers, J. S. Henslow, urged him to read *Principles of Geology*, Charles Lyell's new book that argued effectively for James Hutton's concept of uniformitarianism. Henslow also cautioned Darwin not to believe Lyell's ideas, but Darwin was soon convinced by Lyell's arguments. On the voyage of the *Beagle*, Darwin observed processes in action that clearly could have produced major features of Earth's crust. In Chile, for example, he witnessed violent volcanic eruptions in the Andes, and during the period of eruption he experienced an earthquake and soon found that a strip of seafloor along the coast had abruptly risen to a position above sea level. It seemed obvious to Darwin that such processes, operating over millions of years, could produce a tall mountain range such as the Andes by elevating it a little at a time and occasionally piling volcanic rocks here and there on its surface. Lyell's uniformitarian picture of Earth's history provided Darwin with the vast stretches of geologic time required for the evolution of life.

Although Lyell's concept of uniformitarianism influenced Darwin profoundly, Darwin's concept of biological evolution differed in a significant way from Hutton's and Lyell's concepts of Earth's history. Lyell recognized that the upper part of the planet is locked in an endless rock cycle. Life also changes, Darwin concluded, but not in a cyclical fashion. Instead, it is constantly moving in new directions. These changes have introduced altogether new kinds of organisms over millions of years and have greatly increased the variety of life on Earth.

Physical and chemical features of Earth have also changed

Not only living things, but also physical and chemical features of Earth, have undergone major directional changes in the course of geologic time. Furthermore, biological, physical, and chemical changes have influenced one another.

One of the best examples of a physical directional change for Earth is the cooling of the planet since early in its history. Two factors have caused this decline in average temperature. First, Earth has been continuously losing the heat that was generated when it came into being. Second, because Earth's radioactive furnace has been weakening, it has been generating less and less heat to replace heat that escapes from the planet. As the radioactive materials that fire this furnace decay, fewer and fewer remain to decay at a future time; in effect, the fuel is being used up. As a result of Earth's declining temperature, plate tectonic processes have weakened in the course of geologic time. Fewer spreading and subduction zones exist today than existed early in Earth's history.

The concentration of oxygen in Earth's atmosphere has also undergone long-term shifts. As later chapters will spell out more fully, little oxygen was present in the planet's early atmosphere. The concentration of oxygen increased substantially only after the evolution of bacteria that, like today's green plants, manufactured their own food by the process of photosynthesis. Oxygen is a product of photosynthesis, and in time it built up in the atmosphere to a level that allowed animals to exist. In other words, the evolution of life altered the environment of the entire planet, and the new environmental conditions led in turn to further evolutionary changes. In the interplay between life and environments we see the intimate intertwining of the physicochemical and biological histories of Earth.

Life and environments have changed in concert

To study the history of interactions between the biological and physicochemical aspects of our planet, it is useful to identify ecosystems. An **ecosystem** is an environment together with the group of organisms that live within it. Ecosystems come in all sizes. Earth and all the forms of life that inhabit it constitute an ecosystem, but so does a tiny droplet of water inhabited by only a few microscopic organisms. Obviously, then, large ecosystems can be divided into many smaller ecosystems, and the size of the ecosystem that is treated in a particular study depends on the type of research that is being conducted.

Modern-day ecosystems are the products of billions of years of Earth's history. Even during the past few hundred million years, continents have broken apart and collided, and mountains have risen and succumbed to erosion. The ocean has repeatedly flooded vast areas of continents and receded again, and massive glaciers have spread across broad regions and then melted away. The deep sea, now near freezing, has at times been much warmer. Likewise, climates on land have warmed and cooled. For example, as we will learn in Chapter 18, 40 million years ago warm temperatures extended to Earth's poles, and palm trees grew in the northern United States; then these northern regions cooled, and they have never become so warm again. Environmental change has not only caused animals to migrate, but has also profoundly influenced evolution and, in the course of geologic time, caused many forms of life to disappear.

As we noted at the beginning of this chapter, by reconstructing the patterns of change undergone by ancient ecosystems, geologists learn lessons that will benefit civilization as it confronts future environmental changes.

Episodic Change in Earth's History

Recall that when James Hutton and Charles Lyell described Earth as if it were a huge machine within which materials cycled, they portrayed the cycles as moving continuously and very slowly. Geologists now recognize that sudden, episodic events have actually played large roles in Earth's physical and biological history.

Abrupt events have occurred on many scales. Sudden events of erosion or deposition can last only a few seconds, for example, and be confined to a few square centimeters. In contrast, the asteroid that apparently swept away the dinosaurs and many other forms of life 65 million years ago seems to have produced this catastrophe by altering climates throughout the world within the space of a few days. On a geologic scale of time, even events that have spanned a few hundred thousand years appear abrupt.

Sedimentation occurs in pulses

The stratigraphic record is a great archive of past events, but upon close inspection, most sedimentary rocks reveal a history of discontinuous deposition (Figure 1-20). As we have seen, strata in rocks represent discrete depositional events, and a surface that separates two such layers represents the interval between those events. Such gaps generally include not only intervals during which no sediments were deposited, but also intervals during which strata were built up but then eroded away. It is easy to picture how such a sequence of sedimentary layers can develop along a sandy beach, where the waters of broken waves wash up. Here one broken wave may deposit a new layer consisting of sand that the wave scoured from the seafloor farther offshore. Another broken wave may arrive later with greater force and scour away previously deposited layers. A surface between two strata may represent a gap in deposition of only a few seconds, or if produced by severe scouring, it may represent hundreds or even thousands of years. So substantial are the gaps within most large bodies of rock that the total time required for deposition of the preserved sedimentary layers was quite brief in comparison with the total time required for accumulation of the entire body of rock.

Deposition can be catastrophic

Some episodic deposition can be described as catastrophic. Hurricanes and other large storms produce

Figure 1-20 Episodic deposition. This portion of a sandstone displays four episodes of deposition. The penny rests on beds that represent the first episode; these beds were laid down along a surface that sloped to the left. Scouring of these beds produced the surface, just above the penny, that slopes to the right. Then additional beds accumulated along another surface that sloped to the left. After a second interruption of deposition, a third series of beds accumulated; these beds are nearly horizontal. After another interval of nondeposition, the dark sediment at the top of the photograph accumulated. (S. Stanley.)

Figure 1-21 A sequence of Late Cretaceous turbidite beds separated by thin shale beds in Alaska's Arctic National Wildlife Refuge. The turbidite beds stand out because they consist of coarser sediment than the intervening shales. Each turbidite bed was deposited rapidly by a current that carried sediment from shallow to deep water. (USGS, photo by David W. Houseknecht, Reston, VA.)

strong waves and currents that erode sediment and then deposit it elsewhere. An event of this type can deposit a meter or more of sediment during a few hours or even a few minutes. Catastrophic deposition occurs even in deep water. Along the submarine margins of continents, for example, an event that amounts to a submarine landslide will occasionally send a slurry of dense, sediment-laden water flowing down to the broad floor of the deep sea. There the flow spreads out, slows down, and deposits a *turbidite* bed—a layer of sediment that may be several centimeters thick. Turbidite beds, about which we will learn more in Chapter 5, are often stacked one on top of another to great thicknesses (Figure 1-21).

Unconformities represent large breaks in the rock record

Many large bodies of sedimentary rock, including some formally recognized as formations, have been deposited

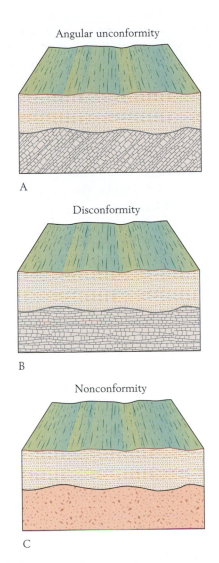

Figure 1-23 An angular unconformity, a disconformity, and a nonconformity. *A.* An angular unconformity separates tilted beds below from flat-lying beds above. *B.* A disconformity separates flat-lying beds below from other flat-lying beds above, but the upper beds rest on an erosion surface that developed after the lower beds were deposited. *C.* A nonconformity separates flat-lying beds from crystalline (igneous or metamorphic) rocks.

Figure 1-22 An angular unconformity in the Grand Canyon. Horizontal Cambrian strata rest on tilted strata of Precambrian age. (Peter Kresan.)

on irregular surfaces produced by the erosion of preexisting bodies of rock. In such cases, the irregular surface between the two bodies of rock is termed an **unconformity**. An unconformity represents a substantial interval of time when erosion, rather than deposition, occurred. There are three kinds of unconformities. When a group of rocks has been tilted and eroded and younger rocks have been deposited on top of them, the eroded surface is termed an **angular unconformity** (Figures 1-22 and 1-23*A*). Other unconformities are less dramatic. Sometimes the beds below an eroded surface are undisturbed, and only the irregular surface between the upper and lower bodies of rock reveals a past episode of erosion. This kind of unconformity is called a **disconformity** (Figure 1-23*B*). An unconformity in which bedded rocks rest on an eroded surface of crystalline rocks (Figure 1-23*C*) is sometimes called a **nonconformity**.

Life on Earth has experienced pulses of change

Like the deposition of rocks, the evolution of life can be interrupted. From time to time, individual species disappear from Earth, or undergo **extinction**. At many times in the history of the planet, extinctions have been clustered: many species have died out during brief intervals of geologic time. The largest of these events, known as **mass extinctions**, have amounted to global catastrophes in which a large percentage of the species on Earth have disappeared. The event that swept away the dinosaurs was only one of several mass extinctions that have occurred during the past half-billion years.

Major episodes of evolution have also occurred during relatively brief intervals of geologic time. Many of the basic kinds of multicellular animals that exist today evolved within just a few million years at the beginning of the Phanerozoic Eon. About half a billion years later, most of the basic groups of mammals that occupy the world today evolved within an interval of about 15 million years. Among them were the group that includes whales, the group that includes bats, and the group that includes humans.

As Chapters 12 through 20 will demonstrate, major pulses of evolutionary diversification and extinction have repeatedly transformed the fabric of life on Earth. One major task of earth scientists is to reconstruct these events and relate them to environmental changes that are recorded within ancient strata.

In Chapters 11 through 20, dramatic changes in the Earth system will be highlighted as Earth System Shifts. Most of these events affected life on Earth profoundly, and many were quite abrupt in the framework of geologic time.

Chapter Summary

What fundamental principles guide geologists as they reconstruct Earth's history?

The principle of uniformitarianism, which is fundamental to natural science, asserts that the laws of nature do not vary in the course of time. Actualism is the uniformitarian principle according to which events and processes of the geologic past are interpreted in light of events and processes observed in the modern world or re-created in the laboratory.

What are the basic kinds or rocks and how are they interrelated?

A rock is an aggregate of mineral grains; a mineral is an inorganic element or compound that is characterized by its chemical composition and internal structure. Igneous rocks form by the cooling of magma; sedimentary rocks are layered rocks that accumulate under the influence of gravity; metamorphic rocks form by the alteration of preexisting rocks at high temperatures and pressures. The rock cycle is the endless cycle in which rocks of various kinds change into rocks of other kinds. Water moves in a complex cycle that is linked to the rock cycle; reservoirs through which it moves include the oceans, lakes and rivers, groundwater, glaciers, and water vapor in the atmosphere.

How do geologists unravel the age relations of rocks?

The relative ages of rocks that come into contact with one another can often be determined by the principles of superposition, original horizontality, and original lateral continuity, as well as by the principle of intrusive relationships and and the principle of components. The scale that is employed to divide the rock record into units representing discrete intervals of geologic time was developed in Europe during the nineteenth century. Earth's history entails not only cyclical changes but also directional changes, including the evolution and extinction of life. Some of these changes are abrupt, even catastrophic. The record of sedimentary rocks forms episodically and contains numerous gaps, the largest of which represent unconformities. Fossils are remains or tangible evidence of ancient life found within rocks. Changes in life on Earth during the course of geologic time are documented in the fossil record. The succession of fossils reveals the relative ages of rocks in different regions. The decay of naturally occurring radioactive materials reveals the actual ages (in years) of some rocks.

How does the lithosphere relate to Earth's inner regions, and how does it move and deform?

Earthquakes create seismic waves that reveal much about Earth's subsurface structure. Earth's interior is divided into concentric layers. A central core of high density is surrounded by a less dense mantle, which is blanketed by a crust. The parts of Earth's crust that form continents are thicker and less dense than the parts that lie beneath the oceans. The crust and upper mantle constitute the rigid lithosphere, which is divided into discrete plates that move laterally over the partially molten lower mantle, termed the asthenosphere. Lithosphere that underlies the ocean forms along narrow ridges, spreads away from them, and descends into the asthenosphere along deep-sea trenches. Heat from radioactive decay drives this movement. Rocks fold and contort where plates converge, sometimes forming mountains.

Review Questions

1. Give general examples of the use of actualism to interpret ancient rocks.

2. In which of the three basic kinds of rocks do nearly all fossils occur? Why?

3. What is metamorphism?

4. Where is lithosphere formed along plate boundaries, and where does it disappear into the asthenosphere?

5. What is isostasy, and how does it explain why mountains have roots?

6. What kinds of features distinguish one geologic system from another?

7. Name three kinds of unconformities. How does each type form?

8. Describe three kinds of relationships between two bodies of rock that indicate which of the two bodies is the younger one.

9. How do sedimentary rocks relate to igneous and metamorphic rocks in the rock cycle?

10. What drives the water cycle, and how does this cycle relate to geologic, biological, and atmospheric processes near Earth's surface?

11. In what ways was Charles Lyell's view of Earth's history flawed?

Evaporite deposits of Oligocene age from Germany. The white and colorless crystals are halite. (Lawrence Hardie.)

CHAPTER 2

Rock-Forming Minerals and Rocks

Rocks, the building blocks of Earth, provide evidence of conditions and events of the geologic past. The composition and configuration of rocks tell stories of mountain-building episodes, changing positions of land and sea, climatic changes, and many other aspects of Earth's history. In this chapter our emphasis is on materials that play prominent roles in the rock cycle: rocks and the minerals that form them.

The Structure of Minerals

Minerals, as we saw in Chapter 1, are naturally occurring solid elements or compounds. In order to understand the properties of minerals, we need to examine the nature of atoms, the fundamental units of elements. Then we can see how atoms combine to form minerals and how minerals combine to form rocks.

Isotopes of an element have distinctive atomic weights

At the center of an atom is a *nucleus*, in which nearly all of the atom's mass resides. Particles called *protons* and *neutrons* form the nucleus. Each of these particles has the same mass of *one atomic mass unit*; but whereas protons each have a positive charge of $+1$, neutrons, as their name suggests, have no charge. In orbit around the nucleus are *electrons*, each having a charge of -1. Electrons move about within zones called *orbitals* or *shells* around the nucleus (Figure 2-1). Their atomic mass is negligible.

Atoms form *elements*, which, being indivisible into simpler substances, are the most basic chemically distinct form of matter. Every chemical element consists of atoms that have a particular number of protons. The number of protons is the element's unique *atomic number*. Hydrogen, with only one proton, has an atomic number of 1; for oxygen the number is 8, for silicon it is 14, and for iron it is 26 (Figure 2-2). A neutral atom has as many

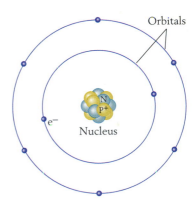

Figure 2-1 **Model of a carbon atom.** The central nucleus is surrounded by orbiting electrons (e^-). Each proton (P^+) has an atomic mass of 1, as does each neutron (N). The mass of an electron is negligible. Protons and electrons have opposite charges.

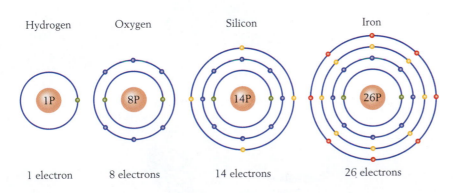

Hydrogen   Oxygen   Silicon   Iron

1P   8P   14P   26P

1 electron   8 electrons   14 electrons   26 electrons

Figure 2-2 Models of several common elements, each with a different number of shells.

Visual Overview
Rocks and Their Origins

The origins of igneous, sedimentary, and metamorphic rocks are interrelated through the rock cycle.

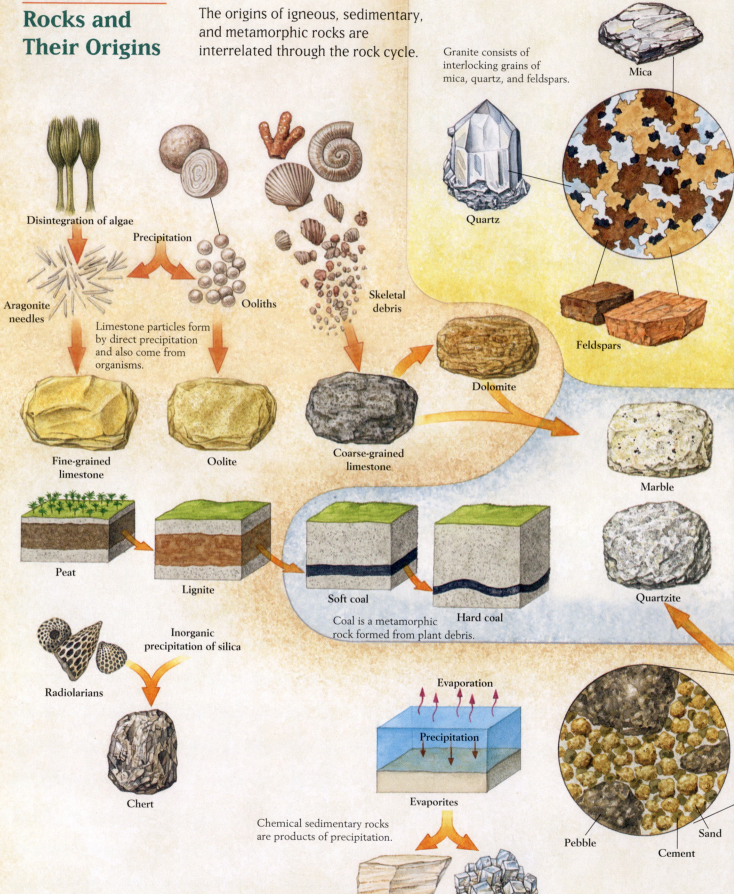

Granite consists of interlocking grains of mica, quartz, and feldspars.

Mica

Quartz

Feldspars

Disintegration of algae

Precipitation

Ooliths

Skeletal debris

Aragonite needles

Limestone particles form by direct precipitation and also come from organisms.

Dolomite

Marble

Fine-grained limestone

Oolite

Coarse-grained limestone

Peat

Lignite

Soft coal

Hard coal

Quartzite

Coal is a metamorphic rock formed from plant debris.

Radiolarians

Inorganic precipitation of silica

Chert

Evaporation

Precipitation

Evaporites

Chemical sedimentary rocks are products of precipitation.

Pebble

Cement

Sand

Gypsum

Halite

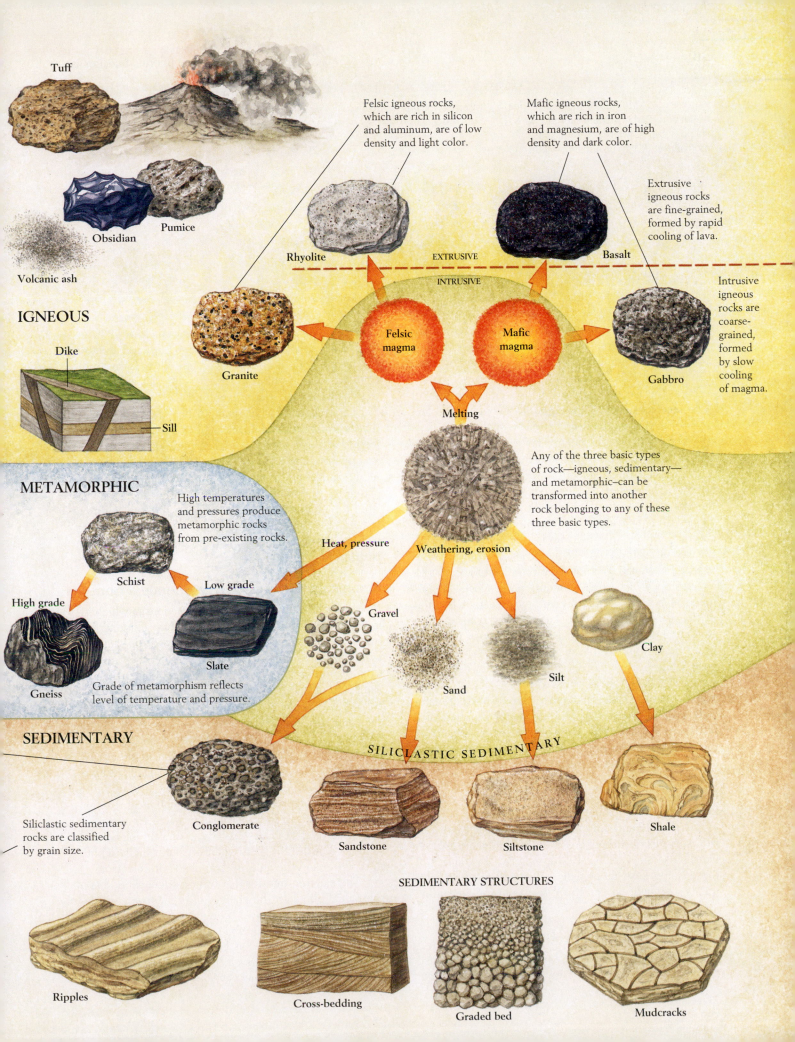

Tuff

Obsidian

Pumice

Volcanic ash

IGNEOUS

Dike

Sill

Felsic igneous rocks, which are rich in silicon and aluminum, are of low density and light color.

Mafic igneous rocks, which are rich in iron and magnesium, are of high density and dark color.

Extrusive igneous rocks are fine-grained, formed by rapid cooling of lava.

Rhyolite

EXTRUSIVE

INTRUSIVE

Basalt

Intrusive igneous rocks are coarse-grained, formed by slow cooling of magma.

Felsic magma

Mafic magma

Granite

Gabbro

Melting

METAMORPHIC

High temperatures and pressures produce metamorphic rocks from pre-existing rocks.

Schist

Low grade

High grade

Slate

Gneiss

Grade of metamorphism reflects level of temperature and pressure.

Heat, pressure

Weathering, erosion

Any of the three basic types of rock—igneous, sedimentary—and metamorphic–can be transformed into another rock belonging to any of these three basic types.

Gravel

Sand

Silt

Clay

SEDIMENTARY

SILICLASTIC SEDIMENTARY

Siliclastic sedimentary rocks are classified by grain size.

Conglomerate

Sandstone

Siltstone

Shale

SEDIMENTARY STRUCTURES

Ripples

Cross-bedding

Graded bed

Mudcracks

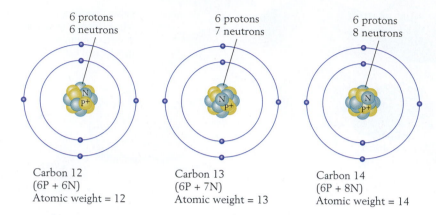

Figure 2-3 Models of three isotopes of carbon. Carbon 12 and carbon 13 are stable isotopes, whereas carbon 14 is radioactive. (P^+ = proton; N = neutron.)

6 protons
6 neutrons

6 protons
7 neutrons

6 protons
8 neutrons

Carbon 12
(6P + 6N)
Atomic weight = 12

Carbon 13
(6P + 7N)
Atomic weight = 13

Carbon 14
(6P + 8N)
Atomic weight = 14

electrons as it has protons. Consequently, as the number of protons increases, so does the number of electrons—and the number of shells also increases, by steps. The innermost shell holds a maximum of two electrons, whereas the second and third shells hold a maximum of eight electrons each. Thus eight of the twenty-six electrons of iron have to occupy a fourth shell.

Unlike protons, the neutrons of a given element may vary in number. Thus carbon, which has six protons, may have six, seven, or eight neutrons (Figure 2-3). The atomic masses of the protons and neutrons in a carbon atom may therefore add up to 12, 13, or 14. These numbers are termed the *atomic weights* of the different kinds of carbon atoms. Each kind of atom, with its unique atomic weight, is called an **isotope** of its element. The three carbon isotopes are designated carbon 12, carbon 13, and carbon 14.

Isotopes of certain elements have special importance in geology. Some are *radioactive*; that is, their nuclei decay at a constant, known rate to form other isotopes, informally termed *daughter isotopes*. By measuring the amount of decay such isotopes have undergone while present in rocks, geologists can date those rocks. The rapid rate at which carbon 14 decays, for example, enables geologists to date carbon-bearing materials, such as wood, that are only a few hundred or a few thousand years old. Most other naturally occurring radioactive isotopes decay much more slowly, and some are widely used to date rocks that are hundreds of millions of years old. The total number of radioactive atoms in Earth is so enormous that their decay releases enough heat to warm the planet significantly (p. 18).

Carbon 12 and carbon 13, the nonradioactive or **stable isotopes** of carbon, are also useful to geologists. As we shall see in Chapter 10, the relative abundances of these isotopes within organic matter and fossils in sediments shed light on aspects of Earth's history, including changes in the composition of Earth's atmosphere.

Chemical reactions produce minerals

Rocks are made of minerals, but for a mineral to exist, elements must combine by way of a *chemical reaction*, in which two or more atoms interact to form a structure called a *molecule*. An atom of sodium and an atom of chlorine, for example, react chemically to form a molecule of sodium chloride (NaCl)—the salt with which we flavor our food. Sodium chloride occurs in nature as the mineral **halite**, informally called rock salt. Halite that was precipitated from ancient seas or salty lakes forms large deposits that we can mine for our use.

Just as an atom is the basic unit of a chemical element, a molecule is the basic unit of a *chemical compound*. Most compounds have very different properties from the elements of which they are formed. Sodium, for example, is a metal, and chlorine is a gas; neither bears any resemblance to sodium chloride, which is a clear, solid compound.

Chemical reactions create chemical bonds

Molecules form through chemical reactions in which interactions between electrons produce attachments between atoms known as *chemical bonds*. In one kind of bond, known as an **ionic bond**, one atom loses an electron to another atom. An ionic bond forms in the chemical reaction that produces a molecule of sodium chloride. Elements having one or two electrons in their outer shell tend to lose electrons to other kinds of atoms. Sodium, for example, tends to lose its lone outer electron to chlorine. Chlorine, with an atomic number of 17, has seven electrons in its outermost (third) shell, just one short of the full complement. If a sodium atom and a chorine atom approach each other, a chemical reaction can occur in which sodium transfers an electron to chlorine (Figure 2-4). The result is a stable molecule of sodium chloride (NaCl), in which the outer shells of both the sodium and the chlorine are filled with electrons.

The electron transfer that forms an ionic bond has a substantial effect on the charges of the atoms: each develops a charge imbalance. In the formation of a sodium chloride molecule, for example, the sodium ends up with a net charge of +1, since its loss of an electron leaves its protons outnumbering its electrons by one. In complementary fashion, chlorine develops a charge of −1 by

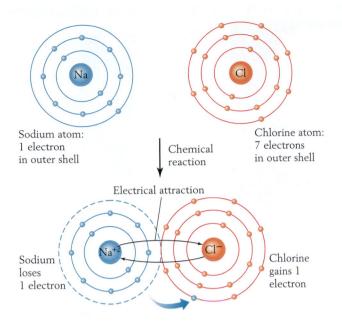

Figure 2-4 A chemical reaction forms sodium chloride. Sodium chloride (NaCl) forms by the transfer of the only electron in the outer shell of sodium (Na) to the outer shell of chlorine (Cl). The sodium ion (Na^+) and chloride ion (Cl^-) thus formed are held together by an electrical attraction resulting from their opposite charges.

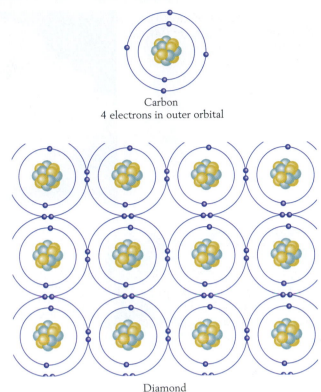

Carbon
4 electrons in outer orbital

Diamond

Figure 2-5 Covalent bonding in diamond. An isolated carbon atom (above) has four electrons in its outer shell. All atoms in diamond (below) share the electrons of their outer shell equally so that each atom ends up with a total of eight electrons in its outer shell instead of four. Because all of the electrons of the outer shell are shared, the atoms are strongly bonded, and diamond is very hard.

gaining an electron. When an atom becomes charged, it becomes an *ion*. It is the attraction between the positively charged sodium ion and the negatively charged chlorine ion that holds the sodium chloride molecule together.

Ions retain their identity when the compounds they form dissolve in water. When sodium chloride dissolves in seawater, for example, Na^+ and Cl^- ions are released to form a salt solution. In fact, the accumulation of these and other, less abundant ions is what makes seawater salty. When seawater evaporates in a dry climate, Na^+ and Cl^- can become too highly concentrated to remain in solution. Then Na^+ and Cl^- ions combine, and salt precipitates. This is how halite deposits form in nature.

Sodium chloride displays the simplest kind of ionic bond, in that a single electron is transferred from sodium to chlorine. Many other kinds of molecules form by the transfer of more than one electron; in fact, many form by the union of three or more ions. Calcium chloride exemplifies both of these complex features; as $CaCl_2$, it consists of a Ca^{2+} ion that is attached to two Cl^- ions.

Ions often group together to form *complex ions*. Examples are carbonate (CO_3^{2-}) and sulfate (SO_4^{2-}). Both of these complex ions are abundant in seawater, from which they can precipitate by combining with positive ions to form minerals.

Atoms can also form **covalent bonds**, in which electrons are shared rather than exchanged. The mineral diamond, which is cut to produce gems, consists entirely

of carbon atoms united by covalent bonds. An isolated carbon atom, with an atomic number of 6, has four electrons in its outermost (second) shell. When carbon forms diamond, each of the electrons in its outer shell is shared by two atoms (Figure 2-5). Thus the number of electrons in the outer shell of each atom is doubled to the full complement of eight. The result is a stable molecular structure in which every carbon atom shares two electrons with each of four neighboring atoms.

Crystal lattices are three-dimensional molecular structures

For ions to form a stable compound, not only must their ionic charges be in balance, but the ions must have relative diameters that allow them to fit together. In the mineral halite, for example, the small sodium ions fit comfortably between larger chloride ions (Figure 2-6A).

The three-dimensional molecular structure of a mineral is known as a *crystal lattice*. The configuration of a particular mineral's crystal lattice reflects the relative sizes and numbers of the various kinds of ions that make up the mineral. The external form of the mineral,

Chloride

Sodium

A   B   C

Cleavage   Sodium
planes   Chloride

Figure 2-6 **Sodium chloride (halite).** *A.* Sodium and chloride ions pack together in equal numbers, forming a cubic structure. *B.* Cubic crystals of halite reflect the crystal structure of sodium chloride. *C.* Shown more widely separated, the sodium and chloride ions can be seen to form a crystal lattice in which ions form parallel layers. These layers are separated by planes of weakness, known as cleavage planes. *D.* Table salt consists of tiny rectangular solids formed when halite is crushed and breaks along cleavage planes oriented at right angles to one another. (*B,* Breck P. Kent; *D,* Chip Clark.)

D

in turn, reflects the configuration of the mineral's crystal lattice. The crystal lattice of halite, for example, consists of equal numbers of Na^+ and Cl^- ions packed closely together in a regular pattern. It is easy to see how this pattern is reflected in the cubic shape of halite crystals (Figure 2-6*B*).

Two or more minerals can have identical chemical compositions but altogether different crystal lattices. **Calcite** and **aragonite**, for example, are common minerals of this kind that both consist of calcium carbonate ($CaCO_3$). Both of these minerals precipitate from watery solutions in nature, and a variety of organisms se-

A   B

Figure 2-7 **Calcite and aragonite, two forms of calcium carbonate ($CaCO_3$).** *A.* Calcite crystals. *B.* Aragonite crystals in the form of very small needles; crystals like these form carbonate mud on shallow tropical seafloors. (*A,* E. R. Degginger/Photo Researchers; *B,* Ian G. Macintyre, National Museum of Natural History, Smithsonian Institution.)

crete one or both to form a skeleton. Calcite forms blocky or tooth-shaped crystals (Figure 2-7*A*). Aragonite precipitates directly from shallow tropical seas as tiny needle-shaped crystals that accumulate as what is termed **carbonate mud** (Figure 2-7*B*). This mud and the aragonitic and calcitic skeletons of marine animals are major components of sediment that hardens to form limestone. Limestones more than a few million years old, however, consist almost entirely of calcite, because aragonite is not stable: over time, it is transformed into calcite. Corals that have lived during the past 250 million years have secreted skeletons of aragonite, but most of these skeletons that have been preserved as fossils were long ago altered to calcite (Figure 2-8). This alteration is often brought about by watery solutions that destroy some of the original features of the coral skeleton.

Ions of an element can substitute for ions of another similar element

By definition, a particular mineral can have a specified range of chemical compositions (p. 6). The reason for this latitude in the definition of a mineral is that many minerals have crystal lattices in which a small quantity of a particular ion can substitute for another ion without significantly altering the mineral's physical or chemical properties. Again, calcium carbonate serves to illustrate the point. The strontium ion sometimes substitutes for the calcium ion in the calcium carbonate crystal lattice, being only slightly larger than the calcium ion and having the same charge of

Figure 2-8 Coral skeletons. On the left is the cross section of a present-day coral skeleton that consists of aragonite. On the right is a piece of fossil coral of the same species that is only a few hundred thousand years old, but has been largely altered to calcite—a change often brought about by watery solutions. Secondary calcite also fills many of the pores of the original skeleton, obscuring some of the original features. (Courtesy Smithsonian Institution, photo by Chip Clark.)

+2. Two stable isotopes of strontium occur in Earth's crust and in the ocean. Because strontium ions are present in seawater, some of them find their way into skeletons of calcium carbonate secreted by marine organisms. The relative amounts of these two isotopes in seawater have changed during Earth's history. Therefore, as Chapter 6 will explain more fully, the isotope ratio of strontium in a fossil can be used to establish the age of the fossil.

The Properties of Minerals

The molecular structure of a mineral determines many properties that influence its physical and chemical behavior in nature. The densities of minerals influence whether they are found in Earth's crust or mantle, for example, and their crystal lattice structures influence how they break and how readily they abrade when they roll about as isolated grains. Table 2-1 shows the chemical and physical properties of the major mineral groups.

Chemical bonds determine hardness

Molecular bonds within minerals vary greatly in strength. The covalent bond structure of diamond is so strong that diamond is the hardest of all minerals (see Figure 2-5). Because the powerful covalent bonding that forms diamond can develop only under very high pressure, diamond forms primarily within Earth's mantle and only rarely is elevated to its surface. Very small diamond crystals can also form when a large meteorite strikes Earth, creating enormous pressures at impact.

Graphite, like diamond, is a mineral consisting of pure carbon, but some of the bonds within the crystal lattice of graphite are very weak. In fact, graphite is so soft that even paper can abrade it; that is why we can use it as what we call "lead" in our pencils. Graphite forms in Earth's crust, at much lower pressures than are required to produce diamond in the mantle.

The weight and packing of atoms determine density

Density is the weight of a given volume of any substance. The density of a mineral—often expressed in grams per cubic centimeter (g/cm^3)—depends on two things: the atomic weights of the atoms that form the mineral and the degree to which those atoms are packed together. Iron, for example, with an atomic number of 26, has a greater atomic weight than many other rock-forming elements; thus minerals that contain iron tend to be relatively dense. This is why such minerals are relatively abundant in Earth's mantle. Iron tended to sink deep within Earth early in its history, when the planet was in a hot, liquid state, and in fact it became the dominant element of Earth's core.

We can see how atomic packing increases a mineral's density by comparing diamond and graphite. The tight packing of diamond's carbon atoms, which results from its formation at high pressures deep within Earth, gives it a density of $3.5 \ g/cm^3$; graphite, which forms under lower pressures, has a density of only $2.1 \ g/cm^3$.

Fracture patterns reflect crystal lattice structure

Weak bonding within a mineral's crystal lattice can create parallel planes of weakness along which the mineral tends to break. If you observe several grains from a salt shaker through a strong magnifying glass, you will see that many are tiny rectangular blocks. These blocks were formed when larger chunks of halite broke along planes of weakness termed *cleavage planes* (see Figure 2-6*D*).

Minerals and rocks form under particular physicochemical conditions

The composition and internal structure of minerals reflect the conditions under which they form. We have seen that diamond, for example, forms at much higher pressures than graphite, although both consist solely of

TABLE 2-1 Major Mineral Groups

	Chemical properties	Physical properties	Rock-forming contribution	Comments
Silicates	SiO_4^{4-} tetrahedra are the basic units	Mostly hard, except for mica and clay minerals; most have a glassy or pearly luster	Dominant mineral group in igneous, sedimentary, and metamorphic rocks	Most crystallize at high temperatures and occur in sediments only as detritus
Carbonates	Positive ions attached to CO_3^{2-}	Soft, light-colored	Mostly sedimentary, but also form marble, a metamorphic rock	Include calcite, aragonite, and dolomite
Sulfates	Positive ions attached to SO_4^{2-}	Soft, light-colored, water-soluble	Most rock-forming varieties are sedimentary	Form large sedimentary evaporite deposits, including gypsum and anhydrite
Halides	Positive ions attached to negative ions of elements such as chlorine (Cl) and bromine (Br)	Soft, light-colored, water-soluble	Most rock-forming varieties are sedimentary	Form large sedimentary deposits, including halite (rock salt)
Oxides	Metallic ions combined with oxygen	Soft to hard	Mostly sedimentary, but many varieties are present in igneous and metamorphic rocks	Some, including the iron materials magnetite and hematite, are major ore minerals
Sulfides	Metallic ions combined with sulfur	Soft to medium hard, often with a metallic luster	Have a minor role in rock forming	Many are important ore minerals that form at high temperatures

carbon. We have also noted that iron-rich minerals are rarer in Earth's crust than in its mantle, where elements of high atomic weight are concentrated. In lakes and the ocean, the only minerals that can precipitate are ones such as halite and aragonite that are soluble in water.

These constraints on mineral origins allow a geologist to identify the setting where a particular body of rock originated. Sometimes an individual mineral tells the story, and sometimes it is the entire suite of minerals within the rock. Minerals within rocks found at the top of a mountain, for example, may reveal that those rocks originated deep below Earth's surface at high temperatures and pressures; such rocks have obviously been uplifted to their present elevation by the forces that formed the mountain. As another example, the oldest

sedimentary rocks that contain abundant iron minerals that are rich in oxygen have been taken to indicate when oxygen first built up to relatively high levels in Earth's atmosphere. These rocks are about 2 billion years old.

A few families of minerals form most rocks

The **silicates** are minerals containing *silica*, which is silicon bonded to oxygen. They are the most abundant group of minerals in Earth's crust and mantle. In silicates, four negatively charged oxygen atoms form a tetrahedral structure around a smaller, positively charged silicon atom. Figure 2-9 illustrates how silicate tetrahedra unite in various ways, usually with other atoms, to form the most prominent silicate minerals of Earth's crust. Most silicate minerals form at high temperatures. Silicates, including

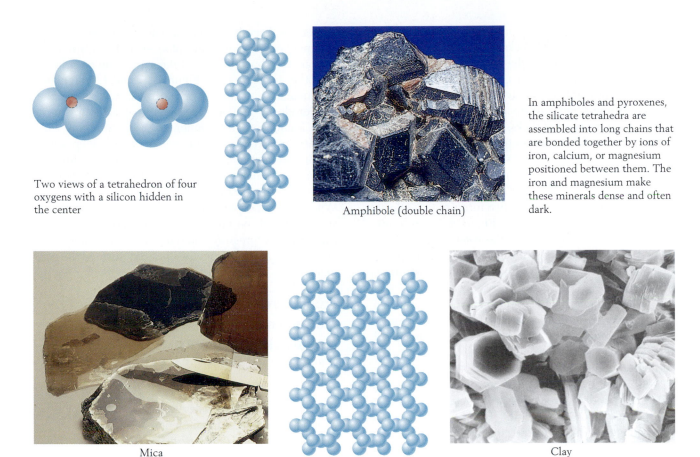

Two views of a tetrahedron of four oxygens with a silicon hidden in the center

Amphibole (double chain)

In amphiboles and pyroxenes, the silicate tetrahedra are assembled into long chains that are bonded together by ions of iron, calcium, or magnesium positioned between them. The iron and magnesium make these minerals dense and often dark.

Mica

Clay

In mica and clay minerals, the silicate tetrahedra are more fully connected to form two-dimensional sheets that are bonded together by sheets of aluminum, iron, magnesium, or potassium. Because the bonds between these sheet silicates are weak, micas and clays cleave into thin flakes. Clay minerals are especially weak and almost always occur naturally as small flakes.

Very complex crystal lattices

Quartz   Pink feldspar

Quartz is the simplest silicate mineral in chemical composition, consisting of nothing but interlocking silicate tetrahedra. Because each oxygen is shared by two adjacent tetrahedra, the mineral as a whole has only two times rather than four times as many oxygens as silicons (the ratio in a single tetrahedron). Hence the chemical formula of quartz is SiO_2. Quartz is very hard because its silicons and oxygens are tightly bonded. Feldspars differ from quartz in that their structure includes both silicate tetrahedra and tetrahedra in which aluminum takes the place of silicon. Ions of one or more additional types (potassium, sodium, and calcium) also fit into the framework in varying proportions. Unlike quartz, feldspars display good cleavage.

Figure 2-9 Rock-forming silicate minerals. Diagrams show the arrangements of silicate tetrahedra (SiO_4) in five rock-forming silicate mineral groups. (Other ions are omitted for simplicity.) (Amphibole, Breck P. Kent; mica, Chip Clark; clay, W. D. Keller; quartz and pink feldspar, Chip Clark.)

feldspar and quartz, are the primary constituents of the igneous and metamorphic rocks of Earth's crust. Because many silicate grains survive the weathering of these rocks and accumulate as sediments, most sedimentary rocks also consist primarily of silicate minerals.

Carbonate and **sulfate minerals** also play large roles in the formation of rocks, but, unlike silicates, most of these minerals form at low temperatures near Earth's surface. Carbonates are constructed of one or more positive ions, such as calcium, magnesium, and iron, bonded to the complex ion CO_3^{2-}. The carbonate minerals calcite and aragonite are two forms of $CaCO_3$ with different crystal structures (see Figure 2-7). **Dolomite** resembles calcite, but half of the calcium ions are replaced by magnesium. Unlike calcite and aragonite, dolomite is not secreted by any organism in the form of a skeleton.

Sulfates are formed of positive ions (such as calcium, iron, or strontium) that are attached to the complex ion SO_4^{2-}. As we shall see later in this chapter, many sulfates also form at low temperatures near Earth's surface through the evaporation of ocean or lake water.

Although **oxides** make up only a small percentage of the large bodies of rock on Earth, these minerals form many important ore deposits. Rocks whose primary components are the oxides magnetite (Fe_3O_4) and hematite (Fe_2O_3), for example, yield most of the iron that is put to human use.

Types of Rocks

Rocks are classified on the basis of their composition and the size and arrangement of their constituent grains. As we saw in Chapter 1, a body of rock belonging to any of the three fundamental types of rocks—igneous, sedimentary, and metamorphic—can be transformed into another body of rock of the same type or either of the other two types (see Figures 1-7 and 1-9). Igneous, sedimentary, and metamorphic rocks can all weather and erode to produce sediment that accumulates to form a new body of sedimentary rock. Similarly, each of the three types of rock, when buried deep within Earth, can melt into magma that later cools to form igneous rock. Finally, each can be altered by high pressures and temperatures to form metamorphic rock. When a metamorphic rock is remetamorphosed, the result will usually be a new kind of metamorphic rock, because the new pressure and temperature conditions will almost always differ from the ones that formed the original rock.

Igneous rocks form when molten rock cools

Igneous rocks are classified according to their chemical composition and grain size, both of which reflect a rock's mode of origin.

Composition and density Recall from Chapter 1 that most igneous rocks of continental crust fall into one of two major groups—felsic or mafic. Felsic rocks, being rich in silicon and aluminum, are generally light-colored and of low density. For this reason, continental crust is predominantly felsic. **Granite** is Earth's most abundant kind of rock (Figure 2-10). Two kinds of feldspar constitute about 60 percent of its volume; one of them gives many bodies of granite a pink color (see Figures 1-2 and 2-9). Felsic rocks also contain large percentages of quartz. Quartz is the silicate mineral with the highest

Figure 2-10 Common igneous rocks.
Basalt (upper left) and rhyolite (lower left) are mafic and felsic extrusive rocks, respectively; they are fine-grained because they cooled rapidly at Earth's surface. Gabbro (upper right) and granite (lower right) are the intrusive equivalents, which are coarser-grained because they cooled slowly at great depth. (Chip Clark.)

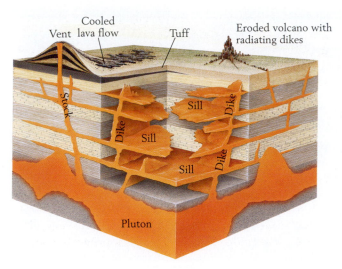

Figure 2-11 labels: Cooled Vent lava flow, Tuff, Eroded volcano with radiating dikes, Stock, Sill, Dike, Dike, Sill, Sill, Dike, Pluton

Figure 2-11 Configurations of extrusive and intrusive bodies of igneous rock on a continent. Magma in large chambers can cool to form massive plutons. Magma that rises through a steeply sloping crack cools to form a dike. Magma that is injected between sedimentary strata cools to form a sill.

concentration of silicon because, as SiO_2, it consists entirely of silicon and oxygen. Feldspar is also relatively rich in silicon, although it also contains aluminum and either potassium, sodium, or calcium, or a mixture of sodium and calcium. Mafic igneous rocks, in contrast, are relatively low in silicon and contain no quartz. An abundance of magnesium and iron makes mafic rocks, such as **gabbro** and **basalt** (see Figure 2-10), darker than felsic rocks, and the iron also makes them denser. Basalt forms most of the oceanic crust, while ultramafic rocks, which are even lower in silicon, form the mantle below the crust. Polished slabs of gabbro are sometimes sold commercially under the misleading name "black granite;" in fact, no granite is black.

Cooling rate and grain size Igneous rocks are also classified according to grain size, which is especially useful because grain size reflects the rate at which igneous rocks cooled from a molten state. If a rock cools slowly, deep within the crust, its crystals can grow large, thus producing a coarse-grained rock. In contrast, rapid cooling, which takes place at Earth's surface, "freezes" molten rock into small crystals that yield a fine-grained rock (see Figure 2-10). Extremely rapid cooling produces glassy volcanic rock known as **obsidian**, one of the kinds of rock that Native Americans and other groups used to form arrowheads and cutting tools.

Most molten rock that cools within Earth's crust or at its surface comes from the mantle; as this molten rock rises, in the form of a blob or plume, it melts crustal rock with which it comes in contact. Molten rock found within Earth is known as **magma**.

The coarse-grained bodies of rock that magma forms when it cools within Earth are referred to as **intrusions**, because they often displace or melt their way into preexisting rocks. They are also called **plutons**. **Sills** are sheetlike or tabular plutons that have been injected between sedimentary layers, and **dikes** are similarly shaped plutons that cut upward through sedimentary layers or crystalline rocks (Figure 2-11).

Molten rock that appears at Earth's surface through an opening, or **vent**, is called **lava**. Lava that has cooled to form solid rock often exhibits "frozen" flow structures on its surface (Figure 2-12). Some lavas erupt from tube-shaped vents to build cone-shaped volcanoes (see Figure 2-11). One such structure is Mount St. Helens, which erupted in Washington State in 1980. A hollow crater forms at the summit of most volcanoes after an eruption, when the unerupted lava sinks back down into the vent and hardens.

Other volcanic rocks form simply by flowing out of cracks, or **fissures**, from which they spread over large areas. These rocks are almost always mafic, because felsic lavas are more viscous (thicker liquids) and do not flow as easily. Mafic extrusive rocks that have flowed widely are often referred to as **flood basalts**. A flood basalt forms the broad Columbia Plateau in the northwestern United States (Figure 2-13). Another, in Siberia, was extruded at virtually the same time as the biggest global extinction of the Phanerozoic Eon. As we shall see in Chapter 15, this volcanic event may have been associated with that extinction.

Lava that emerges from the crust beneath the sea cools rapidly in a way that gives its surface a hummocky configuration, creating rock known as **pillow basalt** (Figure 2-14). Thus pillow structures in ancient basalt indicate that it cooled under water rather than on land.

Volcanic rocks can also form in ways other than by cooling of flowing lava. Some volcanic eruptions—

Figure 2-12 **Recently cooled lava in Hawaii.** The surface of the rock exhibits a ropy structure that reflects the pattern of the lava flow. (Peter Kresan.)

Figure 2-13 Flows of Miocene basalt along the Columbia River near Vantage, Washington. The cliff is about 330 meters (1100 feet) high. (Calvin Larsen/Photo Researchers.)

including that of Mount St. Helens in 1980—are explosive, hurling solid fragments of previously formed volcanic rock great distances. These fragments range from dust size to several meters across. Loose fragments of various sizes settle to form rock known as **tuff**. Although deposited in the same manner as sedimentary rock, tuff is usually classified as volcanic simply because it consists of volcanic particles. In fact, some tuffs form from hot grains that melt together as they settle, but others harden only after water percolating through them precipitates cement. Volcanoes sometimes eject frothy masses of lava that cool quickly to form the glassy rock known as **pumice**. Most pumice is so full of small air pockets that it floats on water.

Rocky material is not all that issues from volcanic vents. Gases of many kinds are also emitted, and in areas such as Yellowstone National Park, steamy geysers shoot up from sites where underground water is heated by magma and vaporized.

Sedimentary rocks form from particles that settle through water or air

Most sedimentary rocks are formed of particles that belong to one of three categories: (1) fragments produced by the weathering and erosion of other rocks, (2) crystals precipitated from seawater, or (3) skeletal debris from organisms.

Fragments of rock produced by destructive processes are termed **clasts**. **Siliciclastic rocks**, then, are sedimentary rocks composed of clasts of silicate minerals.

Erosion and sediment production As we have seen, rocks at or near Earth's surface can be broken down by the processes that constitute *weathering*. There are physical as well as chemical weathering processes. Ice, snow, water, and earth movements are the primary agents of physical weathering. Water expands when it freezes, and when it freezes in cracks and crevices within rocks, it exerts such tremendous pressure that it can split the rocks apart. The products of weathering ultimately move away from the site of their origin under the influence of gravity, wind, water, or ice: the agents of *erosion*.

Destructive chemical processes constitute the most pervasive kind of weathering, and water and watery solutions act as their primary agents. Water at Earth's surface readily converts feldspar to clay, for example, carrying away some ions in the process. Like micas, clay minerals are sheet silicates (see Figure 2-9). Clay differs from mica, however, in that the molecular structure of its sheets is weaker. Thus clay minerals form only very small flakes, instead of large sheets like those that typify mica; this is why clay sediments have fine-grained textures.

Figure 2-14 Pillow basalt. This pillow basalt formed on Española, one of the Galápagos islands, more than 3 million years ago, when lava cooled beneath the sea. Most of the "pillows" are less than 1 meter (3 feet) across. (S. Stanley.)

Quartz, in contrast to feldspar, is quite resistant to weathering. This characteristic, and the resistance of quartz to abrasion, accounts for the abundance of sand on Earth's surface. Most sand consists of quartz grains that are similar to or slightly smaller in size than the quartz grains of granites and other crystalline rocks, from which most of them are derived.

When the feldspar grains of a crystalline rock such as granite weather to clay, the rock crumbles, releasing both flakes of clay and grains of quartz as sedimentary particles. Rainfall or the meltwaters of snow or ice wash many of these particles into streams, from which they are carried to larger bodies of water. Eventually many of the particles settle from the waters of rivers, lakes, or oceans as sediment, which in time may become hard sedimentary rock.

Both water and oxygen take part in the weathering of mafic rocks, converting these iron-rich minerals to clay and iron oxide minerals that resemble rust. In the process, silica is carried away in solution. Mafic miner-als are generally less stable at Earth's surface than felsic minerals and therefore undergo more rapid chemical weathering. Because most mafic minerals weather soon after exposure to air and water, these minerals are not abundant in most beach sands that fringe oceans. Feldspars, too, are rarely found on sandy beaches, because most turn to clay either within or close to their parent rock.

Siliciclastic sedimentary rocks Siliciclastic sedimentary particles and rocks are classified according to grain size (Figure 2-15). The term **clay** is applied to particles smaller than $\frac{1}{256}$ millimeter. Although this quantitative definition may seem redundant and contradictory, because clays also constitute a family of sheet silicate minerals, the use of the term *clay* in reference to both mineralogy and particle size seldom creates confusion: nearly all clay mineral particles are of clay size, and nearly all particles of clay size are clay minerals. **Silt** is the name given to particles in the next size category,

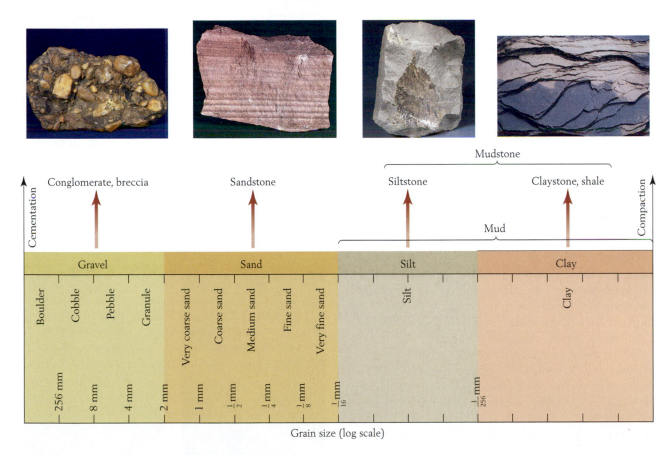

Figure 2-15 Classification of sedimentary rocks according to grain size. Sediments range in size from clay to silt, sand, and gravel; gravel is divided into granules, pebbles, cobbles, and boulders. Gravelly rocks are called conglomerates when their pebbles and cobbles are rounded and breccias when they are angular; these rocks normally contain sand as well. Rocks in which sand dominates are called sandstones.

Rocks in which silt dominates are called siltstones. Rocks formed of clay are called claystone if they are massive and shale if they are fissile (or platy). Siltstones, claystones, and shales are all varieties of mudstone. All of the specimens shown here would fit in the palm of your hand. (Conglomerate, Glenn M. Oliver/Visuals Unlimited; sandstone and siltstone, Breck P. Kent; claystone and shale, D. Cavagnaro/Visuals Unlimited.)

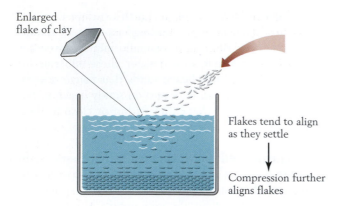

Figure 2-16 Alignment of clay particles. Flakes of clay tend to assume a horizontal orientation as they settle through a body of water and pile up on the bottom. Those that initially lie at an angle tend to be aligned by compaction of the sediment.

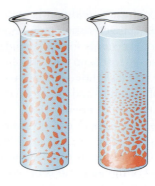

Figure 2-17 The settling pattern of sediment after it is suspended in water. The coarsest sediment settles most quickly and therefore ends up at the bottom of the deposit. The finest sediment settles last.

with diameters between $\frac{1}{256}$ and $\frac{1}{16}$ millimeter. **Mud** is a term that embraces aggregates of clay and silt. Rocks formed largely of mud are termed **mudstones**. Those that consist largely of clay tend to break along bedding surfaces; they are said to be **fissile** and are called **shale**. Fissility results from the tendency of flakes to align horizontally during their deposition (Figure 2-16). Mudstone that contains an abundance of silt is less fissile than shale.

Material of **sand** size ranges from $\frac{1}{16}$ millimeter to 2 millimeters in diameter. When sand is cemented, it becomes sandstone (p. 8). Most sedimentary grains of silt and sand size are composed of quartz. Even though quartz is very hard, quartz grains suffer some abrasion as they bounce and slide downstream along the floors of rivers. In the process, they tend to become smaller and more rounded.

Gravel refers to all particles larger than sand, including **granules**, **pebbles**, **cobbles**, and **boulders**. A rock containing large amounts of gravel is termed a **conglomerate** if the gravel is rounded and a **breccia** if it is angular. In both cases, sand or granules nearly always fill the spaces between the pieces of gravel.

When sediment settles from water, coarse-grained particles settle faster than fine-grained particles, as can be seen when sediments of various grain sizes are mixed with water in a tall glass container and allowed to settle (Figure 2-17). Clay settles so slowly in water that in nature very little of it falls from rapidly moving water, such as that of a stream or wave-ridden shallow sea. Most of it is deposited in calm waters, such as those of lakes, quiet lagoons, and the deep sea. Sand, in contrast, tends to accumulate along beaches and on the bottoms of swiftly flowing streams.

Coarse sediments not only settle more quickly from water than do fine sediments, but they are also less eas-

ily picked up or rolled along surfaces by moving water. Gravel, for example, tends to remain near its source area (the area in which it was originally produced by erosion); thus gravel usually accumulates along the flanks of the mountains from which it has eroded, seldom reaching the center of a large ocean.

Differing degrees of sorting of sediment are seen in Figure 2-18. Grains are said to be *poorly sorted* when they are of mixed sizes; the implication is that moving water did not separate the grains well according to size before they were deposited. Sand along a beach, in contrast, has usually been washed and transported by water currents and waves, and thus it tends to be *well sorted*—that is, its particles tend to fall within a narrow size range. Most of the particles in a handful of beach sand are likely to be either medium- or fine-grained (see Figure 2-15).

Siliciclastic rocks are classified according to composition as well as grain size. Because so many sand-sized

A   B

Figure 2-18 Poorly sorted (A) and well-sorted (B) grains of sand. (A, Rex R. Elliott; B, E. R. Degginger/ Photo Researchers.)

grains are quartz, the word **sandstone** is sometimes automatically interpreted to mean quartz sandstone, but several other kinds of siliciclastic rocks also consist largely of sand-sized grains. One of them is **arkose**, whose primary constituents, grains of feldspar, often give the rock a pinkish color. Because feldspar tends to weather rapidly to clay, arkose usually accumulates only in proximity to its parent rock, soon after the feldspar grains are released by partial weathering and erosion.

Another rock in which sand-sized particles predominate is **graywacke**, so called because it is usually dark gray. Graywacke consists of a variety of sedimentary particles, including sand- and silt-sized grains of feldspar and dark rock fragments as well as substantial amounts of clay. Most of the clay in graywacke was not carried to the environment of deposition in its present state, but was formed by the disintegration of larger grains within the rock.

Thus far we have discussed the nature of siliciclastic sediments, but not the ways in which these loose sediments become hard rock. A variety of processes lithify soft sediments, or transform them into rock. The primary physical process of **lithification** is *compaction*, a process in which grains of sediment are squeezed together beneath the weight of overlying sediment. Muddy sediments usually undergo a great deal of compaction after burial as water is squeezed from them.

The most important chemical process of lithification is *cementation*. In this process, minerals crystallize from watery solutions that percolate through the pores between grains of sediment. The cement thus produced

Figure 2-19 Cement bordering quartz grains in sandstone, as seen in a thin section through a microscope. Here the rounded sand grains appear gray or white in polarized light, depending on their orientation, and the granular calcite cement between the grains is iridescent. (Peter Kresan.)

may or may not have the same chemical composition as the sediment. Sandstone, for example, is often cemented by quartz but more commonly by calcite. Cement can be observed microscopically in slices of rock ground thin enough to transmit light (Figure 2-19). Cementation is less extensive in clayey sediments than in clean sands because after clays undergo compaction they are relatively impermeable to mineral-bearing solutions. Clean sands are initially much more permeable than clays, but their porosity is reduced—and sometimes virtually eliminated—by pore-filling cement.

Cement sometimes gives sedimentary rocks their color. This is often the case for red siliciclastic sedimentary rocks called **red beds**. Some red beds are mudstones, and others are sandstones or conglomerates, but most derive their color from iron oxide, which acts as a cement.

Chemical and biogenic sedimentary rocks The grains of some sedimentary rocks are products of precipitation. **Chemical sediments** result from precipitation of inorganic material in natural waters, sometimes as a result of evaporation. **Biogenic sediments** consist of mineral grains that were once parts of organisms. Some of these grains are pieces of skeletons, such as snail shells or "heads" of coral, and others are the tiny, complete skeletons of single-celled creatures. Most biogenic sediments, however, consist of the skeletal remains of a variety of organisms, rather than just one or two. Because it is not always possible to determine whether a sedimentary rock is of chemical or biogenic origin, we will consider both groups together.

The most common chemical sedimentary rocks are **evaporites**, which form from the evaporation of seawater or other natural water. Many evaporites are massive, well-bedded deposits that consist of vast numbers of crystals. Among the most abundant evaporites are **anhydrite** (calcium sulfate, $CaSO_4$) and **gypsum** (calcium sulfate with two water molecules attached, $CaSO_4 \cdot 2H_2O$). The terms *anhydrite* and *gypsum* refer both to the minerals with these names and to rocks that are composed largely of these minerals. *Halite* is another term that refers both to a mineral and to an evaporite rock. Recall that in regions with high rates of evaporation, evaporites can form even in lakes where ions released from rocks by weathering have become concentrated (p. 20). Salty lakes in Death Valley, California, for example, lie atop halite deposits several kilometers deep that have accumulated over millions of years.

Just as evaporites are readily precipitated from water, they are readily dissolved, so they do not survive long at Earth's surface except in arid climates. Evaporites can survive for long geologic intervals, however, when they are buried deeply enough beneath younger deposits to be protected from groundwater.

Figure 2-20 Chert, a sedimentary rock composed of very small quartz crystals. *A.* Some chert is a chemical sedimentary rock and some is biogenic. *B.* Native Americans made arrowheads from chert, which is known informally as flint. (Breck P. Kent)

A   B

Other types of chemical sediments are less abundant than evaporites. Among the most common of these are chert, phosphate rocks, and iron formations. **Chert,** also called **flint,** is composed of extremely small quartz crystals that have precipitated from watery solutions. Typically, impurities give chert a gray, brown, or black color. Chert breaks along curved, shell-like surfaces (Figure 2-20); Native Americans took advantage of this feature when they fashioned chert into arrowheads. Some cherts occur as scattered, irregular masses in other kinds of sedimentary rocks. These masses, as well as some extensive beds of chert, grow from silica-rich solutions that have moved through rock. Other bedded cherts are thought to have formed by direct precipitation of silica (SiO_2) from seawater. Still others are biogenic deposits that result from the accumulation on the seafloor of the microscopic skeletons of single-celled organisms. These skeletons consist of a type of silica that differs from quartz in that it is amorphous (or noncrystalline). Over long geologic intervals, water percolates through deposits of these skeletons, converting them to very hard chert (Figure 2-21). Cherts older than 100 million years or so have suffered such extensive chemical alteration that many are difficult to identify as biogenic or chemical.

Banded iron formations are complex rocks that consist of oxides, sulfides, or carbonates of iron interlayered with thin beds of chert. Banded iron formations are widespread only in very old Precambrian rocks, and some form large iron ore deposits.

Limestones include both chemical and biogenic bodies of rock. Because they are not as soluble in water as evaporites, limestones are much more common at Earth's surface, where they are quarried extensively for the production of building stone, gravel, and con-

crete. We have already noted that although ancient limestones consist primarily of the mineral calcite, many of their grains were initially composed of aragonite that in time was transformed into calcite (p. 32).

Dolomite is a carbonate mineral that is relatively uncommon in modern marine environments, but is common in many ancient rocks. It differs from calcite in that half of the calcium ions of the crystal lattice are replaced by magnesium ions. In fact, much dolomite has formed by the chemical alteration of calcite. When dolomite is the dominant mineral of rock, the rock, too, is sometimes called dolomite, but is more properly labeled **dolostone.** Because limestones and dolostones are similar rocks and are often found closely associated, they are sometimes referred to collectively as **carbonate rocks.** Similarly, unconsolidated sediments con-

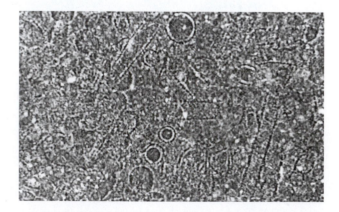

Figure 2-21 Photomicrograph of biogenic chert. "Ghosts" of silicious skeletal elements of sponges and single-celled organisms called radiolarians are still visible in this thinly sliced rock. (E. F. McBride.)

Figure 2-22 **Photomicrograph of limestone.** This rock consists largely of partly dissolved skeletal debris from organisms, together with the calcium carbonate cement that locks the grains together. The skeletal grains are a few millimeters in maximum dimension. (A. J. Copley/Visuals Unlimited.)

sisting of aragonite, calcite, or both are often called **carbonate sediments**.

Carbonate sediments form in two ways: by direct precipitation from seawater and through accumulation of skeletal debris from organisms. Many types of marine life grow shells or other kinds of skeletons that consist of aragonite or calcite. These organisms contribute skeletal material to the seafloor as sedimentary particles. Some of these particles retain their original sizes, while others diminish through breakage or wear. The

product of this biological contribution is an array of clastic particles that are similar to siliciclastic grains and, like them, can be classified according to size. Thus we speak of carbonate sands and carbonate muds.

Although watery solutions tend to dissolve the carbonate grains in limestone, most of the remaining carbonate particles that are sand-sized or larger can be seen to be skeletal particles (Figure 2-22). The origin of the mud-sized material is more difficult to determine. The main constituents of carbonate muds in modern seas are **aragonite needles** (see Figure 2-7B), which are produced by both direct precipitation and the collapse of carbonate skeletons, especially of algae. In ancient fine-grained limestones, aragonite needles have been transformed into tiny calcite grains. The resulting granular texture reveals little about the configuration or mode of origin of the original carbonate particles.

Oolites are sediments consisting of nearly spherical grains (*ooliths*) that can be seen to grow in modern seas by rolling about and accumulating aragonite needles in the same way a snowball rolled about in the snow becomes large enough to make a snowman (Figure 2-23A). Ooliths form in shallow water, where the seafloor is agitated by strong water movements. A cross section of an oolith displays a concentric structure, with thin, dark bands representing breaks in deposition when the grain was at rest and ceased to grow for a time (Figure 2-23B).

Calcium carbonate is precipitated only from seawater that contains relatively little carbon dioxide. Carbon dioxide combines with water to form carbonic acid, thus tending to retain calcium and carbonate ions in solution. Carbon dioxide is less soluble in warm than in cold water, so carbonate sediments accumulate primarily in tropical seas, where winter water temperatures seldom

A

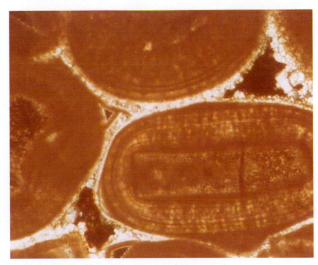

B

Figure 2-23 **Oolite.** *A.* Nearly spherical ooliths, which are the size of sand grains. *B.* A photomicrograph of thinly sliced oolite rock reveals the concentric structure of the ooliths. (*A*, S. Stanley; *B*, A. J. Copley/Visuals Unlimited.)

Figure 2-24 A graded bed. When sediments consist of grains of mixed sizes, the coarser grains settle more rapidly than the finer ones, so that grain size diminishes from bottom to top. (S. Stanley.)

drop below 18°C (64°F). This phenomenon also accounts for the fact that few cold-water organisms secrete massive skeletons of calcium carbonate. Carbonate sediments can also form in freshwater habitats, usually as a result of the carbonate-secreting activities of certain algae.

Unlike siliciclastic clays, carbonate rocks do not compact greatly after burial. Instead, they harden primarily by cementation, in the same manner as siliciclastic sands. Carbonate sediments are nearly always cemented by carbonate minerals simply because rich sources of such cements are close at hand.

Sedimentary structures Distinctive arrangements of grains in sedimentary rocks are termed **sedimentary structures**. Sedimentary structures reflect modes of deposition and provide useful tools for interpreting the environments in which ancient sediments were deposited.

A **graded bed** is a sedimentary structure in which grain size decreases from the bottom to the top (Figure 2-24). This pattern usually results from the normal settling process that characterizes sediments of mixed grain size, with the coarser sediment settling more rapidly than the finer sediment (see Figure 2-17). Most graded beds form in nature when a strong current suddenly introduces a large volume of sediment to a quieter body of water, where it settles.

Ripples formed on a beach have symmetrical cross sections because wave motion oscillates back and forth (see Figure 1-1). Water currents and wind, in contrast, produce ripples and larger ridges—bars or dunes—with asymmetrical cross sections. Sediment accumulates along the lee (downstream) side of such a ridge (Figure 2-25). The accumulation of beds here produces what is termed **cross-bedding** or **cross-stratification** because sets of parallel beds slope at an angle to the horizontal. Although sets of cross-bedding in a single rock unit vary somewhat in their orientation, their average direction of slope indicates the general direction of the prevailing winds when the ancient beds were deposited to form a dune, or of currents when the beds were deposited in a stream.

Mudcracks form as fine-grained, clay-rich sediments dry out and shrink. These structures, which of-

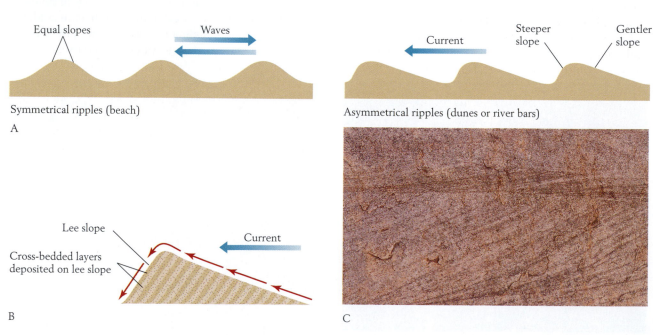

Figure 2-25 Ripples and cross-bedding. *A.* Profiles of ripples produced by waves and by currents. *B.* The formation of cross-bedding on the lee slope of a ripple. *C.* Cross-bedding in coarse sandstone, formed as in *B*, by currents flowing toward the left. (*A* and *B* after F. Press, R. Siever, J. Grotzinger, and T. H. Jordon, *Understanding Earth*, W. H. Freeman and Company, New York, 2004; *C*, Peter Kresan.)

ten form patterns like the hexagonal tiles of a bathroom floor, are visible in ancient rocks even if sediment fills them. Mudcracks indicate deposition in shallow waters, such as those at the edge of a lagoon or lake, which receded and exposed the sediments to the air.

Metamorphic rocks form from other rocks at high temperatures and pressures

Recall that metamorphic rocks form by the alteration of other rocks at temperatures and pressures that exceed those normally found at Earth's surface. Metamorphism alters rocks without melting them, whereas igneous rocks, which also form at high temperatures, are products of the cooling of melted rock. Metamorphism alters both the composition and the texture of all kinds of rocks—igneous, sedimentary, and metamorphic. Mineral assemblages of metamorphic rocks serve as critical "thermometers" and "barometers," varying with the temperature and pressure of metamorphism. **Grade** is the word used to indicate the levels of temperature and pressure of metamorphism. High-grade assemblages of metamorphic minerals form at higher temperatures and pressures than low-grade assemblages.

Regional metamorphism As its name implies, **regional metamorphism** transforms deeply buried rocks over distances of hundreds of kilometers. Igneous activity usually extends along the length of an actively forming mountain chain, supplying heat for metamorphism. Extending outward along each side of such an igneous belt is a zone of regional metamorphism produced by high temperatures and pressures. Most rocks in zones of regional metamorphism display a texture known as **foliation**, which is an alignment of platy minerals caused by the pressures applied during metamorphism.

The grade of metamorphism in a regional metamorphic zone typically declines away from the neighboring belt of igneous activity. The following three types of rocks represent different grades of metamorphism. **Slate** is a fine-grained rock of very low metamorphic grade in which foliation produces fissility much like that of shale (Figure 2-26A). Slate broken into rectangular plates covers the roofs of many houses. **Schist** is a low- to medium-grade metamorphic rock that consists largely of platy grains of minerals, often including mica; because of its strong foliation, schist tends to break along parallel surfaces. **Chlorite** is a micalike green mineral that occurs primarily in schist. When abundant, it gives the rock the informal name *greenschist* (Figure 2-26B). **Gneiss** is a high-grade metamorphic rock whose intergrown crystals resemble those of igneous rock, being granular rather than platy, but whose minerals tend to be segregated into wavy layers (Figure 2-26C).

Not all rocks in regional metamorphic zones are foliated. Some have homogeneous granular textures, which indicate not only that their interlocking mosaics of crystals lack preferred orientations but also that certain minerals are not segregated into bands. **Marble** and **quartzite**, for example, are usually homogeneous granular metamorphic rocks (Figure 2-27). Marble consists of calcite, dolomite, or a mixture of the two minerals, and it forms from the metamorphism of sedimentary carbonates. Marble is popular as a decorative stone, not only because impurities often create attractive patterns within it but also because it is relatively soft and easy to cut and polish. Quartzite, consisting of nearly pure quartz, is much harder than marble; it forms from the metamorphism of quartz sandstone.

Contact metamorphism When an igneous intrusion "bakes" surrounding rock (see Figure 1-9), the result is **contact metamorphism**. High temperature plays a larger role in this process than high pressure. The resulting rocks are usually fine-grained. Contact metamorphism may occur deep within Earth or near the surface. Though a more localized phenomenon, it resembles regional metamorphism in displaying a gradient: the grade of metamorphism declines away from the heat source.

A   B   C

Figure 2-26 Foliated metamorphic rocks. Slate (A) is a low-grade metamorphic rock, schist (B) is a medium-grade rock, and gneiss (C) is a high-grade rock. (A, Breck P. Kent; B, S. Stanley; C, Visuals Unlimited.)

A

B

Figure 2-27 Nonfoliated metamorphic rocks. *A.* A marble quarry in Ireland. *B.* Quartzite from Australia. (*A*, Tom Bean; *B*, Joyce Photographics/Photo Researchers.)

Hydrothermal metamorphism The percolation of hot, watery fluids through rocks can result in **hydrothermal metamorphism**. Fluids of this type escape when magma intrudes continental crust, but most hydrothermal metamorphism takes place along mid-ocean ridges, where seawater circulates through the hot, newly formed lithosphere. Here basalt (see Figure 2-10) is extensively altered to greenschist (Figure 2-26*B*). Many valuable ore minerals, including gold deposits, have been emplaced by hydrothermal fluids.

Burial metamorphism Rocks are altered when they are buried so deeply that they are exposed to temperatures and pressures high enough to change their chemical composition. For siliciclastic rocks, the consequences of this **burial metamorphism** are similar to those of regional metamorphism.

When the original material of burial metamorphism is plant debris, the product can be coal (Figure 2-28). **Coal** is rock formed by the low-grade metamorphism of stratified plant debris. It can be burned because organic carbon compounds account for more than 50 percent of its composition. The starting material for the formation of coal is **peat**, leafy and woody plant tissue that accumulates in oxygen-deprived water where few bacteria are present to cause decay. The heat and pressure of burial turn peat into brown coal called *lignite* through compression and expulsion of water, hydrogen, and nitrogen. This metamorphism increases the carbon content of the organic material, and deeper burial continues the process, producing soft coal and eventually hard coal. The higher the carbon content of coal, the more energy it produces when it burns, because burning is rapid oxidation of carbon (chemical combination of carbon and oxygen).

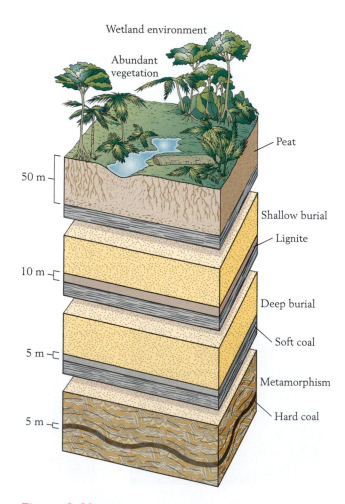

Figure 2-28 The origin of coal. Peat, which accumulates in water where little oxygen is present, becomes compressed and heated through burial, eventually becoming hard coal.

The upper limit of metamorphism When conditions become so hot that a rock melts, metamorphism ceases. Rocks that form when the resulting molten material cools are then classified as igneous. Very high-grade meta- morphic rocks that come close to a molten state before cooling resemble coarse-grained igneous rocks in being granular and lacking foliation. In fact, these metamor- phic rocks are difficult to distinguish from igneous rocks.

Chapter Summary

What traits of minerals determine their physical properties?

Minerals are elements or chemical compounds that are the building blocks of rocks. An isotope is a form of an element that differs in number of neutrons (and, hence, in atomic weight) from other isotopes of the same element; unstable isotopes decay to form other isotopes, making minerals that contain them radioac- tive. A mineral's hardness is determined by the strength of the chemical bonds of its crystal lattice; its density is determined by the weight of its atoms and by their degree of packing within the crystal lattice; and its pattern of fracturing reflects the configuration of its crystal lattice. Most rocks of Earth's crust and mantle are formed by a few families of minerals, of which the most abundant is the silicate family, in which the ba- sic building block is a silicon atom surrounded by four oxygen atoms.

What conditions produce various kinds of igneous rock?

Mafic igneous rocks are denser than felsic igneous rocks, primarily because they are rich in minerals that contain iron, which has a high atomic weight. Intrusive igneous rocks are coarse-grained because they form from magma that cools slowly below Earth's surface, al- lowing time for the growth of relatively large crystals. Extrusive igneous rocks are fine-grained because the molten lava from which they form cools and solidifies quickly at Earth's surface.

What are the ways in which sedimentary rocks form?

Siliciclastic sedimentary rocks consist of silicate grains that are products of weathering and erosion. Weathering of granite, the most abundant kind of rock in continental crust, releases grains of quartz sand, the dominant con- stituent of sandstone, and clay, the dominant constituent of shale. Chemical sediments are formed by precipita- tion of inorganic material from natural waters; for exam- ple, evaporites, or salt deposits, precipitate from water as it evaporates. Biogenic sediments consist of mineral grains that were once parts of the skeletons of organisms. The most abundant biogenic sedimentary rock is lime- stone, which consists largely of calcium carbonate; limestone also forms by precipitation of carbonate from seawater. Some geologically young limestones consist of the chemically unstable mineral aragonite, but old lime- stones consist of the more stable mineral calcite. Arrange- ments of grains, known as sedimentary structures, reflect the mode of deposition of sedimentary rocks.

How do metamorphic rocks form from other rocks?

Metamorphic rocks vary in composition depending on the degrees of heat and pressure responsible for their origin. Regional metamorphism alters rocks over hun- dreds of kilometers, whereas contact metamorphism re- sults when the heat of an igneous intrusion bakes local rock. Hydrothermal metamorphism results when hot, watery fluids percolate through rocks. Burial metamor- phism of plant debris produces coal.

Review Questions

1. How does an ion form from an uncharged atom?

2. What allows one element to substitute for another in the crystal lattice of a particular mineral?

3. How does the orientation of a dike of igneous rock differ from the orientation of a sill?

4. Under what circumstance does the eruption of lava produce a flood basalt instead of a volcano?

5. Why do most sediments that form from weathering consist of silicate minerals?

6. How does the origin of clay particles from igneous rock differ from the origin of grains of quartz sand from the same kinds of rock?

7. How does the process that turns clay into shale differ from the process that turns sand into sandstone?

8. Why do evaporite deposits weather quickly?

9. How do the following sedimentary structures form? (1) Graded beds. (2) Ripples. (3) Cross-bedding. (4) Mudcracks.

10. How does the origin of a metamorphic rock differ from the origin of an igneous rock?

11. Water plays a major role in the origin of sedimentary rocks. Use the Visual Overview on page 28 as a guide to explore the ways in which water serves as a medium for the production, transport, deposition, and lithification of the various kinds of mineral grains that form sediment.

A living chambered nautilus swimming close to the seafloor. It swims by squirting water from a tube that is visible beneath its tentacles. (Douglas Faulkner/Photo Researchers.)

CHAPTER 3

The Diversity of Life

In the course of geologic time, many of the organisms that have inhabited Earth have left a partial record in rock of their presence and their activities. This record reveals that life has changed dramatically since it first arose. At times of sudden environmental change, many forms of life have died out. After each such crisis, evolution has produced many new forms of life. This chapter introduces the major groups of organisms that have evolved on Earth, some extinct and others still flourishing.

It is not easy to define *life* precisely, but two attributes that are generally regarded as essential to life are the capacity for self-replication and the capacity for self-regulation (criteria that viruses do not meet). On Earth today, all entities that are self-replicating and self-regulating are also cellular: they consist of one or more discrete units called cells. A living **cell** is a membrane-bounded module with a variety of distinct features, including structures in which certain chemical reactions take place. The chemical "blueprint" for a cell's operation is encoded in the chemical structure of its genes. Essential to this blueprint is the cell's built-in ability to duplicate its genes so that a replica of the blueprint can be passed on to another cell or to an entirely new organism. The fossil record reveals that life existed 3.8 billion years ago; it may have originated hundreds of millions of years earlier, but the qualities that define it have not changed.

Fossils and Chemical Remains of Ancient Life

Most of our knowledge about the life of past intervals of geologic time is derived from **fossils**, the tangible remains or signs of ancient organisms that died thousands or millions of years ago. Few fossils consist of materials that can survive the high temperatures at which igneous and metamorphic rocks form. Consequently, almost all fossils are found in sediments or sedimentary rocks. Fossils are especially abundant in sedimentary rocks that were formed in the ocean, where animals with skeletons abound.

Hard parts are the most commonly preserved features of animals

The most readily preserved features of animals are "hard parts"—teeth and bones of vertebrate animals and comparable solid mineralized skeletal structures of invertebrate animals. Many groups of invertebrates lack skeletons and therefore have left poor fossil records or none at all. Some,

Visual Overview
The Six Kingdoms

Relationship among taxa that constitute the Archaebacteria, Eubacteria, Protists, Plants, Fungi, and Animals.

PROTOZOANS WITH SKELETONS

Radiolarians

Foraminifera

Protozoans are animal-like protists.

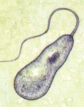

Amoebas

Zooflagellates

Ciliates

Many Eubacteria break down cells and tissues of dead organisms; others cause diseases.

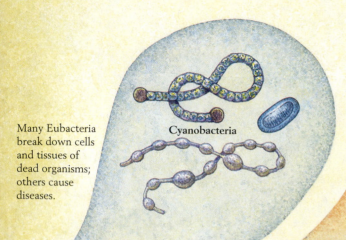

Cyanobacteria

EUBACTERIA

Protozoans *consume* food.

Fungi absorb food from dead organisms.

FUNGUSLIKE PROTISTS

PROTISTS

Plantlike protists *manufacture* food.

UNICELLULAR ALGAE

Dinoflagellates

Nannoplankton

Diatoms

ARCHAEBACTERIA

Deep-sea vents

Many Archaebacteria tolerate extreme environmental conditions.

Heat-tolerant bacteria

MULTICELLULAR ALGAE

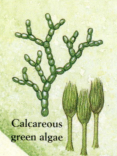

Calcareous red algae

Calcareous green algae

Multicellular algae lack tissues.

Fleshy green algae

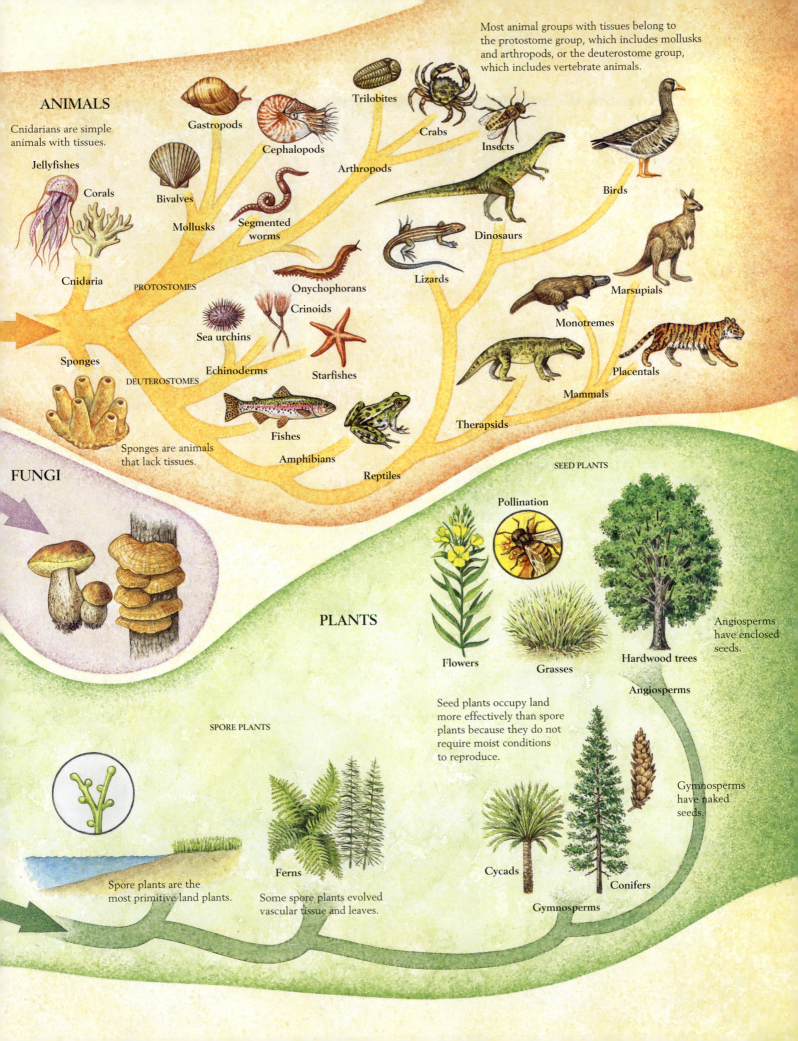

ANIMALS

Cnidarians are simple animals with tissues.

Most animal groups with tissues belong to the protostome group, which includes mollusks and arthropods, or the deuterostome group, which includes vertebrate animals.

Jellyfishes

Corals

Gastropods

Cephalopods

Trilobites

Crabs

Insects

Bivalves

Arthropods

Birds

Mollusks

Segmented worms

Dinosaurs

Cnidaria

PROTOSTOMES

Lizards

Marsupials

Onychophorans

Monotremes

Sea urchins

Crinoids

Placentals

Sponges

DEUTEROSTOMES

Echinoderms

Starfishes

Mammals

Therapsids

Sponges are animals that lack tissues.

Fishes

Amphibians

Reptiles

FUNGI

PLANTS

Pollination

SEED PLANTS

Flowers

Grasses

Hardwood trees

Angiosperms have enclosed seeds.

Angiosperms

Seed plants occupy land more effectively than spore plants because they do not require moist conditions to reproduce.

Gymnosperms have naked seeds.

SPORE PLANTS

Ferns

Cycads

Conifers

Spore plants are the most primitive land plants.

Some spore plants evolved vascular tissue and leaves.

Gymnosperms

Figure 3-1 Fossil crinoids of Paleozoic age. *A.* In life each of these animals was attached to the seafloor by a flexible stalk. *B.* The calcium carbonate skeleton of this specimen has been altered chemically to pyrite, a mineral that consists of iron sulfide and is known as "fool's gold." (Left, Chip Clark; right, courtesy Smithsonian Institution, photo by Chip Clark.)

A   B

however, have internal skeletons embedded in soft tissue; among them are relatives of starfishes called crinoids (Figure 3-1*A*). External skeletons protect other invertebrates, among them bivalve and gastropod mollusks, whose tissues are housed inside skeletons popularly known as seashells. Hard parts are often preserved with only a modest amount of chemical alteration, but at times they are completely replaced by minerals that are unrelated to the original skeletal material (Figure 3-1*B*).

Soft parts of animals are rarely preserved

Fleshy parts of animals, or "soft parts," are occasionally found in the fossil record, but only in sediments that

date back a few millions or tens of millions of years. In older rocks, nothing but chemical residues of organisms remain.

One deposit that is famous for preservation of soft parts is the Geiseltal Formation of Germany, which is more than 40 million years old. In the Geiseltal sediments, which are nearly impermeable to air and water because they are rich in oily plant debris, the skin and blood vessels of long-extinct frogs can still be studied, fossil leaves are still green, and insects retain their iridescent colors. Some animals are preserved here with their last meals remaining only partly decomposed in their stomachs, and some have fur still intact (Figure 3-2). Protection from oxygen is the secret for fossilization of soft

Figure 3-2 Remarkable preservation of soft parts of an Eocene mammal. This creature, which resembled a hedgehog, became buried in the bottom of a lake, where its flesh and fur were partially preserved. This fossil from the Geiseltal Formation is about 40 centimeters (16 inches) long. (Nature Museum, Senkenberg, Germany.)

tissue: it is most likely to be preserved when organisms are buried in fine-grained, relatively impermeable sediment, especially if oily water-repellent organic matter is also present. The kinds of bacteria that cause the decay of soft tissue cannot live in the absence of oxygen.

Permineralization produces petrified wood

Terrestrial plants do not generally have mineralized skeletal structures, but the cellulose walls of their cells are so rigid that woody tissue and even leaves are much more commonly preserved in the fossil record than is the flesh of animals. After plants are buried in sediment, the spaces left inside the cell walls of woody tissue may be filled by inorganic materials—most commonly chert (finely crystalline quartz). Among other kinds of fossils, this process, known as **permineralization**, produces what is informally called petrified wood.

Molds and impressions are imprints

Sometimes solutions percolating through rock or sediment dissolve fossil skeletons, leaving a space within the rock that is a three-dimensional negative imprint of the organic structure, called a **mold** (Figure 3-3). If it has not been filled secondarily with minerals, a paleontologist can fill the mold with wax, clay, or liquid rubber to produce a replica of the original object.

Fossils called **impressions** might be viewed as squashed molds. Impressions usually preserve in flattened form the outlines and some of the surface features of soft or semihard organisms such as insects or leaves (Figure 3-4). A residue of carbon remains on the surface of some impressions after other compounds have been lost through the escape of liquids and gases. This process of carbon concentration is known as **carbonization**.

Figure 3-3 Preservation of brachiopods. Brachiopods are animals with two shell halves that meet along a hinge. This surface of a Paleozoic rock contains fossil brachiopod shells (*S*) as well as molds of the internal surfaces of shells (*M*). (Ken Lucas/Visuals Unlimited.)

Trace fossils are records of movement

Tracks, trails, burrows, and other marks left by animal activity are known as **trace fossils** (Figure 3-5). Trace fossils can reveal a great deal about the behavior of extinct animals, even though the animal that made a particular trace cannot always be identified. Trackways of dinosaurs, for example, show that these animals were very active. Unlike modern reptiles, they customarily moved about at a fast pace. Farther back in the geologic record, preserved tracks and burrows reveal how

Figure 3-4 Carbonized fossil leaf impression from the Jurassic Period. This leaf belongs to the group of plants known as cycads. (Martin Land/Science Photo Library/Photo Researchers.)

Figure 3-5 Dinosaur tracks in Connecticut. A dinosaur formed these tracks by walking across wet mud that subsequently hardened into rock. (Dinosaur State Park, Rocky Hill, CT.)

the earliest animals colonized the floors of ancient seas at a time when very few kinds of animals left fossils that revealed the shapes of their bodies.

The quality of the fossil record is highly variable

Although fossils are found with great frequency in many sedimentary rocks, many kinds of animals and plants that have existed in Earth's history have never been discovered as fossils. Rare species and those that lack skeletons are especially unlikely to be found in fossilized form. Even most forms with skeletons have left no permanent fossil record. A variety of processes destroy skeletons. Animals that scavenge carcasses, for example, may splinter bones in the process. Also, many bones, teeth, and shells are abraded beyond recognition when they are transported by moving water before finally becoming buried. Even after burial, many fossils fail to survive metamorphism or erosion of the sedimentary rocks in which they are embedded. Finally, many kinds of fossils remain entombed in rocks that have never been exposed at Earth's surface or sampled by drilling operations. Because sediments accumulate on nearly all parts of the seafloor, marine organisms are more frequently preserved than terrestrial plants and animals.

Biomarkers are useful chemical indicators of life

A dead organism that decays within sediment can leave a chemical residue behind. Some residues of this kind, known as **biomarkers**, provide key information about the presence of ancient life. Certain biomarkers show, for example, that organisms more complex than bacteria existed more than 1.7 billion years ago.

Dead organisms decay to form fossil fuels

Organisms usually lose their identity in contributing to *fossil fuels*. Coal is formed by metamorphism of plant debris that accumulates in water as peat (p. 46). Petroleum and natural gas form from smaller particles of organic matter, mostly debris from small floating organisms that sinks to the seafloor and becomes buried, thus escaping total destruction by bacteria. Deep burial exposes such material to elevated temperatures that transform it into liquid and gaseous compounds of carbon and hydrogen. These fluids sometimes accumulate in porous reservoirs within soft sediment or hard rock, from which they can be extracted for human use.

Taxonomic Groups

Most biologists divide life on Earth into six kingdoms (Figure 3-6). Two of these large groups contain bacteria; one is the **Archaebacteria** ("old bacteria") and the other is the **Eubacteria** ("true bacteria"). The organisms in these groups are known as **prokaryotes**. The organisms in the four remaining kingdoms, which contain more complex forms of life, are known as **eukaryotes**. The cells of prokaryotes differ from those of eukaryotes in lacking certain internal structures, including a nucleus to house their genetic material.

Three eukaryotic kingdoms—Plantae, Fungi, and Animalia—are distinguished by their nutrition. Members of the **Plantae**, in a process called **photosynthesis**, harness the sun's energy to produce their own food from water and carbon dioxide. Thus we speak of plants as being **producers**. Members of the **Fungi** and the **Animalia**, in contrast, are **consumers**: they obtain nutrition by consuming the organic material of other forms of life. Whereas most animals digest food outside their cells, but inside a body cavity, and then absorb the products, fungi release enzymes to break down food materials in the external environment and then absorb the food into their cells. Both plants and animals are *multicellular* (many-celled), as are nearly all kinds of fungi.

The fourth eukaryotic kingdom is a receptacle for all forms of life that do not fit within any of the other three kingdoms. This hodgepodge, the **Protista**, is an assortment of simple, mostly *unicellular* (single-celled)

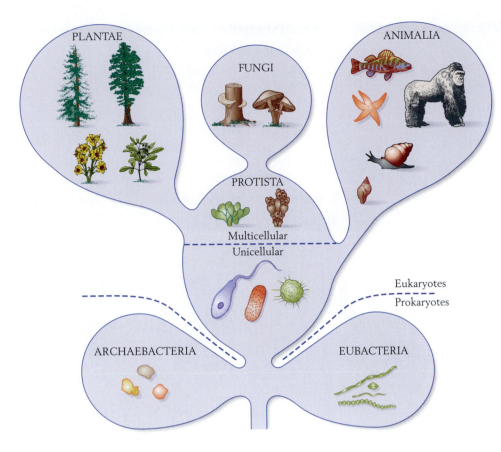

Figure 3-6 The six kingdoms of living things. Archaebacteria and Eubacteria are termed prokaryotes because they lack a nucleus and other cellular features. Those features are shared by the protists, plants, fungi, and animals, which are collectively termed eukaryotes.

species that includes the groups that were ancestral to plants, fungi, and animals. Protists include both producers and consumers.

All six kingdoms—Archaebacteria, Eubacteria, Protista, Plantae, Fungi, and Animalia—are well represented in the fossil record.

The six kingdoms and the numerous subordinate groups into which they are divided are known as **taxonomic groups**, or **taxa** (singular, **taxon**), and the study of the composition and relationships of these groups is known as **taxonomy**.

Taxa within kingdoms range from the broad category known as the phylum (plural, phyla) to the narrowest category, the species (Table 3-1). A **species** is a group of individuals that can interbreed. The basic categories of *higher taxa*—the genus (plural, genera), family, order, class, phylum, and kingdom—are sometimes supplemented by categories such as the subfamily and the superfamily. Names of genera are capitalized, and names of both genera and species are printed in italics. Actually, the name of a species consists of two words, the first of which is the name of the genus to which the species belongs. This scheme of classification, introduced by the Swedish biologist Carolus Linnaeus in the eighteenth century, has the advantage of identifying every species as a member of a particular genus. For

example, the species name of the modern lion is *Panthera leo*. The first name—the genus name—connects this animal to closely related big cats, including *Panthera tigris*, the tiger, and *Panthera pardus*, the leopard. Thus, if we learn that an extinct animal belonged to the genus *Panthera* but was the size of a bear, we know that it was a huge cat. In fact, a lion much larger than today's African lion inhabited North America in the recent geologic past, dying out just a few thousand

---

TABLE 3-1 Major Taxonomic Categories within a Kingdom, as Illustrated by the Classification of Humans

Kingdom: Animalia
　Phylum: Chordata
　　Class: Mammalia
　　　Order: Primates
　　　　Family: Hominidae
　　　　　Genus: *Homo*
　　　　　　Species: *Homo sapiens*

Note: Between these categories, intermediate ones (e.g., superorders, suborders, superfamilies, and subfamilies) are sometimes recognized.

Figure 3-7 The taxonomic position of the human genus, *Homo*, within the order Primates and the family Hominidae. There are four other genera in the superfamily Hominoidea: three ape genera of the family Pongidae and one genus of the family Hylobatidae.

PRIMATES

SUBORDER PROSIMII

Lemurs and their relatives

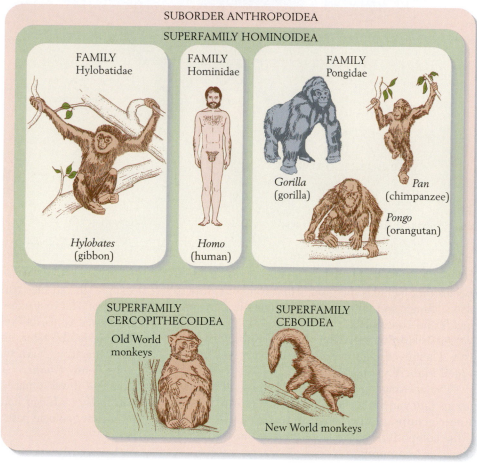

SUBORDER ANTHROPOIDEA

SUPERFAMILY HOMINOIDEA

FAMILY Hylobatidae

Hylobates (gibbon)

FAMILY Hominidae

Homo (human)

FAMILY Pongidae

Gorilla (gorilla)

Pongo (orangutan)

Pan (chimpanzee)

SUPERFAMILY CERCOPITHECOIDEA

Old World monkeys

SUPERFAMILY CEBOIDEA

New World monkeys

years ago. As a very close relative of the modern lion, it is assigned to the genus *Panthera*.

Figure 3-7 illustrates how humans are classified within the order Primates of the class Mammalia. In general, the narrower the taxonomic category, the greater the biological similarity of its members. Humans and gorillas, for example, have enough in common to be assigned to the same superfamily. Monkeys, however, differ sufficiently from these groups to be assigned to other superfamilies. All of these superfamilies are nonetheless similar enough to be united in a single suborder. Often one or a small number of biological features distinguish one higher taxon from closely related taxa of the same rank.

To be united within a higher taxonomic category, a group of species must not only resemble one another, but must also form a discrete portion of the so-called tree of life. The tree of life is actually shaped more like a bush, having many branches, but no central trunk; formally, it is known as the **phylogeny** of life. This structure grows as species arise from others, each of them forming a separate branch. Some of those species die out, but still the phylogeny grows because more species originate than become extinct. Large phylogenies typically display a complex structure. As Figure 3-8 indicates, their species tend to be clustered into groups that share particular traits. Researchers recognize such clusters as higher taxa. Small clusters constitute genera.

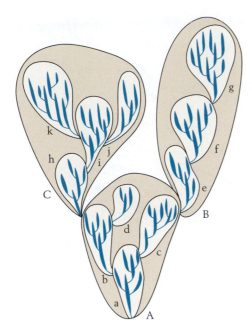

Figure 3-8 Natural clustering of species within phylogenies. Species that form clusters are closely related to one another and have a common ancestry. Such clusters form natural groupings that researchers often designate as higher taxa. The clusters labeled a–k, for example, might be recognized as genera. Those clusters, in turn, form three larger clusters, A, B, and C, each of which might be designated as a family.

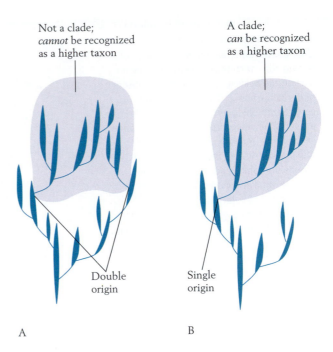

Figure 3-9 A higher taxon must have a single evolutionary origin—it must be a clade. *A.* The group of species in the shaded area does not meet this requirement because not all of the species have a common ancestor. *B.* The group of species in the shaded area does meet the requirement; it is a clade and may be recognized as a higher taxon.

Genera that share particular traits form larger clusters, which may be recognized as families. Even larger groups that are characterized by distinctive features are recognized as still higher taxa, such as orders and classes.

Identifying Clades and Their Relationships

Dividing a phylogeny into genera, families, and even higher taxa is a subjective matter, but one rule must be followed: all of the species within each taxon must be traceable to a common ancestor, as illustrated in Figure 3-9. A cluster of species that share such an ancestry is termed a *clade*.

In the past, efforts to reconstruct phylogenies were based on degrees of resemblance between taxa and on evidence that some taxa were more primitive than others or existed earlier in geologic time. A relatively new technique, now widely employed, instead emphasizes branching events in phylogeny—events that form new clades. A researcher who uses this approach, known as *cladistics*, makes the initial assumption that when two groups share a particular biological trait, both groups

have inherited the trait from a common ancestor. Thus the traits in the two groups are said to be *homologous*.

Certain traits in any biological group are *primitive*, appearing early in the group's evolutionary history. Others are *derived*, having evolved later and occurring in only some of the subgroups. For example, when the six major subgroups of present-day vertebrates are compared (Figure 3-10A), five traits or pairs of traits—jaws, lungs, claws or nails, feathers, and fur together with mammary glands—are found in one or more of these subgroups, but are absent from the hagfishes. These five traits represent derived features, and the hagfishes are a primitive subgroup. A **cladogram** is a diagram depicting the relative phylogenetic positions of the various taxa in a group, as reconstructed from the distribution of derived traits. Figure 3-10A is a cladogram for the major vertebrate subgroups. Jaws are a derived trait that separates all five of the other vertebrate subgroups from the hagfishes. Lungs separate the other four subgroups with jaws from the jawed fishes. The origin of each derived trait, including jaws and lungs, marked a branching point in the evolution of vertebrates. Each of these branching events, by introducing the new trait, produced a new kind of animal.

Figure 3-10 Relationships among groups of vertebrate animals. *A.* This cladogram shows the major subgroups of present-day vertebrates arrayed along an axis on the basis of the presence or absence of derived traits. *B.* The actual, highly generalized phylogeny of mammals. The present-day vertebrate groups are arranged in the order shown in the cladogram (*A*), but this phylogeny also includes two extinct groups, therapsids and dinosaurs, that cladistic analysis shows to occupy intermediate evolutionary positions between surviving groups. Note that the approximate times of origin of the various vertebrate groups, based on fossil occurrences, form a sequence that corresponds to the branching sequence indicated by the cladogram.

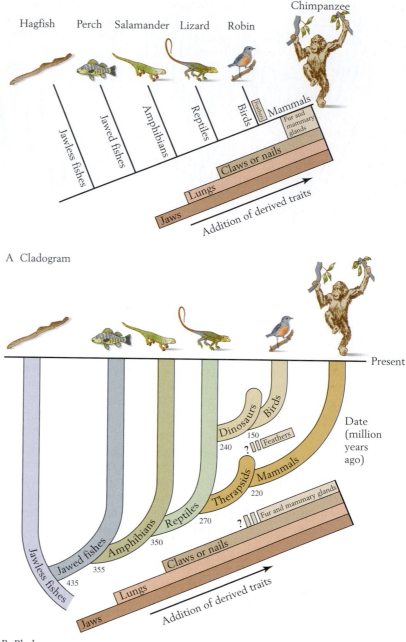

Each of the five kinds of jawed animals shown in Figure 3-10*A* represents a particular class of vertebrates. A cladogram such as this does not depict a complete phylogeny; it simply shows what appear to be the relative evolutionary positions of selected taxa. As a result, a cladogram may not include a clade that is intermediate between two of the clades that it depicts. Figure 3-10*A*, for example, might give one the mistaken impression that mammals evolved directly from reptiles. The fossil record shows that an extinct group of animals known as **therapsids** actually occupied an intermedi-

ate evolutionary position between reptiles and mammals (Figure 3-10*B*). Two observations point to this status: therapsids share with mammals several derived features of the skull, teeth, and limbs that are absent from reptiles; yet they lack other derived features of skull form that are present in mammals.

Anatomical evidence also indicates that birds did not evolve directly from reptiles, as one might conclude from Figure 3-10*A*. As Figure 3-10*B* illustrates, a group of dinosaurs was intermediate between reptiles and birds. Surprising as it may seem, birds evolved from small dinosaurs that apparently used feathers as insulation to retain body heat.

From the intermediate positions of therapsids and dinosaurs depicted in Figure 3-10*B*, we can conclude that one can establish ancestral and descendant relationships in a cladistic analysis only by including all biological groups. Sometimes this task is impossible because a key group is unknown, never having been discovered in the fossil record.

Figure 3-10*B* indicates the geologic age of the oldest known fossil representatives of each of the vertebrate subgroups that are included within it. Note that the relative times of these first appearances are consistent with the relationships indicated by cladistics. Jawed

fishes appear in the fossil record below amphibians, for example; amphibians appear below reptiles, and reptiles below mammals.

Thus far we have examined relationships only among large groups of organisms, such as classes. Thus the phylogeny of vertebrates depicted in Figure 3-10*B* is very generalized. Paleontologists can begin this kind of cladistic analysis at lower taxonomic levels—even at the level of the species. Studies of the evolution of horses provide an illustration.

Figure 3-11 shows the evolutionary relationships among species of the horse genus *Nannippus*. The name of this genus, from the Greek, means "extremely small horse," which is appropriate because typical members

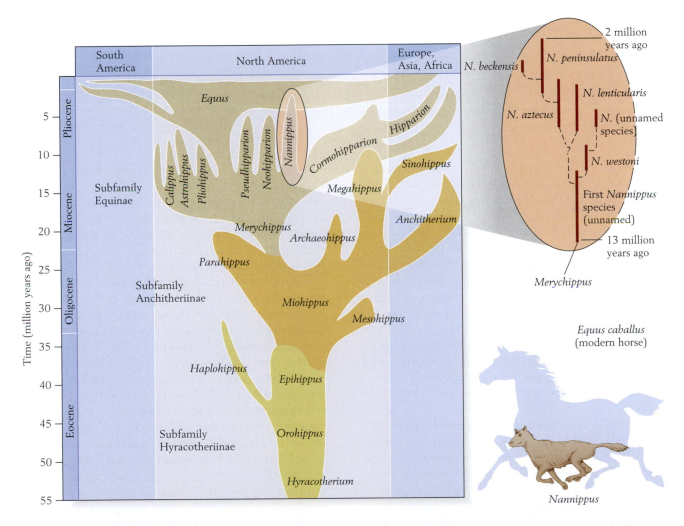

Figure 3-11 The phylogeny of *Nannippus*, a genus of small horses that lived in North America between about 13 million and 2 million years ago. On average, *Nannippus* was about half the height of a modern horse. It was a member of the Equinae, which is the subfamily that includes *Equus*, the genus that includes all living members of the horse family. The fossil record has yielded seven species of *Nannippus*. Cladistic analysis of the bones and teeth of *Nannippus* has produced the phylogeny shown here, in which the ancestry of two species remains uncertain. Dashed lines show inferred evolutionary connections. (After B. J. MacFadden, *Paleobiology* 11:245–257, 1985, and R. C. Hulbert, *Paleobiology* 19:216–234, 1993.

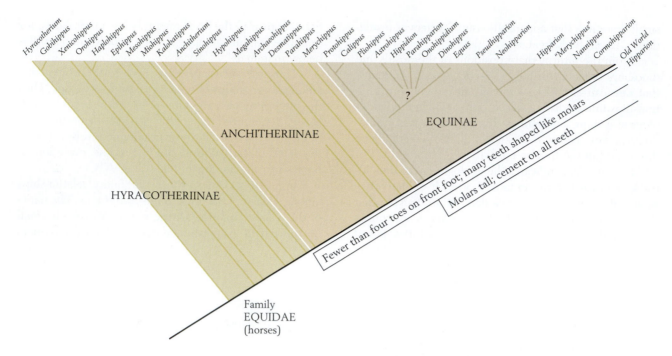

Figure 3-12 Cladogram for genera of the horse family, Equidae. (After B. J. MacFadden, *Fossil Horses*, Cambridge University Press, 1992.)

of *Nannippus* were only about half as tall as a modern horse. They also had three toes on each foot, rather than a single hoof like that of a modern horse.

At the level of the species, as at higher taxonomic levels, cladistic analysis does not always indicate certain ancestry. Thus the question mark beneath *Nannippus aztecus* and *Nannippus lenticulatus* indicates that the ancestry of each of these species within the genus is uncertain. In addition, an imperfect fossil record often leaves us with an underestimation of the total time of a species' existence. Thus one species may have descended from another even though fossils have not yet proved that they overlapped in time. For some evolutionary relationships, the situation is even worse: fossils representing the ancestral species have never been found.

Actually, the fossil record of horses is of relatively high quality and has been so well studied that it effectively illustrates how paleontologists reconstruct phylogenies. The phylogeny of *Nannippus* displayed in Figure 3-11 is part of a much larger phylogeny that has been reconstructed for species of fossil horses. This kind of analysis has clustered horse species into small clades, such as *Nannippus*, that have been designated as genera.

Figure 3-12 is a cladogram for all horse genera, including *Nannippus*, that are recognized by one expert. The phylogeny of the horse family, Equidae, includes three clades that are recognized as subfamilies. The general phylogeny for the Equidae depicted in Figure 3-11 shows the relationships of these three subfamilies.

This figure also shows the positions within the phylogeny of some of the genera, including *Nannippus*. Note that each of the first two subfamilies persisted for a time after giving rise to another subfamily. All present-day species of the horse family, including zebras, wild horses, and wild asses, belong to the genus *Equus*, the name of which is the Latin word for "horse." Figure 3-11 shows that the horse family originated in North America and that most of its members have inhabited that continent. The last of the native horses of North America became extinct a few thousand years ago, however, and horses were absent from the continent until Spanish conquistadors introduced the domestic horse.

Cladistic analysis is not perfectly reliable. Not only does the imperfection of the fossil record produce gaps and uncertainties, but evolution sometimes follows unexpected paths that fool researchers. For example, certain derived traits have actually arisen more than once. In addition, traits have sometimes been lost secondarily and thus are absent from groups of species whose ancestors had possessed them. In general, researchers favor cladograms that minimize the numbers of multiple origins and disappearances of traits. Although gaps remain, we have an accurate picture of the general phylogenies—the relationships among major higher taxa—for many kinds of organisms. In recent years, scientists have used genetic information to construct cladograms in the same way that they construct them from anatomical information. When anatomical and genetic traits yield different cladograms, experts favor the genetic results because an anatomical feature is more likely to have evolved more than once.

In the following sections we will review the basic biological features and interrelationships of major groups of organisms—mainly phyla and classes—that have played large roles in the history of life and left conspicuous fossil records that allow us to assess those roles.

Prokaryotes: The Two Kingdoms of Bacteria

Bacteria gain nutrition in a great variety of ways. Some share with plants the ability to use sunlight to convert chemical compounds into food. Others produce their own food by harnessing chemical energy rather than light. Still others are consumers, absorbing organic compounds from which they gain food and energy.

The fossil record of bacteria extends back more than 3 billion years, which is more than a billion years beyond the record of eukaryotes. Remarkably, many kinds of bacteria found in the modern world appear to differ little from forms that lived early in Earth's history.

Archaebacteria tolerate hostile environments

Several distinctive chemical compounds found within Archaebacteria distinguish this group from Eubacteria, as do certain unique genetic features. Archaebacteria are notable for their tolerance of extreme environmental conditions. Some forms thrive in hot springs (Figure 3-13). One group tolerates only a combination of very high temperatures and extremely acidic conditions. Another lives only in the absence of oxygen. Still another occupies only very salty waters, such as those of the Dead Sea.

Figure 3-13 Archaebacteria that are adapted to high temperatures. A greenish mat of archaebacteria surrounds a steaming hot spring in Yellowstone National Park. (Phil Farnes/Photo Researchers.)

Eubacteria include decomposers, germs, and polluters

Eubacteria are subdivided taxonomically on the basis of the structure of their cell walls. Some cause diseases in plants or animals. Others play a more positive role in the natural world, breaking down the cells and tissues of dead organisms and thus liberating nutrients for the use of other life forms. Many kinds of eubacteria are capable of locomotion.

The **cyanobacteria** are a group of photosynthetic eubacteria that have left an especially important fossil record. Some cyanobacteria are more or less spherical (Figure 3-14A), and others are filamentous (threadlike) in form (Figure 3-14B). Some filamentous types float as greenish scum that pollutes lakes, streams, or the sea.

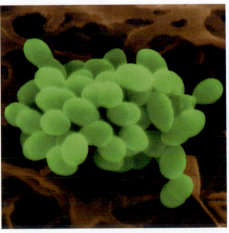

A

B

Figure 3-14 Two genera of eubacteria. A. Nearly spherical cyanobacteria. B. *Oscillatoria* is a photosynthetic form of cyanobacteria. Its filaments, which are about 5 micrometers (μm) in diameter, often form sticky mats. (A, Dennis Kunkel/Visuals Unlimited; B, Sinclair Stamers/Science Photo Library/Photo Researchers.)

Others form mats on the seafloor that can trap sediment to produce distinctive three-dimensional structures. The fossil record of these structures extends back more than 3 billion years.

The Protista: A Kingdom Consisting Mainly of Single-Celled Organisms

Protists include many kinds of single-celled organisms and a few kinds of simple multicellular organisms. Some scientists classify only unicellular organisms as protists, but we will include in this group the multicellular photosynthetic forms informally known as seaweeds. Seaweeds, together with single-celled protists that are also photosynthetic, are known as *algae* (singular, *alga*). Algae are the most prominent producer organisms in lakes and the ocean.

Animal-like protists are known as **protozoans** (Figure 3-15). Among them are amoebas, which have irregular, constantly changing shapes (Figure 3-15A); zooflagellates, which employ a whiplike structure, called a *flagellum*, for locomotion (Figure 3-15B); and ciliates, which move by means of cilia—structures that resemble flagella but are shorter and more numerous, and that beat in unison (Figure 3-15C).

Unicellular algae   Three groups of unicellular algae that float in natural bodies of water are well represented by abundant fossils: dinoflagellates, diatoms, and calcareous nannoplankton. These groups all originated during the Mesozoic Era, and together they are the most prominent producers in modern seas, serving as food for a great variety of animals.

Dinoflagellates employ two flagella for limited locomotion (Figure 3-16A) but are transported chiefly by movements of the water in which they drift. When conditions are unfavorable for survival, some dinoflagellates enter a state of dormancy, armoring themselves in a tough organic structure called a cyst. When conditions improve, they can emerge to resume an active life; some cysts, however, have become fossilized by sinking to the bottom and becoming buried in sediment.

Diatoms are unicellular forms that secrete two-part skeletons of opal, a form of silicon dioxide that differs from quartz in lacking a crystal lattice (p. 42). The two halves of the skeleton fit together like the top and bottom of a cannister (Figure 3-16B). Some diatom species live in lakes and others in the ocean. Most species float, but some lie on the bottom in shallow water where they receive enough light for photosynthesis. At certain times in the geologic past, diatoms have flourished so spectacularly that their skeletons have rained down on the seafloor to produce thick bodies of sediment. Some such sediments have eventually turned into chert; others remain as soft accumulations that are mined to produce the scouring powder we use for cleaning bathrooms.

Calcareous nannoplankton are small, nearly spherical cells that secrete minute, shieldlike plates of calcium carbonate that overlap to serve as armor against small attackers (Figure 3-16C). Nearly all species of this group drift in the ocean, but a few live in lakes. Some concentrated accumulations of plates from calcareous nannoplankton have been weakly lithified to become the fine-grained form of limestone called chalk, which we use to write on chalkboards.

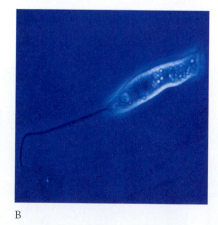
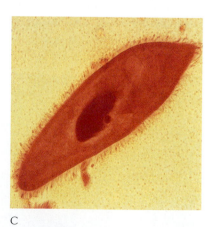

A   B   C

Figure 3-15   **Three types of protozoans.** *A. Amoeba*, which has a variable shape, and as seen here, has a maximum dimension of about 0.5 μm. *B.* A zooflagellate, which (not including the flagellum, which extends to the left) is about 0.1 μm long. *C. Paramecium*, which is surrounded by cilia and is about 0.2 μm long. (*A*, M. Walker/NHPA; *B*, Eric Grave/Photo Researchers; *C*, Michael Abbey/Photo Researchers.)

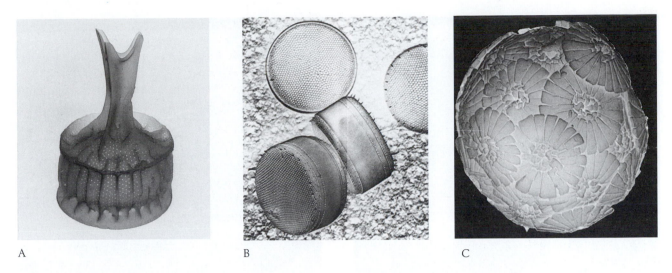

A   B   C

Figure 3-16 Representatives of three major types of single-celled algae found in natural bodies of water today. *A.* A dinoflagellate (about 10 μm in diameter). *B.* A diatom (about 10 μm in diameter). *C.* A cell of calcareous nannoplankton (about 8 μm in diameter). (*A*, Michael Hoban, California Academy of Sciences; *B*, C. L. Stein; *C*, Mitch Covington, Florida Geological Survey.)

Multicellular algae Some multicellular algae drift passively in bodies of water, but most kinds attach to the bottom. Most are soft, fleshy organisms (Figure 3-17A), but some bottom-dwelling members of the groups known as red algae and green algae secrete skeletons of calcium carbonate (Figure 3-17B). Fragments of such skeletons are major constituents of limestones.

Protozoans with skeletons Foraminifera and radiolarians are amoeba-like protozoan groups that secrete skeletons and have fossil records spanning the entire Phanerozoic Eon. **Foraminifera**, nicknamed "forams,"

form a chambered skeleton by secreting calcite or cementing grains of sand together. Long filaments of their protoplasm extend through pores in the skeleton and interconnect to form a sticky net in which they catch food (Figure 3-18). Some forams float in the ocean, and others live on the seafloor. So abundant are their skeletons on shallow seafloors that a few are likely to be present in a handful of beach sand. Foraminifera are widely used to date rocks. Because of their small size, they are especially useful when the samples to be dated are too small to contain many larger fossils. Thus forams are widely used in the search for petroleum, in which the

A   B

Figure 3-17 Green algae. *A.* The fleshy form *Ulva* resembles a bundle of spinach leaves in size. *B.* The calcareous green alga *Halimeda* consists of flat segments less than a centimeter in width. (*A*, Breck P. Kent; *B*, R. J. Goldstein/Visuals Unlimited.)

Figure 3-18 A living planktonic foraminifer. Strands of protoplasm radiate from the skeleton, which is the size of a grain of sand. (Manfred Kage/Peter Arnold.)

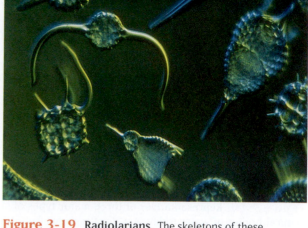

Figure 3-19 Radiolarians. The skeletons of these protozoans, which are made of silica, are the size of a grain of sand. (Eric Grave/Photo Researchers.)

only available rock samples are cuttings from drill cores. Similarly, forams are frequently used to date layers in cores of sediment obtained by drilling into the floor of the deep sea.

Radiolarians are closely related to forams and, like them, capture food with threadlike extensions of protoplasm that radiate from their skeletons. These marine floaters secrete skeletons of noncrystalline silicon dioxide that are perhaps the most beautiful organic structures in the sea (Figure 3-19).

The Fungi: A Kingdom of Decomposers

Fungi absorb most of their food from dead organisms. Most mushrooms, for example, feed on wood, bark, or dead leaves. Fungi typically have filamentous cells. Some groups, including mushrooms, have many cells that are packed tightly into bundles. A few kinds of fungi, including yeasts, have evolved a unicellular condition from multicellular ancestors. Although fungi have a generally poor fossil record because they lack skeletons, those fossils that survive can be quite revealing. An abundance of fungal filaments in sedimentary layers just above those that mark the largest extinction of all time, at the end of the Paleozoic Era, points to a sudden killing off of life across broad regions of Earth: numerous fungi apparently were feasting on the victims.

The Plant Kingdom

The bodies of plants, in contrast to the other kingdoms we have discussed so far, are divided into tissues. A **tissue** is a connected group of similar cells that perform a particular function or group of functions. Plants have a pattern of development that distinguishes them from multicellular green algae, the group from which they evolved (Figure 3-20). Algae shed their eggs and sperm into the water in which they live, so that fertilization is external and offspring develop independently. The egg of a plant, in contrast, is fertilized within the parent plant, and the embryo remains protected there for a time, living within the plant tissue as what amounts to a parasite.

The early evolution of plants from green algae entailed a shift from life in the water to life on land. Early plants were transitional, living in shallows along the margins of bodies of water. Some species became partly emergent, and then others became entirely so. Many basic features of plants relate to problems of living on land. Early plants evolved rigid stems and roots that enabled them to stand upright without the support of water. Roots also provided water and nutrients from the soil, since the entire plant could no longer absorb them from a surrounding watery medium. Some early land plants became *vascular*: they evolved vessels for transport of water, dissolved nutrients, and the food that they manufactured.

Mosses and a few other small, simple plants of the modern world are *nonvascular*: they lack con-

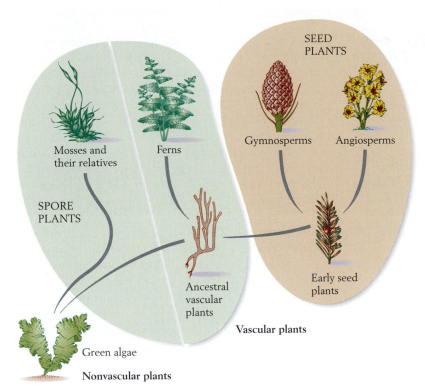

Figure 3-20 Phylogeny of multicellular plants.

ducting vessels and rely on diffusion for transport of materials from cell to cell. Mosses lack multicellular roots; they soak up water and nutrients with tiny hair-like extensions of cells. Such forms have changed little from land plants that lived more than 400 million years ago.

Seedless vascular plants came first

The simplest vascular plant of the modern world is *Psilotum* (Figure 3-21), which lacks roots and leaves and resembles the earliest vascular plants known from the fossil record.

Ferns represent a more advanced level of evolution, having both roots and leaves. Ferns, like other seedless plants, including mosses, reproduce by means of spores. **Spores** are tiny, durable structures that are shed by the conspicuous fern or moss plant that is familiar to us. Whereas this plant has two sets of chromosomes—two copies of genetic material—a spore has just one. As spores drift through the air, they disperse, and, if conditions are favorable, each grows into a tiny adult plant. This tiny plant and the eggs and sperm that it produces each possess a single set of chromosomes (Figure 3-22). A sperm slithers over the moist surface of the tiny plant to fertilize an egg. A fern's reproductive cycle thus entails what is called an *alternation of generations*: a spore-producing generation alternates with one that produces eggs and sperm. Because sperm require moisture to complete their journey, ferns typically occupy moist habitats.

Most groups of spore plants in the modern world are only a few inches tall. Late in the Paleozoic Era, however, several groups grew to the size of trees, and

Figure 3-21 *Psilotum*, **the simplest vascular plant alive today.** This plant, which is the size of a small fern, lacks leaves and roots. The yellow knoblike structures at the ends of stalks are spore-bearing organs. (William Ormerod/Visuals Unlimited.)

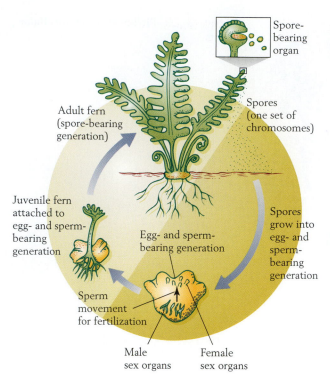

Spore-bearing organ

Adult fern (spore-bearing generation)

Spores (one set of chromosomes)

Juvenile fern attached to egg- and sperm-bearing generation

Egg- and sperm-bearing generation

Spores grow into egg- and sperm-bearing generation

Sperm movement for fertilization

Male sex organs

Female sex organs

Figure 3-22 Life cycle of a fern. The large fern plant with which we are familiar has two sets of chromosomes, but produces spores that have only one. A spore grows into a tiny plant that also has only one pair of chromosomes, but produces eggs and sperm. A sperm fertilizes an egg to produce another large fern plant with two sets of chromosomes, which grows on top of the tiny parent plant.

some of these occupied broad swamps, where their remains accumulated to form peat. Much of this peat turned into the coal that supplies a large part of the electric power that humans use today.

Seed plants invaded dry land

Seed plants include most species of large land plants in the modern world. Seeds are durable structures that disperse the offspring of these plants. Some blow in the wind, and others have barbs that can attach to the fur of mobile animals. Still other seeds are surrounded by fruits that animals can eat, so that the seeds are deposited in feces that serve as fertilizer. The evolutionary origin of the seed during the Paleozoic Era triggered a great ecological expansion of land plants beyond the moist environments required by spore plants.

Gymnosperms The plants known as **gymnosperms** (meaning "naked seeds") produce seeds that are exposed to the environment. Gymnosperms include not only **conifers**, or cone-bearing plants (pine, spruce, and fir trees and their relatives), but also less common

groups such as cycads (see Figure 3-4). The seeds of conifers are exposed on the tops of the scales that together form a cone. After pollen grains, which bear sperm, form on a cone, wind transports them to another cone that has produced eggs, and there, after fertilization, seeds grow and are released. Gymnosperms were the dominant large plants of the Mesozoic Era, when dinosaurs roamed the land. Since early in the Cenozoic Era, however, flowering plants have been more abundant than gymnosperms in most terrestrial regions. Today gymnosperms are the dominant large plants only in cold regions and in some dry, sandy environments.

Angiosperms Flowering plants are more formally known as **angiosperms** (meaning "seeds within a vessel"). Their flowers endow them with special means of pollen transport. The colors and fragrances of flowers attract animals—especially insects and birds—to feed on the nectar they provide and carry away sticky pollen to other plants of the same species, which the animals visit as they continue to feed. Many angiosperms, including hardwood trees, such as oaks and maples, have inconspicuous flowers. Others have showy flowers that attract particular species of pollinators. The seeds of angiosperms have a special advantage over those of gymnosperms: angiosperms engage in double fertilization, which places two eggs in close contact. One egg becomes the seed, and the second grows into a special tissue that nourishes the young plant that grows from the seed. We consume some of this tissue in the form of vegetables such as peas, beans, and corn. Thus the seedling gets off to a good start—one of the reasons that angiosperms greatly outnumber gymnosperms in the modern world.

The Animal Kingdom

Animals are conveniently divided into **vertebrates**, which possess a backbone, and **invertebrates**, which lack one. The bodies of most animals are formed of tissues. Only a few simple invertebrates lack tissues; the largest and most important of these groups is the sponges. All animals more advanced than very simple worms have a body cavity known as a *coelom*, which houses internal organs. These coelomates are divided into two large groups, the *protostomes* and *deuterostomes*, depending on how the mouth forms during embryonic development (Figure 3-23). In protostomes the first opening to form in the embryo becomes the mouth. In most deuterostomes, on the other hand, the first opening to form in the embryo becomes the anus and another opening becomes the mouth; this group includes a few invertebrate groups and all vertebrate animals.

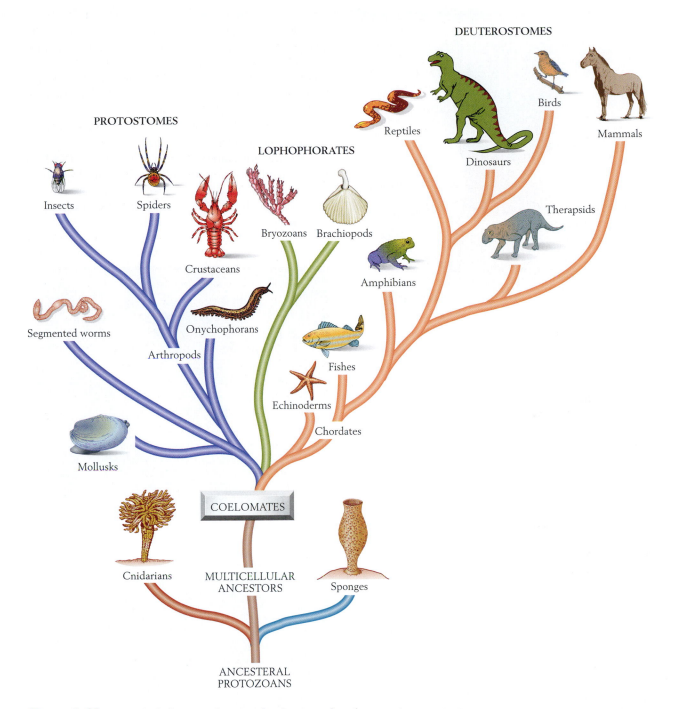

PROTOSTOMES

LOPHOPHORATES

DEUTEROSTOMES

Insects

Spiders

Crustaceans

Bryozoans

Brachiopods

Reptiles

Dinosaurs

Birds

Mammals

Therapsids

Amphibians

Segmented worms

Onychophorans

Arthropods

Mollusks

Fishes

Echinoderms

Chordates

COELOMATES

Cnidarians

MULTICELLULAR
ANCESTORS

Sponges

ANCESTERAL
PROTOZOANS

Figure 3-23 Animal phylogeny, showing the division of coelomates into protostomes, deuterostomes, and lophophorates.

Sponges are simple invertebrates

The simplest animals that are conspicuous in the modern world are the *sponges*. Sponges have irregular shapes and are formed of several cell types that are widely distributed throughout their bodies. Most species live in the ocean, but some live in lakes. All are **suspension feeders**; that is, they strain small particles of food from water. Bacteria are their primary food. Cells that bear flagella pump water into a sponge through numerous pores, and individual cells capture bacteria from currents that pass through internal canals (Figure 3-24). The canals converge, and water exits through one or more central canals. Some sponges have supportive skeletons of tough organic material—the substance of bath

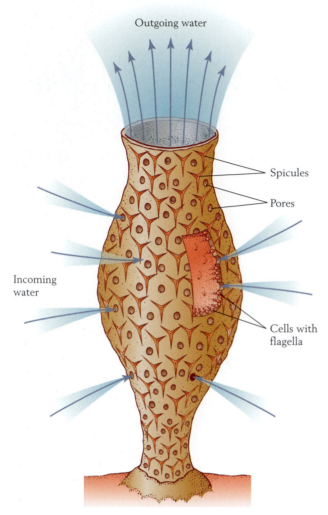

Figure 3-24 Body plan of a simple sponge. A modern sponge is typically a few centimeters tall. (After W. K. Purves, G. H. Orians, and H. C. Heller, *Life: The Science of Biology*, Sinauer Associates, Sunderland, MA, 1995.)

sponges. Others secrete calcium carbonate or silica in the form of small, needlelike or many-pointed elements called *spicules* or as three-dimensional skeletons. Spicules and skeletons give sponges a conspicuous fossil record that extends back to the Cambrian Period.

Cnidarians include the corals

Jellyfishes, corals, and their relatives represent the group known as **cnidarians** (pronounced with a silent c). Most members of this group live in the ocean, but a few occupy freshwater environments. Though they are simple animals, cnidarians exhibit a tissue level of organization (Figure 3-25). Tissues that form an inner and an outer body layer change the shape of a cnidarian's cylindrical body by contracting or relaxing. Between these two layers is a jellylike layer that stiffens the body. Cnidarians are carnivores that catch small animals with

tentacles, which are armed with special stinging cells. The tentacles surround the mouth and pass food through it into a large digestive cavity. Cnidarians have nearly *radial symmetry*—that is, they have no left and right sides or front and back, but face the environment with similar biological features on all sides. Radial symmetry is a typical trait of animals that are immobile or that have no preferred direction of movement.

Corals are cnidarians that have left an excellent fossil record because they secrete skeletons of calcium carbonate. Cnidarians reproduce not only sexually but also asexually, by budding. Some coral species can produce large colonies of interconnected individuals by bud-

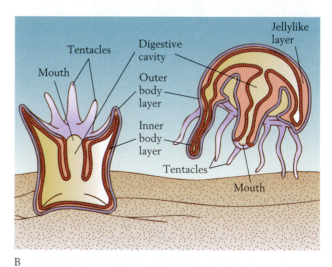

A

B

Figure 3-25 Body plans of cnidarians. *A.* A coral colony, some individuals of which are extending their tentacles from lobes of the colonial skeleton that are the size of a child's finger. *B.* On the left is a cross section of a bottom-dwelling form that resembles a coral but lacks a skeleton. On the right is a cross section of a jellyfish. (*A,* Gerry Ellis/NP Images.)

ding. The colonial skeletons of some corals form reefs that stand above the surrounding seafloor and serve as habitats for a great variety of plants and animals.

Protostomes include most kinds of animals that lack skeletons

Segmented worms **Segmented worms** are complex worms whose bodies are divided into segments. They have a fluid-filled coelom that serves as a primitive skeleton under the pressure of muscular contraction (Figure 3-26). Each segment of the worm has its own coelomic cavity that can expand or contract independently, allowing the animal to move as bulging waves pass from end to end. Among the segmented worms are earthworms, which burrow in soil and feed on organic matter, as well as inhabitants of sand or mud at the bottom of rivers, lakes, or the sea; their burrows are conspicuous in sedimentary rocks hundreds of millions of years old.

Arthropods The **arthropods** include a great variety of living animals, among them insects, spiders, crabs, and lobsters. Although their name means "jointed foot," most arthropods have jointed appendages that serve as legs rather than feet, as well as others that have become modified for such activities as manipulating food, sensing the environment, and swimming. Like segmented worms, which are their ancestors, arthropods have bodies that are divided into segments. The bodies of arthropods are not soft and wormlike, however, but are protected by an external skeleton of stiff organic material

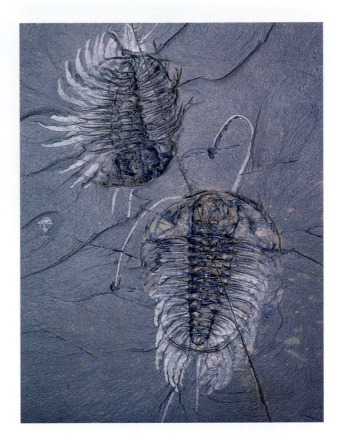

Figure 3-27 **Trilobite fossils.** Not only are the segments and the three lobes of the skeleton visible in these exceptionally well-preserved specimens, but because the animals were buried in sediment that lacked oxygen, their legs (one per segment) and antennae were also preserved. This species is about 10 centimeters (4 inches) long. (Courtesy Smithsonian Institution, photo by Chip Clark.)

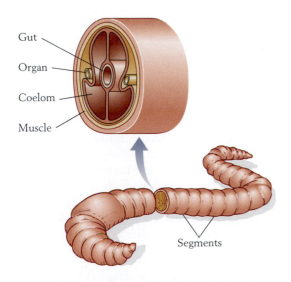

Figure 3-26 **Body plan of a segmented worm.** A fluid-filled coelom lies within the middle body layer. Each segment contains its own coelomic compartments. (After W. K. Purves, G. H. Orians, and H. C. Heller, *Life: The Science of Biology,* Sinauer Associates, Sunderland, MA, 1995.)

that in some forms is strengthened by impregnation with a mineral such as calcite. Like a suit of armor, this jointed skeleton allows for some flexibility of the body inside. Among the many groups of arthropods are the trilobites, crustaceans, and insects.

Trilobites are an extinct group of marine arthropods whose fossils are popular with the general public. They were especially common during the Cambrian Period, the interval during which a great variety of animals with hard parts first appeared on Earth. As their name indicates, trilobites had a body consisting of three lobes, a central lobe and left and right lateral lobes (Figure 3-27). Several of their segments were fused to form a rigid head structure, and several others were fused to form a rigid tail structure. Beneath the heavily calcified external skeleton of most species were many pairs of appendages, each branching to form a gill-like structure for respiration and a leg for locomotion. Trilobites fed on small animals or particles of organic matter. A few types floated in the water or burrowed in sediment, but

Figure 3-28
Onychophorans are intermediate in form between segmented worms and arthropods. *A.* A living onychophoran, which is a terrestrial predator about the length of a pen. *B.* A fossil of a species belonging to this group, from Cambrian strata deposited in a marine environment more than 500 million years ago. (*A*, Michael Fodgen/Bruce Coleman; *B*, courtesy Smithsonian Institution.)

A

B

most crawled over the surface of the seafloor. Many species of trilobites had primitive eyes.

Crustaceans are arthropods with a head formed of five fused segments, behind which are a thorax and an abdomen formed of additional segments. Among the many kinds of crustaceans are lobsters, shrimps, and crabs. Others are small floating creatures. The ocean contains the greatest variety of crustaceans, but many live in lakes or streams, and a few are land dwellers. Because the external skeletons of most of these forms are uncalcified or only weakly calcified, crustaceans have a relatively poor fossil record. Even so, this record extends back to early Paleozoic time.

Insects are a group of arthropods that includes most of the animal species on Earth, yet almost none live in the ocean. Insects breathe air through a system of tubes. Like crustaceans, they have bodies divided into a head, thorax, and abdomen, but nearly all have two pairs of wings (flies have only one). Insects play many roles in nature, but one of the most important is fertilizing plants by transporting pollen from flower to flower. Although insects are preserved in sediments only under unusual circumstances, their fossil record extends far back beyond the first appearance of flowering plants, revealing much about the group's early evolutionary history.

The **onychophorans** contribute greatly to our understanding of invertebrate evolution because they are intermediate in form between segmented worms and arthropods. They are wormlike in shape but, like arthropods, have a series of legs along their bodies (Figure 3-28). These legs are not jointed like those of arthropods, and thus represent an earlier stage of evolution. The fossil record of onychophorans extends back almost to the beginning of the Paleozoic. Early onychophorans lived in the sea. In contrast, modern representatives, which reach about 15 centimeters (6 inches) in length, are terrestrial animals that occupy moist forests, where they feed on other small animals.

Mollusks Snails, clams, octopuses, and their relatives are familiar groups of **mollusks**. Most mollusks have a shell of aragonite, calcite, or a combination of these forms of $CaCO_3$. A *mantle*, which is a fleshy, sheetlike organ, secretes this shell. Most mollusks respire by means of featherlike gills, and many use a filelike structure called a *radula* to obtain food. The major groups of mollusks all have fossil records that extend back to the beginning of the Cambrian. Figure 3-29 shows the anatomical relationships of the various classes of mollusks.

Monoplacophorans are the most primitive mollusks. This group was long considered to be extinct, but in 1952 marine biologists discovered living monoplacophorans on the floor of the deep sea. Members of this class have cap-shaped shells and creep about on a broad foot. Monoplacophorans use the radula to graze on organic matter. It appears that all other mollusks ultimately trace back to monoplacophoran ancestors.

Gastropods, informally termed snails, evolved from monoplacophorans at the beginning of the Paleozoic Era. A snail of the most basic type can be viewed as a monoplacophoran whose body has been twisted so that the digestive tract is U-shaped, with the anus positioned above the head. Snails constitute the largest and most varied class of mollusks. Most are marine, but some have come to live in freshwater environments. Through the evolution of a lung for respiring in air, others have become land dwellers. Some snails, including slugs, have lost their shells. A few types have become suspension feeders rather than grazers, and many marine groups feed on other animals.

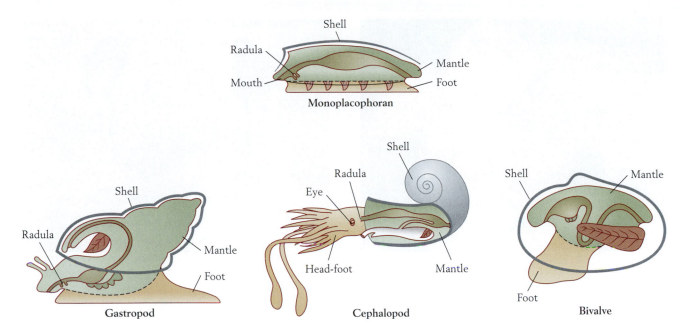

Figure 3-29 Body plans of four groups of mollusks. Monoplacophorans represent the primitive ancestral group from which the other groups were derived.

Cephalopods include squids, octopuses, chambered nautiluses, and their relatives. All members of this molluscan group swim in the sea and feed on other animals. They have eyes and pursue their prey by jet propulsion, squirting water out through a small opening in the body. They capture their prey with tentacles and have a strong, hard beak that they use to eat it. The living chambered nautilus belongs to the most primitive cephalopod group (p. 48). The chambers of its shell are filled partly with water, but also partly with gas, which prevents the animal from sinking to the seafloor. Cephalopods with chambered shells, though not well represented in modern seas, were a diverse group throughout most of the Phanerozoic Eon, and their fossils are widely used to date rocks. Forms known as ammonoids secreted beautiful shells that had more complex partitions between their chambers than those of the living chambered nautilus.

Bivalves, as their name suggests, have a shell that is divided into two halves, known as *valves*. This group includes clams, mussels, oysters, scallops, and their relatives (see Figure 1-10). The head and radula were lost in the evolution of this group, and the shell became folded along a hinge, with each side becoming one valve. One or two short cylindrical muscles pull the two valves together. The single muscle of this type is the part of the scallop that we eat. Some bivalves use the foot to burrow in sediment. Others, including most species of oysters, cement their shell to another shell or a rock or secrete threads to attach themselves to hard objects.

Most bivalves are suspension feeders, but a few kinds of burrowers extract food from sediment.

Coelomates with lophophores are hard to classify

A small number of coelomates feed with a frilly, loop-shaped organ called a *lophophore*. These lophophorate animals have developmental patterns that distinguish them from both protostomes and deuterostomes. Two lophophorate groups, the brachiopods and bryozoans, have extensive fossil records.

Brachiopods Having shells that are divided into two halves, **brachiopods** look superficially like bivalve mollusks, but actually the two groups are not closely related (Figure 3-30; see also Figure 3-3). Brachiopods are sometimes called lamp shells because some resemble artists' portrayals of Aladdin's oil lamp. They employ the lophophore to pump water and sieve small food particles from it. Brachiopods live only in the ocean. They are uncommon today, but they played a major ecological role in ancient seas. In fact, they are the most conspicuous fossils in rocks of Paleozoic age. *Articulate brachiopods* are characterized by teeth that interlock along the hinge between the two valves. Most articulate brachiopods attach to the substratum by means of a fleshy stalk (Figure 3-30A), but some extinct forms lived free on the substratum or cemented their shells to hard objects. *Inarticulate brachiopods* lack hinge teeth. Among

A   B

Figure 3-30 Brachiopods. *A.* Articulate brachiopods attach to hard surfaces with their stalk; most species fall within the size range between a pea and a golf ball. *B. Lingula* is an inarticulate brachiopod that anchors itself in the substratum with its fleshy stalk; it is about the length of a pencil. (*A,* James R. McCullagh/Visuals Unlimited, *B,* A. J. Copley/Visuals Unlimited.)

them is the living genus *Lingula*, which belongs to a group that originated early in the Paleozoic Era. Members of this group live in the sediment, anchored by their fleshy stalk (Figure 3-30*B*).

Bryozoans Although **bryozoans**, or moss animals, form colonies of tiny, interconnected individuals by budding, they are close relatives of brachiopods. Each individual is housed within a chamber in the colonial skeleton (Figure 3-31). Unlike a brachiopod, however, a bryozoan animal extends its lophophore from its skele-

Figure 3-31 Bryozoans. Most of these colonial animals cement their skeletons to hard objects. Individuals within the colony are no more than a few millimeters long; they cannot be distinguished at the scale shown here. (Ed Robinson/Tom Stack & Associates.)

ton in order to feed. Many bryozoans have heavily calcified skeletons, and the group has an excellent fossil record that extends back to the Ordovician Period.

Echinoderms are deuterostome invertebrates

The name **echinoderm** means "spiny-skinned form." The most conspicuous trait of echinoderms, however, is their fivefold radial symmetry—most visible in the five arms that radiate from the central body of a starfish. Echinoderms also possess radial rows of *tube feet*, which are small appendages that terminate in suction cups. All echinoderms are ocean dwellers, and most have an internal skeleton formed of calcite plates that is readily fossilized in marine sediments.

Starfishes With a few exceptions, **starfishes** are flexible animals whose internal skeletal plates are not locked rigidly together. Most starfishes are predators that use their tube feet to grasp their victims, such as bivalve mollusks. When the tension it applies to a bivalve shell spreads the two valves apart slightly, the starfish extrudes its stomach and digests the victim within its own shell. The fossil record of starfishes, though relatively poor, extends back to lower Paleozoic strata.

Sea urchins The rigid skeleton of a **sea urchin** is formed of interlocked plates to which numerous spines attach by ball-and-socket joints. There are two groups of these bottom dwellers. *Regular sea urchins* (Figure 3-32) have radially symmetrical bodies that typically are

Figure 3-32 **A regular sea urchin.** Tube feet that terminate in suction cups extend between the spines. This animal is about the size of an orange. (Brian Parker/Tom Stack & Associates.)

Figure 3-33 **A stalked crinoid feeding in the deep sea.** The body is tilted, and the arms are spread to sieve food from a current. This animal is less than 1 meter (about 3 feet) tall. (David L. Meyer, University of Cincinnati.)

the size of a lemon or orange. By moving their spines, these animals crawl over the substratum with no preferred direction; most graze on algae using a complex feeding apparatus that has five teeth. *Irregular sea urchins*, in contrast, are bilaterally symmetrical, which means that they have a front end and a back end. They are burrowers that have very short spines, and most feed on organic matter in the sediment. Flat irregular urchins that live along sandy beaches are known as sand dollars, and more inflated forms that live farther offshore are called heart urchins.

Crinoids Informally termed sea lilies, **crinoids** sieve food from the water with featherlike arms and pass it to the centrally positioned mouth with tube feet. Most living species are free-living forms that can swim by waving their arms, but some that live in the deep sea resemble fossil forms in being attached to the seafloor by a long, flexible stalk (Figure 3-33). Crinoid stalks are supported by disk-shaped, grooved plates that are stacked together like poker chips (see Figure 3-1).

Chordates include animals with backbones

Vertebrate animals belong to the phylum known as chordates. **Chordates** are defined by their possession of a *notochord*—a flexible, rodlike structure that runs nearly the length of the body and provides support—at least early in the animal's life history, if not in adulthood. The *spinal cord*, a long stem of the nervous system, lies adjacent to the notochord.

Primitive chordates The lancelet is a primitive living chordate in which the notochord is the only skeletal feature (Figure 3-34). It swims like a fish, by flexing its tail back and forth with V-shaped muscles arrayed along its body. Most of the time, however, a lancelet rests partly burrowed into the sediment, sieving small food particles from the water with its gills. In its basic anatomy, the lancelet resembles the ancestors of all modern chordates.

Vertebrates During the maturation of a vertebrate animal, the notochord develops into a vertebral column. This and other skeletal elements are bony in most vertebrates, but in a few groups, including sharks, they consist of cartilage. Because vertebrates are more familiar than many of the animal groups described earlier, they require a less extensive introduction. We have already examined relationships among the major vertebrate

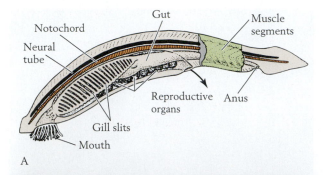

Figure 3-34 A lancelet. *A*. The notochord, adjacent to the neural tube (which resembles a spinal cord), identifies this animal as a chordate. *B*. These two animals, which are about 3 centimeters (>1 inch) long, are feeding while partly buried in sediment. (*A* from W. K. Purves, G. H. Orians, and H. C. Heller, *Life: The Science of Biology*, Sinauer Associates, Sunderland, MA 1995; *B*, Heather Angel.)

Figure 3-35 A conodont tooth and the conodont animal. The tooth (*A*) is just over 1 millimeter long, and the animal (*B*) is about 4 centimeters (2.5 inches) long. Its head faces to the right (*A*, A. J. Copley/ Visuals Unlimited; *B*, Dr. J. K. Ingham, Hunterian Museum & Art Gallery, Glasgow.)

groups (see Figure 3-10). The earliest vertebrates were primitive fishes or fishlike animals that evolved a supple structure to support the body—the jointed vertebral column, to which the skull and skeletal supports for appendages were attached.

For more than a century, paleontologists used small toothlike fossils to date marine rocks of Cambrian through Triassic age without knowing what kind of animal the fossils represented. These fossils were given the descriptive name **conodonts**, meaning "cone-teeth" (Figure 3-35*A*). Conodonts were often found clustered together in rocks in a consistent pattern that had the appearance of a chewing apparatus. Many conodont species were found to have broad geographic distributions, suggesting that the teeth had belonged to animals that moved about effectively in ancient oceans—probably a group of swimmers. For decades paleontologists searched in vain for remains of the conodont animal. Finally, in 1982, a worker who was examining fossil-bearing rocks of Early Carboniferous age from Edinburgh, Scotland, came upon a fossil that represented an elongate, soft-bodied creature about 4 centimeters (about 2.5 inches) in length (Figure 3-35*B*); a conodont tooth apparatus was embedded in one end. These were the fossil remains of an actual conodont animal! The animal's body was formed of a series of V-shaped bundles of muscles, like those of a lancelet (see Figure 3-34) or a modern fish. Two long fins supported by skeletal rays were attached along each side of the body. The conodont animal fitted the description that scientists had previously pieced together for it on the basis of fragmentary and circumstantial evidence: it was a small, predatory swimmer. More recently, scientists have found that the internal structure of the conodont tooth resembles that of the bones and calcified cartilage of vertebrate animals, and conodonts have joined the ranks of the vertebrates.

Fishes were present early in the Paleozoic Era, and partway through that era, they became effective predators through the evolution of jaws. Early jawed fishes, including primitive sharks, had a skeleton composed of cartilage (a trait retained in modern sharks). Others evolved bony external armor. Still others developed a bony internal skeleton, and their descendants include most fishes of the modern world.

Midway through the Paleozoic Era, two groups of bony fishes evolved, each with a distinctive kind of fin. As their name suggests, **ray-finned fishes** have fins that are supported by thin bones that radiate outward from the body. They are the dominant fishes of modern seas, lakes, and rivers and include tuna, barracuda, salmon, and bass. **Lobe-finned fishes**, in contrast, have fleshy fins supported by a complex assembly of heavy bones. Only a few species of lobe-finned fishes survive today. One of the most remarkable of them is the coelacanth, a large, primitive creature that lives off the eastern coast of Africa and was discovered only in 1939 (Figure 3-36).

Lobe-finned fishes played a special role in the history of life. An early group of these fishes evolved into four-legged land animals. Each fin became a limb with toes. The lung, which terrestrial animals use to breathe air, had evolved in the ancestral group of lobe-finned fishes, perhaps as a device for coping at times when the bodies of water in which they lived dried up.

Amphibians were the first four-legged vertebrates to spend their adult lives on land. They laid their eggs in water, however, and spent their early lives there. They grew legs as they matured and then moved onto the land. We see this pattern of development, known as *metamorphosis*, in living amphibians such as frogs, whose juvenile forms are tadpoles that swim with fish-like tails. Most kinds of early amphibians were more similar to salamanders than to frogs, but many were much larger than modern salamanders. Some grew to the size of a large pig.

Reptiles evolved from amphibians by way of another major evolutionary step: the origin of eggs that had protective shells and could survive on dry land. This biological innovation enabled reptiles to invade dry habitats—habitats that had been inaccessible to amphibians because of their dependence on water for reproduction. Surviving reptiles include turtles, lizards, snakes, and crocodiles. Like fishes and amphibians, reptiles are **ectothermic**; that is, the environment exerts control over their internal body temperature.

Dinosaurs were rather closely related to crocodiles, yet many experts do not consider them to have been reptiles. Certain groups of dinosaurs shared many additional anatomical features with birds. In fact, di-

Figure 3-36 A coelacanth. This lobe-finned fish, which lives in moderately deep water, is about 1.5 meters (3 feet) long. (Tom McHugh/Photo Researchers.)

nosaurs have recently been found with feathers attached to their bodies. There is much evidence that most dinosaurs were very active animals, not the slow, lumbering creatures often portrayed.

Birds evolved from a group of dinosaurs during the Mesozoic Era. Birds are **endothermic**; that is, they control their body temperature internally. As we shall see in Chapter 16, dinosaurs probably shared this feature with birds.

Mammals are separated from other vertebrates by a unique set of traits. Not only are mammals endothermic, but they have hair on their bodies for insulation and sweat glands for cooling when they become overheated. They also bear live young and suckle them with sweat glands modified to secrete milk. The teeth of mammals are more highly differentiated than those of reptiles, having a variety of shapes and functions. Mammals also exhibit advanced features for locomotion. In particular, their legs are positioned fully beneath their bodies, rather than extending out from the sides as those of reptiles do (Figure 3-37). Three groups of mammals occupy the modern world, all with histories extending back to the Mesozoic Era.

Monotreme mammals are a small group today, including only the echidna and platypus. These creatures are unusual in retaining the primitive trait of laying eggs, rather than giving birth to live young.

Marsupial mammals bear live young, but their offspring are tiny and immature at birth and grow for a time in a pouch on the mother's abdomen, where they also obtain their milk. Today most marsupials, including kangaroos, inhabit Australia. An opossum is the only marsupial species native to North America.

Placental mammals include the vast majority of living mammal species, including humans. The newborn

A   B

Figure 3-37 **Differences between reptiles and mammals in tooth and limb structure.**
The alligator (*A*) has simple spikelike teeth, whereas the lion (*B*) has long "canine teeth" (the
equivalent of "eyeteeth" in humans) near the front of its mouth, for slashing, and slicing teeth in the
rear, for processing meat. The alligator's legs sprawl outward from its body, whereas the lion's legs
are beneath its body. (*A*, John Green/Visuals Unlimited; *B*, S. R. Maglione/Photo Researchers.)

offspring of placental mammals are much larger and
more mature than those of marsupials and do not spend
their infancy in a pouch. Not until the Cenozoic Era,
when the dinosaurs were gone, did the placentals expand to become the dominant large animals on all continents except Australia, where marsupials came to prevail in the absence of placental competitors.

Therapsids evolved from reptiles and were ancestral to mammals (see Figure 3-10*B*). Living during late
Paleozoic and early Mesozoic time, they were intermediate between reptiles and mammals. They stood more
upright than reptiles, and their teeth were more highly
differentiated, but in neither their posture nor their
tooth patterns were they as advanced as mammals.
Therapsids may have resembled mammals in being
hairy animals that regulated their internal body temperature. Early in the Mesozoic Era, they shared the
world with dinosaurs.

Chapter Summary

What are fossils?

Fossils include hard parts of organisms, which have
sometimes been chemically altered, and molds of those
structures; they also include trace fossils, which are
marks of activity, and impressions of soft parts.

How do scientists arrange organisms in natural groups?

A group of closely related organisms is formally recognized as a taxon. A species is a taxon, and scientists group
species into higher taxa. A phylogeny is a tree of life
produced by evolutionary branching. A species forms a
single branch, and a cluster of branches, traceable to a
single branching event, constitutes a higher taxon.

What is the most fundamental taxonomic division of life?

Organisms are divided into six kingdoms. Two of those
kingdoms, the Archaebacteria and Eubacteria, contain
bacteria, which are prokaryotic, differing from the other
(eukaryotic) kingdoms of organisms in lacking a nucleus
and other structures found in the cells of eukaryotes.
The four eukaryotic kingdoms are the Protista, Fungi,
Plantae, and Animalia.

What kinds of organisms constitute the Protista and Fungi?

Protists include single-celled eukaryotes and fleshy algae,
which are simple multicellular forms. Several groups of

protists have important fossil records. Among them are three groups of floating algae that are major producers in the modern world (dinoflagellates, diatoms, and calcareous algae) and two groups of amoeba-like forms with skeletons (foraminifera and radiolarians). Fungi are simple multicellular forms of life that absorb food from decaying dead organisms.

What are the major features of the Plantae?

The earliest plants to invade the land evolved from algae and, like modern mosses, lacked vessels to conduct fluids through their tissues. Like modern ferns, early vascular plants reproduced by means of spores. Of plants that reproduce by means of seeds, the two largest groups are the gymnosperms, which include conifers (cone bearers) and angiosperms (flowering plants).

What are the major invertebrate members of the Animalia?

Animals that lack backbones are informally termed invertebrates. The simplest of these, including sponges, do not have their cells organized into tissues. Cnidarians, which include corals, are among the simplest animals having tissues. Segmented invertebrates include worms such as earthworms, arthropods, and also onychophorans, which are intermediate in form between the two other groups; of the segmented groups, only the arthropods have an extensive fossil record (including the

jointed skeletons of trilobites and even a wide range of insect remains). The entire Phanerozoic displays a rich fossil record of mollusks—especially of gastropods (snails), cephalopods (relatives of the chambered nautilus), and bivalves (clams, mussels, oysters, scallops, and their relatives). Brachiopods survive in relatively low diversity today, but are the most conspicuous fossils in Paleozoic rocks. Bryozoans, or moss animals, are colonial creatures that have been abundant since early in the Paleozoic Era. Echinoderms—marine animals with radial symmetry—have an excellent fossil record that includes free-living forms (especially sea urchins) and attached forms such as crinoids (sea lilies).

What are the relationships among the major vertebrate groups of the Animalia?

A variety of fishes evolved during the first half of the Paleozoic Era, among them jawless groups as well as jawed groups that included armored forms; ray-finned forms, which include most living fish species; and lobe-finned forms, some of which gave rise to the first land-dwelling vertebrates (amphibians). Reptiles, whose hard-shelled eggs liberated them from reproduction in water, evolved from amphibians late in the Paleozoic Era. Before they died out, the dinosaurs gave rise to birds. Mammals evolved from reptiles by way of the therapsids, which were intermediate in form between the two groups.

Review Questions

1. What conditions favor the preservation of soft parts as fossils within sediment?

2. What are the six kingdoms of organisms, and what traits characterize each one?

3. What is the value of derived traits for the reconstruction of phylogenies?

4. What kinds of organisms are grouped as algae?

5. How did the evolution of certain reproductive features allow early plants to invade the land? Answer the same question for animals.

6. In what ways do animals participate in the reproduction of seed plants?

7. What is a colonial animal?

8. What kinds of animals are included among the arthropods? What kinds are included among the mollusks?

9. How do lobe-finned fishes differ from ray-finned fishes, and why were the ray-finned forms unlikely to give rise to terrestrial animals?

10. How do mammals differ from reptiles?

11. The history of life has been highlighted by evolutionary innovations. Using the Visual Overview (p. 50) and what you have learned in this chapter, identify important evolutionary breakthroughs in the overall phylogeny of life that have distinguished new taxa from their predecessors, and explain the biological significance of each of the features that you identify.

Zebras, springbok antelopes, and wildebeests gather at a water hole in a savannah in southwestern Africa. (Robert W. Hernandez/Photo Researchers.)

CHAPTER 4

Environments and Life

Organisms are able to live only in environments where they can find food, tolerate physical and chemical conditions, and elude natural enemies. These were the requirements for life in the past just as they are in the present. Climate is the environmental factor that exerts the greatest control over the distributions of species throughout the world, influencing conditions not only on land but also in bodies of water. On land, both temperature and moisture strongly affect the distributions of plant species, and plant species in turn influence the distributions of animal species. In this chapter, therefore, we will examine the mechanisms that create the prevailing climatic patterns on Earth today as we explore the relationships between environments and life.

If the planet had undergone little change in the course of geologic time, we could directly apply what we know about life and environments today to ancient fossils and rocks. But Earth has changed dramatically over the course of its history. The planet's materials—including living matter—have changed continually in both composition and location. Life has undergone vast evolutionary changes, and many conspicuous physical features of the modern world—including the polar ice caps of Greenland and Antarctica and the very cold body of water that now forms the deep ocean—were not present 100 million years ago. Although this situation does not violate the principle of uniformitarianism (p. 4), inasmuch as natural laws are not broken, it does require us to take changing environmental conditions into account when we interpret the rock record. What we learn in this chapter will serve as a starting point for our exploration of environments and life through the eons, beginning with the planet's origins and moving forward until we reach the present.

One means of gaining an understanding of environments on Earth is to examine the configuration of the planet's surface. Recall that Earth's crust is divided into the thin, dense oceanic crust and the thicker, less dense continental crust—a distinction that accounts for Earth's external shape, with continental surfaces standing above the seafloor. Currently, about 70 percent of Earth's surface lies below sea level, and most of this area forms the deep-sea floor (Figure 4-1). Those settings on or close to Earth's surface that are inhabited by life are called **habitats**. Nearly all habitats can be classified as terrestrial or aquatic. Aquatic habitats are further divided into marine habitats (e.g., those within oceans and seas) and freshwater habitats (e.g., lakes, rivers, and streams).

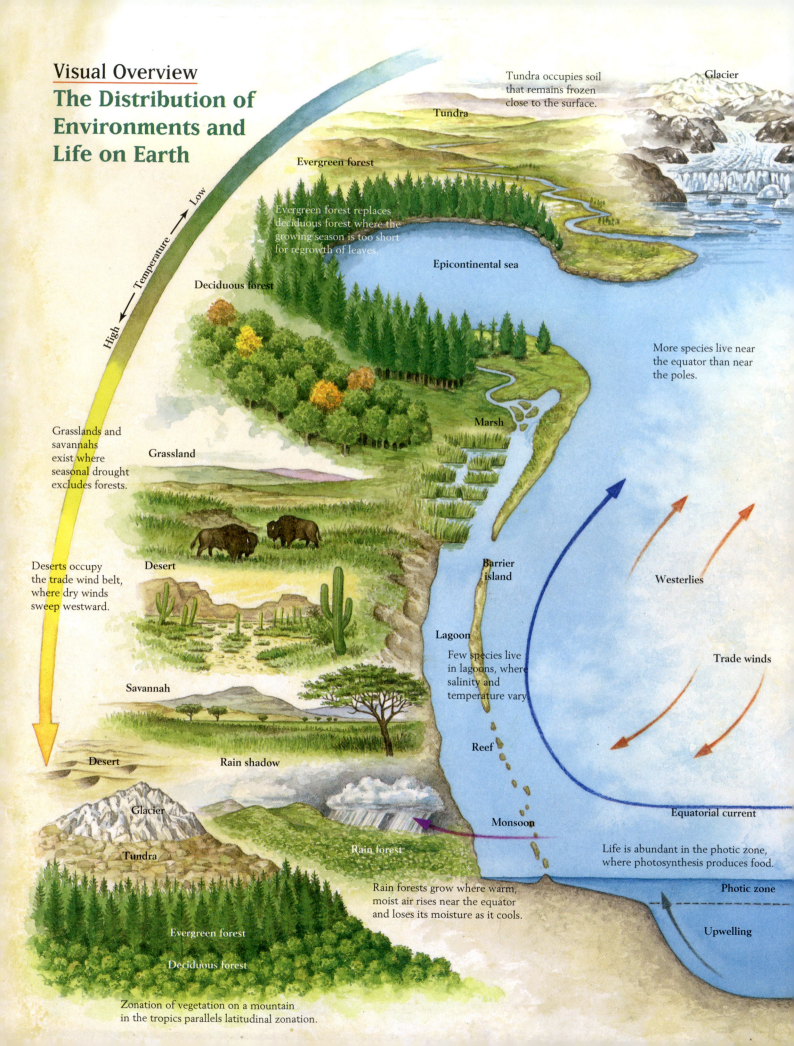

Visual Overview
The Distribution of Environments and Life on Earth

Temperature — Low · High

Tundra occupies soil that remains frozen close to the surface.

Glacier

Tundra

Evergreen forest

Evergreen forest replaces deciduous forest where the growing season is too short for regrowth of leaves.

Epicontinental sea

Deciduous forest

More species live near the equator than near the poles.

Marsh

Grasslands and savannahs exist where seasonal drought excludes forests.

Grassland

Desert

Deserts occupy the trade wind belt, where dry winds sweep westward.

Savannah

Barrier island

Westerlies

Lagoon

Few species live in lagoons, where salinity and temperature vary.

Trade winds

Reef

Desert

Rain shadow

Glacier

Tundra

Monsoon

Equatorial current

Rain forest

Life is abundant in the photic zone, where photosynthesis produces food.

Rain forests grow where warm, moist air rises near the equator and loses its moisture as it cools.

Photic zone

Evergreen forest

Deciduous forest

Upwelling

Zonation of vegetation on a mountain in the tropics parallels latitudinal zonation.

EVERGREEN
FOREST

Fir

Moose

Lynx

TUNDRA

Lichen

Lemming

Caribou or reindeer

GLACIAL
ENVIRONMENT

Polar bear

Lichen

Penguin

DECIDUOUS
FOREST

Maple

Squirrel

Wolf

GRASSLAND

Bison

Snake

Grass

DESERT

Owl

Saguaro

Scorpion

Dry, cool air
descends.

Cold air
descends.

Warm,
moist air
rises.

Tundra

Glacier

Evergreen
forest

Deciduous
forest

Grassland

Desert

Savannah

Rain forest

Westerlies

Trade winds

Equatorial current

Summer

Summer occurs in a
hemisphere when it is
tilted toward the sun.

Winter

RAIN FOREST

Monkey

Gavial

Canopy-forming
tree

SAVANNAH

Acacia

Zebra

Rhino

CORAL REEF

Palm

Organic
matter

Abyssal plain

Many animal species occupy the
deep sea, but the density of life
is low because there is very little food.

SEA

Fish

Killer whale

Shrimp

Figure 4-1 Relative elevations of Earth's surface. The curve shows the relative amounts of land and seafloor that lie at various distances above and below sea level. The plot is cumulative, depicting the total percentage of land that lies above each depth or altitude. About 70 percent of Earth's surface lies below sea level. Continental shelves are borders of continents flooded by shallow seas. As the left side of the diagram shows, mountains account for relatively little of Earth's surface.

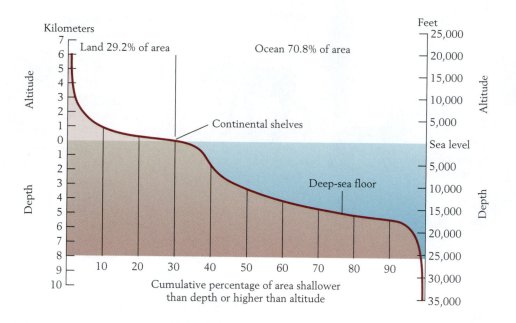

Principles of Ecology

Ecology is the study of the factors that govern the distribution and abundance of organisms in natural environments. Some of these factors are conditions of the physical environment, and others are modes of interaction between species.

A species' niche is its position in the environment

The way a species relates to its environment defines its ecological **niche**. The niche requirements of a species include particular nutrients or food resources and particular physical and chemical conditions. Some species have much broader niches than others. Before human interference, for example, the species that includes grizzlies and brown bears ranged over most of Europe, Asia, and western North America, eating everything from deer and rodents to fish, insects, and berries. The sloth bear, in contrast, has a narrow niche. It is restricted to Southeast Asia, where it feeds mainly on insects, for which its peglike teeth are specialized, and on fruits. The ecological niches of many other closely related species present similar contrasts.

We speak of the way a species lives within its niche as a **life habit**. A species' life habit is its mode of life—the way it obtains nutrients or food, the way it reproduces, and the way it stations itself within the environment or moves about.

Every species is restricted in its natural occurrence by certain environmental conditions. Among the most important of these **limiting factors** are physical and chemical conditions. Most ferns, for example, live only under moist conditions, whereas cactuses require dry

habitats. The salt content of water is a major limiting factor for species that live in the ocean. Few starfishes or sea urchins, for example, can live in a lagoon or bay where normal ocean water is diluted by fresh water from a river.

Almost every species shares part of its environment with other species. Thus, for many species, **competition** with other species for an environmental resource that is in limited supply is a limiting factor as well. Among the resources for which species commonly compete are food and living space. Often two species that live in similar ways cannot coexist in an environment because one species competes so much more effectively than the other that it monopolizes the available resources. Terrestrial plants often compete for water and nutrients; as a result, plant species that live close together often have roots that penetrate the soil to different depths (Figure 4-2).

Predation by one species upon another is another limiting factor. An especially effective predator can prevent another species from occupying a habitat altogether.

A community of organisms and its environment form an ecosystem

A **population** is a group of individuals that belong to a single species and live together in a particular area. Populations of several species living together in a habitat form an ecological **community**. In most ecological communities, some species feed on others. The foundation of such systems consists of producers—photosynthesizing organisms or, less commonly, bacteria that harness the energy of chemical reactions. Consumers consist of **herbivores**, which feed on producers, and **carnivores**, which feed on other consumers. Terrestrial herbivores

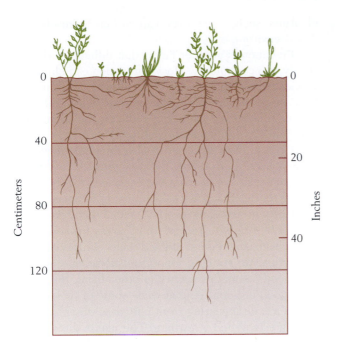

Figure 4-2 Differences in the niches of coexisting species of plants. The roots of different species occupy different depth zones of the soil and thus avoid competing for water and nutrients. (After H. Walter, *Vegetation of the Earth in Relation to Climate and Eco-Physiological Conditions*, Springer-Verlag, Stuttgart, 1973.)

include such diverse groups as rabbits, cows, pigeons, garden slugs, and leaf-chewing insects. Terrestrial carnivores include weasels, foxes, lions, and ladybugs.

The organisms of an ecological community and the physical environment they occupy constitute an **ecosystem**. Ecosystems come in all sizes, and some encompass many communities. Earth and all the forms of life that inhabit it represent an ecosystem, but so does a tiny droplet of water that is inhabited by only a few microscopic organisms. Obviously, then, small ecosystems exist within larger ones. The animals and protozoans of an ecosystem are collectively referred to as its **fauna** and the plants and plantlike protists as its **flora**. A flora and a fauna living together constitute a **biota**.

One of the most important attributes of an ecosystem is the flow of energy and materials through it. When herbivores eat plants, they incorporate into their own tissue part of the food synthesized by those plants. Carnivores assimilate the tissue of herbivores in much the same way. In most ecosystems, carnivores that eat herbivores are eaten in turn by other carnivores; in fact, several levels of carnivores are often present in an ecosystem. A sequence of this kind, from producer to top carnivore, constitutes a **food chain**. Because most carnivores feed on animals smaller than themselves, the body sizes of carnivores often increase toward the top of a food chain (Figure 4-3).

Simple food chains in which a single species occupies each level are uncommon. Most ecosystems are characterized by **food webs**, in which several species occupy each level. Most species below the top carnivore level serve as food for more than one consumer species. Similarly, most consumer species feed on more than one kind of prey.

Parasites and scavengers add further complexity to ecosystems. **Parasites** feed on living organisms, only occasionally causing their death, while **scavengers** feed on organisms that are already dead. A flea that feeds on the blood of a dog, for example, is a parasite, as is a tapeworm that lives within a human. A vulture, in contrast, is a scavenger, as is a maggot, which feeds on dead flesh.

Material is lost in each step of the food web, for two reasons. First, not all organic matter is consumed or digested. Second, some organic matter that is consumed is used to provide energy rather than to form tissue.

Although material flows from one level of a food web to the next, it does not stop at the highest level. In fact,

Figure 4-3 A sequence of species forming a food chain within a stream. Single-celled algae are fed upon by nymphs (i.e., the juvenile stage) of mayfly insects. The nymphs, in turn, are eaten by sunfishes, which are preyed upon by large-mouthed bass. Otters eat the bass. (Organisms are not drawn to scale; the algae and nymph are greatly enlarged.)

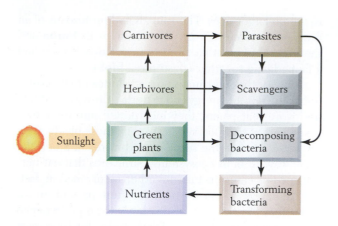

Figure 4-4 The cycle of materials through an ecosystem.

materials are cycled through the ecosystem continuously, with bacteria completing the cycle (Figure 4-4). Some of these bacteria decompose dead animals and plants of all types into simple chemical compounds, while others transform decomposed material, liberating nutrients to be reused by plants.

The term **diversity** is used to designate the number of species that live together within a community. Diversity is normally low in habitats that present physical difficulties for life. For example, because plants require water to make food, deserts contain fewer species of plants than do moist tropical forests. Only a few types of plants, such as cactuses, can sustain themselves in desert environments.

Predation is another factor that influences the diversity of a community. Heavy predation can eliminate species from a community, reducing diversity. In contrast, *moderate* predation can increase the number of species able to live together by reducing the abundance of potentially dominant competitors.

Physical disturbances can also influence diversity. Storm waves, for example, may tear animals and plants from rocky shores, leaving bare surfaces for the invasion of species that are weak competitors. Species that specialize in invading newly vacated habitats—land cleared by fire, say, or new shore areas formed along rivers that change course at flood stage—are aptly called **opportunistic species**. Opportunistic species are seldom good competitors. On the other hand, they tend to be good invaders, so that while some of their populations are disappearing from one area, others are becoming established elsewhere. The plants that we call weeds are opportunistic species par excellence. In gardens and lawns, many types of weeds come and go in the course of a few seasons.

Biogeography concerns broad patterns of occurrence

Distributions and abundances of organisms on a broad geographic scale are studied within the field known as **biogeography**. The greatest limiting factor at this scale

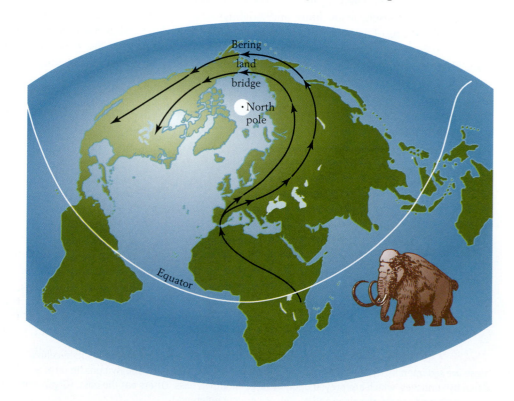

Figure 4-5 Changes in the geographic distribution of mammoths. Mammoths evolved in Africa during the Pliocene Epoch and spread to Eurasia. Later, during the Pleistocene Epoch, when sea level was lowered, they crossed the Bering land bridge, which emerged between Eurasia and North America. (After V. J. Maglio, *Trans. Amer. Philos. Soc.* 63:1–149, 1973.)

is temperature, with some species being restricted to polar regions, others to tropical regions near the equator, and still others to temperate regions in between. In general, communities increase in diversity toward the equator, because many more kinds of animals and plants can survive there than in the harsher conditions at higher latitudes.

Temperature, however, is not the only control of biogeographic patterns of occurrence, as evidenced by the fact that most species are not found in every habitat that meets their particular ecological requirements. Dispersal of most species is also restricted by barriers, the most obvious of which are land barriers for aquatic forms of life and water barriers for terrestrial forms of life. Of course, these barriers change over time, and the geographic distributions of species shift accordingly. *Mammuthus*, the genus that includes the extinct members of the elephant family known as mammoths, evolved in Africa about 5 million years ago, during the early part of the Pliocene Epoch. Blocked by northern oceans, mammoths were unable to migrate to North America until the Pleistocene Epoch. During the Pleistocene Epoch, however, large volumes of water were locked up on land as glaciers, and sea level fell throughout the world. Consequently, a land bridge formed between Siberia and Alaska, allowing mammoths to invade the Americas, where they survived until several thousand years ago (Figure 4-5).

The survival of mammoths in the Americas represents an interesting biogeographic phenomenon—the development of a *relict distribution*, or the presence of a taxonomic group in one or two locations after it has died out elsewhere. By late in the Pleistocene Epoch, mammoths remained in only a relatively small area of North America.

The Atmosphere

The **atmosphere**, the envelope of gases that surrounds Earth, affects life primarily by regulating Earth's temperature and by constituting a reservoir of chemical compounds that are used within living systems. The atmosphere has no outer boundaries, thinning gradually into interplanetary space. More than 97 percent of the mass of the atmosphere lies within 30 kilometers (19 miles) of Earth's surface; thus it is comparable to the continental crust in thickness.

Nitrogen, oxygen, and carbon dioxide constitute most of the atmosphere

Nitrogen (N_2) is the most abundant component of the atmosphere, making up about 78 percent of the total volume of atmospheric gas. It is a major constituent of

proteins, which regulate chemical reactions within cells and serve as building blocks of all living things. Second in abundance is oxygen (O_2), which forms about 21 percent of the volume of the atmosphere.

Atmospheric nitrogen and oxygen are maintained at consistent levels by being cycled through living organisms continuously and returned to the air. Most oxygen enters the atmosphere from plants, which produce it as a by-product of photosynthesis. A smaller amount of oxygen comes from the upper atmosphere, where sunlight splits water vapor (H_2O) into oxygen and hydrogen.

Carbon dioxide (CO_2), from which plants produce oxygen, is contributed to the atmosphere by the respiration of organisms—the process in which they burn food—and in the modern world by the burning activities of humans. It forms only about 1 percent of the atmosphere's volume, but plays a major role in trapping the sun's heat. In Chapter 10 we will see how oxygen and carbon dioxide are cycled through the atmosphere as part of large-scale geochemical cycles.

Temperature variations and Earth's rotation govern circulation in the atmosphere

Without the atmosphere, Earth's average temperature would be about $-18°C$, and there would be no life because it would be too cold for liquid water to exist. The energy that warms Earth from outer space comes from the sun, but the atmosphere helps trap solar heat and, in concert with the ocean, circulates it around the globe. Large-scale movements of air and water that arise from unequal solar heating of Earth's surface create regional climates and thus influence the distributions of organisms.

When solar energy reaches Earth, much of it is absorbed and turned into heat energy. The amount of solar radiation that is absorbed varies from place to place according to the nature of Earth's surface. Sunlight generates less heat when it strikes ice, for example, than when it strikes water, soil, or vegetation, because ice reflects more radiation. The percentage of solar radiation reflected from Earth's surface, called the **albedo**, ranges from 6 to 10 percent for the ocean; from 5 to 30 percent for forests, grassy surfaces, and bare soil; and from 45 to 95 percent for ice and snow.

Earth's polar regions receive just as many hours of sunlight as the equatorial region, but their sunlight is heavily concentrated in the summer. This happens because the tilt of Earth's rotational axis causes the angle and daily duration of sunlight to vary from season to season as Earth moves around the sun. Whichever hemisphere is tilted toward the sun experiences summer because the sun's rays strike it more directly and

Figure 4-6 **Sunlight and temperatures on Earth.** A ray of sun strikes Earth at a lower angle near the poles than near the equator and is therefore spread over a broader area, so that it heats the surface with less intensity. The tilt of Earth's axis creates the seasons by aiming one pole toward the sun while the other is aimed away from it. This orientation increases the angle at which the sun's rays strike near one pole (concentrating them to intensify heating) and reduces this angle near the other pole.

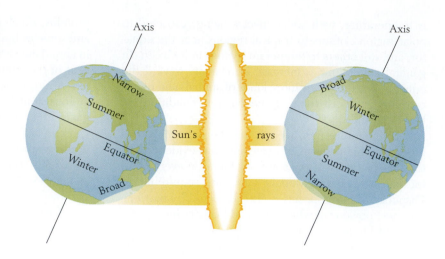

for more hours every day than during other seasons of the year (Figure 4-6).

Absorption of solar radiation warms land and water, and they, in turn, warm the atmosphere. Warming of fluids causes them to move. Much of the transfer of heat from place to place in the ocean and atmosphere occurs by *convection*, which results from the fact that a liquid or gas is less dense when it is warm than when it is cool (see Figure 1-16).

Let us first consider what happens to the atmosphere where Earth is warmest and coldest: warm air rises near the equator, while cool air sinks near the poles. To understand what happens in between, it is necessary to take into account the **Coriolis effect**,

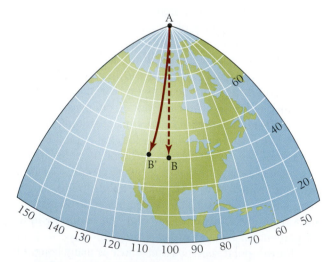

Figure 4-7 **The Coriolis effect.** If Earth did not rotate, an air current flowing from the north pole toward the equator would move directly along a line of longitude (from A to B). Because Earth rotates from west to east, however, the current bends to the west with respect to features of Earth's surface (moving from A to B'). All horizontal currents of air and water in the Northern Hemisphere tend to bend toward the right (clockwise).

which results from Earth's rotation. As Earth rotates, air currents are deflected clockwise in the Northern Hemisphere and counterclockwise in the Southern Hemisphere. The Coriolis effect is most easily envisioned for an air current moving from one of Earth's poles toward the equator (Figure 4-7). Instead of moving directly along a line of longitude toward the equator, such an air current will bend to the west with respect to landmarks on the surface of the rotating planet and will continue to do so because the surface speed of the planet increases toward the equator. More generally, the Coriolis effect causes an air current flowing in any direction in the Northern Hemisphere to flow toward the right, or clockwise. Similarly, it bends all air currents in the Southern Hemisphere toward the left, or counterclockwise.

By preventing Earth's atmosphere from circulating in the simple pattern shown in Figure 4-8, the Coriolis effect creates important wind and rainfall patterns. Consider what happens near the equator. Although air in this region rises from the warm surface of Earth and spreads poleward at high elevations, it does not move *directly* to the north and south because the Coriolis effect deflects it toward the east. As a result, the air piles up north and south of the equator more rapidly than it can escape toward the poles, producing belts of high atmospheric pressure about 30° from the equator. This high pressure bears down on Earth's surface. Air that rose near the equator and cooled aloft descends in this zone, pushing winds at the surface both poleward and equatorward. These winds are also deflected by the Coriolis effect. The ones that flow equatorward—the well-known **trade winds**—bend westward (Figure 4-9). Note that the trade winds replace the warm air rising near the equator; thus, they complete a cycle, or **gyre**, of airflow on either side of the equator.

The northern and southern trade winds converge in the **intertropical convergence zone**. Because of Earth's axial tilt, this warm tropical zone is usually not

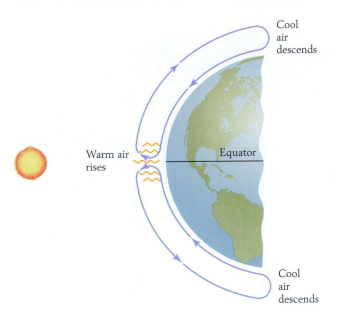

Figure 4-8 Atmospheric circulation of an imaginary Earth that receives equal amounts of sunlight on all sides from a source in the plane of the equator. Air warms and rises over the equator, and cools and descends at the poles. Therefore, one gyre occupies the Northern Hemisphere and another the Southern Hemisphere.

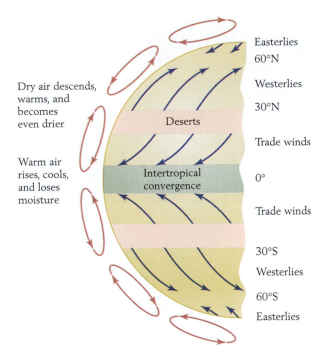

Figure 4-9 The major gyres of Earth's atmosphere. The lower segments (blue arrows) of these gyres represent the major wind systems, labeled at the right.

positioned on the equator, but shifts a few degrees north or south seasonally with the location of maximum solar heating (see Figure 4-6). Another gyre is positioned poleward of the trade winds. Here, because of the Coriolis effect, the **westerlies** bend toward the east instead of flowing directly poleward. Like the trades, these winds originate from a high-pressure zone where air piles up about 30° from the equator. Still farther from the equator in each hemisphere is yet another gyre, which originates where cool air descends near the poles and twists westward to produce **easterlies**. These major air movements in the atmosphere influence the distributions of climates and organisms.

The Terrestrial Realm

Currently, the continents of the world stand relatively high above sea level (see Figure 4-1). Thus the expanses of land are broader today than during most of Phanerozoic time. Another unusual feature of the modern world is a steep temperature gradient between each pole and the equator. Whereas tropical conditions prevail near the equator, the average summer temperature near the poles is well below freezing. At certain times in the geologic past, polar regions were warmer than they are today and the pole-to-equator temperature gradient was gentler. Because climatic conditions have a profound

effect on the distribution of organisms on land, climates will be our first consideration as we discuss the distribution of life on the broad continental surfaces of the modern world.

Vegetation patterns parallel climatic zones

It is remarkable how closely the distribution of terrestrial vegetation corresponds to the geographic pattern of climates. This correspondence, coupled with the fact that plants are the dominant producers of the food web and thus strongly affect the distribution and abundance of animals, makes climate an especially significant factor in terrestrial ecology. Plants, in fact, serve not only as sources of food but also as habitats for many animals; numerous insects, for example, spend their entire lives on certain types of trees, and insects account for most species of organisms living in the world today.

Tropical climates—those in which the average air temperature ranges from 18° to 20°C (64° to 68°F) or higher—are usually found at latitudes within 30° of the equator. These climates are not only very warm, but often very moist as well. When air in this region rises after being heated at Earth's surface, it cools (see Figure 4-9); because cool air cannot hold as much water vapor as warm air, it loses moisture in the form of rain. In South America and Africa, the only large continents

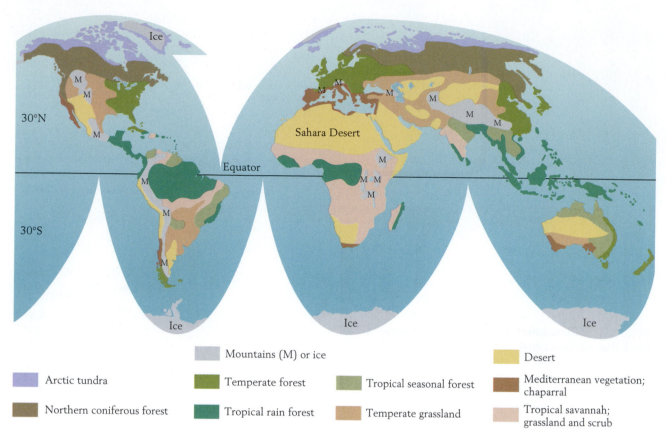

	Mountains (M) or ice		Desert
Arctic tundra	Temperate forest	Tropical seasonal forest	Mediterranean vegetation; chaparral
Northern coniferous forest	Tropical rain forest	Temperate grassland	Tropical savannah; grassland and scrub

Figure 4-10 **The major terrestrial communities.** Each kind of community is characterized by a particular association of plants adapted to particular climatic conditions. (After C. B. Cox and P. D. Moore, *Biogeography: An Ecological and Evolutionary Approach*, John Wiley & Sons, New York, 1980.)

with equatorial regions (Figure 4-10), the warm, moist conditions that characterize tropical **rain forests** allow so many kinds of plants to thrive (Figure 4-11) that they

Figure 4-11 **A tropical rain forest in Peru.** (Andre Bartschi.)

form what are informally called jungles. These plants provide food and shelter for a wide variety of animals.

Recall that about 30° north and south of the equator, the trade winds form (see Figure 4-9). The air of these winds is dry, having cooled at high altitudes and dropped much of its moisture as rain. As a result, the trade winds drop little rain; instead, they pick up moisture from the surface of Earth, creating **deserts** in broad continental areas that lie 20° to 30° north and south of the equator. The Sahara is the largest of these deserts, but broad deserts also occupy southern Africa, central Australia, and southwestern North America (see Figure 4-10).

Most deserts receive less than 25 centimeters (10 inches) of rain per year. Because only a few types of plants can live under such conditions, deserts are characterized by sand and bare rock rather than by dense vegetation (Figure 4-12). Some desert plants, such as cactuses, are able to store water, and nearly all have small leaves to minimize water loss by evaporation. Few animal species can survive in this environment, and a large percentage of those that do are nocturnal; many

Figure 4-12 A desert landscape in Arizona. (Peter Kresan.)

are small rodents that find refuge from the hot sun by remaining in burrows during the day.

Savannahs and **grasslands** form in areas where rainfall is sufficient for grasses—but not forests—to thrive. Many savannahs and grasslands are, in fact, positioned between dry deserts and wet woodlands. Savannahs differ from grasslands in containing scattered trees, which sometimes form small groves. Along their moister margins, savannahs grade into woodlands. The Great Plains of North America constitute a grassland, and savannahs are well represented in Africa.

In Africa and elsewhere, broad savannahs lie between the equatorial zone, where it is rainy year-round, and the deserts that occupy the trade wind belt. In the Northern Hemisphere, such savannahs receive most of their rain during the northern summer, when the intertropical convergence zone has shifted north of the equator, bringing heavy tropical rainfall with it. Similarly, savannahs of the Southern Hemisphere receive most of their rain during the southern summer.

Savannahs are noted for their populations of large animals (see p. 78). Most of the herbivores found in savannahs and grasslands—including bison, antelopes, zebras, and wildebeests—are relatively large animals that

graze on grasses and have enough stamina to flee from carnivores—such as lions, hyenas, and wolves—that are large enough to capture them.

In sharp contrast to the rich biotas of tropical rain forests and warm savannahs are the meager biotas of the regions near the poles. Today large ice caps cover Greenland and Antarctica, the two large landmasses of polar regions. Such ice caps have been absent from Earth in past times, when no large landmasses have occupied polar regions or when the polar regions have been warmer. The ice caps of Greenland and Antarctica are now so heavy that they actually depress the continental crust (Figure 4-13). These ice caps are continental glaciers.

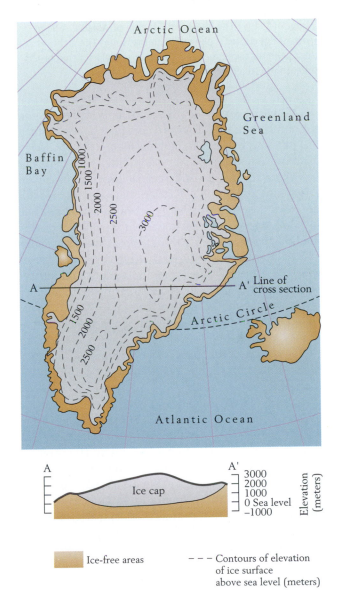

Figure 4-13 The thick glacial ice cap of Greenland depresses the continental crust.

Figure 4-14 **The Hubbard Glacier of Alaska.** A large piece of the glacier is plunging into the water. The icebergs floating in the foreground are pieces of the glacier that broke off earlier. (Tom Bean.)

Glaciers are among the most impressive physical structures on Earth. They are not simply masses of ice; they are masses of ice in motion (Figure 4-14). Glaciers form from snow that accumulates until it is so thick that the pressure of its weight recrystallizes the individual flakes into a solid mass. Glaciers slide slowly downhill, but they also spread over horizontal surfaces. This movement resembles the flowing of any tall pile of solid material that is near its melting point. Glaciers form not only at high latitudes but also at high elevations, even near the equator, where the atmosphere and surface of Earth are cool. Small glaciers occupy mountain valleys, through which they flow downhill. When they encounter warmer temperatures closer to sea level, they usually melt, though near the poles they may reach the sea before they can do so. Large chunks of ice then break from their terminal portions and float off in the form of icebergs.

Nearly all of Antarctica is covered by ice, but no other large continental area in the Southern Hemisphere is close enough to the south pole to experience very cold conditions year-round. In the Northern Hemisphere, however, cold conditions exist year-round in many regions near glaciers, and they support a subarctic ecosystem known as **tundra**. Tundra exists in areas where a layer of soil beneath the surface remains frozen during the summer, even though air temperatures rise above freezing. Under these conditions, water is never available in abundance. The dominant plants in tundra communities are not grasses and tall trees, but rather plants that need little moisture, such as mosses, sedges, lichens (associations of algae and fungi), and low-growing trees and shrubs. A broad belt of tundra stretches across the northern margins of North America and Eurasia, sup-

porting a low diversity of animal life (Figure 4-15). Rodents and snowshoe hares are present in tundras today, and the dominant herbivores of larger size are the caribou (called reindeer in Europe and Asia) and musk ox. Foxes, wolves, and bears are the primary hunters.

South of the tundra in the Northern Hemisphere, in areas where moisture is sufficient, forests, rather than deserts or grasslands, are found. The cold regions adjacent to tundra are cloaked in **evergreen coniferous forests**, dominated by such gymnosperm trees as spruce, pine, and fir. These trees are successful in very cold regions because they retain their needles year-round. The summers in such regions are too short to permit deciduous trees to regrow their leaves and still have enough warm days ahead to produce the food they need. The diversity of animals is also low, partly because relatively few plant species are present to support them and partly because many groups of animals simply cannot survive the cold winters.

To the south, in slightly warmer climates with longer summers, **temperate forests** replace evergreen forests. There deciduous trees such as maples, oaks, and beeches are usually present in greater abundance than evergreen trees. Deciduous trees are angiosperms, or flowering plants (p. 66). Where conditions are favorable for their growth, angiosperms defeat conifers, which are gymnosperms, in competition for resources. Ground animals are more diverse in temperate forests than in the colder evergreen forests, and birds are especially well represented.

Mediterranean climates, which are characterized by dry summers and wet winters, often prevail along coasts that lie about 40° from the equator. During the summer the land in these areas is warmer than the ocean, so that moist air coming off the ocean is warmed over the land and retains its moisture. In the winter the land is cooler than the ocean, so that moist sea air cools over the land and drops its moisture. This type of climate

Figure 4-15 A caribou grazing in North American tundra. (Michael Francis/Wildlife Collection.)

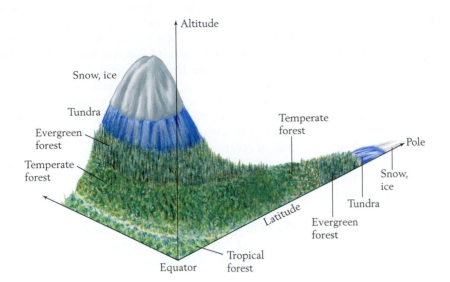

Figure 4-16 The similarity between altitudinal and latitudinal zonation of terrestrial vegetation. The zonation from the bottom to the top of a mountain is shown for a region near the equator, where a tropical forest occupies the lowland in the foreground.

characterizes much of California and southeastern Australia as well as the Mediterranean region. Mediterranean climates support *chaparral* vegetation, which consists primarily of shrubby plants with waxy leaves that retain moisture during summer droughts. Such climates have attracted large human populations, which have altered them greatly by decimating the native biotas.

Because oceans, continents, and continental topography are not distributed uniformly over Earth's surface, regional climates and vegetation vary from place to place even within a single broad climatic zone. Earth's surface features exert strong influences on climates and thus on forms of life.

Climates change with altitude

The temperature gradient between the base of a mountain and its top resembles the latitudinal gradient between a warm climatic zone and a polar region. As a result, the peaks of tall mountains are very cold, even in the equatorial region, and the zonation of vegetation on a mountain resembles the broader geographic zonation

between low and high latitudes (Figure 4-16). Thus a mountain in the temperate zone will typically have deciduous forest at its base, evergreen forest higher on its slopes, and tundra above the limit for tree growth. Snow and glaciers cap tall mountains, just as they cover polar regions. Near the equator, tropical forest may occupy the base of a mountain, below deciduous forest. At times in the past when global climates have been warmer than they are today, vegetation adapted to very cold climates may have been restricted to mountains.

Mountains not only offer plants growing conditions cooler than those of surrounding lowlands, they also alter rainfall patterns. A mountain range deflects prevailing winds upward, causing them to cool and drop their moisture on the windward side of the mountain (the side from which the wind is blowing). The result is relatively heavy rainfall on the windward side, where the winds rise, and dry conditions on the leeward side (the side that is sheltered from the wind), where the winds descend (Figure 4-17). On the leeward side, in what is termed the **rain shadow**, the climate may be so arid as to produce a desert.

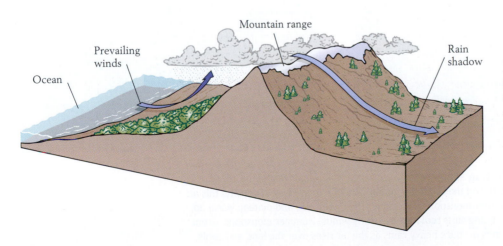

Figure 4-17 A mountain range forming a rain shadow. Air loses its moisture as it rises over the mountains and cools, producing rain on the windward side and dry conditions on the leeward side. (From W. K. Purves, G. H. Orians, and H. C. Heller, *Life: The Science of Biology*, Sinauer Associates, Sunderland, MA, 1995.)

For example, the Great Basin of North America, a desert that includes most of Nevada, lies in the rain shadow of the Sierra Nevada range. Winds carry moisture eastward from the nearby Pacific Ocean, but most of it falls on the western side of the Sierra Nevada, supporting abundant vegetation. Dry air descends into the Great Basin, which also receives little moisture from the trade wind belt to the south. The Rocky Mountains, though much farther from the ocean, create a modest rain shadow immediately to their east, where there is only enough moisture to support short grasses. The so-called tallgrass prairie grows farther to the east, where winds bring more moisture northward from the Gulf of Mexico.

Land and water influence seasonal temperature change

The ocean tends to undergo less extreme temperature shifts than do interior regions of continents. Large bodies of water also tend to stabilize the climates of neighboring land areas. These relationships result from the high *heat capacity* of water: its ability to absorb or release a great deal of heat without changing its temperature very much. Land (rock and sediment) has a much lower heat capacity than water, and air has a lower heat capacity still. Thus, outside the equatorial zone, land areas and air masses tend to become warmer in summer and colder in winter than oceans at the same latitude. In fact, pronounced seasonality is the most conspicuous trait of continental climates.

At latitudes where temperatures change from season to season, temperatures tend to be less extreme near the margins of a continent than within its interior. The reason for this difference is that a neighboring sea shares its thermal stability with coastal regions. Having a low heat capacity, the land cools more quickly than the ocean when winter comes. Winds transfer some heat from the ocean to the nearby land, however, reducing the impact of winter. Similarly, the ocean remains relatively cool in summer and substantially lowers the temperature of nearby land areas through the transfer of heat by winds.

Winter
Summer

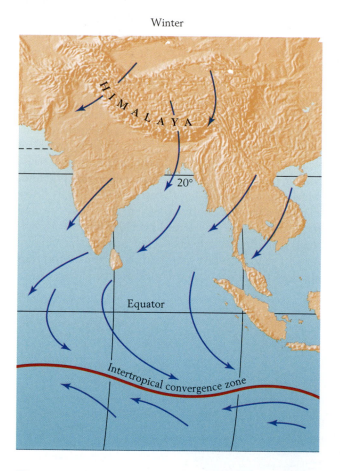
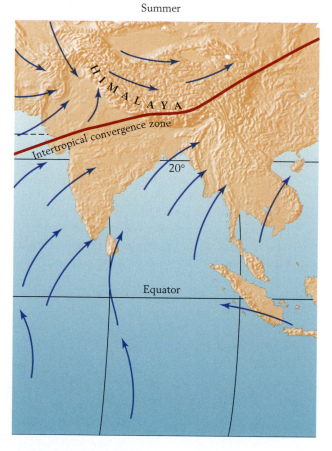

Figure 4-18 Monsoonal winds in southern Asia. In winter the Himalaya and the Tibetan plateau to their north become cold, and they cool the air above them. Increasing in density, this air descends and flows seaward as winter monsoons. In summer the mountains and plateau warm up, and the air above them rises. The rising air is replaced by strong winds—summer monsoons—from the Indian Ocean. These moist winds produce heavy rains as the air rises over the land and cools.

Large lakes can stabilize the climate of a continental interior in the same way that an ocean can stabilize a terrestrial climate near the coast. Thus the Great Lakes of North America ameliorate climates within several kilometers of their shores. Lakes are so much smaller than the ocean, however, that winds from the land can significantly cool them in winter and warm them in summer. For the same reason, in regions with seasonal climates, bays and lagoons at the margin of the ocean display summer and winter temperatures that are much more extreme than those of the open ocean nearby.

Monsoons are strong onshore and offshore winds that are caused by the difference in heat capacity between land and water. When winter comes to a nonequatorial region at the margin of a continent, the land cools more rapidly than the neighboring ocean. The cool, dense air mass that forms over the land then pushes seaward, producing winter monsoons as it displaces warmer air above the ocean. Summer monsoons blow in the opposite direction as the land heats up more rapidly than the ocean and the air above it warms and rises, to be replaced by air that flows in from the ocean. Summer monsoons carry a great deal of moisture, releasing it as rain if they rise over elevated terrain and cool.

Today the most powerful monsoons occur in southern Asia (Figure 4-18). Here the effect of heating and cooling of the land is intensified by the massive Himalaya and neighboring plateaus. A pronounced change in the temperature of this great body of uplands has a powerful effect on the atmosphere. The uplands cause onshore summer winds to rise abruptly, cool markedly, and release torrential rain. So strong are these monsoonal winds that they cause the intertropical convergence zone in this region to shift northward and southward annually through a distance of about 5000 kilometers (about 3000 miles).

Less powerful monsoons occur in other regions of the world, including the central United States, where summer heating brings winds laden with moisture inland from the Gulf of Mexico. Without these winds, summer droughts would be more severe in the eastern half of the United States.

Fossil plants reflect ancient climatic conditions

Because plants are so sensitive to environmental conditions, the fossils of many plants can be used to infer climatic conditions of the past. The **cycads**, for example, are an ancient group of plants that today grow only in the tropics and subtropics (Figure 4-19). Because this distribution seems to reflect a fundamental physiological limitation of this plant group, it is assumed that cycads of the past also lived in warm climates. Thus the finding of fossil cycads indicates the presence of warm climates in the geologic past.

Flowering plants are valuable indicators of climates of the past 80 or 90 million years, the interval during which they have been abundant on Earth. Angiosperms include not only plants with conspicuous flowers but also hardwood trees as well as grasses and their relatives. Because thick, waxy leaves help plants retain moisture, fossil leaves of this type that represent hardwood plants indicate that those plants lived in warm climates. The margins of leaves provide an even more useful means of assessing temperatures of the past. Leaves with smooth rather than jagged margins are especially common in the tropics. In fact, if a large flora occupies a region characterized by moderate or abundant precipitation, the percentage of plant species with smooth

A

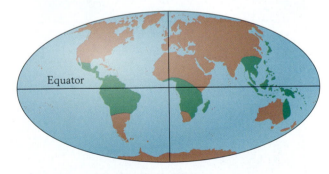

B

Figure 4-19 The cycad, a plant that was especially common during the Mesozoic Era. *A.* Living cycads. Today few cycads are found outside the tropics, and it appears that ancient cycads were also restricted to warm climates. *B.* The distribution of cycads today is shown in green. (*A*, Gerald Cubitt; *B* after C. B. Cox and P. D. Moore, *Biogeography: An Ecological and Evolutionary Approach*, John Wiley & Sons, New York, 1980.)

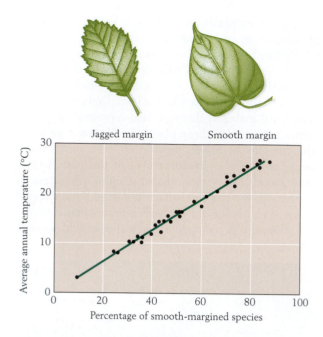

Figure 4-20 **Relation between climate and the shapes of leaves of flowering plants.** In the modern world, the average annual temperature of a region is closely correlated with the percentage of plant species with smooth margins. (After J. A. Wolfe, *Amer. Sci.* 66:994–1003, 1978.)

leaf margins provides a remarkably good measure of the region's average annual temperature (Figure 4-20). The application of this relationship to fossil faunas has revealed a considerable amount of information about temperatures on Earth during the past 100 million years.

Inasmuch as certain groups of terrestrial animals are also restricted to warm regions, some animal fossils, too, serve as indicators of ancient climates. Reptiles, which do not maintain constant warm body temperatures, are among the animals that cannot live in very cold climates.

The Marine Realm

The ocean floor is a vast basin in which sediments have accumulated over the course of Earth's history. For this reason, and because so many types of organisms with readily preserved skeletons live in the ocean, the seafloor is also where most species of the fossil record have been preserved.

We have seen how the geographic distribution of terrestrial species reflects broad patterns of air movement in the atmosphere. In a similar way, the distribution of marine species reflects large-scale movements of water in the ocean.

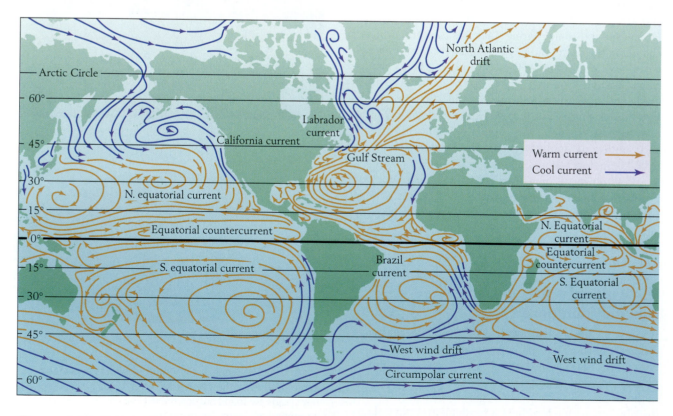

Figure 4-21 **Major surface currents of the ocean.** Note that large gyres north of the equator move clockwise, while those south of the equator move counterclockwise. (After P. R. Ehrlich, A. H. Ehrlich, and J. P. Holdren, *Ecoscience: Population, Resources, and Environment*, W. H. Freeman and Company, New York, 1977.)

Winds drive ocean currents

Most ocean currents at Earth's surface owe their existence to large-scale wind patterns and the Coriolis effect. The trade winds blow toward the equator from the northeast and southeast (see Figure 4-9), pushing equatorial water westward to form the north and south **equatorial currents** (Figure 4-21). These currents pile water up on the western sides of the major ocean basins, where some of the water flows backward under the influence of gravity as **equatorial countercurrents**.

Within each major ocean there is a full clockwise gyre north of the equator and a counterclockwise gyre south of the equator. The Indian Ocean, most of which lies south of the equator, has only a counterclockwise gyre. The Gulf Stream, a famous segment of the North Atlantic gyre, carries warm water from low latitudes to the shores of Great Britain, where palm trees survive more than 50° north of the equator. An eastern segment of the Pacific gyre has the opposite effect, bringing cool water to the coast of California.

Strengthened by westerly winds, the southern segments of the three gyres in the Southern Hemisphere join to form the Antarctic **circumpolar current** (Figure 4-22). Water trapped in this current becomes very cold because of the high latitude and contributes to the frigid condition of Antarctica. The landmasses of North America and Eurasia prevent the development of a comparable circumpolar current in the Northern Hemisphere.

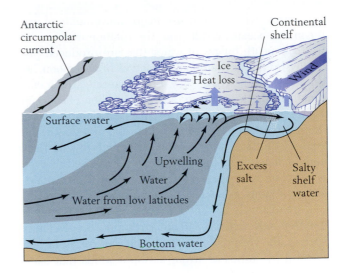

Figure 4-23 Formation of cold bottom water around Antarctica. Water flowing below the surface of the ocean from low latitudes rises up against the continental shelf all around Antarctica. Here it becomes more dense, for two reasons. First, it loses heat to the cold atmosphere. Second, freezing of some of the water to form sea ice leaves excess salt behind, so that the remaining water is slightly hypersaline. The cold, salty shelf water formed in this way sinks to the deep-sea floor, where it spreads throughout the world. (After A. L. Gordon and J. C. Comiso, *Sci. Amer.* June 1988. © 1988 by Scientific American, Inc. All rights reserved.)

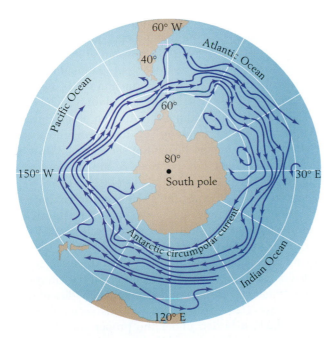

Figure 4-22 The eastward-flowing circumpolar current around Antarctica. Note in Figure 4-21 how this current is formed by the three large counterclockwise gyres of the southern oceans. (After A. N. Strahler, *The Earth Sciences,* Harper & Row, New York, 1971.)

Near the poles, water loses heat to the cold air and gains salt from water that freezes to form sea ice. This water is especially dense because it is colder and more saline than normal, so it sinks to great depths (Figure 4-23). When this water reaches the deep-sea floor, it spreads toward the equator. Antarctic water that descends in this manner is denser than the water from the north, so it hugs the bottom of the sea and flows well into the Northern Hemisphere. Above this water, which remains at near-freezing temperatures, is slightly warmer water that descends near the north pole, north of Iceland. These descending currents supply the deep sea with oxygen from Earth's atmosphere. This oxygen permits a wide variety of animals to live in the bottom water despite its near-freezing temperatures. As we shall see in later chapters, there have been times in the geologic past when the deep sea was warmer and poorly supplied with oxygen.

Waves are yet another important form of water movement. Surface waves result from the circular movement of water particles under the influence of the wind. Because this movement decreases with depth, wave motion has only a weak effect several meters below the surface. Waves far from shore form swells that lack sharp crests, but close to shore, the seafloor impedes the forward movement of waves, which steepen and break.

Tides, which also cause major movements of water in the oceans, result from the rotation of the solid Earth beneath bulges of water that are produced primarily by the gravitational attraction of the moon. As tides approach a coast, they often generate strong currents. A tide flows toward a coast and then ebbs again within a few hours as Earth rotates, moving the coast away from the tidal bulge. As tides ebb and flow at the edge of the sea, they cause the shoreline to shift back and forth across an **intertidal zone**. Because this zone is alternately exposed to air and water and also experiences pronounced changes in temperature, it is inhabited by relatively few species. Landward of this zone is one that is even harsher to life and supports even fewer species: the **supratidal zone**, which is dry except when flooded by storms or strong onshore winds that coincide with high tides.

Marine life varies with water depth

The depth of the sea varies from the thickness of a watery film at the shoreline to more than 10 kilometers (6 miles) in the deep sea. Water depth by itself has little effect on the distribution of marine life. Nonetheless, many species are restricted to particular depth zones because important limiting factors vary with depth, among them light conditions, temperature, and water movements.

The configuration of the seafloor is shown in Figure 4-24. A **continental shelf** is nothing more than

Figure 4-25 An intertidal marsh along the coast of Virginia. (Bates Littlehales.)

the submarine extension of a continental landmass. Many continental shelves, including the one fringing eastern North America, are broad expanses of shallow seafloor that settle gradually under the weight of great thicknesses of sediment. The **shelf break** marks the edge of the shelf; seaward of it, the continent pinches out along the steeper **continental slope**. Today, throughout the world, the shelf break of continents is about 200 meters (650 feet) below sea level. This consistency reflects the fact that all continents are roughly the same thickness, so that their surfaces stand at similar heights above the intervening oceanic crust (see Figure 1-13). Near the base of the continental slope, continental crust gives way to oceanic crust. Just seaward of this juncture is the **continental rise**, consisting of sediment that has been transported down the continental slope. Beyond the continental rise lies the **abyssal plain**, which is the surface of a layer of sediment resting on oceanic crust. When we speak of the **deep-sea floor**, we are usually referring to the region below the shelf break. Most of the deep-sea floor lies between 3 and 6 kilometers (about 2 and 4 miles) below sea level (see Figure 4-1). The water above it constitutes the **oceanic realm**.

Along the margin of the sea, **barrier islands** of sand heaped up by waves and wind often parallel the shoreline. In the protection of these elongate islands are relatively quiet lagoons or bays. **Marshes**, which are formed by low-growing plants that inhabit the intertidal zone, fringe the margins of these ponded bodies of water (Figure 4-25). Here plant remains accumulate as peat, which, if buried under the proper conditions of temperature and pressure, can turn to coal. In places the sea spreads farther inland over a continent, forming a broad, semi-isolated **epicontinental sea**. At present, seas stand lower in relation to continental surfaces

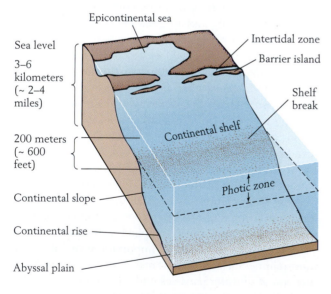

Figure 4-24 The aquatic environments at the edge of a continent. When the water is clear, the continental shelf—the submerged margin of a continent—usually lies within the photic zone, where enough sunlight penetrates to permit photosynthesis.

than they have at most times during the past 600 million years. For this reason, epicontinental seas are not well developed. As we journey through Phanerozoic time in later chapters, we will examine many ancient epicontinental seas and the life they harbored.

How is the life of the ocean related to water depth? The upper layer of the ocean, where enough light penetrates the water to permit plants to conduct photosynthesis, is known as the **photic zone**. The base of this zone varies from place to place, depending on the clarity of the water and the angle of the sun's rays, but it usually lies between 100 and 200 meters (300 to 600 feet) below sea level. It is sheer coincidence that 200 meters is also the approximate depth of the shelf break in most areas. Animals that live below the photic zone feed on organic matter that settles from above, or they feed on other animals.

For life on the seafloor, the most profound environmental change associated with depth takes place along the margins of the ocean. In contrast to the life of the adjacent **subtidal zone**, which is never exposed to air, the biota of the intertidal zone must endure large, often rapid fluctuations in environmental conditions. At some latitudes, hot, dry conditions prevail in this zone at low tide during the summer, yet winter chills drop temperatures below freezing. Furthermore, in the *surf zone*, where waves break along a beach, the constant movement of the sand permits only a few species to survive—such as small crabs that are exceptionally mobile and can quickly reestablish themselves in the sand if they are dislodged by a wave.

Although there is little food in the deep sea, the environment is relatively stable: temperatures remain slightly above freezing year-round. In fact, all ocean waters at depths below about 500 meters (1600 feet) are cool and support a unique fauna adapted to their low temperatures and weak food supply.

Marine life floats, swims, or occupies the seafloor

Most of the photosynthesis that takes place in the ocean is conducted by single-celled drifting algae. Organisms that drift in water are known as **plankton**, and plantlike organisms that belong to this group constitute the **phytoplankton** (Figure 4-26). The most important protists that constitute phytoplankton are *dinoflagellates, diatoms*, and (in warm regions) *calcareous nannoplankton* (see Figure 3-16). All three of these groups have extensive fossil records—the diatoms and calcareous nannoplankton because they have hard skeletons, and the dinoflagellates because they form cysts that have durable cell walls (p. 62). Also abundant as phytoplankton in the open ocean (far from shore) are very small cyanobacteria.

Feeding on the phytoplankton are drifting animals—the **zooplankton**—among which are small shrimplike crustaceans and other animals that spend their full lives adrift. Also included are the floating larvae of some invertebrate species (e.g., snails, bivalves, and starfishes) that spend their adult lives on the seafloor. These larvae, which may drift long distances, eventually settle to the seafloor and develop into adults. Some members of the zooplankton are carnivores that feed on other zooplankton.

Although many planktonic species can swim, they move primarily by drifting passively along. Animals that move through the water primarily by swimming are termed **nekton**, the most important being fishes. Both the plankton and the nekton include not only herbivores, which feed on phytoplankton, but also carnivores, which feed on other animals. Planktonic and nektonic organisms together constitute **pelagic life**, or oceanic life that exists above the seafloor.

Both immobile and mobile organisms also populate the seafloor, and these organisms are known as **benthos**, or **benthic life**. The seafloor itself is often referred to as the **substratum**. Some substrata are formed of rock, but they are more likely to be composed of soft substances such as loose sediment. Some benthic organisms live on top of the substratum, while others live within it.

Just as producers and consumers float in the water, both nutritional groups are found on the seafloor. Benthic producers include certain kinds of single-celled algae as well as multicellular plants. Because they require light, these photosynthetic forms must live on or close to the surface of the substratum.

Some benthic herbivores graze on plantlike forms, especially algae growing on hard surfaces. Forms known as **suspension feeders** strain phytoplankton and plant debris from the water. Others, known as **deposit feeders**, consume sediment and digest organic matter mixed in with the mineral grains. Bottom-dwelling carnivores of modern seas include crabs and starfishes as well as several kinds of snails and worms. In addition, many fishes swim close to the sea bottom and feed on bottom-dwelling animals.

Figure 4-26 depicts the basic features of the marine food web. The phytoplankton occupy the photic zone above both the continental shelves and the deep sea. Joining them as producers on continental shelves, where the photic zone reaches the seafloor, are bottom-dwelling plants. Another important food supply in shallow water is organic debris transported to the ocean by rivers or washed in from marginal marine marshes. The high concentration of phytoplankton in the photic zone causes zooplankton to be concentrated there as well. Some herbivorous zooplankton can also be found at

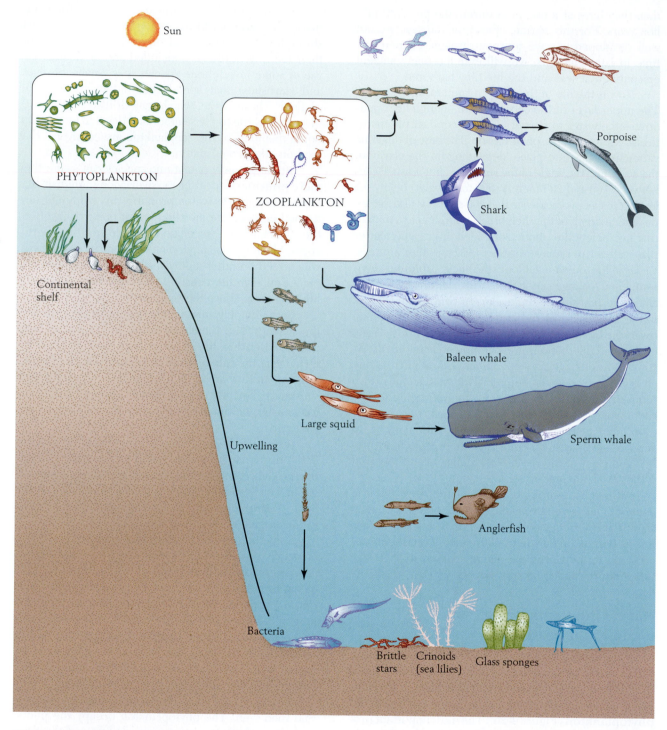

Figure 4-26 **The marine food web.** (The various forms of life are not drawn to scale.) Phytoplankton occupy the photic zone of the ocean, and thus most zooplankton, which feed on phytoplankton, also live there. On continental shelves, especially near the shore, bottom-dwelling plants also contribute food to the marine ecosystem. Most large marine carnivores are fishes. The whales are mammals that include carnivorous porpoises and sperm whales, which feed on large animals, as well as baleen whales, which strain tiny zooplankton from the water. As the amount of plant material diminishes with depth, the abundance of animal life diminishes as well. A few herbivores that feed on plankton suspended in the water, such as sponges and crinoids (sea lilies), live on the deep-sea floor, but most herbivores there extract their food from the sediment. Bacteria in the deep sea turn dead organic matter into nutrients, which upwelling currents carry to the surface for use by phytoplankton and other photosynthetic life.

greater depths, where they feed on the algal cells and plant debris that rain slowly down from the photic zone.

Different kinds of swimmers occupy different depth zones in the ocean. Some, such as herring, feed on zooplankton. So do the great baleen whales, which strain zooplankton through a sievelike bony structure. Other fishes, including nearly all sharks, are carnivorous, as are many kinds of whales. Carnivores are found at all depths of the ocean, although some species are restricted to narrow depth zones.

In the deep sea, however, where suspended food is scarce, only a few kinds of benthos strain food from the water. Most herbivores are deposit feeders, and there are many types of carnivorous fishes. In fact, a remarkably large number of species live along the cold, dark abyssal plain. Organic debris arrives here from shallow waters at a very slow rate, however, so the density of animals is low. Thus a survey of 1000 square meters of deep-sea floor might uncover dozens of species, each represented by a small number of individuals.

Bacteria live throughout the open ocean but are most abundant in the deep sea, where much organic debris accumulates. Some bacteria decompose this debris, while others transform some of the products of decay into compounds of nitrogen and phosphorus. Phytoplankton use these compounds to make food, thereby cycling the materials back through the ecosystem. A crucial step in this recycling is the physical process known as **upwelling**—the movement of cold water upward from the deep sea to the photic zone. Upwelling tends to occur along the margins of continents, where the large oceanic gyres drag water away from the land (see Figure 4-21); this water is replaced by water that wells up from the deep sea. Upwelling often brings nutrients to the photic zone in large quantities, producing an unusually rich growth of phytoplankton. The phytoplankton support large populations of zooplankton, which in turn support large populations of fishes.

Water temperature influences biogeographic patterns

In the marine realm, as in the terrestrial realm, temperature plays a major role in the geographic distribution of species. The calcareous nannoplankton, for example, live primarily in warm waters (Figure 4-27). Most species of planktonic diatoms, in contrast, live in cool waters at high latitudes.

The geographic distributions of benthic species are also limited by temperature, with some large groups being restricted to certain latitudes. The reef-building corals, which are restricted largely to the tropics, are among the most prominent of these groups (Figure 4-28). Corals are cnidarians that feed on smaller animals, which they

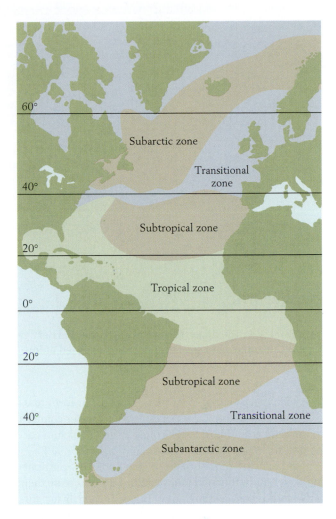

Figure 4-27 **Latitudinal zones of calcareous nannoplankton in the modern Atlantic ocean.** Although some species are found in cold waters, fewer species live there than in the tropical and subtropical zones. (After A. McIntyre and A. W. H. Be, *Deep Sea Res.* 14:561–597, 1967.)

capture by means of stinging cells on their tentacles. Reef-building corals are colonial animals that grow as clusters of connected individuals, or *polyps*. A colony forms from a single polyp, which develops from a larva that settles on the seafloor. The original polyp gives rise to a colony by budding off additional polyps, which in turn bud off others. Each polyp secretes a cup of calcium carbonate, and the adjacent cups are fused to form a large composite skeleton (see Figure 3-25A). Aiding corals in forming their large colonies are single-celled algae that live and multiply in the coral tissues. The corals supplement their diet by digesting organic matter produced by these algae. Before they are digested, however, the algal cells also remove carbon dioxide from the corals for use in photosynthesis. In this way, the algae facilitate the corals' secretion of their calcium carbonate skeleton. Species that

Figure 4-28 The upper surface of the Great Barrier Reef of Australia, showing coral colonies exposed at low tide. (Roger Steene/ENP Images.)

lack such algae cannot form reefs. Corals can form massive reefs only in tropical seas because cool water contains too much carbon dioxide for corals to grow calcium carbonate skeletons rapidly.

Other types of reef dwellers join corals in contributing skeletons to the solid reef structure. Reefs create their own fossil record as they grow. Beneath their living surface they consist of the remains of the dead animals and plants that were responsible for their construction.

Oxygen isotopes of fossil skeletons have been used in efforts to determine the temperatures at which animals lived millions of years ago. Oxygen occurs naturally in two isotopic forms: oxygen 18 has two more neutrons than the more common oxygen 16, and is thus the heavier isotope. The two isotopes have the same chemical properties, but marine organisms that secrete skeletons incorporate the isotopes in slightly different proportions, depending on the temperature of the environment: as the temperature decreases, the percentage of oxygen 18 in skeletons increases. Care must be taken in using oxygen isotopes to measure ancient temperatures, however, because some ancient shells have suffered chemical alteration after burial; as a result, temperature estimates based on oxygen isotopes are sometimes inaccurate.

Salinity is an important limiting factor near shore

The saltiness of natural water is called **salinity**. Oceanic seawater contains about 35 parts of salt per 1000 parts of water, and is said to have a salinity of 35 parts per 1000 (3.5 percent). Salinities of 30 to 40 parts per 1000 lie within the normal range for seawater; water of lower salinity is called **brackish**, while water of higher salinity is termed **hypersaline**.

Brackish conditions are most commonly found in bays and lagoons along the margins of the ocean. Brackish conditions result from an influx of fresh water from rivers into bays or lagoons that are partially isolated from the open ocean. Hypersaline conditions also develop in bays and lagoons, but only in those whose waters evaporate rapidly—usually in hot, arid climates.

The salinity of brackish and hypersaline waters typically changes frequently in response to rainfall and evaporation. Because animal life evolved in the ocean, the salt concentration in the tissues of most animals is similar to that of normal seawater. The maintenance of that concentration is essential for proper functioning of cells, so marine animals find it difficult to move into a habitat where the salinity is abnormal or fluctuating. It is hardly surprising that most bays and lagoons contain fewer species of animals than normal marine habitats. Many marine animals migrate into marshes to breed, however, because the scarcity of predators there and the abundance of organic matter from decaying vegetation improve the chances that their offspring will survive.

Freshwater Environments

Rivers and lakes contain **fresh water**, which by definition has a salinity below 5 parts per 1000 (0.5 percent). The difficulty of living in freshwater habitats is a major reason the faunas of rivers and lakes are not very diverse. Most freshwater animals must have ways of excreting the excess water that enters their tissues from the environment. Phytoplankton and zooplankton similar to those of the ocean also inhabit lakes, but in reduced variety. Because streams and rivers are constantly in motion, they do not sustain a planktonic community as complex as that of the ocean. Most producers that occupy rivers live on the bottom, and a large proportion of consumers are immature growth stages of terrestrial insects (see Figure 4-3). Both lakes and rivers also differ from the ocean in their low diversity of larger animal species. Unfortunately, relatively few kinds of these animals are readily preserved as fossils; fishes and shelled mollusks are the primary exceptions. As the next chapter will illustrate, ancient oceans have also accumulated a much larger volume of sediment than have lakes and rivers, so fossils of aquatic life are not only more varied in oceans, but also relatively more abundant.

Chapter Summary

What factors determine the ecological niches of species, and by what means do species obtain nutrition?

The way a species interacts with its environment defines its ecological niche. The distribution and abundance of any species are governed by a number of limiting factors: the availability of food, the physical and chemical conditions in the environment, and the presence of other species that are potential predators or competitors. Communities are groups of coexisting species that form food webs. Plants, which are at the base of most food webs, are fed upon by herbivores, which are fed upon in turn by carnivores. Communities and the environments they occupy constitute ecosystems. The diversity of a community is the variety of species it encompasses. Bacteria decompose dead organisms and transform the products of decay into compounds that serve as plant nutrients; thus materials are cycled through the ecosystem continuously.

What factors govern the geographic distributions of species?

Physical barriers to dispersal and changes in environmental temperature are the most important factors that limit geographic distributions of species. Many more species exist in warm climates than in cold climates. Tropical rain forests develop near the equator, where warm air rises and loses its moisture. The dried air descends north and south of the equator and circles back toward the equator as trade winds. In many areas that lie between 20° and 30° from the equator, trade winds produce deserts and dry grasslands. Continental glaciers (ice caps) exist near Earth's poles in Greenland and Antarctica, excluding all but a few species from these regions. Mountains provide cool habitats even in the tropics. They also alter climates in neighboring areas by intensifying monsoons and creating rain shadows. Because land plants are highly sensitive to climatic conditions, fossil land plants are useful indicators of ancient climates.

What factors govern the distribution of aquatic life?

In the oceans, prevailing winds and the Coriolis effect create huge gyres of water movement, influencing temperature patterns that govern the occurrence of organisms. The deep sea is cold because its waters come from near the poles, where frigid water sinks to great depths and flows throughout the deep sea. Many species of animals inhabit the deep sea, but their populations are small because little food reaches this environment. The salinity of bays and lagoons near the margin of the ocean differs from that of normal seawater and fluctuates greatly; relatively few species are able to live in these environments. Freshwater environments, such as lakes and rivers, usually harbor relatively few species because life in fresh water poses physiological problems for many kinds of animals.

Review Questions

1. Sometimes the species of a community are described as forming a food chain. Why is it usually more appropriate to speak instead of a food web?

2. Which terrestrial and marine environments characteristically contain few species? Explain why each of these environments has a low diversity.

3. How do the main kinds of producers (photosynthesizers) in the ocean differ in mode of life from those on land?

4. What can fossil plants tell us about ancient environments?

5. How does water depth in the ocean relate to the distribution of seafloor environments?

6. What produces the intertropical convergence zone?

7. What conditions create monsoons?

8. How do winds affect the ocean on a large scale?

9. Rain forests are sometimes likened to coral reefs, in that both support communities that include large numbers of species. Why are both communities restricted to the tropics?

10. What kinds of salinities characterize lagoons along the margin of the ocean? What causes these salinity conditions?

11. Identify the limiting factors that play important roles for various communities of marine organisms (from the deep sea to the intertidal zone and from the tropics to polar regions) and for various communities of nonmarine organisms (from lowland equatorial regions to mountaintops and polar regions). How do the compositions of various communities reflect the presence of the limiting factors? Use the Visual Overview on page 80 as a guide.

Aerial view of oolite shoals along the Grand Bahama Bank, with deep water on the left. (Paul M. [Mitch] Harris with permission of Chevron Petroleum Technology Company.)

CHAPTER 5

Sedimentary Environments

The stratigraphic record reveals remarkably detailed pictures of the settings where sediments accumulated long ago, enabling geologists to reconstruct ancient environments and, more broadly, **paleogeography**—the geography of the past. The goal is not only to learn about the distribution of land and sea at a particular time, but also to identify localized environmental features of the past, such as deserts, lakes, river valleys, lagoons, and submarine shelf breaks. In most instances, geologists can learn not only where a valley was located, for example, but also what kind of river occupied it—perhaps one consisting of many small, intertwining channels choked with bars of gravel and sand, or one flowing along a single broad, winding channel. Frequently geologists can also "read" from the sedimentary record whether the terrain that once bordered an ancient river was a dry, sparsely vegetated plain or a swamp densely populated by water-loving trees and undergrowth. In this and other ways, geologists reconstruct ancient climates.

The identification of ancient sedimentary environments also provides geologists with a framework within which to interpret life of the past. Although we can learn some aspects of how an organism lived by studying its fossil remains alone, a fuller understanding of that species can come only when its habitat is taken into consideration. It was once widely believed, for example, that the largest dinosaurs were too big to be fully terrestrial and therefore must have spent much of their time in water, like hippopotamuses. The stratigraphic record contradicts this idea: the fossilized bones of these giant creatures are frequently found in sedimentary rocks that represent nonaquatic environments.

Furthermore, by understanding the environmental relationships of sedimentary rocks, geologists can often predict where sedimentary deposits may harbor such natural resources as petroleum and natural gas or where economically valuable sedimentary bodies, such as deposits of halite and gypsum, may exist. Petroleum and natural gas tend to accumulate in porous sediments, such as clean sands deposited along ancient shorelines or rivers, and in ancient limestone reefs constructed in shallow seas by corals or other organisms.

Geologists rely heavily on actualism to reconstruct the environments in which sedimentary rocks accumulated. In other words, they study patterns of deposition in modern settings in order to recognize ancient deposits that formed in similar environments. Because sediments deposited in modern settings tend to be buried quickly beneath younger strata, geologists often have to excavate, tunnel, or core before they can observe those sediments. *Coring*—whereby a tube is driven into a sedimentary deposit and then withdrawn—is especially useful for examining deposits at the center of a lake, lagoon, or deep ocean. A core of sediment thus extracted can indicate

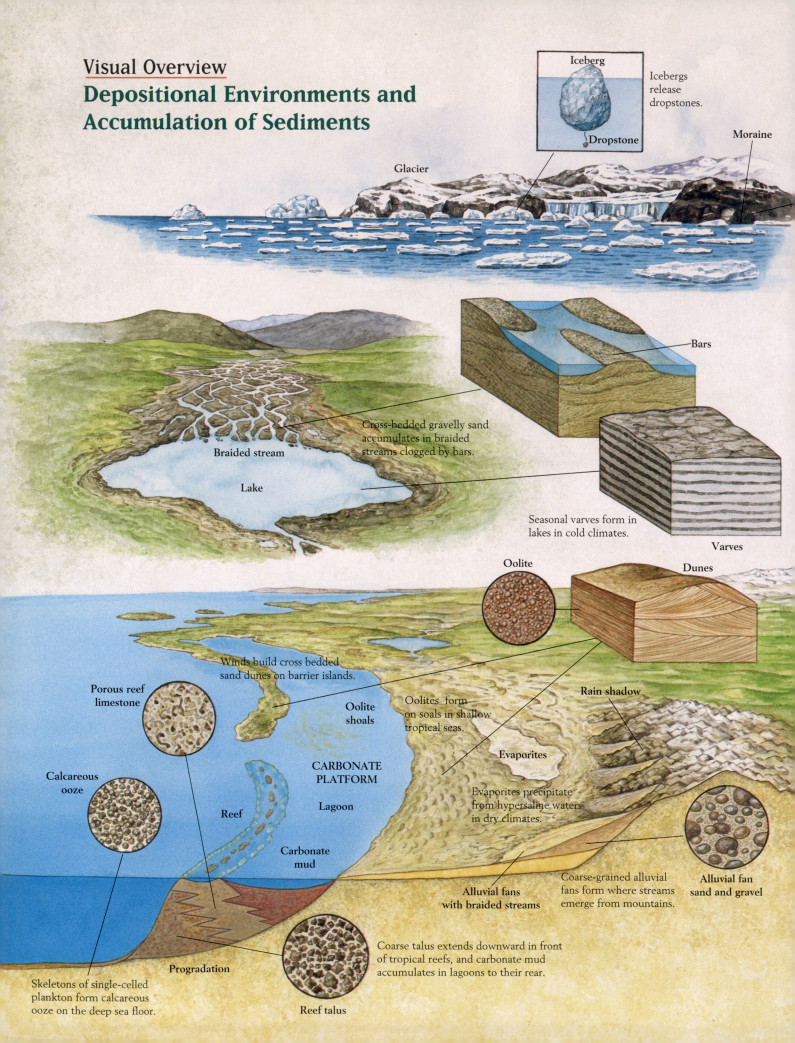

Visual Overview
Depositional Environments and Accumulation of Sediments

Iceberg

Icebergs release dropstones.

Dropstone

Moraine

Glacier

Bars

Cross-bedded gravelly sand accumulates in braided streams clogged by bars.

Braided stream

Lake

Seasonal varves form in lakes in cold climates.

Varves

Oolite

Dunes

Winds build cross bedded sand dunes on barrier islands.

Rain shadow

Porous reef limestone

Oolite shoals

Oolites form on soals in shallow tropical seas.

Evaporites

Calcareous ooze

CARBONATE PLATFORM

Lagoon

Reef

Carbonate mud

Evaporites precipitate from hypersaline waters in dry climates.

Alluvial fan sand and gravel

Coarse-grained alluvial fans form where streams emerge from mountains.

Alluvial fans with braided streams

Progradation

Skeletons of single-celled plankton form calcareous ooze on the deep sea floor.

Coarse talus extends downward in front of tropical reefs, and carbonate mud accumulates in lagoons to their rear.

Reef talus

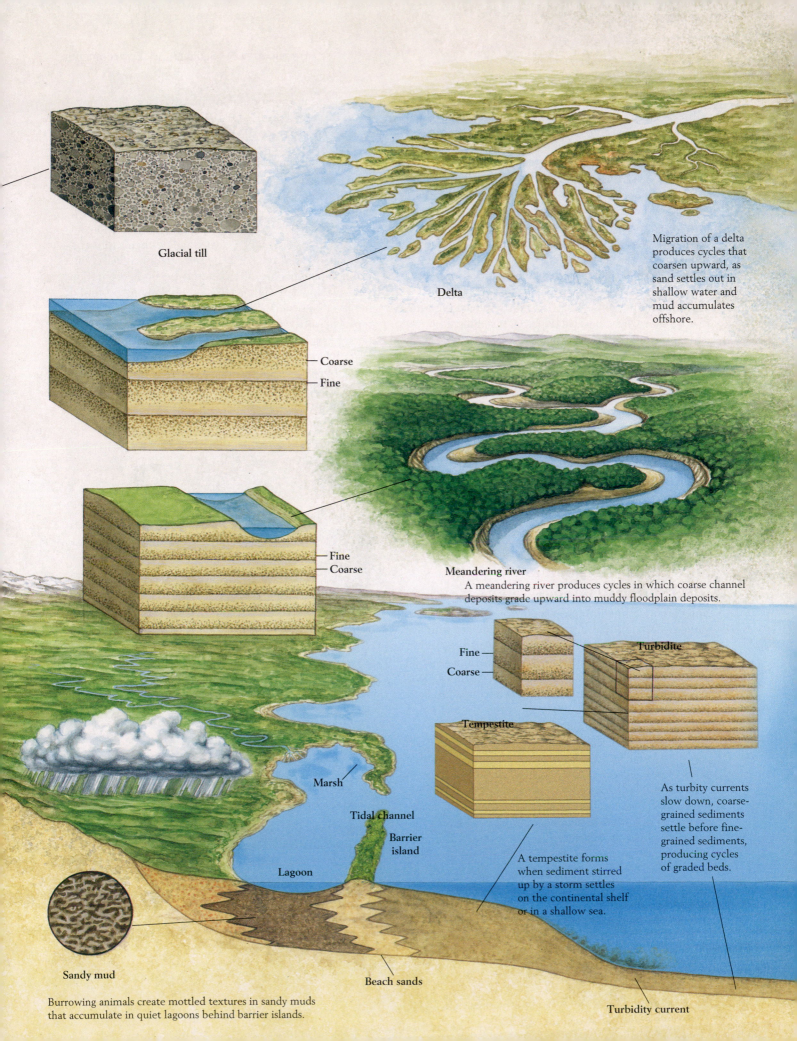

Glacial till

Delta

Migration of a delta produces cycles that coarsen upward, as sand settles out in shallow water and mud accumulates offshore.

Coarse
Fine

Meandering river
A meandering river produces cycles in which coarse channel deposits grade upward into muddy floodplain deposits.

Fine
Coarse

Fine
Coarse

Turbidite

Tempestite

As turbity currents slow down, coarse-grained sediments settle before fine-grained sediments, producing cycles of graded beds.

A tempestite forms when sediment stirred up by a storm settles on the continental shelf or in a shallow sea.

Marsh

Tidal channel

Barrier island

Lagoon

Sandy mud

Beach sands

Turbidity current

Burrowing animals create mottled textures in sandy muds that accumulate in quiet lagoons behind barrier islands.

the sequence of deposition at the site and can provide a three-dimensional picture of the deposit. Similarly, geologists who wish to examine a meandering river's depositional record must either dig one or more pits in the valley floor adjacent to the channel or sample the floor by means of coring.

While some sedimentary features provide highly reliable information about the nature of the environments in which they formed, many others offer ambiguous testimony. Most coal deposits, for example, represent swamps choked with vegetation—but such swamps are typical of both the banks of rivers and the shores of marine lagoons. To determine which environment pertains, geologists must evaluate the beds that lie above or below coal deposits. If sediments containing fossils of marine animals are found, it is likely that a marine lagoon, not a river, lay adjacent to the swamp or marsh in which the coal formed. Such vertical stacking of sedimentary features often provides the key to recognizing ancient environments.

Nonmarine Environments

We will begin our exploration of sedimentary clues to ancient environments by considering depositional settings on the surfaces of continents. Then our focus will shift to the margins of continents, and finally to the marine realm.

Ancient soils can point to past climatic conditions

Soil can be defined as loose sediment that contains organic matter and has accumulated in contact with the atmosphere rather than under water. Soil rests either on sediment of some sort—such as sand or mud—or on rock. It supplies essential nutrients and provides physical support for plants, and it serves as a habitat for many organisms. Soils develop largely through weathering processes and decay of plant material. **Topsoil**, the upper zone of many soils, consists primarily of sand and clay mixed with humus. **Humus**, the organic matter that gives topsoil its dark color, is derived from the decay of plant debris by bacteria and in turn supplies nutrients for other plants; thus it has an important position in the cycling of materials through terrestrial ecosystems.

Soils form in a variety of environments throughout the world—in tropical rain forests, in arid regions, and even on mountaintops—and can often be found buried, albeit chemically altered, within thick sequences of ancient sediment. The type of soil that forms depends in part on climatic conditions. For example, in warm climates that are dry part of the year, evaporation of groundwater causes calcium carbonate

to precipitate as nodular or massive deposits known as **caliche** (Figure 5-1). In moist tropical climates, warm waters percolate through the soil, destroying humus by *oxidizing* its components (combining them with oxygen). Silicate minerals in such areas also break down quickly, producing **laterites**: soils rich in iron oxides, which give them a rusty red color, as well as aluminum oxides.

Most soils are destroyed by erosion. In addition, ancient, buried soils can be exceedingly difficult to recognize and to interpret, partly because their original chemical components are often altered beyond recognition. One place where an ancient soil is likely to be found, however, is beneath an unconformity. Ancient soils are most likely to be found beneath unconformities that have been recognized on the basis of other criteria. Plant roots also provide clues to the presence of ancient soils. Similarly, burrows made by animals such as insects and rodents are diagnostic features. Certainly the most unusual of these excavations are the structures known as "devil's

A

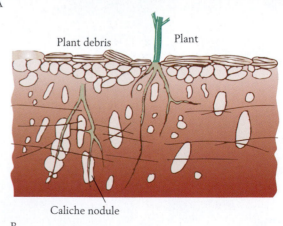

B

Figure 5-1 Nodules of caliche in soil. The white nodules of calcium carbonate (*A*) formed around plant roots, as shown in the drawing (*B*). (*A*, Lawrence Hardie, Johns Hopkins University.)

A

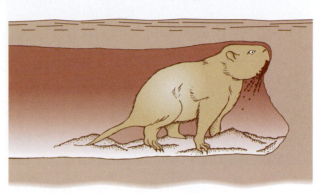

B

Figure 5-2 Fossil burrows made by beavers in soils more than 20 million years old. *A.* These "devil's corkscrews" have marks showing that beavers excavated them. Unlike modern beavers, which live near water, these animals lived in the grasslands of what is now Nebraska, where they formed colonies like those of prairie dogs. The burrows, which the beavers excavated with their front teeth (*B*), extend downward as far as 10 meters (33 feet) into the soil. (*A,* Carnegie Museum of Natural History; *B* after L. D. Martin and D. K. Bennett, *Palaeogeogr. Palaeoclimatol. Palaeoecol.* 22:173–193, 1977.)

corkscrews," which are actually burrows that beavers of an extinct species dug with their teeth in the Oligocene and Miocene soils of Nebraska (Figure 5-2). Skeletons of these beavers have been found in the burrows, and scratches on the burrow walls match their front teeth! The fact that these animals lived as far as 10 meters (33 feet) below ground level indicates that the level of standing water in the ancient soil stood below this depth; had it been higher, the beavers would have drowned.

Freshwater lakes and glaciers leave clues to ancient climates

Freshwater lakes and glaciers serve as reservoirs of water on land. Both leave distinctive features in the geologic record and provide indications of ancient climatic conditions.

Freshwater lake environments At no time in geologic history have lakes occupied a large fraction of Earth's surface. However, because lakes form in basins that lie at lower elevations than most soils, lake deposits are much more likely than soils to survive erosion.

Evidence of large freshwater lakes in the geologic past indicates abundant precipitation in a region because such lakes must be fed by substantial runoff from the land. In addition, because of the high heat capacity of water, large lakes tend to stabilize the temperature of nearby land areas, reducing annual fluctuations.

The sediments around the margins of a freshwater lake tend to be coarser than those at its center, partly because the current of a stream or river slows when it meets the waters of a lake, dropping its load of coarse sediment near the shore. Furthermore, the wind-driven waves on a deep lake's surface touch bottom only where they approach the shore, winnowing the sediment there and driving clay-sized particles into suspension. These particles later settle to the bottom toward the lake's center. Some lakes receive almost all of their coarse sediment during a moist season, and only fine-grained sediment accumulates during the rest of the year. The resulting pairs of coarse-grained and fine-grained layers are known as **varves**.

Fossils are valuable tools for distinguishing lake sediments from marine sediments. Although fish fossils are found in both kinds of deposits, the presence of exclusively marine fossils, such as corals or echinoderms, provides strong evidence that ancient sediments did not originate in a lake. Because burrowing animals are not as abundant in lakes as they are in many marine environments (and because waves and currents in lakes are generally weak), the fine-grained sediments that accumulate in the centers of lakes are likely to remain well layered—just as well layered as when they were laid down (Figure 5-3).

Another clue in the identification of lake deposits is close association with other nonmarine deposits, such as river sediments. It would be highly unusual to find lake deposits directly above or below deep-sea deposits, or even above continental-shelf sediments, unless the two types of deposits were separated by an unconformity.

Figure 5-3 Lake deposits from the Eocene Green River Formation of Wyoming. The even layers are typical of sediments that have accumulated in lakes. The alternating coarse-grained and fine-grained layers in this rock may represent varves. (S. Stanley.)

Glacial environments Sedimentary features associated with continental glaciers are excellent indicators of cold climates. Glaciers that form in mountain valleys seldom leave enduring geologic records because the mountains stand above the surrounding terrain and therefore tend to be eroded rapidly. Continental glaciers, however, leave legible records that survive for hundreds of millions of years. The geologic record reveals that ice sheets have spread over broad geographic areas several times during Earth's history. In fact, we are living during an interval of continental glaciation. Modern examples of continental glaciers include the one now occupying most of Greenland (see Figure 4-13) and the even larger one covering nearly all of Antarctica.

As glaciers move, they erode rock and sediment, transporting both in the direction of flow. Rocks embedded in the base of a glacier commonly leave deep scratches in the underlying rock that serve to record the direction of the glacier's movement. When such scratches are found in an area now free of glaciers, it is safe to assume that a glacier passed that way in the distant past (Figure 5-4).

Glaciers leave records not only of erosion but also of deposition. Some of the mixture of boulders, pebbles, sand, and mud scraped up by a moving glacier is deposited as the glacier advances, and some is deposited when it melts. This unsorted, heterogeneous material is called **till** (Figure 5-5); when lithified, it is known as **tillite**. At the farthest reach of a glacier's advance, till plowed up in front of the glacier is left standing in ridges known as **moraines**. In front of a moraine, sediment from a melting glacier is often deposited by streams of **meltwater** issuing from the front of the retreating mass of ice. Here the sediment tends to be sorted by size into layers of gravel, cross-bedded sand, and mud, forming well-stratified glacial material known as **outwash**.

Figure 5-4 A rocky surface in eastern Canada scoured by glaciers. (Peter Kresan.)

Figure 5-5 Glacial till. A mountain glacier deposited this bouldery sediment on the eastern side of the Sierra Nevada in California during the recent (Pleistocene) Ice Age. (Martin G. Miller/Earth Lens.)

In many instances, streams of meltwater converge to form a lake in front of the glacier. The alternating layers of coarse and fine sediment that typically accumulate in such a lake constitute varves (Figure 5-6). Each coarse layer forms during the summer months, when meltwater carries sand into the lake, whereas each fine layer is formed during the winter, when the surface of the lake is sealed by ice and clay and organic matter settle slowly from suspension in the still water. Typically the abundant organic matter makes the winter layer darker than the summer layer. Each pair of layers of varved sediment thus represents a single year's deposition, so geologists can count the number of years represented by a series of layers. In some areas, thousands of years of deposition have been tallied in this manner.

When a glacier encroaches on a lake or ocean, pieces of it break loose and float away as icebergs (see

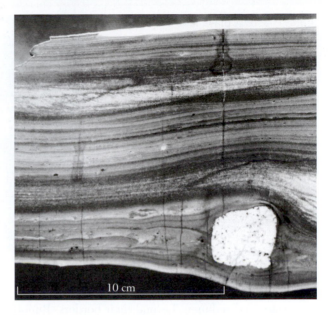

Figure 5-7 Fine-grained sedimentary rock of the Proterozoic Gowganda Formation in southeastern Canada. An igneous pebble apparently dropped from floating ice. This formation is about 2 billion years old. The layers in the sedimentary rock are varves. (D. A. Lindsey, U.S. Geological Survey.)

Figure 4-14). As these chunks of glacial ice melt, the sediment they carried sinks to the bottom of the lake or ocean, creating a highly unusual deposit in which pebbles or even boulders rest in a matrix of finer sediments (Figure 5-7). Unlike the tightly packed, coarse material that characterizes glacial till, these **dropstones** occur either singly or scattered throughout the matrix. Very few natural mechanisms other than this *ice rafting* bring large stones to the middle of a lake or to a seafloor far from land.

Deserts and arid basins accumulate salt and sand

Like ancient glacial environments, dry environments of the distant past can be identified from diagnostic traits of sedimentary rocks. Desert soils contain little organic matter because dry conditions support little vegetation, the source of the organic matter in soils. The rain that occasionally falls in deserts leads to erosion and deposition of sediment, and temporary streams carry the chemical products of weathering to desert basins. The subsequent precipitation of evaporite minerals in these basins sets arid regions apart from those with moist climates. In moist regions, permanent streams flow great distances without being absorbed into the soil or evaporating into the air, so they are frequently capable of reaching the ocean. As a consequence, we speak of most

A

B

Figure 5-6 Varved clays in Canada. The varves are layers of clay and silt deposited in quiet lake waters. Varved clays produced during the Pleistocene Epoch at Toronto (*A*) are remarkably similar to Precambrian varves in southern Ontario that are nearly 2 billion years older (*B*). (*A*, P. F. Karrow, Ontario Department of Mines; *B*, F. J. Pettijohn.)

Figure 5-8 A playa lake in Death Valley, California. (Peter Kresan.)

humid regions as having **exterior drainage**, meaning that water and sediment beyond their borders. Runoff from arid regions, in contrast, is too sparse and intermittent to form permanent streams and rivers, and often the result is **interior drainage**—a pattern in which streams die out through evaporation, seepage into the dry terrain, or drainage into lakes. Lakes in areas with interior drainage, known as **playa lakes** (Figure 5-8), also tend to be temporary, and when they shrink by evaporation, evaporites precipitate from them.

In dry, sparsely vegetated regions, the wind may pile loose sand into hills of sand called **dunes**. Dunes occupy less than 1 percent of some deserts, but sometimes form magnificent landscapes (Figure 5-9). Where the wind blows across loose sand, a dune can begin to form over any obstacle that creates a wind shadow in which sand can accumulate. Figure 5-10A shows how a dune tends to "crawl" downwind as sand from the upwind side moves over the top and accumulates on the downwind side. As the prevailing wind direction shifts back and forth, the direction in which the dune migrates shifts as well. A shift in wind direction usually leads to the truncation of preexisting deposits, which often causes a new set of beds to accumulate on a curved surface cut through older sets. The result is called **trough cross-stratification**. Figure 5-10B shows an idealized cross section through a dune, and Figure 5-10C shows a real section through an enormous lithified dune more than 200 million years old.

Dunes are familiar sights not only in desert terrain but also landward of the sandy beaches that border oceans and large lakes. For this reason, a geologist must consider other indicators to determine the setting in which lithified sand dunes formed. However, even after a geologist identifies an arid region as the setting for ancient dunes, there is work to be done. Dry climates are widespread in the trade wind belt but are also found in the rain shadows of mountains and in inland regions, far from oceans. To determine which of these settings was the site of an arid environment recorded in ancient rocks, geologists must reconstruct the geographic features of neighboring regions.

Many of the features typical of an arid basin can be seen in Death Valley, California (Figures 5-8, 5-9, 5-11, and 5-12). Temporary streams carry sediment down valleys incised into the naked rocks of nearby highlands to form low cone-shaped structures called **alluvial**

Figure 5-9 Sand dunes in Death Valley. (Tom Bean.)

A

Windstream lines

Erosion   Deposition

Movement of dune

B

C

Figure 5-10 The internal structure of a sand dune.
A. The windstream becomes compressed just above a dune and consequently increases in velocity. The dune ceases to grow taller when its height becomes so great that it causes the windstream to move rapidly enough to transport sand. As sand passing over the dune accumulates on the steep leeward slope, the dune begins to "crawl" in that direction. *B.* This cross section of a dune shows the cross-bedding that results from shifts in the wind. As the wind direction changes, the shape of the dune is altered by removal of sand, and a new leeward slope forms. *C.* Cross-stratified dune deposits of the Jurassic Navajo Sandstone in Arizona. (*C*, Tom Bean.)

fans, which spread out onto the floor of Death Valley (Figure 5-11). These structures form where a mountain slope meets the valley floor, causing streams to slow down and drop much of their sediment. Alluvial fans consist of poorly sorted sedimentary particles that range from boulders to sand near the source area and from sand to mud on the lower, gentler slopes. Most alluvial fans include broad deposits of coarse, cross-stratified sediments laid down by a complex network of channels, or **braided streams**, that carry water and sediment during the infrequent rainy intervals. After a heavy rain or snowmelt in the mountains, water gushing onto a fan from a steep valley carries so much sediment that sand and gravel clogs portions of the initial channel, dividing it into numerous smaller channels that connect and divide in complex patterns.

Figure 5-11 Overlapping alluvial fans in Death Valley, California. These alluvial fans (A, B, and C) have formed at the mouths of narrow valleys between mountains, where flowing streams slow down and deposit the sediment they have carried. Braided streams can be seen on the surfaces of the fans. (Michael Collier.)

Figure 5-12 Large mudcracks in Death Valley, California. (Peter Kresan.)

The braided streams formed in this way lead to the center of the basin, where their occasional flow may form a temporary playa lake (see Figure 5-8). As the lake waters dry up, evaporite minerals accumulate. Similar minerals also accumulate on those parts of the basin floor where groundwater seeps to the surface and evap-

orates. The evaporites of Death Valley are composed primarily of halite, gypsum, and anhydrite. Alternate wetting and drying in this basin produce large polygonal mudcracks in many areas (Figure 5-12). Around the margins of these "salt pans," calcium carbonate deposition forms caliche.

Braided and meandering rivers deposit sediment in moist regions

In areas that receive abundant rainfall, the precipitation normally creates exterior drainage: small streams meet to form larger streams, which in turn flow into still larger rivers. Many small rivers terminate at lakes, but most large rivers reach the sea. Figure 5-13 summarizes the sequence of aquatic environments through which water typically passes as it moves from the headwaters of river systems in hills or mountains to the deep sea, transporting and depositing sediment in the process. In the remainder of this chapter we will investigate the diagnostic features of ancient deposits formed in each of these environments—features that enable us to recognize each type of deposit as far back in the geologic record as a few hundred million or even 2 or 3 billion years.

Alluvial fans and braided-stream deposits Sediment in the form of alluvial fans accumulates at the feet of mountains and steep hills in moist regions, just as it does in arid regions. In moist climates, however, water often spreads these coarse sediments farther from their source, producing fans that slope more gently than those of dry basins. Alluvial fans in moist regions nonetheless slope steeply enough to remain poorly vegetated, and the large volumes of coarse sediment that arrive at times of heavy rainfall or snowmelt produce braided streams.

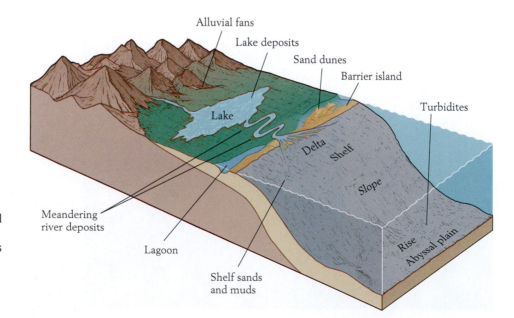

Figure 5-13 The downhill course of sediment transported from mountains to the sea. Along the way, sediment is trapped in a wide variety of depositional environments.

Figure 5-14 A braided stream in Alaska. Bars of sediment divide the flow into numerous winding channels. The water and sediment emerge from the base of a melting glacier that is out of view. (Tom Bean.)

Braided streams also form in front of mountain glaciers that experience substantial melting in summer, as the meltwaters are choked with sand and gravel plowed up by the flowing ice (Figure 5-14).

Meandering rivers In gently sloping terrain far from uplands, rivers generally occupy solitary channels that wind back and forth like ribbons (Figure 5-15). Unlike braided streams, these **meandering rivers**—such as the Mississippi or Thames—are not choked with sediment, because sediment is supplied to them slowly in relation to the rate at which the water flows. Any irregularity in the local terrain causes the river's path to curve. Because the water flows least rapidly near the inside of a bend and most rapidly near the outside, the river tends to cut into the outer bank of the bend, where

the current is swiftest. On the inside of the bend, the current is so weak that sediment is deposited rather than eroded, and it accumulates there to form a **point bar** (Figure 5-16A).

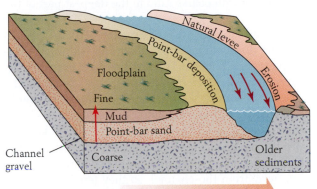

A

B

Figure 5-16 Deposition of sediment by a meandering river. *A.* At each bend, the river migrates outward. The current flows most rapidly on the outside of a bend (long arrow), so the water cuts into the outer bank. On the inner side of the bend, where the current moves more slowly, sand accumulates to form a point bar. As the channel migrates outward, the point-bar sands advance over the coarser, cross-bedded sands deposited within the original channel. Pebbles often accumulate at the base of the channel. Muddy floodplain deposits, which form when the river floods its banks, migrate in their turn over the point-bar sands. As a result of this shifting of depositional environments, coarse sediments at the base grade upward into fine sediments at the top. *B.* Cyclical meandering-river deposits of Devonian age in the Catskill Formation, south of Harrisburg, Pennsylvania. In each cycle, coarse-grained sandstone (light-colored, thickly bedded deposits), which formed in the river channel, grades upward into fine-grained sediments, which formed in floodplains. (*B* from F. J. Pettijohn and P. E. Potter, *Atlas and Glossary of Primary Sedimentary Structures,* Springer-Verlag Publishing Company, New York, 1964.)

Figure 5-15 A meandering river. Sand forms point bars on the insides of the bends. (Peter Kresan.)

Point-bar deposits usually consist of sand. Most of this sand is cross-bedded, because large ripples migrate along the riverbed in the course of time. In deeper water, where the current is stronger, the sediment in the river channel is coarser (Figure 5-16B); gravel is often found along with coarse sand in the deepest part of the channel.

Because mud consists of very fine particles, it tends to move downstream without settling within a meandering river's channel. When the river overflows its banks, however, it carries fine sediment laterally to the adjacent lowlands, known as **floodplains** or **backswamps**. Here, the spreading floodwaters flow slowly, allowing mud to settle before they recede. In keeping with the normal pattern of sediment deposition, these floodwaters move progressively more slowly as they flow away from the channel, so they tend to drop the coarser portion of their suspended sediment before they spread far from the channel. Sand is therefore dropped first, followed by silt, and together these deposits form a gentle ridge or **natural levee** alongside the channel. Because natural levees and floodplains are inundated only periodically, their surfaces tend to dry out and crack. Many mudcracks formed in this way have been preserved in the stratigraphic record. Levees and floodplains also become populated by moisture-loving plants, which may leave traces of their roots in the rock record or even form deposits that eventually turn into coal.

The vertical sequence in which sediments are deposited by a meandering river is seen in Figure 5-16— from coarse channel deposits at the base to cross-bedded point-bar sands in the middle to muddy backswamp deposits at the top. Levee sediments sometimes lie between the point-bar and backswamp deposits. This coarse-to-fine sequence can be thought of as forming a composite depositional unit. It forms because, as the channel migrates laterally, the point bar builds out over deeper gravels and the floodplain shifts over older point-bar deposits. The meandering-river sequence shown in Figure 5-16 illustrates **Walther's law**, which states that when depositional environments migrate laterally, sediments of one environment come to lie on top of sediments of an adjacent environment.

In a broad basin that happens to be *subsiding* (sinking relative to surrounding terrain), a river may migrate back and forth over a large area many times, piling one coarse-to-fine composite depositional unit on top of another as it goes (5-16B). Each of these composite units, or *sedimentary cycles*, lies unconformably on the one beneath it, because the channel in which the basal deposits accumulate removes the uppermost sediments of the preceding cycle as it migrates. Sometimes a channel cuts deeply, removing not only the uppermost deposits of the preceding cycle but some lower ones as well. Thus many preserved cycles of migrating channels are really only partial cycles.

Two final points deserve mention. First, not every river can be assigned to the braided or meandering category. Some rivers have segments or branches that are braided and others that are meandering. Second, no river in a moist region deposits its entire load of sediment in its valley. A river ultimately discharges its water and remaining sediment into a lake or the sea (see Figure 5-13).

Marginal Marine and Open-Shelf Environments

Along the edge of the sea, sediment accumulates under the influence of both fresh water and seawater. Thus, freshwater plants may form peat deposits in a marsh near the coast, while waves that break nearby deposit sand and bury shells of marine animals along a beach exposed to the open ocean.

A delta forms where a river meets the sea

Where a river empties into either a lake or the sea, its current dissipates, and it often drops its load of sediment in a fanlike pattern. The depositional body of sand, silt, and clay that is formed in this way is called a **delta** because of its resemblance to the Greek letter Δ. Most of the large deltas that have been well preserved in the geologic record formed in areas where sizable rivers emptied into ancient seas.

Like all moving water, as river water mixes with standing water and begins to slow down, it loses sand first. Silt, being finer, spreads farther from the mouth of the river, and clay is carried even farther. The typical result is a delta structure that includes **delta-plain**, **delta-front**, and **prodelta** deposits (Figure 5-17).

Delta-plain beds, which consist largely of sand and silt, are nearly horizontal except where they are locally cross-bedded. Some delta-plain deposits accumulate within river channels. As a river slows on the surface of the delta, sand builds up on the bottom, causing the channel to branch repeatedly into smaller channels that radiate out from the mainland. These **distributary channels** are floored by cross-bedded sands. Sand also spills out from the mouths of the channels, forming shoals and sheetlike sand bodies along the delta front. Between the distributaries and separated from them by natural levees are swamps, which are sometimes dotted by lakes. Here, as in the backswamps of meandering

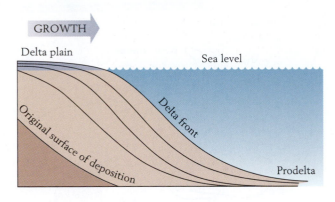

Figure 5-17 **Cross section of a delta.** As river water flows into the sea, it slows down. First sand drops from suspension on the delta plain, then silt and clay settle on the delta front (the slope is greatly exaggerated in the diagram), and finally clay settles on the prodelta. As the delta grows seaward, the sandy shallow-water sediments build out over the finer-grained sediments deposited in deeper water.

rivers, muds accumulate and marsh plants often grow, contributing to future coal deposits.

Delta-front beds slope seaward from the delta plain, usually lying in waters too deep to be agitated by wind-driven surface waves. These beds consist largely of silt and clay, which can settle under these quiet conditions. Because they lie fully within the marine system, delta-front muds harbor marine faunas that often leave fossil records, but these muds usually contain fragments of waterlogged wood as well. In fact, the presence of abundant fossil wood in ancient marine muds testifies to the presence of both land and a river system near the site, and such ancient subtidal muds usually represent deltaic deposits.

Spreading seaward at a low angle from the lowermost delta-front deposits are the prodelta beds, which consist of clay. Even during floods, the fresh water that spreads from distributary channels slows down so abruptly that it loses its silt on the delta front. It is partly because fresh water is less dense than seawater that clay is carried far from the mouths of distributary channels. Because the fresh water floats on top of the denser seawater, it does not mix in quickly; instead, it spreads seaward for some distance, still carrying much of its clay.

As a delta **progrades**, or grows seaward, the relatively coarse deposits of the delta plain build out over the finer-grained delta-front beds in accordance with Walther's law (see Figure 5-17), and the delta-front beds build out over the still finer-grained prodelta beds. The result is a sequence of deposits that coarsens toward the top.

The famous Mississippi River delta spills into the Gulf of Mexico in an area that is protected from strong wave action. As a result, the delta projects far out into the sea. Because construction of the delta from river-borne sediment has prevailed decisively over the destructive forces of the sea, this type of delta is sometimes called a *river-dominated delta*. The growing portion, or **active lobe**, of such a delta is the site of the functioning distributary channels (Figure 5-18). Many previously active lobes can be identified in the delta-plain portion of the Mississippi delta, and these lobes, which have been dated by the carbon 14 method (described in Chapter 6), provide a history of deltaic development during

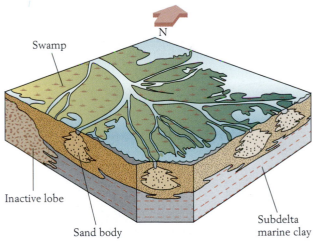

Figure 5-18 **Structure of the active lobe of the modern Mississippi River delta.** At this lobe (identified by the pink arrow on the map), sediments are being deposited in the shallow water where the river meets the sea. Here, as shown in the block diagram below, the river channel divides into many distributary channels. Swamp deposits accumulate between the distributaries. Bodies of sand accumulate in front of the distributaries where the river water meets the ocean and slows down. As the active lobe builds seaward, the distributaries extend over these sand bodies. (After H. N. Fisk et al., *J. Sediment. Petrol.* 24:76–99, 1954.)

Figure 5-19 Lobes of the Mississippi delta. The active lobe is numbered 16. Older lobes that are now inactive are numbered 1 through 15, with 1 being the oldest. These older lobes are now subsiding, and some are already partially or entirely submerged. (After T. Elliott, in H. G. Reading, ed., *Sedimentary Environments and Facies*, Blackwell Scientific Publications, Oxford, 1978.)

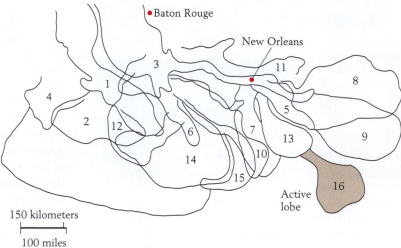

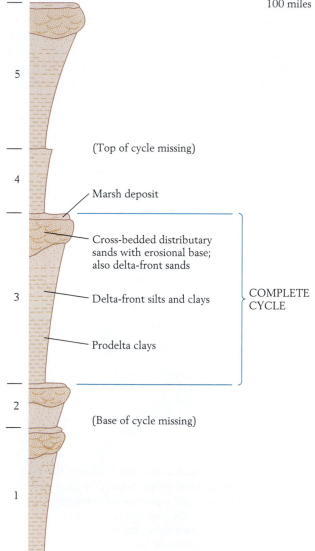

Figure 5-20 **Five deltaic cycles.** Cycle 3 is a complete cycle; within it prodelta clays grade upward to delta-plain sands. Each cycle represents an accumulation of sediments resulting from the seaward growth of a deltaic lobe.

the past few thousand years (Figure 5-19). Depositional activity (or lobe growth) periodically shifted when floods caused the river to cut a new channel and to abandon the previously active channel and its distributaries.

The fate of an abandoned delta lobe is the key to the stratigraphic sequence of a river-dominated delta. An abandoned lobe gradually sinks, for two reasons: first, the sediments of which it is formed compact under their own weight, and second, the lobe is part of the entire delta structure, which is constantly sinking as a result of the isostatic response of the underlying crust to the weight of the continually growing mass of sediment. A younger lobe eventually grows on top of an abandoned lobe, with each consisting of the typical upward-coarsening sequence. The result is an accumulation of sedimentary cycles that differ markedly from those of meandering rivers, which, as you will recall, become finer-grained toward the top.

Some deltaic cycles in the rock record lack tops because sediment was eroded away before another cycle was superimposed (Figure 5-20). Because the building of a river-dominated delta is limited to the active lobe at any given time and because older exposed beds erode easily, a delta can seldom be traced very far laterally in the rock record. When preserved, the porous sand bodies in the upper parts of deltaic cycles may serve as reservoirs for petroleum or natural gas.

Changes in the rate at which a delta sinks or is supplied with sediment can alter the size of the active lobe. Today such changes are causing the active lobe of the Mississippi delta to shrink rapidly, with alarming consequences. The building of levees along the Mississippi River and of dams along its tributaries have reduced the rate of deposition of sediment on the Mississippi delta. In addition, the removal of groundwater for human use in the region of the delta has increased the rate of subsidence of the delta fivefold. Because of these two factors, the Louisiana coast is now losing about 100 square kilometers (40 square miles) of land every year. Furthermore, as the marine waters of the Gulf of Mexico encroach farther and farther inland over the delta, they are drowning wetlands that provide nearly 30 percent of the United States' annual seafood harvest and support the nation's largest population of waterfowl.

Lagoons lie behind barrier islands of sand

Deltas of large rivers occupy only a small percentage of the total shoreline of the world's oceans. Where they are absent, long stretches of shoreline are fringed by **barrier islands**, composed largely of clean sand piled up by waves. Although some barrier islands extend laterally from deltas, most derive their sand not from neighboring rivers, but from the marine realm. They are built up as waves and the shallow currents that flow along the coast, called **longshore currents**, winnow sediments and sweep sand parallel to the shoreline. Where the beach of a barrier island is washed by breaking waves, deposits tend to have nearly horizontal bedding, but often dip gently seaward. Cross-bedding develops in areas where the beach surface is gently irregular and changes from time to time. Wind-blown sand often accumulates behind the beach as sand dunes, but in time these dunes are often eroded away.

Lagoons lie behind long barrier islands, such as those that border the Texas coast (Figure 5-21). Protected from strong waves, lagoons trap fine-grained sediment and are usually floored by muds and muddy sands. Small rivers often build deltas along the landward margins of lagoons. A barrier island and the lagoon behind it form a **barrier island–lagoon complex**.

Barrier islands often form chains with tidal channels separating adjacent islands. Tidal currents pass through these channels and deposit cross-bedded *tidal*

deltas within the lagoon. Other depositional environments are also found along the shores of lagoons. Among them are **tidal flats**—formed of sand or muddy sand—whose surfaces are alternately exposed and flooded as the tide ebbs and flows. High in the intertidal zone, above the barren tidal flats, marshes fringe one or both margins of many lagoons. Here plant debris accumulates rapidly and decomposes to form peat or, after long burial, coal.

Fresh water from rivers and streams tends to remain trapped in coastal lagoons for some time, so the waters of lagoons in moist climates are often brackish. The salinity of these waters at any given time depends on the rate of freshwater runoff from the land, which varies during the course of the year. Laguna Madre of Texas is typical of lagoons found in warm, arid climates (see Figure 5-21). The ponded waters of this long lagoon are hypersaline because they receive little fresh water from rivers and suffer a high rate of evaporation. Whether lagoons are brackish or hypersaline, their abnormal and fluctuating salinity excludes many forms of marine life. As a result, the fossil faunas in the ancient sediments of lagoons are not very diverse; those species that are present, however, often occur in large numbers. Usually among them are burrowers, such as segmented worms, that disturb the muddy lagoonal sediments, leaving them mottled and largely devoid of bedding structures.

When a barrier island–lagoon complex receives sediment at a sufficiently high rate, it progrades—that is, it migrates seaward—like the active lobe of a delta. Unlike the migration of a delta, however, this progradation takes place along a broad belt of shoreline (Figure 5-22). As the shoreline of a sea migrates seaward, marsh and

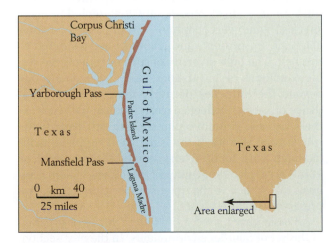

Figure 5-21 Barrier islands along the Texas coast. Tides here are weak, and there are few tidal channels or passes, so large lagoons lie behind these Texas barrier islands.

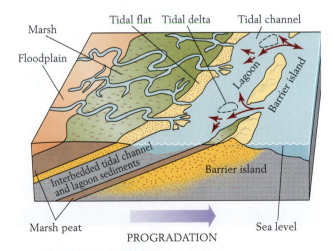

Figure 5-22 The stratigraphic sequence produced when a barrier island–lagoon complex prograde. Sediments of the lagoon and of the adjacent marshes and tidal flats are superimposed on beach sands of the barrier island.

tidal-flat deposits prograde over sediments of the lagoon and its associated tidal channels. All of these sediments in turn build out over deposits of the barrier islands and over the tidal deltas and marshes behind them. Thus the horizontal sequence of depositional environments (barrier island, marsh or tidal delta, lagoon, tidal flat, and marsh) comes to be represented by a corresponding vertical sequence of sedimentary deposits, in accordance with Walther's law.

Open-shelf deposits include tempestites

Seaward of barrier islands, continental shelves display a variety of physical conditions and therefore produce a variety of sedimentary deposits. On open shelves where tides produce strong currents and sand is abundant, the currents may pile the sand into large ridges or dunelike structures. On shelves where waves have a stronger effect than tidal currents, wave motion tends to flatten the bottom, and the sand spreads out in sheets.

On quieter shelves, mud or muddy sand accumulates most of the time, but storms occasionally produce **tempestites**, which are sandy beds that are usually a few centimeters thick. A storm that pounds a coastline may produce waves that pile up water carrying sand that they have scoured from the shallow seafloor. After the storm passes inland, the piled-up water flows seaward, and as it loses velocity, deposits the suspended sediment on the shelf in the form of a tempestite. The occasional deposition of tempestites on a normally quiet shelf pro-

Figure 5-23 Tempestite beds of Ordovician age in Kentucky. The coarse-grained sediment in the lower part of this photograph was deposited by strong water movements produced by a storm along the shelf of an inland sea. (Frank R. Ettensohn.)

duces a succession of thin sandy beds in a body of finer-grained sediment (Figure 5-23). Individual tempestites are commonly graded, having formed as sand settled before silt or mud.

Animal burrows are abundant in most shelf sediments, and as we will soon see, particular skeletal fossils also point to an offshore, open marine environment.

Fossils serve as indicators of marine environments

Ancient sediments deposited within and seaward of barrier island–lagoon complexes often yield fossils that help geologists to recognize particular depositional environments. Figure 5-24 depicts an example in Wales, west of central England. Here, in rocks of mid-Silurian age, fossil communities of marine invertebrates are arrayed roughly parallel to the ancient shoreline. Adjacent to that shoreline is a narrow zone of fine-grained sediments, in which the inarticulate brachiopod *Lingula* is especially abundant, but only a few other species are present (Figure 5-24A). Presumably the water here was brackish: *Lingula*, a living fossil genus (see Figure 3-30B), today tolerates nearshore environments of brackish and variable salinity where few other species are able to live.

Seaward of the zone where *Lingula* predominates in the Silurian deposits is a more diverse fossil community, adapted to the more stable conditions of the center and seaward margin of a lagoon (Figure 5-24B). Sandy deposits representing a barrier island have not been preserved, but we can infer that one was present because seaward of the lagoonal deposits are sandy, often cross-stratified marine deposits in which the most common fossil is a type of brachiopod that was adapted to such agitated conditions (Figure 5-24C).

In finer-grained sediments deposited farther offshore is a fossil community that includes many species, none very common. Many types of brachiopods are present, and trilobites are restricted to this belt. The high diversity of species reflects the stable conditions of an offshore shelf environment, one beyond the influence of river water; the low abundance of species reflects a weak food supply, far from the algae and primitive plants that must have flourished in the vicinity of the lagoon and supplied its inhabitants with food. Where muddy sediments occur in this area, planktonic graptolites—fragile colonial animals that settled to the quiet seafloor after death—are preserved. (Figure 5-24D).

Fossils have played a key role in the reconstruction of this set of Silurian environments. In the next section, we will examine deposits that are composed entirely of recognizable fossils and fossil debris.

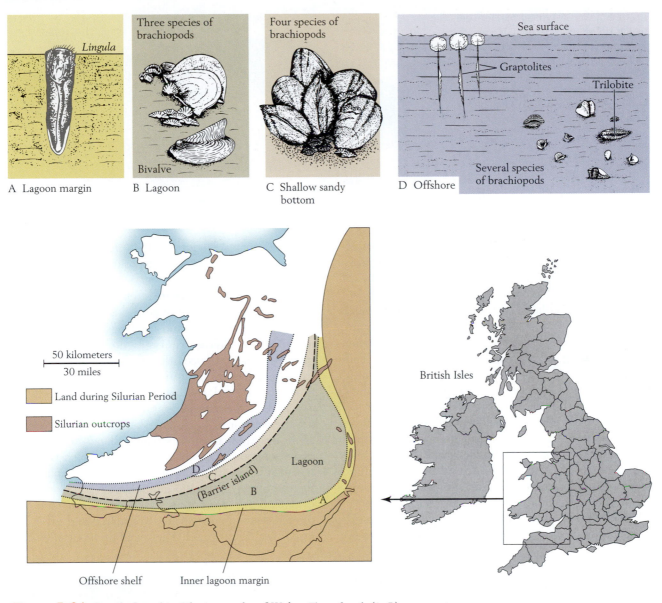

Figure 5-24 Fossils found in Silurian rocks of Wales. These fossils (A–D) represent successive marine habitats from an inner lagoon margin to an offshore shelf. (After A. M. Ziegler, in W. S McKerrow, ed., *The Ecology of Fossils*, MIT Press, Cambridge, MA, 1978.)

Organic reefs are bodies of carbonate rock

In tropical shallow marine settings where siliciclastic sediments are in short supply, carbonate sedimentation usually prevails. Here coral reefs are often prominent, rising above the seafloor as rigid structures. Because they are produced largely by organisms that secrete calcium carbonate, organic **reefs** form their own distinct depositional records—as bodies of limestone. Although some ancient reefs were formed by organisms other than corals, they, like their modern counterparts, grew in shallow waters of high clarity and normal marine salinity.

The basic framework of a reef consists of the calcareous skeletons of organisms, primarily corals. This framework is strengthened by cementing organisms that encrust the surface of the reef. Carbonate sediment, composed of fragments of the skeletons of reef-dwelling organisms, is trapped within the porous framework, filling some voids. With their complex internal structure, reef limestones are typically either unbedded or only poorly bedded. Even with the presence of infilling debris, reef limestone is so porous that many ancient buried reefs serve as traps for petroleum, which migrates into them from sediments rich in organic matter.

Figure 5-25 A typical barrier reef. The reef grows up to sea level, so the reef flat is exposed at low tide. Waves break across the reef flat, losing energy in the process and leaving a quiet lagoon behind. Sediment accumulates in the back-reef area and in the lagoon, and here and there patch reefs rise from the lagoon floor.

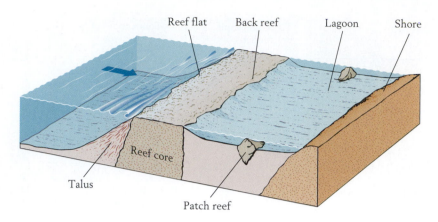

Because living reefs stand above the neighboring seafloor, they alter patterns of sedimentation nearby. On the leeward side of an elongate reef—the side nearest the land—there is often a relatively calm lagoon, especially if the reef has a typical **reef flat**, or horizontal upper surface, that stands close to sea level (Figure 5-25). Below the living surface of the reef is a limestone core consisting of a dead skeletal framework and trapped sediment. A pile of rubble called **talus**, which has fallen from the steep, wave-swept reef front, often extends seaward from the living surface.

Reefs build upward rapidly enough to remain near sea level even when the seafloor around them is becoming deeper. Many reefs, in fact, grow so rapidly and are so durable that they build seaward in the manner of a prograding delta. Figure 5-26 shows a spectacularly exposed cross section of a Devonian reef in Australia. Although it was built by organisms that have been extinct for hundreds of millions of years, this reef closely resembles many modern reefs in its basic structure—it displays both seaward talus deposits and leeward back-reef strata.

Isolated **patch reefs** are often found in lagoons behind elongate reefs (see Figure 5-25). Elongate reefs that face the open sea and have lagoons behind them are known as **barrier reefs** (Figure 5-27). Reefs that grow right along the coastline without a lagoon behind them are known as **fringing reefs**. Some fringing reefs grow seaward and eventually become barrier reefs.

Perhaps the most curious reefs in the modern world are the circular or horseshoe-shaped structures known as **atolls**. Atolls form on volcanic islands and thus are quite common in the tropical Pacific, which is dotted by many such islands. Charles Darwin's explanation for the origin of Pacific atolls is still accepted today (Figure 5-28). According to Darwin, each atoll was formed when a cone-shaped volcanic island was colonized by a fringing reef. The island then began to sink,

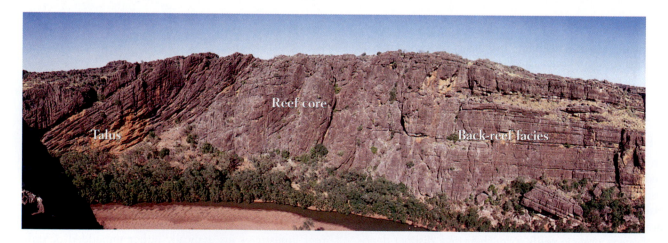

Figure 5-26 Outcrop along Windjana Gorge, northwestern Australia, revealing the internal structure of a Devonian reef. The reef core consists of unbedded limestone. The talus is crudely bedded, and the beds slope away from the reef core. The back-reef strata are also crudely bedded, but the beds are approximately horizontal. (P. E. Playford, Geological Survey of Western Australia.)

Figure 5-27 A barrier reef.
The open sea is on the right, the
lagoon on the left. (Ralph and Daphne
Keller/NHPA.)

turning the reef into a barrier reef, with a lagoon separating it from the remnant of the volcano. The island eventually sank beneath the sea, leaving a circular reef standing alone with a lagoon in the center, where limestone now accumulates in quiet water. Often the reef does not quite form a full circle but is broken by a channel on the leeward side, where food supplies are low

and reef-building organisms do not thrive. Horseshoe-shaped atolls range up to about 65 kilometers (40 miles) in diameter; during World War II their lagoons served as natural harbors for ships.

Ancient atolls that lie buried beneath younger sediments can be identified by studying cores of rock brought up from drilling operations—and because

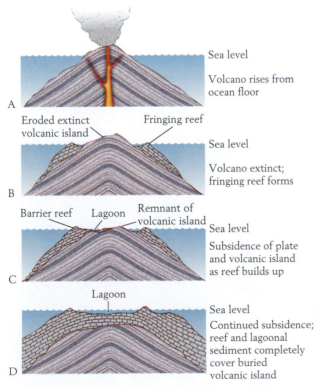

A. Sea level

Volcano rises from ocean floor

B. Eroded extinct volcanic island / Fringing reef

Sea level

Volcano extinct; fringing reef forms

C. Barrier reef / Lagoon / Remnant of volcanic island

Sea level

Subsidence of plate and volcanic island as reef builds up

D. Lagoon

Sea level

Continued subsidence; reef and lagoonal sediment completely cover buried volcanic island

E

Figure 5-28 **Development of a typical coral atoll in the Pacific.** *A–D.* The stages of development of an atoll, as proposed by Charles Darwin. *E.* A coral reef encircling a volcanic island in the Pacific Ocean, representing the stage of atoll formation illustrated in *C.* (*E,* Truchet/Tony Stone Worldwide.)

Figure 5-29 A late Paleozoic horseshoe-shaped atoll lies almost a kilometer (0.6 mile) below the surface of the land in Texas. The atoll was discovered when rocks of the region were drilled for petroleum. The reef appears to have faced prevailing winds from the south. (After P. T. Stafford, U.S. Geological Survey Professional Paper no. 315-A, 1959.)

porous reef rocks often serve as traps for petroleum, drilling in the vicinity of these atolls is often profitable. Figure 5-29 shows the outline of a subsurface atoll of late Paleozoic age that has yielded considerable quantities of petroleum in the state of Texas.

Carbonate platforms form in warm seas

An organic reef commonly forms part of a **carbonate platform**, which is a broad structure that consists of

calcium carbonate and stands above the neighboring seafloor on at least one of its sides. Organic reefs often grow along the windward margins of carbonate platforms, where food is plentiful. The calcium carbonate of carbonate platforms precipitates from shallow tropical waters at or near the site where it accumulates.

Reefs and carbonate platforms are largely restricted to tropical seas because carbon dioxide is less soluble in warm water than in cold water. (You may have noticed that a warm bottle of carbonated soda fizzes readily when shaken, releasing bubbles of carbon dioxide.) Removal of carbon dioxide from seawater favors the precipitation of calcium carbonate because carbon dioxide dissolves in seawater to form carbonic acid:

$$CO_2 + H_2O = H_2CO_3$$

The cooling of water in contact with the atmosphere increases the concentration of dissolved carbon dioxide in the water, driving this chemical reaction to the right and increasing the concentration of carbonic acid. Similarly, warming the water removes carbon dioxide from solution, decreasing the concentration of carbonic acid.

Carbonic acid breaks down calcium carbonate to form ions of calcium and bicarbonate, which remain in solution:

$$H_2CO_3 + CaCO_3 = Ca^{2+} + 2HCO_3^-$$

An increase in the carbonic acid concentration in water causes more calcium carbonate to break down. In complementary fashion, a decrease in the carbonic acid concentration favors the precipitation of calcium

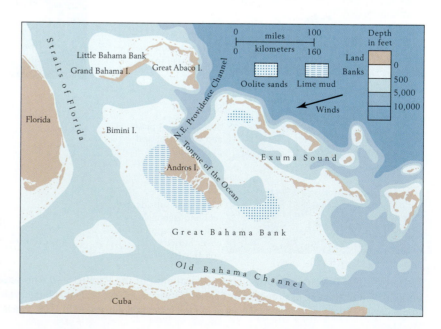

Figure 5-30 The Bahama banks, which are now separated from Florida by the Straits of Florida. Here great thicknesses of sediment have accumulated over 170 million years. (After N. D. Newell and J. K. Rigby, Society of Economonic Palaeontologists and Mineralogists, Special Paper no. 5, 1957.)

carbonate. Thus, low levels of carbonic acid in warm tropical seas favor the growth of organic reefs and carbonate platforms.

In times past, carbonate platforms that are now buried have stretched along most of the eastern margin of the United States, but because climates are cooler today than they have been during most of Earth's history, carbonate platforms in the modern world are restricted to low latitudes. In the western Atlantic and Caribbean region, a large carbonate platform extends seaward from the Yucatán Peninsula of Mexico. Smaller platforms border the Antilles, and the platforms known as the Little Bahama Bank and Great Bahama Bank lie to the east and southeast of Florida (Figure 5-30).

The varied sediments currently accumulating on the Bahama banks resemble those of many ancient carbonate platforms. Here, as on carbonate platforms generally, sediments accumulate rapidly. Since mid-Jurassic time, about 170 million years ago, some 10 kilometers (6 miles) of carbonates have accumulated both on the Bahama banks and in southern Florida, which was part of the same carbonate platform during Cretaceous time and earlier. This heavy buildup of sediments has caused the oceanic crust to subside, so that shallow-water Jurassic deposits now lie up to 10 kilometers below sea level.

Among the accumulating carbonate sediments are oolites, which are composed of ooliths—spherical grains that consist of aragonite needles precipitated from seawater (see Figure 2-23). Strong currents pile ooliths into spectacular shoals in some shallow subtidal areas of the Bahama banks (p. 102). Oolites formed in this manner display conspicuous cross-bedding. This feature and the need for individual ooliths to roll around in order to form makes cross-bedded oolites in the rock record reliable indicators of shallow seafloors swept by strong currents.

Bordering tidal channels in some areas of the Bahama banks are knobby intertidal structures known as **stromatolites**. They are produced by threadlike cyanobacteria. As Figure 5-31 indicates, these organisms form sticky mats by trapping carbonate mud. After forming a mat, they grow up through it to produce another one. Repetition of this process on an irregular surface forms a cluster of stromatolites. Each is internally layered: organic-rich layers alternate with organic-poor layers. The fossil record of stromatolites is unusually ancient, extending back more than 3 billion years.

There is a simple reason that stromatolites are found almost exclusively in supratidal and high intertidal settings. Because these environments are exposed above sea level much of the time, they become hot and dry, and relatively few marine animals can survive in them. Thus there is little to interfere with the tendency of the mats created by cyanobacteria to form layered

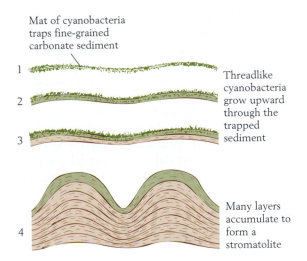

Figure 5-31 **The growth of stromatolites.** A mat of cyanobacteria traps sediment, and the cyanobacteria grow through it to form another layer. The accumulation of several layers leads to the formation of a stromatolite.

structures. Such mats can grow under water, but they are quickly eaten by grazing marine animals and damaged by burrowers, so they seldom accumulate to form stromatolites or well-layered limestones. Exceptions are large column-shaped stromatolites that grow in subtidal channels in the Bahamas where tidal currents are very strong. Few animals can survive in these current-swept areas, so stromatolites flourish there. Stromatolites also flourish in Shark Bay, Western Australia, where the waters are hypersaline and animals are very rare (Figure 5-32). As we shall see in later chapters, very ancient stromatolites formed beneath the sea before the origin of marine animals that feed on cyanobacteria.

Figure 5-32 **Living stromatolites.** These forms, exposed at low tide, occupy the margin of Shark Bay, a lagoon in Western Australia where hypersaline waters allow few animals to survive. (Paul F. Hoffman.)

Stromatolites are not the only indicators of extreme environments near shore. Mudcracks are as telling here as they are on continental interiors. Ridges of sediment that build up along tidal channels in the Bahama banks dry out after occasional flooding by storm-driven seas. As a result, the surface of the sediment here is broken by mudcracks resembling those that often form when a mud puddle on land dries up. Mudcracks associated with ancient marine deposits usually represent intertidal or supratidal environments that were alternately wetted by the sea and dried by the sun.

Deep-Sea Environments

To reconstruct the distribution of continents and ocean basins for any time interval in the past, geologists must identify not only nearshore deposits of the various kinds just described, but also deposits of deep-water environments beyond continental shelves. Coarse clastic deposits derived from a continental shelf accumulate along its base, but fine-grained sediments predominate in the middle of a huge ocean, far from sources of clastic sediments, and they accumulate very slowly.

Turbidites flow down submarine slopes

One of the most remarkable advances in sedimentology took place in the middle of the twentieth century with the recognition that certain sedimentary rocks have been produced by turbidity currents. A **turbidity current** is a flow of dense, sediment-charged water moving down a slope under the influence of gravity.

Turbidity currents were first noticed in clear lakes, where flows formed from muddy river water that hugged the lake floor. These currents flowed for a considerable distance, slowing and dropping their sediment only when they reached gentler slopes and spread out far from shore. In the 1930s the Dutch geologist Philip Kuenen demonstrated in the laboratory that turbidity currents can attain great speed, especially when they are heavily laden with sediment and moving down steep slopes. Sediment suspended in a turbidity current behaves as part of the moving fluid, and its presence increases the density of the fluid by as much as a factor of 2.

When the slope beneath a turbidity current begins to flatten out, the current slows and spreads out, dropping its sediment in the general sequence that we have seen again and again: first the coarse sediment falls from suspension, and then, much later, the fine material follows. The result is a graded bed of sediments, with poorly sorted sand and granules at its base and mud at the top. Such a graded bed is known as a **turbidite** (Figure 5-33A).

Large turbidity currents flow down continental slopes and deposit turbidites along continental rises and on the abyssal plain. Turbidity currents that originate near the edge of the continental shelf not only carry sediment to the deep sea but also erode both the continental slope and part of the continental rise. Thus,

A

B

Figure 5-33 Ordovician turbidites in New York State. *A.* A turbidite bed that grades from coarse at the bottom to fine at the top; the upper surface is irregular because it was disturbed by the succeeding turbidity current and distorted by subsequent compaction. *B.* The bottom surface of a turbidite bed, showing "sole marks" produced when the sediment forming the bed filled depressions that the turbidity current depositing it scoured in the preexisting sediment surface. Most scour marks taper toward the upper left, indicating that the current flowed in that direction. (Earle F. McBride.)

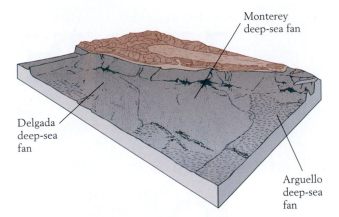

Figure 5-34 Deep-sea fans spreading from submarine canyons along the coast of California. The fans consist primarily of turbidite deposits. (After H. W. Menard, *Geol. Soc. Amer. Bull.* 71:1271–1278, 1960.)

such currents are largely responsible for carving great submarine canyons in many parts of the slope. The turbidity currents slow down at the mouths of these canyons, dropping part of their sedimentary load to form deep-sea fans that superficially resemble alluvial fans (Figure 5-34). In fact, much of the continental rise actually consists of coalescing submarine fans.

Ancient turbidites normally are stacked one on top of another in groups. The result is that the deposits they form are cyclical (see Figure 1-21). The bottom of each cycle consists of coarse material at the base of the turbidite, and the top consists of mud that accumulated in the quiet depths before the next turbidite formed.

The sandy portions of lithified turbidites are typically graywackes, which are quite unlike the clean sands of meandering-river deposits. This is only one of several ways in which a cyclical sequence of turbidites differs from a meandering-river cycle, although both show a grading of sediment from coarse to fine within each complete cycle. Another difference is that a single turbidite is usually only a few centimeters and seldom as much as a meter thick (see Figure 1-21), whereas most complete meandering-river cycles measure at least two or three meters from bottom to top. In addition, turbidites lack the large-scale cross-bedding characteristic of meandering river channels. Furthermore, the base of a turbidite is often irregular because the earliest, most rapidly moving waters of a turbidity current scour depressions in the sedimentary surface laid down earlier. These scours are subsequently filled in by the first sediments that settle as the current slows. When the base of a lithified turbidite is turned over for inspection, its irregularities, or "sole marks," can reveal the direction of water flow (Figure 5-33*B*).

Pelagic sediments are fine-grained and accumulate slowly

Turbidity currents and other bottom flows carry mud to the abyssal plain of the ocean well beyond the continental rise. None of these flows, however, contributes sediment to the deep sea at a sustained high rate; indeed, sediment in most areas of the abyssal plain accumulates at a rate of about a millimeter per thousand years! In the deep sea, most sediments are clays, which come from two sources. One source is the weathering of rocks produced by oceanic volcanoes. Such clays are less abundant in the Atlantic than in the central Pacific, where volcanoes are common. Clays also reach the deep sea by settling from the water above, having traveled through the air as wind-blown dust or having drifted seaward from the land at very low concentrations in surface waters of the ocean. Just as organisms that occupy the water above the deep-sea floor are termed pelagic forms of life, fine-grained sediment that settles from these waters to the deep-sea floor is called **pelagic sediment**.

In addition to clays, pelagic sediment includes skeletal material contributed by small pelagic organisms. Whether clays predominate in any area of the deep sea depends on the extent to which they are diluted by the more rapid accumulation of biologically produced sediments. Some of the latter consist of calcium carbonate, while others consist of silica.

Where deposition of calcium carbonate predominates in the deep sea, its fine grain size has led oceanographers to refer to it as **calcareous ooze** (Figure 5-35). This sediment consists of skeletons of single-celled planktonic organisms, including planktonic foraminifera, which are amoeba-like protozoans (see Figure 3-18). Other important constituents are the armorlike plates that surround calcareous nannoplankton, the single-celled floating algae that are major components of tropical phytoplankton (see Figure 3-16*C*).

Calcareous ooze is abundant in the modern ocean only at depths of less than about 4000 meters (about 13,000 feet) because calcium carbonate begins to dissolve as it settles to great depths. As pressure increases and temperature declines with depth, the concentration of carbon dioxide increases—and so, therefore, does the concentration of carbonic acid. As a result, most small particles of calcium carbonate dissolve by the time they sink to a depth of 4000 meters.

In many regions at high latitudes, as well as in tropical Pacific regions characterized by strong upwelling, biologically produced **siliceous ooze** carpets the deep-sea floor. This sediment consists of the skeletons of two groups of organisms that thrive where upwelling supplies nutrients in abundance: diatoms, a highly productive phytoplankton group found in nontropical waters

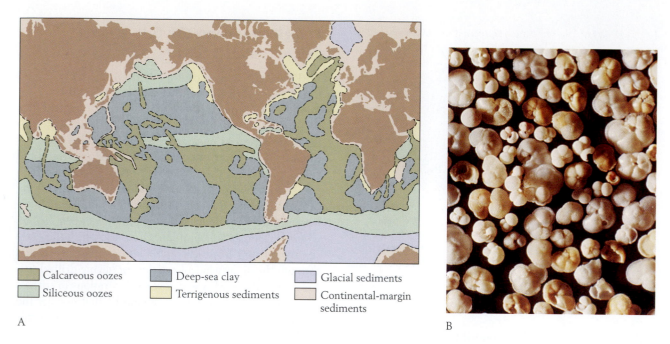

■ Calcareous oozes	■ Deep-sea clay	■ Glacial sediments
■ Siliceous oozes	■ Terrigenous sediments	■ Continental-margin sediments

A   B

Figure 5-35 The global pattern of deep-sea sediments. *A.* Calcareous oozes are restricted to low latitudes. Most areas of siliceous ooze lie closer to the poles, although some occur close to the equator in areas of the Pacific and Indian oceans. *B.* Calcareous ooze composed of planktonic foraminifera. (*A* after T. A. Davies and D. S. Gorsline, in J. P. Riley and R. Chester, eds., *Chemical Oceanography*, Academic Press, London, 1976; *B*, Heather Angel/Biofotos.)

(see Figure 3-16*B*), and radiolarians, which are single-celled planktonic protozoans related to foraminifera (see Figure 3-19). Recall that the skeletons of both diatoms and radiolarians consist of a soft form of silica called opal, which tends to recrystallize so that individual skeletons cannot be discerned (see Figure 2-21). In the process, the rock that they form becomes a dense, hard chert (see Figure 2-20). Before recrystallization, the soft sediment, consisting largely of diatoms, is known as *diatomaceous earth*—the abrasive component of many scouring powders.

Thick bodies of diatomaceous sediment have formed in marine areas of strong upwelling. Diatoms did not exist until late in Mesozoic time, however, and here we come to an important point: the composition of pelagic sediments has changed markedly in the course of geologic time as groups of sediment-contributing organisms have waxed and waned within the pelagic realm.

Chapter Summary

How does vertical stacking of distinctive types of strata provide clues to environments of deposition?

Sometimes one kind of rock alone serves to identify an ancient environment. Usually, however, suites of closely associated rock types are required for this purpose. These deposits are commonly organized into cycles in which, in accordance with Walther's law, one kind of sediment tends to lie above another that accumulated in an adjacent environment.

What sedimentary features result from deposition in particular nonmarine environments?

Ancient soils resembling those of the modern world are sometimes found beneath unconformities, although they may be hard to identify because of chemical alteration. Some types of ancient soils reflect the climatic conditions under which they formed. Lake deposits, which are much less common than marine deposits, are characterized by thin horizontal layers, few burrows, and an absence of marine fossils. Glaciers, which plow over the surface of the land, often leave a diagnostic suite of features, including scoured and scratched rock surfaces, poorly sorted gravelly sediment, and associated lake deposits that exhibit annual layers. In hot, arid basins, erosion of the surrounding highlands creates gravelly alluvial fans. Braided streams flowing from the fans toward the basin center deposit cross-bedded gravels and sands. Beyond these deposits there may be shallow lakes and salt flats where evaporites accumulate. Some arid basins also contain dunes of clean, cross-bedded, wind-blown sand. In moist climates, braided streams also form on alluvial fans, and in lowland areas meandering rivers leave characteristic deposits in which channel sands and gravels grade upward through point-bar sands to muddy backswamp sediments.

What are the distinctive features of marginal marine and continental-shelf deposits?

Where a river meets a lake or an ocean, it drops its sedimentary load to form a delta. Deltaic deposition typically produces an upward-coarsening sequence as shallow-water sands build out over deeper-water muds. More widespread than deltas along the margin of the ocean are muddy lagoons bounded by barrier islands formed of clean sand. Coral reefs border many tropical shorelines. A typical reef stands above the surrounding seafloor, growing close to sea level and leaving a quiet lagoon on its leeward side. Most reef limestones are supported by rigid internal organic frameworks. Coral reefs form parts of many carbonate platforms, although these platforms contain a number of other deposits as well. On continental shelves, storms occasionally produce thin sandy beds known as tempestites.

What are the characteristics of deep-sea sediments?

Beyond the edge of the continental shelf, turbidity currents intermittently sweep down continental slopes to the continental rise and abyssal plain, where they spread out, slow down, and deposit graded beds of sediment, known as turbidites. Still farther from continental shelves, only fine-grained sediments accumulate. Clay reaches these deep-sea areas very slowly. In some areas the deposition of clay is far surpassed by the accumulation of minute skeletons of planktonic marine life, which settle to the seafloor to form calcareous or siliceous oozes.

Review Questions

1. What kinds of nonmarine sedimentary deposits reflect arid environmental conditions?

2. What kinds of nonmarine sedimentary deposits reflect cold environmental conditions?

3. What kinds of deposits indicate the presence of rugged terrain in the vicinity of a nonmarine depositional basin?

4. In what nonmarine settings do gravelly sediments often accumulate?

5. Contrast the patterns of occurrence of sediments and sedimentary structures in the following three kinds of depositional cycles: the kind produced by meandering rivers, the kind produced by deltas, and the kind produced by turbidity currents.

6. Draw a profile of a barrier island–lagoon complex, and label the various depositional environments.

7. What features typify sediments that accumulate in the centers of lakes?

8. How do stromatolites form?

9. Describe the kind of rock found in a typical organic reef.

10. Where is a lagoon in relation to a barrier reef? Where is it in relation to an atoll?

11. Which features of carbonate rocks suggest intertidal or supratidal deposition? Which features suggest subtidal deposition?

12. What types of sediments and sedimentary structures usually reflect deposition in a deep-sea setting?

13. What are the important depositional environments of the basic kinds of sediment (such as mud, well-sorted sand, gravel, evaporites, and various kinds of limestone) and of particular sedimentary structures (such as cross-bedding, graded beds, mudcracks, and ripples)? How are the kinds of sediment and sedimentary structures found within each environment related to processes operating within it? Use the Visual Overview on page 104 for reference.

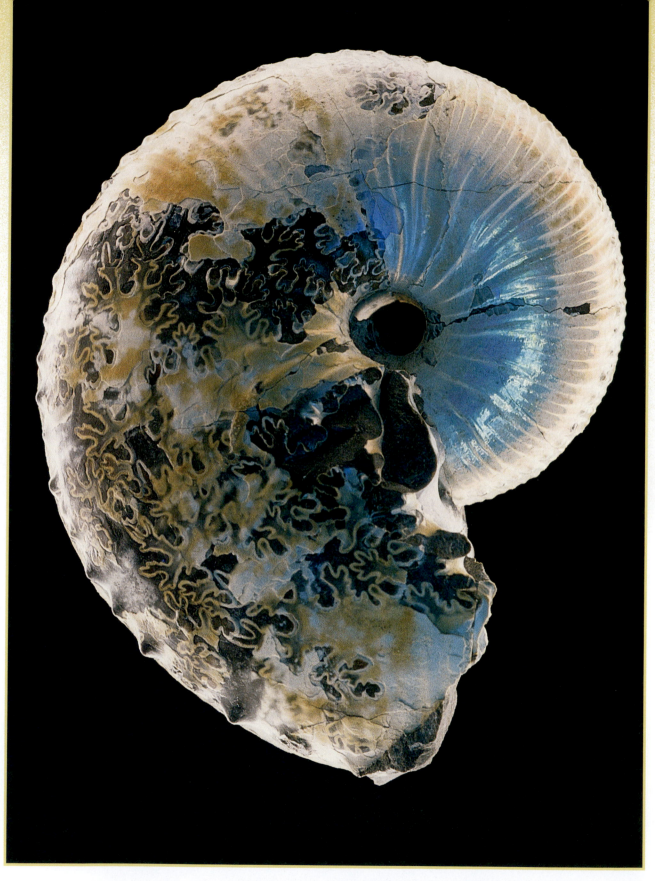

A distinctive Upper Cretaceous ammonoid, displaying some original brownish shell material on the upper left and pearly inner shell material on the right. Dark sediment below fills chambers between convoluted interior partitions of the shell called septa. Ammonoids are valuable index fossils for dating rocks. (David M. Dennis/Tom Stack & Associates.)

CHAPTER 6

Correlation and Dating of the Rock Record

By the 1830s, through the writings of James Hutton and Charles Lyell, most scientists were convinced of the immensity of geologic time. The challenge of piecing together the chronological details of Earth's history, however, remained. First, the *relative* ages of bodies of rock needed to be established: was a body of rock younger, older, or the same age as another? Second, the *absolute* ages of bodies of rock needed to be established: how long ago, in years, or thousands or millions of years, did they form?

The Geologic Time Scale

Geologists have made great progress since the 1830s in determining relative and absolute ages of rocks throughout the world. This progress has been crucial to unraveling the history of Earth and its life. Only after placing rocks and fossils in their relative positions in geologic time can scientists trace the history of environmental change and reconstruct the comings and goings of species within this environmental context. Absolute dating provides a true time scale for the chronological framework, revealing just how rapidly ice ages have begun and ended, how suddenly mass extinctions have occurred, and how quickly mountains have risen and eroded away.

Fossil succession reveals the relative ages of rocks

In the seventeenth century, as noted in Chapter 1, Nicolaus Steno established ways of showing that some rocks were older than others, but his laws applied only to bodies of rock that were in contact with one another. Then, late in the eighteenth century, William Smith discovered that fossils could often be used to establish the relative ages of strata that lay far apart; he introduced the concept of fossil succession. Smith showed that fossils could be used to determine the relative ages of certain bodies of rock in England. Having established the sequence of rocks—from lowest to highest—in which various kinds of fossils occurred, he showed that he could place an isolated outcrop of rock in its proper position in this stratigraphic sequence. In addition, he concluded that the entire sequence represented a very long interval of time. William Smith was a man of modest background and limited education, but his studies of fossil succession served a practical purpose. As a surveyor and engineer who planned the siting of canals, he had much to gain by learning how to recognize particular strata easily, on the basis of their fossils, and to predict what other strata would be found below or above them.

Baron Georges Cuvier, of the National Museum of Natural History in France, added new understanding to William Smith's discovery of fossil succession. A highly educated aristocrat dedicated to academic pursuits,

Visual Overview
Methods of Stratigraphic Correlation

Scientists use fossils, radioactive dating, unconformities, isotope stratigraphy, and patterns of transgression and regression to correlate rocks with respect to time.

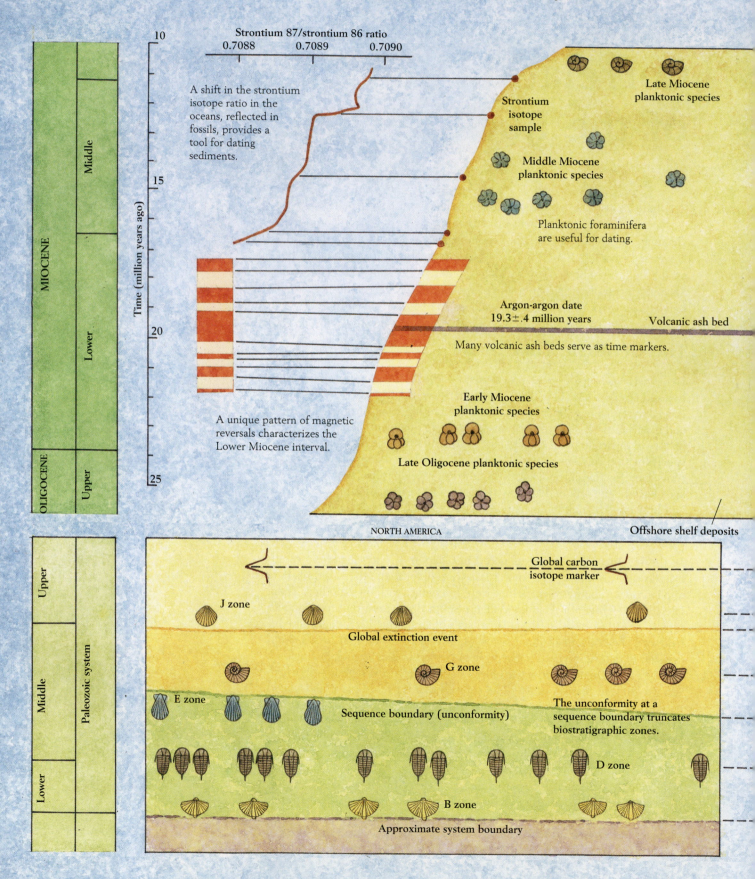

Strontium 87/strontium 86 ratio

A shift in the strontium isotope ratio in the oceans, reflected in fossils, provides a tool for dating sediments.

Strontium isotope sample

Late Miocene planktonic species

Middle Miocene planktonic species

Planktonic foraminifera are useful for dating.

Argon-argon date 19.3±.4 million years

Volcanic ash bed

Many volcanic ash beds serve as time markers.

A unique pattern of magnetic reversals characterizes the Lower Miocene interval.

Early Miocene planktonic species

Late Oligocene planktonic species

NORTH AMERICA

Offshore shelf deposits

Time (million years ago)

MIOCENE — Middle, Lower
OLIGOCENE — Upper

Global carbon isotope marker

J zone

Global extinction event

G zone

E zone

Sequence boundary (unconformity)

The unconformity at a sequence boundary truncates biostratigraphic zones.

D zone

B zone

Approximate system boundary

Upper, Middle, Lower

Paleozoic system

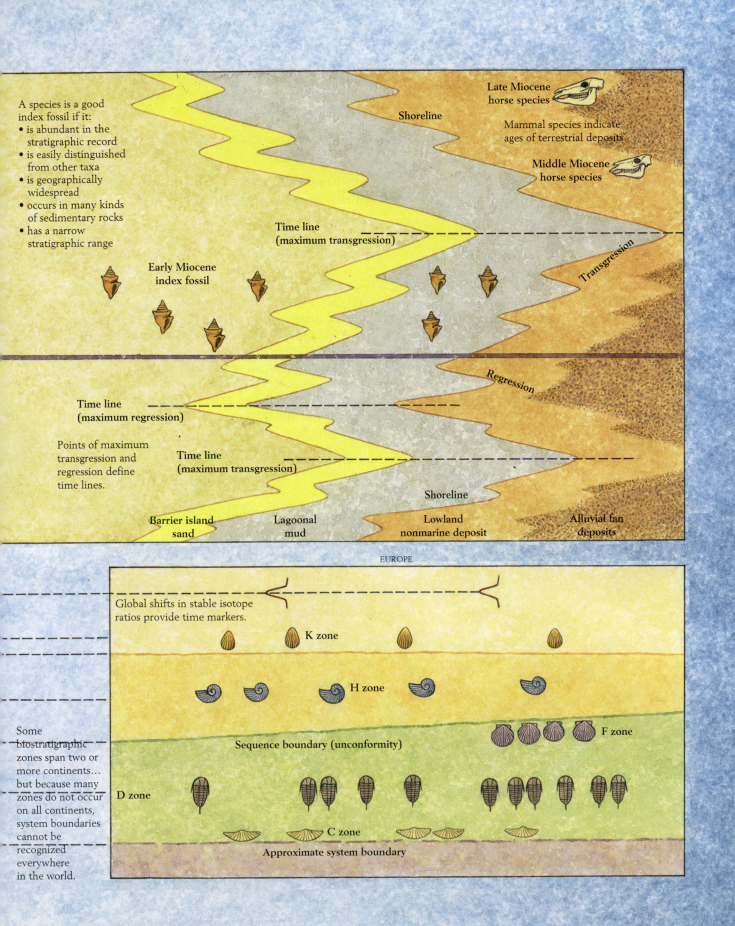

A species is a good
index fossil if it:
• is abundant in the
  stratigraphic record
• is easily distinguished
  from other taxa
• is geographically
  widespread
• occurs in many kinds
  of sedimentary rocks
• has a narrow
  stratigraphic range

Shoreline

Late Miocene
horse species

Mammal species indicate
ages of terrestrial deposits.

Middle Miocene
horse species

Time line
(maximum transgression)

Early Miocene
index fossil

Transgression

Time line
(maximum regression)

Regression

Points of maximum
transgression and
regression define
time lines.

Time line
(maximum transgression)

Shoreline

Barrier island
sand

Lagoonal
mud

Lowland
nonmarine deposit

Alluvial fan
deposits

EUROPE

Global shifts in stable isotope
ratios provide time markers.

K zone

H zone

Sequence boundary (unconformity)

F zone

Some
biostratigraphic
zones span two or
more continents…
but because many
zones do not occur
on all continents,
system boundaries
cannot be
recognized
everywhere
in the world.

D zone

C zone

Approximate system boundary

Cuvier was the first to conclude that species have become extinct in the course of Earth's history. From the Paris Basin, where sediments were being quarried extensively, Cuvier collected the fossils of many species of mammals unknown in the present world. We now know that these sediments were deposited early in the Cenozoic Era (Figure 6-1). Cuvier identified a succession of seven distinct fossil faunas of terrestrial mammals, each of which disappeared abruptly from the stratigraphic record, to be followed by another. In between any two successive terrestrial fossil faunas, however, were strata that yielded marine mollusks. Cuvier concluded that cataclysmic environmental changes had punctuated the history of life in this region. The sea must occasionally have swept inland from the northeast, wiping out terrestrial mammals and introducing marine life. When the waters receded again, a new terrestrial fauna appeared. Nonetheless, Cuvier—who died in 1832, long before Charles Darwin published his influential book—did not conclude that new species had arisen on Earth by evolving from others. He believed that species of each new mammalian fauna had migrated to the Paris Basin from elsewhere after a crisis, when conditions once again became favorable there.

Certainly, the first members of each new fauna to occupy the Paris Basin immigrated to that region from other regions, but we now know that many other species arose in that region or elsewhere by evolving from other species. What Cuvier established, however, was that entire biotas had become extinct because of environmental change and had then been replaced by other biotas.

Since Cuvier's time, scientists have established the relative positions of many kinds of fossils in strata, and these fossil occurrences have provided the most important evidence of the relative ages of strata around the world. Assembly of the geologic time scale that we still employ began shortly after Cuvier's death.

Geologic systems were founded in the nineteenth century

The modern science of geology was born in Great Britain, and it was there and in nearby areas of Europe that early geologists divided bodies of rock that contain fossils of distinctive animal or plant life into **geologic systems**, each of which corresponds to a geologic period (see Figure 1-11). The oldest of these fossil-rich systems is the Cambrian, and thus the label Precambrian was applied to all older rocks, which we now know represent the first 4 billion years or so of Earth's history. Because early scientists had no way of knowing the relative ages of rocks separated by great distances, they founded the various systems haphazardly. They could establish the relative ages of systems with certainty only by finding two of them in the same area and observing that one was positioned above the other.

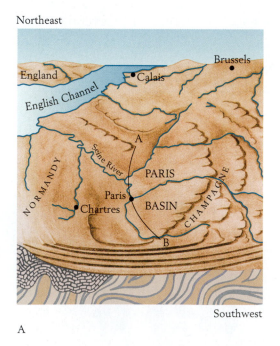

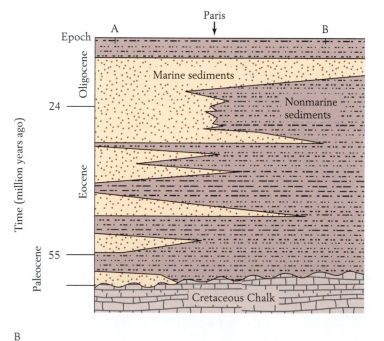

Figure 6-1 Cenozoic strata of the Paris Basin, where Cuvier discovered successive faunas of mammals. A. A schematic view of the basin structure. Erosion of weak strata around the basin margins has left durable strata standing as ridges. B. Schematic cross section along the line AB shown in A. Marine strata interfinger with nonmarine strata, having been laid down during episodic incursions of the sea from the northeast; the nonmarine strata yield mammalian fossils.

The British geologists Adam Sedgwick and Roderick Murchison named the Cambrian and Silurian systems in a joint paper published in 1835, principally on the basis of geologic studies in Wales. (The distribution of Silurian rocks in Wales is shown in Figure 5-24.) Sedgwick, a professor at Cambridge University, derived the name Cambrian from Cambria, the Roman name for Wales, and Murchison, a wealthy landowner who made geology his full-time hobby, named the Silurian for the Silures, an ancient Welsh tribe. Murchison defined the Silurian primarily on the basis of its fossils—including taxa of trilobites, brachiopods, and crinoids—as well as corals that formed small reefs. Sedgwick, finding no diagnostic fossils in his Cambrian System, simply defined the Cambrian as a body of sedimentary rocks that rested directly on ancient crystalline rocks.

By 1839, Silurian fossils had been found throughout Europe and even in the Americas and South Africa, allowing Murchison to proclaim the Silurian System a global entity. In time, he went too far, claiming that the Silurian included the earliest fossil-bearing rocks on Earth. He argued that the Welsh rocks on which Sedgwick had founded the Cambrian System actually belonged to the lower portion of the Silurian and simply happened to contain few fossils. Murchison's expanded version of the Silurian System incurred Sedgwick's bitter enmity. Eventually, distinctive fossils of Sedgwick's Cambrian System came to light not only in Wales but throughout the world. The Cambrian is now universally recognized as the oldest geologic system to contain a great variety of fossil shells and other skeletal remains of invertebrate animals.

In 1879 Charles Lapworth, a British schoolmaster, showed that in many parts of the world the rocks assigned to the Cambrian and Silurian systems actually displayed a succession of three distinct groups of fossils. He proposed that the Cambrian be retained as the system harboring the oldest group and the Silurian be retained as the system harboring the youngest group. For the intervening rocks, with their own distinctive fossils, he erected the Ordovician System.

In the twentieth century, geologists came to recognize that many taxa fossilized in uppermost Cambrian rocks became extinct abruptly, before the earliest Ordovician strata were laid down. Thus the Ordovician System records the evolution of many new forms of life. Uppermost Ordovician rocks document another episode of widespread extinction—along with an expansion of glaciers near the south pole that we now know was related to that extinction. Silurian rocks, in turn, contain many new kinds of fossils that represent another evolutionary expansion of life.

Unlike the Cambrian, Ordovician, and Silurian, which were founded on the basis of distinctive fossils, some other systems were founded to embody distinctive strata. The Cretaceous System, for example, was erected for a body of rocks whose upper portions include great thicknesses of chalk in Britain, France, Texas, Kansas, and many other areas. The Belgian Omalius d'Holloy established the Cretaceous System in France in 1822, aptly deriving its name from *creta*, the Latin word for chalk. Not until later in the century did geologists learn that Cretaceous chalk is formed of the very small external plates of calcareous nannoplankton that sank to the seafloor after the deaths of these single-celled floating algae. When producing the Cretaceous chalk, these algae were more productive than at any other time in their history.

Stratigraphic Units

What Sedgwick, Murchison, Lapworth, and d'Holloy, among others, were occupied in doing came to be known as stratigraphy. Broadly defined, **stratigraphy** is the study of stratified rocks, especially their geometric relations, compositions, origins, and age relations. **Stratigraphic units** include strata, or groups of adjacent strata, that are distinguished by physical, chemical, or paleontological properties; they also include units of time that are based on the ages of such strata. Geologic systems, such as the Cambrian, Silurian, and Cretaceous, represent one kind of stratigraphic unit.

Correlation is the procedure of demonstrating correspondence between geographically separated parts of a stratigraphic unit. Recall that a distinctive body of rock may be formally recognized as a *formation* on the basis of its composition and appearance. After a formation is established in one region, outcrops at other locations may be correlated with it and assigned to it, in accordance with Steno's principle of original lateral continuity. This or any other correlation based on rock type is termed **lithologic correlation**. The term *correlation* is more commonly used to indicate that widely separated bodies of rock are the same *age*, a procedure known as **temporal correlation**.

The rock record is divided into time-rock units and geologic time, into time units

A **time-rock unit**, formally termed a *chronostratigraphic unit*, includes all the strata in the world that were deposited during a particular interval of time. A geologic system is a time-rock unit. A **time unit**, formally termed a *geochronologic unit*, is the interval during which a time-rock unit formed. Thus a time-rock unit, such as the Silurian System, is defined in the field, and we refer to the time interval that this system represents as the Silurian Period.

TABLE 6-1   Geologic Time Units and Time-Rock Units

Time unit	Example	Time-rock unit	Example
Eon	Phanerozoic	None	
Era	Paleozoic	Erathem	Paleozoic
Period	Devonian	System	Devonian
Epoch	Late Devonian	Series	Upper Devonian
Age	Famennian	Stage	Famennian

Note: Time-rock units are bodies of rock that represent time units bearing the same formal name. When an epoch (the time unit) is designated by the term *Early, Middle,* or *Late,* the corresponding series (the time-rock unit) is identified by the adjective *Lower, Middle,* or *Upper.* For example, the Upper Devonian Series of rocks represents Late Devonian time.

Geologic systems are grouped into **erathems**. Eras are the time units that correspond to erathems; for the Phanerozoic Eon, these are the Paleozoic, Mesozoic, and Cenozoic erathems. Systems are also subdivided into **series**, which are further subdivided into **stages**; **epochs** and **ages** are the corresponding time units for these smaller time-rock units (Table 6-1).

A boundary between two systems, series, or stages is formally defined at a single locality, by means of a body of rock known as a **boundary stratotype**. Often the lower boundary of a time-rock unit is defined in one region and the upper boundary in another. The challenge is to extend each boundary throughout the world by means of temporal correlation, a task geologists undertake by a variety of methods.

Biostratigraphic units are based on fossil occurrences

Fossil occurrences are the most widely used means of extending the boundaries of systems, series, and stages around the globe. The results are far from perfect, however, for reasons that will soon be evident.

Stratigraphic units of the rock record that are defined and characterized by their fossil content are termed **biostratigraphic units**. These units are based on the stratigraphic ranges of fossil taxa. The *stratigraphic range,* or simply *range,* of a species is the total vertical interval through which that species occurs in strata, from lowermost to uppermost occurrence.

The most fundamental biostratigraphic unit is the zone, more formally termed a *biozone.* A **zone** is a body

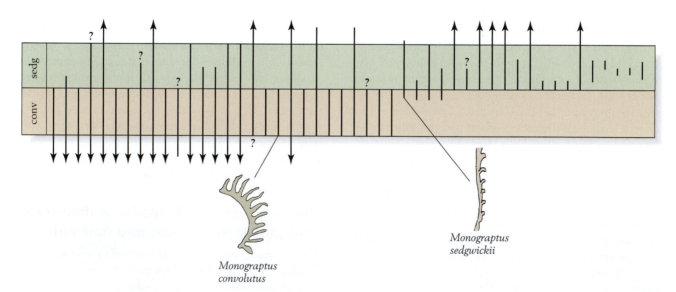

Monograptus convolutus

Monograptus sedgwickii

Figure 6-2   Zones based on the presence of graptolite fossils. The two zones represented in this bar graph are situated in the lower part of the Silurian System in the British Isles. Vertical bars represent the known ranges of graptolite species; the species for which the two zones are named are illustrated. The stalks of these animals are about 2 or 3 millimeters ($\frac{1}{10}$ inch) wide. All ranges shown are for species that first appear in the *convolutus* or *sedgwickii* zone. An arrow indicates that a range continues beyond the figured interval; a question mark indicates that a range may be longer than that shown here. (After R. B. Rickards, *Geol. J.* 11:153–188, 1976.)

of rock whose lower and upper boundaries are based on the ranges of one or more taxa—usually species—in the stratigraphic record. A zone may be defined by the range of a single taxon, although most zones are more complex than this. Many have a lower boundary defined by the lowermost or uppermost occurrence of one taxon and an upper boundary defined by the lowermost or uppermost occurrence of another taxon. Other zones consist of the stratigraphic interval within which two or more taxa occur together. Every zone is named for a taxon that occurs within it. Figure 6-2 shows how the stratigraphic ranges of a number of graptolite species in the Silurian System of Britain are used to define two graptolite zones. The graptolites are extinct colonial planktonic animals whose fossils are extremely useful for correlation. One of the zones shown is defined by the range of a single graptolite species, *Monograptus convolutus*, and is named for that species. The other is named for *Monograptus sedgwickii*, a species named in honor of Adam Sedgwick, but this zone is defined by the co-occurrence of several graptolite species.

Unfortunately, no zone represents exactly the same time interval everywhere it occurs. For one thing, the members of an extinct taxon will not have appeared or disappeared simultaneously in all the areas they inhabited. A species will commonly have originated in a small area and later greatly expanded its geographic range. Furthermore, a species or genus—the mammoth, for example—will often have persisted in a restricted region after having died out elsewhere (see Figure 4-5). In fact, many taxa have had complex histories of migration, largely as a result of changing environmental conditions. Another problem in defining biostratigraphic zones is that the fossil record is incomplete: a taxon may have existed at a given time and place without leaving a fossil record—or its fossils may remain undiscovered.

Although fossil species and genera do not provide for perfect correlation, some are reliable enough to be designated **index fossils** or **guide fossils**. A taxon of this kind has some or all of the following desirable characteristics:

1. It is abundant enough in the stratigraphic record to be found easily.

2. It is easily distinguished from other taxa.

3. It is geographically widespread and thus can be used to correlate rocks over a large area.

4. It occurs in many kinds of sedimentary rocks and therefore can be found in many places.

5. It has a narrow stratigraphic range, which allows for precise correlation if its mere presence, rather than its lowermost or uppermost occurrence, is to be used to define a zone.

Unfortunately, few taxa exhibit all of these traits as strongly as we might wish. Consider the planktonic foraminiferal fossils found in late Mesozoic and Cenozoic sediments (p. 63; see also Figure 5-35B). They meet two criteria for index fossils: they are easily identified under the microscope, and having been floaters in the sea, they settled in a wide variety of sedimentary environments. They are not ideal index fossils, however, because they lived in offshore areas and are therefore seldom found in nearshore sediments. Moreover, some extinct species of planktonic foraminifera lived for such lengthy intervals—for as long as 15 or 20 percent of the Cenozoic Era—that only their earliest and latest appearances provide useful information for correlation. Other foraminiferal species, however, lived for such short intervals that their mere occurrence permits quite precise correlation.

Although the identification of biozones is of immense value in stratigraphy, the accuracy of correlations can sometimes be improved by evaluating the magnetic properties of rocks. Such use of rock magnetism constitutes **magnetic stratigraphy**.

Magnetic stratigraphy identifies polarity time-rock units

As we have seen, Earth's core consists of dense material made up of iron and other heavy substances. In the outer part of the core, this material is in a liquid state, and its motion generates a **magnetic field**. As a result, the planet behaves like a giant bar magnet, with a north and south pole (Figure 6-3). When iron-containing minerals form sedimentary or igneous rocks at or near

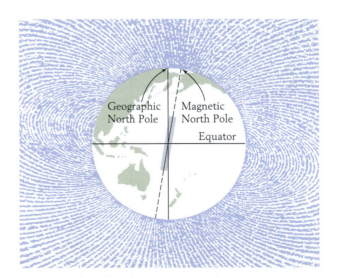

Figure 6-3 Earth's magnetic field. The magnetic field, represented by the lines of force surrounding the planet, resembles the one that would be produced by a bar magnet located within the planet (at the position shown in gray) with its long axis inclined slightly (11°) from Earth's axis of rotation.

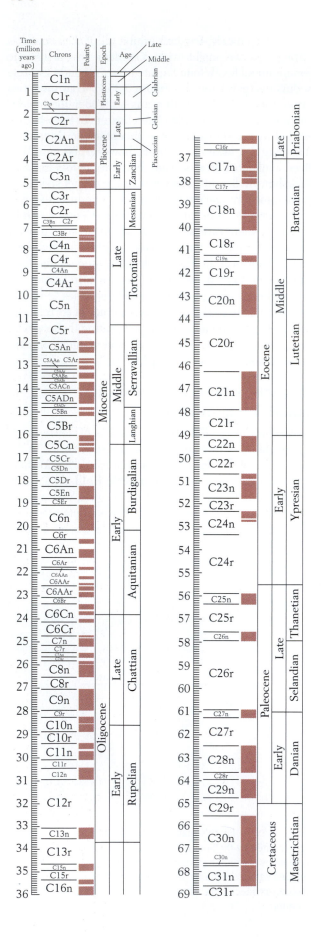

Earth's surface, they often become aligned with Earth's magnetic field, just as a compass needle does when it is allowed to rotate freely. As small grains of iron minerals settle from water to become parts of sedimentary rocks, they often rotate so that their magnetism becomes aligned with that of the planet. Similarly, iron minerals that crystallize from lava or magma automatically become magnetized by Earth's magnetic field as they cool.

It is a startling fact that Earth's north and south magnetic poles occasionally switch positions. No one knows why. Intervals between such magnetic reversals vary considerably, but during the Cenozoic Era they have averaged about a half-million years. Sequences of magnetized rocks that can be dated with radioactive materials have revealed the history of magnetic reversals during the Cenozoic Era (Figure 6-4) and during much of the Mesozoic Era as well. These reversals provide for accurate correlation of iron-containing rocks throughout the world. Periods when the polarity was the same as it is today are known as *normal intervals*, and periods when the polarity was the opposite of what it is today are called *reversed intervals*. Each of these, sometimes in conjunction with one or more neighboring intervals, is designated a chron, which is a **polarity time-rock unit**. A **chron** is either assigned a number or formally named for a geographic locality where it is well represented by magnetized rocks.

Magnetic stratigraphy generally provides for more accurate correlation than biostratigraphy. The primary difficulty in using the magnetic record for correlation lies in assigning strata of known (normal or reversed) polarity to a particular polarity time-rock unit. Especially helpful in this respect is a *signature*—that is, a distinctive sequence of reversals that, when plotted on a stratigraphic column, produces a unique pattern resembling the bar code for an item in a store. In rocks known to be of Eocene age, for example, a pattern that indicates a long normal interval flanked by two long reversed intervals can only be early middle Eocene in age; furthermore, the base of this first reversal interval coincides with the base of the middle Eocene (see Figure 6-4).

Unfortunately, magnetic stratigraphy has not yet been extensively employed for temporal correlation of Paleozoic and Mesozoic strata.

◀ **Figure 6-4** Magnetic polarity scale for the Cenozoic Era. Numbers at the left represent millions of years. About half of the time the magnetic field has had normal polarity (polarity like that of the present), as indicated by the colored segments of the scale, and about half of the time the polarity has been reversed, as indicated by white space. (After W. B. Berggren et al., *Geol. Soc. Amer. Bull.* 96:1407–1418, 1985.)

Rock units are defined by lithology, not age

As we saw in Chapter 1, geologists divide the stratigraphic record into local three-dimensional bodies of rock known as *formations*. Formations are sometimes united into larger divisions known as *groups* and sometimes include smaller units called *members*. Groups, in turn, may be united into *supergroups*. All these entities are known as **rock units**, or more formally, *lithostratigraphic units*. Formations are delineated not by age, but on the basis of **lithology**, or the physical and chemical characteristics of rock. Many are relatively homogeneous bodies, consisting of a single rock type. Others, as we have seen, consist of two or more rock types in alternating layers; formations of this sort include the sedimentary rock cycles produced by deltas and by meandering rivers.

Rock units are observable only in segments. Typically, some portions of a formation have been eroded away and others lie buried in the subsurface. A local outcrop of a formation that displays a vertical sequence constitutes a **stratigraphic section**. Every formally recognized rock unit is assigned a **type section**—designated at a particular locality where the unit is well exposed—which serves to define the unit. A rock unit is given the name of a local geographic feature, such as a river or town, where its type section is located, followed by the word "Formation," "Group," or "Member" or by the name of a specific rock type, such as Sandstone, Shale, or Limestone.

Lithologic correlation Geologists construct diagrammatic cross sections of strata to establish their geometric relationships and interpret their modes of origin. Such cross sections are constructed from local stratigraphic sections (Figure 6-5). Typically a stratigraphic section is measured from bottom to top, and the positions of various types of rock are recorded, as are the locations of all the kinds of fossils that can be collected and identified. The strata of two or more sections in a given region can then be correlated. Figure 6-5 depicts the correlation of fifteen local sections located along the walls of the Grand Canyon, Arizona.

A key aspect of rock units is that their boundaries are defined without reference to biostratigraphic units or time. In fact, the ages of the upper and lower boundaries of many rock units vary widely from place to place. This point is well illustrated by three Cambrian formations of marine origin exposed along the walls of the Grand Canyon (Figure 6-6A). To early geologists it appeared that the Bright Angel Shale rested on the Tapeats Sandstone, and was therefore the younger unit; the Muav Limestone appeared to bear a similar relation to the Bright Angel Shale. In the 1940s, however, a geologist named Edwin McKee showed that this idea of a layer-cake pattern was in error. By carefully measuring the sections illustrated in Figure 6-5 and collecting fossils from them, McKee showed that a trilobite zone that belongs to the Early Cambrian Series passes through the Tapeats Sandstone and into the lower part of the Bright Angel Shale (Figure 6-6A). He further discovered that a second trilobite zone, belonging to the *Middle* Cambrian Series, passes through the *upper* part of the Bright Angel Shale and into the Muav Limestone. This pattern shows that the eastern portion of each formation is younger than the western portion.

Facies The strata represented in Figures 6-5 and 6-6 accumulated along the western margin of North America, which in Cambrian time lay far to the east of its present location. The Tapeats Sandstone is a marine unit that was deposited above an unconformity on crystalline rocks of Precambrian age. This means that it was deposited along the shoreline. The eastward migration of the base of the Tapeats Sandstone during Cambrian time indicates that the shoreline shifted in this direction: the sea migrated over the land. This encroachment of the sea onto the land occurred because of a global rise in sea level, which also caused a shift of the Cambrian shoreline inland along the east coast of North America.

Recall that when a sandy beach progrades over muddy offshore deposits, the process of progradation pushes the shoreline seaward (see Figure 5-22). This seaward migration of a shoreline is known as **regression**. The Cambrian strata depicted in Figure 6-6 display the opposite pattern, termed **transgression**, in which the sea spreads over the land. When a shoreline migrates inland as sea level rises, so do the environments seaward of it, in accordance with Walther's law (p. 114). Thus, during Cambrian time in the Grand Canyon region, muddy offshore sediments came to rest on top of nearshore sands (Figure 6-6B). Similarly, carbonate sediments deposited even farther offshore came to rest on the muddy sediments.

The pattern of sediment deposition illustrated in Figure 6-6 reflects the fact that no particular type of environment stretches infinitely far in any direction. One environment of deposition inevitably gives way to another, and when it does so, the transition between the two modes of deposition is either abrupt or gradual. The set of characteristics of a body of rock that represent a particular depositional environment is called a **facies**. Accordingly, lateral changes in the characteristics of ancient strata, which reflect lateral changes in the depositional environment, are known as facies changes. The three formations displayed in Figure 6-6A represent three distinct facies.

In another portion of the geologic record, two or more interfingering facies might be assigned to a single formation because they are similar in general lithology.

For example, an ancient reef might be grouped in a single formation with the rubbly rock formed by talus in front of it and the well-bedded strata that formed in the

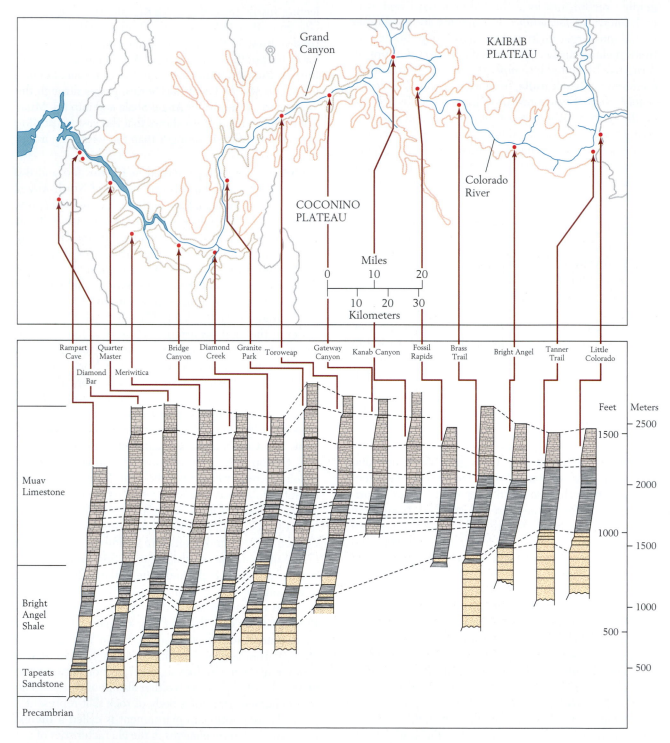

Figure 6-5 Local stratigraphic sections measured along the walls of the Grand Canyon where Cambrian strata are exposed. The strata here are divided into three formations; the lowermost one consists primarily of sandstone, the middle one primarily of shale, and the uppermost one primarily of limestone. Dashed lines between sections indicate lithologic correlation.

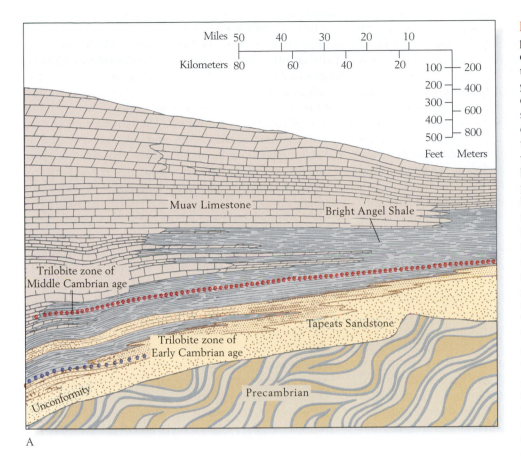

Miles 50 40 30 20 10

Kilometers 80 60 40 20

100	200
200	400
300	600
400	800
500	

Feet Meters

Muav Limestone

Bright Angel Shale

Trilobite zone of Middle Cambrian age

Tapeats Sandstone

Trilobite zone of Early Cambrian age

Unconformity

Precambrian

A

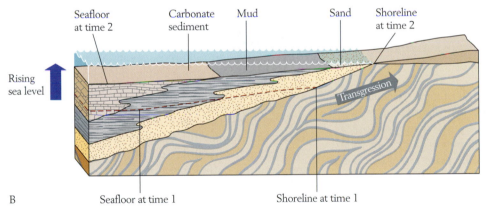

Seafloor at time 2   Carbonate sediment   Mud   Sand   Shoreline at time 2

Rising sea level

Transgression

B   Seafloor at time 1   Shoreline at time 1

Figure 6-6 Stratigraphic pattern of Cambrian rocks exposed along the walls of the Grand Canyon. *A.* A graphic cross section constructed from the stratigraphic sections depicted in Figure 6-5. Fossil trilobites collected from the various sections show that biostratigraphic zones do not parallel formation boundaries, but intersect them, passing from one formation into another. Thus the formations represent deposits that were forming simultaneously within neighboring environments that shifted through time. The basal formation, the Tapeats Sandstone, was deposited along the shoreline that bordered western North America, and this shoreline shifted eastward through time. *B.* The eastward shifting of the shoreline as sea level rose between two "moments" of Cambrian time. The result was a transgressive pattern of deposition, in which offshore muddy sediments came to overlie nearshore sand deposits, and carbonate sediments deposited even farther offshore came to overlie the muddy sediments. (After E. D. McKee, Carnegie Institution of Washington Publication 563, 1945.)

lagoon behind it because all three facies consist of limestone (see Figure 5-26). In contrast, the three facies illustrated in Figure 6-6, being lithologically very different, are assigned to separate formations.

Earth's Absolute Age

Thus far in this chapter, we have discussed only the relative ages of rocks. Until late in the nineteenth century, when radioactivity was discovered, geologists could make only crude estimates of the absolute ages of bodies of rock or of Earth itself. Today, dates obtained

through the use of radioactive materials show that Earth is about 4.6 billion years old—even older than many uniformitarian geologists of the nineteenth century believed.

Early geologists underestimated Earth's antiquity

Prior to the twentieth century, most attempts to establish Earth's absolute age were based on rates of accumulation of materials from weathering and erosion or on ideas about the rate at which the planet had cooled from its fiery beginning. These early estimates led to erroneous suggestions that the planet was younger than the hundreds of millions of years that the conventional uniformitarian view seemed to demand.

Salts in the ocean In the eighteenth and nineteenth centuries, some scientists estimated Earth's age by calculating how long it should have taken for the ocean to accumulate its salts. They assumed that the ocean's waters had originally been fresh and that runoff from the land had progressively increased their salt content, and was still doing so. In 1899, the Irish geologist John Joly estimated that, at the rate at which rivers were then contributing salts to the ocean, about 90 million years would have been required to produce its current salinity. Joly thereby greatly underestimated Earth's age. In fact, we now recognize that the salinity of the ocean may not have changed greatly since early in Earth's history: the removal of salts by deposition of evaporites has approximately balanced the addition of salts by rivers and other sources.

Accumulation of sediment Near the beginning of the nineteenth century, some geologists attempted to use rates of sediment accumulation to estimate Earth's age. First they estimated rates of deposition of sediment in various modern settings; then they estimated the thickness of sedimentary rocks in Earth's crust. Multiplying these two numbers gave estimates for the total time of sediment accumulation—typically 100 million years or less, a small fraction of Earth's actual age. We now know that these estimates were inaccurate for several reasons:

1. The stratigraphic record is full of gaps. In many depositional settings, sediments are deposited in pulses. Within just a few seconds, a current may deposit a bed of sediment a few centimeters thick, but then there may be no net accumulation on top of that bed for a few years, or even hundreds of years. Moreover, scouring by currents may remove the bed that was laid down so rapidly, along with other beds below it (p. 22).

2. Unconformities in the rock record represent even larger breaks in sediment accumulation—times of erosion instead of accumulation (p. 23). Some unconformities represent millions of years of time unrecorded in the rock record.

3. Not recognizing that many metamorphic rocks had once been sedimentary rocks, early geologists failed to recognize that these rocks once formed a significant part of the stratigraphic record.

Earth's temperature The most formidable challenge to the uniformitarian view that Earth must be very old came in 1865 from Lord Kelvin, the eminent British physicist. He presented an address with the presumptuous title "The Doctrine of Uniformity in Geology Briefly Refuted." By "uniformity" Kelvin meant uniformitarianism. His argument was simply that Earth had been very hot when it formed and had been cooling ever since. It was known that the temperature rises as one descends a mine shaft, and calculations based on such temperature increases showed that the interior of the planet is still very hot. Kelvin argued that Earth retained so much of its original heat that it could be only about 20 million or at most 40 million years old. This seemed much too brief a history for the uniformitarian interpretation of Earth's history—for processes such as erosion, deposition, igneous activity, and mountain building to have produced the complex array of rocks displayed at Earth's surface. Thus Lord Kelvin and his followers went so far as to challenge uniformitarianism, which had become the foundation of geologic science.

The discovery of radioactivity finally beat back Lord Kelvin's challenge, providing an explanation for the persistence of high temperatures within Earth. Soon thereafter, radioactivity came to play a different and much larger role in the evaluation of geologic time. Scientists developed methods of using radioactive materials to determine absolute ages of rocks and of Earth itself.

Radioactive decay provides absolute ages of rocks

In 1895 Antoine-Henri Becquerel discovered that the element uranium undergoes spontaneous **radioactive decay**; that is, its atoms change into atoms of another element by releasing subatomic particles and energy. Geologists soon recognized that radioactive elements and the products of their decay could be used as geologic clocks to measure the ages of rocks.

Only a few naturally occurring chemical elements are useful for dating rocks by means of radioactive decay. Recall that *isotopes* are forms of an element that differ in the number of neutrons in their nuclei; furthermore, some isotopes are unstable, or radioactive (p. 30). Atoms of radioactive isotopes decay spontaneously, changing into atoms of a different element. The isotope that undergoes decay is known as the *parent isotope*, and the product is known as the *daughter isotope*. There are three modes of radioactive decay:

1. *Loss of an alpha particle.* An alpha particle consists of two protons and two neutrons (a helium ion, He^{2+}). Loss of an alpha particle converts the parent isotope into the element whose nucleus contains two fewer protons (Figure 6-7).

2. *Loss of a beta particle.* A beta particle is an electron, which has a negative charge but a negligible mass.

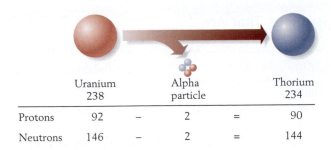

	Uranium 238		Alpha particle		Thorium 234
Protons	92	–	2	=	90
Neutrons	146	–	2	=	144

Figure 6-7 **Decay of an atom of uranium 238 to thorium 234 by loss of an alpha particle.** This loss reduces the atomic number by 2 and the atomic weight by 4.

Its loss turns a neutron into a proton, changing the parent isotope into the element whose nucleus contains one more proton.

3. *Capture of a beta particle.* Addition of a beta particle turns a proton into a neutron, changing the parent isotope into the element whose nucleus has one less proton.

Some radioactive isotopes decay into other isotopes that are also radioactive. In fact, several steps of decay are required to yield a stable isotope from some parent isotopes.

Radioactive elements are useful for dating rocks because each radioactive element decays at its own nearly constant rate. Once scientists have measured this rate in the laboratory, they can calculate the length of time over which decay in a natural body of rock has been proceeding by measuring the amounts of both the radioactive parent isotope and the daughter isotope in the rock. This procedure is known as **radiometric dating**. Radioactive isotopes decay at a constant geometric rate (not at a constant arithmetic rate). This means that, no matter how much of the parent isotope is present when it begins to decay, after a certain amount of time half of that amount will survive (Figure 6-8). After another interval of the same duration, half of the surviving amount (one-fourth of the original amount) will remain, and so on. This characteristic interval is known as the **half-life** of a radioactive element. Thus, in the course of four successive half-lives, the number of atoms of a radioactive element will decrease to one-half, one-fourth, one-eighth, and one-sixteenth of the original number of atoms. The number of atoms of the daughter isotope will increase correspondingly.

As Table 6-2 indicates, several radioactive isotopes are abundant in rocks and are therefore useful for geologic dating. Most of these isotopes occur in igneous rocks; thus, if we know the amounts of parent and daughter isotopes currently present in an igneous rock, we can calculate the time that has elapsed since the par-

ent isotope was trapped—the date when magma cooled to form the rock. A few minerals on the seafloor incorporate radioactive elements as well, and radiometric dates obtained for these minerals represent the interval of time that has elapsed since they formed. Unfortunately, radiometric clocks are reset by metamorphism. Metamorphic processes reposition radioactive isotopes in new minerals that contain none of their decay products. As a result, subsequent decay gives the age of the metamorphic event, not that of the original rock.

Naturally occurring radioactive isotopes vary greatly in their half-lives (see Table 6-2), and these differences have a bearing on their use in dating. More specifically, isotopes with short half-lives are useful for dating only very young rocks, while those with long half-lives are best used to date very old rocks. In essence, isotopes that have short half-lives, such as carbon 14, decay so quickly that their quantities in old rocks are too small to be measured. By the same token, isotopes that have long half-lives, such as rubidium 87, decay so slowly that the quantities of their daughter isotopes in

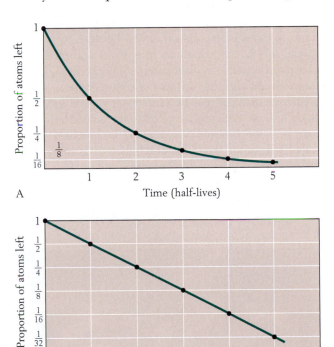

Figure 6-8 **Arithmetic and geometric patterns formed by loss of atoms through radioactive decay.** *A.* When plotted on a standard arithmetic scale, the number of atoms of a radioactive element can be seen to decrease more slowly with each successive interval of time. *B.* When the number of atoms is scaled as a geometric progression, the plot forms a straight line. Half of the atoms present at the beginning of each interval (or half-life) survive to the beginning of the next interval.

TABLE 6-2 Properties of Some Radioactive Isotopes That Are Commonly Used to Date Rocks

Radioactive isotope	Approximate half-life (years)	Product of decay
Rubidium 87	48.8 billion	Strontium 87
Thorium 232	14.0 billion	Lead 208
Potassium 40	1.3 billion	Argon 40
Uranium 238	0.7 billion	Lead 206
Carbon 14	5730	Nitrogen 14

Note: The number after the name of each element signifies the atomic weight of that element and serves to identify the isotope.

very young rocks are too small to be measured accurately. Let us examine the radiometric utility of the various radioactive isotope decay systems listed in Table 6-2. The element most often associated with radioactivity—uranium—is also one that is used extensively to date rocks.

Uranium and thorium The useful radioactive isotopes uranium 238 and thorium 232 decay to different isotopes of lead. The most useful mineral for dating by means of these isotopes is the silicate zircon because it is widespread in low concentrations in igneous and metamorphic rocks as well as in siliciclastic sediments derived from them. By abrading the surfaces of zircon grains to obtain unaltered material and by employing several grains to study a single rock unit, geologists are now able to date even very old (Precambrian) rocks quite precisely. Some calculated ages in the range of 2 billion to 3 billion years are considered to be within just a few million years of the actual ages! Uranium and thorium isotopes have also been used to date rocks from the moon. The ages of the oldest dated moon rocks—slightly more than 4.6 billion

years—closely approximate the age of Earth and its solar system, as estimated from other evidence.

Several radiometric techniques employ uranium and its daughter products to determine the ages of rocks or fossils a few million years old or younger. One of these techniques is **fission-track dating**, a more accurate method for measuring the decay of uranium 238 than measuring the daughter isotope (lead 206) when a rock is too young to have accumulated much lead. When uranium 238 decays, it emits subatomic particles that fly apart with so much energy that they penetrate the surrounding crystal lattice, producing fission tracks (Figure 6-9). These tracks can be enlarged in the laboratory by acid etching and counted under a microscope. The remainder of the uranium is then subjected to a neutron field, which causes it to decay completely. The number of tracks thus produced is compared to the number that formed naturally, and the resulting numerical ratio reveals the age of the mineral.

Additional methods of radiometric dating apply to reef-building corals, which incorporate a small amount of uranium into their skeletons. The fact that uranium 234 decays rapidly to thorium 230 allows accurate dating of corals that range in age from a few thousand years to about 300,000 years. Other radioactive isotopes of uranium yield helium as one of their final daughter products. Thus, by measuring the amount of helium and the amount of undecayed uranium trapped in well-preserved coral skeletons, geochemists can date corals several million years old.

Rubidium-strontium dating Rubidium occurs as a trace element in many igneous and metamorphic rocks and even in a few sedimentary rocks. Nonetheless, because of the long half-life of rubidium 87, the parent isotope, the rubidium-strontium system is generally useful only for dating rocks older than about 10 million years.

Potassium-argon and argon-argon dating Argon, the daughter isotope of potassium 40, is an inert (chemically nonreactive) gas. Argon becomes trapped in the crystal lattice of some minerals that form in igneous and metamorphic rocks. The potassium-argon method is

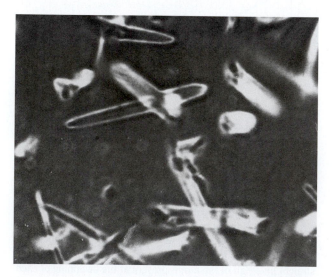

Figure 6-9 Fission tracks in a mineral grain enlarged by etching. The complete track oriented almost horizontally in the center is 13.5 μm long. (Rosen and Miller, Rensselear.)

used to date materials as young as 50,000 years and as old as 4.6 billion years. Generally, more precise dating results from a procedure called argon-argon dating: a sample is bombarded with neutrons in a nuclear reactor to convert potassium 40 to argon. The amount of argon gas produced in this way can be measured more precisely than the amount of potassium in a rock. One deficiency of the potassium-argon and argon-argon methods is that argon can leak from the lattice of a crystal, making it appear that less than the actual amount of decay has occurred. An error of the opposite kind will occur if a sample has absorbed argon from the atmosphere. Nonetheless, these methods have provided the dates of important events in the evolution of humans in Africa.

Radiocarbon dating Radiometric dating that makes use of carbon 14 is known as **radiocarbon dating**. It is the best known of all radiometric techniques, but because the half-life of carbon 14 is only 5730 years, this method can be used only with materials younger than about 70,000 years, and it is sometimes used to date materials younger than a few hundred years. Objects of biological origin, such as bones, teeth, and pieces of wood, make up the bulk of the materials dated by this method. Despite its limitations, radiocarbon dating is of great value for dating materials from the latter part of the Pleistocene Epoch—an interval so recent that most other radioactive materials found in its sediments have not decayed sufficiently to permit their products to be measured accurately. Fortunately, the useful range of carbon 14 encompasses much of the time interval during which modern humans have existed, as well as the interval during which glaciers most recently withdrew from North America and Europe at the close of the recent Ice Age (the Pleistocene Epoch). Thus radiocarbon dating plays a valuable role in the study of human culture.

Carbon 14 is a rare isotope of carbon that forms in the upper atmosphere, about 16 kilometers (10 miles) above Earth's surface, as a result of the bombardment of nitrogen by cosmic rays. The atmospheric ratio of carbon 14 to the stable isotope carbon 12 has nonetheless been fairly uniform over the past thousand years. Both isotopes are assimilated by plants, which turn them into tissue. Once a plant dies, carbon is no longer incorporated into its tissues, but the stored carbon 14 continues to decay to nitrogen 14. Thus the percentage of carbon 14 in dead plant tissue declines in relation to the percentage of carbon 12, and the ratio of the two can be used to determine when the tissue died.

Fossils often provide more accurate correlation than radiometric dating

The fact that the half-lives of radioactive isotopes are well established does not imply that radiometric dating permits more accurate correlation of sedimentary rocks than

fossils do. For one thing, dating stratigraphic boundaries is problematic when appropriate radioactive isotopes are absent from rocks positioned close to them. In addition, most minerals that can be dated radiometrically are of igneous origin, and the dating of igneous rocks often yields only a maximum or a minimum age for associated sedimentary rocks (Figure 6-10). Clasts of igneous rocks or minerals found in sedimentary rocks can also be dated radiometrically, but this yields only estimates of the maximum ages of sedimentary rocks. Another problem is that rocks that are dated radiometrically have not necessarily remained intact in nature: they may have gained or lost atoms of the parent or the daughter isotope. As noted earlier, this is a particularly common source of error for the potassium-argon decay system.

Even more fundamental uncertainties are inherent in radiometric dating. Most published radiometric dates, for example, are followed by a symbol and a number representing a smaller interval of time (for example, 47 ± 3 million years). The plus-or-minus sign indicates the range of uncertainty attributed to the date, based on possible errors in the measurement of the parent and daughter isotopes.

These various types of errors sometimes add up to sizable total errors, especially when very old rocks are being dated. This point is well illustrated by past estimates made for the beginning and end of the Silurian Period (Figure 6-11). In evaluations made between 1959 and 1968 alone, the duration of the Silurian was halved and then doubled and then halved again. It now seems likely that the Silurian began about 440 million years ago and that it ended about 417 million years ago, but each of these dates may be adjusted slightly in the future.

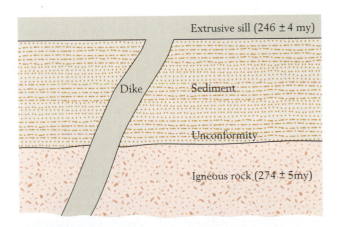

Figure 6-10 The age of a sedimentary unit can be bracketed by radiometric dating of associated igneous rocks. In this cross section, the sediments lie on top of a body of igneous rock dated at 274 ± 5 million years, so they must be younger than that. The sediments are also cut by a dike and covered by a sill of igneous rock dated at 246 ± 4 million years; they are therefore older than that. We can conclude that this sedimentary unit may be as old as 279 million years or as young as 242 million years.

Figure 6-11 Various estimates of the interval of time represented by the Silurian Period. The date when each estimate was made and the author or authors responsible for each estimate are listed. (After N. Spjeldnaes, American Association of Petroleum Geologists, *Studies in Geology* 6:341–345, 1978.)

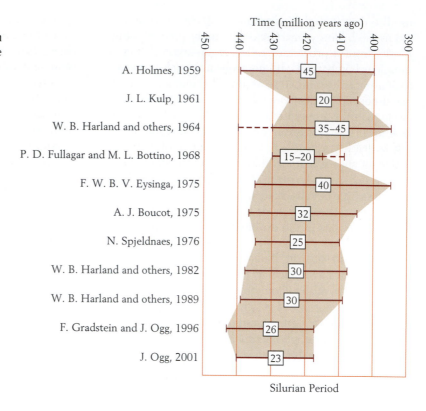

Silurian Period

Fossil graptolites (see Figure 6-2) of Silurian age allow for more accurate correlation of sedimentary rocks within particular regions than do radiometric dates. Many kinds of graptolites floated in the sea and thus spread quickly over large areas. In addition, graptolites evolved rapidly. These two factors together make graptolites highly useful index fossils. Many individual species of graptolites existed for about 1 million or 2 million years; thus correlations based on such species cannot be inaccurate by a larger interval than that.

Although radiometric dating provides an absolute geologic time scale—one based on years—most geologic correlations are still based on fossils. Not only are fossils more common than radioactive elements in sedimentary rocks, but the analysis of fossils often allows for greater accuracy. However, the geologic intervals at the upper and lower ends of the geologic time scale represent striking exceptions to this general rule. At the lower end, most rocks older than about 1.4 billion or 1.5 billion years contain few fossils that are well enough preserved and easily enough identified to serve as guides. Thus radiometric dates serve as a primary basis for correlation of these early rocks, especially from continent to continent. And at the upper end of the time scale, extending from the present back about 60,000 years, radiocarbon dating offers estimates of age that commonly have only small ranges of uncertainty. Because relatively few species of animals or plants have appeared or disappeared during this brief interval, fossils found in the corresponding rocks have much less value for correlation.

Changes in stable isotopes permit global correlation

Within a chemical reservoir such as the ocean, the ratio between two stable isotopes of an element may vary through time. If organisms incorporate the two isotopes into their skeletons, the fossilized skeletons will record changes in the ratio of the two isotopes in the ocean. Distinctive patterns of change can then be used for correlation. One correlation method of this kind employs the element strontium. Two stable isotopes of strontium that occur in all modern seas provide a special opportunity to date relatively young strata containing fossils that have not been altered appreciably since their burial. These two isotopes, strontium 86 and strontium 87, occur in the same relative abundance in all modern seas. The ratio of their abundance has changed through time, however, for many reasons, including changes in the rates at which rocks have yielded their strontium to the ocean as a result of exposure and erosion. During the past 25 million years, the relative abundance of strontium 87 has increased slightly. Small amounts of strontium take the place of calcium in the calcareous skeletons of marine organisms. This strontium has the same isotope ratio as the seawater in which an animal lives. Thus well-dated fossils provide a record of changes in the ratio of strontium 87 to strontium 86 in seawater (Figure 6-12). Once a record of such changes has been established, geologists can date fossil skeletons by measuring their precise isotopic composition. This valuable dating method, known

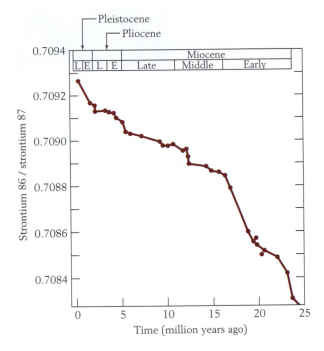

Figure 6-12 The changing ratio of strontium isotopes in marine fossils composed of calcium carbonate during the past 25 million years. (After D. J. DePaolo, *Geology* 14:103–106, 1986.)

as **isotope stratigraphy**, has only recently been put to use. It promises to resolve many stratigraphic problems, especially with respect to Cenozoic fossils, which often have undergone relatively little alteration since burial and retain their original strontium isotope ratios.

Event Stratigraphy

Imagine if someone with magical powers were to spray paint instantaneously over a large region of Earth—ocean floors, lake bottoms, and terrestrial lowlands. In keeping with the law of superposition, this layer of paint would separate deposits that had formed before it was laid down from accumulations that were laid down afterward. Similarly, if a magical force were suddenly to lower sea level throughout the world by 200 meters, the deposition of sediments at the shoreline would shift seaward for great distances, and the first of the new shoreline deposits thus created would be of nearly the same age on all continents, so that these deposits could be

accurately correlated throughout the world. These two hypothetical scenarios—the instantaneous formation of a surface and a sudden relocation of shorelines—are not far removed from actual sudden events whose geologic records have enabled us to correlate rocks of widely separated regions. The use of geo-logic records of this kind for correlation is termed **event stratigraphy**.

Marker beds allow correlation over wide areas

A **marker bed** is a bed of sediment that resembles our hypothetical layer of paint: all parts of it are virtually the same age. Widespread layers of volcanic ash, for example, often function as marker beds. Sometimes an ancient ash fall, which may represent either a single volcanic eruption or a series of nearly simultaneous eruptions, can be traced for thousands of square kilometers, and so represents a brief moment of geologic time. The Bishop Tuff, for example, is a bed of volcanic ash that was emitted from a volcano in eastern California slightly more than 700,000 years ago and spread halfway across the United States (Figure 6-13). Figure 6-14 shows a

A

B

Figure 6-13 The Bishop Tuff, which formed when the Long Valley volcano in eastern California erupted slightly more than 700,000 years ago. *A.* A geologist examines the cream-colored ash bed at Shoshone, where it is very thick, having been deposited close to the source. *B.* The areal extent of recognized occurrences of the tuff; most of the ash spread eastward, blown by the prevailing winds. (*A,* Breck P. Kent; *B* after G. A. Izett and C. W. Naeser, *Geology* 4:587–590, 1976.)

Figure 6-14 A thick band of volcanic ash (the beige bed just above the hammer head) overlain by two similar but thinner ash beds. This conspicuous group of Cretaceous ash beds permits correlation over a large area of Colorado. (Erle Kauffmann, University of Colorado.)

Figure 6-15 Two cores taken from the Castile Evaporites, which were precipitated in western Texas near the end of the Permian Period. The cores are from localities 14.5 kilometers (9 miles) apart, yet their laminations match almost perfectly, allowing for precise correlation. The alternating dark and light bands, which range up to a few millimeters in thickness, probably represent seasonal organic-rich (winter) and organic-poor (summer) layers. If this is the case, each pair of bands represents one year, and the 200,000 or so paired bands of the Castile Formation represent about 200,000 years of deposition. (R. Y. Anderson et al., *Geol. Soc. Amer. Bull.* 83:59–86, 1972.)

much older ash bed positioned within nonvolcanic marine strata. This bed formed during Cretaceous time in Colorado, when a volcano erupted in a region of mountain building to the west and emitted ash that fell into a large inland sea.

Some marker beds are global in distribution. For example, the top of the Cretaceous System can be identified by a thin layer of sediment that is rich in iridium, an element that is generally rare on Earth. The iridium came from fallout from the explosion of an asteroid that struck Earth, resulting in the extinction of the dinosaurs and many other forms of life. The iridium-rich layer also contains sand-sized grains that exhibit features reflecting the heat and pressure generated by the object's impact. These grains were blown high into the atmosphere and spread throughout the world.

Glacial tills such as those described in Chapter 5 also serve as marker beds; even tills formed by glaciers in different parts of the world can be useful if they represent a brief interval of global cooling.

Certain events produce marker beds that allow for correlation within particular basins of deposition. The character of evaporite deposits in a deep basin, for example, can reflect aquatic conditions throughout the basin. Because such evaporites typically form widespread horizontal beds (Figure 6-15), an individual bed that differs in chemical composition from beds above and below it can often be traced over thousands of square kilometers.

Back-and-forth shifting of facies boundaries creates marker beds

When a transgression is followed by a regression (or vice versa), the resulting stratigraphic pattern is useful for correlation (Figure 6-16). A point along the boundary between two facies marks the time of maximum transgression. Two or more such points that represent the same time of maximum transgression can therefore be connected to form a line of temporal correlation. Most time lines of this type are useful only for correlating stratigraphic sections that represent single depositional basins—for example, sections that represent different parts of a lake or shallow sea. The unconformities discussed in the next section can provide evidence for correlation over longer distances.

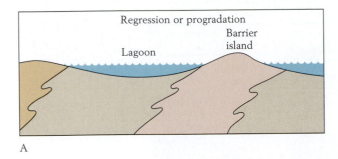

A

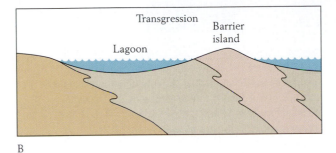

B

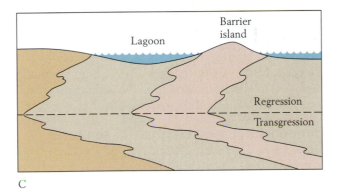

C

Figure 6-16 Correlation based on a stratigraphic pattern in which a regressive depositional sequence follows a transgressive sequence. *A.* The pattern of a regressive (progradational) sequence. *B.* The pattern of a transgressive sequence. *C.* When a regression follows a transgression, the points of maximum transgression of various facies can be connected to form a line of correlation, as indicated by the broken line in the diagram. This was the "moment" in geologic time when the sea shifted farthest inland. A similar line can be constructed for a stratigraphic pattern in which a transgression follows a regression.

Unconformities can be detected by seismic stratigraphy

An unconformity formed during a single interval of tectonic uplift or nondeposition may be of more or less the same age wherever it occurs. Such an unconformity may truncate rocks of many ages, yet the sediments resting directly on top of the erosional surface are often nearly the same age in all parts of a depositional basin. The ages of these superimposed deposits are seldom *pre-*

cisely the same everywhere, however, and may vary greatly from place to place. A sea that has deserted an area, for example, may later invade it again slowly, after a period of erosion, reaching different parts of the area at different times. On the other hand, if sea level drops suddenly throughout the world and then rises again rapidly, the resulting global unconformities may represent fairly accurate time markers.

Global unconformities that occur within Mesozoic and Cenozoic sediments lying along continental shelves have been used with great success as time markers. Most continental shelves have remained below sea level for long intervals after forming, but occasionally have been exposed to erosion through sea-level lowering. Unconformities produced at such times of exposure have been examined by **seismic stratigraphy**, which entails the interpretation of seismic reflections generated when artificially produced seismic waves bounce off physical discontinuities within buried sediments (Figure 6-17). Some of these discontinuities have been found to be approximately the same age everywhere they occur. Many of them reflect changes in lithology

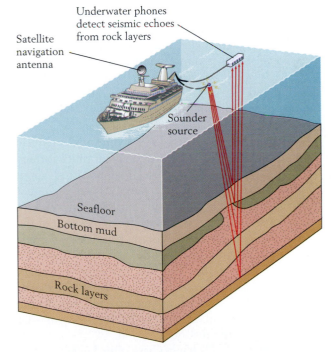

Figure 6-17 The use of seismic reflections to study sediments and rocks buried beneath the seafloor. Sound waves are generated by a sounder that makes a pneumatic explosion like that of a bursting balloon. The sound waves bounce off surfaces of discontinuity, which include bedding surfaces and unconformities. Underwater phones then pick up the reflections, allowing marine geologists to determine the configurations of buried features.

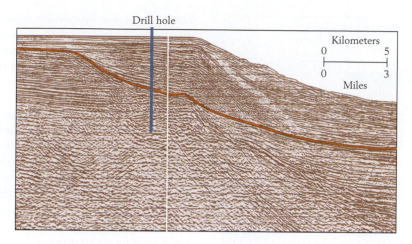

Figure 6-18 **Seismic section of the continental shelf and slope east of Massachusetts.** The heavy colored line marks an unconformity that formed during the Oligocene Epoch. This unconformity separates the Zuni and Tejas sequences (see Figure 6-21). The drill hole represents the location of a sediment core used to date the individual sedimentary beds. (From L. F. Jansa, *AAPG Memoir* 56:111–126, 1993.)

between depositional units, but others represent unconformities. (Figure 6-18).

When compilations of many local seismic profiles dated with fossil evidence from sediment cores reveal that sea level rose or fell in widely separated areas at the same time, the change must have occurred on a global scale. Such evidence has yielded a global curve of sea-level changes during the Cenozoic Era (Figure 6-19). Less precise information for earlier intervals, based in part on rocks, fossils, and unconformities visible on the continents, has yielded a less detailed sea-level curve for the entire Phanerozoic Eon.

Global changes in sea level are termed **eustatic changes**. Figure 6-18 shows an unconformity formed by a eustatic drop in sea level that occurred about 30 million years ago. However, eustatic changes are not reflected in seismic profiles for localities where there was an offsetting tectonic change in the position of the land. If the land in one area rises in pace with a eustatic rise, for instance, the sea-level rise will not be recorded as it would be if the land had remained stationary (Figure 6-20A). Moreover, if the tectonic uplift exceeds the eustatic rise, sea level in that area will actually fall in relation to the land (Figure 6-20B), regardless of what is happening in the rest of the world.

Vertical sea-level changes have often been mistaken for transgressions and regressions, which are lateral shifts in the position of the shoreline. Of course, a strong connection exists both between transgressions and eustatic rises and between regressions and eustatic falls, but there is a complicating factor that makes transgressions and regressions unreliable indicators of eustatic changes. As Figure 6-20C indicates, the rate at which sediment accumulates strongly influences whether transgression or regression will occur in a particular area. Thus, regressions have often occurred locally during global intervals of rising sea level simply because sediment has been supplied from the land at such a high rate that it has pushed the sea back from the land.

Sequences record changes in sea level

The widespread unconformities that represent eustatic falls have been used to divide the stratigraphic record into units called sequences (Figure 6-21). **Sequences** are large bodies of marine sediment deposited on continents when the ocean rose in relation to continental surfaces and formed extensive epicontinental seas (see Figure 4-24). Individual sequences represent tens of millions of years during which sea level rose quite high and then receded again. During most of the time when sequences were deposited—which means during most of Phanerozoic time—the sea stood higher than it does today.

Young sequences also include bodies of sediment that now lie below the sea, having accumulated along continental borders relatively recently. Geologists use seismic stratigraphy to study these bodies of sediment, which underlie continental shelves and slopes. The seismic section of the continental shelf and slope east of Massachusetts shown in Figure 6-18, for example,

▶ **Figure 6-19** **Estimates, based primarily on seismic stratigraphy, of relative changes in sea level during the Cenozoic Era.** Horizontal segments of the sea-level curve represent sudden drops in sea level. (After B. Haq et al., *Science* 235:1156–1167, 1987.)

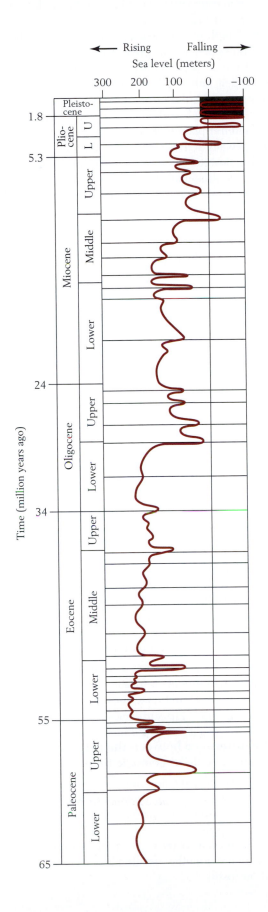

depicts the unconformity between the Zuni and Tejas sequences (see Figure 6-21).

Most long-term changes in sea level that have produced sequences separated by unconformities have resulted from changes in the rate at which new lithosphere has formed along mid-ocean ridges and on submarine plateaus. Mid-ocean ridges, as we know, are great swellings of the seafloor where new lithosphere rises up and remains swollen because of its high temperature. The lithosphere cools and shrinks as it moves away from a ridge axis. Consequently, the seafloor subsides on either side of a ridge, eventually leveling out to form the

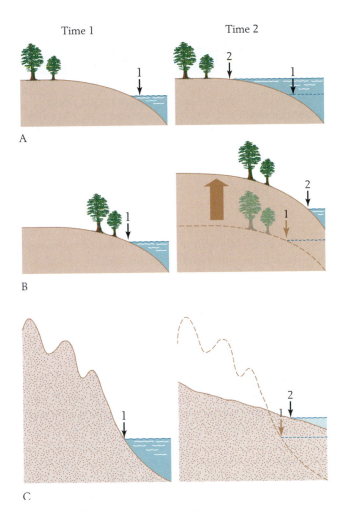

Figure 6-20 A rise in sea level does not necessarily result in a transgression. In each of the three pairs of diagrams, the initial and final positions of sea level are numbered 1 and 2. *A.* The land remains unchanged, and a rise in sea level causes a transgression. *B.* A rise in sea level is accompanied by regression rather than by transgression because the land rises tectonically (broad arrow) more than sea level does. *C.* A rise in sea level is accompanied by regression (progradation) rather than by transgression because sediment eroded from nearby highlands pushes the shoreline seaward.

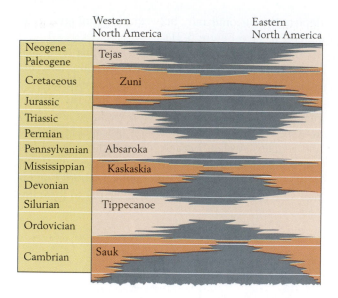

Figure 6-21 Phanerozoic sequences on the North American continent. Sequences are shown in tan and brown. These sequences resulted from transgressions that occurred on both sides of the continent as global sea level rose, and they ended with regressions as global sea level fell. The blue-gray areas are gaps represented by unconformities. (From L. L. Sloss, *Geol. Soc. Amer. Bull.* 74:93–114, 1963.)

abyssal plain (see Figure 1-17). In addition, the swelling of oceanic lithosphere and extrusion of submarine lava that form submarine plateaus elevate global sea level.

Plate tectonic activity has not been uniform during Earth's history. At times of intense plate tectonic activity, the total length of mid-ocean ridges has been relatively great. Rates of seafloor spreading have also tended to be high at these times, and so much heat has flowed to the ridges that individual ones have stood relatively tall. The large total volume of ridges has pushed sea level upward, causing broad continental areas to flood. At times of intense plate tectonic activity, plumes have also been especially active, creating submarine plateaus that have also elevated sea level. At such times, marine deposition on continents has formed sequences. Then, when plate tectonic activity has become less intense again, total ridge volume has shrunk (over millions of years) and sea level has declined correspondingly. Unconformities have then formed as seas have receded from continents, producing sequence boundaries.

Changes in the elevation of the seafloor have moved sea level up and down at rates on the order of 10 meters (33 feet) per million years. Expansion and contraction of continental glaciers have caused much more rapid and dramatic changes in sea level. Many times during the modern Ice Age, sea level has fallen by as much as 100 meters (330 feet) within just a few thousand years as glaciers have expanded over the land, "locking up" water and thus removing it from the global water cycle. Expansion of continental glaciers much earlier in Earth's history also lowered sea level dramatically, draining shallow seas and shifting shorelines thousands of kilometers toward the centers of ocean basins.

Chapter Summary

What are the basic units of stratigraphy?

Geologic systems were erected haphazardly in the nineteenth century and pieced together to form the geologic time scale. Along with erathems, which are larger units, and series and stages, which are smaller units, systems are considered time-rock units. The upper and lower boundaries of time-rock units are called boundary stratotypes. Time units—eras, periods, epochs, and ages—correspond to erathems, systems, series, and stages. Biostratigraphic units are defined by fossil occurrences; the zone is the basic unit of this type. The most useful species for correlation are called index or guide fossils. Ideally, these species are easily identified, widely distributed, abundant in many kinds of rock, and restricted to narrow vertical stratigraphic intervals. Polarity time-rock units are based on periodic reversals in the polarity of Earth's magnetic field, which are recorded in magnetized rocks. A body of rock characterized by a particular lithology or group of lithologies is often recognized as a formal rock unit (lithostratigraphic unit: a member, formation, or group). Seldom is either the upper or the lower boundary of a rock unit the same age everywhere; therefore, rock units are often transected obliquely by biostratigraphic units.

How do facies differ from rock units?

A facies is the set of characteristics of strata that formed in a particular environment. A formation, or any other rock (lithostratigraphic) unit, may consist of a single facies, or of two or more adjacent facies.

What is the difference between the relative scale and the absolute scale of geologic time?

The relative scale of geologic time is based on superposition and fossil succession. Bodies of rock are simply ordered from oldest to youngest. Radiometric dating, in contrast, provides an absolute scale in which events are measured in years. Correlations based on radiometric dates are nonetheless often less accurate than ones based on fossils.

How are stable isotopes, marker beds, and unconformities used for correlation?

Global shifts in the ratios of stable isotopes of certain elements preserved in sediments allow for correlation of strata. Event stratigraphy employs sedimentary layers that form almost simultaneously over large areas—such as ash falls, glacial tills, and evaporite beds—as useful time markers. Unconformities resulting from a sudden, major global lowering of sea level can also mark brief moments in Earth's history. The analysis of seismic reflections allows geologists to recognize these unconformities deep within Earth. A sequence is a body of rock situated between two such unconformities. Sequences have formed at times when the overall rate of plate tectonic activity has been high, elevating the deep-sea floor and the ocean above it. Unconformities between sequences represent times of reduced plate tectonic activity, when the floor of the deep sea has stood at a relatively low level.

Review Questions

1. What kinds of stratigraphic correlation do geologists undertake?

2. What is the difference between the relative and absolute ages of rocks?

3. To what interval of geologic time is radiocarbon dating applicable?

4. How are strontium isotopes useful in dating rocks?

5. Why does a geologic formation not necessarily have an upper or lower boundary that is the same age everywhere?

6. What factors prevent biostratigraphic zones from having either an upper boundary or lower boundary that is the same age everywhere?

7. What factors lead to imperfections in radiometric dating?

8. Construct a diagram resembling Figure 6-16*C* to depict the depositional history of a barrier island–lagoon complex that has undergone a regression followed by a transgression.

9. Construct a diagram resembling Figure 6-20*B* showing that a lowering of sea level need not always lead to regression.

10. How are seismic profiles used to study regional stratigraphy?

11. What is the scientific contribution of event stratigraphy?

12. Using the Visual Overview on page 130 as your guide, evaluate the various methods of correlating rocks with regard to age. Which methods are especially useful for intercontinental correlation?

Megatherium americanum was the largest of the ground sloths, all of which are
long extinct. This Pleistocene species, which was about 6 meters (20 feet) long
from head to tail, migrated to North America from South America before it died out.
(Field Museum of Natural History, Chicago, Neg. GEO 8460c.)

CHAPTER 7

Evolution and the Fossil Record

The central concept of modern biology is that living species have come into being as a result of the evolutionary transformation of quite different forms of life that lived long ago. Indeed, it is often maintained that very little of what is now known about life would make sense in any context other than that of organic evolution. It is important to remember, however, that while the broad definition of evolution is "change," organic evolution does not encompass every kind of biological change. The term refers only to changes in populations, which consist of groups of individuals that live together and belong to the same species.

Adaptations

When we examine the broad spectrum of organisms that inhabit our planet, we cannot help being impressed by the success with which each form functions in its own particular circumstances. Members of the cat family, for example, have sharp fangs at the front of the mouth for puncturing the flesh of prey and bladelike molars in the rear for slicing meat. Horses, for their part, are equipped with chisel-like front teeth for nipping grass and broad molars for grinding it up (Figure 7-1). Thus the teeth of each animal equip it to process the food it eats.

Plants, too, exhibit a variety of forms and features that vary with their ways of life. The leaves of most tree species native to tropical rain forests, for example, are waxier than those of plants found in cool regions, and a typical tropical leaf terminates in an elongate tip called a drip point. The drip point and the waxy surfaces help the leaves shed the rainwater that falls on them daily in a rain forest; the waxy surfaces also keep them from drying out in the tropical heat. In contrast, the leaves of another rain-forest plant group, the bromeliads, form a cup that acts as a private reservoir for rain. Without this feature, a bromeliad would dry up and die, because it lives high above the moist forest floor, attached to a tree. Yet another rain-forest plant, the Venus's-flytrap, secretes a sweet nectar that lures insects to the midrib of its leaf (Figure 7-2). On the margins of its leaves are rows of spines that mesh when the leaves snap shut around an unsuspecting insect. In a reversal of the normal roles of plant and animal, the Venus's-flytrap then digests the insect.

These specialized features of animals and plants that perform one or more useful functions are known as **adaptations**. Each individual organism possesses many adaptations that function together to equip it for its particular way of life. Before the middle of the nineteenth century, adaptations were not well understood, or even recognized. It was assumed that all features of a species were perfect mechanisms that had been specially designed to allow the species to function optimally within its ecological niche. Since then it has become widely acknowledged that the features we now recognize as adaptations are fraught with imperfections, many of which stem from

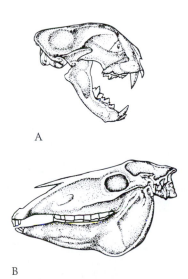

A

B

Figure 7-1 The teeth of (*A*) a cat and (*B*) a horse, both of Pliocene age.

153

Visual Overview
The Evolution of Life

EVIDENCE FOR EVOLUTION

Human arm

Bat wing

Darwin concluded that similar configurations of organs in distantly related animals often reflect a common ancestry: they represent homology.

Living

Armadillos

Sloths

Fossil

Darwin observed that living and fossil members of some animal groups are known only from the Americas and must have originated there by some natural process.

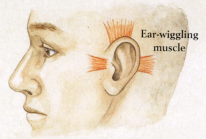

Ear-wiggling muscle

Darwin concluded that some anatomical features that lack functions are vestigial: they are remnants of features that were functional in evolutionary ancestors.

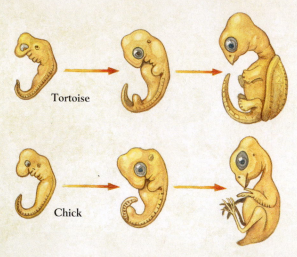

Tortoise

Chick

Darwin concluded that the similarity of embryos of animals that are quite different as adults must reflect a common ancestry.

Wolf

Darwin noted that artificial selection by animal breeders serves as a model for natural selection.

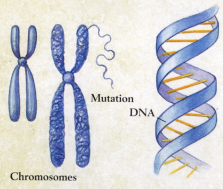

Mutation

DNA

Chromosomes

Genetic changes—mutations in the chemical code of DNA or rearrangements of segments of chromosomes—produce the variability upon which natural selection operates.

The geographic distributions of organisms, such as the restriction of particular species of giant tortoises to single islands in the Galápagos, suggested to Darwin that species arise from isolated populations.

Aardvark

Alligator

Dawn redwood

Ancient taxa that that have survived to the present with little speciation have undergone little evolutionary change and are called living fossils.

TRENDS AND PATTERNS

Axolotl

Salamander

(Adult)

(Larval stage)

The axolotl, which has the form and aquatic habits of a larval salamander, originated rapidly from a salamander species by way of a single genetic change.

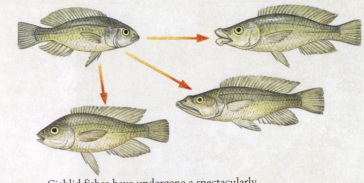

Cichlid fishes have undergone a spectacularly rapid evolutionary radiation in Lake Victoria, a relatively young body of water.

The fossil record documents some gradual evolutionary trends, such as the enlargement and flattening of coiled oysters.

Mammals

Dinosaurs

Mass extinction of the dinosaurs permitted the great evolutionary radiation of the mammals, illustrating how the disappearance of one taxon can trigger the expansion of another.

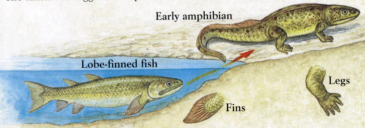

Early amphibian

Lobe-finned fish

Legs

Fins

The origin of the legs of early amphibians from the fins of fishes was the adaptive breakthrough that triggered the evolutionary radiation of terrestrial vertebrates.

Like many other large-scale evolutionary trends, increase in the average body size of horses followed a complex pattern with some small species originating long after horses first appeared.

Thylacine (marsupial)

Wolf (placental)

Evolutionary convergence between distantly related taxa—the evolution of similar shapes associated with a similar mode of life—is powerful evidence that evolution is adaptive.

Figure 7-2 A Venus's-flytrap has lured a bee onto a gaping pair of leaves, which will snap shut. (Michael Viard, Auscape International.)

their evolutionary heritage. In other words, a species may develop a useful new feature with which to perform a function, but the evolution of that feature will sometimes be constrained by the structure of the ancestral organism. Evolution can operate only by changing what is already present; it cannot work with the freedom of an engineer who is designing a new device from raw materials. The business of evolution, in other words, is and always has been remodeling rather than new construction.

It becomes obvious that evolution is a remodeling process when we observe that certain organs, such as the teeth of mammals and the leaves of plants, serve different functions in different species but nonetheless have a common "ground plan," or fundamental biological architecture. All mammals, for example, possess teeth that are rooted in bone and consist of both dentin and enamel. Similarly, certain types of cells and tissues form the leaves of nearly all flowering plants. Common ground plans suggest common origins, and they are one of the many pieces of evidence indicating that groups of species of the modern world have a common evolutionary heritage: no matter how greatly the species of a given order or class may differ, they share certain basic features that reflect their common ancestry.

Evolutionary divergence of groups of organisms with common ground plans first became a widely accepted idea because of the powerful arguments of Charles Darwin.

Charles Darwin's Contribution

Few biologists gave serious consideration to the idea of organic evolution until 1859, when Charles Darwin (Figure 7-3) published his great work, *On the Origin of Species by Natural Selection.* You can appreciate the power of the basic evidence that evolution has occurred by putting yourself in Darwin's position when, in 1831, at the age of 22, he set sail as a naturalist aboard the *Beagle* on a surveying voyage that took him around the world. Through his observations on this trip, Darwin became convinced of the workings of evolution and also accumulated much of the evidence that later enabled him to convince others of the validity of his ideas.

The voyage of the *Beagle* provided geographic evidence

Recall from Chapter 1 that Darwin read Charles Lyell's *Principles of Geology* during the voyage and became convinced that the uniformitarian approach to geology was valid.

While Darwin's adherence to the uniformitarian view of Earth's history provided a framework for his acceptance of evolution, it was his observation of the geographic distributions of living things that ultimately led him to conclude that many different forms of life possess a common biological heritage. Darwin was surprised to find South America inhabited by animals that

Figure 7-3 Charles Darwin in old age. (Painting by John Collier, National Portrait Gallery, London.)

▶ **Figure 7-4** The rare flightless bird *Rhea darwinii*, named after Charles Darwin. (Painting by John Gould, from C. Darwin, *Zoology of the Voyage of H.M.S. Beagle*, Smith and Elder, London, 1838–1843.)

differed substantially from those of Europe, Asia, and Africa. The large flightless birds of South America, for example, were species of rheas (Figure 7-4), which belonged to a different family than the superficially similar birds of other continents—the ostriches of Africa and the emus of Australia. Among other unique South American creatures were the sloths and the armadillos. Not only was South America the home of living representatives of these groups, but it was there that Darwin dug up the fossil remains of extinct giant relatives of the living forms (Figure 7-5, see p. 152). Why, he asked, were all rheas, as well as all living and extinct members of the sloth and armadillo families, found nowhere but in the Americas?

Darwin was also intrigued to find that species of marine life on the Atlantic side of the Isthmus of Panama differed from those on the Pacific side. In places the isthmus is only tens of miles wide, and it struck

Figure 7-5 Unusual South American mammals. *A.* A living three-toed sloth, hanging upside down in its normal mode of life. This animal is the size of a small dog. *B.* A reconstruction of two of the Pleistocene mammals of South America that Darwin unearthed as fossils. *Megatherium*, the giant ground sloth, was larger than an elephant. It ranged northward into the United States. The giant armadillo is *Glyptodon*. (*A*, Jerry Ellis/The Wildlife Collection; *B*, Field Museum of Natural History, Chicago, Neg. CK20T, painting by C. R. Knight.)

A

B

Figure 7-6 Three tortoises, each of which inhabits a different island of the Galápagos. *Testudo abingdonii*, which inhabits Pinta Island, has a long neck and a shell that is elevated in the neck region; these features represent adaptations for reaching tall vegetation. The shells of these animals can exceed 1 meter (3 feet) in length. (After T. Dobzhansky et al., *Evolution*, W. H. Freeman and Company, New York, 1977.)

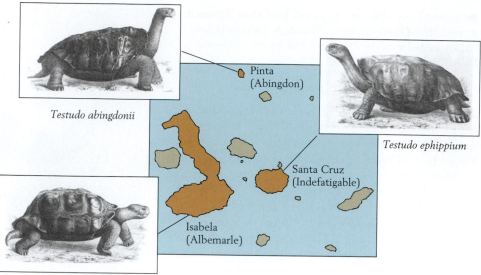

Testudo abingdonii

Testudo ephippium

Testudo microphyces

Darwin as strange that the marine creatures on opposite sides of this narrow neck of land should differ from each other—unless the various species had somehow come into being where they now lived. If the species had instead been scattered over the planet by an external agent, he reasoned, many should have landed on both sides of the isthmus.

Perhaps Darwin's most striking observations concerned life forms on oceanic islands. Darwin noted that no small island situated more than 5000 kilometers (3000 miles) from a continent or from a larger island was inhabited by frogs, toads, or land mammals unless they had been introduced by human visitors. The only mammals native to such islands were bats, which could originally have flown there. This observation led Darwin to suspect that species could originate only from other species. Otherwise, why would isolated areas of land be left without forms of life that were prominent elsewhere?

The Galápagos Islands, which lie astride the equator about 1100 kilometers (700 miles) from South America, played an especially large part in the development of Darwin's new ideas. Darwin found the Galápagos to

be inhabited by huge tortoises, and he thought it curious that the native people could look at a tortoise shell and immediately identify the island from which it had come. The fact that different species of giant tortoises occupied different islands (Figure 7-6) led Darwin to

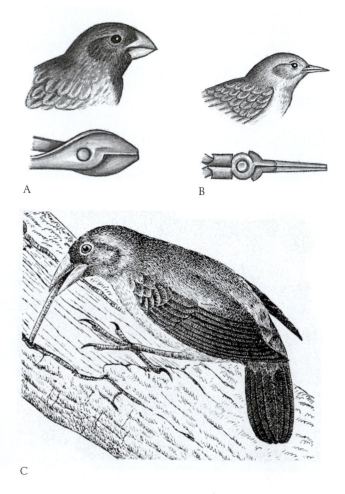

A

B

C

▶ **Figure 7-7** Three of the finch species that Darwin observed in the Galápagos Islands. *A.* The large tree finch's parrotlike beak operates like heavy pliers to crush fruits and buds. *B.* The warbler finch's beak operates like needle-nose pliers to catch insects. *C.* The woodpecker finch, which excavates tree bark with its chisel-like beak, uses a cactus needle as a tool to probe for insects. (*A* and *B* from P. R. Grant, "Natural Selection and Darwin's Finches," *Sci. Amer.*, October 1991. © 1991 by Scientific American, Inc. All rights reserved. *C* from D. Lack, "Darwin's Finches," *Sci. Amer.*, April 1953. © 1953 by Scientific American, Inc. All rights reserved.)

suspect that these distinctive populations of tortoises had a common ancestry, but had somehow become differentiated in form as a result of living separately in different environments.

Even more striking were the various kinds of finches that Darwin found in the Galápagos. Some types had slender beaks, others had somewhat sturdier ones, and still others had very heavy beaks, which they used for breaking seeds (Figure 7-7). One kind of finch behaved like a woodpecker, using a cactus spine as a woodpecker uses its long beak to probe for insects in wood. Furthermore, all of the finches in the Galápagos resembled a species of finch on the South American mainland (the closest large landmass) rather than finches found in other regions of the world. Darwin began to ponder whether a population of finches from the South American mainland might have reached the islands and become altered in some way to assume a wide variety of forms. It seemed that the finches had somehow differentiated in such a manner that they were able to pursue ways of life that on the mainland were divided among several different families of birds. As Darwin put it in the journal in which he described his scientific work on the voyage:

> Seeing this gradation and diversity of structure in one small, intimately related group of birds, one might really fancy that from an original paucity of birds in this Archipelago, one species had been taken and modified for different ends.

Darwin's anatomical evidence was broadly based

When Darwin returned to England and weighed other evidence indicating that one type of organism has evolved from another, he found that certain anatomical relationships seemed to build an especially compelling case. One such piece of evidence was the remarkable similarity of the embryos of all vertebrate animals. Darwin was intrigued by the admission of Louis Agassiz, a noted American scientist, that he could not distinguish an early embryo of a mammal from that of a bird or a reptile. This, Darwin reasoned, was exactly what could be expected if all vertebrate animals had a common ancestry: although adult animals might become modified in shape as they became adapted to different ways of life, early embryos were sheltered from the outer world and would thus undergo less change.

Equally convincing to Darwin was the evidence of **homology**—the presence in two different groups of animals or plants of organs that have the same ancestral origin but serve different functions. The principle of homology is illustrated by the variations in teeth and leaves discussed earlier in this chapter. Another example is the common origin of the toes of land-dwelling mammals,

the fingers of humans, and the wings of bats. Human fingers and bats' wings are actually formed of four toes whose external appearance and bone configuration resemble those of four-legged mammals. If human fingers and bats' wings did not both originate from toes, why should the two types of organs have similar bone configurations? Such evidence of common origins abounds in both the animal world and the plant world.

The existence of **vestigial organs**—organs that serve no apparent purpose but resemble organs that do perform functions in other creatures—further supported Darwin's argument in favor of evolution. We humans, for example, retain muscles that other mammals use to prick up their ears in order to catch sounds more effectively. Some people can wiggle their ears slightly with these muscles, but this ability serves no function for us: in humans the muscles are vestigial structures.

Natural selection is the mechanism of evolution

Darwin also recognized a different type of evidence that pointed to the validity of biological transformation in nature: animal breeders had produced major changes in domesticated animals by means of selective breeding. If wolves could be modified into greyhounds, Saint Bernards, and Chihuahuas under domestic conditions, Darwin saw no reason why animals should be anatomically straitjacketed in nature. The question was, what could bring about such biological changes under natural conditions?

This question led to Darwin's second great contribution. The first, of course, had been his amassing of an enormous amount of evidence indicating that species had evolved in nature. The second was his conception of a mechanism through which evolution could have taken place. The mechanism Darwin proposed was **natural selection**—a process that operates in nature but parallels the artificial selection by which breeders develop new varieties of domestic animals and plants for human use.

Essentially, artificial selection in domestic breeding involves the preservation of certain biological features and the elimination of others. A breeder simply chooses certain individuals of one generation to be the parents of members of the succeeding generation. Darwin recognized that in nature, many more individuals of a species are born than can survive. Accordingly, he reasoned that success or failure of individuals here, as in selective breeding, would not be determined by accident. In nature it would be determined by advantages that certain individuals had over others—a greater ability to find food, for example, or avoid predators, or resist disease, or deal with any of a number of environmental conditions. By virtue of their longevity, these individuals would tend to produce more offspring than others.

Darwin also recognized, however, that survival was not the only factor influencing success in nature, because some individuals with only average life spans were capable of producing more offspring than others simply because they bore large litters or shed large numbers of seeds. Thus, as long as the members of a breeding population varied substantially in either longevity or rate of reproduction, certain individuals would pass on their traits to an unusually large number of members of the next generation. The kinds of individuals that came to predominate as generation followed generation could then be said to be favored by natural selection.

Alfred Russel Wallace, another British scientist who had traveled extensively in far-off lands, conceived of natural selection after Darwin had, but before Darwin had published his ideas. Spurred into action by a letter from Wallace, Darwin wrote *On the Origin of Species by Natural Selection* within just a few months. Acknowledging the great length of time Darwin had spent developing his theory and the large variety of evidence he had amassed to support it, Wallace graciously yielded priority to Darwin for recognizing the process of natural selection. If these two men had not brought the concept to light, some other scientist would have done so before long, because the evidence for evolution by natural selection is so vividly displayed in nature.

Genes, DNA, and Chromosomes

Darwin faced a major obstacle in his efforts to convince others that natural selection could operate effectively to produce evolution. Because he lived before the birth of modern genetics, Darwin was not familiar with the mechanisms of inheritance and thus could not explain how an organism could pass along a favorable genetic trait to its offspring. Although the Austrian monk Gregor Mendel outlined the basic elements of modern genetics only a few years after Darwin published *On the Origin of Species*, Mendel's work was not acknowledged until the turn of the century, two decades after Darwin's death.

Mendel's most significant contribution to modern genetics was the concept of **particulate inheritance**, which explains how certain hereditary factors, which we now call **genes**, retain their identities while being passed on from parents to offspring. Mendel's experiments with pea plants demonstrated that individuals possess genes in pairs, with one gene of each pair coming from each parent. In one of his experiments, Mendel employed a true-breeding white-flowered strain of pea plants and a true-breeding red-flowered strain. (In a *true-breeding* strain, descendants resemble parents throughout a long series of generations.) Mendel's first step was to cross plants of the white-flowered strain with those of the red-flowered strain. The surprising result was that all of the offspring had red flowers. When these red-flowered plants were crossed with one another, however, they pro-

duced both red-flowered and white-flowered offspring. These experiments showed that the effect of the gene for white color could surface in the third generation even after its presence had been masked in the second generation. Preservation of genes in this manner—that is, as discrete entities that maintain their identity from generation to generation—constitutes the basis of particulate inheritance, the cornerstone of modern genetics.

More than 30 years passed before additional breakthroughs in science began to unlock the secrets of inheritance. One important discovery was that genes can be altered. It is now understood that genes are, in fact, chemical structures that can undergo chemical changes, and these changes, or **mutations**, provide much of the variability on which natural selection operates. Genes are now known to be segments of long molecules of deoxyribonucleic acid, or **DNA**—a compound that transmits chemically coded information from generation to generation, providing instructions for the growth, development, and functioning of organisms.

Gene mutations can result from imperfect replication of the DNA strand during cell division. They can also occur when an already existing DNA strand is chemically altered by an external agent, which may be a chemical substance or a dose of radiation such as cosmic radiation or ultraviolet light. Some mutations produce changes in the structure of the proteins that are encoded by the mutated segments of DNA. Proteins are the basic elements in the structure and functioning of organisms. Some are key constituents of tissues, and others serve as enzymes, which trigger chemical reactions in

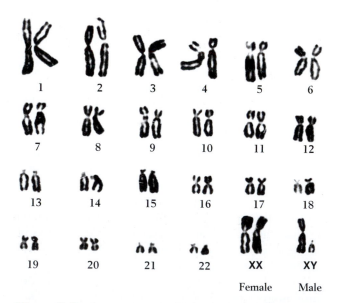

Figure 7-8 **The complete set of human chromosomes.** One member of each of a person's 23 pairs of chromosomes comes from each parent. In pair 23, the presence of two X chromosomes indicates that the set of chromosomes represents a female; the alternative male condition is determined by the presence of one X and one Y chromosome. (M. Grumbach and A. Morishima.)

the physiology of organisms. Some genes that do not encode proteins play a regulatory role, governing the pattern of growth and development by switching other genes on or off. Mutation of a regulatory gene can profoundly modify an organism.

In eukaryotes, such as plants and animals (see p. 54), DNA is concentrated within **chromosomes**, which are elongate bodies found in the nucleus of the cell. Most organisms have chromosomes that are paired, one having been inherited from each parent (Figure 7-8). Chromosomal mutations take the form of changes in the number of chromosomes or in the positions of segments of individual chromosomes. Such changes cause segments of DNA to move with respect to one another, and such changes in relative positions sometimes alter the way one segment affects the operation of another segment. The result can be a change in the way an organism develops and functions.

Populations, Species, and Speciation

When two organisms breed, each normally contributes half of its chromosomes to each offspring by way of a **gamete**, a special reproductive cell that contains only one member of each chromosome pair. The female transmits this set of chromosomes to her offspring by way of an egg cell, which is the female gamete, and the male by way of a sperm cell, the male gamete. Similarly, the offspring, if it mates, combines half of its chromosomes with half of those of a member of the opposite sex in order to produce still another generation. The mixing that takes place in this manner, which is known as **sexual recombination**, continually yields new combinations of genes and chromosomes. The process of sexual recombination, in conjunction with occasional mutations, is responsible for the variability among organisms that provides the raw material for natural selection. Unfortunately, Darwin and his contemporaries had no knowledge of these sources of variability.

In the study of evolution today, the sum total of the genetic components of a population, or group of interbreeding individuals, is referred to as a **gene pool**. And as we have seen (p. 55), populations belong to the same species if their members can interbreed. Reproductive barriers between species keep their gene pools separate and thus prevent interbreeding. These barriers include differences in mating behavior or habitat preference, incompatibility of egg and sperm, and failure of offspring to develop into fully functioning adults.

A species as a whole can not only evolve in the course of time, but can also give rise to one or more additional species. The origin of a new species from two or more individuals of a preexisting one is called **speciation**. Because species are kept separate from one an-

other by reproductive barriers, speciation, by its very definition, entails evolutionary change that produces such barriers. It is widely believed that most instances of speciation result from the spatial isolation of one population from the remaining populations of the parent species. This isolated population then follows an evolutionary course that causes it to diverge from the parent species in both physical form and way of life. Its divergence may result from natural selection operating on unique mutations or gene combinations under unusual environmental conditions. The development of distinct species of finches on the various Galápagos islands (see Figure 7-7) illustrates this principle.

Rates of Origination

One unique contribution of fossils to biological science is the ability they afford us to assess rates of evolution and extinction. It is only through data derived from the fossil record, for example, that we have been able to measure the rates at which new species, genera, and families have appeared within large groups of animals and plants.

At many times in Earth's history, groups of organisms have undergone remarkably rapid evolutionary expansion—that is, one or more phyla, classes, orders, or families have produced many new genera or species during brief intervals of time. Rapid expansions of this kind are known as **evolutionary radiations**. The word *radiation* refers to the pattern of expansion from some ancestral adaptive condition to the many new adaptive conditions represented by the descendant taxa.

Figure 7-9 shows how the fossil record permits us to measure the rate at which evolutionary radiation has taken place. Here we can see that the number of families of corals increased rapidly during the Jurassic Period. The coral families depicted belong to the order known as the *hexacorals*, which first appeared in Middle Triassic time, about 235 million years ago. Living species of hexacorals form the beautiful coral reefs of the modern world (see Figure 4-28).

Evolutionary radiation often occurs in groups of plants or animals within just a few million years of their origin. This pattern is common because the modes of life of recently formed groups often differ from those of other groups from which they originated. Because the old and new groups occupy different niches, ecological competition does not restrain the diversification of the new group. In addition, when a group first evolves, predators may not yet have developed efficient methods for attacking its members. Until they do, the new group is free to form many new species in a short period of time.

Sometimes the extinction of one biological group has allowed for the evolutionary radiation of another even though the radiating group was not a new one on Earth. Mammals, for example, inhabited Earth during

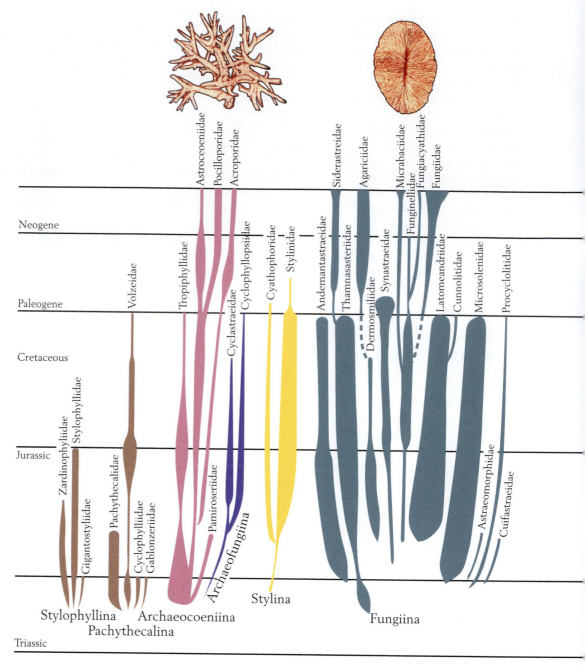

Figure 7-9 The pattern of adaptive radiation of the hexacorals since the order originated in Middle Triassic time. The hexacorals include the builders of modern coral reefs. All but one of the suborders (color groups) originated before the end of Jurassic time, but families (individual bars) continued to proliferate during Cretaceous and Paleogene time. Widths of bars indicate relative diversity. (After J. E. N. Veron, *Corals in Space and Time*, Cornell University Press, Ithaca, NY, 1993.)

almost all of the Mesozoic Era, but they remained small and relatively inconspicuous until the close of that era, when the dinosaurs suffered extinction. Unrestrained by competition or predation by dinosaurs, mammals then underwent a spectacular radiation. Their rise to dominance on land has led paleontologists to label the Cenozoic Era the Age of Mammals. Most of the living orders of mammals, including the order that comprises bats and the one that comprises whales, came into existence within only about 12 million years after the start of the Cenozoic Era. This interval represents only about 2 percent of all Phanerozoic time.

Many episodes of evolutionary radiation have been triggered by **adaptive breakthroughs**—the appearance of key features that, along with ecological opportunities, have allowed the radiation to take place. The porous skeletons of hexacorals, for example, have often allowed these creatures to crowd out other animals that inhabit hard surfaces in shallow marine environments (see Figure 2-8). Because their skeletons are porous,

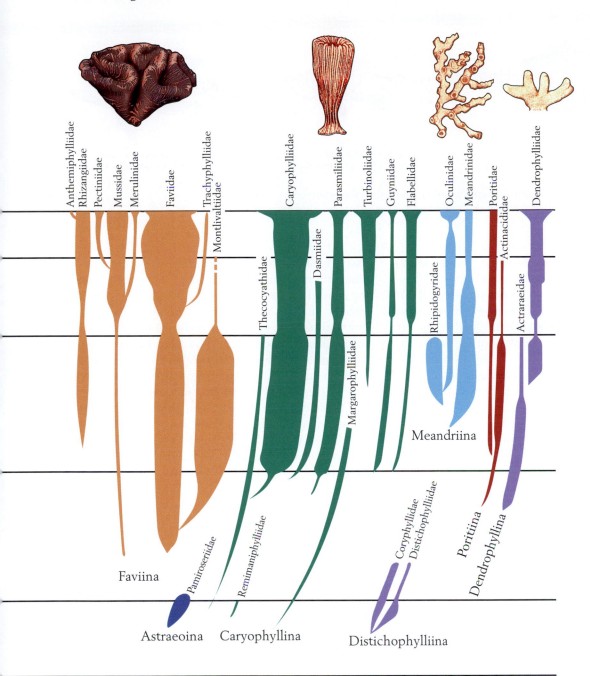

hexacorals can quickly assume large proportions without the need for large volumes of calcium carbonate. Thus they have had an edge over slowly growing organisms with which they compete for space on the seafloor. Another adaptive breakthrough for the hexacorals was the development of a **symbiotic** (mutually beneficial) relationship with algae that live in the tissues of the reef-building species and help the hexacorals secrete their skeletons (p. 99).

A variety of adaptive breakthroughs played major roles in the early Cenozoic radiation of mammals. The key feature for rodents, for example, was continually growing front teeth, which most rodents use for eating nuts and other hard seeds. Beavers, however, use them to cut down trees, and you will recall that one extinct

beaver species even used them to gnaw deep burrows in the ground (see Figure 5-2). Farther back in geologic time, near the end of the Devonian Period, the paddlelike fins of certain fishes evolved toes that served not only for swimming but also for walking in shallow water. The origin of legs and feet from these dual-purpose limbs was the adaptive breakthrough that allowed vertebrate animals to invade and diversify on dry land.

The pattern of evolutionary radiation seen in Figure 7-9 for the hexacorals is typical: early evolution of the order produced large-scale evolutionary divergence at a very early stage. Note that all but one of the suborders that exist today were already present during the Jurassic Period, shortly after the evolutionary radiation

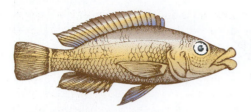

Haplochromis chilotes,
a specialized insect eater (46% of actual size)

Haplochromis estor,
a fish eater (16% of actual size)

Haplochromis sauvagei,
a mollusk eater (44% of actual size)

Figure 7-10 **Three of the roughly 500 species of cichlid fishes that evolved in Lake Victoria, Uganda, within the last thirteen thousand years.** These fishes exhibit a wide variety of adaptations. (After P. H. Greenwood, *Brit. Mus. [Nat. Hist.] Bull.* Suppl. 6, 1974.)

of hexacorals began. Then, however, evolutionary change became more restricted. Several new families arose after Jurassic time, but no new suborders, and no new families have arisen during Neogene time—only new genera and species. The appearance of this kind of pattern again and again in the history of life seems to indicate that as a phylum or class of animals or plants begins to expand, it quickly exploits any adaptations that its body plan allows it to develop with ease. Later, however, evolution is restricted to the development of variations on the basic adaptive themes that evolved early on. In time, few new orders or families evolve, primarily just new genera and species.

Local radiations of the recent past offer special insights into the fundamental nature of evolutionary radiation. Many of these local events occurred in isolated environments, such as islands or lakes, whose well-defined boundaries prevented the escape of the species within them. When these sites of radiation are of recent origin, they can provide evidence of the remarkably rapid diversification of life. This diversification, in turn, usually reflects the fact that the site of radiation was uninhabited territory and thus lacked the predators and competitors that might have inhibited evolutionary diversification.

The Galápagos, where Darwin studied unique groups of tortoises and finches (see Figures 7-6 and 7-7), are the islands most famous as sites of recent evolutionary radiation. The Galápagos originated as a result of volcanic eruptions a few million years ago. Many large lakes have also been sites of radiations in the recent past. Cores taken from the center of Lake Victoria in Uganda have yielded a layer of fossil grass that radiocarbon dating has shown to be only about 13,000 years old. This evidence provides the maximum age for the lake, which evidently formed on top of a broad grassland. Lake Victoria now contains about 500 fish species of the cichlid group, and all but three of them can be found nowhere else in the world. The brevity of the lake's history indicates that the radiation of these fishes was spectacularly rapid. Despite the relative youth of the lake, many of its cichlid fishes have highly distinc-

tive adaptations: some are specialized for eating insects, others for attacking other fish, and still others for crushing shelled mollusks (Figure 7-10). Several species of cichlids in Lake Victoria look very much like the ancestral fishes that gave rise to the great evolutionary radiation that has occurred in the lake. In other words, the original species, or descendants very much like them, remain in the lake along with much more distinctive products of the evolutionary radiation.

In view of such dramatic evolutionary radiations in lakes and on islands, it is interesting to consider what may have happened in the aftermath of major extinctions of the past. When the dinosaurs disappeared from Earth at the end of the Mesozoic Era, for example, the great continents of the world must have been the equivalents of large vacant islands that mammals could colonize on a grander scale than had been possible when dinosaurs ruled. For the mammals, which had remained small and in the shadow of the dinosaurs, the opportunity to undergo evolutionary radiation must have resembled the evolutionary opportunity that was available to the first cichlids that arrived in Lake Victoria or the first finches that landed on a Galápagos island. The entire world, not just a small area, became available for mammals to occupy.

Thus small-scale evolutionary radiations of the recent past seem to offer useful models for understanding larger radiations of the more distant past. The most basic lesson is that when preexisting species do not interfere, a small number of founder species can rapidly produce many new species, some of which differ substantially from the original forms.

The Molecular Clock and Times of Origination

Fossils are rare or unknown for some taxa. There is nonetheless a way of estimating when such taxa originated in the geologic past. Genetic data from living species provide a useful, though imperfect, means of estimating when two related groups of organisms began to diverge from one another. Any two such groups will have accumulated genetic differences through natural selection, and the rate at which these differences developed will have varied, depending on the intensity of natural selection. There are, however, some genes that are unaffected by natural selection—for example, genes that code for amino acids that are interchangeable with other amino acids in proteins. These "neutral" genes undergo mutation without any effect on adaptation. In the absence of natural selection, segments of DNA containing neutral genes will simply change at rates determined by rates of mutation. Mutations occur by chance and are therefore sporadic, but the average rate of mutation for a given kind of DNA segment unaffected by natural selection will be relatively constant over long stretches of geologic time. The accumulation of mutations at a fixed rate provides what is termed a **molecular clock**. To read this clock, however, we need information from the fossil record.

The first step is to measure the percentage of genes that differ between two living taxa in one or more appropriate segments of DNA. The taxa must, however, have fossil records that are complete enough to provide an accurate estimate as to how many million years ago they diverged from one another along separate evolutionary pathways. It is then a simple procedure to calculate the pace of the molecular clock for the DNA segments—the average percentage of genes that changed every million years. Once the molecular clock has been calibrated for two related taxa that have good fossil records, it can be applied to two related taxa with poor fossil records. We can tell how long ago those taxa diverged from one another by determining the percentage of their genes that differ in the calibrated DNA segments. For example, if the segments differ by 10 percent and the mutation rate is 2 percent per million years, then the taxa diverged 5 million years ago. If one of the taxa evolved by branching off from the other, then that taxon arose 5 million years ago.

Unfortunately, the molecular clock is far from perfect. One problem is that rates for the clock are often based on genetic data for taxa that arose relatively recently—and then this rate is extrapolated over long intervals of geologic time. Any inaccuracy in the clock then leads to a large error in the estimated time of origin of a taxon. Another problem is that the clock does not tick at a single rate: its rate varies among taxa and even among different types of genes and segments of DNA in the same taxon. Thus, to apply the clock to a particular fossil group, it is important to calibrate it by means of living organisms that are closely related to that fossil group. It is nonetheless never certain that the clock ticked at exactly the same rate for the living and fossil taxa.

Another problem is that the molecular clock can be erroneously set to run too fast if paleontologists have underestimated the age of the taxon used to calibrate it, which happens when the earliest members of the taxon have not been discovered in the fossil record. This problem can be avoided by calibrating the clock using times of divergence of taxa that are well dated by some physical event. The uplift of the Isthmus of Panama provides such an opportunity, having separated the Caribbean Sea from the Pacific Ocean slightly more than 3 million years ago. Recall that Charles Darwin noted that differences between the modern marine life on the two sides of the isthmus suggested evolutionary divergence (p. 157). As it turns out, calibration of the molecular clock for sea urchin species on both sides of the isthmus gives rates of genetic change of about 2 percent per million years. In fact, these rates happen to resemble the ones calculated for various groups of mammals using fossil data.

Although the molecular clock is imperfect, especially when its rates are extrapolated over hundreds of millions of years, it has provided useful information about certain groups with poor fossil records. For example, in the 1970s, when relatively poor fossil data suggested that the human family diverged from ancestral apes about 15 million years ago, the molecular clock indicated that this important evolutionary event occurred closer to 5 million years ago. This latter estimate is closer to the truth, according to more recent fossil evidence.

Evolutionary Convergence

As a group of organisms undergoes an evolutionary radiation, some of the taxa that arise may come to resemble taxa that evolved separately, in other radiations. **Evolutionary convergence**—that is, the evolution of similar forms in two or more different biological groups—offers convincing evidence that biological form is adaptive. This principle is strikingly illustrated by the similarity between many of the marsupial mammals of Australia and the other kinds of mammals that live in similar ways on other continents (Figure 7-11). The marsupials of Australia, which carry their immature offspring in a pouch, are the products of a radiation that took place on this isolated island continent during the Cenozoic Era. That the marsupials' radiation has been adaptive is indicated by the fact that they have *diverged* from one another, but simultaneously have *converged*, both in way of life and in body form, with one or more groups of placental mammals living elsewhere. (Nearly all nonmarsupial mammals are placentals, as we saw in Chapter 3.) The strong similarities between many Australian marsupials and placental mammals of other regions must partially reflect the basic evolutionary limitations of the mammals in general. In other words, it

MARSUPIALS   PLACENTALS

Tasmanian wolf   Wolf
(*Thylacinus*)   (*Canis*)

Flying phalanger   Flying squirrel
(*Petaurus*)   (*Glaucomys*)

Marsupial anteater   Anteater
(*Myrecobius*)   (*Myrmecophaga*)

Marsupial mole   Mole
(*Notoryctes*)   (*Talpa*)

Figure 7-11 Evolutionary convergence between marsupial mammals of Australia and nonmarsupial mammals of other continents. Although each of the marsupials is more closely related to a kangaroo than to its placental counterpart in the other column, these pairs of mammals have converged in body form and way of life. (After G. G. Simpson and W. S. Beck, *Life*, Harcourt, Brace & World, Inc., New York, 1965.)

would appear that certain adaptations are likely to develop in mammals under a variety of circumstances, while others are never likely to evolve.

Almost all adaptive radiations, however, have produced some surprises. Judging from what we see on other continents, we might have predicted, for example, that hoofed, four-footed herbivores resembling deer, cattle, and antelopes would populate the continent of Australia. As it turns out, the Australian equivalents of these large, fast-moving herbivores are kangaroos—animals that hop around on two legs. Apparently

the breakthrough represented by the kangaroos' hopping adaptation happened to evolve in Australian herbivorous marsupials before an adaptive breakthrough could produce an efficient running apparatus. Once the kangaroos had occupied many regions of Australia, there was simply little opportunity for animals resembling deer, cattle, or antelopes to evolve.

Extinction

Fossils provide the only direct evidence that life has changed substantially over long spans of geologic time. They also offer the only concrete evidence that millions of species have disappeared from Earth, or suffered **extinction**.

The idea that a species could become extinct was not widely accepted until late in the eighteenth century. Before that time, fossil forms that seemed no longer to inhabit Earth were thought to live in unexplored regions. In 1786, however, Georges Cuvier, a French naturalist, pointed out that fossil mammoths were so large that any living mammoths could not possibly have been overlooked. Cuvier thus concluded that mammoths were extinct. His argument was well received, and soon the extinction of many species was accepted as fact.

In general, extinction results from extreme impacts of the limiting factors that normally hold populations in check. Limiting factors, as we saw in Chapter 4, include predation, disease, competitive interactions with one or more other species, and restrictive conditions of the physical environment. They also include chance fluctuations in the number of individuals in a population. Population declines resulting from one or more of these factors have led to the extinction of most of the species of animals and plants that have inhabited Earth; in fact, of all the species that have existed in the course of Earth's history, only a tiny fraction remain alive today.

Species have also disappeared by evolving to the point at which they have been formally recognized as different species. In this process, known as **pseudo-extinction**, a species' evolutionary line of descent continues, but its members are given a new name. The point at which the new species comes into being is often arbitrarily designated because there is no way of determining precisely when members of an evolving group lost the ability to interbreed with its original members.

Rates of extinction vary greatly

Extinction rates have varied greatly within most large groups of animals and plants in the course of geologic time, and they have varied just as greatly from taxon to taxon. Mammalian species, for example, have survived, on average, for just 1 million to 2 million years, which means that the extinction rate for mammals has exceeded 50 percent per million years. In contrast, within

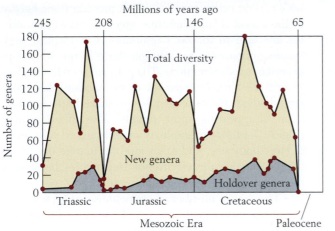

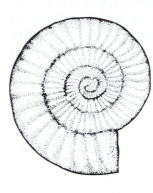

Figure 7-12 The appearance and disappearance of ammonoid genera through time. The turnover rate of genera was high throughout the ammonoids' history. Data plotted for the many ages of the Mesozoic Era show that few genera present during one age were still present in the next. (After W. J. Kennedy, in A. Hallam, ed., *Patterns of Evolution*, Elsevier, Amsterdam, 1977.)

many groups of marine life, an average species has existed for 10 million years or more; among these groups are the bivalve mollusks (clams, scallops, oysters, and their relatives). Under ordinary circumstances, then, only a small fraction of the species within such a group has disappeared every million years.

Groups of animals and plants that are well represented in the fossil record and that have experienced high extinction rates tend to serve well as index or guide fossils (p. 135). The *ammonoids*, an order of swimming mollusks related to the chambered nautilus, meet both requirements (p. 128). The fossil record of species is so incomplete that geologists often use fossil genera to tally rates of extinction; even if only a modest percentage of the species that belonged to a genus are known from the fossil record, they can provide a good estimate of the time when that genus became extinct. As Figure 7-12 indicates, few ammonoid genera found in any stage of the Mesozoic Erathem are also found in the next stage—

and yet these genera are usually succeeded by a large number of new ones. Thus the rate at which genera of ammonoids died out was high, but so was the rate at which new genera formed. In other words, the *turnover rate* of ammonoid genera was high.

Different groups of organisms may exhibit different rates of extinction, but when many groups are viewed together, global trends in rates of extinction become evident. Figure 7-13 shows that the average rate of extinction for marine animals has declined since the Cambrian Period. One explanation for this pattern is that taxa whose genera have tended to suffer high rates of extinction have tended to disappear altogether, while groups characterized by low rates of extinction have tended to accumulate in the course of the Phanerozoic. These resistant taxa make up a relatively large percentage of the taxa of the modern world.

Figure 7-13 also shows that extinctions of species have sometimes been clustered within brief intervals of

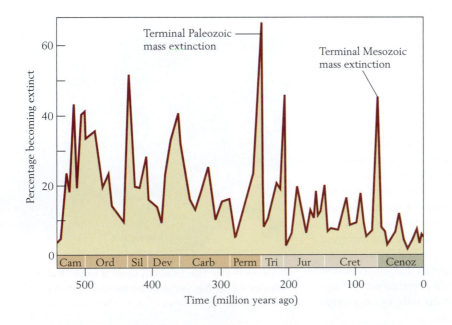

Figure 7-13 Percentages of extinction for genera of marine animals for stages of the Phanerozoic. Conspicuous peaks represent mass extinctions. (After J. J. Sepkoski, *Geotimes*, March 1994.)

time. During several intervals of a few million years or less, large numbers of genera, families, and even higher taxonomic groups have vanished in **mass extinctions**. In Figure 7-13 the largest mass extinctions are represented by peaks that rise above the 40 percent level for extinction of genera. There is no convention specifying how severe a crisis must be to constitute a mass extinction, but even events represented by small peaks (losses of 10 to 20 percent of genera) have sometimes been described as mass extinctions.

Geologists are now studying mass extinctions quite intensively. Although few of these biotic crises are well understood, it is clear that they did not all result from a single kind of cause. As noted earlier, it now appears almost certain that the one that swept away the dinosaurs resulted from the impact on Earth of a large asteroid or comet (see p. 22). As we will also see in later chapters, other mass extinctions had earthly causes, including climatic changes.

After every mass extinction, the number of species on Earth has increased again. A mass extinction affects some groups of living things more severely than others, and some taxonomic groups that have suffered huge losses have failed to regain their previous diversity. Conversely, some taxa that previously were not very diverse have come to flourish in the aftermath of a biotic crisis.

Unfortunately, the world is entering a unique interval of mass extinction today: one that we humans are bringing about.

A mass extinction is occurring today

It appears that modern human activities are triggering a mass extinction that will rival any that the world has known in the past 600 million years. Rates of extinction have varied throughout geologic time, but calculations suggest that during most intervals, only about one species has died out each year on the average. These days several species are believed to be dying out somewhere every day. Experts believe that the rate of extinction may climb to several hundred species a day within 20 or 30 years.

The fossil record alerts us to look for particular patterns in the crisis we are entering. Large animals, for example, have always tended to suffer relatively high rates of extinction, apparently because their typically small populations leave them constantly at risk. The dinosaurs and mammoths spring most readily to mind. The fossil record also shows that when the environment deteriorates on a global scale, the percentage of species that are lost tends to be especially great in tropical regions. Tropical species are particularly vulnerable because the great diversity of life in the tropics is packed into complex communities in which many species have specialized ecological requirements and small populations. A large proportion of these species can exist only in association

with certain other species, which provide their habitat or their food. Thus when one species goes extinct, others are sure to follow. Communities of tropical reef-building organisms, for example, have frequently disappeared in mass extinctions. The patterns seen in the fossil record are already apparent in the modern-day crisis: large-bodied and ecologically specialized species have been disappearing most rapidly. The Siberian tiger (Figure 7-14), for example, is an endangered subspecies of large body size, and many kinds of marine organisms that build tropical reefs are now endangered because of their narrow ecological requirements.

The largest number of extinctions can be traced to our destruction of habitats. Of all our depredations, the destruction of tropical forests—most of them rain forests—has the most dire consequences, for two reasons. First, even though these lush habitats occupy less than 10 percent of Earth's land area, they contain most of the world's species. Second, the total expanse of tropical forests being destroyed every year constitutes an area about the size of West Virginia, and only about 5 percent of the remaining area is protected in parks and preserves. Because their populations fall below critical levels as their habitat is lost, many species die out in a shrinking rain forest long before the forest has disappeared altogether. Rain forests can grow again in areas from which they have disappeared, but recovery requires more than a century.

Rain forests are disappearing so rapidly today that even a massive restoration program would be too late to preserve vast numbers of species of great beauty, scientific interest, and possible value to humankind as sources of medical drugs and other useful products. Evolution would require several million years to restore this lost diversity through the natural speciation process. Even then, the process would not precisely duplicate

Figure 7-14 The Siberian tiger, a large-bodied subspecies on the verge of extinction. (Jim Brandenburg/ Minden Pictures.)

any lost species, or even produce species remotely resembling those that belonged to genera or families that had vanished altogether.

Direct exploitation of animals and plants by humans is another cause of extinction. The African black rhinoceros, for example, will be lucky to survive hunters catering to people in all parts of the world who cling to the old fantasy that its horn enhances sexual prowess.

After earlier mass extinctions, a few kinds of surviving organisms inherited a world all but free of competitors and predators, and their populations exploded. These species were ecological opportunists—species capable of invading vacant terrain readily and then multiplying rapidly. Weeds are conspicuous opportunists in today's world; so are noisy, aggressive birds, such as starlings, and so are the cyanobacteria that form scummy masses in polluted water and then decay, robbing their environment of oxygen. Thus we not only face the prospect of losing vast numbers of species that are important to us—as well as many entire ecological communities—in the next few decades; we must also expect their places to be filled by existing species that impair the quality of human life.

Evolutionary Trends

By examining the evolutionary history of any higher taxon that has left an extensive fossil record, we can observe long-term evolutionary trends—general changes that developed over the course of millions of years. Some of these changes affected form, but others simply affected body size.

Some overall trends for taxonomic groups resulted from evolution, but extinction contributed to some, by preferentially eliminating certain kinds of species to change the groups' composition.

Animals tend to evolve toward larger body size

A general tendency for body size to increase during the evolution of a group of animals is known as **Cope's rule**, after Edward Drinker Cope, a nineteenth-century American paleontologist who observed this phenomenon in his studies of ancient vertebrate animals.

Numerous factors may cause a group of animals to become larger in body size as the group evolves, but all have to do with the tendency of large individuals to produce more offspring than smaller ones. Within species in which males fight for females, for example, larger males tend to win battles and hence to produce a disproportionate number of offspring. Within other species, larger animals may produce more offspring because they are better equipped to obtain food or to avoid predators.

This evolutionary trend cannot continue indefinitely in any animal group, however, because at some point a further increase in body size will inevitably cease to be advantageous. A four-legged animal the size of a large building, for example, could not run or even stand, because its weight would greatly exceed the strength of its limbs. Indeed, many animals could not gather sufficient food or move efficiently if they were appreciably larger than they are.

Given the fact that increases in body size are advantageous only within limits, the great number of animal groups that have evolved toward larger size suggests that most animal orders and families have evolved from relatively small ancestors. Large size tends to impose many adaptive problems for animals, and the specialized adaptations associated with these problems are not easily altered to produce entirely new adaptations. Thus large, highly specialized animals tend to represent evolutionary dead ends.

Some of the problems associated with large size can be seen in the physical adaptations of the elephant. An elephant has such a huge body to feed that it must spend most of its time grinding up coarse food with its molars. The need for constant chewing dictates that an elephant's teeth and jaws must be quite large in relation to its overall size. The elephant's head is therefore large as well—so large that the neck must be quite short to support it (Figure 7-15A). To compensate for their

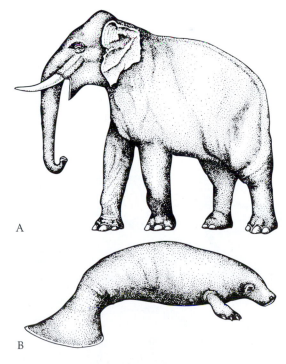

A

B

Figure 7-15 **Two large-bodied mammals.** The elephant (A) and the manatee (B) differ in form but have a common ancestry within the class Mammalia.

consequent short reach, members of the elephant family evolved an enormous trunk from an originally short nose, together with long tusks from originally short teeth. These and other unusual features make it unlikely that modern elephants will ever evolve into very different types of animals.

Manatees (or sea cows), which are blubbery swimmers, are relatives of the elephants (Figure 7-15*B*). The evolution of both groups illustrates Cope's rule. Like elephants, manatees are highly specialized animals with limited potential to give rise to substantially different types of animals. The two groups have common ancestors of early Cenozoic age that were small and rodent-like in their general form and adaptations. These small, relatively unspecialized forms easily evolved in a variety of directions, most of which led to larger animals. Elephants and manatees are among the largest.

Whales are the largest of all mammals; the water that surrounds them supports their massive bodies. The fossil record shows that modern whales evolved from terrestrial animals (Figure 7-16). The earliest known whales, which lived about 50 million years ago during the Eocene Epoch, were four-legged terrestrial animals about 2 meters (6 feet) long with large jaws and teeth. Slightly later whales had shorter legs and long toes that probably belonged to feet that were webbed for swimming. Like seals, they probably spent some of their time on the shore and some in the water. By 40 million years ago, more advanced whales existed that had no hind limbs and only small pelvic bones. Some of the most advanced Eocene whales, measuring more than 20 meters, were as long as good-sized modern whales, though they were more serpentlike in form (Figure 7-16*D*).

Figure 7-16 depicts a general trend, showing stages in the evolutionary transition from four-legged, land-dwelling whales to fully marine modern whales. Because not all extinct whale species are known from fossils, we cannot be sure that any of the whales depicted here actually descended from any of the earlier forms pictured below them—only that they descended from these *kinds* of animals. Now let us examine how fossil data depict more precise trends for some taxonomic groups.

Evolutionary trends can be simple or complex

Trends in evolution occur on both small and large scales. A transition from one species to another, for example, represents a simple trend on a small scale. A complex,

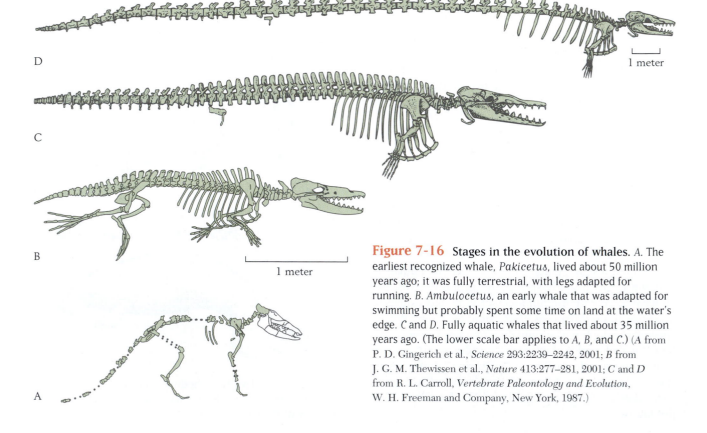

Figure 7-16 Stages in the evolution of whales. *A.* The earliest recognized whale, *Pakicetus*, lived about 50 million years ago; it was fully terrestrial, with legs adapted for running. *B. Ambulocetus*, an early whale that was adapted for swimming but probably spent some time on land at the water's edge. *C* and *D.* Fully aquatic whales that lived about 35 million years ago. (The lower scale bar applies to *A, B,* and *C*.) (*A* from P. D. Gingerich et al., *Science* 293:2239–2242, 2001; *B* from J. G. M. Thewissen et al., *Nature* 413:277–281, 2001; *C* and *D* from R. L. Carroll, *Vertebrate Paleontology and Evolution*, W. H. Freeman and Company, New York, 1987.)

large-scale trend is one that occurs within a branching limb of the tree of life—a *phylogeny* (p. 56). In considering the structure of evolutionary trends, we will focus first on simple trends by which one species is transformed into another by way of a single thread of evolution. We will then examine complex trends involving many species—trends that consist of many threads of evolution and can thus be viewed as having a fabric.

Some species have arisen by the gradual transformation of another species—that is, an entire species has changed sufficiently in the course of many generations to be regarded as a new species. Figure 7-17 illustrates this type of evolutionary change in a group of coiled

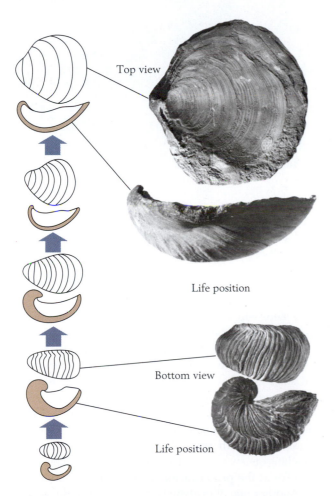

Top view

Life position

Bottom view

Life position

Figure 7-17 An apparently gradual trend in a lineage of coiled oysters of the genus *Gryphaea* during an Early Jurassic interval of about 12 million years. During the interval shown here, the shell became larger, attaining the diameter of a small saucer, but it also became thinner and flatter. These animals rested on the seafloor in the orientation labeled "life position." Perhaps the flatter shell was more stable against potentially disruptive water movements. (After A. Hallam, *Phil. Trans. R. Soc. Lond. B* 254:91–128, 1968.)

oysters during the Jurassic Period. Recall that the "disappearance" of a species because of such evolutionary change is termed pseudoextinction (p. 166).

Like many biologists and paleontologists of the twentieth century, Darwin believed that gradual trends, such as those evident in the evolution of the Jurassic coiled oysters, produced most large-scale evolutionary trends. Recently, however, it has been observed that such gradual trends are relatively rare. Oysters are bivalve mollusks, and hundreds of bivalve mollusks have been identified from Jurassic rocks of Europe—yet very few of these species exhibit gradual trends like the one illustrated in Figure 7-17. In fact, it is estimated that an average bivalve species living in Europe during the Jurassic Period existed without appreciable change for about 15 million years, or for about one-quarter of the entire Jurassic Period.

Many other animal and plant species have also survived for long geologic intervals. Species of benthic foraminifera, for example, are estimated to have survived, on average, 30 million years, diatoms 25 million years, mosses and their relatives more than 15 million years, seed plant species 6 million years, freshwater fishes 6 million years, beetles more than 2 million years, and mammals nearly 2 million years. These various estimates suggest that most species evolve very slowly; they indicate that an average animal or plant species is not likely to evolve sufficiently to be regarded as a new species even after it has passed through about a million generations. Note that the oyster species shown in Figure 7-17 changed enough to be regarded as a new species within 2 million to 4 million years; even though an individual oyster becomes reproductively mature in 2 or 3 years, evolution required a million generations or so to produce a new species.

All of these species longevities must be viewed in the light of the length of time it has taken for new higher taxa of the same group to develop. Early in the Cenozoic Era, bats evolved from small rodentlike mammals in perhaps 12 million years. A typical survival time of 2 million years for a single mammal species seems quite long in comparison. Thus it appears that many new species have arisen from others by way of a rapid step of speciation, with a descendant species evolving from the parent species in a relatively short span of time, rather than by gradual alteration of an entire preexisting species.

An example of a rapid speciation event is the origin of the axolotl, an amphibian species whose members remain aquatic throughout life (Figure 7-18). The axolotl evolved from a population of a normal salamander species, one that is still extant—an amphibian that undergoes metamorphosis from an aquatic juvenile form to a terrestrial adult form. The axolotl becomes

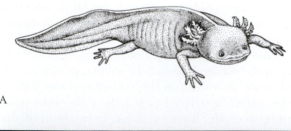

A

B

C

Figure 7-18 **Aquatic and terrestrial amphibians.** *A.* The axolotl, which lives its entire life in fresh water. *B* and *C.* Two stages in the ontogeny of a typical salamander. The aquatic larval stage (*B*) closely resembles the adult axolotl (*A*), which in effect never grows up. (*A* after J. Z. Young, *Life of Vertebrates,* Oxford University Press, London, 1962; *B,* Stephen Dalton/Photo Researchers; *C,* Michael Durham/ENP Images.)

reproductively mature even though it retains the juvenile body form of its ancestors and continues to live in water throughout its life. The evolutionary transition that produced the axolotl was genetically simple. An axolotl can be artificially forced to metamorphose into a terrestrial animal if it is injected with thyroxine, a substance normally produced by the thyroid gland but missing in the axolotl. Thus the axolotl must have been produced by a speciation event consisting of a simple genetic change that impeded the normal development of the thyroid gland. This change must have occurred quite rapidly on a geologic scale of time—probably as the genetic change spread throughout a population of the ancestral species that was confined to a single pond or lake. Presumably, most species arise by way of many such mutations and recombinations of genes, but the case of the axolotl illustrates that just one or a few genetic changes can produce a new species.

According to the traditional, gradualistic model of evolution, most evolutionary change takes place in very small steps within well-established species. Some paleontologists have come to oppose the gradualistic model on the basis of the very slow rates of evolution that have characterized many well-established species. They note that such rates have been too slow to account for the many large evolutionary changes that have occurred quite rapidly on a geologic scale of time. These paleontologists conclude that such evolutionary changes must be associated with speciation—that is, with the rapid branching of new species from existing species. This view is known as the **punctuational model** of evolution.

Another line of evidence cited in favor of the punctuational model is the evolutionary history that typifies long, narrow segments of phylogeny—segments that undergo little branching but span long intervals of geologic time. If speciation were indeed the site of most evolution, such segments of phylogeny would be expected to exhibit little evolution for the simple reason that they have experienced very little speciation. This is exactly the pattern exhibited by the bowfin fishes

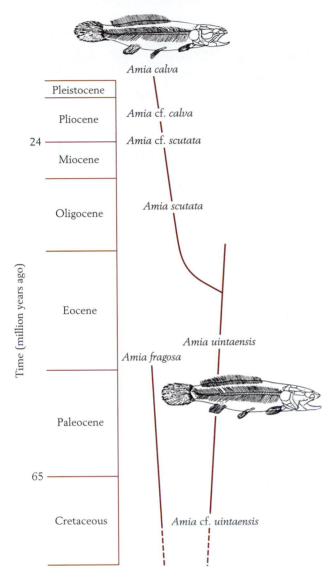

Time (million years ago)

Pleistocene	
Pliocene	
24	
Miocene	
Oligocene	
Eocene	
Paleocene	
65	
Cretaceous	

Amia calva

Amia cf. *calva*

Amia cf. *scutata*

Amia scutata

Amia uintaensis

Amia fragosa

Amia cf. *uintaensis*

Figure 7-19 Living and fossil bowfin fishes. The living bowfin, *Amia calva*, is the product of a long evolutionary history in which there has been little speciation (branching) and little evolutionary change. *Amia* grows to a length of more than a meter (3 feet). (After J. R. Boreske, *Bull. Mus. Comp. Zool.* 146:1–87, 1974.)

(Figure 7-19), a group that has experienced very little speciation and very little evolution during the last 60 million years. The single living bowfin species so closely resembles those of early Cenozoic time that it has been labeled a *living fossil*. Contrasting with the sluggish evolution of the bowfin fishes is the dramatically divergent evolution of the ciclid fishes in Lake Victoria during the past 13,000 years, in association with many speciation events.

As it turns out, all of the living species that we know to be at the end of long, narrow segments of phylogeny are living fossils. Among other living fossils are the alligator, the snapping turtle, and the aardvark. A well-publicized living fossil plant is the dawn redwood, which was thought to be extinct until it was discovered living in a small area of China in the 1940s.

Even when individual species have evolved slowly, large-scale trends have sometimes developed through multiple steps of change in particular evolutionary directions that have led to distinctive new species. Trends have also developed when species of certain types have proliferated while species of other types have died out. Complex trends of this type are evident in the evolution of horses (Figure 7-20). The oldest known horse had four toes on each forefoot and three on each hind foot, had stubby, simple molars, and was the size of a small dog. The modern horse, in contrast, has a single hoofed toe on each foot, has tall, complex molars, and is a relatively large mammal. Many speciation events separated the ancestral kind of horse from the modern kind, however. Both the ancestral genus and the modern one have included several species, as have most intermediate genera, and two or more genera have existed at most times. Thus horse evolution has not entailed a single line of descent, but a complex phylogeny.

The increase in average body size for horses exemplifies Cope's rule. It also illustrates the complex pattern of horse evolution. This and other aspects of horse evolution have been partly a response to environmental change. As we shall see in Chapters 17 and 18, forests cloaked much of the world early in the Cenozoic Era, and grasslands expanded as the era progressed. As grasslands spread, horses that came to occupy them had to be relatively large-bodied to survive because small animals lack the endurance to outrun predators in open terrain, where there is no place to hide. Although most horse species were relatively large by 10 million years ago, several small-bodied species of the genus *Nannippus* evolved after that time (see Figure 3-11). Small horses were probably able to exist during late Cenozoic time only in areas where woodlands served as a refuge.

The evolutionary shift from short, simple molars to tall, complex molars also reflected the spread of grasslands. The molars of early horses were suited to feeding on soft leaves, whereas the taller, more complex molars of all living species are adapted for chewing harsh grasses (see Figure 7-20). Here too the pattern of evolutionary change was complex. Tall molars originated several times in horse evolution, adapting various lineages for feeding on grass. Some species continued to feed partly on soft, leafy vegetation into Neogene time, but an increase in the percentage of species that specialized on grasses led to an increase in average molar height for the horse family.

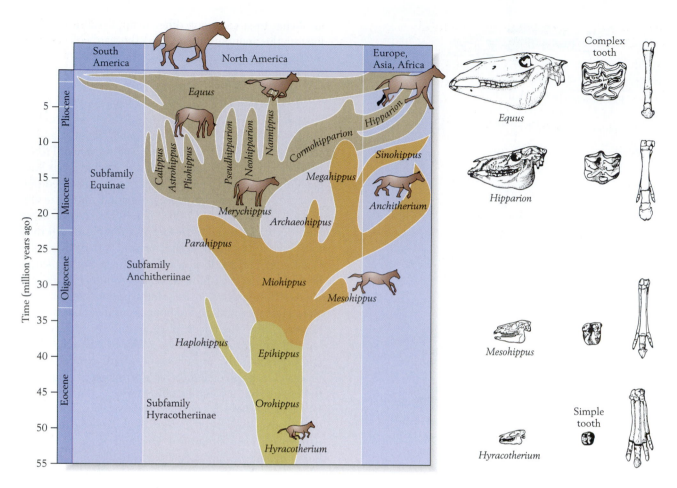

Figure 7-20 The general pattern of the evolution of the horse family. The molars of some members became taller and developed complex grinding surfaces that were associated with a transition from browsing on soft leaves to grazing on tough grasses. The number of toes on the front foot was reduced from four to one. Heads and teeth (but not feet) are drawn to the same scale to show the general increase in size. (After B. J. MacFadden, *Paleobiology*, 11:245–257, with modifications by D. R. Prothero; and G. G. Simpson, *Horses*, Oxford University Press, New York, 1951.)

Evolution is irreversible

A complex evolutionary transition that has resulted from several genetic changes is highly unlikely to be reversed by subsequent evolution. This principle is called **Dollo's law**, for Louis Dollo, the Belgian paleontologist who proposed it early in the twentieth century. Dollo's law reflects the fact that it is extremely unlikely that a long sequence of genetic changes in a population will be repeated in reverse order. Thus evolution occasionally produces a species that crudely resembles an ancestor, but it never perfectly duplicates a species that has disappeared. In other words, once a species has evolved into another or has been eliminated by extinction, it is gone forever.

Chapter Summary

What lines of evidence convinced Charles Darwin that organic evolution produced the vast number of species that inhabit the modern world?

- Many closely related groups of species are restricted to discrete geographic regions separated by barriers.
- Many groups of animals that are dissimilar as adults have similar embryos.
- Animals that live in quite different ways often have similar anatomical "ground plans."
- Some animals possess vestigial organs that serve no apparent function but resemble functioning organs in other species.
- Humans have drastically altered domestic animals and plants through selective breeding.

What are the two components of natural selection?

Natural selection is the process by which certain kinds of individuals become more numerous in a population because they produce an unusually large proportion of the members of the next generation. They manage to do so either by surviving a long time or by reproducing at a high rate.

What is the source of the variability that is the basis of natural selection?

This variability is generated by two mechanisms: mutation of genes and chromosomes and the generation of new gene combinations by sexual recombination.

What role does geography play in speciation, the process by which an existing species gives rise to an additional species?

In most cases, the population that becomes the new species is first geographically isolated from the remainder of the parent species so that, through evolution, it can become reproductively isolated from the parent species.

What factors lead to evolutionary radiation, the proliferation of many species from a small ancestral group?

The evolution of an adaptive breakthrough can trigger an evolutionary radiation, as can the appearance of new ecological opportunities—for example, when another group of organisms becomes extinct or a new habitat appears.

Why is convergence one of the most convincing kinds of evidence that evolutionary changes are adaptive?

It is not likely to be an accident that two unrelated species resemble one another in form and also live in the same way.

Why do species become extinct?

The primary agents of extinction are the ecological limiting factors that normally govern the sizes of populations in nature. Pseudoextinction is the disappearance of a species through its evolutionary transformation into another species. Rates of extinction, like rates of speciation, vary greatly in nature.

What is mass extinction?

Mass extinction is the disappearance of many species during a geologically brief interval of time.

In what ways can evolutionary trends develop?

An evolutionary trend can develop through the gradual transformation of a species. On a larger scale, a trend can result within a higher taxon from the extinction of species of a particular kind or the proliferation of species of a particular kind.

Review Questions

1. What geographic patterns suggested to Charles Darwin that certain kinds of species descended from others?

2. What characteristics make a particular kind of individual successful in the process of natural selection?

3. What conditions make it likely that a small group of closely related species will increase to a large number of species by means of rapid speciation?

4. How can evolution proceed by a change in the growth and development of a species?

5. In what ways can an evolutionary trend develop during the history of a genus or a family?

6. What kinds of environmental change can lead to the extinction of a species?

7. What is a mass extinction?

8. How is pseudoextinction related to gradual evolutionary change?

9. Give an example of evolutionary convergence.

10. What general trends has the evolution of horses displayed?

11. Using the Visual Overview on page 154 and what you have learned in this chapter, compare the kinds of evidence that living organisms on the one hand and fossils on the other hand contribute to our understanding of the evolution of life. What kinds of evidence does each of these two bodies of evidence contribute that the other does not?

Thingvellir Graben in Iceland. This portion of the Mid-Atlantic Ridge is dramatically exposed above sea level. As the rift separates, lava squeezes upward to form new basaltic crust. (Danilo G. Donadoni/Bruce Coleman.)

CHAPTER 8

The Theory of Plate Tectonics

The emergence of the theory of plate tectonics during the 1960s fostered a revolution in the science of geology. **Tectonics** is a term that has long been used to describe movements of Earth's crust. Accordingly, **plate tectonics** refers to the movements of discrete segments of Earth's crust in relation to one another. Whereas continents were once thought to be locked in place by the oceanic crust that surrounds them, the theory of plate tectonics holds that continents move over the surface of Earth because they represent parts of moving plates. Moreover, continents occasionally break apart or, alternatively, fuse together to form larger continents. The theory of plate tectonics explains why most volcanoes and earthquakes occur along curved belts of seafloor, why mountain belts tend to develop along the edges of continents, and why the present ocean basins are very young from a geologic perspective. Most kinds of large-scale rock deformation also result from the movements of plates.

The History of Continental Drift Theory

When the concept of plate tectonics emerged quite suddenly in the 1960s, it resolved many long-standing disputes. The idea that continents move horizontally over Earth's surface—an idea labeled **continental drift**—had been proposed previously, but it had failed to receive general support in Europe or North America. In 1944 one prominent geologist went so far as to assert that the idea of continental drift should be abandoned outright because "further discussion of it merely encumbers the literature and befogs the minds of students." Although many geologists may not have read this comment, most agreed with it in spirit, and during the 1950s, little attention was given to the possibility that continental drift was a real phenomenon.

When the idea of continental drift first emerged, however, it had attracted considerable attention, primarily as a result of the arguments of two scientists: Alfred Wegener of Germany and Alexander Du Toit of South Africa. We will briefly examine the case that these two men and their followers made and the reasons their arguments were rejected by most of their contemporaries.

Some early observations were misinterpreted

Centuries ago, map readers noted with curiosity that the outline of the west coast of Africa seemed to match that of the east coast of South America. The observation that the coasts on the two sides of the Atlantic Ocean fit together like separated parts of a jigsaw puzzle (Figure 8-1) provided the first evidence that continents might once have broken apart and moved across Earth's surface.

Figure 8-1 Computer-generated "best-fit" repositioning of continents that now lie on opposite sides of the Atlantic. This fit was calculated by Sir Edward Bullard and his co-workers at the University of Cambridge. The fit was made along the 500-fathom line of each continental slope. Continental shelves are shown in light green. (After P. M. Hurley, *Sci. Amer.*, April 1968. © 1968 by Scientific American, Inc. All rights reserved.)

Visual Overview
Elements of Plate Tectonics

Plate tectonics provides a unifying picture of the dynamic features of Earth's lithosphere, accounting for the distributions of plants and animals, volcanoes, earthquakes, midocean ridges, and the three basic kinds of faults.

Distributions of fossils and glacial deposits form coherent patterns if landmasses are reassembled to form Gondwanaland.

Gondwanaland

Glacial deposits

● *Lystrosaurus*

● *Mesosaurus*

● *Glossopteris* flora

Trench

Thrust fault

Magma from partial melting of a slab rises to form a volcanic arc parallel to a subduction zone.

Thrust faults form along subduction zones, along with jumbled masses of rock known as mélanges.

Mélange

Drag of convection

Pull of slab

Earthquakes

Suction

Plate movement results from
1. drag from convection in the asthenosphere
2. push from an elevated mid-ocean ridge
3. pull of a cold, dense slab sinking along a subduction zone
4. suction by a sinking segment of a slab that has broken from a plate

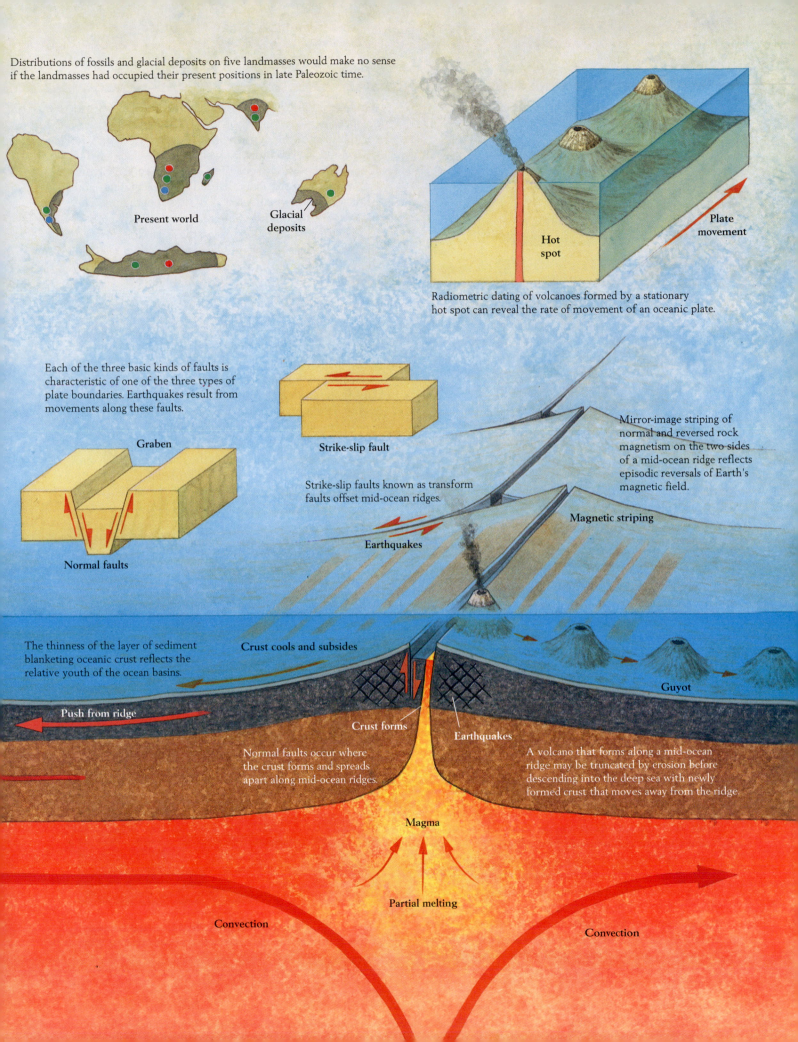

Distributions of fossils and glacial deposits on five landmasses would make no sense if the landmasses had occupied their present positions in late Paleozoic time.

Present world

Glacial deposits

Radiometric dating of volcanoes formed by a stationary hot spot can reveal the rate of movement of an oceanic plate.

Hot spot

Plate movement

Each of the three basic kinds of faults is characteristic of one of the three types of plate boundaries. Earthquakes result from movements along these faults.

Strike-slip fault

Graben

Strike-slip faults known as transform faults offset mid-ocean ridges.

Normal faults

Earthquakes

Mirror-image striping of normal and reversed rock magnetism on the two sides of a mid-ocean ridge reflects episodic reversals of Earth's magnetic field.

Magnetic striping

The thinness of the layer of sediment blanketing oceanic crust reflects the relative youth of the ocean basins.

Crust cools and subsides

Guyot

Push from ridge

Crust forms

Earthquakes

Normal faults occur where the crust forms and spreads apart along mid-ocean ridges.

A volcano that forms along a mid-ocean ridge may be truncated by erosion before descending into the deep sea with newly formed crust that moves away from the ridge.

Magma

Partial melting

Convection

Convection

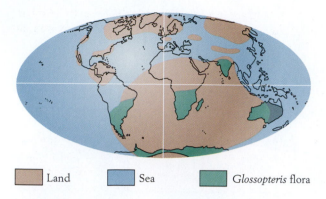

Figure 8-2 Distribution of land and seas during the Carboniferous Period as perceived by nineteenth-century geologists. It was widely believed that large areas of the present-day ocean floor then stood above sea level. The distribution of the *Glossopteris* flora is consistent with this idea, but it is also consistent with the opposing idea that the southern continents were once united.

Nonetheless, nineteenth-century geologists clung to the idea that large blocks of continental crust could not move over Earth's surface. Thus, when the distributions of certain living and extinct animals and plants began to suggest former connections between landmasses now separated, most geologists tended to assume that great corridors of felsic rock (the most abundant material in continental crust) had once formed land bridges that connected continents, but later subsided to form portions of the modern seafloor (Figure 8-2). Today it is recognized that this scenario is not realistic, because felsic crust is of such low density that it cannot possibly sink into the mafic rocks that underlie the oceans. Nonetheless, many prominent geologists of the late nineteenth century presented schemes of Earth history that included the concept of felsic corridors.

One phenomenon that led scientists to speculate about ancient land bridges was the similarity between the fauna of the island of Madagascar and that of India, a land separated from Madagascar by nearly 4000 kilometers (2500 miles) of ocean. Madagascar's mammals are primitive; altogether missing are the zebras, lions, leopards, gazelles, apes, rhinoceroses, giraffes, and elephants that inhabit nearby Africa. In contrast, some of the native animals of India closely resemble those of Madagascar. Some scientists consequently favored the idea that a now-sunken land bridge had once spanned the western part of the Indian Ocean, connecting Madagascar to India.

A second line of evidence for ancient land connections was found in the fossil record. During the nineteenth century, late Paleozoic coal deposits of India, South Africa, Australia, and South America were found to contain a group of fossil plants that were collectively

designated the *Glossopteris* flora, after their most conspicuous genus, a variety of seed fern (Figure 8-3). After the turn of the century, the *Glossopteris* flora was discovered in Antarctica as well. The presence of this fossil flora on widely separated landmasses (see Figure 8-2) was one of the facts that led to the idea that land bridges had once connected all of these continents. The name **Gondwanaland** came to connote the hypothetical continent that consisted of these landmasses and the land bridges that were believed to have connected them. (Gondwana is a region of India where seams of coal yield fossils of the *Glossopteris* flora.)

Not until early in the twentieth century was it hypothesized that, rather than being connected by land bridges, the continents had once actually lain side by side, as components of very large landmasses, and that these supercontinents eventually broke apart and moved across Earth's surface to their present positions. This general idea is central to modern plate tectonic theory. One early surmise, now generally accepted, was that the immense submarine mountain chain called the Mid-Atlantic Ridge, which is one of several ridges of the present-day seafloor, marks the line along which one such ancient landmass ruptured to form the Atlantic Ocean (Figure 8-4).

Figure 8-3 A fossil *Glossopteris* leaf from the Permian of India. The leaf is about 12 centimeters (5 inches) long. (Jim Frazier/Mantis Wildlife Films Pty Ltd.)

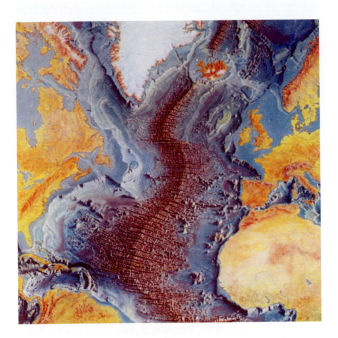

Figure 8-4 Artist's representation of the Mid-Atlantic Ridge in the North Atlantic region. (Painting by Heinrich C. Berann; Bruce C. Heezen and Marie Tharp, *World Ocean Floor*, 1977, © Marie Tharp.)

Alfred Wegener was a twentieth-century pioneer

In 1915 Alfred Wegener, a German meteorologist, presented evidence that virtually all of the large continental areas of the modern world were united late in the Paleozoic Era as a single supercontinent, which he labeled **Pangaea**. Wegener reasoned that Pangaea had broken apart, and that the fragments had drifted away from one another. He cited the great rift valleys of Africa as possible newly forming or failed rifts. We-

gener's insight was correct. As we shall see, the African rift valleys are now regarded as examples of continental rifting at an early stage.

Wegener supported his theory with several additional arguments. He noted numerous geologic similarities between eastern South America and western Africa, for example, and he also called attention to many similarities between the fossil biotas of these two widely separated continents. Several extinct groups of animals and plants, in addition to members of the *Glossopteris* flora, had been found in the fossil records of two or more Gondwanaland continents, and these discoveries led Wegener to argue that these continents must once have lain close together (Figure 8-5). Of the 27 species of fossil land plants recognized within the *Glossopteris* flora of Antarctica, 20 have now been found as far away as India. It might be suggested that winds could have spread the plants this far by carrying their seeds, but in fact the seeds of the genus *Glossopteris* are several millimeters in diameter—much too large to have been blown across wide oceans.

Since the development of plate tectonic theory, professional geologists have accepted the former existence of Pangaea—although, as we shall see, they have found major errors in Wegener's proposed chronology for the fragmentation of this supercontinent (see Figure 8-5).

Alexander Du Toit focused on the Gondwana sequence

Wegener's arguments were more fully developed by the South African geologist Alexander Du Toit. Du Toit and others introduced a wealth of circumstantial evidence in support of the idea of continental drift—evidence that was publicized both before Wegener's death in 1930 and during the three decades of controversy that

Late Carboniferous
(300 million years ago)

Eocene
(50 million years ago)

Early Pleistocene
(1.5 million years ago)

Figure 8-5 Alfred Wegener's reconstruction of the map of the world for three past times. Africa is placed in its present position as a point of reference. Light shading represents shallow seas. Wegener erred in suggesting that Pangaea, the supercontinent shown in the left-hand map, did not break apart until the Cenozoic Era (center and right-hand maps). (After A. Wegener, *Die Entstehung der Kontinents und Ozeane*, Friedrich Vieweg und Sohn, Brunswick, Germany, 1915.)

followed. Du Toit noted, for example, that fossils of the small reptile *Mesosaurus* (Figure 8-6) occurred at or near the boundary between the Carboniferous and Permian systems in both Brazil and South Africa. On both continents, fossils of *Mesosaurus* occur in dark shales along with fossil insects and crustaceans. *Mesosaurus* occupied freshwater and perhaps brackish habitats, and most paleontologists found it difficult to imagine that the animal had somehow made its way across an ocean as broad as the present Atlantic and had then found freshwater depositional settings that were nearly identical to its former habitat.

Even *living* animal and plant groups were shown to exhibit a "Gondwanaland" pattern: a number of individual species and genera were found to be distributed among the southern continents. One genus of earthworm, for example, was found to live only at the southern tips of South America and Africa, which lay close together in Wegener's Gondwanaland reconstruction. Another genus was encountered only in southern India and southern Australia.

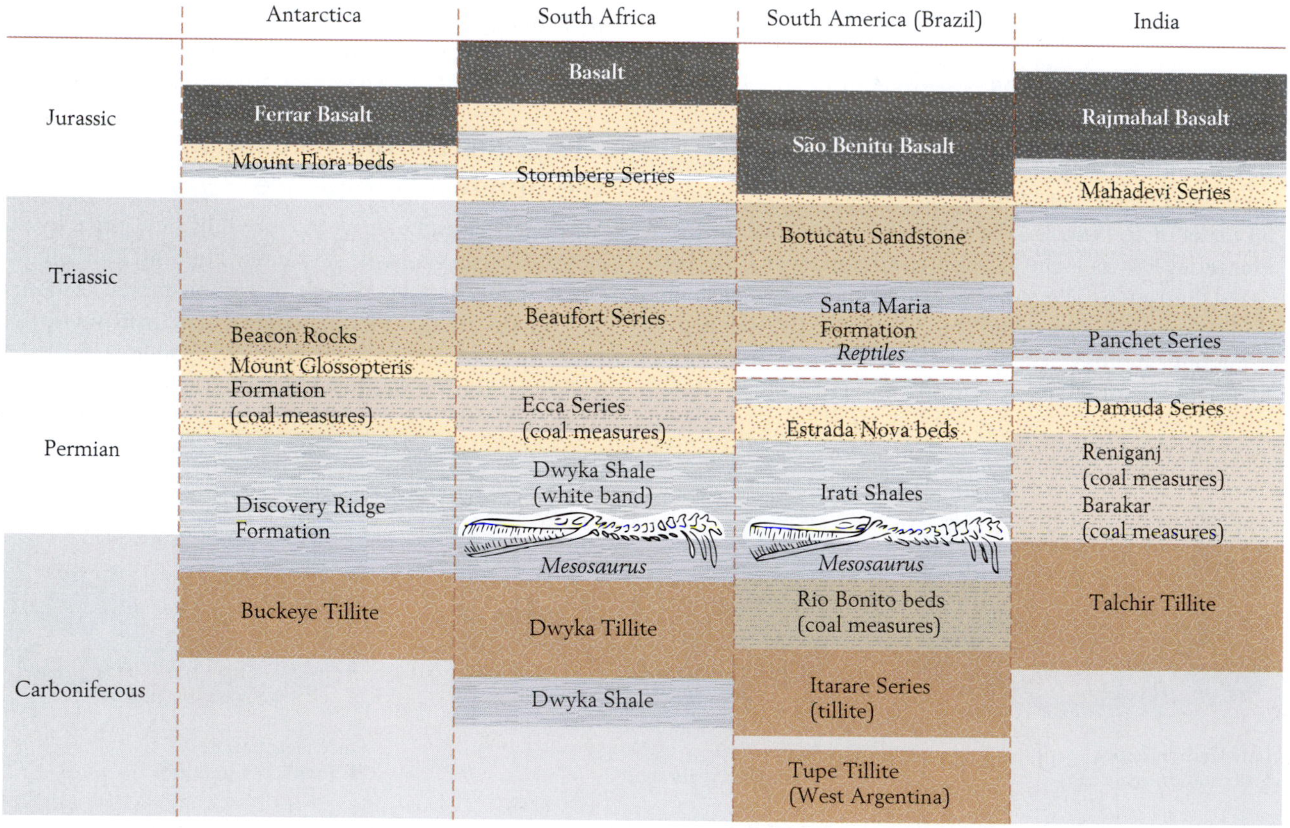

Figure 8-6 *Mesosaurus*, a small Early Permian reptile found in freshwater sedimentary deposits in both South Africa and southern Brazil. The animal was about 0.6 meter (2 feet) long. (After A. Hallam, *Sci. Amer.*, November 1972. © 1972 by Scientific American, Inc. All rights reserved.)

	Antarctica	South Africa	South America (Brazil)	India
Jurassic	Ferrar Basalt / Mount Flora beds	Basalt / Stormberg Series	São Benitu Basalt	Rajmahal Basalt / Mahadevi Series
Triassic	Beacon Rocks / Mount Glossopteris Formation (coal measures)	Beaufort Series	Botucatu Sandstone / Santa Maria Formation *Reptiles*	Panchet Series
Permian	Discovery Ridge Formation / *Mesosaurus*	Ecca Series (coal measures) / Dwyka Shale (white band) / *Mesosaurus*	Estrada Nova beds / Irati Shales	Damuda Series / Reniganj (coal measures) / Barakar (coal measures)
Carboniferous	Buckeye Tillite	Dwyka Tillite / Dwyka Shale	Rio Bonito beds (coal measures) / Itarare Series (tillite) / Tupe Tillite (West Argentina)	Talchir Tillite

Figure 8-7 Correlation of the stratigraphic sequences of four continents. In each sequence, glacial tillites are followed by shales and coal beds containing the *Glossopteris* flora. (After G. A. Duamani and W. E. Long, *Sci. Amer.*, September 1962. © 1962 by Scientific American, Inc. All rights reserved.)

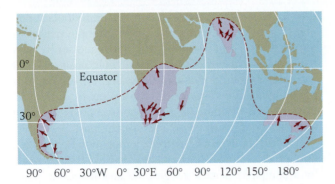

Figure 8-8 Locations of late Paleozoic glaciation and the directions in which glaciers flowed. (After A. Holmes, *Principles of Physical Geology*, Ronald Press Company, New York, 1965.)

The general stratigraphic context of the *Glossopteris* flora and *Mesosaurus* offered further support for the existence of Gondwanaland. Specifically, Carboniferous and Permian rock units that yield the *Glossopteris* flora form what is known as the Gondwana sequence, which occurs with remarkable similarity in South America, South Africa, India, and Antarctica.

The Gondwana sequence of Brazil bears an uncanny resemblance to a sequence representing the same geologic interval in South Africa. At the bases of both sequences (Figure 8-7) are glacial tillites that are coarsest at the base and alternate with interglacial sediments, including coals that yield members of the *Glossopteris* flora. As in South Africa, *Mesosaurus* is found near the

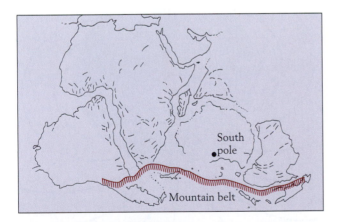

Figure 8-9 Alexander Du Toit's reconstruction of Gondwanaland. The short lines show that regional patterns of faulting and other kinds of rock deformation align well when the continents are assembled to form Gondwanaland. The Andes mountain chain of South America aligns with mountain systems of South Africa, Antarctica, and Australia. (After A. L. Du Toit, *The Geology of South Africa*, Oliver & Boyd Ltd., Edinburgh, 1937.)

base of the Permian in dark shales. Much of the Triassic record consists of dune deposits, which, like similar dune deposits of South Africa, are succeeded by Jurassic lava flows. Similar Gondwana sequences occur in Antarctica and India.

When Du Toit and other followers of Wegener measured the orientations of features scoured into underlying bedrock by glaciers, they found telltale patterns (Figure 8-8). The glacial movement in eastern South America, for example, had been primarily from the southeast, where today no landmass exists that might support large glaciers. Likewise, in southern Australia there was evidence of glacial flow from the south, where there is now only ocean. Obviously, it would not be at all difficult to account for such movement if the continents had been united at the time the glaciers were flowing. Ice flow would then have radiated from the center of a large continent that could have supported large glaciers under cold climatic conditions.

Du Toit correctly deduced from geologic evidence that Pangaea did not form until late in the Paleozoic Era. Before Pangaea was formed, Gondwanaland existed as a distinct supercontinent, and the northern continents were united as a second supercontinent, called **Laurasia**.

Du Toit also recognized that if South America, Antarctica, and Australia were assembled to form Gondwanaland, the mountain belts along their margins would line up, as would regional trends of rock deformation (Figure 8-9).

Continental drift was widely rejected

Despite mounting evidence supporting Wegener's and Du Toit's ideas, geologists of the Northern Hemisphere continued to view the theory of continental drift with considerable skepticism. The primary source of their dissatisfaction lay in the apparent absence of a mechanism by which continents could move over long distances. Geophysicists knew that both continental crust and oceanic crust were continuous above the Mohorovičić discontinuity (or Moho), so they could not imagine how continents could be made to move laterally—to plow through oceanic crust (see Figure 1-13). In addition, some fossil evidence seemed to contradict the notion of continental drift. Wegener had suggested a brief timetable for drift: he proposed that Pangaea, which incorporated virtually all the modern continents, had survived into the Cenozoic Era (see Figure 8-5). But when paleontologists looked for evidence that the world's biotas had evolved into increasingly distinctive geographic groupings since the start of the Cenozoic Era, they found none.

Burdened by so many apparent problems, the idea of continental drift remained highly unpopular in the

Figure 8-10 *Lystrosaurus*, a mammal-like reptile now known from Antarctica as well as from Africa and Southeast Asia. *Lystrosaurus* was a herbivorous animal about a meter (3 feet) long, with short legs, beaklike jaws, and a pair of short tusks. (Painting by Mark Hallett. © 1999, Mark Hallett Illustrations, Salem, OR.)

United States and Europe for decades. The idea was more popular among geologists who lived on fragments of Gondwanaland and recognized the strong geologic similarities among them.

We now know that Wegener made a dating error that misled paleontologists of his time. The rifting of Pangaea had actually begun near the start of the Mesozoic Era—much earlier than Wegener believed. Continents could not have moved far enough during the relatively brief Cenozoic Era to have allowed biotas to diverge greatly; instead, continents have been widely dispersed since the very beginning of this era.

Ironically, after new data supporting plate tectonics had finally brought continental drift back into favor, an exciting fossil find was made in Antarctica. This was the discovery in 1969 of the genus *Lystrosaurus*, an animal classified as a member of the therapsids, the group that gave rise to mammals (p. 76). *Lystrosaurus* was a heavyset herbivorous animal with beaklike jaws (Figure 8-10) that lived in marshy environments. Antarctica is so distant from the other fragments of Gondwanaland that discovery of *Lystrosaurus* fossils there a decade earlier might have revived enthusiasm for continental drift before the advent of plate tectonic theory.

Paleomagnetism showed puzzling patterns

Interest in continental movements was renewed in the late 1950s as a result of new evidence derived from **paleomagnetism**, or the magnetization of ancient rocks at the time of their formation. We have already seen

that Earth's magnetic field has reversed its polarity on many occasions. During the 1950s, geophysicists attempted to ascertain whether the north and south magnetic poles had not only reversed their positions, but had also wandered about periodically. To explore this possibility, these researchers attempted to determine the previous positions of the magnetic poles by using magnetized rocks as compasses for the past.

As we learned in Chapter 6, a magnetic field frozen into a rock is similar to the magnetic field that is "read" by a compass. Earth's geographic and magnetic poles are not in exactly the same place. The angle that a compass needle, which points to the magnetic north pole, makes with the line running to the geographic north pole is called the **declination**. Today the magnetic pole lies about 11° from the geographic pole, so the declination is small (Figure 8-11). A compass needle not only points to the north magnetic pole but, if allowed to tilt in a vertical plane, also dips at a particular angle. The **inclination** of the needle varies with the distance of the compass from the magnetic pole. The inclination is lowest near the equator, where the lines of force of Earth's magnetic field intersect Earth's surface at the lowest angle, as Figure 8-11 shows.

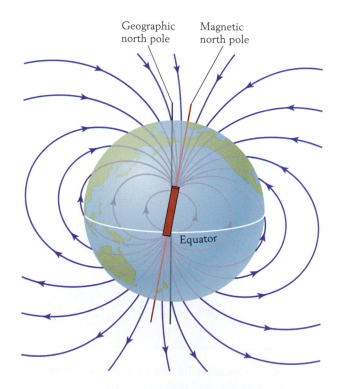

Figure 8-11 The structure of Earth's magnetic field. Earth's core has north and south poles and thus behaves like a bar magnet. At the present time, the north–south axis has a declination of 11° from Earth's north–south geographic axis. Curved lines represent magnetic lines of force. These lines of force intersect Earth's surface at high angles near the poles and low angles near the equator.

Paleomagnetism in a rock also has a declination, which indicates the apparent direction of the north magnetic pole at the time when the rock was first magnetized. In addition, paleomagnetism has an inclination, which indicates the distance between the rock, when it formed, and the geographic pole. It is important to understand, however, that neither a compass needle nor the magnetism of a rock reveals anything about longitude (position in an east–west direction).

When geologists first began to measure rock magnetism, they found that the magnetism of recently magnetized rocks was consistent with Earth's current magnetic field. The magnetism in older rocks, however, had different orientations. As data accumulated, it began to appear that Earth's magnetic north pole had wandered. A plot of the pole's apparent positions, indicated by rocks of various ages in North America and in Europe, showed that the pole seemed to have moved to its present position from much farther south, in the Pacific Ocean. However, the path obtained from European rocks differed in detail from that obtained from North American rocks (Figure 8-12A). It was recognized that this pattern might actually reflect a history in which the north pole had not wandered at all; instead, as Wegener had suggested, the continents of Europe and North America might have moved in relation to the pole and to each other, carrying with them rocks that had been magnetized when the continents were in different positions. This possibility led to the use of the cautious term *apparent polar wander* to describe the pathways that geophysicists plotted.

Tests were conducted to examine the possibility that continents, rather than magnetic poles, had moved. It was hypothesized, for example, that if North America and Europe had once been united and had drifted over Earth's surface together, they should have developed identical paths of apparent polar wander during their joint voyage. The test, then, was to fit the outlines of North America and Europe together along the Mid-Atlantic Ridge to determine whether, with the continents in this position, their paths of apparent polar wander coincided. As Figure 8-12B shows, North America and Europe, when united and then moved to their present positions, had apparent polar-wander paths that coincided almost exactly for both Paleozoic and early Mesozoic time. This evidence strongly suggested that the continents had indeed drifted apart after that time, carrying their magnetized rocks with them.

The Rise of Plate Tectonics

During the late 1950s, these new paleomagnetic data generated widespread discussion of continental drift, but most geologists continued to doubt its validity. There were two reasons for this continuing skepticism: first, many paleomagnetic measurements were imprecise; and second, the belief persisted that no natural mechanism could move continents through oceanic crust. Then, in 1962, the American geologist Harry H. Hess published a landmark paper proposing a novel solution to this problem.

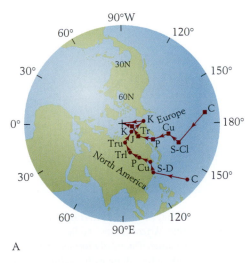

A

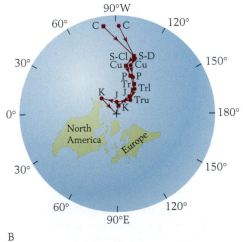

B

Figure 8-12 **Apparent polar-wander paths for North America (circles) and Europe (squares).** *A.* Plot of polar-wander paths based on the assumption that the continents have remained in their present positions. *B.* Plots for North America and Europe juxtaposed, as postulated for Paleozoic time by Wegener and his followers. Here the Paleozoic and Mesozoic apparent polar-wander paths of the two continents

nearly coincide, suggesting that the continents were united during the Paleozoic Era. The time-rock units represented are Cretaceous (K), Triassic (Tr), Upper Triassic (Tru), Lower Triassic (Trl), Permian (P), Upper Carboniferous (Cu), Siluro-Devonian (S-D), Silurian to Lower Carboniferous (S-Cl), and Cambrian (C). (After M. W. McElhinny, *Paleomagnetism and Plate Tectonics*, Cambridge University Press, London, 1973.)

Seafloor spreading explained many phenomena

In essence, Hess suggested that the felsic continents had not plowed through the dense mafic crust of the ocean at all but that instead, the *entire crust had moved.* Hess's ideas were highly unconventional (he labeled his contribution "geopoetry"), but the manner in which he compiled his facts exemplifies the way geologists assemble circumstantial evidence to construct theories. In the following summary of Hess's paper, the critical facts and inferences appear in italics.

One piece of evidence that Hess pondered was the apparent youth of the ocean basins. At the time, it was estimated that sediment was being deposited in the deep sea at a rate of about 1 centimeter (about four tenths of an inch) per thousand years. At that rate, 4 billion years of Earth history would theoretically produce a layer of deep-sea sediment 40 kilometers (25 miles) thick. In fact, *the average thickness of sediment in the deep sea today is only 1.3 kilometers* (less than 1 mile). Thus, allowing for some compaction, Hess estimated that *the layer of sediment existing in the deep sea represents only about 260 million years of accumulation*—a figure that might therefore approximate the average age of the seafloor. (It turns out that Hess's calculation was of the right order of magnitude.)

Like many earlier investigators, Hess noted the central location of the Mid-Atlantic Ridge. He also noted that other mid-ocean ridges tend to be centrally located within ocean basins. (A "best-fit" repositioning of continents along the Mid-Atlantic Ridge, calculated after Hess's paper was written, is shown in Figure 8-1.) Four other curious facts about these ridges seemed significant to Hess:

1. The ridges are characterized by a high rate of upward heat flow from the mantle to neighboring segments of seafloor.

2. Seismic waves from earthquakes move through the ridges at unusually low velocities.

3. A deep furrow runs along the crest of each ridge.

4. Volcanoes frequently rise up from mid-ocean ridges.

Hess developed a hypothesis that seemed consistent with all of these observations. Essentially, he suggested that mid-ocean ridges represent narrow zones where oceanic crust forms as material from the mantle moves upward and undergoes chemical changes. Hess further maintained that as this material rises, it carries heat from the mantle to the surface of the seafloor. *The expanded condition of the warm, newly forming crust thus accounts for the swollen condition of the crust at these locations—that is, for the presence of a ridge.*

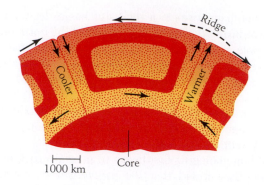

Figure 8-13 Convective motion within the mantle as envisioned by Harry H. Hess. Mid-ocean ridges form where the upward-flowing limbs of two adjacent convection cells approach the surface. (After H. H. Hess, in *Petrologic Studies: A Volume in Honor of A. F. Buddington*, Geological Society of America, 1962.)

Hess then revived a geophysical concept that other researchers had discussed earlier—that the slushlike material that makes up Earth's mantle rotates by means of large-scale thermal convection (see Figure 1-16). Hess proposed that Earth's semi-molten mantle is divided into **convective cells** (Figure 8-13), whose low-density material forms crust as it rises and cools. This crust then bends laterally to become one flank of a ridge (Figure 8-14). *The furrow down the center of many*

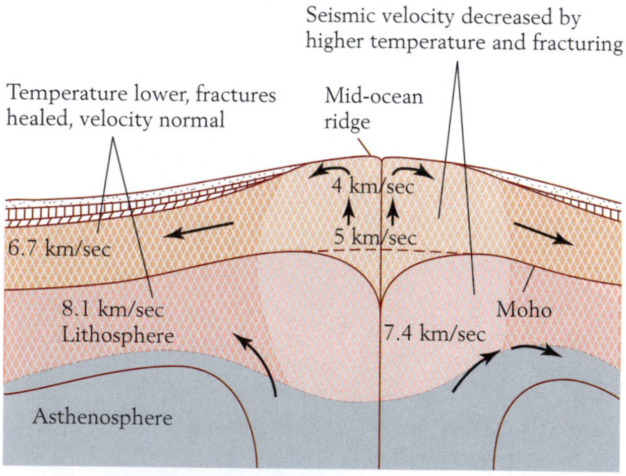

Figure 8-14 Hess's model of the structure of a mid-ocean ridge. Arrows show the flow of new crust derived from the rising mantle. The newly formed crust carries heat from the mantle. This factor, along with fracturing of the rock as it bends laterally, results in low velocities (shown in kilometers per second) for seismic waves passing through the ridge. The elevation of the ridge results from the hot, swollen condition of the newly formed crust. (After H. H. Hess, in *Petrologic Studies: A Volume in Honor of A. F. Buddington*, Geological Society of America, 1962.)

ridges could then be explained as the site at which newly formed crust separates and flows laterally in two directions. Similarly, *volcanoes along mid-ocean ridges could represent the rapid escape of mantle material here and there.* Furthermore, *the low velocity of seismic waves passing through a ridge would result from the fact that the rocks of the ridge are extremely hot and are extensively fractured where they bend laterally to form the basaltic seafloor.*

During World War II, Hess had commanded an American naval vessel in the Pacific, and he had seized upon this opportunity to pursue his interest in geology. In order to study the configuration of the ocean floor, Hess kept his ship's echo-sounding equipment operating for long stretches of time. While profiling the bottom in this way, he discovered curious flat-topped seamounts rising from the floor of the deep sea, which he named **guyots** after the nineteenth-century geographer Arnold Guyot. On the basis of their size and shape, Hess concluded that *guyots were volcanic islands that had been eroded by the action of waves near sea level.* Two decades later, shallow-water fossils of Cretaceous age were recovered from the tops of some guyots, proving that the guyots had indeed once stood near sea level. How the ocean floor on which they sat had subsided to such great depths, however, remained a mystery.

How do the guyots that Hess discovered fit into plate tectonic theory? According to his scheme, the seafloor adjacent to a mid-ocean ridge (together with anything attached to that seafloor) moves laterally, away from the spreading center. The volcanoes that frequently form along mid-ocean ridges sometimes grow upward to sea level, as is the case with Ascension Island in the Atlantic. As a volcano moves laterally away from the ridge along with the crust on which it stands, it moves away from the source of its lava. It then becomes an inactive seamount, and its tip is quickly planed off by erosion and wave action. Recall that the seafloor gradually deepens away from a mid-ocean ridge because newly formed crust cools and therefore shrinks as it moves laterally away from the ridge (see Figure 1-17). Thus *a truncated seamount is gradually transported out into deep water as if it were on a descending conveyor belt, and it then becomes a guyot* (Figure 8-15). Assuming that the Atlantic Ocean had developed by seafloor spreading since the end of the Paleozoic, Hess calculated a spreading rate of about 1 centimeter per year.

Hess found support for the idea of a youthful seafloor in his observation that there are only about 10,000 volcanic seamounts (volcanic cones and guyots) in all the world's oceans. Hess knew that when a volcano has been eroded to the lowest level at which waves can act on it, it undergoes little further erosion. Thus

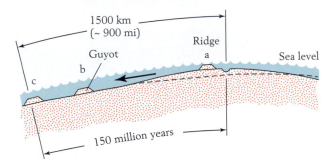

Figure 8-15 Hess's interpretation of guyot formation. First, a volcano builds a cone along a mid-ocean ridge. The cone initially stands partly above sea level, but its tip is soon planed off by wave erosion (a). The resulting flat-topped structure moves laterally with the spreading crust and is carried gradually downward (b and c) as the newly formed crust beneath it cools and therefore shrinks as it moves away from the ridge. (After H. H. Hess, in *Petrologic Studies: A Volume in Honor of A. F. Buddington,* Geological Society of America, 1962.)

he assumed that if the oceans were permanent features, their oldest volcanic seamounts should still be extant. Given the fact that there are only 10,000 volcanic seamounts in modern oceans, Hess further reasoned that if the oceans were nearly as old as Earth—say, 4 billion years old—then an average of only one volcano would have formed every 400,000 years or so. The existence of so many obviously young volcanoes indicated to Hess that new volcanoes appear much more frequently—perhaps at a rate of one every 10,000 years. Thus *the relatively small number of volcanic seamounts in modern oceans suggested to Hess that current ocean basins are much younger than Earth.*

Continents can be viewed as enormous bodies that float in the Earth's mantle by virtue of their low density; thus they would be expected to ride passively along like guyots. Here, then, was Hess's explanation for the fragmentation of continents: he reasoned that when convective cells in the mantle change their locations, the upwelling limbs of two adjacent cells must sometimes come to be positioned beneath a continent. Convective spreading should then rift the continent into two fragments and move them apart from the newly formed spreading center. New ocean floor should subsequently form at the same rate on each side of the spreading center. Hess further maintained that the spreading center would continue to operate along the midline of the new ocean basin—and thus persist as a mid-ocean ridge—as long as the convective cells remained in their new location.

If oceanic crust forms and flows laterally without an enormous change in thickness, however, it must disappear somewhere. Hess postulated that it must be

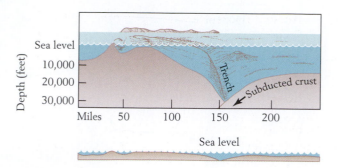

Figure 8-16 Section through the Tonga trench in the Pacific Ocean. The view is northward. In the upper diagram, vertical distances are exaggerated by a factor of 10; the lower diagram is drawn without vertical exaggeration. The island toward the left is Kao, a dormant volcano. The oceanic crust to the right (or east) of the trench is moving down into the mantle beneath the oceanic crust to the left. (After R. L. Fisher and R. Revelle, *Sci. Amer.*, November 1955. © 1955 by Scientific American, Inc. All rights reserved.)

swallowed up again by the mantle along the great **deep-sea trenches** that exist at certain places on the ocean floor (Figure 8-16). Movement of crust into the mantle along one side of a trench provided a ready explanation for the observation that *Earth's gravitational field is unusually weak along deep-sea trenches*; the presence of low-density crustal rock and deep-sea sediment in the trenches in place of dense mantle rock would be expected to weaken the gravitational force exerted by Earth on objects at or above its surface. Hess estimated that *the formation of new crust along mid-ocean ridges and the simultaneous disappearance of crust into the deep-sea trenches would produce an entirely new body of crust for the world's oceans every 300 million or 400 million years.*

Hess's hypothesis of seafloor spreading had two great strengths. First, by asserting that continents move along *with* oceanic crust, it overcame the objection that continents could not move *through* that crust. Second, the hypothesis was consistent with a variety of facts, the most important of which have been italicized above. Most of these facts had not previously seemed to make sense.

Paleomagnetism provided a definitive test

Despite the strong circumstantial evidence in support of Hess's hypothesis, its publication in 1962 created no great stir within the geologic profession. What was needed was a really convincing test of the basic idea. Such a test was soon found. It was based on the well-known fact that Earth's magnetic field has periodically reversed its polarity (p. 136). In 1963 the British geophysicists Fred Vine and Drummond Matthews re-

ported that newly formed rocks lying along the axis of the central ridge of the Indian Ocean were magnetized while Earth's magnetic field was polarized as it is now. This finding came as no surprise, because it was known that other mid-ocean ridges also exhibited "normal" magnetization. It turned out, however, that seamounts on the flanks of the Indian Ocean ridge were magnetized in the reverse way. Vine and Matthews concluded that this pattern might confirm Hess's seafloor-spreading model. They reasoned that if crust is now forming along the axis of any mid-ocean ridge, it must become magnetized with the present polarity of Earth's magnetic field as it crystallizes from the molten mantle. In older crust lying at some distance from the ridge, however, reversed polarity should be encountered, and in even older crust farther from the ridge, the polarity should be normal again (Figure 8-17).

Magnetic "striping" had, in fact, recently been observed on many parts of the seafloor, but the striping patterns were soon put to a more rigorous test. During the 1960s a time scale was developed for late Cenozoic magnetic reversals (see Figure 6-4). This scale was based on measurements of the magnetic polarity of terrestrial rocks of known age. It was assumed that the

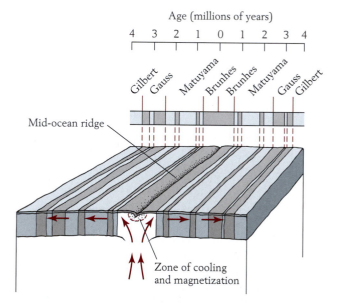

Figure 8-17 Magnetic anomaly patterns of the seafloor fit the prediction that they represent magnetic reversals. The time scale shows known magnetic reversals of the past 4 million years. The labels (Gilbert through Brunhes) represent intervals that are characterized by either normal or reversed polarity, dated by the polarity of terrestrial rocks whose ages are known. The relative lengths of these intervals are remarkably similar to those of the magnetic-anomaly stripes on either side of a mid-ocean ridge. (After A. Cox et al., *Sci. Amer.*, February 1967. © 1967 by Scientific American, Inc. All rights reserved.)

spreading rate for each mid-ocean ridge had remained reasonably constant over the past 4 million or 5 million years. The relative widths of seafloor stripes then turned out to be proportional to the polarity time-rock intervals that these stripes were thought to represent—that is, long intervals of normal polarity were represented by broad stripes and short intervals by narrow stripes. Furthermore, the striping patterns on the two sides of a ridge were mirror images of each other. Thus the detailed patterns of striping were found to match the known timing of magnetic reversals (see Figure 8-17).

An interesting story concerns the misfortunes of a Canadian geologist named L. W. Morley. Morley developed the same model for magnetic anomalies that Vine and Matthews published, but the manuscript in which he outlined his model was rejected by the two journals to which he submitted it in 1963. One reviewer of the manuscript cynically commented that "such speculation makes interesting talk at cocktail parties." Because Vine and Matthews were fortunate enough to have had their paper accepted for publication, they were the ones who ultimately received recognition in the scientific world. Radically new ideas are not easily established.

Faulting and Volcanism along Plate Boundaries

Since the acceptance of plate tectonics, geologists have learned how lithospheric plates move in relation to one another along their mutual boundaries—and how they fracture and experience igneous activity there as well. To comprehend these dynamic features, we must first understand the kinds of faults along which rocks move.

Recall that faults are surfaces along which bodies of rock break and move past each other (p. 14). Faults are classified into three basic types, according to their orientations and the directions in which rocks move along them (Figure 8-18). Each of the three kinds of plate boundaries is characterized by one of the three basic types of faults.

Normal faults take their name from the fact that usually nothing more than gravity accounts for the direction of movement along them. They result from tension: in effect, two blocks of rock are pulled apart along the fault. The plane of a typical normal fault lies at an angle of more than 45° to Earth's surface, and the block of rock above the fault slides downward in relation to the one below (Figure 8-18A). Geologists discuss motion along faults as "relative movement" because any type of fault can result from movement of the rocks on both sides of the fault plane or of those on only one side. A normal fault, for example, occasionally entails uplift of the block of rock below the fault without downward movement of the upper block.

Thrust faults display movement that is opposite in direction to movement along normal faults: the relative movement of the upper block is uphill along the fault surface (Figure 8-18B). Thrust faults occur in areas where opposing horizontal forces compress the lithosphere enough to fracture it and to cause the rocks on the two sides of the fracture to slide past each other.

Strike-slip faults are nearly vertical, and movement along them is nearly horizontal. This movement results from shearing stress that causes the rocks on opposite sides of the fault to move in opposite directions (Figure 8-18C). The most famous strike-slip fault in North America is the San Andreas fault of California. Earthquakes caused by movement along the San Andreas have caused considerable damage in California. Los Angeles and San Francisco lie on opposite sides of this fault. At the present rate of movement along it, the two cities will lie alongside each other in about 30 million years.

Oceanic crust forms along mid-ocean ridges

At places in Iceland, which sits atop the Mid-Atlantic Ridge, the furrow down the center of the ridge is visible as a structural graben. A **graben** is a valley bounded by normal faults along which a central block of rock has

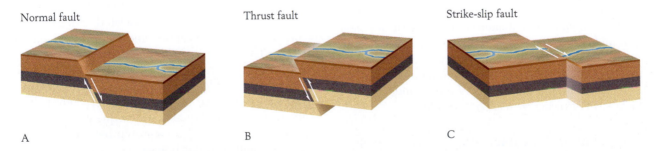

Normal fault   Thrust fault   Strike-slip fault

A   B   C

Figure 8-18 Basic types of faults. Geologists use arrows with barbs on one side to indicate relative movements of the blocks on the two sides of a fault.

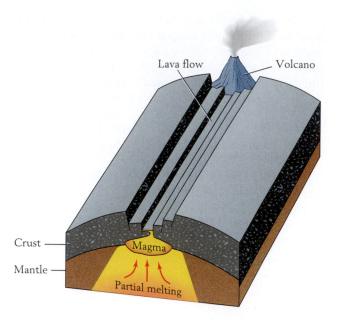

Figure 8-19 A graben forms along a mid-ocean ridge. As tension breaks the crust and spreads it laterally, a block of crust sinks between normal faults, forming a graben. Lavas that emerge along the faults fill in the space produced by rifting and also flow laterally and solidify on the floor of the graben. In places, the lavas erupt through small vents to form volcanoes.

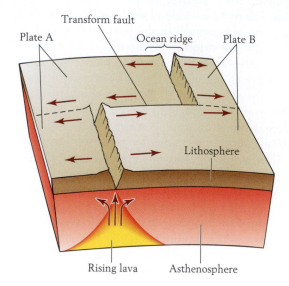

Figure 8-20 A transform fault. The central part of this fault is a plate boundary along which two plates slide past each other. Here plate A and plate B are separating at a rift, and the transform fault offsets this ridge. Arrows indicate the opposite directions of plate motion between the two segments of the ridge.

slipped downward (Figure 8-19). Grabens form where the crust is extending—where it is forming and moving laterally along a mid-ocean ridge. As the crust periodically breaks apart along the ridge, lava moves upward to fill the space thus vacated, producing new oceanic crust. Extruded lavas have also been recorded along submarine mid-ocean ridges. In some places the lavas spread out to form broad flows, but occasionally they build volcanoes (see Figure 8-19). When the lavas cool underwater, they form **pillow basalt** (see Figure 2-14).

It is now recognized that the boundary between the crust and the mantle—the Moho (see Figure 1-13)—is not the surface along which Earth's "skin" moves. That surface, which lies well below the Moho, is the boundary between the plastic asthenosphere (the semi-molten part of the mantle) and the more rigid lithosphere (the uppermost mantle and the crust). The asthenosphere-lithosphere boundary is situated closer to the surface beneath mid-ocean ridges, where upward heat flow keeps mantle material molten even at very shallow depths. Figure 1-17 shows the current configuration of the crust and upper mantle in the vicinity of the Atlantic Ocean, which is still growing by seafloor spreading.

Transform faults offset mid-ocean ridges

Mid-ocean ridges are frequently offset along **transform faults**, which are enormous strike-slip faults (Figure

8-20; see also Figure 8-4). Transform faults form because pressures are uneven along ridges, and some segments of newly formed crust break away from others that move less rapidly away from the ridge axis. Earthquakes emanate from transform faults episodically, when movement takes place along them. The San Andreas fault is a transform fault that happens to cut across the edge of the North American continent.

Lithosphere is subducted along deep-sea trenches

Deep-sea trenches are the sites where slabs of lithosphere descend into the asthenosphere—a process called **subduction**. Most of the trenches of the modern world encircle the Pacific Ocean (Figure 8-21). In the decade before Hess developed his ideas, geologists had noted that trenches are associated with two other geologic features: volcanoes and **deep-focus earthquakes**. The latter are earthquakes that originate more than 300 kilometers (190 miles) below Earth's surface. In areas far from deep-sea trenches, deep-focus earthquakes are rare. Near trenches, however, both shallow- and deep-focus earthquakes are frequent.

The typical spatial relationship between trenches, volcanoes, and deep earthquake foci is shown in Figure 8-22. The earthquake foci fall along a narrow, nearly planar zone that lies along or within a descending slab.

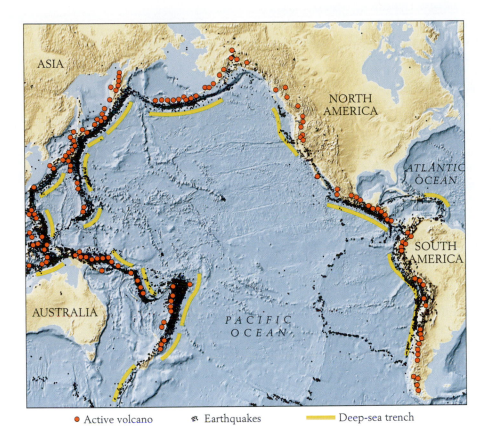

Figure 8-21 Distribution of deep-sea trenches, deep-focus earthquakes, and volcanoes near the margins of the Pacific Ocean. Note that when trenches are viewed from above, many can be seen to curve, and that deep-focus earthquake centers are concentrated along trenches. The volcanoes form the Pacific "ring of fire."

● Active volcano   ❋ Earthquakes   ▬ Deep-sea trench

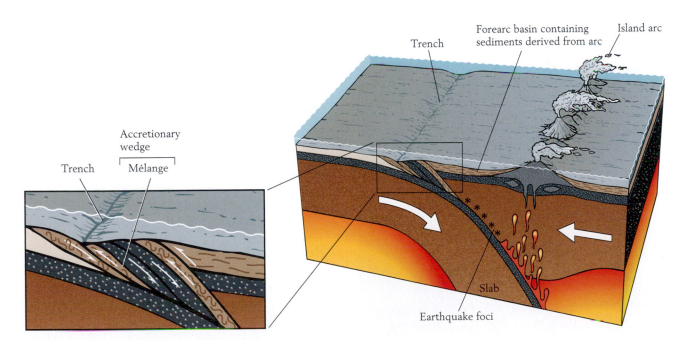

Figure 8-22 Major features of a subduction zone. A subducted plate partially melts after it reaches a critical depth. Magma of low density rises to form intrusions in the crust and volcanoes on the surface. Deep-focus earthquakes are caused by episodic downward movement of the slab. Compression along the trench forms a mélange and piles up slices of seafloor along thrust faults. (After S. Uyeda, *The New View of the Earth*, W. H. Freeman and Company, New York, 1978.)

The slab descends because it is cooler and therefore denser than the semi-molten asthenosphere, and it produces earthquakes by occasionally taking a sudden step downward.

Chains of volcanic islands often parallel deep-sea trenches (see Figure 8-21). They are positioned in this way because the descending slab undergoes partial melting (see Figure 8-22), and any of the molten material that is less dense than the asthenosphere rises toward the surface as magma. Some of this magma solidifies within the crust to form intrusive igneous bodies. The rest reaches the surface to emerge in volcanic eruptions.

A band of subducted lithosphere demarcates a **subduction zone**. Subduction zones border much of the Pacific Ocean (see Figure 8-21). The volcanoes associated with these zones form what is known as the "ring of fire" around the Pacific. When deep-sea trenches and the chains of volcanoes associated with them are viewed from above, many can be seen to have the shape of an arc. The volcanoes that rise above sea level form curved arrays of volcanic islands termed **island arcs**.

Plate subduction typically creates a zone of intensely deformed rocks in the belt between the zone called the **forearc basin** and the deep-sea trench (see Figure 8-22). Most of the rocks in this deformed belt are deep-ocean sediments such as dark muds and graywackes, with bits of oceanic crust mixed in. Some have been scraped from the descending plate. Rocks of subduction zones are characteristically metamorphosed at low temperatures (because the depth at which they

are deformed is not great) and at high pressures (because the plates converge with great force). This chaotic, deformed mixture of rocks is called a **mélange** (the French word for mixture). The mélange shown in Figure 8-23 formed near San Francisco when a subduction zone passed beneath the western margin of North America.

We have seen that spreading zones display normal faults and that the transform faults that offset spreading zones are strike-slip faults. In contrast, subduction zones exhibit thrust faults (see Figure 8-18B); under enormous compressive forces, huge slices of mélange and adjacent seafloor break from a descending lithospheric plate and pile up along thrust faults (see Figure 8-22). The entire body of rocks formed in this way along a subduction zone constitutes what is descriptively termed an **accretionary wedge**. Between the accretionary wedge and the igneous island arc lies the forearc basin, where turbidites and other sediments accumulate in moderately deep water.

Plate Movements

Several forces contribute to plate movements, and they move plates at different rates. It is not clear, however, why some plates move faster than others, because the relative importance of the driving forces is difficult to assess. Nonetheless, there are remarkably accurate techniques for measuring the rates at which plates are moving over Earth today.

Plates move for four reasons

Four driving forces cause plates of lithosphere to move away from spreading zones toward subduction zones (Figure 8-24):

1. Convective motion in the asthenosphere applies drag to the base of a plate.

2. The ascent of magma at a spreading zone pushes the lithosphere upward, and the weight of the elevated ridge then causes the lithosphere to spread laterally in both directions, pushing the plate on each side ahead of it.

3. At the other end of a plate, the cold, relatively dense slab sinks into the hot asthenosphere, dragging the rest of the plate toward the subduction zone.

4. If a segment of a slab breaks loose and sinks into the asthenosphere, it ceases to drag its parent plate downward, but produces other forces that suck the parent plate downward.

Whatever the relative importance of these various forces may be, their combined strength moves not only the thin oceanic crust but also the immense continents that float on Earth's mantle.

Figure 8-23 A mélange of the Franciscan sequence in California. Large blocks of exotic material are visible in the dark, metamorphosed deep-sea sediment. This mélange formed during the Mesozoic Era, when deep-sea sediments were pushed against the margin of the continent along a subduction zone. (John Wakabayashi, Hayward, CA.)

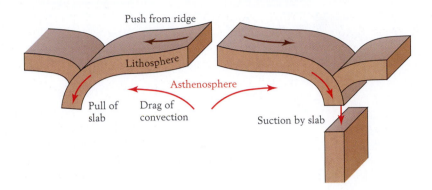

Figure 8-24 Forces that drive plate movements.

Free slabs sink deep into the mantle

A slab eventually breaks away from the plate that is being subducted. The free slab absorbs heat very slowly from the hot asthenosphere around it. Remaining cold and dense, it sinks deep into the asthenosphere. In fact, the behavior of earthquake waves indicates that some slabs sink all the way to the boundary between Earth's core and mantle (see Figure 1-16). Eventually, perhaps only after hundreds of millions of years, a slab lodged deep in the mantle must melt. The mafic magma thus formed, being less dense than the ultramafic asthenosphere, must then rise through convective movements. The particulars of this process remain unclear, but the overall pattern amounts to an enormous version of the rock cycle (p. 10).

Plate movements are measurable

Because all plates are moving, no piece of lithosphere represents a perfectly immobile block against which the movement of all others can be assessed. When two plates are in contact, their relative movements determine whether the boundary between them is a spreading zone, a subduction zone, or a transform fault.

Today Earth's lithosphere is divided into eight large plates and several small ones (see Figure 1-15). The configuration of lithospheric plates and plate boundaries has changed dramatically throughout Earth's history. From time to time, new ridges and subduction zones have formed and old ones have disappeared. Although we cannot reconstruct the history of plate movements in detail for all of geologic time, this history is moderately well known for the Paleozoic Era and very well known for the Mesozoic and Cenozoic eras.

Cores obtained by drilling the floor of the deep sea from ships have shown that all segments of the deep-sea floor are of Mesozoic or Cenozoic age. Radiometric dating gives absolute ages for the oceanic crust beneath the sediments. The resulting patterns reveal rates of seafloor spreading. Australia and Antarctica have been moving apart much more rapidly, for example, than the Americas and Africa (Figure 8-25).

The complex history of continental movement described by plate tectonics differs in an important way from the pattern that Wegener envisioned. Wegener thought that the enormous landmass of Pangaea existed as a stable crustal feature for hundreds of millions of years and that it fragmented in a single event. Plate tectonics, however, entails continuous movement of landmasses in relation to one another.

How can we measure the *absolute* movement of a plate? Absolute movement is defined as any movement in relation to a fixed feature, such as an immobile point at the surface of Earth's mantle. Nearly immobile points appear to have been discovered in the form of hot spots. A **hot spot** is a small geographic area where heating and igneous activity occur within the crust. A search of the entire globe has turned up many hot spots that have been active within the last 10 million years (Figure 8-26). One hot spot is located at Yellowstone National Park in Wyoming, where geysers and volcanoes have been present for millions of years. Some hot spots result from the arrival at Earth's surface of a **thermal plume**, or a column of molten material that rises from the mantle. Plumes appear to arise near the border between Earth's core and mantle and melt their way upward through the convecting mantle. Often a large volcano forms at the surface above a plume that rises through thin oceanic crust. As oceanic lithosphere moves over a plume, its successive positions are commonly recorded as a chain of volcanoes such as the Hawaiian Islands (Figure 8-27).

Radiometric dating tells us that Hawaii, the largest and easternmost of the islands, is less than 1 million years old, while the small northwestern island of Kauai is about 5.6 million years old. A long train of even older submarine seamounts extends northwestward beyond Kauai. This age pattern seems to indicate that the Pacific plate is moving in a west–northwest direction over a stationary hot spot. This direction approximates the one in which the Pacific plate is moving in relation to the plates that border it on the north and northwest, where it is being subducted (see Figure 8-21).

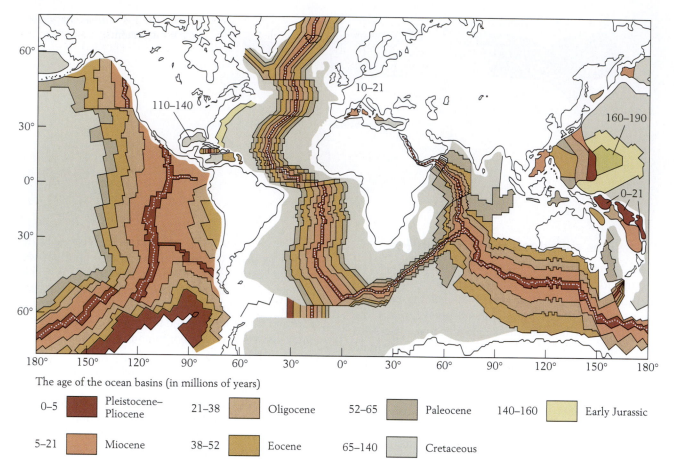

The age of the ocean basins (in millions of years)

0–5	Pleistocene–Pliocene	21–38	Oligocene	52–65	Paleocene	140–160	Early Jurassic
5–21	Miocene	38–52	Eocene	65–140	Cretaceous		

Figure 8-25 **The ages of the world's ocean basins.** No part of the seafloor is older than Mesozoic age. (After W. C. Pitman et al., Geological Society of America Map and Chart MC-6, 1974.)

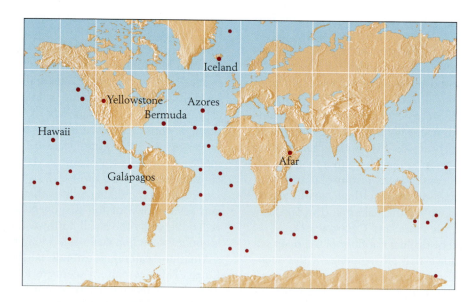

Figure 8-26 **The world's identified hot spots.** Many hot spots are positioned on or close to mid-ocean ridges. Note that hot spots are located at the Hawaiian Islands, Yellowstone National Park, Iceland, and the Afar region of northeastern Africa. (After S. T. Crough, *Annu. Rev. Earth Planet. Sci.* 11:165–193, 1983.)

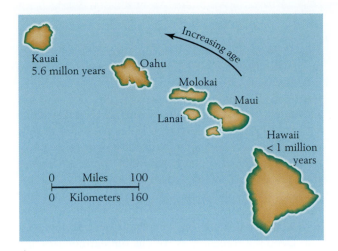

A

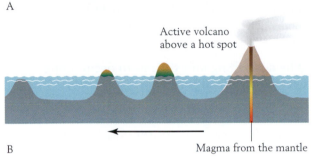

B

◀ **Figure 8-27** Formation of a chain of islands as an oceanic plate moves over a hot spot. *A.* The major Hawaiian islands. The islands increase in age toward the northwest. They have formed, one after the other, as the Pacific plate has moved northwestward over a hot spot in the mantle. *B.* Volcanic islands form, one after the other, as an oceanic plate moves over a hot spot.

Since the mid-1980s, geologists have employed a more precise tool for measuring plate velocities in real time. This tool is the Global Positioning System (GPS), which makes use of artificial Earth-orbiting satellites that serve as known reference points. Transmission of radio waves between a satellite, a small reference point on a plate, and a ground-based receiver makes it possible to measure the absolute location of the earthly reference point with remarkable accuracy. Repetition of these measurements over several years has yielded measurements of motion for all of Earth's plates. The average rate of movement is about 5 centimeters per year, but plates of the Pacific region north of the Antarctic plate are moving faster than this, and the North American plate is rotating relatively slowly toward the Pacific plate (Figure 8-28).

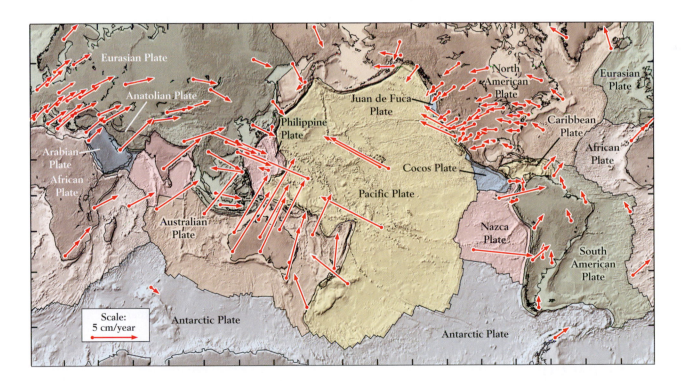

Figure 8-28 Movements of plates on Earth today. These movements were measured by use of the Global Positioning System. (Courtesy of Michael Heflin.)

Chapter Summary

What rocks and fossils in Africa and South America suggest that those two continents were connected to each other as parts of Gondwanaland in late Paleozoic time?

Fossils of many kinds of nonmarine organisms, including members of the *Glossopteris* flora and the lake-dwelling reptile *Lystrosaurus*, are found on both continents. Furthermore, the Gondwana sequence of late Paleozoic strata in South Africa is virtually identical to that in Brazil. In addition, mountain belts and regional patterns of rock deformation in the parts of those continents that border the Atlantic Ocean line up if the continents are repositioned to form parts of Gondwanaland. Glacial scours on ancient bedrock also show that glaciers would have had to move from the ocean onto the land—an impossible occurrence—if Africa and South America had been at their present locations in late Paleozoic time.

How does paleomagnetism demonstrate that continents have moved in the course of time?

Magnetism frozen into ancient rocks is not aligned with Earth's present magnetic field, indicating that continents have rotated relative to Earth's magnetic poles. In addition, the inclination of this magnetism does not align with Earth's present magnetic lines of force, which indicates that the continents have moved in a north–south direction.

How was paleomagnetism used in the 1960s to show that lithosphere forms along a mid-ocean ridge and migrates away from the ridge in both directions?

It was found that reversals in the polarity of Earth's magnetic field had produced striped patterns of normal and reversed magnetism on the two sides of present-day mid-ocean ridges—and that the patterns on the two sides are mirror images of each other. Thus oceanic lithosphere has been forming along mid-ocean ridges and spreading outward in both directions while Earth's magnetic field has undergone reversals.

How did features of the seafloor engender the concept of plate tectonics?

The thinness of the layer of sediment that carpets the deep sea indicates that the present-day oceanic crust cannot be nearly as old as Earth. Given the number of active volcanoes on Earth today, if the ocean basins were 4 billion years old, they should contain many more old, inactive volcanoes than they actually do. Upward movement of magma, faulting, and earthquakes along mid-ocean ridges could finally be explained as results of formation of oceanic crust in these narrow zones of ocean basins. Flat-topped seamounts on the deep-sea floor that were truncated by wave action could be explained as volcanoes that grew above the sea surface at mid-ocean ridges, were truncated by erosion and wave action, and rode along on a moving plate into the deep sea. Deep-sea trenches, as well as the earthquakes and volcanoes associated with them, could be attributed to subduction of oceanic lithosphere into the asthenosphere.

Why do faulting and volcanism occur along oceanic plate margins?

Spreading along mid-ocean ridges produces normal faults that bound grabens, and magma rises along these faults to form volcanoes. Plates move laterally past each other along strike-slip faults. Convergence of two plates causes one to be subducted; thrust faults develop in the subduction zone, and volcanic activity results from partial melting of the subducted slab of lithosphere.

What causes lithospheric plates to move?

Four kinds of force are responsible: mantle convection applies drag to the base of a plate; the elevated mid-ocean ridge pushes the lithosphere away from it; a subducted slab, being cold and dense, drags its parent plate into the subduction zone; and a slab that becomes separated from the lithosphere continues to sink and sucks its parent plate downward.

How can geologists measure rates of plate movement?

Radiometric dating of seafloor that extends away from a mid-ocean ridge shows how rapidly that seafloor has moved away from the ridge. Movement of oceanic lithosphere over a stable hot spot produces a chain of volcanoes that can be dated to give the velocity of the lithospheric plate. The satellite-based Global Positioning System permits geologists to measure present-day plate motions in millimeters per year.

Review Questions

1. List as many pieces of evidence as you can in support of the idea that continents have moved over Earth's surface.

2. What is the geographic extent of the lithospheric plate on which you live?

3. What is apparent polar wander? Draw pictures of Earth showing how the movement of a continent can produce apparent wander of the north magnetic pole.

4. Why are most volcanoes that have been active in the last few million years positioned in or near the Pacific Ocean?

5. Why do mid-ocean ridges form?

6. Why do deep-focus earthquakes occur along subduction zones?

7. Compare the three basic kinds of faults.

8. What are the multiple driving forces of plate movement?

9. What happens to slabs of subducted lithosphere?

10. How does the Global Positioning System allow geologists to measure the velocity at which a plate is moving in the present world?

11. Suppose that you were to encounter a well-trained geologist who was unfairly imprisoned in 1955 and was deprived of reading materials until he was released last week. He entered prison firmly opposed to the idea that continents have moved large distances across Earth's surface. Given an hour of time, how would you convince this unfortunate geologist that continents have actually moved thousands of kilometers? Use the Visual Overview on page 178 as a guide to develop your argument.

Mount Everest, at 8853.5 meters (29,028 feet), is the tallest mountain on Earth. Standing along the border between Tibet and Nepal, it is part of the Himalaya, an immense mountain chain that formed when a fragment of Gondwanaland collided with Asia to become the peninsula of India. (Jock Montgomery/Bruce Coleman.)

CHAPTER 9

Continental Tectonics and Mountain Chains

Plate tectonic forces modify the thick, felsic crust of continents as well as the thin, mafic crust beneath the deep ocean. These forces break continents apart, weld them together, and build mountain chains along their margins. Continental crust undergoes changes along all three kinds of plate boundaries: spreading zones, subduction zones, and transform faults.

Continents break apart when spreading zones propagate from oceanic crust to continental crust. First valleys form in the zone of continental rifting, and then a new seaway forms between the separated landmasses.

Collision along a subduction zone may suture two continents together, or it may suture a small crustal fragment to a large continent. A mountain chain forms along a zone of suturing. A mountain chain also forms along the margin of a continent beneath which oceanic lithosphere is being subducted.

Sometimes transform faults—the strike-slip faults that offset spreading zones (p. 190)—intersect the margins of continents. The San Andreas fault, for example, slices through western California, where in some places movements along it have shattered rocks and accelerated erosion to create a narrow, scarlike valley (Figure 9-1). San Francisco sits at the western

Figure 9-1 The San Andreas fault in California. The Pacific plate is on the left, and the North American plate is on the right. The Pacific plate periodically slides northwestward in relation to the North American plate. (Wallace/USGS.)

Visual Overview
Formation and Deformation of Continental Margins

Continents rift apart, forming passive margins that become the sites of mountain building along subduction zones, with or without collision with other continents.

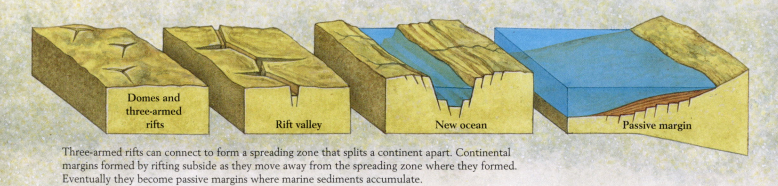

Domes and three-armed rifts | **Rift valley** | **New ocean** | **Passive margin**

Three-armed rifts can connect to form a spreading zone that splits a continent apart. Continental margins formed by rifting subside as they move away from the spreading zone where they formed. Eventually they become passive margins where marine sediments accumulate.

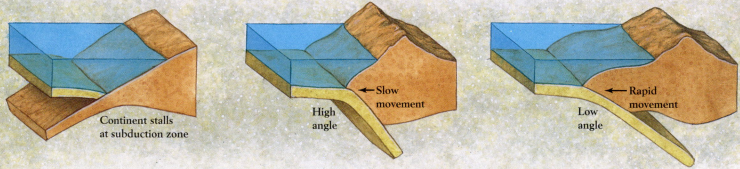

Continent stalls at subduction zone | **Subduction reverses** | **Continents collide**

When a continent arrives at a subduction zone, its low density prevents its subduction; the oceanic plate colliding with it then begins to be subducted instead. If a continent within the second plate arrives at the subduction zone, its collision with the first continent results in mountain building.

Continent stalls at subduction zone | ← **Slow movement** **High angle** | ← **Rapid movement** **Low angle**

When a continent arrives at a subduction zone and the direction of subduction is reversed, igneous activity and the pressure of collision build mountains along the margin of the continent even if it does not collide with another continent. Slow movement against a subducted plate permits the slab to rotate to a high angle.

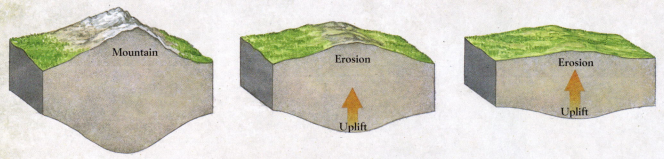

Mountain | **Erosion** **Uplift** | **Erosion** **Uplift**

Continental crust floats on the denser mantle. In keeping with the principle of isostasy, a mountain chain is balanced by a root of felsic material that extends downward into the mantle.

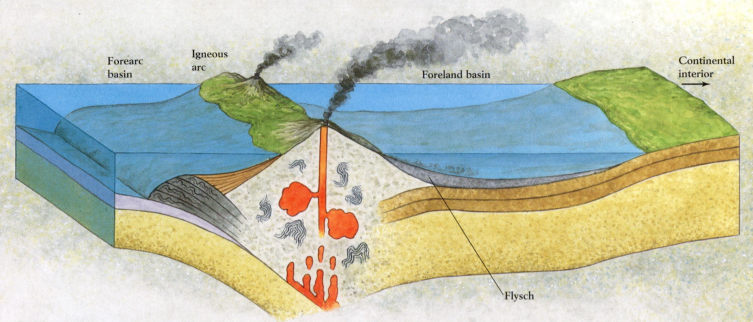

Forearc basin   Igneous arc   Foreland basin   Continental interior

Flysch

1 When a mountain chain begins to form along a continental margin through igneous activity and compression, it weighs down the neighboring continental crust to form a deep foreland basin where flysch accumulates in an arm of the ocean.

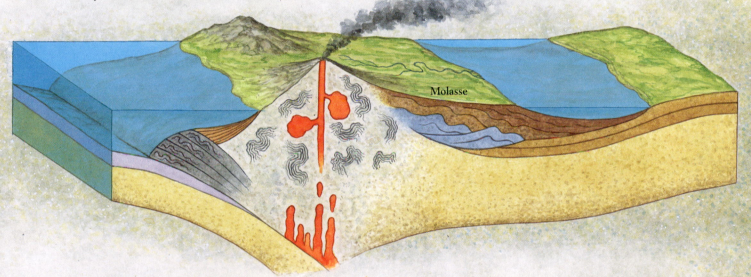

Molasse

2 As mountains continue to grow, sediments they shed push back the marine waters of the foreland basin, and deposition of flysch gives way to accumulation of shallow marine and nonmarine deposits known as molasse.

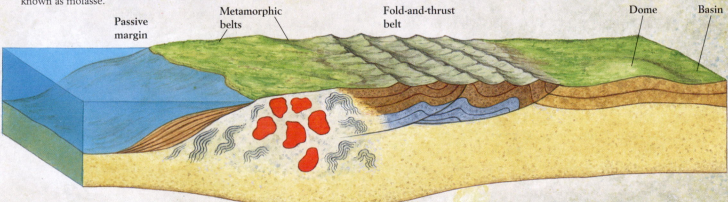

Passive margin   Metamorphic belts   Fold-and-thrust belt   Dome   Basin

3 An older mountain chain shows a zone of metamorphic rocks on either side of the core of igneous rocks. A fold-and-thrust belt extends farther into the continental interior. Weak deformation still farther inland produces domes and basins.

edge of the North American plate, whereas Los Angeles occupies a sliver of continental crust that is part of the Pacific plate. Movements along the San Andreas fault are bringing these two cities closer together at the rate of 5.5 centimeters (about 2 inches) per year. In several tens of millions of years, the sliver that Los Angeles occupies may move beyond San Francisco to end up as a slender island in the Pacific Ocean. Just as movement along a transform fault can slide a segment of a continent away from that continent, it can bring two bodies of continental crust into contact with each other.

The Rifting of Continents

We saw in Chapter 8 that spreading zones occupy the central portions of ocean basins, where oceanic lithosphere forms along a medial ridge and spreads in either direction. Continental crust, too, can spread apart. A new ocean then forms between the two continental fragments that remain. Because continental crust is about five times thicker than oceanic crust, it does not break apart as easily. In fact, a rift may begin to fracture a continent and then fail to complete the break, leaving telltale scars of its activity.

Hot spots give rise to three-armed rifts

Africa and the adjacent seas provide the best example of active continental rifting in the modern world. A system of rift valleys, formed during the Cenozoic Era, extends southward through Africa from the Red Sea and the Gulf of Aden (Figure 9-2). The central rift valleys and the broader basins harboring the Red Sea and Gulf of Aden are grabens formed by extension and breaking of the continental lithosphere.

When rifts develop, they often begin as three-armed grabens at plate boundaries known as **triple junctions**. Long before the advent of plate tectonics, geologists noticed that at the locations of three-armed grabens, continental crust is frequently elevated into a dome. In the context of plate tectonics, this elevation represents the development of a hot spot. In the area of Ethiopia called the Afar Triangle, the Red Sea, the Gulf of Aden, and the north end of the African rift system form a triple junction. Such junctions are common features of Earth's crust (see Figure 1-15). More than one kind of plate boundary can meet at a triple junction. Each boundary may consist of a spreading zone, a subduction zone, or a transform fault. At the Afar Triangle, the junction happens to involve three spreading zones.

When a large continent breaks apart, the jagged line along which it divides often represents a composite structure formed from arms of several three-armed rifts. This phenomenon is seen in the rifting that formed

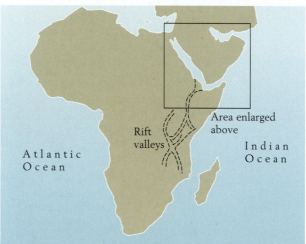

Figure 9-2 A three-armed rift along the northeastern margin of Africa. Two of the arms represent new oceans: the Red Sea and the Gulf of Aden. The third is beginning to break the continent of Africa apart along Africa's famous rift valleys. The Afar Triangle (cross-hatched area) is a small region of oceanic crust that has been elevated to become land.

the Atlantic Ocean (Figure 9-3). A three-armed rift usually contributes two of its arms to the composite rift, while the third arm becomes a **failed rift**—a plate tectonic dead end. Before it ceases to be active, this third arm forms a graben or a system of grabens that projects inland from the new continental margin formed by the other two arms. Some of the world's largest rivers, including the Mississippi and the Amazon, flow through valleys located in failed rifts that border the Atlantic basin.

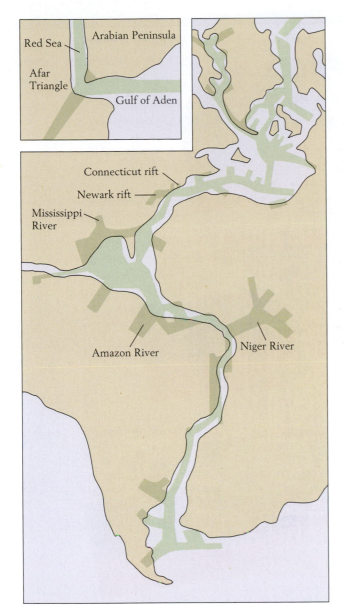

Figure 9-3 Ancient three-armed rifts become apparent when we reassemble the continents now bordering the Atlantic Ocean. Many of these rifts contributed two arms to the spreading zone that became the Mid-Atlantic Ridge. (After K. C. Burke and J. T. Wilson, *Sci. Amer.*, August 1976. © 1976 by Scientific American, Inc. All rights reserved.)

It is not uncommon, however, for all three arms of a three-armed rift to develop into segments of plate boundaries. Note that the Mid-Atlantic Ridge, for example, terminates at a triple junction in the South Atlantic. Similarly, although the rift arm that projects into Africa has not yet divided that continent, it may do so in the future.

It is not surprising that many hot spots are situated on or very near mid-ocean ridges (see Figure 8-26). These may be the only surviving hot spots of a larger

number that produced the three-armed rifts that contributed one or two arms to the formation of actively spreading ridges. The southern portion of the Mid-Atlantic Ridge has shifted to a position slightly to the west of three surviving hot spots that appear to have played a role in its origin.

Rift valleys form when continental breakup begins

A mid-ocean ridge is often associated with *block faulting*, which is normal faulting that produces blocks of crust. As we have seen, this block faulting is expressed by a graben running along the ridge's midline (see Figures 8-4 and 8-19). When a spreading zone first encounters continental crust, however, it seldom produces faults that cut cleanly through. Instead, its extension along a rift tends to break the thick continental crust into a complex band of fault blocks. Today this process is under way in Africa, where a system of rift valleys passes southward from the Red Sea (see Figure 9-2). Each of these valleys, or **fault block basins**, is a long, narrow, downfaulted block of crust associated with mafic volcanoes that have welled up from the mantle along the faults; the rifting has also produced mafic dikes and sills and flood basalts. Some of the rift valleys cradle great lakes such as Lake Tanganyika. These rift valleys have been in existence only since Early Miocene time (less than 20 million years).

The rapid subsidence of a fault block basin creates a rugged landscape that is subject to rapid erosion, so that clastic sediments often accumulate quickly to great thicknesses. The sediments typically include conglomerates derived from the steep valley walls as well as red beds—sediments whose reddish color is usually attributable to iron oxide cement—and alluvial-fan deposits. Lakes that form within the elongate valleys also leave a sedimentary record. In arid climates, temporary lakes leave accumulations of nonmarine evaporites.

If rifting continues long enough, a rift valley becomes so wide and so extended that it opens up to the sea. Because inflow from the sea tends at first to be restricted or sporadic, saline waters within the rift valley evaporate more rapidly than they are renewed. Under such circumstances, marine evaporites, such as halite and gypsum, form. Waters from the Indian Ocean, for example, only recently gained full access to the currently widening Red Sea. Beneath a thin veneer of marine sediment in the Red Sea, geologists have found evaporites that formed during an earlier time, when there was only a weak connection to the larger ocean, or perhaps even earlier, when the basins were nonmarine but arid.

The margins of the Red Sea also exhibit geologic features that typify the early stages of continental rifting.

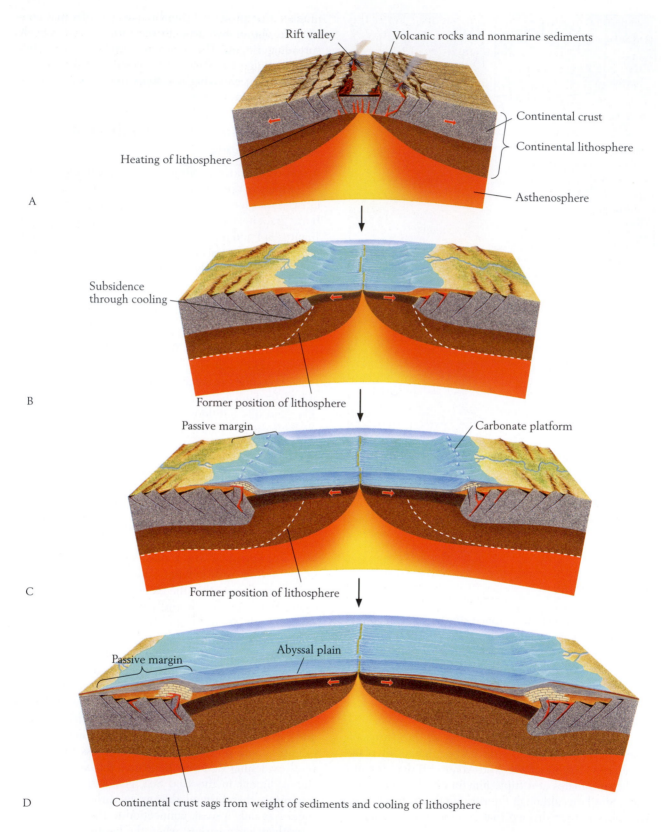

A

Rift valley

Volcanic rocks and nonmarine sediments

Continental crust

Continental lithosphere

Heating of lithosphere

Asthenosphere

B

Subsidence through cooling

Former position of lithosphere

C

Passive margin

Carbonate platform

Former position of lithosphere

D

Passive margin

Abyssal plain

Continental crust sags from weight of sediments and cooling of lithosphere

Figure 9-4 **The rifting of Pangaea that produced the Atlantic Ocean.** (*A*) Rifting began early in the Triassic Period, forming nonmarine basins along block faults. (*B*) The sea spread into the zone of rifting during the Jurassic, and passive margins formed on either side of the new Atlantic Ocean. (*C* and *D*) The Atlantic has continued to expand, and the passive margins have persisted to the present day.

The Afar Triangle (see Figure 9-2) was once part of the Red Sea floor. Most of the rocks in this area are basalts similar to those that form oceanic crust. Much of the topography is the product of block faulting and uplift produced by the high rate of heat flow from the mantle, and in places there are great thicknesses of evaporite deposits.

In summary, then, regions of continental rifting are characterized by block faults, mafic dikes and sills, and thick sedimentary sequences within fault block basins that often include lake deposits, coarse terrestrial deposits, and evaporites followed by oceanic sediments. This suite of features can be found in fault block basins of the eastern United States. Two of these basins are labeled in Figure 9-3 as the Newark and Connecticut rifts. They are actually failed rifts—rift arms that never became part of the composite rift that formed the Atlantic Ocean. These rifts extended far inland and never opened wide enough to allow the sea to invade, although they contain other sedimentary sequences that typify incipient continental rifts.

Rifting creates passive margins

When continental rifting does not fail, one continent becomes two, and a narrow ocean forms between them. Eventually the two new continental margins move far from the spreading zone where they formed. Soon they are likely to be flooded by shallow seas, because they move laterally away from the ridge axis, down the slope of the asthenosphere's surface, to regions where heat flow is lower and the asthenosphere is not so swollen (Figure 9-4).

Thus the continental borders, which were tectonically active when they were still close to the spreading zone, become what are termed **passive margins**. Having descended below sea level, these tectonically inactive areas of continental crust accumulate sediment along shallow shelves. Thus, after Pangaea broke up early in the Mesozoic Era, the newly formed Atlantic margin of the United States migrated away from Africa. It soon

subsided below sea level and began to accumulate great thicknesses of marine sediment (see Figure 9-4). As later chapters will describe in greater detail, to the present day this passive margin has continuously subsided under the weight of added sediment, making way for still more to be laid down.

As the term "passive margin" suggests, continental margins can also be tectonically active. Indeed, **active margins** are zones of tectonic deformation and igneous activity. Simply put, they are sites of mountain building. Before we can investigate how tectonic forces and igneous activity form mountain chains, however, we must consider how bodies of rock bend and flow under stress from tectonic forces, in addition to moving past one another along faults.

Bending and Flowing of Rocks

Geologic outcrops reveal that some rocks have been warped, twisted, and folded; some have even flowed. When forces are applied to a rock, the component grains may be affected in various ways; they may slide past one another, change shape, or break along parallel planes. (In the last case, a grain deforms in the way a deck of cards on a table changes shape when we push on one end of the deck near the top.) Internal deformation of a large body of rock by any of these mechanisms takes place very slowly, but when many of the grains are affected, the entire body of rock can undergo radical changes in shape in the course of millions of years. Such changes usually take place at great depths within Earth's crust.

When rocks are deformed at high temperatures and pressures, they also undergo high-grade metamorphism (p. 45). Under less extreme conditions, sedimentary rocks and the fossils they contain may be deformed with little or no metamorphism.

One common type of large-scale rock deformation is referred to as **folding**. Compressive forces can shorten Earth's crust, creating folds (Figure 9-5). Folds

Figure 9-5 Folded rocks in the Andes. (Richard W. Allmendinger, Cornell University.)

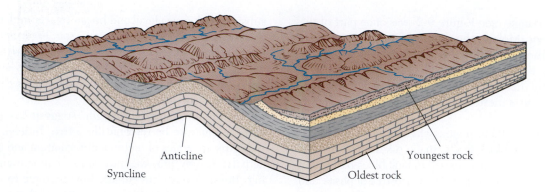

Figure 9-6 Humplike folds, or anticlines, and troughlike folds, or synclines.

come in many sizes. When folded sedimentary rocks are viewed with their oldest beds at the bottom and their youngest beds on top, they display two kinds of folds. A **syncline** is concave in an upward direction, having its vertexes at the bottom, like the letter U. An **anticline**, in contrast, is concave in a downward direction, having its vertexes at the top, like an arch (Figure 9-6).

Many rocks do not simply bend when they are folded; instead, material is displaced from one part of a bed toward another. When a large, complex body of rock is subjected to an external force, weak beds tend to become more intensely folded than durable beds. Shales, for example, tend to be weak and to deform much more severely than massive sandstones and limestones that are subjected to the same forces. Igneous and metamorphic rocks are relatively strong, but they, too, can be folded.

Geologists have developed special terminology to describe shapes of folds in detail. A tilted bed, for ex-

ample, is said to have a **dip**—a term that describes the angle that the bed forms with the horizontal plane. In other words, the dip is the direction in which water would run down the surface of the bed. The **strike** of a bed, in contrast, is the compass direction that lies at right angles to the dip (Figure 9-7); strikes are always horizontal. It is sometimes said that the **regional strike** of a given area is in a particular geographic orientation—north–south, for example. This does not mean that every strike in this area has the same orientation, only that most of the fold axes trend north–south, so that the strikes of most beds do too.

A fold is said to have an **axial plane**, which is an imaginary plane that cuts through the fold and divides it as symmetrically as possible. In actuality, many folds are asymmetrical, with one limb (or flank) dipping more steeply than the other. If either limb is rotated more than 90° from its original position, so that older strata overlie younger strata, the fold is said to be **overturned** (Figure 9-8).

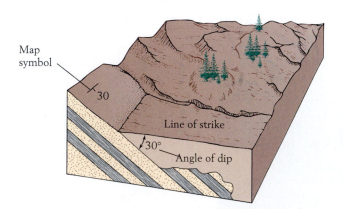

Figure 9-7 The strike and dip of inclined beds. The map symbol indicates the geographic orientation of the strike and dip.

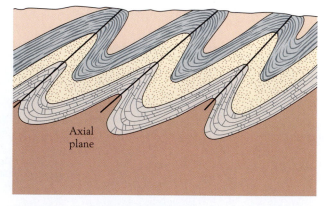

Figure 9-8 Overturned folds in cross section. Both limbs of an overturned fold dip in the same direction. One limb of such a fold is upside down, with younger strata lying below older strata.

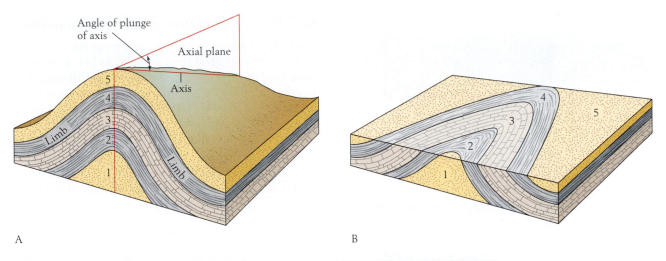

A

B

C

Figure 9-9 Plunging folds. A plunging fold (*A*) before and (*B*) after it has been truncated by erosion; note how erosion produces a curved outcrop pattern. *C*. A plunging fold seen from the air; the view is along the axis of the Virgin Anticline of southwestern Utah. (*C* from J. S. Shelton, *Geology Illustrated*, W. H. Freeman and Company, New York, 1966.)

The **axis of a fold** is the line of intersection between the axial plane and the beds of folded rock. Often the axis plunges—that is, it lies at an angle to the horizontal (Figure 9-9*A*). When a **plunging fold** is truncated by erosion, its beds form a curved outcrop pattern (Figures 9-9*B* and *C*). A series of plunging folds then produces a scalloped surface pattern. Because most folds plunge (we could not expect many to be perfectly horizontal), this scalloped pattern is characteristic of regions where sedimentary rocks have been extensively folded.

Most large folds in continental crust form where two plates converge, applying compressive forces to the rocks adjacent to the juncture. Folding in such regions thickens the crust and contributes to the growth of mountain chains.

Mountain Building

Before the advent of plate tectonic theory, the process of mountain building, or **orogenesis**, was a subject of debate. North American geologists had long been struck by the fact that one chain of mountains, the Cordilleran system, parallels the west coast of their continent and another, older chain, the Appalachian system, parallels the east coast. Before the 1960s, this symmetrical pattern led some North American geologists to conclude that mountains tend to form along the margins of relatively stable continental masses, known as **cratons**. Most European geologists, however, pointed to the Ural Mountains, standing between Europe and Asia within the largest landmass on Earth, as evidence that long mountain chains could indeed rise up in the center of a continent. Plate tectonic theory explains both of these patterns. It is true that continuous long mountain chains form only along continental margins. Continents can unite, however, and when they do, a mountain chain forms where the margins are welded together within the newly formed large landmass. The unification of two continents along a subduction zone is termed *suturing*.

According to the theory of plate tectonics, subduction zones are the key to the origin of mountain chains

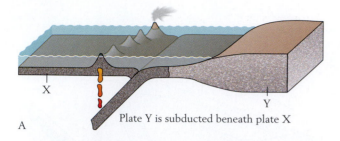

A   Plate Y is subducted beneath plate X

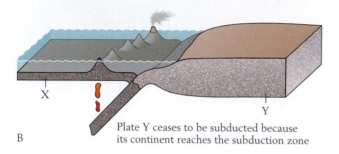

B   Plate Y ceases to be subducted because its continent reaches the subduction zone

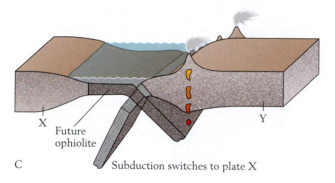

C   Future ophiolite   Subduction switches to plate X

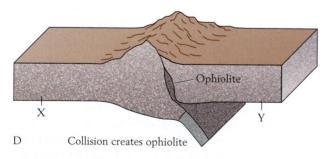

D   Collision creates ophiolite   Ophiolite

Figure 9-10 The processes of suturing and mountain building where two continents meet along a subduction zone.

on continents. But not all mountain-building events—called **orogenies**—result from continental suturing; a mountain chain also forms when an oceanic plate descends beneath the margin of a solitary continent. Furthermore, not all mountain chains are formed of continental crust. Rifting and swelling of oceanic crust create the great mid-ocean ridges, which are, in effect, submarine mountain chains (see Figure 8-4).

Continental collision produces orogenies

When two continents collide, two properties of continental crust—its great thickness and its low density relative to that of the asthenosphere—lead to mountain building. Continents are simply too thick and too buoyant to be subducted. When a continent encounters a deep-sea trench, its resistance to subduction forces a reversal in the direction of subduction (Figure 9-10). As a result, the oceanic plate opposite the continent is forced to descend into the mantle. If a second continent is riding on the newly subducting plate, it will eventually collide with the first continent along the subduction zone, and the two continents will be welded together. The juncture formed in this way is called a **suture**.

Continental suturing creates mountain chains because subduction along a trench causes the margin of one continent to wedge beneath the margin of the other (Figure 9-10D). The forces of collision cause both continental margins to thicken through deformation, and a mountain chain is uplifted along the suture. Often remnants of seafloor are pinched up along the suture, forming an ophiolite. **Ophiolites** typically include two components: ultramafic rocks from the upper mantle and basalts, including pillow basalts (see Figure 2-14), that formed oceanic crust. Often present as well are deep-sea sediments, such as turbidites, black shales, and cherts. In effect, ophiolites are samples of ancient ocean basins that have been conveniently elevated for our study. Because ophiolites mark the positions of vanished oceans that once lay between continents, they are key features in our recognition of plate convergence along subduction zones.

Orogenies can occur without continental collision

Mountains can form along the margin of a continent that is resting against a subduction zone even when that continent does not collide with another. Figure 9-11 shows what happens when a continent encounters a subduction zone. Magma produced by the partial melting of the subducted slab rises into the margin of the continent positioned above the slab. Some of the magma reaches the surface and forms a chain of volcanoes that elevate the crust, forming mountain peaks. Some magma also cools within the crust, forming plutons, or massive intrusions of granitic rock. Such a zone of volcanoes and intrusions constitutes an **igneous arc**, which is a belt of igneous activity equivalent to the volcanic arc that forms through subduction beneath oceanic lithosphere (see p. 192).

In keeping with the principle of isostasy (p. 14), the addition of large volumes of low-density igneous rock to the base of the crust causes the crust along the

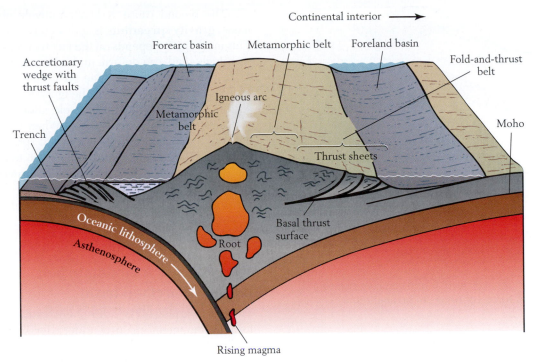

Figure 9-11 An idealized mountain chain forming where an oceanic plate is being subducted beneath the edge of a continent. This cross section illustrates the general symmetry of the mountain chain. Metamorphism dies out both toward the sea and toward the land from the central igneous arc. Beyond the metamorphic belt, in the direction of the continental interior, is a fold-and-thrust belt. Beyond the inland fold-and-thrust belt, the crust is warped downward to form a foreland basin, where sediments from the mountain system accumulate.

igneous arc to bob upward. This vertical movement contributes to the elevation of the mountain chain. At the same time, a root of crustal rock produced by the igneous arc extends downward beneath the mountain chain, balancing the weight of the mountains by displacing dense rocks in the asthenosphere.

Mountain belts have a characteristic structure

An igneous arc like the one just described forms the core of a typical mountain chain. Extending along either side of this core is a belt of regional metamorphism (see p. 45). This belt consists of crustal rocks that have been metamorphosed by heat from the core and locally intruded by magma issuing from the core. Rocks of this **metamorphic belt** (see Figure 9-11) are also deformed by processes that will be described shortly.

Metamorphism dies out in both directions from the igneous core. Toward the continental interior, the metamorphic belt gives way to a **fold-and-thrust belt** (see Figure 9-11). The folds of the fold-and-thrust belt are typically overturned away from the core of the mountain chain, reflecting the fact that the prevailing forces of deformation come from the direction of the core. Be-

cause of their distance from the igneous core, the sedimentary rocks of the fold-and-thrust belt are largely unaffected by metamorphism and are folded less severely than the rocks in the metamorphic belt. As a result, the behavior of these sedimentary rocks during the deformation process is more brittle and less plastic than that of the rocks in the metamorphic belt. Thus the folds are broken by enormous thrust faults along which large slices of crust, known as **thrust sheets**, have moved away from the core. The thrust sheets usually slide along a *basal thrust surface*. A look at Figure 9-11 reveals how the telescoping action of folding and thrusting above the lowermost, or basal, thrust shortens and thickens the crust.

Figure 9-12 shows the relationship between folds and thrust faults. After an overturned fold forms, continued stress along it can break the folded rock, forming a thrust fault. Figure 9-13, a cross section of the fold-and-thrust belt of the Rocky Mountains, shows thrust faults that slice through previously folded rocks.

Along the subduction zone seaward of an igneous arc, rocks are also deformed by folds and thrust faults. This is the position of the accretionary wedge, in which thrust sheets are piled up along the continental margin (see Figures 8-22 and 9-11).

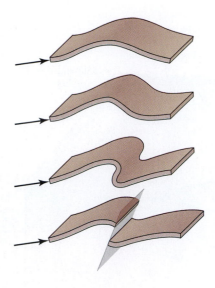

Figure 9-12 An overturned fold can give rise to a thrust fault as force continues to be applied.

Compressive forces cause deformation

It is easy to understand how deformation occurs in the accretionary wedge along a deep-sea trench. Here compression breaks off slices of material and piles them up along thrust faults (see Figure 9-11). The compression also folds material within individual thrust slices. Deformation within the metamorphic belts and the fold-and-thrust belt on the continental side is more complex. It has two primary causes. The first cause is simply pressure that the subducted plate applies to the mountain chain, pushing it laterally toward the interior of the continent. The resulting compression takes the form of folding near the igneous arc and folding and thrusting in the more brittle terrain farther toward the continental interior.

The second cause is less intuitively obvious. Its name, **gravity spreading**, is aptly descriptive, however. This mechanism depends on the fact that rock, although seemingly rigid, can deform under its own weight when its height and weight become great enough. The necessary weight can develop as the igneous arc builds the core of the mountain to a high elevation while a root develops by isostatic adjustment. When the mountain chain becomes tall enough, it tends to spread laterally, like a mound of pudding heaped too high to remain stable. A mountain chain spreads by deformation along folds and thrust faults. This process can even cause thrust sheets to move uphill along the basal thrust, just as pudding that is heaped too high can spread up the gently sloping sides of a dish.

The weight of a mountain belt creates a foreland basin

The downwarping of the lithosphere beneath an actively forming mountain chain continues for some distance beyond the fold-and-thrust belt. This activity produces an elongate **foreland basin**, whose long axis lies parallel to the mountain chain (Figure 9-14; see also Figure 9-11). The foreland basin forms rapidly and is usually so deep initially that the sea floods it either through a gap in the mountain chain or through a passage around one end of the chain (Figure 9-14C).

The foreland basin typically subsides so quickly that the first sediments to accumulate in it are deepwater deposits, especially muds. Turbidites also form later if the slope from the foot of the mountain is steep enough to send turbidity flows out into the basin. Figure 1-21, for example, displays turbidite deposits that accumulated in front of Cretaceous mountains in

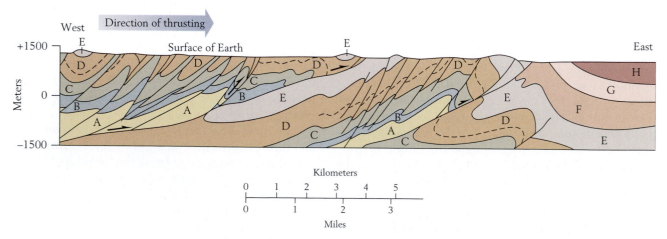

Figure 9-13 A fold-and-thrust belt of the Rocky Mountains southwest of Calgary, Alberta, Canada. Thrust faults, shown as dark lines, are intimately associated with overturned folds. Thrusting has been toward the east, and folds are overturned in the same direction. The oldest beds (A) are Lower Carboniferous, and the youngest (H) are Paleogene. (After P. B. King, *The Evolution of North America*, Princeton University Press, Princeton, NJ, 1977.

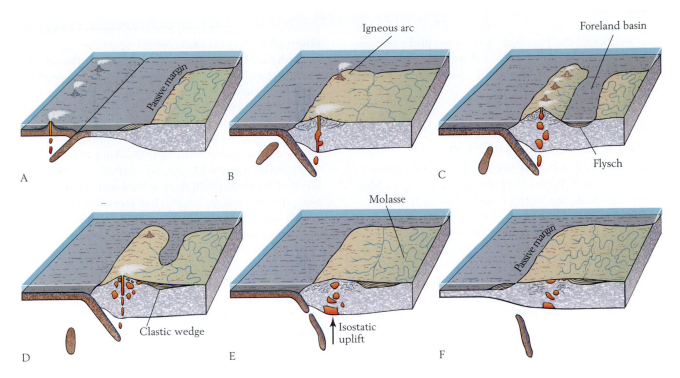

Figure 9-14 History of a typical mountain belt and foreland basin. *A.* A continent bounded by a passive margin approaches a subduction zone. *B.* The continent stalls at the subduction zone, forcing a reversal in the direction of subduction, and mountain building begins. *C.* A deep foreland basin forms and accumulates flysch. *D.* As mountain building progresses, the accumulation of sediment pushes the sea from the foreland basin. *E.* The foreland basin accumulates nonmarine sediment (molasse) until the mountain chain wears down, but its root remains, and isostasy causes the mountain belt to bob up again. *F.* Isostatic uplift and erosion continue until no root remains, but geologic features of the mountain system remain in the bedrock of the region.

northern Alaska. When both shales and turbidites accumulate in foreland basins, they are collectively known as **flysch**.

As a mountain system evolves, folding and thrust faulting move progressively farther inland, and mountain building proceeds toward the continental interior. In the process, the shale or flysch deposits become folded and faulted. At the same time, the mountain core rises and sheds sediments more and more rapidly. Eventually sediment chokes the foreland basin, pushing marine waters out and leaving nonmarine depositional settings in their place (see Figure 9-14D and E). These settings include alluvial fans along the mountain chain as well as riverbeds, floodplains, and other lowland environments. The resulting nonmarine sediments are collectively termed **molasse**. Molasse deposits can accumulate to great thicknesses; as deformation continues, they too can sometimes be folded and faulted. During molasse deposition, the foreland may no longer be a topographic basin, but may appear instead as a broad depositional surface sloping away from the mountain front (Figure 9-14E). As the foreland subsides beneath the accumulating sediments, however, it remains a *structural basin*, which is a circular or elongate depression of stratified rock. Molasse deposits pinch out from the margin of a mountain belt toward the interior of the craton. Because of its prismlike configuration, a thick body of molasse is sometimes referred to as a **clastic wedge**.

The depositional transition from deep-water sediments to nonmarine sediments occurs during the evolution of most foreland basins. Even molasse deposition comes to an end after orogenic activity stops. Not only does igneous activity eventually cease in the core of the mountain chain, but so do folding and thrusting along the margin. Erosion soon subdues the mountainous terrain, and the source of the molasse sediment disappears (Figure 9-14E).

After a mountain system is initially leveled by erosion, some of its root remains. The remnant of a root creates gravitational instability, so that the thick mass of felsic rocks in the mountain belt tends to rise up, just as a block of wood floating in water will do if its top is sliced off (see Figure 1-14). Orogenic belts bob up sporadically long after subduction has ceased, and each time they rise, erosion temporarily subdues them. This process can continue over hundreds of millions of years, until no root is left (Figure 9-14F). Even after the root is gone and the land is level, folds and faults remain in the bedrock of the region, as do igneous and metamorphic rocks—and often flysch and molasse of the foreland basin. These rocks are the marks of an ancient mountain belt.

The Andes exemplify mountain building without continental collision

The Andes of South America are still in the process of forming. In fact, Figure 9-11 may be viewed as an idealized representation of the Andean belt as it is today.

The Andes are associated with the Pacific "ring of fire." This composite feature, consisting of volcanoes that encircle much of the Pacific Ocean, is in most areas a product of subduction zones where oceanic plates are colliding (see Figure 8-21). In certain segments of the ring, however, subduction is instead occurring along blocks of continental crust. One of these segments is the coast of South America, where the Andes continue to form.

The Andean system is the longest continuous mountain chain in the world. Its history extends well back into the Paleozoic Era, but the present pattern of mountain building began early in the Mesozoic Era, when a subduction zone came to lie along the margin of South America (Figure 9-15). Enormous volumes of igneous rock have since risen from the subducted oceanic plate and have been added to the Andean crust, thickening it in some places to more than 70 kilometers (45 miles).

When Charles Darwin sailed around the world on the *Beagle*, he noted the presence of Cenozoic marine fossils at high altitudes in the Andes. These fossils offered proof that the Andes had been greatly elevated rather recently in geologic time. Darwin also saw firsthand that the movements occurred in pulses. He witnessed earthquakes during which land along the seacoast was suddenly raised several feet, leaving marine

animals rotting in the sun. From a distance Darwin also observed Andean volcanoes erupting on the night of the earthquakes. We now know that for about the last 200 million years the Andean crust has not only been thickened by the addition of igneous material below but has also been bobbing up isostatically. At the same time, volcanic rocks have been piled on top. We also understand why most major continental earthquakes originate in major orogenic belts such as the Andes.

Igneous activity has steadily shifted toward the interior of South America during Mesozoic and Cenozoic time; in other words, magma has ascended at positions farther and farther inland (see Figure 9-15). Probably because today the subducted plate descends at a low angle, the zone where volcanoes are formed and magma cools below the surface to form intrusive rocks is now centered about 200 kilometers (125 miles) inland from the coast. Earlier, the subducted plate probably descended at a steeper angle, thus reaching the depth of partial melting nearer the coast.

A change in the angle of subduction usually reflects a change in the rate of plate movement. When the plate that is not being subducted is moving rapidly toward the subduction zone, it overrides and bears down on the upper part of the subducted slab. This pressure "rolls back" the plate, much as you might roll back a wrinkle in a carpet by pushing on it with a stick (Figure 9-16). Migration of the subduction zone causes the subducted slab to dip at a low angle. In contrast, when the plate that is not being subducted is moving too slowly to roll the subduction zone back rapidly, the subducted slab is free to rotate toward a vertical position under the influence of gravity.

As recently as 10 million years ago, during the Miocene Epoch, a long seaway occupied the foreland basin to the east of the Andes (Figure 9-17). Though relatively shallow, it stretched along most of the length of the continent. Interestingly, a smaller seaway connected this inland sea to the Atlantic Ocean along the

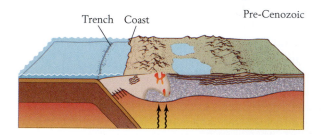

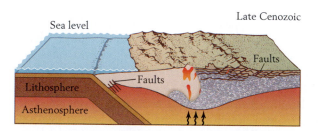

Figure 9-15 **Formation of the Andes.** During pre-Cenozoic time, igneous material was added to the crust from the oceanic plate descending along the marginal trench. During Cenozoic time, igneous activity shifted farther east. Thrust faulting has occurred both east and west of the area of igneous activity. (After D. E. James, *Sci. Amer.*, August 1973. © 1973 by Scientific American, Inc. All rights reserved.)

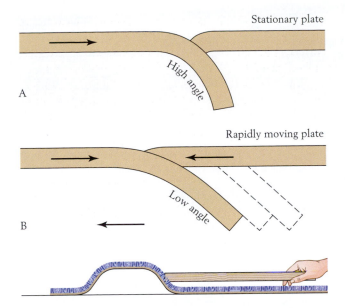

Figure 9-16 **The rate of plate movement and the angle of subduction.** *A.* The plate not being subducted moves slowly, and the weight of the subducted slab rotates the slab to a high angle. *B.* The plate not being subducted moves rapidly against the subducted slab, rolling it back and forcing it to maintain a low angle of descent. *C.* A situation analogous to the "rollback" phenomenon in *B*: the stick rolls back the wrinkle in the carpet.

axis of a failed rift that today marks the path of the lower Amazon River (see Figure 9-3). As the foreland basin filled in and its marine waters receded, freshwater runoff that had flowed into the shallow inland sea coalesced to form the Amazon River. The Amazon is still home to dolphins, stingrays, and manatees—freshwater species descended from marine ancestors that occupied the foreland basin sea millions of years ago.

The Andes orogenic belt has matured to the degree that large volumes of sediment shed from the mountains now keep the sea from flooding the foreland basin (see Figure 9-14E and F). Today only nonmarine molassic sediments accumulate there.

The Andes exemplify mountain building along a single continent bordered by a subduction zone. In contrast, the Pyrenees—the second youthful mountain range we will examine—have been created by the collision of two continents.

The Pyrenees exemplify mountain building by continental collision

The Pyrenees, which separate France from Spain, stand tall above the surrounding terrain because they are a relatively youthful mountain system. They formed during Cretaceous and Paleogene time, when Iberia, the peninsula occupied by Spain and Portugal, collided with Eurasia (Figure 9-18).

Iberia had actually been attached to Eurasia previously, but it broke away during Early Cretaceous time when a spreading zone branched from the Mid-Atlantic Ridge and propagated across the neck that connected the two landmasses (Figure 9-18A). Later in the Cretaceous Period, this spreading zone ceased to function and was replaced by a subduction zone that extended westward from the Mediterranean Sea (Figure 9-18B). Subduction was toward the north, beneath France, and it soon led to the reattachment of Iberia to Eurasia. Today ophiolites in the northern Pyrenees mark the site of suturing (Figure 9-18E). Soon after the collision, compressive forces began to fold rocks and push slabs of crust to both the north and south along thrust faults. By Paleocene time, flysch was accumulating in foreland basins on both sides of the newly forming mountain chain (Figure 9-18C). In typical fashion, marine deposition in the foreland basin

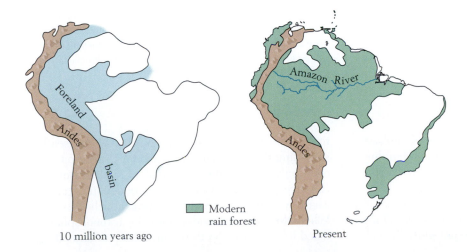

Figure 9-17 **A shallow sea occupied the foreland basin of the Andes during the Miocene Epoch.** (After S. D. Webb, *Science* 269:361–362, 1995.)

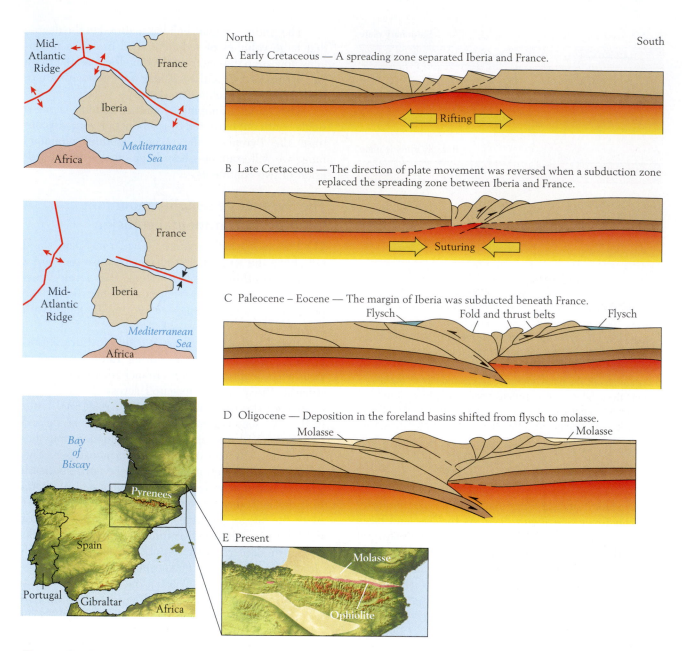

Figure 9-18 Origin of the Pyrenees through the collision of Iberia and France. Iberia had previously been attached to France, but a rift zone spread across the juncture between the two in Early Cretaceous time, separating them (*A*). They were sutured together again soon afterward, when a subduction zone replaced the rift zone (*B*) and mountain building began, producing flysch (*C*) and then molasse (*D*) in foreland basins. Ophiolites in the modern Pyrenees mark the zone of suturing (*E*). (After F. Roure and P. Choukroune, *Mém. Géol. Soc. France*, new series, 173:37–52, 1998.)

then gave way to nonmarine deposition of molasse (Figure 9-18*D*). Subduction and mountain building continued throughout the Oligocene Epoch, ending about 20 million years ago.

Curiously, the Pyrenees orogeny produced no igneous rocks, either intrusive or extrusive. Why did the slab that descended beneath France not partially melt to liberate magma? The answer is that the seafloor spreading that separated Iberia from France operated over only a very short interval of Cretaceous time (Figure 9-18*A*). As a result, when subduction began, only a narrow segment of oceanic crust was attached to Iberia. When this slab descended into the asthenosphere, it failed to reach the depth required for partial melting.

Apart from its lack of igneous activity, the Pyrenees orogeny was a typical episode of mountain building: a fold-and-thrust belt formed on either side of the axis of the mountain chain, and flysch and then molasse accumulated in foreland basins beyond the fold-and-thrust belts.

Suturing of Small Landmasses to Continents

Iberia, which became attached to Eurasia during the Cretaceous Period, is a relatively small landmass, slightly larger in area than California. Even smaller pieces of continental lithosphere, known as *microcontinents*, also become sutured to large continents, as do island arcs that have ceased to be active but remain as large bodies of volcanic rock and sediment.

Later chapters will describe the attachment of several small landmasses to the eastern and western margins of North America during the past 500 million years. The addition of these **exotic terranes** has expanded the size of the continent considerably, especially in the west, where they constitute large areas of the United States and Canada from Alaska to California.

Tectonics of Continental Interiors

Powerful forces can deform Earth's crust or create faults even far inland from continental margins. The causes of tectonic activity in continental interiors are poorly understood, but it is evident that the building of both the Appalachian and Rocky mountains caused mild deformation in neighboring areas of the North American craton. Evidently, other forces were also at work.

Gentle vertical movements of continental crust produce structural basins and domes. Recall that a structural basin is a circular or oval depression of stratified rock. A *structural dome* is a comparable uplift—a blisterlike structure (Figure 9-19). Vertical forces can form these structures: local depression can create a basin and local uplift, a dome. Both basins and domes can also result from compressive forces of the sort that form anticlines and synclines when they are applied along a broader zone (see Figure 9-6). In fact, an extremely elongate dome amounts to an anticline that plunges in two directions. Similarly, a very long basin amounts to a syncline that has upturned ends.

Both basins and domes, when eroded, form outcrops with concentric, circular bands of stratified rocks. There is a simple way to distinguish the outcrop pattern of a dome from that of a basin (see Figure 9-19). In a

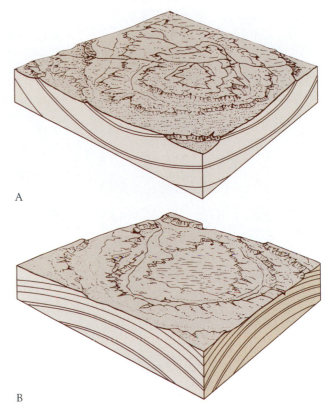

A

B

Figure 9-19 The concentric outcrop pattern of a structural basin and of a structural dome. The rocks in the center of this structural basin (*A*) are relatively durable and thus have remained at a high elevation; those in the center of this structural dome (*B*) are relatively weak and have been deeply eroded to form a topographic basin. (After W. K. Hamblin and J. D. Howard, *Exercises in Physical Geology*, Burgess Publishing Company, Minneapolis, 1975.)

dome, the oldest beds lie in the center, whereas in a basin, it is the youngest beds that are centrally positioned. Figure 9-19 illustrates how rocks within a structural basin that are resistant to erosion can form topographic ridges; similarly, erosion of weak rocks in the center of a structural dome can create a central basin.

The Black Hills of South Dakota are the surface expression of an oblong dome (Figure 9-20). Although they lie about 250 kilometers (150 miles) from the Rockies, they rose up with those mountains early in the Cenozoic Era, and erosion has since exposed ancient rocks in the center of the dome.

As we have seen, a massive mountain chain typically depresses adjacent continental crust to produce a foreland basin. The Appalachian Basin, lying just to the west of the Appalachian Mountains (Figure 9-21), is a remnant of the Appalachian foreland basin that was too far from the axis of Appalachian mountain building to have been strongly deformed and uplifted. The Denver

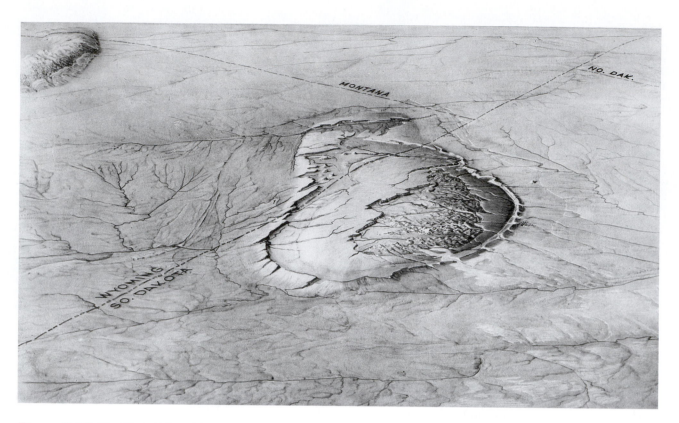

Figure 9-20 The Black Hills, a blisterlike structural dome to the east of the Rocky Mountains. Paleozoic and Mesozoic strata flank crystalline Archean rocks that erosion has exposed in the center of the dome. (From J. S. Shelton, *Geology Illustrated*, W. H. Freeman and Company, New York, 1966.)

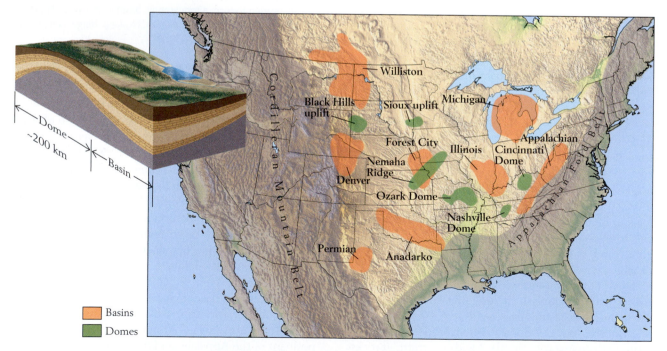

Figure 9-21 Structural basins of the United States. (From F. Press, R. Siever, J. Grotzinger, and T. H. Jordon, *Understanding Earth*, W. H. Freeman and Company, New York, 2004.)

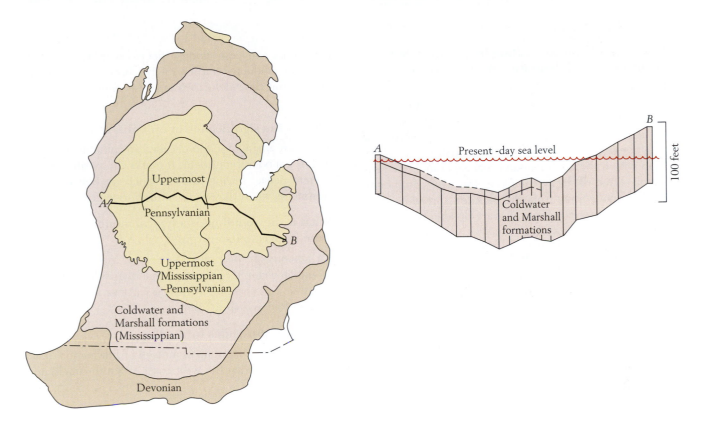

Figure 9-22 **A simplified geologic map of the state of Michigan.** The fact that the youngest rock units are in the center indicates that this is a structural basin. The cross section from A to B illustrates the configuration of the Coldwater and Marshall formations, which occur throughout most of the state but are buried beneath younger deposits in the center. This cross section has been constructed from information obtained by drilling. (After V. Brown Monnett, *Amer. Assoc. Petrol. Geol. Bull.* 32:629–688, 1948.)

and Williston basins bear a similar relationship to the Rocky Mountains.

Large domes and basins also originate deeper within the interiors of cratons and probably result from forces not directly related to mountain building. Most of the state of Michigan, for example, constitutes the central part of a structural basin that formed during the Paleozoic Era through a series of episodes of regional subsidence (Figure 9-22). It is not fully understood what caused Earth's crust to subside to form either the Michigan Basin or the Illinois, Anadarko, and Permian basins farther south (see Figure 9-21). The subsidence that formed these basins may have resulted from stretching and weakening of the crust as North America moved over the lithosphere or as a hot spot briefly heated the crust. It is likely that some of the domes on the North American craton—for example, the Black Hills in the west and the Cincinnati and Nashville domes in the east—were elevated by the arrival of deep crustal material that was squeezed out from beneath basins that were subsiding nearby (see Figure 9-21).

Many basins and domes of eastern North America continued to form over long stretches of Paleozoic time. When actively forming, the basins accumulated great thicknesses of sediment and the domes formed islands surrounded by shallow seas.

Occasionally rocks of continental interiors undergo small movements along faults that lie far from zones of rifting or mountain building. Most movements of this kind probably result from slight amounts of compression or extension of the continental lithosphere as it moves over the asthenosphere. In 1811 and 1812, powerful earthquakes with epicenters in Missouri issued from movements of this kind along faults that had formed long ago, during Proterozoic time, and had apparently long been dormant. At the time of these huge earthquakes, there were no buildings close enough to their epicenters to suffer severe damage, but as far away as the District of Columbia, the earthquakes cracked sidewalks and rang church bells.

Chapter Summary

How does continental rifting begin, and what environments of deposition does it produce?

Fracturing of continents often begins with the doming of continental crust in several places. Each dome then fractures to form a three-armed rift system. The joining of some of the rift arms may produce a fracture that cuts across the entire continent. Initially, rift valleys form within the continent and receive nonmarine sediment. After rifting is complete, the new continental borders move away from the elevated spreading zone to become passive margins that are flooded by seas and accumulate marine sediment.

How do rocks become folded?

Compressive forces associated with mountain building cause rocks buried deep within Earth to flow and bend to form folds.

How does a mountain chain form when a continental margin encounters a subduction zone?

The igneous arc moves onto the land, producing tall volcanoes. Granitic plutons also form at depth, and their low density leads to isostatic uplift of the mountain chain. In addition, compressive forces of collision and gravity spreading form fold-and-thrust belts on either side of the crystalline core of the mountain chain.

Why does a foreland basin form and accumulate large volumes of sediment on the continent, just inland from the mountain chain?

The weight of the mountain chain and sediment eroded from it causes the continental lithosphere to subside.

Why, when two continents collide, is one not subducted beneath the other?

Continental material is of such low density that it cannot descend into the mantle. As a result, when two continents converge along a deep-sea trench, neither becomes subducted; rather, the two become sutured together, and their deformed crust becomes a mountain chain.

What is the significance of ophiolites?

When a remnant of seafloor, called an ophiolite, is found within a modern continent, it marks the position of an ancient ocean, having been pinched up between two continents when they were sutured together.

What is the zonation of a typical mountain chain, from its axis to its margins?

An igneous arc occupies the central axis of a typical mountain chain. It is bordered on either side by a metamorphic belt, beyond which lies a less intensely deformed fold-and-thrust belt.

How have the Andes formed?

The Andes have risen up as a result of subduction of an oceanic plate along the west coast of South America.

How did the Pyrenees form?

Unlike the Andes, the Pyrenees are the result of collision between two continental landmasses, Iberia and Eurasia.

What is an exotic terrane?

An exotic terrane is a landmass that has become attached to the margin of a continent through suturing of a microcontinent or island arc.

What broad features does rock deformation create in continents far from their margins?

Basins and domes are the most common large structural features of continental interiors.

Review Questions

1. What geologic features enable us to recognize ancient continental rifting? What features enable us to recognize ancient subduction zones?

2. What are failed rifts, and how are they important to our understanding of the breakup of continents? (Hint: Refer to Figure 9-3.)

3. What is flysch and where does it form?

4. What is molasse? Why does it normally accumulate after flysch?

5. How can mountain chains form without continental collision?

6. How does the angle of subduction beneath a mountain chain relate to the rate at which the plate and the continent are moving toward each other?

7. Why do mountains have roots?

8. Are the rocks that become ophiolites within a mountain chain older or younger than molasse deposits that form along the mountain chain?

9. Examine a world map or globe to locate mountain chains that are not discussed in this chapter. Then locate those chains on the plate tectonic map of the world (see Figure 8-21). See if you can figure out how

the presence of each mountain system might relate to plate tectonic processes. (Some of the answers appear in the chapters that follow.)

10. Using the Visual Overview on page 200 and what you have learned in this chapter, trace the history of a continental margin that experiences the following events: (a) it originates by rifting and becomes a passive margin along which sediment accumulates, (b) stalls at a subduction zone, where subduction reverses, (c) grows a mountain chain, (d) becomes a passive margin again when the igneous arc that formed it ceases to function, and (e) eventually loses its root through erosion and isostatic uplift.

A moist temperate forest in Oregon, in which tall, deeply rooted Douglas firs and rapid recycling of water produce rapid weathering. (From David Middleton, *Ancient Forests* © 1992, published by Chronicle Books, San Francisco.)

Major Chemical Cycles

Many environmental changes that affect broad regions of Earth are of a chemical nature—or they are the physical results of chemical changes. The most widely publicized of these environmental changes are the increases in so-called **greenhouse gases**—atmospheric gases that trap warming solar radiation near Earth's surface—caused by human activities. This chapter describes evidence that the atmospheric concentration of carbon dioxide, the most important greenhouse gas, has undergone major changes during the Phanerozoic Eon. During long stretches of geologic time, Earth's climate has been even warmer than it soon will become as a result of human-induced greenhouse warming. These ancient "greenhouse" intervals serve as models for the future Earth system. Humans evolved in a much cooler, "icehouse" world for more than 100,000 years before we began to burn fossil fuels at a rate that now threatens our habitat. Will human activities eventually turn this icehouse world into a greenhouse world?

Just as chemical cycles have affected the levels of greenhouse gases in the atmosphere, they have influenced the chemistry of the ocean in ways that affect marine organisms and the sediments they produce. This chapter explores how changes in rates of plate tectonic activity have altered the chemistry of the ocean, with profound effects on the kinds of organisms that have produced reefs and carbonate sediments.

Chemical Reservoirs

The chemical changes that we will review in this chapter are quite simple. Most are changes in the rates at which key chemical elements and compounds of the Earth system move in huge cycles that carry them through two or more vast reservoirs. These **reservoirs** are bodies of chemical entities that occupy particular spaces. Examples are the total volume of carbon dioxide in the atmosphere and the total volume of glacial ice on land. Reservoirs also include the *biomass* of living organisms (the tissue volume of a group of organisms living in the same area). These reservoirs expand and contract through changes in the rates at which elements or compounds flow to or from them.

Chapter 4 described the cycling of nitrogen and phosphorus—key nutrients for producer organisms—through marine ecosystems. Here we focus on cyclical pathways of two other elements essential to life: carbon and oxygen. All of these large-scale chemical cycles move materials through portions of the water cycle (discussed in Chapter 1).

We will also examine how two isotopes of a particular element can differ in their movements through global chemical cycles. Scientists use these differences as indicators of past conditions and events. The distribution of carbon isotopes in sedimentary rocks and in the fossils they contain, for

Visual Overview
Key Chemical Cycles in Earth System History

Chemical cycles influence Earth's climate as well as the composition of its atmosphere and natural waters.

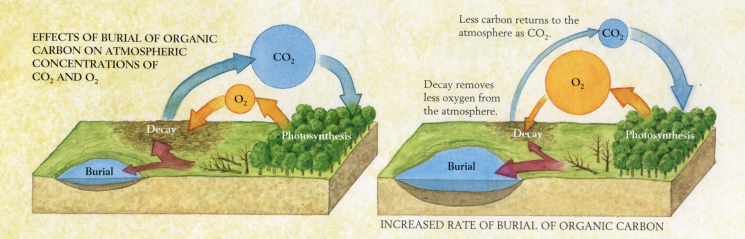

EFFECTS OF BURIAL OF ORGANIC CARBON ON ATMOSPHERIC CONCENTRATIONS OF CO_2 AND O_2

CO_2

O_2

Decay

Photosynthesis

Burial

Less carbon returns to the atmosphere as CO_2.

CO_2

O_2

Decay removes less oxygen from the atmosphere.

Decay

Photosynthesis

Burial

INCREASED RATE OF BURIAL OF ORGANIC CARBON

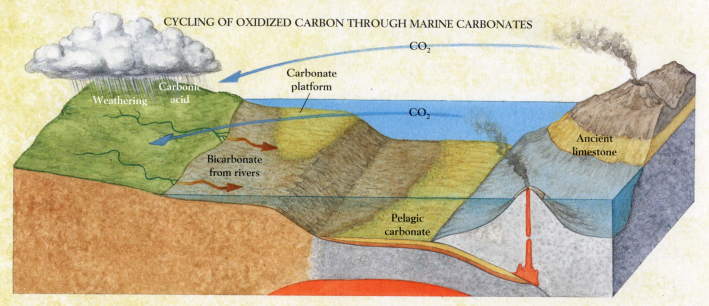

CYCLING OF OXIDIZED CARBON THROUGH MARINE CARBONATES

CO_2

Carbonate platform

Weathering

Carbonic acid

CO_2

Bicarbonate from rivers

Ancient limestone

Pelagic carbonate

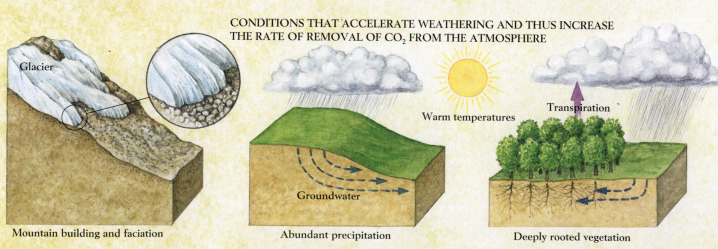

CONDITIONS THAT ACCELERATE WEATHERING AND THUS INCREASE THE RATE OF REMOVAL OF CO_2 FROM THE ATMOSPHERE

Glacier

Mountain building and faciation

Warm temperatures

Groundwater

Abundant precipitation

Transpiration

Deeply rooted vegetation

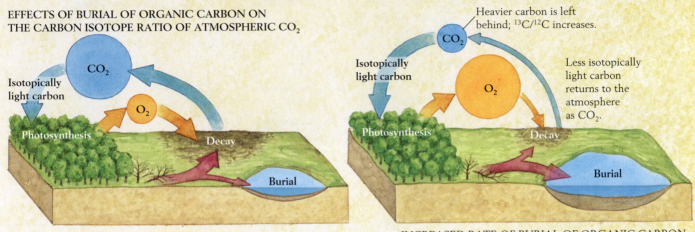

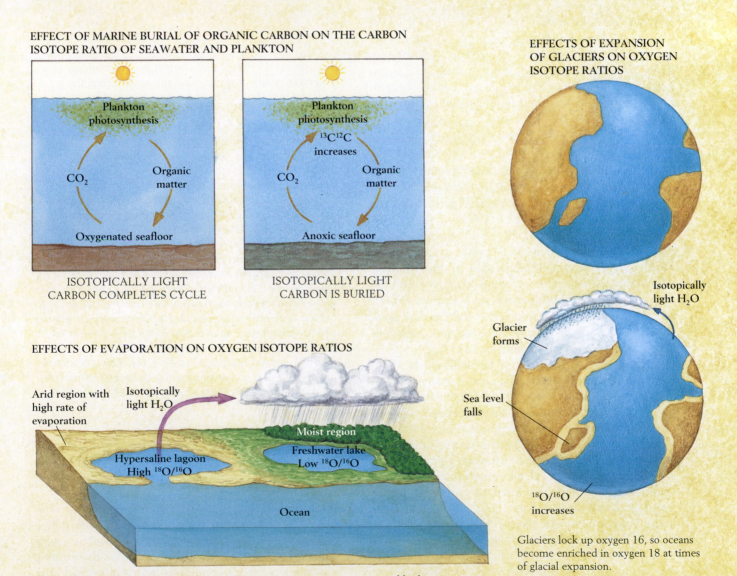

example, gives evidence of changes in the abundance of carbon dioxide (CO_2) and oxygen (O_2) in Earth's atmosphere over the course of hundreds of millions of years. The distribution of oxygen isotopes in marine fossils offers testimony about the temperatures and salinities of ancient seas.

The fundamental aspects of chemical cycles presented in this chapter provide a framework for understanding many environmental and biological events described in later chapters.

Fluxes are rates of movement between reservoirs

When one reservoir for a chemical entity increases in size, one or more other reservoirs for that material must shrink. In the water cycle, for example, the expansion of glaciers on land robs the ocean of water, and sea level drops (p. 18). Reservoirs expand or contract because of changes in the rate at which they gain or lose their contents. In a global chemical cycle, a rate of this kind is termed a **flux**. An everyday example of a flux is the number of gallons of water per minute that a pump draws from a well and sends through a hose to a swimming pool. Another is the number of bushels of corn per week harvested from a field and fed to a herd of farm animals.

Feedbacks affect fluxes

To envision how reservoirs and fluxes operate, imagine a balloon with two openings in it, so that water can flow in at one end and out at the other (Figure 10-1). The water in the balloon increases in volume until its pressure forces water out as rapidly as it is flowing in. The pressure exerted by the balloon operates as a **negative feedback**, opposing the expansion of the reservoir within the balloon more and more strongly, until the fluxes to and from this reservoir are in balance. At this point the volume of the reservoir is stabilized. A **positive feedback** operates in the opposite way, accelerating a change instead of braking it.

A variety of factors operate as feedbacks in global chemical cycles. Forests may be providing an important negative feedback in the global carbon cycle today. During recent decades, humans have increased the concentration of carbon dioxide in the atmosphere by burning wood and fossil fuels. Plants use carbon dioxide in photosynthesis, the process by which they produce their own food. An increase in the concentration of atmospheric carbon dioxide in effect fertilizes plants. If they have not already done so, forests of middle and high latitudes may soon increase their biomass as a result of rising levels of atmospheric carbon dioxide. The biomass of plants is a reservoir for carbon, and an increase in the size of this reservoir will reduce the rate of buildup of

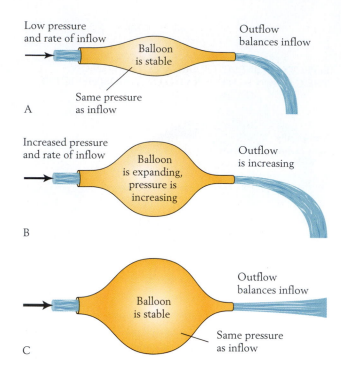

Figure 10-1 How fluxes control the volume of a reservoir. In this imaginary laboratory experiment, the reservoir is a balloon with a tube-shaped opening at each end. Water flows into the balloon through one tube and out of it through another, smaller tube. (*A*) Initially the pressure and rate of inflow are low. Even so, the water cannot escape from the small exit tube as rapidly as it enters the balloon, so the balloon begins to fill with water and expand. As the balloon expands, its rubber stretches and applies increasing backpressure to the water it contains. The balloon continues to expand until its constantly increasing backpressure balances the pressure of the inflowing water. At this point the balloon stabilizes, and the outflow balances the inflow. (It is as if the balloon simply became an extension of the inflow tube.) *B* and *C* show what happens when the pressure and rate of inflow are then increased. (*B*) At first, the newly applied higher pressure is greater than the opposing backpressure from the balloon. The balloon therefore begins to expand, and it continues to expand until its backpressure again equals that of the water inside (*C*). Then the balloon (reservoir) is stabilized at a larger volume than in *A*, and the inflow (flux) to it and the outflow (flux) from it are again equal, but larger than in *A*. Backpressure from the balloon is the negative feedback that has stabilized the system twice (in *A* and *C*).

carbon dioxide in the atmosphere. Thus the fertilization effect of carbon dioxide on plants is a negative feedback against the buildup of this gas in the atmosphere.

As an example of a positive feedback, we can consider what will result from changes in vegetation at high northern latitudes if humans' burning of fossil fuels produces global warming in the future. Climatic warming will cause evergreen coniferous forests to expand their range northward, so that they replace tundra (p. 90). Forests have a lower albedo than tundra, so they absorb more heat from the sun. Thus, as global warming causes these forests to expand, they will create additional global warming.

Carbon Dioxide, Oxygen, and Biological Processes

Living things play major roles in chemical cycles within the Earth system. We have already seen that land plants form an integral part of the water cycle, soaking up moisture with their roots and transpiring large volumes of water vapor to the atmosphere (p. 19). Water is also one of the two raw materials for photosynthesis, the most important process providing food for both plants and animals. The other raw material for photosynthesis is carbon dioxide, whose two constituent elements, carbon and oxygen, move back and forth between plants and animals by way of the *photosynthesis-respiration cycle*. First we will examine the factors that control the abundances of oxygen and carbon dioxide in the atmosphere. Then we will consider how these abundances may have changed in the course of geologic time.

Plants employ a photosynthesis-respiration cycle

The biological processes of photosynthesis and respiration are opposites. Whereas photosynthesis employs energy from the sun, **respiration** releases energy (Figure 10-2). In photosynthesis plants combine carbon dioxide and water to form what we can loosely call *sugars*—compounds

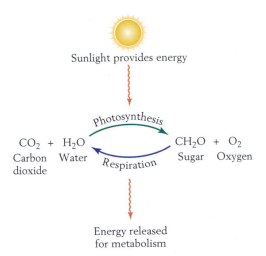

$$CO_2 + H_2O \quad\xrightarrow{\text{Photosynthesis}}\quad CH_2O + O_2$$

Carbon dioxide   Water   Respiration   Sugar   Oxygen

Sunlight provides energy

Energy released for metabolism

Figure 10-2 The photosynthesis-respiration cycle in plants. This cycle turns energy from sunlight into energy that can be used in metabolism. The cycle requires no input of carbon, hydrogen, or oxygen.

of carbon, hydrogen, and oxygen. Having consumed plenty of sugars, we are all familiar with the large amount of energy they contain. When plants conduct photosynthesis, sunlight provides the energy for the formation of sugars, and this energy ends up stored within the sugars. Oxygen is a by-product of photosynthesis.

Respiration moves in the opposite direction, combining sugars and oxygen to release the sugars' energy. Organisms use this energy to fuel their metabolism. Carbon dioxide and water, the raw materials of photosynthesis, are the products of respiration. Because oxygen is added to the elements of sugars in respiration, the process is an example of *oxidation*. Fire, which is very rapid oxidation, illustrates how this process releases energy. *Organic compounds* contain relatively small amounts of oxygen, or none at all, and are referred to as "reduced carbon" compounds; the carbon atoms within them are described as **reduced carbon**. Sugars are reduced carbon compounds, and *reduction* is the process of oxygen removal that produces them.

Plants employ the photosynthesis-respiration cycle because they are unable to use sunlight directly for their metabolism. They are able to store the energy of sunlight in sugars, however, and, through respiration, they can later release it in a usable form. Although energy is released less rapidly in respiration than in fire, we can refer to it as *burning* of sugars.

Photosynthesis produces tissue growth

Note that the photosynthesis-respiration cycle in a plant harnesses energy from the sun and releases it for the plant's use without gaining or losing chemical components (see Figure 10-2). Thus there is no net exchange of CO_2 or O_2 between the plant and the atmosphere. The plant has to conduct more photosynthesis than it needs for energy, however. It needs sugars to build tissues while growing and to produce spores or seeds. If some of the sugars it manufactures are to become plant tissue—a reservoir for carbon—the plant cannot destroy those sugars through respiration. When the plant builds sugars and does not burn them, the CO_2 and water that it uses in forming them are not returned to the atmosphere, but are stored within the plant's tissues as sugars and other compounds derived from sugars that contain carbon, hydrogen, and oxygen. In addition, because plants do not use O_2, the by-product of photosynthesis, to burn these stored sugars, they release it to the atmosphere. Thus it is because plants grow leaves, stems, roots, and reproductive structures that they remove CO_2 from the atmosphere and contribute O_2 to it.

Almost every bit of plant tissue has one of three final destinies: it can be eaten by an animal, it can decompose, or it can be buried in sediment.

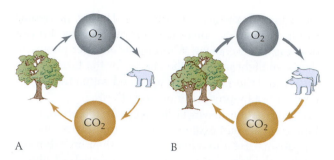

Figure 10-3 The photosynthesis-respiration cycle involving both plants and animals. *A*. In this cycle, oxygen and carbon dioxide move through atmospheric reservoirs (balloons) between plants and animals. *B*. Doubling the biomass of both plants and animals doubles the volume of all of the fluxes (arrows) both to and from each atmospheric reservoir, but leaves the volume of each reservoir unchanged.

Respiration releases energy

For humans, as for all other animals, respiration is a process in which gases are exchanged with the environment. We take in O_2 when we inhale and release CO_2 when we exhale.

Animals employ respiration to gain energy from the sugars of the plants they eat. Thus, as Figure 10-3*A* illustrates, animals form a photosynthesis-respiration cycle with plants like the one that plants themselves employ in manufacturing and burning sugars. Plants use as much CO_2 and H_2O to make sugars eaten by animals as the animals release in the respiration that liberates the sugars' energy. Similarly, animals use the same amount of O_2 to burn the sugars that the plants liberate in producing them. In this cycle, unlike the comparable one that occurs entirely within plants, CO_2 and O_2 pass through the atmospheric reservoir as they move between plants and animals. Even so, because the exchange between plants and animals is in balance, the volume of O_2 and CO_2 in the atmospheric reservoir is unaffected by the cycle.

If the number of plants increases, providing more food for animals, the number of animals will increase in proportion—as will all the fluxes of O_2 and CO_2 between the organisms and the atmosphere. If the biomasses of plants and animals double, for example, so will the various fluxes of O_2 and CO_2 to and from the atmosphere. The sizes of the atmospheric reservoirs of these gases will therefore remain unchanged (Figure 10-3*B*).

Of course, the food web extends beyond herbivores to carnivores. Carnivores consume the protein and fat of other animals, rather than sugars, but these compounds have ultimately been manufactured from sugars, and carnivores oxidize them to obtain metabolic energy in the same way that herbivores oxidize sugars. Thus the photosynthesis-respiration cycle extends to the top of the food web without essential modification.

Bear in mind that some animals—endothermic forms (p. 75)—normally maintain a body temperature higher than the temperature of their environment. To accomplish this feat, mammals and birds must consume more food and respire at rates higher than the normal rates of ectothermic animals, such as fishes, amphibians, and reptiles.

Decomposers employ respiration

To this point we have ignored an important component of ecosystems: the *decomposers*, which break down dead organic debris not consumed by animals. The most important decomposers are bacteria and fungi. Decomposers, like animals, use respiration to break down tissues. They simply consume dead rather than living organic matter. Thus decomposers too extract O_2 from the atmosphere and release CO_2 into it (Figure 10-4).

There is plenty of room in the ecosystem for decomposers of dead plant tissue because animals do not harvest the entire biomass of living plants. Animals' consumption of plants is limited by their lack of access to certain plant structures and their inability to eat and

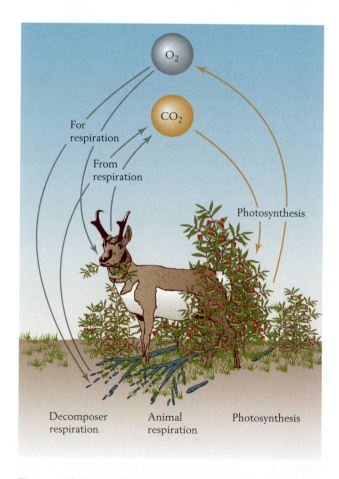

Figure 10-4 Respiration of plant material. The total amount of consumption and respiration in nature is divided in variable proportions between animals and decomposers.

digest some parts of plants. In addition, predation or limiting factors imposed by the physical environment often keep animal populations below the levels that the biomass of plants can potentially support.

Simply put, some plant tissues are consumed by animals and others are consumed by decomposers when the plants die. Both kinds of consumption entail respiration. Therefore, if populations of herbivorous animals decline, more plant tissue is left for decomposers, and these organisms contribute more total respiration. Conversely, if herbivore populations expand, they assume a larger portion of the total amount of respiration. All of the plant material *that is not somehow removed from the ecosystem* ends up as part of the photosynthesis-respiration cycle that includes plants and the organisms that destroy their tissues.

Burial of plant debris alters atmospheric chemistry

The italicized portion of the previous sentence provides the key to understanding long-term changes in the abundances of CO_2 and O_2 in Earth's atmosphere. Some dead plant debris escapes from the photosynthesis-respiration cycle through burial—burial in swamps, for example, or in sediments along the margins of continents. Buried organic matter, including that contained in coal, constitutes a reservoir for reduced carbon compounds (Figure 10-5). Just as sedimentation is always burying reduced carbon, however, erosion is always exposing it to the atmosphere, where it is soon oxidized by decomposers or inorganic processes. Over long

stretches of geologic time, the rate of burial of organic carbon has nearly balanced its rate of exposure by erosion, and the reservoir of buried carbon has undergone little net change in volume. In other words, the cycle of burial and erosion has been in balance, like the photosynthesis-respiration cycle of plants, animals, and decomposers. As a result, levels of CO_2 and O_2 in the atmosphere have been relatively stable.

Occasionally the system is thrown out of balance, however, when the overall rate of carbon burial increases. Atmospheric CO_2 was used to produce any reduced carbon compounds that are buried. Thus, in effect, excess burial of these compounds removes CO_2 from the photosynthesis-respiration cycle—specifically from the atmospheric reservoir—and stores it in a subterranean reservoir. As a result, the concentration of CO_2 in Earth's atmosphere declines.

What happens to oxygen when the overall rate of carbon burial increases? It becomes more abundant in the atmosphere (Figure 10-6). Had the large volume of reduced carbon compounds not been buried, decomposers would have oxidized it through respiration. Because these carbon compounds have been buried, the oxygen that would have oxidized them remains in the atmosphere.

A large global change in the rate of burial of organic matter powerfully alters the concentration of atmospheric CO_2 and O_2. Where do reduced carbon compounds accumulate in large quantities? The most important sites are anoxic bodies of water. An **anoxic** condition (or **anoxia**) is the virtual absence of O_2. Anoxia allows debris from dead plants to survive on the

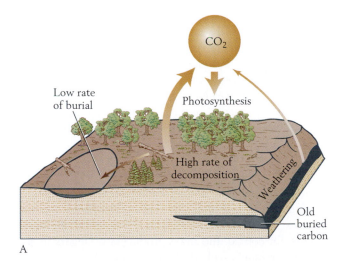
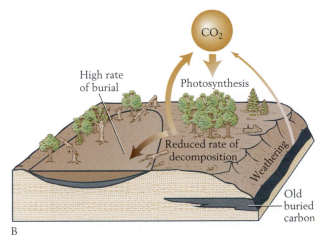

A

B

Figure 10-5 **The effect of the global rate of organic carbon burial on the size of the atmospheric reservoir of CO_2.** *A.* Initially, the rate of carbon burial (left) balances the rate of weathering of buried carbon (right). *B.* Later, swamps expand and vegetation adapted to them flourishes. The increased rate of burial of dead plant material in swamps leaves less carbon to be returned to the atmosphere, through decomposition, as CO_2. As a result, the atmospheric reservoir of CO_2 shrinks.

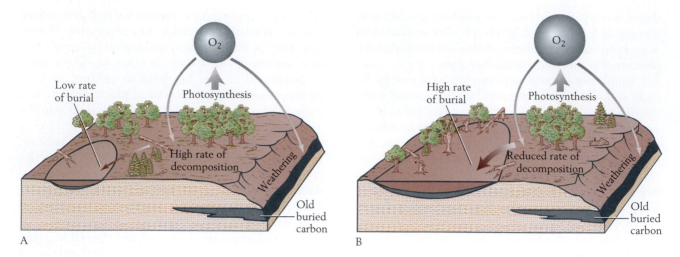

Figure 10-6 The effect of the global rate of organic carbon burial on the size of the atmospheric reservoir of O_2. *A.* Initially, the rate of carbon burial (left) balances the rate of weathering of buried carbon (right). *B.* Later, as in Figure 10-5, swamps expand and vegetation adapted to them flourishes. An increased rate of burial leaves behind less dead plant material to use up oxygen through decomposition. As a result, the atmospheric reservoir of O_2 grows.

floor of a body of water, and eventually to become deeply buried, without having been fully decomposed. The reason is quite simple: the organisms—mostly bacteria—that cause decomposition by respiration require oxygen to do so. In environments where these organisms cannot live, organic matter survives and accumulates with silt and clay. Conditions of this kind have provided for exceptional preservation of fossils, sometimes allowing remnants of soft tissue to become fossilized (p. 52). Today debris from sphagnum moss and other plants accumulates in bogs in cold climates. The bottom waters of these bogs, being not only anoxic but also highly acidic, are inhospitable to the kinds of bacteria that cause decomposition. If deeply buried, the peat that accumulates within them will eventually turn to coal (p. 46). If they remain near the surface, they can also be exploited for use as fuel, or they can instead be used for enrichment of garden soils.

When anoxic conditions become widespread on Earth, burial of plant debris can form large reservoirs of reduced carbon compounds within Earth's crust. At times when deep portions of the oceans became anoxic or when swamps with stagnant, anoxic bottom waters expanded to occupy large areas of continents, these swamps became burial sites for large volumes of organic matter from trees, some of which has become coal.

The Carboniferous Period was a time when primitive trees were buried extensively in what have come to be called *coal swamps*. The remains of the trees accumulated on the anoxic bottoms of the swamps, where they turned into peat. This peat eventually became coal,

which gave the Carboniferous its name and which provides humans with large quantities of fossil fuel.

At times when organic matter is buried rapidly, there is no reason why the rate at which older buried carbon is exposed should increase. In fact, because erosion occurs throughout the world and uncovers buried carbon of many ages, different rates in different areas tend to average out: there is little overall change through time. The rate of burial is more unstable because a global change in climate can alter floras and thereby expand swampy areas, drastically changing rates of burial of organic matter throughout the world. Thus, given little change in the rate at which buried carbon is exposed to oxidation, a large, long-term increase in the flux of organic debris to the buried carbon reservoir increases the size of this reservoir substantially. The result is a substantial reduction of the CO_2 in Earth's atmosphere and a corresponding increase in O_2 (see Figures 10-5 and 10-6).

Marine cycles resemble terrestrial cycles

Our discussion thus far has focused on nonmarine plants and animals, but the same principles hold for the oceans, where dissolved CO_2 is available for photosynthesis. Of course, the main producers of sugars in the oceans are single-celled planktonic algae.

Animals consume a percentage of the biomass of marine plankton that is larger than the percentage of plant biomass that animals are able to consume on land. (Consider, for example, how difficult it is for animals to eat wood or roots and how many leaves fall to the forest floor uneaten.) Thus a smaller proportion of the sugars produced by marine photosynthesis is available for either decomposition or burial. Terrestrial plants, however, contribute additional dead tissue to marine environments. Rivers carry large amounts of partly decomposed plant tissues from land to lagoons, deltas, or the continental shelf and slope. Those partly decomposed plant materials that are not eaten by marine animals or totally decomposed by bacteria accumulate in muddy sediments.

Organic matter can also become buried farther from shore. In the modern oceans, cold, dense surface water sinks near both of Earth's poles, spreading throughout the deep sea and supplying it with oxygen (p. 95). At certain times in the past, however, polar regions were relatively warm, and water in their vicinity did not descend. At these times, the deep sea was relatively stagnant and anoxic. With few bacteria to cause decomposition, much of the organic matter from dead phytoplankton that settled on the deep-sea floor was buried. This burial, like that of terrestrial plant debris in swamps, transferred carbon from the atmosphere to the reservoir of buried carbon.

In fact, the abyssal plain of the ocean is never the site of rapid carbon burial. It lies far from the zones of upwelling that lie along continental margins, where the productivity of phytoplankton is highest and where plant debris from the land also accumulates in the sediments

of lagoons and deltas. At certain times, however, low-oxygen conditions extended upward to such shallow depths that even the floors of large epicontinental seas became largely anoxic. Organic-rich muds accumulated in these settings. These muds eventually became shales that were black because they contained a large amount of carbon. Figure 10-7 shows the locations of deposits of this type that formed in mid-Cretaceous time, slightly more than 100 million years ago. High temperatures and pressures later altered some of the organic matter buried in these Cretaceous settings to liquid and gaseous compounds. These fluids migrated through porous rocks and became trapped in underground reservoirs that have provided humans with large quantities of petroleum and natural gas.

Use of Carbon Isotopes to Study Global Chemical Cycles

Carbon isotopes serve as tools for tracing the history of some aspects of atmospheric chemistry. Recall that isotopes are varieties of chemical elements whose atoms differ only in the number of neutrons they contain. Thus an atom of carbon 13 has one more neutron, and is heavier, than an atom of carbon 12. These two isotopes of carbon are *stable isotopes*—they do not spontaneously decay in the manner of radioactive isotopes, such as those used to date rocks (p. 30). The relative proportion of one isotope, such as carbon 13 (abbreviated ^{13}C), in a specimen is usually specified by the symbol δ, which relates the isotopic composition of the specimen to that of a standard specimen to which all others are compared. This relative proportion is expressed as parts per thousand (‰).

Two molecules of a particular compound that contain different isotopes of a given element will behave slightly differently as they flow through the Earth system. Our first concern will be with carbon. Carbon isotopes in sedimentary materials offer evidence of changes in the composition of Earth's atmosphere. To understand why, we must understand how carbon 12 and carbon 13 move through carbon reservoirs in slightly different ways.

Most molecules of CO_2 in the atmosphere contain carbon 12; a smaller proportion contain carbon 13. When organisms of any kind conduct photosynthesis, they employ a proportion of carbon 12 slightly larger than that found in the atmosphere because this isotope is relatively light. As a result, all plant tissue contains a disproportionately large percentage of carbon 12. Therefore, the distribution of carbon isotopes in organic sedimentary materials allows scientists to identify past changes in the concentrations of CO_2 and O_2 in Earth's atmosphere.

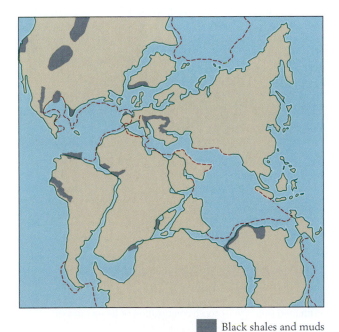

■ Black shales and muds

Figure 10-7 Locations of black shales and muds deposited in anoxic marine waters of mid-Cretaceous epicontinental seas. (After A. G. Fischer and M. A. Arthur, *Soc. Econ. Paleontol. Mineral. Spec. Publ.* 25:19–50, 1977.)

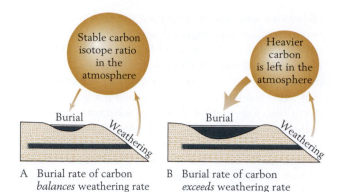

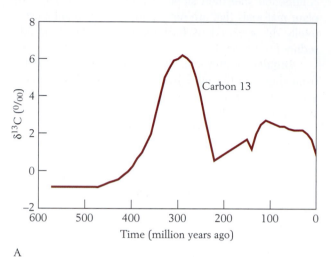

A   Burial rate of carbon *balances* weathering rate   B   Burial rate of carbon *exceeds* weathering rate

Figure 10-8 The effect of the rate of carbon burial on the isotope ratio of carbon in atmospheric CO_2. *A.* When the rate at which organic carbon is buried equals the rate at which weathering returns organic carbon to the atmosphere as CO_2, burial of carbon does not alter the isotope ratio of carbon the atmospheric CO_2. *B.* When organic carbon is buried at a much higher rate, however, the total reservoir of buried organic carbon, which is isotopically light, expands; conversely, the atmospheric reservoir of CO_2 shrinks and becomes isotopically heavier.

Carbon isotopes record the cycling of organic carbon

The isotopically light carbon (carbon 12) that plants preferentially extract from the atmosphere is returned to the atmosphere through the respiration of animals and decomposers (as in Figure 10-4). Thus the photosynthesis-respiration cycle has little effect on the isotope ratio of CO_2 in the atmosphere. The same is true for burial of isotopically light organic carbon, as long as the carbon is returned to the atmosphere by weathering of organic matter as rapidly as it is added to the underground reservoir by burial—that is, as long as the system is in balance.

The system is *not* always in balance, however. At certain times carbon can be buried rapidly even though weathering rates remain unchanged (see Figure 10-5). At these times, a relatively large proportion of the carbon 12 from the atmosphere becomes locked up in the reservoir of buried organic matter, leaving the atmosphere with an elevated ratio of carbon 13 to carbon 12 (Figure 10-8).

At any time in Earth's history when the rate of carbon burial increases on continents or beneath the sea, both CO_2 in the atmosphere and CO_2 dissolved in the ocean become depleted of carbon 12 and therefore enriched in carbon 13. Both reservoirs are affected because the movement of carbon between them is rapid on a geologic time scale. Exchange between these two reservoirs results largely from movement of CO_2 back and forth between the atmosphere and surface waters of the ocean. Changes in the abundance and isotopic composition of CO_2 are spread rapidly through the atmosphere by winds and vertical air movements and through the upper ocean by large-scale water currents.

Isotope ratios in limestones and deep-sea sediments record changes in rates of carbon burial

Because of the rapid mixing between atmospheric and oceanic carbon reservoirs, carbon isotope ratios in

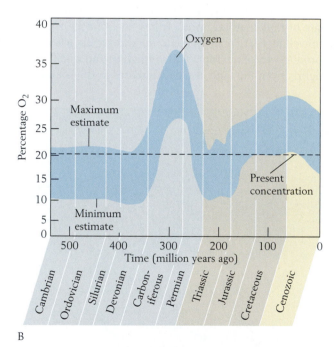

A

B

Figure 10-9 Estimated changes in the size of the atmospheric reservoir of oxygen during the Phanerozoic. *A.* Changes in the relative percentage of carbon 13 in seawater, estimated from the isotopic composition of limestones of various ages. *B.* Estimated changes in the fraction of Earth's atmosphere that consisted of free oxygen. This estimate is based on the carbon content of sediments; an increase in the rate of carbon burial causes oxygen to build up in the atmosphere. The broad band depicts uncertainties in calculations. (*A* from R. A. Berner, *Amer. J. Sci.* 287:177–196, 1987; *B* from R. A. Berner and D. E. Canfield, *Amer. J. Sci.* 289:333–361, 1989.)

shallow-water limestones provide a record of changes in carbon isotope ratios in the atmosphere as well as in the ocean.

Figure 10-9A is a plot of the carbon isotope ratios in the calcium carbonate ($CaCO_3$) of limestones over the entire Phanerozoic Eon. The most conspicuous feature of this plot is the large increase in the relative percentage of carbon 13 in the latter part of the Paleozoic. Limestones with the highest values of carbon 13 formed in Late Carboniferous time, when large volumes of carbon accumulated in coal swamps. Because the buried carbon was isotopically light, an excess of carbon 13 was left in the atmosphere. As a result, $CaCO_3$ precipitated in the ocean became isotopically heavy.

The carbon isotope ratios of phytoplankton, the primary photosynthesizers of the open ocean, follow a similar pattern. When high burial rates of organic matter, which is isotopically light, leave behind an excess of carbon 13 in the CO_2 of the atmosphere and ocean, marine phytoplankton that use this CO_2 for photosynthesis assimilate a higher proportion of isotopically heavy carbon. When waters of the deep sea become anoxic, for example (Figure 10-10), the carbon in organic matter accumulating on the seafloor becomes increasingly heavy. In other words, a sudden pulse in the carbon isotope ratio toward heavy values in the marine stratigraphic record can indicate an episode of increased burial of isotopically light organic carbon.

Carbon burial enlarges the atmosphere's oxygen reservoir

We have seen that an increase in the rate of organic carbon burial causes oxygen to build up in the atmosphere (see Figure 10-6). Conversely, a decrease in the rate of burial leaves more carbon behind to be oxidized by decomposers; increased respiration then pulls oxygen out of the atmosphere.

Because changes in carbon isotope ratios in marine carbonates provide a continuous record of overall rates of carbon burial, these ratios provide a general picture of rising and falling oxygen in the atmosphere (Figure 10-9B). A shift of the carbon in limestones toward heavier values, for example, reflects burial of more organic carbon, which is isotopically light. The fact that this carbon was buried rather than decomposed must have caused oxygen to accumulate in the atmosphere.

We cannot reconstruct oxygen levels in the atmosphere precisely from the graph displayed as Figure 10-9A because secondary factors also come into play. We can nonetheless conclude that atmospheric oxygen reached its highest Phanerozoic level during the late Paleozoic interval, when so much carbon was buried in coal swamps (see Figure 10-9B). Some estimates suggest that the level of oxygen then was about twice what it is now.

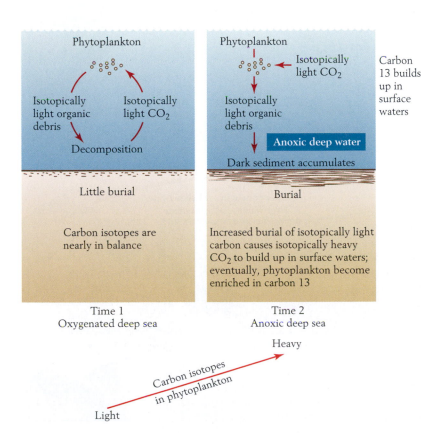

Figure 10-10 Influence of the rate of organic carbon burial on carbon isotope ratios in phytoplankton and sedimentary organic matter. Debris from phytoplankton is isotopically light. When deep water becomes anoxic at time 2, most of the organic debris that sinks is buried. As a result, decomposition and upwelling no longer return its light carbon to the upper ocean, as they did at time 1. Thus the CO_2 that remains in the upper ocean becomes enriched in carbon 13. Because phytoplankton conduct their photosynthesis there, they become isotopically heavier, as does the organic debris that sinks to accumulate in sediments when the phytoplankton die. Thus an increase in the proportion of carbon 13 in the geologic record may reflect an increased rate of carbon burial.

Carbon dioxide is removed from the atmosphere by weathering and ends up in limestone

Of course, burial of organic carbon not only enlarges the atmospheric reservoir of O_2, but also shrinks the atmospheric reservoir of CO_2. It does so by shifting reduced carbon to the reservoir of buried organic carbon (see Figure 10-5). But whereas the level of O_2 in the atmosphere depends largely on the rate of burial of organic carbon, other factors exert a strong influence over levels of atmospheric CO_2. Especially large changes in the concentration of CO_2 result from chemical reactions that create and destroy minerals in soils and rocks near Earth's surface.

Figure 10-11 shows how chemical processes apart from photosynthesis and respiration (p. 225) cycle oxidized carbon through the atmosphere, ocean, and solid Earth. These processes form part of the rock cycle (outlined in Chapter 1). Chemical weathering processes on land use up CO_2 from the atmosphere to break down rocks. A primary agent in these processes is CO_2 that has combined with water in Earth's atmosphere to form carbonic acid (H_2CO_3). Carbonic acid attacks both limestone ($CaCO_3$) and silicate rocks, releasing both positively charged ions of calcium (Ca^{2+}) and negatively charged combinations of carbonate and hydrogen known as bicarbonate ions (HCO_3^-). Dissolved in water, these chemical species travel in rivers to the ocean, where they recombine to form $CaCO_3$. As we have seen, some of this $CaCO_3$ is precipitated by inorganic processes in warm seas, and some is secreted by organisms in the form of skeletons. Thus some limestones are bioclastic sediments and some are chemical sediments (p. 42).

Eventually metamorphism and melting complete the cycle by breaking down calcium carbonate sedi-

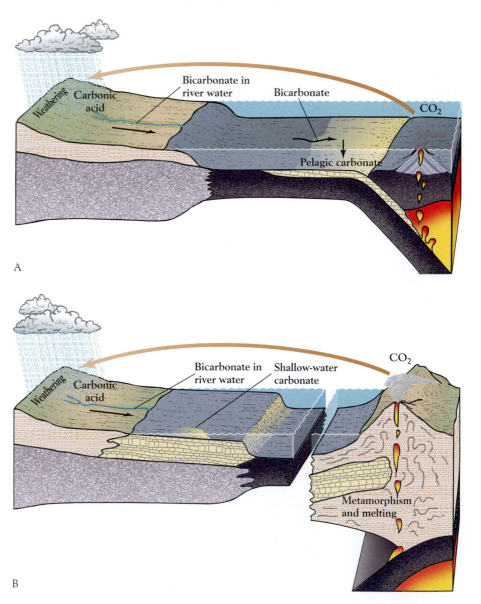

Figure 10-11 The cycling of oxidized carbon through marine limestones. Bicarbonate formed by weathering is carried by rivers to the ocean, where it can become pelagic carbonate that accumulates offshore (A) or shallow-water carbonate, most of which accumulates on nearshore platforms (B). Metamorphism or melting of these carbonate sediments releases CO_2, which completes the cycle by entering into weathering processes and forming bicarbonate. Pelagic carbonates are readily destroyed by subduction along trenches near their place of origin (A). Shallow-water carbonate platforms, in contrast, release their CO_2 only if they are caught up in mountain building (B); this may not happen for hundreds of millions of years after they form and has never happened to many very old shallow-water carbonates.

ments, extracting CO_2 and releasing it to the atmosphere. Calcium carbonate sediments accumulate mainly in two distinct settings: carbonate platforms, such as the Great Bahama Bank (p. 122), and deep-sea settings, where minute skeletons of plankton accumulate as fine-grained pelagic carbonates (p. 125). The two separate reservoirs of carbonate sediment that form in this way—one in deep water and one in shallow water—are subject to different kinds of metamorphism. Many pelagic carbonates are eventually subducted into the asthenosphere, riding on the surface of oceanic crust. Their destruction liberates CO_2, which escapes into the atmosphere through volcanoes that form along subduction zones (see Figure 10-11A). Carbonate platforms, in contrast, are thick bodies of low density that cannot be subducted. Only by becoming caught up in mountain building are they likely to undergo metamorphism and contribute CO_2 to the atmosphere (Figure 10-11B).

In short, most carbonate sediments and rocks ultimately undergo metamorphism and return their CO_2 to the atmosphere along subduction zones. This liberated CO_2 is available to take part in weathering, which produces HCO_3^- that returns to the ocean. There, $CaCO_3$ forms from the HCO_3^- through direct precipitation and skeletal secretion by organisms, completing the global cycle for oxidized carbon. In contrast to bodies of pelagic carbonate, shallow-water carbonate platforms amount to reservoirs where carbon extracted from Earth's atmosphere is stored for long geologic intervals.

Changes in rates of weathering affect the atmospheric carbon reservoir

We have traced the movement of oxidized carbon through a great cycle, but an important question remains: What causes one key segment of this cycle—the atmospheric reservoir—to expand and contract? As we have seen, CO_2 and other greenhouse gases trap solar radiation near Earth's surface, in a manner similar to the glass in a greenhouse. This **greenhouse effect** makes changes in the atmospheric concentration of CO_2 a prime factor in the warming and cooling of Earth's climate. Weathering of calcium and magnesium silicate rocks is the primary process that removes CO_2 from the atmosphere. As we shall see shortly, change in the global rate of weathering appears to have been the most important factor in the largest drop in atmospheric CO_2 concentration of the entire Phanerozoic Eon.

Mountain building and weathering The elevation of a large mountain range is one geologic change that accelerates weathering. The steep slopes of a mountain range undergo rapid weathering and erosion because of the strong influence of gravity and the activity of mountain glaciers (Figure 10-12). Glaciers accelerate chem-

Figure 10-12 **A glacier in the Alps.** The glacier is grinding up the rock beneath it, accelerating the process of chemical weathering. (Farrell Grehan/Photo Researchers.)

ical weathering by grinding up rock so as to increase the surface area exposed to weathering. Chemical weathering extracts a large volume of CO_2 from the atmosphere. As we have seen, most of this CO_2, along with calcium released by the same weathering process, ends up forming marine carbonate sediments. Tens or hundreds of millions of years must pass before these sediments reach the metamorphic settings of mountain belts where their gases can be released. Thus CO_2 from Earth's atmosphere remains bound for a considerable time in vast reservoirs of shallow-water carbonate sediments, to be released only sporadically by mountain building.

Temperature and weathering Like many other chemical processes, weathering speeds up as the temperature rises. As a result, chemical weathering occurs more rapidly, on average, in tropical climates than in cold climates. Thus rates of weathering have increased at times when Earth's climates have been warm.

Precipitation and weathering Water flowing through soils and porous rock is responsible for most chemical weathering. In any region, then, the rate of chemical weathering will vary with the amount of precipitation that falls, including snow that melts. The dry climate of the Colorado Plateau, for example, produces such low rates of chemical weathering that limestone forms broad upland surfaces and steep cliffs, such as those that rim the Grand Canyon (see Figure 1-5). In the moist Appalachian region of the eastern United

States, in contrast, bodies of limestone tend to weather preferentially, so limestone floors valleys instead of capping uplands.

At some times in the geologic past moist conditions have been widespread on Earth and arid environments have been confined to small areas. At other times, moist climates have been more restricted and arid environments more widespread. The overall rate of weathering of Earth's surface has been rapid when moist conditions have extended over much of the globe.

What factors influence global patterns on so large a scale? One is the distribution of continents. When large continental areas occupy the central tropics, for example, moist terrestrial climates are relatively widespread (p. 88). On the other hand, when large continental areas lie in the trade wind belt, arid terrestrial climates are relatively widespread.

The size of continents also plays an important role in the weathering process. Much of the precipitation that falls on continents returns to the ocean by way of rivers, and continents rely on wind-borne moisture from the ocean to replenish this runoff. For this reason, regions of large continents that lie far inland tend to be dry. It follows that a large portion of a huge continent is likely to be arid. When a continent is so small that all portions of it lie close to the ocean, it tends to receive a large amount of precipitation for its size.

At certain times in Earth's history, most of its continental lithosphere was united to form one or a very few large continents. When such continents were not extensively invaded by shallow seas, arid conditions were widespread and global rates of weathering were relatively low.

Vegetation and weathering The effects of plentiful precipitation on weathering are greatly amplified by a key biological factor. Abundant moisture allows forests to occupy the land, and the roots of large plants accelerate chemical weathering. Through their roots, land plants secrete acids and other compounds that break down minerals. Furthermore, forests trap water in a local cycle that moves it through the soil repeatedly. This cycle results from plants' transpiration of water into the atmosphere—photosynthesis uses only a small percentage of the water that their roots absorb. Much of the transpired moisture forms clouds that hover above the forest. Eventually the clouds release the moisture as rain, which soaks into the forest soil, completing the local cycle. This cycling results in the rapid dissolution of soil minerals. Studies of the chemical products of weathering in modern streams reveal that the rate of weathering is typically about seven times higher in a forested area than in a nearby barren area.

Extensive vegetation on Earth's surface has accelerated weathering and extracted CO_2 from Earth's atmosphere at some times in the geologic past. At other times, the aridity of much of the terrain has reduced rates of weathering.

Phanerozoic Trends in Atmospheric CO_2

By taking into account factors of the kinds we have just discussed, scientists have created computer models to estimate historical levels of CO_2 in Earth's atmosphere. Figure 10-13 depicts the results of such a model. Let us examine the major features of this model, recognizing that it may require changes as new information appears. By assessing the net effects of a variety of factors that have influenced atmospheric CO_2, the model provides an estimate of CO_2 levels in Earth's atmosphere since the start of the Cambrian.

The most conspicuous feature of Figure 10-13 is a sharp decline in CO_2 during the latter part of the Paleozoic Era, beginning in the Devonian Period. This decline results largely from a substantial increase in the rate of weathering estimated for terrestrial environments. Chemical weathering is believed to have intensified during Devonian time because of evolutionary changes in land plants that allowed them to inhabit upland environments, rather than simply occupying swamps and fringes of lakes and rivers. This evolutionary step,

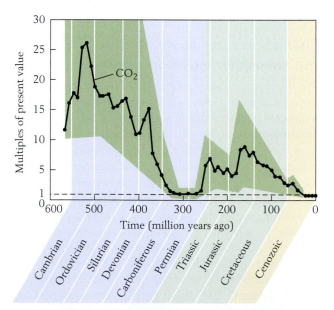

Figure 10-13 Plot of results of a computer model created to estimate changes in atmospheric concentrations of CO_2 during Phanerozoic time. The plotted values are multiples of the present value, which is designated as 1. The shaded area shows the estimated range of uncertainty. The severe decline during Devonian and Carboniferous time is the most conspicuous feature of the plot. (After R. A. Berne and Z. Kothavala, *Amer. J. Sci.* 301:182–204, 2001.)

which we will examine in greater detail in Chapter 14, allowed forests to spread over broad areas of Earth's surface for the first time. In newly forest-clad regions, weathering rates should have increased to about seven times their previous levels (p. 234). Incorporation of this change into the model for a time when other factors underwent little change produced the dramatic drop in atmospheric CO_2 estimated for the Devonian Period.

Note, however, that Figure 10-13 shows the late Paleozoic decline in atmospheric CO_2 continuing into the Carboniferous Period, after upland forests had become widespread. In this model, the continued decline results from another change that we have already discussed: an increase in the rate of burial of organic carbon when coal swamps became widespread (p. 225). Burial of organic material that was produced by photosynthesis depleted the atmospheric reservoir of CO_2 by transferring carbon to a buried reservoir rather than allowing respiration to return it to the atmosphere as CO_2 (see Figure 10-5). Recall that the resulting shift toward heavier isotope ratios for CO_2 in the atmosphere and oceans is recorded in marine limestones (see Figure 10-9A).

The large total decline in atmospheric CO_2 calculated for the Devonian and Carboniferous would have greatly weakened the greenhouse warming of Earth's surface and lower atmosphere. Reduced temperatures resulting from this change probably contributed to another phenomenon of the Carboniferous: the expansion of glaciers across broad areas of the Southern Hemisphere.

Let us step back for a moment and note the very high estimate of atmospheric CO_2 for the early portion of the Paleozoic Era. Might such a high concentration of CO_2, through its powerful greenhouse effect, have produced exceedingly warm temperatures on Earth? Indeed, climates may have been relatively warm, on average, but an opposing factor was the relative weakness of solar warming. The sun's output of radiation has increased progressively since early in the history of the solar system. Weaker solar warming during the first half of the Paleozoic may have more or less offset the powerful greenhouse effect operating at that time. As a result, the mean annual temperature of the early Paleozoic Earth may not have differed greatly from that of the Mesozoic Earth.

The model produces a modest rise in atmospheric CO_2 after the late Paleozoic decline. This rise results largely from two components of the model. One is the rate of mountain building. Because mountain building was not very extensive on a global scale during the Mesozoic Era, the overall rate of weathering on Earth was low—and this reduced the rate at which weathering removed CO_2 from the atmosphere.

The second factor that elevates the Mesozoic level of atmospheric CO_2 in the model is an increase in the amount of calcium carbonate that accumulated on the deep-sea floor relative to the amount that accumulated on nearshore carbonate platforms (see Figure 10-11). This increase resulted from the evolutionary origin of calcareous nannoplankton and planktonic foraminifera, which have formed calcareous oozes in the deep sea only during the past 150 million years or so (see Figure 5-35). Once these oozes became plentiful, their subduction increased the rate at which metamorphism released CO_2 to the atmosphere.

Feedbacks in the Carbon Cycle

Positive feedbacks within the carbon cycle accentuate changes in the atmospheric reservoir of carbon, and negative feedbacks stifle such changes.

Release of frozen methane provides positive feedback

In a molecule-for-molecule comparison, methane (CH_4) is about sixty times as effective as CO_2 in warming Earth's atmosphere. At certain times in Earth's history, methane appears to have warmed the global climate briefly but substantially.

Metabolic activity of archaebacteria is the primary source of atmospheric methane, some of it coming from the flatulence of large hoofed herbivores that harbor methane-generating archaebacteria in their digestive tracts. Although the archaebacteria's overall rate of methane production is quite high, the concentration of methane in the atmosphere is normally very low because nearly all of it becomes oxidized to CO_2 after less than a decade of exposure to atmospheric oxygen. However, some methane produced by archaebacteria becomes stored in the absence of free oxygen (O_2), and the sudden release of large quantities of this stored methane has occasionally created a brief pulse of greenhouse warming in the course of Earth's history. This methane is stored in an icy state as what is called **methane hydrate**—methane frozen within a cage of water molecules.

Masses of methane hydrate constitute the largest fossil fuel reservoir on Earth—one that will never be easy for humans to exploit because of its frozen state. Methane freezes through a combination of low temperature and high pressure, and the resulting frozen masses survive only in the absence of O_2. On land at high latitudes, methane hydrate occupies the permanently frozen zone below tundra, extending down to depths of about 2 kilometers. These nonmarine methane hydrates are too deeply buried to be readily melted, however.

Masses of methane hydrate also occur profusely in marine sediments of continental slopes throughout the world (Figure 10-14). These methane hydrates are more easily melted by environmental change. Outside polar regions, they are restricted to sediments beneath portions of the continental slope that lie between about 400 and 1000 meters below sea level, where the temperature

Figure 10-14 A mass of methane hydrate from the continental slope. (I. R. MacDonald, Texas A&M University; J. Kennett, University of California, Santa Barbara.)

of seawater ranges from 13°C in the shallower regions to 4° at depth. Because of the high heat capacity of water, warming of the seas at the depth of the upper continental shelf can cause large volumes of methane hydrates to melt during brief intervals of geologic time. The result is a positive feedback because global warming releases methane, which accentuates greenhouse warming to cause further global warming. There is evidence that pulses of greenhouse warming of this kind may have occurred several times in the course of Earth's history. The evidence comes from sedimentary organic matter that is isotopically very light.

The melting of methane hydrates ultimately produces reduced carbon that is very light because this carbon is the product of three steps, each of which favors the lighter isotope, carbon 12. Much of the methane in methane hydrates is produced by archaebacteria that break down organic matter. Because this organic matter is produced by photosynthesis, it is isotopically light to start with. The archaebacteria's generation of methane, by metabolizing this carbon further, favors carbon 12. Then, after the methane is oxidized in the environment to form CO_2, the carbon becomes lighter still when cyanobacteria or algae or plants make use of that CO_2 in their photosynthesis. It is the resulting organic matter that is preserved in sediments.

So light is the carbon of methane hydrates that their massive release along continental margins will cause carbon isotopes in newly forming organic matter to shift suddenly toward much lighter values throughout the world. Through isotopic study of carbon in the stratigraphic record, geologists have identified several such shifts that were so pronounced that they cannot easily be attributed to any other cause. These shifts appear as sharp deflections in plots of isotope ratios against time.

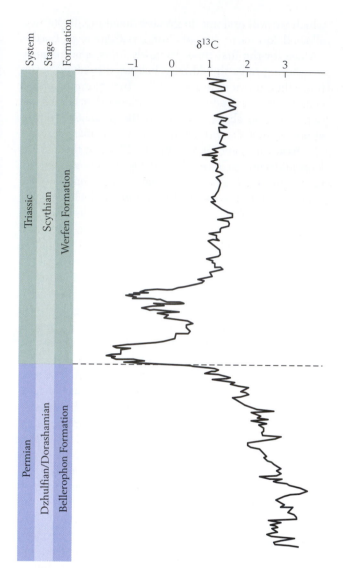

Figure 10-15 Carbon isotope ratios shift toward negative values in strata that span the boundary between the Permian and Triassic systems in the Austrian Alps. (After W. T. Holser et al., *Abj. Geol.* Bundesanstalt 45:213–232, 1991.)

One of these shifts is recorded throughout the world at the end of the Paleozoic Era, when the greatest mass extinction of all time occurred (Figure 10-15). We shall examine this event and the evidence that it coincided with a pulse of pronounced global warming in Chapter 15.

Negative feedbacks hold CO₂ levels in check

Could the atmospheric reservoir of CO_2 grow to the point at which severe greenhouse warming would elevate temperatures at Earth's surface far above present levels? Similarly, could greenhouse cooling plunge the entire world into frigid conditions? In fact, neither drastic heating nor drastic cooling of Earth's climate appears to have been possible in the Phanerozoic world. The factors responsible for this stability are examples of neg-ative feedbacks, which stifle change (p. 224). One of these entails temperature and the other, precipitation.

Warmth and weathering Recall that chemical weathering speeds up as the temperature rises. Con-sider, then, what must happen when for some reason CO_2 begins to build up in the atmosphere and, through the greenhouse effect, increases the average tempera-ture of Earth's surface (Figure 10-16). The warming ef-fect accelerates chemical weathering, increasing the rate at which this process extracts CO_2 from the at-mosphere. The more the CO_2 builds up, the more rapid the weathering. This increase in weathering slows the rate of CO_2 buildup until finally the system is in bal-ance. In this way, barring additional changes, the at-mospheric reservoir is stabilized.

Chemical weathering operates as a negative feed-back in the opposite way if the atmospheric reservoir of CO_2 begins to shrink. Then greenhouse warming weak-ens and global temperatures decline. As temperatures fall, the overall rate of weathering declines, and so does the rate at which weathering extracts CO_2 from the at-mosphere. The result is that the atmospheric reservoir of CO_2 shrinks less rapidly. The more slowly weather-ing proceeds, the more slowly the CO_2 concentration falls, until a stable level is reached.

Precipitation and weathering As we have seen, the configuration of continents influences patterns of pre-cipitation. The temperature of the ocean also affects rates of precipitation on a global scale. The surface wa-ters of warm oceans evaporate more rapidly than those of cold oceans and tend to supply nearby land areas with more abundant moisture. This relationship between temperature and precipitation gives us a second nega-tive feedback for changes in atmospheric CO_2.

Consider what happens when increasing atmo-spheric CO_2 warms Earth considerably. As we have seen, the first negative feedback is acceleration of the chemical reactions of weathering by the higher temperatures. But, in addition, the climatic warming increases precipitation on a global scale. As a result, more watery fluids—the agents of weathering—move through soil and rock every year, and weathering accelerates, depleting the atmo-sphere of CO_2. Forests expand as a result of the increase in both of the materials they employ for photosynthesis: water in the soil and CO_2 in the atmosphere (p. 225). This expansion of forests further accelerates weathering, for reasons described earlier. All of this weathering uses up CO_2 from the atmosphere (see Figure 10-11). The result is less CO_2 in the atmosphere and a reduced greenhouse effect. Thus the increased global precipitation, like the in-creased temperatures that produce it, acts as a negative feedback for greenhouse warming.

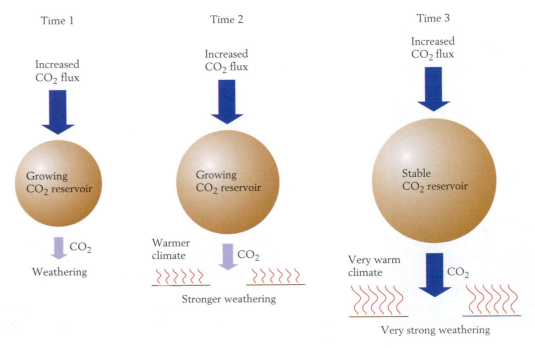

Figure 10-16 **Weathering as a negative feedback opposing an increase in atmospheric CO_2.** An increase in the flux of CO_2 to the atmosphere causes the atmospheric reservoir to grow between time 1 and time 3. The strengthening greenhouse effect that results, however, causes the rate of weathering to increase. These trends continue until the rate at which CO_2 is removed from the atmospheric reservoir by weathering balances the increased flux of CO_2 to the atmosphere (time 3).

Presumably, increases in weathering as climates have become warmer and moister have prevented Earth from undergoing what might be called a runaway greenhouse effect: a persistent buildup of atmospheric CO_2 leading to extraordinarily high temperatures on Earth.

Oxygen Isotopes, Climate, and the Water Cycle

Oxygen, like carbon, exists in two stable isotopic forms. Oxygen 16 is more common than oxygen 18. The slightly different behavior of these two isotopes as oxygen moves through the Earth system provides information about the temperatures of ancient environments, the volumes of glacial ice on Earth, and the salinities of ancient oceans.

Oxygen isotope ratios in skeletons reflect temperatures

Just as photosynthesizing plants slightly favor CO_2 molecules that contain carbon 12 over those that contain carbon 13, organisms that secrete skeletons of calcium carbonate ($CaCO_3$) incorporate oxygen 18 and oxygen 16 into those skeletons in ratios that differ from those found in the environment. In addition, the ratio of the two isotopes incorporated into skeletons varies with temperature. At low temperatures, organisms incorpo-

rate a relatively large proportion of oxygen 18 because ions containing this heavier isotope tend to be more sluggish than ions containing oxygen 16, and so more readily combine with other ions. Once scientists have determined the pattern of variation in oxygen isotope ratio with temperature for the skeletons of a particular group of organisms, they can use the fossil skeletons of those organisms as what amount to paleothermometers.

The fossil record of planktonic foraminifera extends back only to the Cretaceous Period, and fossils of these organisms have been widely used to assess temperatures of ancient oceans for Cretaceous and Cenozoic time. The utility of these organisms stems from their widespread occurrence in the open ocean and from the abundance of their skeletons in deep-sea sediments (p. 63). The oxygen isotope technique for determining temperatures can be applied to fossils older than Cretaceous, but the older the fossil, the greater the possibility that some of the material within it has been altered after burial. Watery solutions may have recrystallized the skeletal material, removing some oxygen atoms and replacing them with others that have different isotope ratios. The skeletal material in question has to be examined through a microscope and by other means to establish that it has undergone little recrystallization.

Figure 10-17 depicts a remarkably detailed record of oxygen isotope ratios for a rudist bivalve from Late

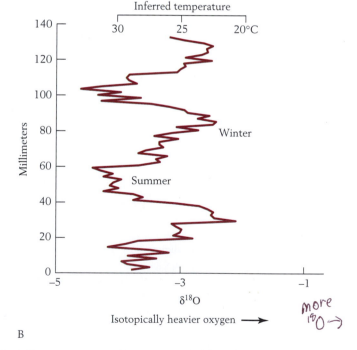

Figure 10-17 **Seasonal temperature shifts estimated from oxygen isotope ratios in a reef-building rudist bivalve from Greece.** *A.* Samples for isotopic analysis were taken along the length of a shell similar to this cucumber-sized specimen. *B.* The plot depicts about 3 years of upward growth. (*A,* T. Steuber, Institut für Geologie und Mineralogie, Der Universität Erlangen, Nürnberg; *B* from T. Steuber, *Geology* 24:315–318, 1996.)

Cretaceous deposits in Greece. Rudists are an extinct group of large bivalve mollusks that built reefs during the Cretaceous Period. The specimen in Figure 10-17A comes from the Tethys Seaway, which at that time connected the Pacific and Atlantic oceans by way of the Mediterranean region. Figure 10-17B is based on samples from a vertical section through a rudist shell. As would be expected, these samples reveal that the oxygen isotope composition of the calcium carbonate secreted by the animal was heavier in winter than in summer. Comparison with isotope ratios in living species indicates an annual temperature range between about 22°C and 32°C (72°F and 88°F) for the rudist's environment. These temperatures are much warmer than those of the Mediterranean waters near Greece today, resembling the present temperatures of seas off Miami, Florida. In other words, the isotope ratios show that the Mediterranean region was remarkably warm during the Late Creta-

ceous—probably because warm waters were flowing through the Tethys Seaway from the tropical Pacific. The isotope data also show that under these conditions rudists grew about 3.0–4.5 centimeters (1.2–1.8 inches) per year—more rapidly than most modern reef corals.

Oxygen isotope ratios vary with salinity

Additional factors complicate the use of oxygen isotopes to estimate past temperatures. One is the fact that the oxygen isotope ratio of ocean water varies from time to time and from place to place. The most important cause of this phenomenon is variation in the rate at which surface waters evaporate. Because H_2O molecules containing oxygen 16 are lighter than those containing oxygen 18, they evaporate more readily. As a result, high rates of evaporation leave behind seawater that is enriched in oxygen 18 (Figure 10-18A). This pattern is evident in modern seas, in which hypersaline waters are

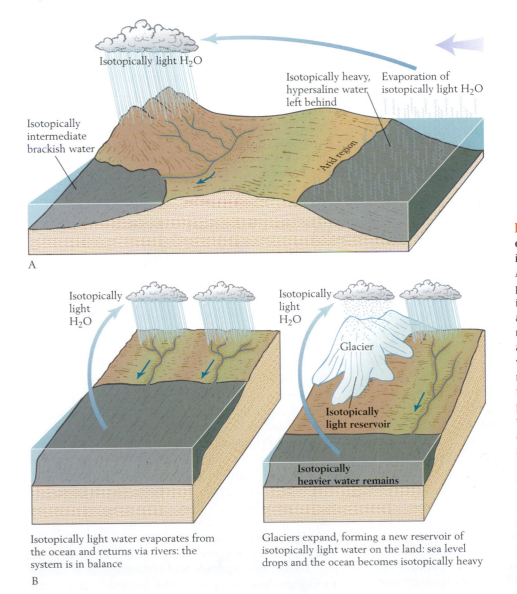

A

Isotopically light H_2O

Isotopically heavy, hypersaline water left behind

Evaporation of isotopically light H_2O

Isotopically intermediate brackish water

Arid region

Isotopically light H_2O

Isotopically light water evaporates from the ocean and returns via rivers: the system is in balance

B

Isotopically light H_2O

Glacier

Isotopically light reservoir

Isotopically heavier water remains

Glaciers expand, forming a new reservoir of isotopically light water on the land: sea level drops and the ocean becomes isotopically heavy

Figure 10-18 Effects of evaporation on the oxygen isotope ratio of seawater. *A.* A high rate of evaporation preferentially removes isotopically light H_2O from one arm of the sea, leaving the remaining water hypersaline and isotopically heavy. The water vapor thus produced moves through the atmosphere to an area of abundant precipitation, where it moves via rainfall and river flow to another arm of the sea. As a result, the waters of this second body of water are brackish and isotopically intermediate. *B.* Growth of continental glaciers locks up isotopically light atmospheric H_2O in a new reservoir on land. As a result, throughout the world, sea level drops and the H_2O of the ocean becomes isotopically heavy.

isotopically heavier than waters of normal salinity (about 35 parts per thousand). Given the difference in evaporation rates, it is easy to see why brackish waters are isotopically lighter than waters of normal salinity. Because H$_2$O molecules that contain oxygen 16 evaporate more readily than those containing oxygen 18, atmospheric moisture is isotopically light. When this moisture reaches Earth's surface as precipitation and returns to the sea in rivers, the brackish waters that result near the margin of the ocean are isotopically lighter than normal seawater.

The variation of oxygen isotope ratios with salinity complicates efforts to derive accurate paleotemperatures from these ratios. Sometimes, however, salinity can be shown to be a relatively minor source of error. This is the case with the rudist depicted in Figure 10-17A, because reefs grow only in seawater close to normal marine salinity. The waters in which this rudist grew probably underwent only minor isotopic shifts as a result of high evaporation rates in summer or seasonal increases in freshwater runoff to the sea.

Glaciers lock up oxygen 16

Because atmospheric moisture is isotopically light, so are glaciers, which form from snow that precipitates from clouds. For this reason, expansion and contraction of large glaciers affect the oxygen isotope ratio of seawater (Figure 10-18B). Recall that glaciers have at times locked up a large portion of the H$_2$O in the water cycle (p. 19). At many times during the past 2.5 million years, expansion of glaciers has temporarily lowered sea level by about 100 meters (330 feet). At such times, the storage of relatively large amounts of oxygen 16 in glacial ice has left the oceans enriched in oxygen 18. During the intervening intervals, when glaciers have shrunk, the ratio of oxygen 18 to oxygen 16 has fallen again. We live at such a time today. Figure 10-19 shows a plot of oxygen isotope ratios for bottom-dwelling deep-sea foraminifera over the past 2.5 million years.

Like changes in salinity, changes in the volume of glacial ice create problems for the scientist who wants to use oxygen isotopes to estimate ancient ocean temperatures. Nonetheless, the effects of ice volume and temperature combine to give a clear picture of the timing of expansion and contraction of large glaciers. When glaciers have expanded, foraminifera in most regions have secreted isotopically heavy skeletons because of the effects of both ice volume and temperature. At these times, the oceans in which they lived have been enriched in oxygen 18 because more oxygen 16 has been locked up in glaciers. In addition, the organisms have preferentially taken up this isotope because the seawater has been cooler. Similarly, when glaciers have contracted, the effects of ice volume and temperature have both contributed to lower oxygen 18/oxygen 16 ratios.

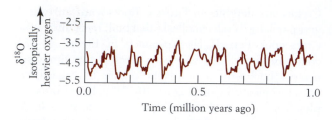

Figure 10-19 Fluctuations of oxygen isotope ratios in the skeletons of foraminifera resulting from the expansion and contraction of continental glaciers during the past million years. The foraminifera sampled here lived on the Pacific deep-sea floor. Peaks on the graph depict relatively high isotope values, which represent times when glaciers expanded, locking up large volumes of isotopically light H$_2$O. The lowest isotope values represent times when glaciers melted, returning isotopically light H$_2$O to the oceans. (After R. Tiedmann, M. Sarnthein, and N. J. Shackleton, *Paleoceanography* 9:619–638, 1994.)

However, because temperature changes in the deep sea have been minor during the modern ice age, the isotopic shifts depicted in Figure 10-19 are largely the result of changes in ice volume.

Shifts in oxygen isotope ratios in fossils have also been related to episodes of glaciation in pre-Cenozoic intervals. At times when few glaciers have existed, ice volume can be ignored in the analysis of oxygen isotopes; isotope ratios can be assumed to reflect temperature and salinity alone. The Late Cretaceous, when the reef-building rudist depicted in Figure 10-17A lived, was such a time.

Ocean Chemistry and Skeletal Mineralogy

A variety of evidence suggests that the general chemical composition of seawater has not changed greatly for hundreds of millions of years. Sodium and chloride, which combine to form halite when seawater evaporates (p. 32), have remained the most abundant dissolved ions, and other components have also varied only modestly. The abundances of some chemical components of seawater have nonetheless fluctuated enough to bring about changes in the types of minerals that have precipitated to form sediments.

For our purposes, the fluctuations of two ions, calcium (Ca^{2+}) and magnesium (Mg^{2+}), in seawater have been especially important. These two ions are chemically similar, having the same charge and differing only modestly in diameter. Magnesium can substitute for calcium in the crystal lattice of calcite but is too small to lodge in the lattice of aragonite, the other form of CaCO$_3$ that precipitates from seawater (see p. 32).

Today both aragonite and high-magnesium calcite precipitate from seawater. The latter is calcite with magnesium substituting for several percent of the calcium ions. Most oolites today are composed of aragonite, but both minerals fill voids in coral reefs and cement sediments here and there along tropical seashores. Aragonite and high-magnesium calcite also formed oolites and seafloor cements on the seafloor during two earlier geologic intervals: early in the Cambrian and again from late in the Paleozoic Era until well into the Mesozoic Era. Between these intervals were ones during which oolites and seafloor cements were composed of ordinary calcite—calcite containing little magnesium. Thus geologists speak of the Phanerozoic Earth as three times having had "aragonite seas" and twice having had "calcite seas."

It turns out that the relative proportions of Ca^{2+} and Mg^{2+} in seawater have determined which minerals have precipitated from warm seas. Laboratory experiments show that relatively high Mg^{2+}/Ca^{2+} ratios result in precipitation of aragonite and high-magnesium calcite. Relatively low Mg^{2+}/Ca^{2+} ratios result in precipitation of normal calcite.

What causes the Mg^{2+}/Ca^{2+} ratio of seawater to rise and fall in the course of geologic time? The primary cause is change in the volume of mid-ocean ridges. Seawater circulates through the sediments and fractured rocks that flank mid-ocean ridges and returns to the ocean. In the process, the water heats up and reacts chemically with the rock of the newly forming lithosphere. These chemical reactions transfer Ca^{2+} from the rocks to the seawater, while also extracting Mg^{2+} from the seawater and locking it up in rocks of the newly forming oceanic crust. Similar chemical exchanges occur during the formation of large submarine plateaus of basalt, which may be studded with active volcanoes even after they form. In other words, mid-ocean ridges and submarine plateaus amount to ion-exchange systems, extracting Mg^{2+} from seawater and releasing Ca^{2+} to it. Variations in the volume of submarine volcanism throughout geologic time have therefore exerted strong influences on seawater chemistry. When the total volume of ridges and volcanically active submarine plateaus increases, the Mg^{2+}/Ca^{2+} ratio falls because the ion exchange system speeds up (Figure 10-20). The result is precipitation of calcite from seawater. At other times, such as the present, when the total volume of mid-ocean ridges is smaller, the Mg^{2+}/Ca^{2+} ratio of seawater is higher and high-magnesium calcite and aragonite precipitate from seawater.

How do we know how deep-sea volcanism has changed through time? The record of global sea level change provides the evidence. When ridges and submarine plateaus increase in volume, they push the oceans upward, and shallow seas expand over continents. Likewise, a decrease in the total volume of ridges and submarine plateaus causes seas to recede from continents. The sedimentary record of movements of shorelines back and forth over continental surfaces (see Figure 6-21) makes it possible to estimate sea level changes through time. By using sea level as an indicator of global submarine volcanism, one can calculate the amount of Mg^{2+} and Ca^{2+} present in the ocean at any time in the Phanerozoic Eon.

More important than the changes in the mineralogy of inorganically precipitated carbonates during transitions between calcite and aragonite seas are corresponding changes in the kinds of marine organisms that have secreted large volumes of calcium carbonate. The organisms that form chalk are a prime example. **Chalk** is a soft, white, fine-grained limestone that we use to write on blackboards. Grains of chalk are minute plates of calcite that once covered the spherical cells of single-celled floating algae called calcareous nannoplankton (p. 62). (The prefix *nanno* means "very small.")

The Upper Cretaceous Series contains a vastly larger volume of chalk than any other portion of the geologic record (Figure 10-21). Upper Cretaceous chalk is conspicuous throughout the world. Calcareous nannoplankton were so prolific when this chalk was deposited that their minuscule plates often accumulated at a rate of about a millimeter per year (an inch every 25 years). At no other time during its history has this group approached such productivity, despite the fact that it has been represented by many species at all times since the Jurassic Period (nearly 400 species inhabit modern seas). Here is where seawater chemistry comes in. During the Cretaceous interval of massive chalk deposition, the Mg^{2+}/Ca^{2+} ratio in seawater was lower than at any other time in the Phanerozoic. Thus, calcite precipitated from seawater more easily at this time than at any other, and chalk-producing plankton flourished as never before or after. Laboratory experiments confirm that modern species of calcareous nannoplankton grow more prolifically in artificial seawater of Cretaceous composition than in modern seawater, which has a much higher Mg^{2+}/Ca^{2+} ratio.

Calcareous nannoplankton suffered heavy extinction at the end of the Cretaceous, when the dinosaurs died out, yet they flourished enough to form moderate volumes of chalk again at the beginning of the Cenozoic Era, when the Mg^{2+}/Ca^{2+} ratio remained low. As the ratio then rose substantially during the Cenozoic Era, the productivity of calcareous nannoplankton declined, even at times when warm global climates should have favored their growth.

Seawater chemistry has also influenced which groups of organisms have formed reefs at particular times and which have contributed large volumes of carbonate sediment. Because organisms need ideal conditions to form large reefs (p. 99), we might expect

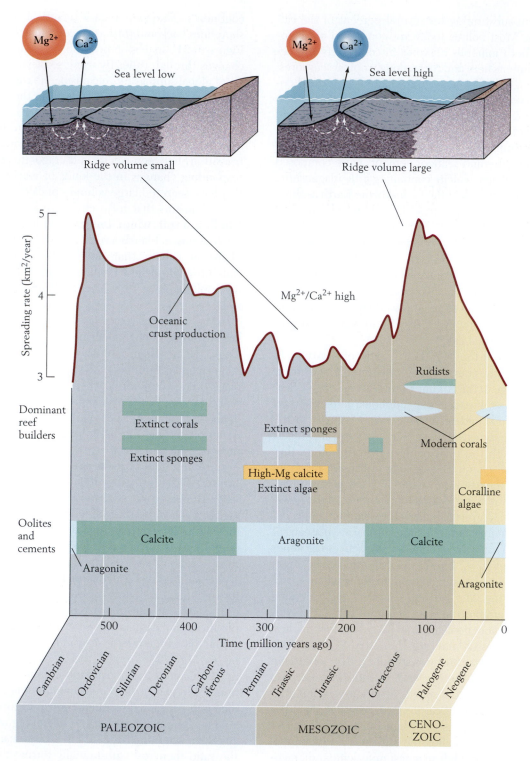

Figure 10-20 Changes in spreading rates along mid-ocean ridges and corresponding changes in the mineralogy of oolites and carbonate cements and in dominant reef-building organisms. The curve shows changes in the rate of oceanic crust production during Phanerozoic time. This curve is based on evidence of long-term changes in global sea level. Seawater that circulates through mid-ocean ridges gives up Mg^{2+} and carries away Ca^{2+}. When spreading rates are high, the volume of mid-ocean ridges increases, causing a global rise in sea level as well as a decrease in the Mg^{2+}/Ca^{2+} ratio of seawater. Low Mg^{2+}/Ca^{2+} ratios produce calcite seas, in which calcite forms oolites and marine cements and organisms with calcite skeletons become successful reef builders. Aragonite and high-magnesium calcite play these roles instead when the total volume of mid-ocean ridges is low and the Mg^{2+}/Ca^{2+} ratio of seawater is high. (After S. M. Stanley and L. A. Hardie, *Palaeogeogr. Palaeoclimatol. Palaeoecol.* 144:3–19, 1998.)

the success of reef builders to reflect seawater chemistry. Modern taxa of reef-building corals can exist outside the tropics and can survive without symbiotic algae, but under neither condition are they able to build large reefs. Warm temperatures and symbiotic algae facilitate modern reef-building corals' secretion of calcium carbonate. These corals, which secrete aragonite, also seem to benefit from the high Mg^{2+}/Ca^{2+} ratio of modern seawater. The same condition accounts for the abundance on modern reefs of calcareous red algae, which secrete skeletons of high-magnesium calcite. Aragonitic corals were major reef builders in aragonite seas early in the Mesozoic Era, but relinquished that role after calcite seas developed (see Figure 10-20). Not surprisingly, groups of calcitic corals and sponges that are now extinct were the primary reef builders in the calcite seas that existed during a long segment of Paleozoic time.

Figure 10-21 Cliffs of chalk in Denmark. The lower, white portion is of Late Cretaceous age; the darker, upper portion was deposited in earliest Cenozoic time, when the productivity of calcareous nannoplankton was lower and clay constituted a larger percentage of the sediment. (Anthony Ekdale, University of Utah.)

Chapter Summary

What are chemical reservoirs in the Earth system?

Chemical reservoirs are bodies of key elements and compounds in the Earth system that shrink or expand as fluxes between them change.

What is the difference between photosynthesis and respiration?

Photosynthesis is the process by which plants use the energy of sunlight to produce sugars from carbon dioxide and water; oxygen is a by-product of this process. Respiration entails the opposite chemical reaction and is used by organisms to oxidize sugars in order to release their energy.

What happens to sugars that plants produce?

Plants use some sugars for their respiration and some to form their tissues, a portion of which animals and decomposers consume for their respiration.

What would happen to atmospheric CO_2 and O_2 if no dead plant tissue were buried?

If no dead plant tissue were buried, it would be decomposed, and the CO_2 thus produced would return to the atmosphere. The photosynthesis-respiration cycle—involving plants, animals, and decomposers—would be roughly in balance. Both CO_2 and O_2 would cycle through these organisms and the atmosphere without much change in the volume of the atmospheric reservoir of either gas.

How does burial of dead plant tissue affect atmospheric CO_2 and O_2?

Burial of dead plant tissue, which contains reduced carbon, upsets the balance of the global photosynthesis-respiration cycle. Because it prevents some reduced carbon from decomposing and returning to the atmosphere as CO_2, the atmospheric reservoir of CO_2 shrinks. At the same time, oxygen that would have been used to decompose the organic matter if it had not been buried remains in the atmosphere, and the atmospheric reservoir of O_2 expands.

Where is organic carbon buried in large quantities?

Most burial of organic carbon takes place in anoxic swamps or marine environments that exclude most kinds of bacteria that decompose organic matter. The spread of these environments in the geologic past has caused the atmospheric reservoir of CO_2 to shrink and the atmospheric reservoir of O_2 to expand.

How can carbon isotopes reveal the geologic history of carbon burial?

Photosynthesis employs a disproportionate percentage of relatively light carbon 12. Therefore, at times in the geologic past when much organic carbon has been buried, the relative abundance of carbon 13 has increased in the atmosphere and ocean. Shifts of this kind toward heavier carbon are reflected in the carbon isotope composition of limestone and of sedimentary organic matter produced by photosynthesis.

How does weathering affect atmospheric CO_2?

Weathering of minerals removes CO_2 from the atmosphere, so factors that intensify weathering can deplete the atmospheric reservoir of CO_2. These factors include mountain building, warm climates, high rates of precipitation, and deeply rooted vegetation.

What caused a decline in the concentration of atmospheric CO_2 during the latter part of the Paleozoic Era?

The initial spread of forests on Earth in the Devonian Period intensified weathering and depleted the atmospheric reservoir of CO_2. This depletion continued as coal swamps became widespread sites of carbon burial during Carboniferous time. It reduced greenhouse warming and contributed to the cooler climatic conditions of a major ice age.

How do methane hydrates on the seafloor create a positive feedback during global warming?

Methane is a powerful greenhouse gas. When global warming melts masses of methane hydrate on the seafloor, the addition of methane to the atmosphere produces further global warming.

What negative feedbacks have prevented a buildup of CO_2 from creating extreme greenhouse warming of Earth?

When the climate warms, chemical weathering accelerates, extracting CO_2 from the atmosphere and thus weakening greenhouse warming. Warm climates also increase the rate of evaporation of water from the oceans; the resulting increase in precipitation on land accelerates chemical weathering, weakening greenhouse warming by extracting CO_2 from the atmosphere.

How do oxygen isotopes in skeletons of marine organisms provide a record of past ocean temperatures?

Organisms incorporate more of the heavier oxygen isotope (oxygen 18) in cool waters than in warm waters. In addition, because water molecules containing the lighter oxygen isotope (oxygen 16) evaporate more readily from the surface of the ocean than water molecules containing oxygen 18, atmospheric moisture is enriched in oxygen 16, and this lighter isotope becomes locked up preferentially in glaciers. Thus, when the global climate cools and glaciers expand, seawater becomes enriched in oxygen 18, and so do the skeletons of organisms.

Why is the water of hypersaline bodies of seawater enriched in oxygen 18?

The high rate of evaporation that produces hypersaline waters preferentially removes water molecules containing oxygen 16.

How have changes in the overall volume of submarine volcanism altered the chemistry of seawater, and what are the important consequences of these changes?

Changes in submarine volcanism have altered the Mg^{2+}/Ca^{2+} ratio of seawater. This ratio governs the kinds of carbonate minerals that have precipitated from seawater at different times in the geologic past, and it has also strongly influenced which kinds of organisms have functioned as major reef builders and sediment producers.

Review Questions

1. What are the two possible fates of plant material not eaten by animals?

2. What is a negative feedback? What is a positive feedback? Give an example of each kind of feedback.

3. How can carbon isotopes in limestones provide evidence about the history of atmospheric oxygen? (Hint: Refer to Figure 10-9.)

4. Draw sketches illustrating how increased burial of carbon reduces the atmospheric reservoir of CO_2 and enlarges the atmospheric reservoir of O_2.

5. In what kinds of marine environments can carbon be buried in large quantities?

6. Draw a diagram depicting the cycle of oxidized carbon that includes limestone, atmospheric CO_2, and weathering.

7. Why are pelagic carbonates more likely than shallow-water carbonates to melt and return CO_2 to the atmosphere?

8. How do glaciers promote chemical weathering?

9. Why does the influence of moist climates on vegetation accelerate weathering?

10. What can the study of oxygen isotopes tell us about ancient oceans?

11. What controls the ratio of magnesium to calcium in the ocean? How have changes in this ratio influenced the mineralogical composition of limestone during Phanerozoic time?

12. Using the Visual Overview on page 222 and what you have learned in this chapter, (a) summarize how burial of organic carbon, alteration of carbonates at high temperatures, and changes in rates of weathering alter greenhouse warming by Earth's atmosphere, and (b) explain how carbon isotopes are used to assess rates of burial of organic carbon.

PART II

The Story of Earth

Armed with basic Earth science knowledge from the first ten chapters, we can now start at the beginning of Earth's story and travel forward. Chapters 11 through 20 portray key periods in the history of environments and life on Earth. Each chapter highlights one or two global changes of great magnitude and special significance under the title "Earth System Shift." Chapters 13 through 19 introduce the cast of characters—the life forms—of an interval of geologic time and then reviews important global and regional events for that interval. Together, the following ten chapters trace the evolution of continents, climates, and biological communities that led to the present world.

In northwestern Canada, the Canadian Shield is a largely barren Precambrian terrane that was scoured by glaciers during the recent Ice Age and is now dotted by lakes. Numerous dikes that formed when magma forced its way into giant cracks in the Precambrian crust now stand above the rocks they intruded. (R. S. Hildebrand, Geological Survey of Canada.)

CHAPTER 11

The Archean Eon of Precambrian Time

Since the nineteenth century, the interval of Earth history that preceded the Phanerozoic Eon has been known as the Precambrian. Although the term "Precambrian" has no formal status in the geologic time scale, it has traditionally been employed as though it did. The Precambrian includes nearly 90 percent of geologic time, ranging from 4.6 billion years ago, when Earth formed, to the start of the Cambrian Period, about 4 billion years later. Two eons are formally recognized within the Precambrian: the Archean and the Proterozoic. The Archean Eon includes about 45 percent of Earth's history. During Archean time, Earth underwent enormous physical changes and life developed on its surface. Many details of Archean history, however, remain unknown or poorly understood. Some scientists recognize a third Precambrian eon, the Hadean, representing the earliest segment of Earth's history (4.6–4 billion years ago). No bodies of rock this old have yet been discovered on Earth.

One reason for our limited understanding of Precambrian history is that, although this history spans about 4 billion years, Precambrian rocks form less than 20 percent of the total area of rocks exposed at Earth's surface (Figure 11-1). Erosion has destroyed many Precambrian rocks, and metamorphism has so altered others that they can no longer be dated and therefore cannot be recognized as Precambrian. Still other Precambrian rocks lie buried beneath younger sedimentary and volcanic rocks. All of these problems are especially profound for the Archean record. In addition, the simple unicellular fossils of Archean rocks are uncommon and difficult to assign to species and genera, so that few are recognized as index fossils. Stratigraphic correlation of these rocks has therefore been based largely on radiometric dating. The richer fossil record of younger Proterozoic strata includes more advanced life forms and provides many index fossils. In the absence of useful biostratigraphic data, the boundary between the Archean and Proterozoic intervals is defined by its absolute age of 2.5 billion years before the present. Despite its deficiencies, the Archean geologic record offers important evidence about the first half of Earth's history.

Most geologic information about the Precambrian is derived from **cratons**, those large portions of continents that have not undergone tectonic deformation since early Paleozoic time. All the continents of the present world include cratons that consist primarily of Precambrian rocks (Figure 11-2). A **Precambrian shield** is a largely Precambrian portion of

Major Events of the Archean Eon

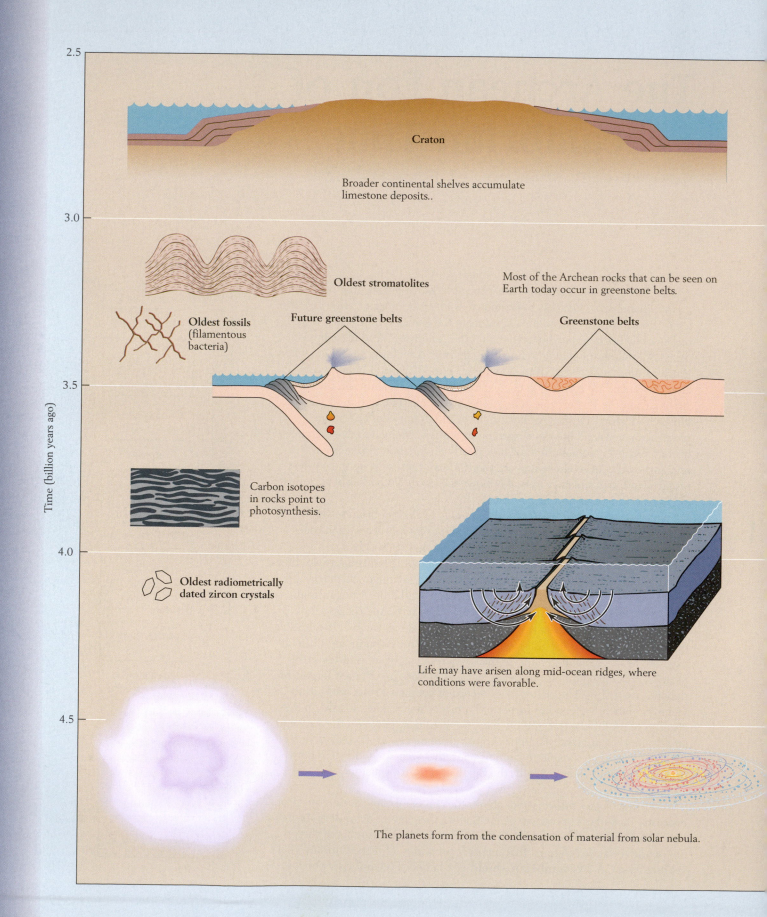

Time (billion years ago)

2.5

3.0

3.5

4.0

4.5

Craton

Broader continental shelves accumulate limestone deposits..

Oldest stromatolites

Oldest fossils (filamentous bacteria)

Future greenstone belts

Most of the Archean rocks that can be seen on Earth today occur in greenstone belts.

Greenstone belts

Carbon isotopes in rocks point to photosynthesis.

Oldest radiometrically dated zircon crystals

Life may have arisen along mid-ocean ridges, where conditions were favorable.

The planets form from the condensation of material from solar nebula.

Origin of the early atmosphere

Huge impact creates Earth's moon

Formation of Earth's layers

Differentiation of materials according to density

Molten Earth

Larger continents form

Frequency of impacts declines

Earth's interior continues to cool

THE FIRST 100 MILLION YEARS

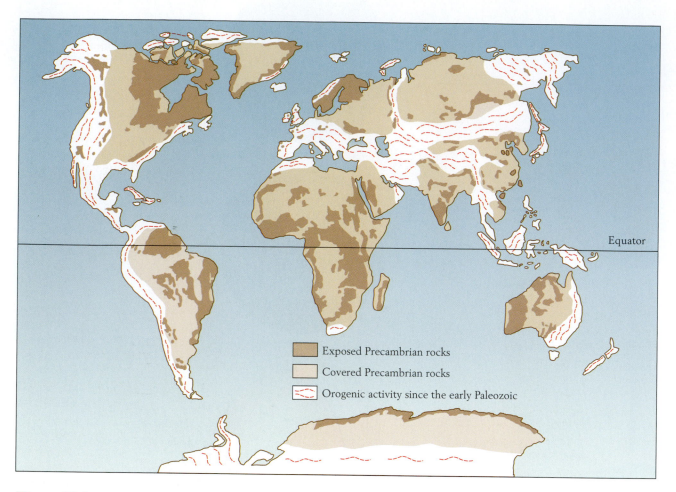

Figure 11-1 **Distribution of Precambrian rocks in the modern world.** Note that these rocks form or underlie most of the cratonic area of the modern world. (After A. M. Goodwin, in B. F. Windley, ed., *The Early History of the Earth*, John Wiley & Sons, New York, 1976.)

a craton that is exposed at Earth's surface. The largest is the vast Canadian Shield, which has become more fully exposed during the past 3 million years by the action of glaciers. Although shields contain some sedi-

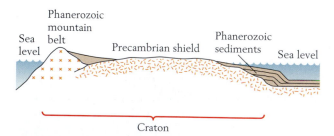

Figure 11-2 **A Precambrian shield flanked by younger (Phanerozoic) mountain belts and sediments.** A shield and those solid rocks that surround it that have not been extensively deformed since early Paleozoic time constitute a craton.

mentary rocks, they consist primarily of crystalline (igneous and metamorphic) rocks. As we shall see, mountain belts that formed during Precambrian time left recognizable traces within Precambrian shields, but erosion long ago destroyed their elevated topography. Today the only Precambrian rocks that stand at high elevations in mountain ranges are those that have been uplifted by Phanerozoic orogenies.

It was during the Archean Eon that Earth acquired its basic configuration, with the mantle and crust surrounding the core. Continental crust also formed during Archean time, but the flow of heat from Earth's interior was greater than it is today, and continents remained small. Near the beginning of Proterozoic time, however, large cratons began to form.

We will begin our review of Archean events by considering when and how Earth and other planets of the solar system came into being. We will then review evidence that suggests how Earth changed during the remainder of Archean time.

The Ages of the Planets and the Universe

The planets of our solar system all rotate around the sun in the same direction, and in orbits that lie in nearly the same plane. This pattern is strong evidence that the planets formed simultaneously from a single disk of material that rotated in the same direction as the modern planets. Precisely when the planets came into being has been a difficult issue to resolve.

Astronauts report that the most beautiful object they see from space is Earth, whose surface is partially blanketed by swirling white clouds set against the blue background of extensive oceans (Figure 11-3). Although Earth's water is aesthetically pleasing and necessary for life, its abundance near the planet's surface makes rapid erosion inevitable. Continuous alteration of Earth's crust by erosion as well as by igneous and metamorphic processes makes unlikely any discovery of rocks nearly as old as Earth. Thus geologists have had to look beyond our planet in their efforts to date Earth's origin. Fortunately, we do have samples of rock that appear to represent the primitive material of the solar system. These samples are **meteorites**—extraterrestrial objects that have been captured in Earth's gravitational field and have crashed into our planet.

Some meteorites consist of rocky material and, accordingly, are called **stony meteorites** (Figure 11-4A). Others are metallic and have been designated **iron**

Figure 11-3 Earth viewed from space. The blue regions are oceans, and the swirling white masses are clouds. (NASA.)

meteorites, even though they contain small amounts of elements other than iron. Still others consist of mixtures of rocky and metallic material and thus are called **stony-iron meteorites** (Figure 11-4B). Meteorites range in size from tiny particles to the small planetlike bodies known as **asteroids**; fortunately, no asteroid has struck Earth during recorded human history. Many

A

B

Figure 11-4 Meteorites. *A.* A stony meteorite that fell in a meteorite shower near Plainville, Texas. *B.* A polished surface of a stony-iron meteorite. The silver-colored iron portion, representing material from the core of a planet or asteroid, contrasts with the stony portion, which consists of greenish olivine that formed the mantle of a planet. The two materials were mixed together in a collision of two large bodies. (*A,* R. A. Oriti; *B,* J. Beckett, American Museum of Natural History Library.)

Figure 11-5 The configuration of a spiral galaxy. This is galaxy NGC 6744. Its arms spiral outward from the central bulge. (Anglo-Australian Telescope Board.)

meteorites appear to be fragments of larger bodies that have undergone collisions and broken into pieces. Among the most interesting meteorites are stony-iron forms in which the stony portion consists primarily of olivine, the dominant mineral of Earth's mantle, and the metallic portion consists primarily of iron, the dominant element of Earth's core (Figure 11-4*B*). These meteorites apparently formed when one or more asteroids collided and their mantle and core materials, which resembled those of Earth, became fused together.

Presumably, some **comets**, which are icy bodies that contain rock fragments, have also struck Earth to become meteorites, but have then quickly melted. Modern humans have never been able to examine the remains of a comet on Earth, however, because most comets are so small that if they approach Earth, they melt and burn from frictional heating as they pass through the atmosphere. Most so-called shooting stars that we see in the night sky are comets that are being destroyed in this way.

Meteorites have been radiometrically dated by means of several decay systems, including rubidium-strontium, potassium-argon, and uranium-thorium. The dates thus derived have tended to cluster around 4.6 billion years, suggesting that this is the approximate age of the solar system. After many meteorites had been dated, it was gratifying to find that the oldest ages obtained for rocks gathered on the surface of the moon in 1969 also approximated 4.6 billion years. This must, indeed, be the age of the solar system. Ancient rocks can be found on the moon because the lunar surface, unlike that of Earth, has no water to weather and erode rocks and is characterized by only weak tectonic activity.

Determining the age of the universe, which turns out to be more than three times as old as our solar system, is more complicated. Most stars in the universe are clustered into enormous spiral galaxies (Figure 11-5). The distance between our galaxy, known as the Milky Way, and all others is increasing. In fact, all galaxies are moving away from one another, giving evidence that the universe is expanding. It is actually not the galaxies that are expanding, but the space between them. What is happening is analogous to inflation of a balloon with small coins attached to its surface (Figure 11-6). The coins behave like galaxies: although they do not expand, they become farther and farther apart. Before the galaxies formed, matter that they contain was concentrated, with infinite density, at a single point. From this point, matter exploded in an event irreverently called the **big bang**. Calculations indicate that protons and neutrons were forming about one ten-thousandth of a second after the start of the big bang, but that atoms would not have appeared until about a half billion years later. It took about a billion years for galaxies and 10 billion years for planets to form. Even after matter became assembled into galaxies, the galaxies spread in all directions from the site of the big bang.

The evidence that the universe is expanding makes it possible to estimate its age. This evidence, called the **redshift**, is an increase in the wavelengths of light waves traveling through space—a shift toward the end of the spectrum of wavelengths where visible light is red. Expansion of the space between galaxies causes this shift by stretching light waves as they pass through it. The farther these light waves have traveled through space, the greater the redshift they have undergone. For this reason, light waves that reach Earth from distant galaxies have larger redshifts than those from nearby galaxies. Calculations based on redshifts indicate that about 15 billion years ago all of the galaxies would have been at one spot—the site of the big bang. This, then, is the approximate date of the big bang and the age of the universe.

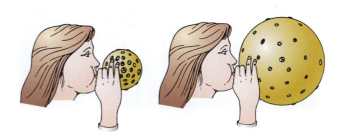

Figure 11-6 An analogy for the expansion of the universe. The space between galaxies expands, like the balloon, but the galaxies themselves, like the coins attached to the balloon, do not expand, but only move farther apart.

The Origin of the Solar System

We know more about the origin of distant stars than we do about the origin of our own solar system. Our solar system formed long ago, and we have no younger solar systems to study at close range.

Galaxies form by the gravitational collapse of dense clouds of gas (mainly hydrogen) into stars. Our galaxy, which originated less than 10 billion years ago, is made up of approximately 250 billion stars. Powerful telescopes reveal that long after galaxies form, secondary stars continue to be born within their spiral arms, where galactic matter is concentrated (Figure 11-7). The sun is a star that formed in such a setting.

The sun formed from a nebula

Our sun formed from material remaining from another star that collapsed violently, forming heavy elements. After this collapse, a **supernova**—an exploding star that casts off matter of low density—was formed. What remained was a dense cloud that condensed as it cooled. This dense cloud, or **solar nebula** (Figure 11-8), is assumed to have had some rotational motion when it formed and must then have rotated more and more rapidly as it contracted, just as ice skaters automatically spin more rapidly (conserving angular momentum) as they pull in their arms.

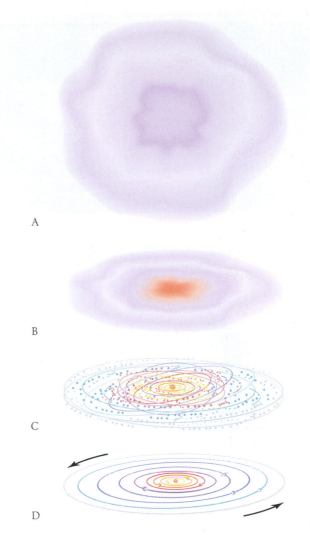

Figure 11-8 The origin of the solar system from a nebula. *A.* A nearly spherical young nebula rotates slowly. *B.* The nebula rotates faster and contracts to form a disk. *C.* As contraction continues, rings of material separate from the ancestral sun. *D.* The material in the rings condenses to form planets.

Figure 11-7 **The birth of stars.** In this photograph, taken by the Hubble Space Telescope, stars emerge from bright terminations of huge columns of hydrogen and dust in the constellation Serpens. (J. Hester and P. Scowen [AZ State Univ.], NASA, Hubble Space Telescope.)

The planets formed from a rotating dust cloud

The planets formed either during or soon after the birth of the sun. In fact, the sun and the planets originated during an interval no longer than 50 million to 100 million years, roughly the length of the Cenozoic Era (the Age of Mammals). This has been deduced from the high concentration in some meteorites of the stable isotopes xenon 129 and plutonium 244. Relatively large amounts of these isotopes in a meteorite indicate that some of their short-lived parent isotopes were originally present in the meteorite material and then decayed to xenon 129 and plutonium 244. Thus the planet or planetlike body of which such a meteorite is a fragment must have

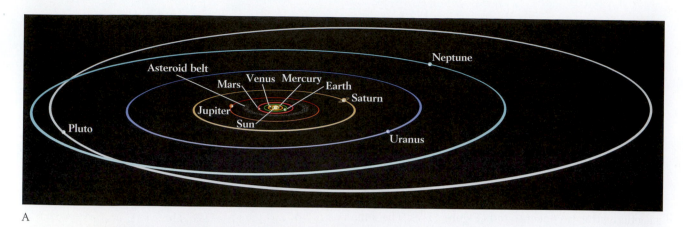

A

B

Figure 11-9 **The solar system.** *A.* Relative sizes of the orbits of the planets. *B.* Relative sizes of the planets. The inner planets, including Earth, are small rocky bodies. The asteroid belt lies between them and the outer planets, which consist largely of volatile elements.

formed soon after the elements of the solar system came into being; otherwise, very little of the short-lived parent isotopes of xenon 129 and plutonium 244 would still have existed to become incorporated into the material of the meteorite. The conclusion is that the sun cannot be very much older than the meteorites and the planets, which are about 4.6 billion years old.

Planets that are positioned far from the sun formed largely from *volatile* (i.e., easily vaporized) elements, which were frozen to form the icy materials of which these planets are composed. These elements were expelled from the hot inner region of the solar nebula to solidify in colder regions far from the sun. Materials that are denser and less volatile tended to be left behind, and these materials formed the inner rocky planets, including Earth (Figure 11-9).

Several steps led to the formation of the planets. When the solar nebula reached a certain density and rate of rotation, it flattened into a disk, and the material of the disk then segregated into rings, which later condensed into the planets (see Figure 11-8). Each

planet began to form by the aggregation of material within one of these rings. The aggregates eventually reached the proportions of asteroids, which are commonly about 40 kilometers (25 miles) in diameter, and some of these coalesced to form planets.

After the planets formed, the rocky debris that we refer to as asteroids remained in orbit around the sun (Figure 11-10). Although some of these asteroids survive to this day, most have collided with larger planets to become part of them. A large swarm of asteroids remains in orbit, forming the asteroid belt between Mars and Jupiter, not far from Earth in the solar system (see Figure 11-9*B*). The motions of these asteroids are perturbed by the gravitational attraction of nearby planets—especially Jupiter because of its large size—and occasionally one leaves its orbit. Most asteroids that have struck Earth during the Phanerozoic Eon have escaped their orbits in this way.

Today comets are concentrated in the Oort cloud, which encircles the solar system far from Pluto, the outermost planet. Despite the great distance of the Oort

Figure 11-10 An asteroid. This false-color picture of 243 IDA, taken by the Galileo spacecraft, reveals many craters. (Jet Propulsion Laboratory, California Institute of Technology.)

cloud from the sun, a comet occasionally plunges into the inner solar system, having been dislodged from the Oort cloud by a passing star or cloud of dust.

The Origin of Earth and Its Moon

As Earth accreted, the impacts of giant bodies, some the size of Mercury or Mars (see Figure 11-9B), are thought to have contributed between half and three-quarters of its mass. These giant bodies must have been quite hot initially and would have become even hotter as their energy of motion turned into heat at impact. Decay of radioactive isotopes within early Earth generated additional heat.

Early melting produced a layered Earth

The heat from these various sources produced a molten planet, in which the most dense material sank toward the center and the least dense material rose toward the surface (Figure 11-11). The result was a predominantly iron core and a mantle of dense silicate minerals. Less dense silicates must have floated to the surface to form what has been termed a **magma ocean**. Eventually this liquid surface layer cooled to form a feldspar-rich crust, which was the precursor of the oceanic crust of the modern world. As we shall see shortly, the continental crust originated later.

Heavy metals were virtually absent from the mantle of the molten planet because they tended to sink to the core. The impacts of large meteorites over hundreds of millions of years increased the concentrations of elements such as gold and platinum in the upper Earth.

The moon formed from a collision

For many decades, scientists debated the origin of Earth's moon. Was it an asteroid captured by Earth's gravitational field, a body that accreted separately from numerous small bodies that once orbited Earth, or a chunk of Earth knocked loose by collision with an asteroid? A variation of the third alternative is now widely accepted.

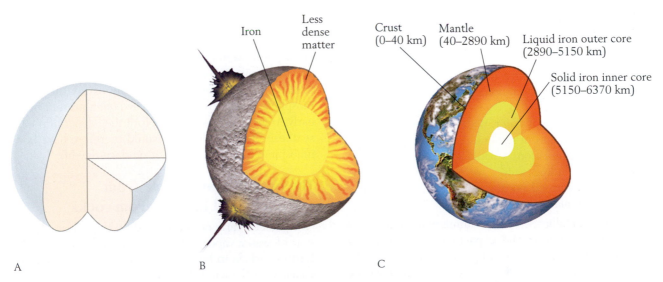

Figure 11-11 The origin of Earth's layers. A. Early Earth accreted as a generally homogeneous body. B. Iron sank to the center of the liquid Earth, and less dense material rose to the surface. C. Eventually the planet became fully differentiated into core, mantle, and crust.

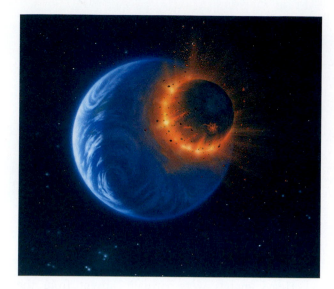

Figure 11-12 **The origin of the moon.** This artist's rendition shows a Mars-sized object striking Earth a glancing blow. The moon is thought to have formed from the impactor's mantle, which broke free. (Painting by Alfred T. Kamajian, *Sci. Amer.*, July 1994, cover. © 1994 by Scientific American, Inc. All rights reserved.)

It now appears that a body about the size of Mars (one-tenth of Earth's mass) formed the moon by striking Earth a glancing blow shortly after our planet's initial accretion (Figure 11-12). The moon is not a chunk of Earth, however; it formed almost entirely from the mantle of the impacting body. This origin accounts for the fact that the proportions of two elements abundant in lunar rocks, iron and magnesium, differ from their proportions in Earth's mantle, from which the moon would have been largely derived if it had broken loose from Earth.

A computer simulation of the kind of glancing impact now thought to have formed the moon has predicted features that the moon actually displays:

1. A virtual absence of water. Water is absent even from the crystal lattices of minerals in lunar rocks. In the computer simulation, volatile elements and compounds, including water, were expelled from the moon as it formed because, owing to its small size, its gravitational attraction is weak.

2. A small metallic core. The simulation showed the core and mantle of the impacting body separating during the impact. The impactor's dense metallic core sank into Earth, joining its core, whereas the impactor's mantle exploded to form a disk of material that encircled Earth, held in orbit by the planet's gravitational field. This orbiting material soon coalesced to form the moon.

3. A feldspar-rich outer layer. The astronauts who reached the moon in 1969 found such a layer everywhere except where volcanoes and asteroid impacts had allowed other material to rise from the moon's mantle. This outer layer resulted from the heat of the moon's formation, which in the simulation produced a magma ocean resembling that of early Earth.

The model also indicated that the heat generated by the moon-forming impact should have been intense enough to melt newly formed Earth and disrupt its layering. The result would have been a new magma ocean—and when it cooled, an entirely new crust on Earth. The impact of a Mars-sized object also explains why Earth rotates more rapidly than it should for its size and position in the solar system. The glancing blow that formed the moon knocked Earth into a faster rate of spin.

Earth's early atmosphere came from within

The asteroids that coalesced to form Earth were too small to permit their gravitational fields to hold gases around them as atmospheres. We can thus conclude that Earth did not inherit its atmosphere from those ancestral bodies. Instead, while it was in a liquid state so that gases could easily escape to its surface, Earth itself must have emitted and retained the gases that formed its atmosphere.

Weaker degassing has continued to the present by way of volcanic emissions. The chemical composition of gases released from modern volcanoes indicates what gases the early atmosphere contained: primarily water vapor, hydrogen, hydrogen chloride, carbon monoxide, carbon dioxide, and nitrogen. Chemical reactions in the atmosphere would also have produced methane (CH_4) and ammonia (NH_3). In the modern world, photosynthesis is responsible for most of the oxygen in the atmosphere. Less oxygen would have been present in the early Archean atmosphere, before the advent of photosynthesis. This atmosphere would therefore have been inhospitable to most forms of modern-day life.

The ocean's water came from volcanoes and its salts from rocks

Our planet's early ocean formed by the volcanic emission of water vapor, which cooled and condensed at Earth's surface to form liquid water. The rain that fell on early Earth would have contained almost no salts because it was derived from the water vapor of volcanic emissions. Salts were brought to early seawater by rivers that carried the products of weathering on land. At the same time, the seawater that seeped into mid-ocean ridges and returned to the ocean carried with it ions of elements such as calcium and potassium that were liberated in the chemical alteration of newly formed oceanic crust. Calculations show that seawater should have become approximately as saline early in Archean time as it is now. Since then, chemical components of seawater have precipitated as carbonate and evaporite sediments about as rapidly as they have been added to the oceans. Thus, seawater's total salinity has varied only modestly, even though the relative proportions of dissolved ions have varied significantly (see Figure 10-20). At the same time, the global water cycle has moved water, but not salts, from the oceans to the atmosphere and back again (see Figure 1-19).

Early Earth experienced many meteorite impacts

Even after a giant impact formed the moon, additional large impacts must have liquefied Earth's crust and mantle repeatedly. For about 800 million years after Earth accreted, large bodies continued to strike it and the other planets of the solar system at a high, though declining, rate. In this way, the planets were sweeping up debris left over from the formation of the solar system. Given the rarity of early Archean rocks, our only evidence that this extraterrestrial shower took place comes from other planets and the moon, which have not undergone the kind of weathering and erosion that constantly alter Earth's surface.

Earthbound observers have long commented on the moon's pockmarked appearance (Figure 11-13).

Figure 11-13 The arrangement of the moon's maria on the side that faces Earth. (Lick Observatory.)

The moon's enormous craters, known as *maria* (singular, *mare*; from the Latin word for "seas"), were first sketched by Galileo. It is only from manned lunar exploration and from photographs provided by artificial satellites that we have gained detailed knowledge of these maria. Those that face Earth have an average diameter of about 200 kilometers (125 miles). At first it was not known whether the maria were craters produced by volcanoes or were the impact scars of huge asteroids. Detailed study of the moon's surface established that asteroid impacts formed the maria. The lunar highlands surrounding the maria consist of rock fragments that testify to the pulverization of the lunar crust by the impacts of falling asteroids. The maria facing Earth are floored by immense flows of dark volcanic basalts, which give the maria their dusky appearance. These basalts, which formed as a result of the heat generated by impacts, are themselves scarred by smaller craters that indicate their relative ages: the older the basalt, the more cratering it has undergone.

Dating of associated rocks has shown that most large lunar craters are quite old, ranging in age from about 3.8 billion years to about 4.6 billion years. Estimates indicate that during the early cataclysmic interval of lunar history, meteorite impacts were more than a thousand times more frequent than they are now. Craters of all sizes are also abundant on planets of the solar system whose surfaces have not been as heavily altered as Earth's. The conclusion seems inescapable that Earth was subjected to the same kind of asteroid bombardment as its moon and nearby planets.

The frequent occurrence of impacts early in the history of the solar system explains why Earth and other planets rotate on tilted axes. Earth's axis tilts at an angle of 23.5° from the plane of its orbit. The planet Uranus must have been struck a more powerful blow, because it lies on its side, with its axis aimed almost directly at the sun (see Figure 11-9B). Venus rotates about a more upright axis but has a retrograde motion, an indication that a large asteroid struck it off center and reversed its direction of spin from that of the other planets.

During the cataclysmic period before 3.8 billion years ago and ever since, the huge planet Jupiter has afforded our planet considerable protection from bombardment. Acting as an immense shield, Jupiter's powerful field of gravity has pulled in many asteroids that would otherwise have struck Earth. Without Jupiter's presence, our planet would have suffered about a thousand times more impacts than it has actually received, and life on Earth would have been devastated by frequent major impacts even after 3.8 billion years ago. Despite the presence of Jupiter, one such crisis did occur 65 million years ago, when the impact of a large asteroid wiped out the dinosaurs and many other forms of life on Earth.

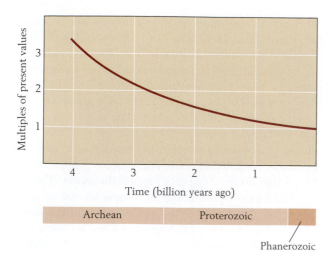

Figure 11-14 The decline of heat production by radioactive decay in Earth's interior. Heat production values are relative to that of the present, which is assigned a value of 1. (After F. M. Richter, *Earth Planet. Sci. Lett.* 68:471–484, 1984.)

Early Earth was hot and its plates were small

During Archean time, heat must have flowed upward through Earth's lithosphere more rapidly than it does today, because Earth's radioactive "furnace" was hotter. (Recall from Chapter 1 that Earth's heat source is constantly diminishing as radioactive isotopes decay without being renewed.) Because particular isotopes decay at constant rates, geologists can calculate the approximate difference between the rate at which Earth produces heat today and the rates of times past. The total rate of heat production was perhaps twice as high near the end of Archean time as it is now (Figure 11-14), and earlier it was even higher. As a consequence, hot spots should have been numerous during the Archean, and the lithosphere should have been fragmented into small plates separated by numerous rifts, subduction zones, and transform faults. As we shall see, the nature of Archean rocks confirms this prediction.

The Origin of Continents

Originally, the many plates into which Earth's crust was divided consisted of basaltic material, like that of modern ocean basins, which formed when the final magma ocean cooled. Only later did felsic material begin to segregate from this mafic crust and the mantle to form nuclei of continental crust. Igneous activity in the modern world reveals how the earliest continental crust probably formed.

Felsic crust formed at hot spots

For continental crust to form, felsic components must be extracted from mafic rocks. Although igneous mate-

rial is extracted from such rocks along subduction zones, the resulting volcanic rocks—those extruded along island arcs—are also mafic or intermediate in composition between mafic and felsic. In contrast, igneous processes associated with hot spots in oceanic crust can produce felsic volcanic rocks—rocks with the chemical composition of granite. Hot spots, then, appear to have been the sites where continental crust came into being.

Iceland is an island formed entirely by volcanic activity at a hot spot along the Mid-Atlantic Ridge (see Figure 8-26). Here the generation of large bodies of felsic igneous rock begins in the lower crust, where small pods and lenses of felsic composition segregate from igneous material derived from the mantle (Figure 11-15). Recall that normal faults run parallel to a mid-ocean ridge, where two plates are diverging (see Figure 8-20). Mafic magma flows along the normal faults beneath Iceland, and its great heat melts the small felsic bodies that the faults intersect. The resulting felsic magma rises readily because of its low density, and some of it forms volcanoes that have the chemical composition of granite. Basaltic volcanoes are more common than felsic ones, but at least 10 percent of Iceland's upper crust is felsic.

As volcanic rocks build up on the surface of Iceland, a second process produces additional felsic magmas. This process is the isostatic sinking of the pile of volcanic rocks to form a root, like that of a mountain. Partial melting of these volcanics extracts additional felsic components, forming magma that later cools to form felsic rocks. A special feature of Iceland allows felsic material to form rapidly: the low rate of crustal spreading here—about 2 centimeters per year—allows many areas of crust to undergo repeated melting and concentration of felsic components before they move away from the zone of igneous activity.

Today, because of its high rate of formation, the crust of Iceland is about twice the average thickness of oceanic crust. Iceland has been forming for only about 16 million years and is still accumulating felsic volcanics.

Presumably, numerous small continents resembling Iceland emerged in the course of Archean time and had time to grow somewhat larger than Iceland before the hot spots that formed them died out when plate tectonic patterns shifted. Such small Archean continents are termed **protocontinents**. Weathering and metamorphism would have generated additional felsic crystalline rocks. First, weathering would have removed iron and magnesium from the mafic rocks of protocontinents, leaving behind clays of felsic composition that would have eventually formed shales. Then metamorphism would have converted some of these shales into more durable felsic metamorphic rocks. In addition, small protocontinents would have become sutured together to form larger protocontinents.

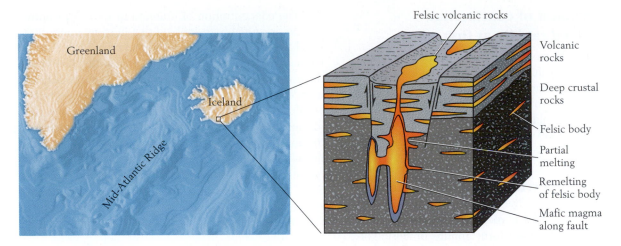

Felsic volcanic rocks

Volcanic rocks

Deep crustal rocks

Felsic body

Partial melting

Remelting of felsic body

Mafic magma along fault

Greenland

Iceland

Mid-Atlantic Ridge

Figure 11-15 The origin of felsic volcanic rocks in Iceland, which is situated over a hot spot along the Mid-Atlantic Ridge. Small felsic bodies that have formed in deep ocean crust are melted by mafic magma moving along faults. The resulting felsic magma rises to the surface. Additional felsic magma forms by the isostatic sinking and partial melting of the pile of volcanics. (After B. Gunnarsson, B. D. Marsh, and H. P. Taylor, *J. Volcanol. Geotherm. Res.* 83:1–45, 1998.)

Archean continents remained small because of Earth's hot interior

It appears that protocontinents failed to produce huge continents on Earth during most of Archean time because broad blocks of crust as old as 3 billion years are absent from Precambrian shields. Archean crustal areas embedded in modern continents are shown in Figure 11-16. Some represent Archean protocontinents sutured together during Proterozoic time. Despite partial destruction by erosion and subduction, some large Archean shield areas would remain today if Archean cratons had been as large, on average, as those of the present world.

It was heating from below that prevented Archean protocontinents from coalescing. Just as numerous rifting events maintained plates of oceanic lithosphere at small sizes, they prevented blocks of continental crust from coalescing to form large continents.

Strong heat flow from the mantle not only prevented continents from becoming large, but also prevented them from attaining the thickness of modern continents. The thinness of the Archean continental crust is preserved in the modern world: areas of existing cratons that are Archean in age are thinner, on average, than are younger portions of cratons.

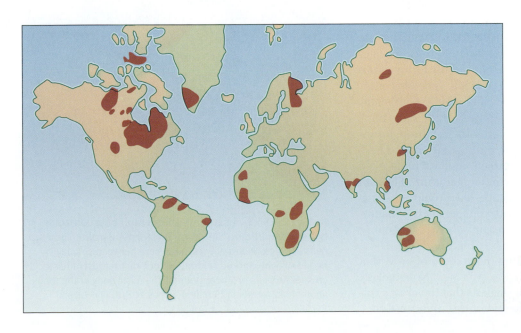

Figure 11-16 The locations of Archean shields in the modern world.

The total amount of continental crust increased rapidly at first

Because most early Archean crustal rocks have been destroyed by metamorphism and erosion, we have no direct evidence of the rate at which continental crust approached its present volume. Some crustal materials are about 4 billion years old, however, and indirect evidence indicates that the total volume of crust approached that of the present by the end of Archean time.

The oldest known crustal materials Small grains of the mineral zircon are the only identifiable remnants of very early Archean felsic rocks. Zircon crystals form when molten rock cools to form granite or felsic volcanic rock. Clastic grains of zircon are so resistant to heat and weathering that they often survive unaltered in metamorphic and sedimentary rocks formed from the igneous rocks in which they crystallized. Fortunately, zircon grains also contain small quantities of uranium, so their ages can be measured. The oldest zircon grains yet dated, from weakly metamorphosed sandstone in Western Australia, give ages of 4.1 billion to 4.2 billion years. These grains are the oldest known remnants of continental crust, having weathered from felsic igneous rock—and perhaps having spent time in various other kinds of rock—before finally lodging in sandstone.

The oldest body of actual continental crust yet discovered is in the northwestern corner of the Canadian Shield. Here radiometric dating of a body of felsic rocks of the Acosta Formation gives ages that range from 3.8 billion to 4.0 billion years. A block of continental crust in Western Australia is also older than 3.5 billion years. Given that continental crust began to form early in Archean time and resided in protocontinents, the question remains: how rapidly did this crust grow in total volume?

The overall growth of continental crust The rate at which the total volume of continental crust changes at any time is the rate at which new crust forms minus the rate at which old crust disappears. Igneous contributions from the oceanic crust and mantle continually add to the total volume of felsic crust. At the same time, felsic material is lost through erosion and through subduction of small amounts of continental crust. These processes of addition and subtraction slowly replace old continental crust with new continental crust. Through the ages, erosion and subduction have removed much Archean crust, replacing it with younger crust. What remains from the Archean represents about 7 percent of the modern continental crust (see Figure 11-16). There is no direct way of measuring the volume of crust that existed at any time during the Archean. Nonetheless, the distribution of uranium suggests that the crust's volume was quite large by the end of the Archean Eon.

The concentration of uranium in Earth's primitive mantle presumably resembled that in stony meteorites, which is well known. As continental crust has grown, it has accumulated uranium from the mantle. The larger the crust has grown, the more depleted of uranium the mantle has become. Mafic rocks 2.7 million years old, which reflect the composition of the mantle when they formed, are just as depleted of uranium, in comparison to stony meteorites, as the modern mantle is. This means that by the end of Archean time continental crust had grown to approximately its present total volume. Since that time, destruction of crustal material has approximately balanced additions of material to the crust from the mantle.

Greenstone Belts

Archean rocks reveal a world that differed in interesting ways from that of the Proterozoic and Phanerozoic. The rocks themselves differ in average composition from younger rocks. For example, during Archean time, numerous volcanic arcs produced large volumes of dark igneous rocks. In addition, large bodies of dark sedimentary rocks formed from the erosion of these volcanic rocks.

The typical configuration of Archean terranes is evident in satellite photographs of shield areas (Figure 11-17). A **terrane** is a geologically distinctive region of

Figure 11-17   A satellite photograph of greenstone belts in the Pilbara Shield of Western Australia. These belts are dark bodies of rock, seen here between circular bodies of light-colored crystalline rock that represent felsic crust of Archean protocontinents. The felsic body at the top is about 40 kilometers (25 miles) across. (CSIRO Division of Exploration Geoscience, Australia.)

Earth's crust that has behaved as a coherent crustal block. Over broad regions, elongate bodies known as **greenstone belts** sit in masses of high-grade metamorphic rocks of felsic composition (gneisses, for example). Rocks of the greenstone belts themselves are generally weakly metamorphosed; in fact, the green metamorphic mineral chlorite gives them their name.

The igneous rocks of greenstone belts, before they were metamorphosed, were mostly mafic and ultramafic volcanic rocks of the kind extruded along volcanic arcs, with felsic volcanic rocks present in smaller volumes. Many of these volcanics display pillow structures, which indicate that the lava that formed them was extruded under water (Figure 11-18; see also Figure 2-14). Many sedimentary rocks of greenstone belts, though now metamorphosed, can be seen to have originally formed from detritus eroded from dark volcanic rocks. In the absence of large continental shelves, it is not surprising that most Archean sedimentary rocks formed in deep water. They include graywackes, mudstones, iron formations, and sediments derived from volcanic activity. Metamorphosed turbidites are common, as are dark mudstones, now mostly metamorphosed to slate. Many of these sediments were deposited in forearc basins and other environments situated along subduction zones.

Cherts and the iron-rich sedimentary rocks known as **banded iron formations** are also found in the Archean sedimentary belts. Dated at about 3 billion

Figure 11-19 **Some of the oldest known rocks on Earth.** These banded iron formations from the Isua area of southern Greenland were deposited as sediments at least 3.8 billion years ago. Individual layers, which are deformed and weakly metamorphosed, are a few millimeters thick. Layers of chert (fine-grained quartz) alternate with darker layers that contain iron minerals. (Preston Cloud and David Pierce.)

years, the banded iron formation at Isua, in southern Greenland, represents one of the oldest known bodies of sedimentary rock on Earth. In fact, the Isua rocks are weakly metamorphosed (Figure 11-19). Like other Precambrian banded iron formations, these rocks consist of iron oxide-rich layers that alternate with quartz layers. The Isua rocks are believed to have originated by precipitation in marine basins, and the quartz within them is thought to have existed initially as chert precipitated from seawater. Submarine volcanic eruptions were the source of most of the dissolved silica that was precipitated as Archean chert. Bacteria may have played a major role in the precipitation of the iron in banded iron formations. Some bacteria in the modern world oxidize iron much more rapidly than it oxidizes inorganically in natural waters.

The rocks that today form greenstone belts came to lie within Archean protocontinents through continental accretion (Figure 11-20). Having formed along subduction zones, they were wedged between protocontinents that collided and became sutured together. Because the protocontinents were so small and the subduction zones so abundant, greenstone belts formed frequently and came to constitute a large proportion of Archean terranes.

Major metamorphic episodes occurred in many parts of the world between about 2.7 billion and 2.3 billion years ago. Why this widespread metamorphism occurred

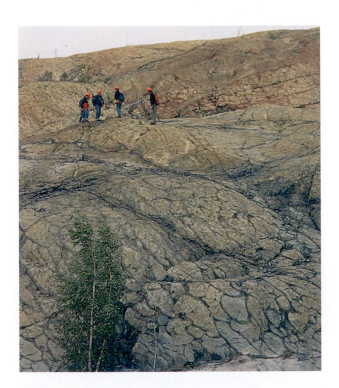

Figure 11-18 **Archean pillow basalts in the Yellowknife region of Canada.** The pillows have been planed off by erosion. (Paul F. Hoffman.)

Figure 11-20 The formation of greenstone belts. Forearc basin sediments, deformed oceanic crust, and arc volcanics along the margins of protocontinents (*A*) became squeezed between protocontinents and were metamorphosed during suturing, to become podlike greenstone belts (*B*) in a larger protocontinent.

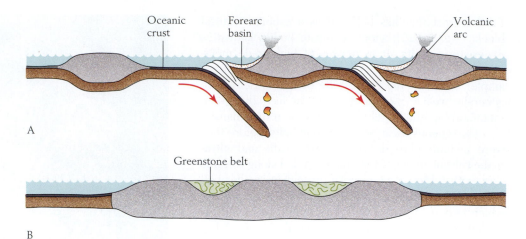

is unclear, but it reset many radioactive clocks and consolidated many small crustal elements into sizable cratons. By this time, it was finally possible for large continents to form because Earth's interior was generating less heat than it had before, and rifting zones, which tear apart continental crust, were less extensive (see Earth System Shift 11-1).

Even at the close of Precambrian time, large continents remained different from those of the present in one important respect: they were barren of advanced forms of life. Long before the first large cratons existed, however, living cells had begun to populate the marine realm, where they remained at a primitive stage of development for a billion years or more.

Evidence of Archean Life

Of all the planets in our solar system, only Earth is well suited to life as we know it. One of the reasons is that its size is right. A much larger planet's gravitational pull on its atmosphere would be so great that the atmosphere would be too dense to admit sunlight, which is the fundamental source of energy for life. A much smaller planet would lack sufficient gravitational attraction to retain an atmosphere with life-giving oxygen. In addition, Earth's temperatures are such that most of its free water is liquid, the form of H_2O that is essential to life. Even Venus, our nearest neighbor closer to the sun, is much too hot for water to survive in a liquid state. Mars, our nearest neighbor farther from the sun, has a cooler surface, but its atmosphere is so thin that liquid water would evaporate from the planet's surface almost immediately. Evidence indicates, however, that water once flowed over the surface of Mars (Figure 11-21), and it is speculated that life may have evolved independently there long ago. If so, we may someday find fossils in Martian rocks.

Life originated shortly after 4 billion years ago

It is estimated that living things could not have survived on Earth until about 4.2 billion years ago. Before that time, giant asteroid impacts would have sterilized the planet by generating enormous amounts of heat. For example, the impact of an asteroid 500 kilometers in diameter would have created a hot atmosphere of vaporized rock and caused the ocean to evaporate temporarily. This leaves a window of at most 700 million years for the appearance of life, because Western Australian

Figure 11-21 Branching valleys that were formed by flowing water on Mars, long ago in the planet's history. (NASA/JPL/Malin Space Science Systems.)

rocks about 3.5 billion years old contain abundant organic compounds containing a relatively large percentage of carbon 12. Such isotopically light carbon compounds result only from photosynthesis. Among prokaryotic organisms, photosynthesis is confined to certain eubacteria. Thus, we have indirect evidence that eubacteria, if not archaebacteria, were present close to a billion years after Earth's origin.

Presumably, then, life evolved on Earth between 4.2 and 3.5 billion years ago, but it remains possible that the planet's first forms of life somehow arrived from outer space aboard an asteroid. Such an introduction of early life may never be documented with geologic evidence.

Prokaryotes left a fossil record

Archean fossils appear to represent only prokaryotes, and most major eubacterial and archaebacterial groups of the modern world seem to have appeared before the end of Archean time.

In the 1950s paleontologists were astounded by the discovery of molds of individual prokaryotic cells in Precambrian cherts. Chert forms by hardening of gelatinous silicon dioxide (SiO_2), and the finely crystalline quartz formed in this way can faithfully preserve the shapes of individual cells. This kind of preservation has shed much light on the Precambrian evolution of unicellular organisms. South African cherts 3.4 billion years old have yielded spherical structures that may be fossilized cells. They consist of organic matter, and some appear to have been preserved in the act of cell division. These ancient structures are too small to have been

Figure 11-23 **Examples of the oldest known stromatolites.** These fossils, from Western Australia, are 3.45 billion years old. They include cone-shaped structures, and their layers extend across the depressions that separate these elevated features. (H. J. Hofmann.)

eukaryotes, and it is likely, but not certain, that they are fossil prokaryotes. The oldest unquestioned thread-shaped fossils of single-celled organisms come from 3.2-billion-year-old cherts from Australia (Figure 11-22). These fossils are intertwined filaments that must have formed tangled mats on the seafloor.

Stromatolites in Western Australia suggest that photosynthesis was occurring on Archean seafloors by about 3.5 billion years ago (Figure 11-23). Probably cyanobacteria played a major role in the production of these ancient layered structures, as they do today. It remains possible, however, that nonphotosynthetic bacteria produced the structures.

Biomarkers in strata about 2.7 billion years old leave little question that cyanobacteria were present at that time. These biomarkers are organic compounds that strengthen the membrane bounding the cell wall of a cyanobacterium. Stromatolites become increasingly abundant toward the top of the Archean rock record. This change may in part reflect an increase in the area of shallow seafloor—the habitat where stromatolites grow—as continents increased in size.

Some rocks about 2.8 billion years old contain carbon that is isotopically very light, even for organic material. Carbon so enriched in the light isotope carbon 12 is produced only by methane-producing archaebacteria (p. 236). We can therefore conclude that these particular forms were also present by late Archean time.

Archean rocks have yielded no fossils that appear to represent organisms with *eukaryotic cells*—cells that contain chromosomes and nuclei. Even so, steranes, which are organic compounds formed only by eukaryotes today, occur in rocks about 2.7 billion years old. Apparently eukaryote ancestors, if not eukaryotes themselves, were present late in Archean time.

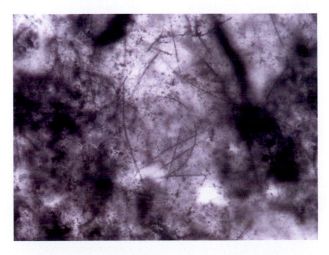

Figure 11-22 **The oldest known fossils?** These threadlike structures, about 2 μm across, are preserved in 3.2-billion-year-old rocks from northwestern Australia. They have been altered to pyrite and may represent archaebacteria or eubacteria. (Birger Rasmussen, University of Western Australia.)

Earth System Shift 11-1 | Large Cratons Appear near the End of Archean Time

Throughout Phanerozoic time, quartz sands have accumulated extensively on continental lowlands and—along with carbonate sediments—in shallow seas. It is striking that comparable bodies of sediment are uncommon in the Archean record. The apparent reason for their sparse occurrence is that no large continents existed early in Earth's history. Archean protocontinents provided only narrow lowlands and continental shelves for sediment accumulation. Many were bordered by subduction zones.

Only after heat flow from Earth's interior diminished substantially as the planet's radioactive furnace wound down did protocontinents coalesce to form large continents. There is evidence that this "cratonization" did not occur simultaneously throughout the world, however. In most areas, typical Archean greenstone belts formed until approximately 2.5 billion years ago, but in southern Africa a large craton was already present about a half billion years earlier. Here, between about 3.1 and 2.7 billion years ago, a large body of sedimentary rocks formed in what are known as the Pongola and Witwatersrand basins.

The great extent of these strata indicates that the sand and mud that formed them were eroded from a sizable continental area. Some deposits of the Pongola basin are strikingly similar to intertidal sequences of younger portions of the stratigraphic record. The Witwatersrand strata accumulated in nonmarine environments to the west. They have yielded abundant small clasts of detrital gold, whose great density caused it to become concentrated with larger silicate pebbles in braided-steam deposits that now form conglomerates. The gold clasts were eroded from rocks that formed at high temperatures within the new continent from magma that rose from the mantle.

Among the Pongola rocks are glacial tillites and dropstones. Having formed about 2.9 billion years ago, these are the oldest known glacial deposits on Earth. Obviously the Archean craton where they accumulated was large enough to experience severe winters inland from the ocean. More generally, these deposits provide evidence that in late Archean time Earth's climate resembled that of today.

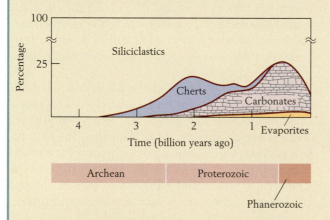

Figure 1 Relative percentages of various kinds of sedimentary rocks have changed in the course of Earth's history. Cherts are relatively abundant in late Archean and early Proterozoic rocks; most were deposited in relatively deep water, commonly in banded iron formations. Carbonates became much more abundant during the Proterozoic, as continents and continental shelves expanded. Evaporites appear to have been rare before mid-Proterozoic time, but their rarity in Archean rocks is partly a matter of preservation because these rocks dissolve readily in water. (After A. B. Ronov and A. A. Yaroshevskiy, *Geochim. Int.* 4:1041–1069, 1969.)

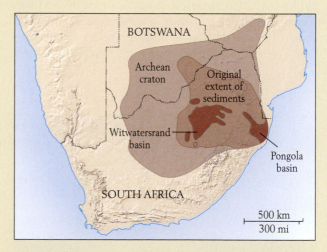

Figure 2 Sedimentary rocks about 3 billion years old in southern Africa document the existence of the oldest recognized continent of substantial proportions. Deposits of the Witwatersrand and Pongola basins accumulated in nonmarine and shallow marine environments from the erosion of large bodies of rock. (After C. R. Anhaeusser, *Phil. Trans. R. Soc. Lond.* A 273:359–388, 1973.)

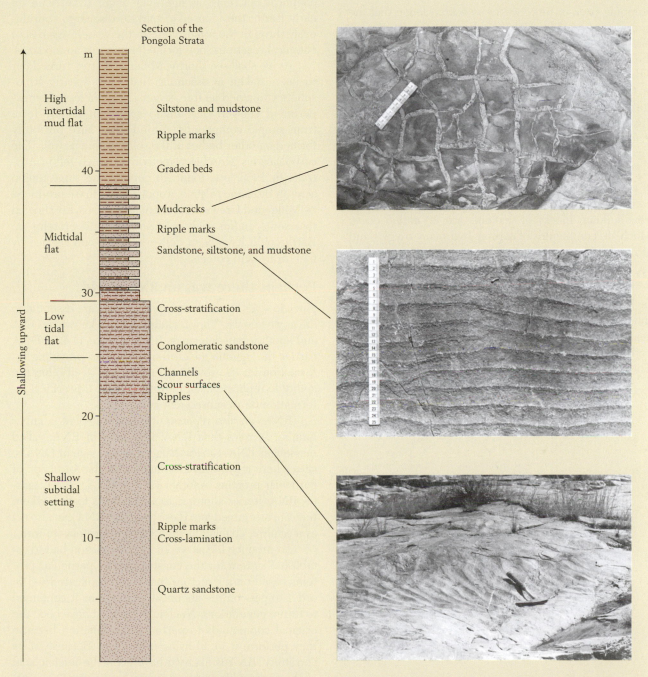

Figure 3 The Pongola strata of southern Africa displays extensive shallow marine deposits. The section depicted here includes a late Archean regressive sequence in which a tidal flat prograded seaward over subtidal environments. In the lower, sandy portion, cross-stratification and symmetrical ripples are common. Some of these ripples are double-crested, reflecting the ebb and flow of tides. Tidal channel deposits floored by pebbles are present, particularly in the upper part. Above them, the ripples and mudcracks of the sand-mud member point to a shallower, midtidal environment. The uppermost mud member bears smaller mudcracks as well as mud chips and represent a high intertidal mud flat. (After V. von Brunn and D. K. Hobday, *J. Sediment. Petrol.* 46:670–679, 1976.)

Chemical Evidence Bearing on the Origin of Life

Stepping back in time, we must acknowledge that the rock record provides little evidence concerning the actual origin of life. For clues we must instead consider the likely nature of primitive life. Recall from Chapter 3 that two essential attributes of life are self-replication, or the ability to reproduce, and self-regulation, or the ability to sustain orderly internal chemical reactions. Sustaining chemical reactions requires energy, such as that provided by respiration.

Many of the compounds that life requires for self-replication and self-regulation are *proteins*. Some kinds of proteins form physical structures, and others enable particular chemical reactions to take place within cells. The building blocks of proteins are 20 *amino acids*, which are compounds of carbon, hydrogen, oxygen, and nitrogen.

Amino acids formed easily

In 1953 Stanley Miller and Harold Urey reported on a simple laboratory experiment in which they produced nearly all of the amino acids found in proteins. The experiment was designed to mimic the conditions under which life arose on Earth. In a closed vessel, above a pool of boiling water, the researchers created a primitive "atmosphere" of hydrogen, water vapor, methane (CH_4), and ammonia (NH_3) (Figure 11-24). To trigger chemical reactions, as lightning might have done on early Earth, they caused a spark to discharge continuously through the atmosphere in the vessel. A series of chemical reactions soon formed numerous amino acids.

As we shall see shortly, Miller and Urey turned out to be mistaken in assuming that Earth's early atmosphere contained no free oxygen. Nonetheless, their experiment showed that amino acids can readily form from simple compounds. Amino acids have obviously formed on other bodies of the solar system as well. The carbonaceous Murchison meteorite, which fell in Australia in 1969, was found to contain the same amino acids in the same relative proportions as those produced by Miller and Urey's experiment. This discovery showed that some amino acids incorporated into proteins on Earth could have been delivered from outer space by meteorites and comets.

Perhaps there was an RNA world

Nucleic acids are other compounds that are essential to life as we know it. They include one type known as DNA and another type known as RNA (Figure 11-25). DNA carries the genetic code of an organism, providing information for its growth and regulation. It also has the ability to replicate itself in order to pass this critical information on to subsequent generations.

RNA can also replicate itself, and it plays a larger number of roles than DNA. One kind of RNA, called messenger RNA, carries the genetic message of DNA to sites where it provides information for the formation of particular proteins. Another kind of RNA, called transfer RNA, ferries appropriate amino acids to sites where they are assembled into these proteins. RNA can also act as a catalyst, enabling certain kinds of proteins to form.

To produce life as we know it, evolution had to establish a system for the construction of particular proteins based on a structure that could replicate itself in order to pass on the chemical instructions it contained to future generations. Nucleic acids must have been the original compounds to perform this function because they perform it in all living organisms. Because of its versatility, RNA is likely to have been the nucleic acid of the earliest life forms. It could have served as a catalyst for the production of key proteins, and it could also have replicated itself in order to pass its coded message on to descendants. Thus, most experts now envision an early global ecosystem known as the *RNA world*. Once the RNA system was in place, Darwinian evolution was possible, with natural selection operating on occasional mutations of the RNA molecule. Eventually DNA, a more stable molecule, evolved to replace RNA as the genetic code.

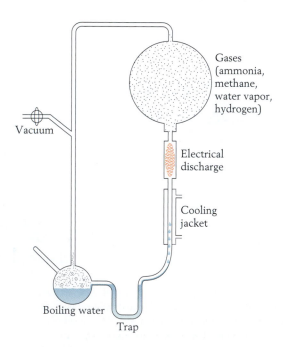

Gases (ammonia, methane, water vapor, hydrogen)

Vacuum

Electrical discharge

Cooling jacket

Boiling water

Trap

Figure 11-24 **The laboratory apparatus in which Miller and Urey produced amino acids.** They circulated an "atmosphere" of ammonia (NH_3), methane (CH_4), water vapor (H_2O), and hydrogen past an electrical discharge. Amino acids accumulated in the trap. (After G. Wald, *The Origin of Life*.)

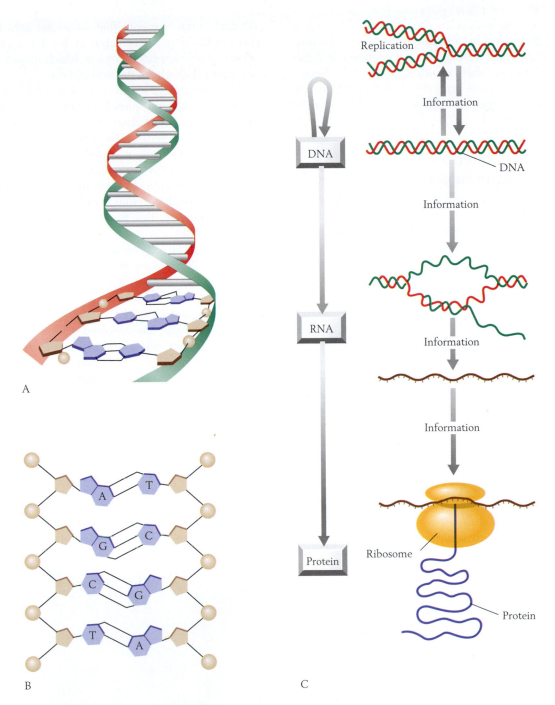

Figure 11-25 **The structure and function of nucleic acids.** *A.* Each strand of the double helix of DNA consists of a chain of nucleotide units. A nucleotide unit includes a phosphate group (sphere), a sugar (pentagon), and a nitrogenous base (blue). The nitrogenous bases bond the two strands of the helix together. *B.* Of the four kinds of bases, adenine (A) and thymine (T) are mutually attached by a double bond, and cytosine (C) and guanine (G) are mutually attached by a triple bond. *C.* For replication, the two strands of DNA separate, and each then duplicates itself. The double helix also separates to allow messenger RNA to translate a portion of the code carried within the bases of DNA. Messenger RNA carries this translation to a site called a ribosome, where it specifies a sequence of amino acids that will form a particular protein. Transfer RNA brings the appropriate amino acids to assemble the protein.

At a very early stage in the history of life, organisms must have evolved a protective external structure. This structure must have been a semipermeable membrane like the one that bounds modern cells. Such a membrane would have protected the chemical system of the primitive organism, allowing only a few kinds of compounds to pass in and out.

Life may have originated along mid-ocean ridges

Whether life arose on Earth or arrived from outer space, it could not have survived until a few hundred million years after Earth's formation, when bombardment by asteroids subsided substantially. Even after Earth's surface had stabilized, however, certain conditions were required for the origin of life.

Miller and Urey's experiment produced amino acids. As a result, they concluded that precursor compounds and ultimately life itself arose in small, ponded bodies of water that were struck by lightning and turned into what is sometimes referred to as the "primordial soup." Even Charles Darwin speculated that life may have first arisen in "a warm little pond." The problem with this idea is that it would have required an atmosphere lacking free oxygen, because even a small amount of O_2 would have oxidized, and thereby destroyed, the chemical raw materials necessary for the production of essential organic compounds.

Knowing that photosynthesis produces the preponderance of oxygen in Earth's atmosphere today, scientists once assumed that the atmosphere lacked free oxygen before the origin of photosynthetic organisms.

We now know, however, that ultraviolet light from the sun breaks down water vapor in Earth's upper atmosphere, slowly liberating oxygen, which spreads in small quantities throughout the atmosphere.

Life must have originated not in a small pond that was exposed to atmospheric oxygen, but in some environment that was isolated from Earth's atmosphere. The most likely setting was a warm area beneath the seafloor in the vicinity of a mid-ocean ridge.

Heat that rises from Earth's mantle along mid-ocean ridges warms seawater that has percolated into the crust through pores and cracks. Because heating reduces its density, this water rises back to the ocean (Figure 11-26A). In some areas it flows from the seafloor through large vents as columns of very hot water (Figure 11-26B). Many kinds of eubacteria and archaebacteria inhabit the warm water of ridge environments, occupying pores, cracks, and vents. They live in a variety of ways, but most of them make use of chemicals that the hot water has dissolved while moving through the ridge system. Some of these simple organisms live in water warmer than 100°C, which remains in a liquid state because of the great pressure applied by the ocean above. Others live in lukewarm water farther from ridge axes. These high-temperature prokaryotes may be inhabiting the kind of setting where life originated.

The principle that most warm-adapted prokaryotes put into practice to obtain energy is quite simple: they harness the energy of naturally occurring chemical reactions. Many of the chemical entities that emerge from deep within mid-ocean ridges are not in chemical equilibrium after the rising water in which they are dissolved cools and mixes with seawater; as a result, these chem-

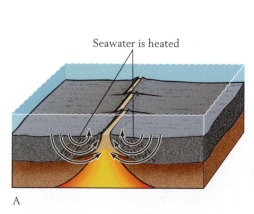

Seawater is heated

A

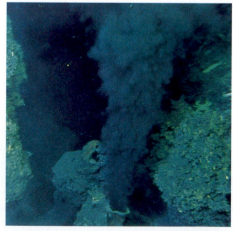

B

Figure 11-26 **The setting where warm-adapted eubacteria and archaebacteria flourish along mid-ocean ridges.** Water that penetrates the crust is heated, dissolves minerals, and rises to return to the ocean (*A*). Some of this warmed water simply seeps out of the seafloor, but some of it spouts out of chimneylike structures formed by sulfides and other minerals that precipitate as the rising water cools (*B*). A wide range of thermal and chemical conditions beneath the seafloor along mid-ocean ridges offer varied habitats for life. (*B*, Woods Hole Oceanographic Institute.)

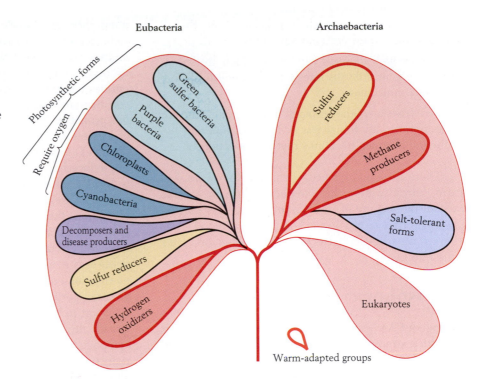

Eubacteria   Archaebacteria

Figure 11-27 Phylogenetic relationships of major groups of prokaryotes to one another and to the eukaryotes. Note the basal positions of warm-adapted eubacteria and archaebacteria and the origin of the eukaryotes from the archaebacteria. No archaebacteria are known to be photosynthetic.

ical entities enter into chemical reactions. Many of these reactions do not occur quickly, however, and warm-adapted prokaryotes take advantage of this situation. These simple organisms consume the chemical compounds and employ enzymes to speed up the chemical reactions, which release energy that the organisms harness for their metabolism.

Some of the warm-adapted prokaryotes are producers, but unlike photosynthetic organisms, they do not use light as an energy source. The processes they employ are collectively termed **chemosynthesis**. Many chemosynthetic reactions are quite simple. Here are some examples of these reactions, for which Figure 11-27 illustrates some of the performers:

Hydrogen oxidation:

$$2H_2 + O_2 \rightarrow 2H_2O + energy$$

hydrogen oxygen   water

Sulfur reduction:

$$S + H_2 \rightarrow H_2S + energy$$

sulfur hydrogen hydrogen sulfide

Methane production:

$$CO_2 + 4H_2 \rightarrow CH_4 + 2H_2O + energy$$

carbon dioxide hydrogen methane water

Biologists have studied the DNA and RNA of many kinds of eubacteria and archaebacteria to reconstruct the prokaryotes' evolutionary tree (see Figure 11-27).

Features of DNA and RNA shared by many prokaryotes are regarded as primitive features that were inherited from very ancient ancestors. It turns out that the warm-adapted prokaryotes that live in the vicinity of mid-ocean ridges are the most primitive living prokaryotes on Earth today. This finding suggests that prokaryotes originated in such habitats, perhaps in the vicinity of mid-ocean ridges.

In fact, mid-ocean ridges exhibit several features that would have made them likely sites for the evolution of very primitive organisms—and even for the origin of life:

1. Their enormous size offered a large range of temperatures, which provided many opportunities for key evolutionary events to take place.

2. Organic compounds of the kind required for the origin of early life readily dissolve in their warm waters. Furthermore, many of these waters would have been anoxic, so that they could have protected those compounds essential to life that are destroyed by free oxygen.

3. They offer protection from the sun's ultraviolet radiation, which was very intense at Earth's surface before there was an abundance of atmospheric O_2, which produces ozone (O_3) in the upper atmosphere. Ozone shields modern Earth from ultraviolet radiation.

4. They are unusual environments in offering an abundance of phosphorus, an element that all organisms require in substantial quantities.

5. They contain metals, such as nickel and zinc, that all organisms require in trace quantities.

6. They are well supplied with clays, which are known to serve as useful substrates for the assembly of large organic molecules.

7. As we have seen, they provide simple organisms with the opportunity to harness a variety of naturally occurring chemical reactions that release energy.

There is a good chance, then, that life evolved in warm, anoxic waters that circulated through oceanic crust in the vicinity of mid-ocean ridges. This may be where the RNA world began, and in all likelihood it was where the earliest eubacteria and archaebacteria later came into being.

Atmospheric Oxygen

Throughout Archean time the concentration of oxygen in Earth's atmosphere remained far below that of the present. Late in Archean time, O_2 was being released by cyanobacteria through photosynthesis and by other bacteria that broke down water molecules. These sources greatly augmented the small amount of oxygen that had been liberated from water vapor in the upper atmosphere since early in Earth's history. Nonetheless, as bacteria began to liberate oxygen, reduced chemical entities—ones with the potential to be oxidized—soaked up oxygen about as rapidly as it appeared. Thus, these entities constituted sinks for oxygen. A **sink** is a reservoir that grows so as to take up a chemical as rapidly as it is produced. The largest of these sinks in Earth's early crust were reservoirs of reduced iron and sulfur. Not enough O_2 was liberated in Archean time to oxidize all the reduced iron and sulfur that were exposed to air and water. Therefore O_2 failed to build up in the atmosphere toward its present level. As evidence of this low O_2 concentration, the mineral pyrite (FeS_2), which contains both reduced iron and reduced sulfur, moved great distances in streams without being oxidized. As a result, pyrite was commonly deposited with siliciclastic grains in Archean sediments. In the present, oxygen-rich world, grains of pyrite weather so rapidly that very few survive transport to be deposited as detrital sediment; they are quickly altered at Earth's surface to form highly oxidized compounds or complex ions of iron and sulfur. A sharp reduction in the deposition of pyrite as sediment shortly after the end of the Archean Eon forms part of the evidence for a significant buildup of oxygen in Earth's atmosphere at that time. This important change will be a central topic of the next chapter.

Chapter Summary

When and how did Earth and its moon come into being?

Radiometric dating has revealed that stony meteorites, which represent the primitive material of early bodies of the solar system, are 4.6 billion years old, as are the most ancient moon rocks. This, then, is the apparent age of Earth and the other planets of the solar system. Earth originated by condensation of material that had been part of a rotating dust cloud. Shortly thereafter, the moon originated when a body the size of Mars struck Earth a glancing blow; the moon formed from the mantle of the impacting body. Between the time it formed and slightly later than 4 billion years ago, Earth was pelted by large numbers of meteorites. During the same interval, meteorites produced most of the large craters that are still visible on the moon, whose surface is less active than Earth's.

How did the core, mantle, oceanic crust, and continental crust form?

Earth was liquefied by the impact that formed the moon, and probably by later giant impacts. Earth became stratified into core, mantle, and oceanic crust because material of high density sank toward its center. Continental crust formed at hot spots, where felsic components were extracted from mafic and ultramafic materials. The formation of large continents was inhibited in early and middle Archean time by the abundance of radioactive elements, whose decay produced heat at a high rate. Under conditions of high heat flow, Earth's crust was divided into small protocontinents, which probably formed by the accumulation of felsic material above hot spots.

Where did Archean rocks form, and what is their nature?

Most Archean rocks occur in greenstone belts. Rocks of these belts consist largely of metamorphosed dark volcanic rocks and the sedimentary rocks that formed from them, along with banded iron formations, in deep waters alongside small continents.

When and why did large continents begin to form?

Large continental landmasses apparently did not form until late in Archean time. The oldest of these landmasses now recognized are in South Africa, where shallow marine and nonmarine siliciclastic sediments were spread over sizable areas about 3 billion years ago. Large continents were able to form only after Earth's radioactive furnace had reduced its rate of heat production to the degree that the number of rift zones declined markedly.

Where did life arise, and what kinds of life existed at the end of Archean time?

Life may have arisen along mid-ocean ridges, where temperatures were warm, free oxygen was nonexistent, and other conditions were favorable. All known Archean fossils appear to represent prokaryotes, which are the most primitive forms of cellular life on Earth today, lacking cell nuclei and chromosomes. Prokaryotes are preserved as outlines of fossil cells, and cyanobacteria also formed stromatolites.

Why did relatively little free oxygen accumulate in Earth's atmosphere throughout Archean time?

Chemical sinks, mainly in the form of reduced ions and compounds of sulfur and iron, were so abundant that they soaked up nearly all oxygen released by photosynthesis and other chemical processes. As a result, the concentration of oxygen in the Archean atmosphere remained low.

Review Questions

1. What is a Precambrian shield? Where is one located in North America?

2. What reasons are there to believe that Earth was pelted by vast numbers of meteorites early in its history?

3. Why might we expect Earth to be nearly the same age as its moon and the material that forms meteorites?

4. What geologic features characterize greenstone belts, and how did these belts form?

5. What types of sedimentary rocks were rare in the Archean Eon? What does this suggest about the nature of cratons during Archean time?

6. Why did magma rise from the mantle to Earth's surface at a higher rate during Archean time than it does today?

7. What features make Earth a more hospitable place than other planets for life as we know it?

8. Why is it likely that life arose in the vicinity of mid-ocean ridges?

9. What are stromatolites? From what we know of their formation today, why might we expect them to have been present early in Earth's history?

10. How do the modes of life of certain living eubacteria and archaebacteria shed light on early evolution?

11. Why did O_2 remain at a low concentration in the atmosphere until long after photosynthesizing cyanobacteria were very abundant on Earth?

12. The composition and configuration of Earth's crust changed more profoundly in the course of Archean time than during any later interval of Earth's history. Using the Visual Overview on page 248 and what you have learned in this chapter, describe major changes in the Archean crust and explain how they relate to one another and to changes in Earth's deep interior.

Stromatolites along Great Slave Lake, Canada. These fossil structures have been planed off by glaciers. (Ron Redfern, author/photographer, "Origins." Reproduced with permission of Orion Publishing Group, London, and Oklahoma University Press, USA.)

CHAPTER 12

The Proterozoic Eon of Precambrian Time

The Proterozoic Eon, which succeeded the Archean Eon 2.5 billion years ago, was in many ways more like the Phanerozoic Eon, in which we live. We have already seen a foreshadowing of this difference between the Proterozoic and Archean eons in the origin of large cratons late in Archean time. The persistence of large cratons throughout the Proterozoic Eon produced an extensive record of deposition in broad, shallow seas—a pattern that differed substantially from the Archean record of deep-water deposition, which is now confined largely to greenstone belts and adjacent areas. In addition, more Proterozoic than Archean sedimentary rocks remain unmetamorphosed and are therefore accessible for study.

The extensive deposits of Proterozoic age document ancient mountain-building events that are strikingly similar to those of the Appalachian Mountains and other younger orogenic belts, and they reveal records of major intervals of glaciation, at least two of which affected most of the world. Proterozoic rocks also harbor a fossil record of organic evolution that reveals a transition from the simplest kinds of single-celled organisms at the start of the eon to more advanced single-celled forms and finally to multicellular plants and animals, some of which belonged to modern phyla. This fossil record provides one of the methods by which geologists divide the Proterozoic into three eras: the Paleoproterozoic; the Mesoproterozoic, which began 1.6 billion years ago; and the Neoproterozoic, which began 1 billion years ago.

Global events are one of the subjects of this chapter, but we will also view the Proterozoic world on a regional scale and learn how the modern continents began to take shape.

A Modern Style of Orogeny

As we saw in Chapter 11, cratons of modern proportions first began to form about 3 billion years ago, late in Archean time. This was when the oldest sedimentary deposits to be laid down over broad continental areas accumulated in southern Africa (p. 261). Although mountain-building processes resembling those of the Phanerozoic world were undoubtedly in operation by this time, it is in Canadian rocks about 1 billion years younger that geologists have found the oldest well-displayed remains of a mountain system that is thoroughly modern in character. This Wopmay system is a body of deformed rocks that has been leveled by erosion. Because there is no longer any part of this system that is mountainous, it cannot be termed a mountain belt. Instead, it is labeled an **orogen**, a term for any body of rocks

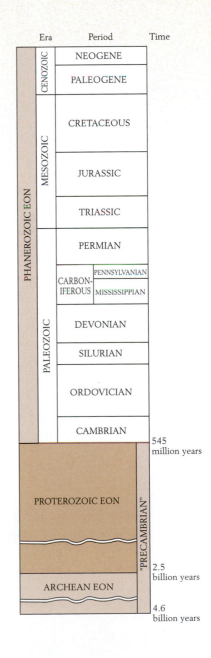

Era	Period	Time
CENOZOIC	NEOGENE	
	PALEOGENE	
MESOZOIC	CRETACEOUS	
	JURASSIC	
	TRIASSIC	
PALEOZOIC	PERMIAN	
	CARBON-IFEROUS — PENNSYLVANIAN / MISSISSIPPIAN	
	DEVONIAN	
	SILURIAN	
	ORDOVICIAN	
	CAMBRIAN	545 million years
PROTEROZOIC EON "PRECAMBRIAN"		2.5 billion years
ARCHEAN EON		4.6 billion years

PHANEROZOIC EON

Visual Overview

Major Events of the Proterozoic Eon

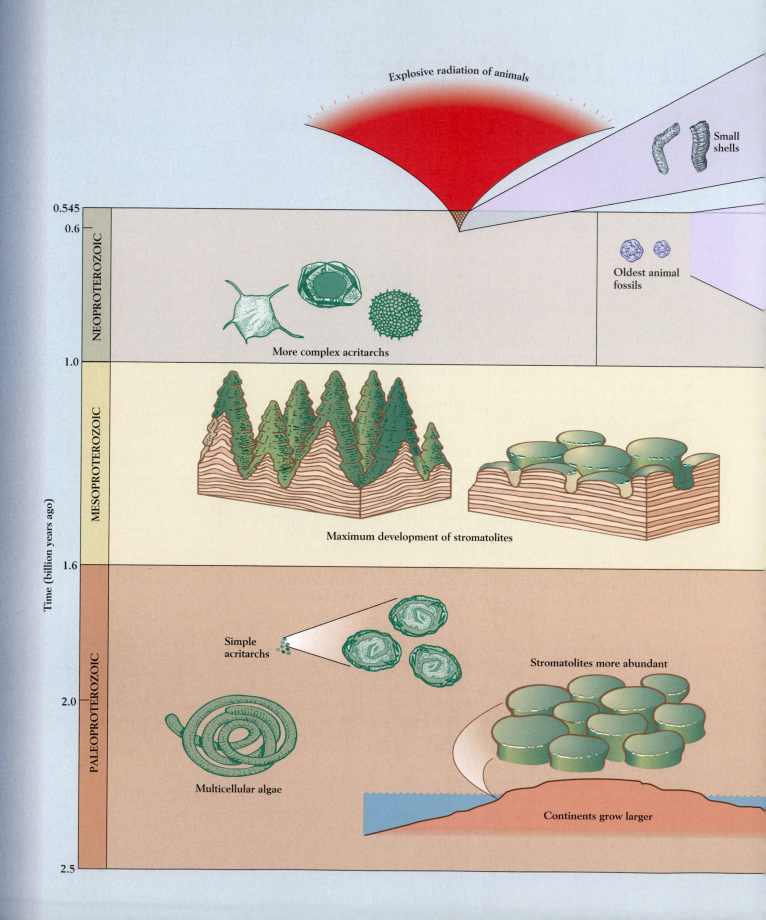

Explosive radiation of animals

Small shells

Oldest animal fossils

More complex acritarchs

Maximum development of stromatolites

Simple acritarchs

Stromatolites more abundant

Multicellular algae

Continents grow larger

NEOPROTEROZOIC

MESOPROTEROZOIC

PALEOPROTEROZOIC

Time (billion years ago)

0.545
0.6
1.0
1.6
2.0
2.5

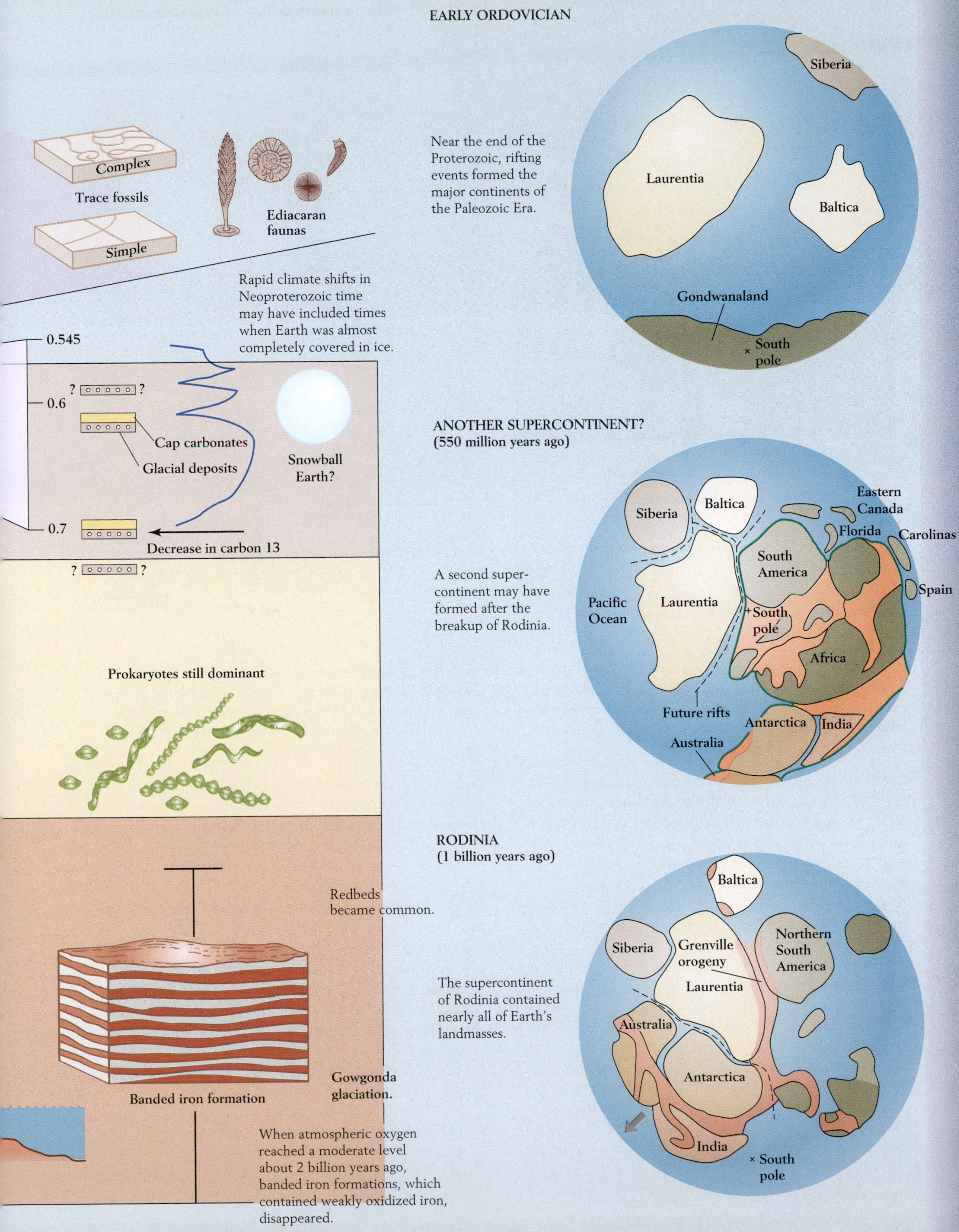

EARLY ORDOVICIAN

Near the end of the Proterozoic, rifting events formed the major continents of the Paleozoic Era.

Complex

Trace fossils

Simple

Ediacaran faunas

Rapid climate shifts in Neoproterozoic time may have included times when Earth was almost completely covered in ice.

— 0.545

? ⬚⬚⬚⬚⬚ ?

— 0.6

Cap carbonates

Glacial deposits

Snowball Earth?

— 0.7

Decrease in carbon 13

? ⬚⬚⬚⬚⬚ ?

Prokaryotes still dominant

Redbeds became common.

Gowgonda glaciation.

Banded iron formation

When atmospheric oxygen reached a moderate level about 2 billion years ago, banded iron formations, which contained weakly oxidized iron, disappeared.

ANOTHER SUPERCONTINENT?
(550 million years ago)

A second supercontinent may have formed after the breakup of Rodinia.

RODINIA
(1 billion years ago)

The supercontinent of Rodinia contained nearly all of Earth's landmasses.

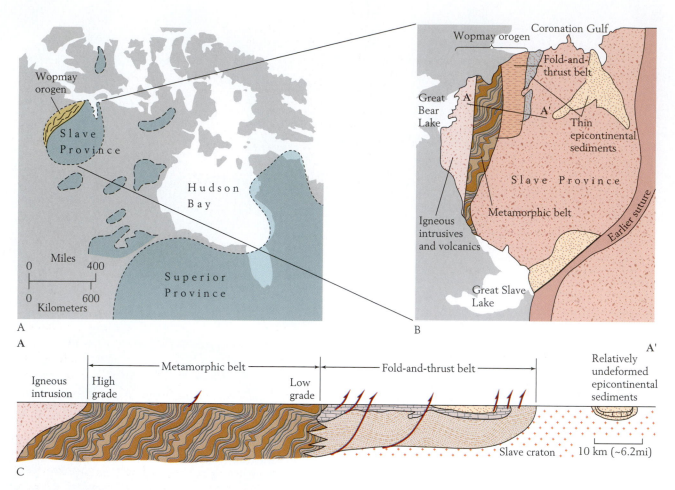

Figure 12-1 The Wopmay orogen, which formed nearly 2 billion years ago along the margin of the Slave Province of northwestern Canada. *A.* The Slave Province and other Archean terranes are shown in blue. *B.* The igneous, metamorphic, and fold-and-thrust belts of the Wopmay orogen. *C.* Cross section of the Wopmay orogen along the line A–A′ shown in *B* (depicted with vertical exaggeration). (After P. F. Hoffman, *Phil. Trans. R. Soc. Lond. A* 233:547–581, 1973.)

deformed by an orogeny. The Wopmay orogen, which formed along the margin of an early continent, developed slightly after 2 billion years ago, over a large area that is now approximately 1000 kilometers (600 miles) to the west of Hudson Bay. Today remarkably well-preserved sedimentary rocks of this orogen are exposed along the low-lying surface of the Canadian Shield as a result of continental glaciation that has repeatedly scoured the orogenic belt over the past 3 million years or so.

The Wopmay orogen lies along the western margin of the geologic region known as the Slave Province and displays an ancient fold-and-thrust belt (Figure 12-1). Although it has long been planed off by erosion, this zone of deformed rocks bears a striking resemblance to the younger fold-and-thrust belts described in Chapter 9. In the Wopmay orogen, thrusting was toward the east, and igneous intrusions associated with the deformation now lie primarily within the Bear Province, to the west. A belt of metamorphism lies between the ig-

neous belt and the fold-and-thrust belt. To the east, epicontinental sedimentary rocks continuous with those of the fold-and-thrust belt are relatively undeformed.

Near the end of Archean time, before the Wopmay orogen was formed, most of what is now called the Slave Province existed as a discrete craton. The rocks to the west of the orogen are those of an island arc. The Wopmay orogeny occurred when the Slave craton collided with this island arc slightly after 1.9 billion years ago and thick shelf deposits accumulated in a foreland basin along its western margin. Like sedimentary deposits of younger fold-and-thrust belts, those of the Wopmay belt show a clear relation to its tectonic history. As in younger mountain belts, the shelf deposits were succeeded by flysch and then molasse deposits (p. 211). The Wopmay sequence has the following characteristics:

1. The first thick deposit, which formed along the passive margin of the Slave craton, is a quartz sandstone that prograded toward the basin (Figure 12-2). This quartz sandstone grades westward into deep-water mudstones and turbidites that now lie within the metamorphic belt.

2. Carbonate rocks that contain abundant stromatolites accumulated along the passive margin on top of the quartz sandstone. These rocks formed a carbonate platform. Sedimentary cycles in these platform deposits record repeated progradation of tidal flats across a shallow lagoon. Laminated dolomite that formed in the lagoonal environment is at the

Figure 12-3 Stromatolites within shelf deposits of the Wopmay orogen. (Paul F. Hoffman, Geological Survey of Canada.)

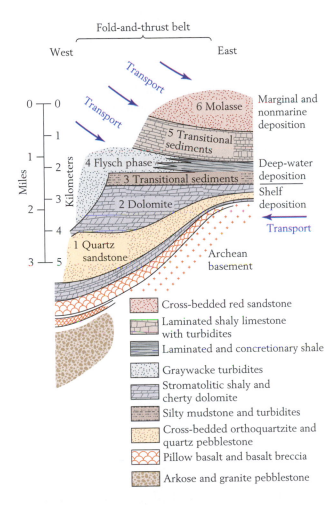

Figure 12-2 The sequence of deposition of sediments in the fold-and-thrust belt of the Wopmay orogen. The units numbered 1 and 2 represent marine deposition along a shallow continental shelf. Units 3 and 4 are deep-water deposits, including flysch, that accumulated when the shelf foundered as mountain building began to the west. Unit 5 consists of shallow-water deposits transitional between flysch below and molasse above. Unit 6, the molasse phase of deposition, followed the exclusion of marine waters by a heavy influx of sediment from the west. (After P. F. Hoffman, in M. R. Walter, ed., *Stromatolites*, Elsevier, Amsterdam, 1976.)

base of each cycle, while at the top are oolitic or stromatolitic deposits that must have formed in environments fringing the lagoon on its landward side (Figure 12-3). Enormous stromatolite mounds grew to the west, along the shelf margin. The fine-grained deposits of the lagoon were trapped behind the persistent barrier formed by these mounds. Thus stromatolites bounded the lagoon on both its landward and seaward margins. The present metamorphic zone consists of a thinner sequence of mudstones that represent deeper environments beyond the shelf edge, together with beds of dolomite breccia that contain blocks as long as 50 meters (165 feet). These blocks were transported down the steep slope in front of the shelf edge by catastrophic flows of submarine debris.

3. The carbonate platform deposits give way to transitional mudstones, which reflect a downwarping of the platform as a foreland basin was formed.

4. As is typical of foreland basin sequences of Phanerozoic age, flysch deposits (turbidites) follow the mudstones. The Wopmay flysch thickens to the west, and includes particles derived from uplifted plutonic rocks to the west. This pattern is also typical of Phanerozoic foreland basins: the source area of siliciclastics was seaward of the foreland basin (see Figure 9-14).

5. The deep-water turbidites grade upward into beds containing mudcracks and stromatolites, both of which formed in shallow-water environments and thus point to a shallowing of the foreland basin.

6. The influx of sediments eventually pushed marine waters from the Wopmay foreland basin, and the Wopmay cycle, like tectonic cycles of the Phanerozoic,

ended with an interval of molasse deposition (p. 211). The Wopmay molasse consists largely of river deposits in which cross-bedding is conspicuous.

In summary, two kinds of evidence suggest that the Proterozoic Wopmay orogen had the same pattern of formation as a modern orogenic system. First, the parallel igneous, metamorphic, and fold-and-thrust belts resemble similarly arranged belts of younger mountain ranges (see Figure 12-1). Second, within the fold-and-thrust belts, shallow-water shelf deposits are succeeded by flysch deposits that give way to molasse deposits. The rocks to the west of the Wopmay orogen are those of an island arc. The Wopmay orogeny occurred when the Slave craton collided with this island arc slightly after 1.9 billion years ago.

Global Events of the Paleoproterozoic and Mesoproterozoic

The fact that the extensive Slave terrane along the Wopmay orogen behaved like rigid continental crust when it was deformed indicates that by about 2.3 billion years ago the deep Earth was much cooler than it had been a billion years earlier, when magmas were pushing up from the mantle to the surface throughout much of what is now the Canadian Shield. That climates in this region were also quite cool approximately 2 billion years ago is shown by evidence that glaciers spread over the land. After reviewing the evidence for this glacial interval, we will examine other remarkable changes that took place in Earth's environments and life early in the Proterozoic Eon.

Glaciation was widespread early in Proterozoic time

Just to the north of Lake Huron in southern Canada are some of the most spectacularly exposed ancient glacial deposits in the world: those of the Gowganda Formation. Well-laminated mudstones in this formation consist of varves that formed in the standing water of a lake or ocean in front of glaciers. In Chapter 5 these ancient deposits were compared to the strikingly similar glacial varves that formed nearby, where Toronto is now located, just a few thousand years ago (see Figure 5-6). Some of the laminated Gowganda mudstones contain dropstones—pebbles and cobbles that appear to have fallen from ice that melted as it floated out from a glacial front (see Figure 5-7). These mudstones alternate with tillites, which were deposited when glaciers encroached on the body of water. Some of the

pebbles and cobbles of these tillites are faceted or scratched from having slid along at the bases of moving glaciers.

The Gowganda deposits cannot be dated directly, but the best estimate of their age is about 2.3 billion years, because they rest on 2.6-billion-year-old crystalline rocks and are intruded by igneous rocks that are 2.1 billion years old. Tillites of similar age are found elsewhere in Canada and in Wyoming, Finland, southern Africa, and India. These widespread glacial deposits testify to extensive continental glaciation not long after the transition from Archean to Proterozoic time.

Cyanobacteria flourished in the oceans and eukaryotes joined them

No abrupt change in life on Earth marked the Archean-Proterozoic transition. There is good evidence that eukaryotic algae existed early in Proterozoic time, but cyanobacteria remained more abundant than algae in Earth's oceans.

Stromatolites Beginning in strata about 2.2 billion years old, stromatolites become increasingly abundant in Proterozoic rocks. Their proliferation probably reflects an increase in the size of continents and therefore in the breadth of continental shelves, where stromatolites flourished.

Stromatolites also began to grow into a greater range of shapes, attaining their greatest diversity about 1.2 billion years ago (Figure 12-4). Distinctive cone-shaped stromatolites, which must have been produced

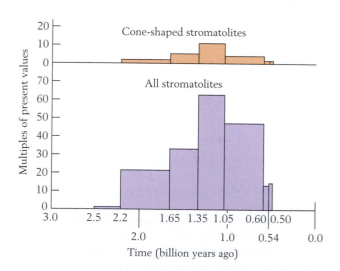

Figure 12-4 Form and distribution of stromatolites. Stromatolites, cone-shaped forms among them, reached their maximum diversity about 1.2 billion years ago. (After M. R. Walter, pp. 270–286 in S. Bengtson, ed., *Early Life on Earth*, Columbia University Press, New York, 1992.)

by a particular kind of cyanobacteria, lived primarily on offshore continental shelves. Stromatolites became less diverse during the Neoproterozoic Era, and cone-shaped forms were the first to decline.

Early eukaryotes Recall from Chapter 11 that chemical biomarkers reveal that eukaryotes existed before the end of Archean time. Presumably these primordial forms lacked important features of advanced eukaryotes, however. To understand what they probably lacked, let us review the steps by which the advanced eukaryotic cell arose.

It now appears that the more advanced cell that characterizes all living eukaryotes arose from the union of two prokaryotic cells, one of which came to reside within the other. The cell that lived within the other was altered in minor ways to form a structure called a **mitochondrion** (Figure 12-5). Mitochondria are the structures that allow cells to derive energy from their food by means of respiration (p. 225), and one or more are present in nearly all eukaryotic cells today. Evidence of this curious origin of mitochondria is the presence

within them of both DNA and RNA that differ from the genetic material of the surrounding cell. It is assumed that the smaller cell that became a mitochondrion, complete with its own DNA and RNA, was eaten by the larger one, but proved resistant to the digestive processes of the predator cell. Such a union of two cells resulted in the first protozoans.

It is widely agreed that plantlike protists later evolved as a result of another union of two kinds of cells. In this major evolutionary step, a protozoan consumed and retained a cyanobacterial cell. This cell then became an intracellular body known as a **chloroplast** (see Figure 12-5), which serves as the site of photosynthesis both in plantlike protists and in plants, which evolved from them. The similarity between cyanobacteria and chloroplasts is striking. In both, for example, the pigment chlorophyll, which absorbs sunlight and permits photosynthesis, is located on layered membranes. Furthermore, chloroplasts, like mitochondria, contain their own distinctive DNA and RNA.

Photosynthesis is conducted within chloroplasts by many kinds of protists. It is generally believed that plantlike protists evolved several times, when protozoans retained within their cells cyanobacteria they had eaten. Thus mobile protozoans may have evolved into mobile photosynthetic forms, while immobile protozoans evolved into immobile photosynthetic forms.

Those living eukaryotes that, according to genetic data, represent the lowermost branches of the eukaryotic family tree exist as parasites in animals. They have nuclei, but live without oxygen and lack mitochondria, obtaining energy directly from their hosts. These forms seem to represent an early stage of eukaryotic evolution. Possibly, like them, all Archean eukaryotes had nuclei but in other ways resembled prokaryotes.

Perhaps eukaryotes are not found in the Archean fossil record because they continued to resemble prokaryotes, which were rarely preserved except indirectly by way of the stromatolites that they produced. Eukaryotes may have continued to resemble prokaryotes because they had not yet taken one key evolutionary step that was necessary for the origin of mitochondria. Before some of them could develop the habit of eating prokaryotic cells, including the cell that became the first mitochondrion, they had to evolve a structure that prokaryotes lack. That structure is the *cytoskeleton*—a dynamic set of fibers that underlie the outer membrane of the cell and allow the cell to change its shape for various purposes. The presence of a cytoskeleton allows one cell to engulf another. The origin of this structure may have triggered an evolutionary expansion of eukaryotes to include a variety of new species, some of which were larger and more readily preserved as fossils than most kinds of bacteria.

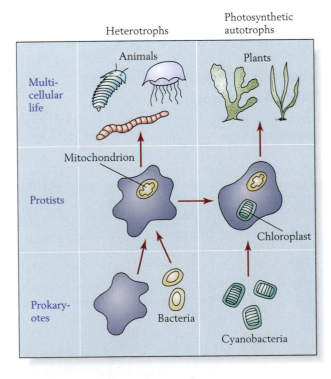

Figure 12-5 The probable sequence of major events leading from prokaryotes to multicellular animals and plants. The first protist apparently evolved when one prokaryote engulfed, but failed to digest, another, which then became a mitochondrion. The first plantlike protist evolved when an animal-like protist engulfed, but failed to digest, a cyanobacterium, which then became a chloroplast.

Figure 12-6 *Grypania*, a genus of coiled multicellular algae. This fossil was found in 2.1-million-year-old rocks in Michigan. (Bruce Runnegar, University of California.)

Algae Having formed by the ingestion of one cell by another, the earliest eukaryotes with mitochondria must have been unicellular. Multicellular plantlike protists—seaweedlike algae—may have arisen soon after the evolution of fully developed eukaryotic cells with mitochondria and chloroplasts. In fact, the oldest fossil eukaryotes now recognized are algal ribbons, commonly wound into loose coils, that date to about 2.1 billion years ago (Figure 12-6). Even after 2 billion years ago, however, prokaryotes greatly outnumbered eukaryotes in floras of single-celled organisms. The Gunflint flora of the Lake Superior region, for example, includes only prokaryotic forms (Figure 12-7A).

Nonetheless, single-celled algae termed **acritarchs** become increasingly conspicuous in Proterozoic rocks younger than 2 billion years (Figure 12-7B–C). These nearly spherical or many-pointed forms are the dominant type of algal plankton found in Paleozoic as well as Precambrian strata. They may include more than one taxonomic group of algae. Some acritarchs are believed to have been the resting stages (or cysts) of dinoflagellates, which are one of the most prominent groups of planktonic algae today (p. 62). The size and complexity of many Proterozoic acritarchs, together with the chemical composition of their cell walls, indicate a eukaryotic level of organization. All living prokaryotes are smaller than Proterozoic acritarchs, and many have simpler wall patterns.

Distinctive organic compounds known as *biomarkers* also indicate the presence of eukaryotes early in Proterozoic time. For example, compounds called steranes, which are present in eukaryotic but not prokaryotic cell membranes, have been detected in rocks about 1.7 billion years old.

Fossil microbiotas indicate that until some time after 2 million years ago, bacteria, including cyanobacteria, continued to play a more important role as producers in marine ecosystems than did acritarchs or other eukaryotes. Photosynthesis by these groups released large amounts of oxygen.

A buildup of atmospheric oxygen favored eukaryotes

It is widely agreed that photosynthesis caused atmospheric oxygen to build up during Precambrian time.

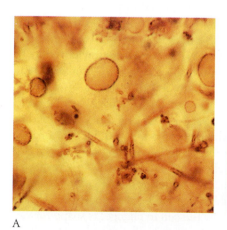

A

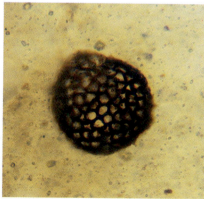

B

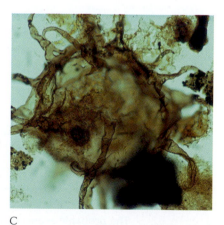

C

Figure 12-7 Fossil prokaryotic and eukaryotic cells of Proterozoic age. Eukaryotic cells were larger and more complex in form than prokaryotic cells. *A.* Fossil prokaryotic cells of the Gunflint Formation, which is about 1.9 billion years old. Both filamentous and spheroidal forms are present. The large spherical forms are about 10 μm in diameter. *B.* The acritarch *Dictyotidium*, which has a thick, complex wall structure that is unknown in prokaryotes. This fossil is about 750 million years old and has a diameter of about 35 μm. *C.* Another acritarch of similar size but different form. (*A*, Andrew H. Knoll; *B*, Nicholas Butterfield; *C*, Department of Industry and Resources/Geological Survey of Western Australia.)

Before oxygen could build up in the atmosphere, however, natural reservoirs known as chemical sinks had to be filled. Oxygen sinks, as we saw in Chapter 11, are chemical elements and compounds that combine readily with oxygen and isolate it from the atmosphere and living things. Sulfur and iron were two of the most important oxygen sinks present in Earth's crust and oceans immediately after Earth formed. (Note how iron that we extract from naturally occurring compounds rusts, or oxidizes to form iron oxide, when exposed to the oxygen in the atmosphere.) Several lines of evidence indicate that the concentration of oxygen in Earth's atmosphere increased significantly slightly after 2 billion years ago (see Earth System Shift 12-1).

The Paleoproterozoic buildup of atmospheric oxygen was one of the most important events in Earth's history. It may, indirectly, have promoted the success of eukaryotic algae. As oxygen built up in the Paleoproterozoic atmosphere, the concentration of dissolved oxygen inevitably increased in the upper ocean. As a result, more nitrogen must have been oxidized to form nitrate (NO_3^-), which is an important nutrient for eukaryotic algae (p. 99). Cyanobacteria, in contrast, do not require nitrate from their environment because they can use pure nitrogen (N_2), which is abundant in the atmosphere and waters of shallow seas. Because cyanobacteria could prosper even before nitrates became abundant in their environment, they had a temporary advantage over eukaryotic algae. The Paleoproterozoic oxygen buildup must have resulted in fertilization of eukaryotic algae, however, by increasing their supply of nitrate. This phenomenon may partly explain the Proterozoic expansion of acritarchs and multicellular algae.

The Beginnings of Modern Life

Animal life may or may not have originated during Neoproterozoic time, but it certainly diversified spectacularly to produce a conspicuous fossil record during the last 30 million years of this era, some 4 billion years after Earth's origin. Even early in the Neoproterozoic there were stirrings of change. Fossil seaweeds reveal that multicellular green and red algae became abundant in Neoproterozoic ecosystems, and distinctive biomarkers confirm this conclusion. In the planktonic realm, acritarch species evolved that were larger and had more complex shapes than species of earlier times. Thus it appears that a large adaptive radiation of eukaryotes began slightly before a billion years ago.

Animals burst on the scene

The evolutionary radiation of animal life during the last 30 million years of Proterozoic time has been aptly described as explosive. Three kinds of fossils contribute to our understanding of this spectacular expansion: trace fossils, soft-bodied fossils, and skeletal fossils. Many of these fossils are markings or remains of creatures that cannot be assigned to any known phylum or class with certainty. Others can be confidently assigned to well-known taxa.

Trace fossils Tracks, trails, and burrows—what are called trace fossils (p. 53)—have provided special evidence of the early diversification of animal life. Even during the nineteenth century it was acknowledged that fossil skeletons appear in the stratigraphic record quite suddenly near the base of the Cambrian System. The complexity and variety of fossilized Cambrian life gave rise to speculation that multicellular animals had a long Precambrian history, during which they lacked hard parts and therefore left no fossil record.

One effective test of this hypothesis was based on the assumption that early multicellular animals that were soft-bodied—those that lacked hard parts—crawled over the seafloor or burrowed into it throughout their existence, and in doing so produced trace fossils in sedimentary rocks. If soft-bodied invertebrate animals had existed for a long interval of Proterozoic time, scientists reasoned, some of them would have left such trace fossils. A search for trace fossils in Precambrian rocks has since turned up a striking pattern: such fossils have been found only in rocks about 570 million years old or younger. For example, 1.3-billion-year-old sedimentary rocks of the Belt Supergroup of Montana exhibit no tracks or trails. As Figure 12-8 indicates, Belt

Figure 12-8 Laminated siltstone from the 1.3-billion-year-old Greyson Shale (Belt Supergroup) in Montana. Here, as in other rocks older than the Neoproterozoic, we see no evidence of burrowing by invertebrate animals. (Charles Byers, University of Wisconsin, Madison.)

Earth System Shift 12-1 | Buildup of Atmospheric Oxygen

There is abundant evidence that until about 2.3 billion years ago, chemical sinks were soaking up oxygen so effectively that the concentration of oxygen in the atmosphere remained at only 1 or 2 percent of its modern level. Some of this evidence comes in the form of uranium and iron minerals in early Proterozoic rocks.

Recall from Chapter 11 that in the modern world, the uranium oxide mineral uraninite (UO_2) and the iron sulfide mineral pyrite (FeS_2, known as "fool's gold") disintegrate readily by oxidation when exposed to the atmosphere. Nonetheless, these minerals are relatively common in Archean sedimentary rocks. In contrast, they are rare in sandstones deposited about 2.3 billion years ago. This pattern suggests that atmospheric oxygen had by then risen above its Archean level. Rocks containing an abundance of iron oxide offer similar testimony: weakly oxidized banded iron formations rarely formed after

Figure 1 Banded iron formations ceased to form about 1.9 billion years ago. This weakly metamorphosed banded iron formation, in northern Michigan, is about 2 billion years old. Banded iron formations are among the oldest known rocks on Earth and are quite common in Archean terranes. Most of them accumulated between about 3.5 billion and 1.9 billion years ago. The term "banded iron formation" refers to a bedding configuration in which layers of chert, often contaminated by iron that gives them red or brown color, alternate with layers of other minerals that are richer in iron than the chert. The iron in these formations may occur in a variety of minerals, and in many cases the mineralogy of the iron has altered over time in ways that cannot be reconstructed. Banded iron formations account for most of the iron ore mined in the world today. Those with great economic value contain iron in the form of magnetite (Fe_3O_4), whose oxygen-to-iron ratio is lower than that of hematite (Fe_2O_3). In most of these rocks the average composition of iron oxide is intermediate between hematite and magnetite.

Banded iron formations accumulated in offshore waters. Many are associated with turbidites. Both the iron and the silica in these sediments appear to have come from hot, watery emissions from the seafloor associated with igneous activity. The kind of layer that was deposited at any time probably depended on the chemical composition of nearby watery emissions. The weak oxidation of the iron indicates that deep and even moderately deep waters of the ocean were poorly supplied with oxygen. Banded iron formations apparently ceased to form about 1.9 billion years ago because the concentration of oxygen built up in the waters of the deep ocean, reflecting a buildup of oxygen in Earth's atmosphere. (Bruce Simonson, Oberlin College.)

1.9 billion years ago, and highly oxidized red beds never formed before that time.

Precambrian soils, though rarely preserved, offer a more detailed picture of change in atmospheric oxygen. These thin units reveal the chemical nature of weathering during the time when the soils formed. In moist soils, iron exposed to abundant oxygen precipitates as hematite and other highly oxidized minerals that are relatively insoluble in water. In contrast, when little oxygen is present, iron that weathers from rocks remains in solution and is carried away by moving water. It is striking that soils that formed on basaltic rocks before about 1.9 billion years ago lost nearly all of their abundant iron. Thus there was still not enough oxygen in the atmosphere to precipitate the iron in these soils as highly oxidized minerals. In contrast, all Proterozoic soils younger than 1.9 billion years accumulated highly oxidized iron. The heavy oxidation of one well-preserved soil that formed 1.9 billion years ago in South Africa indicates that by this time atmospheric oxygen had built up to at least 15 percent of its present level.

Presumably part of the reason for the buildup of atmospheric oxygen about 2 billion years ago was that chemical sinks, including reduced iron compounds, were filling up. As they did so, more of the oxygen produced by photosynthesis accumulated in the atmosphere. In addition, however, a large amount of organic carbon was apparently buried in marine sediments. Oxygen that would have been used up in the decomposition of this organic matter was left to accumulate in the atmosphere (see Figure 10-6). The evidence for this burial of organic carbon is a marked shift toward heavier carbon isotope ratios in limestones throughout the world between about 2.2 billion and 2.0 billion years ago. Recall that the burial of large volumes of organic carbon, which is isotopically light, leaves isotopically heavy carbon behind in the ocean. This heavy carbon ends up in limestones that are precipitated from the seawater (see Figure 10-9).

The buildup of atmospheric oxygen early in Proterozoic time had major ramifications for life on Earth, setting the stage for animals to evolve.

Figure 2 Red beds are never found in terranes much older than 2 billion years. In other words, their pattern is the opposite of that displayed by banded iron formations. These highly oxidized iron-rich sediments are in the Proterozoic Hamersley Group in Western Australia. Hematite, a highly oxidized iron mineral, gives red beds their color. Often the hematite found in red beds has formed secondarily by oxidation of other iron minerals that accumulated with the sediments. Oxygen has been plentiful in Earth's atmosphere during Phanerozoic time, so that this secondary oxidation has often occurred within a few millions or tens of millions of years after the sediments were deposited. It would appear that oxidation of this type did not occur early in Earth's history. (Cornelius Klein, The University of New Mexico)

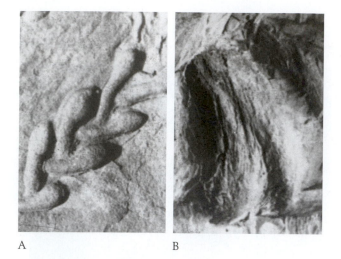

◀ **Figure 12-9** Undersurfaces of sandstone with fillings of Proterozoic burrows in Norway. *A.* Filling of a feeding burrow. *B.* Filling of a shallow burrow on which can be seen scratch marks left by the legs of the animal that dug it. (N. L. Banks.)

A

B

mudstones are often strikingly well layered in comparison with younger deposits, in which burrowing animals have often disrupted or destroyed layers of sediment.

Neoproterozoic trace fossils display a general evolutionary pattern. The oldest ones are simple tubes made by wormlike animals that burrowed through the sediment. In several regions of the world, as stratigraphic sections progress up toward and into the Cambrian

A

B

C

D

E

F

System, trace fossils become increasingly complex and varied (Figure 12-9). This increase in both complexity and variety seems to represent the initial evolutionary diversification of mobile animals in the world's oceans.

Fossils of soft-bodied animals Bodies of animals trapped beneath sediment also left conspicuous imprints in Neoproterozoic sediments younger than about 570 million years. Many of these fossils have been thought to represent jellyfishes or sea pens, both of which belong to the phylum Cnidaria, which contains modern corals (p. 68). Whereas jellyfishes float in the water, sea pens are stalked creatures that stand upright on the seafloor. The assignment of these Proterozoic fossils to the Cnidaria has recently been questioned, however, and they may actually represent groups without living representatives. Because the Ediacara fauna of Australia, named for the region where it was discovered, is the most famous of these late Precambrian "soft-bodied" faunas, all of the Neoproterozoic faunas that resemble it are termed *Ediacaran faunas* (Figure 12-10).

Assemblages of soft-bodied animals were seldom preserved in a similar way on sandy, well-oxygenated seafloors of Phanerozoic age. Probably the Ediacaran animals were preserved only because they lived before the evolution of predators or scavengers capable of readily devouring their carcasses. A few Ediacara-type fossils are known from Cambrian rocks. These occurrences indicate that Ediacaran animals did not become extinct at the end of Proterozoic time, but apparently were seldom preserved once numerous animals capable of destroying their carcasses were on the scene.

Whatever kinds of animals the Ediacaran fossils represent, it is clear that animals more advanced than cnidarians were well established before the end of Proterozoic time. Among these advanced groups were the segmented worms known as annelids, which today include not only modern earthworms but also many kinds of marine and freshwater species (see Figure 3-26). Annelids undoubtedly formed many of the tube-like fossil burrows of Neoproterozoic age. Also present were early members of the phylum Arthropoda,

Figure 12-11 Animal embryos consisting of a few cells. These remarkable fossils consist of phosphate, which replaced the cells before they decomposed. Their configuration resembles that of some living arthropod embryos. (Courtesy of Shuhai Xiao, Virginia Tech, and Andrew H. Knoll, Harvard University.)

which includes modern crabs, lobsters, and insects. Arthropods have external skeletons and jointed legs, a set of which presumably made the scratches visible in the burrow shown in Figure 12-9*B*. One Ediacaran form appears to have been a primitive mollusk that crept over the seafloor like a snail (see Figure 12-10*F*). Tiny embryos that are spectacularly preserved through replacement of their individual cells by phosphate also point to the existence of bilaterally symmetrical animals—perhaps arthropods—about 570 million years ago (Figure 12-11).

Skeletal fossils For many decades, the oldest known fossil shells and other hard parts of animals were from the base of the Cambrian System. Although abundant and varied shelly faunas make their earliest appearance at that level, scientists have recently found small shells throughout the Neoproterozoic interval that contain fossils left by soft-bodied animals. These shells are vase-shaped and tubular structures made of calcium carbonate (Figure 12-12). Although these fossils cannot be assigned with assurance to any previously recognized taxonomic group, they provide further evidence of the broad scope of the Proterozoic radiation of animals.

Skeletal elements (spicules) of sponges are also known from rocks younger than about 580 million years. This is no surprise because sponges are among the simplest of animals.

Evidence of the molecular clock on the timing of animal origins The molecular clock has been used to estimate when various types of animals diverged from one another—that is, when their last common ancestor lived. Unfortunately, as explained in Chapter 7, the clock can be calibrated only for organisms that lived after Proterozoic time, and it also keeps

◀ **Figure 12-10** Representatives of the late Precambrian Ediacara fauna of Australia. *A*. Simple disks that represent the oldest known animal fossils. *B*. An animal that may be a sea pen (0.6 life size). *C*. A problematic flat, segmented form (life size). *D*. An animal that may be a jellyfish (0.7 life size). *E*. An animal that appears to be intermediate in form between a segmented worm and an arthropod (magnified 1.7 times). *F*. Imprint of the foot of an early mollusk or mollusk relative that crawled over the seafloor. (*A*, H. J. Hoffman; *B–E*, M. F. Glaessner; *F*, Courtesy of the Smithsonian Institution.)

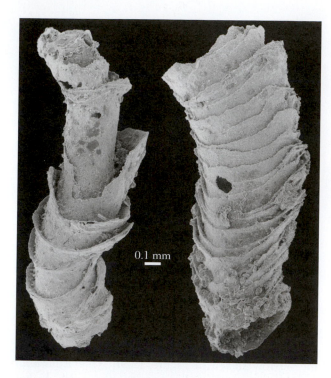

0.1 mm

Figure 12-12 *Cloudina,* **one of the oldest known skeletal fossils.** This tubelike form was slightly less than 1 millimeter in diameter. (S. Bengtson and Yue Zhao, *Science* 257:367–369, 1991.)

time imprecisely. Using the molecular clock, some scientists have estimated that protostome and deuterostome animals diverged about 670 million years ago (recall that the protostomes include groups such as mollusks and arthropods and the deuterostomes include groups such as echinoderms and vertebrates). This timing is in accord with the known fossil record, which documents the existence of a large variety of animals about a hundred million years later. On the other hand, other applications of the molecular clock have yielded estimates of a billion or a billion and a half years ago for the initial divergence of animal groups. If protostomes and deuterostomes arose as early as that, they must for some reason have remained soft-bodied and very small, because they left no known fossil record.

Did physical events trigger the explosive evolution of life?

In Neoproterozoic time, Earth experienced major ice ages. At least twice—slightly before 700 million years ago and again slightly before 600 million years ago—continental glaciers spread throughout the world, even to tropical latitudes (see Earth System Shift 12-2). The second—Marinoan—ice age preceded the initial radia-

tion of conspicuous animals by only a few million years. Possibly the two phenomena were causally related. There is, however, fossil evidence that small, inconspicuous animals existed prior to the Marinoan ice age. Simple, coin-sized disks resembling members of the Ediacaran fauna are known from strata slightly below Marinoan tillites. Possibly only very small, simple animals existed before 600 million years ago, but then ice age cooling caused the extinction of most of them. The origins of larger, more complex animals might have been part of the evolutionary recovery that followed, as all sorts of new anatomies arose.

Interestingly, acritarchs also experienced a dramatic change about 650 million years ago. Astoundingly, individual acritarch species before that time typically survived for about a billion years. After the Marinoan ice age, Neoproterozoic acritarch species survived for only a few tens of millions of years. Thus fundamental changes must have occurred in the marine ecosystem at the time of the Marinoan (or Varangian) ice age. One possibility is that grazing by newly evolved zooplankton accelerated the origin and extinction of phytoplankton species. Before animals began to graze effectively on algae, the availability of nutrients such as nitrates and phosphates may have imposed limits on the population sizes of phytoplankton species. Under such circumstances, the oceans might have been saturated with floating algae. Having large, stable populations, algal species would have been resistant to extinction, and by monopolizing nutrients, existing species would have suppressed the origins of new species. Once advanced zooplankton arose, however, they would have grazed down populations of algae, increasing their extinction rates and opening up ecological space so that new species could arise more easily.

Why did animal life diversify dramatically after the Marinoan ice age? Possibly the great radiation occurred because oxygen rose to a new level in the atmosphere. Recall that atmospheric oxygen rose to about 15 percent of its present level about 2 billion years ago, but possibly after this rise it underwent little increase until about 600 million years ago. Extremely small animals can exist at oxygen concentrations far below that of the modern world, but larger animals require higher oxygen levels. As we have noted, very small animals might have existed well before 600 million years ago without leaving a fossil record. Unfortunately, we have no strong evidence for an increase in atmospheric oxygen near the end of Proterozoic time.

What triggered the initial expansion of animal life remains one of the most intriguing questions in all of Earth's history. Paleontologists have learned much about the pattern of this expansion. They continue to search for its cause.

The Expansion and Contraction of Continents

Although geologists have long attempted to determine how the continents of the modern world originated, they have managed to trace back the histories of these continents only into Neoproterozoic time. Uncertainties remain about the histories of older Proterozoic cratons, and the configurations and relative positions of Archean microcontinents will probably never be known. The difficulty is that the depositional patterns and structural trends of rocks more than half a billion years old are often obscured by erosion, metamorphism, or burial; paleomagnetic data are also sparse and difficult to interpret. Nonetheless, geologists have reconstructed partial histories for most large blocks of Proterozoic crust and have thus gained some knowledge about how they became part of the Phanerozoic world.

In the next section we will learn how North America grew episodically during Precambrian time and how that continent became part of a vast supercontinent that fragmented late in the Proterozoic Eon. We will also review the origins of the landmasses that came to constitute the continents of the Paleozoic Era, including Gondwanaland. Before we discuss the histories of individual cratons, however, let us consider how cratons in general become larger or smaller.

As we have seen, cratons increase greatly in size when they become sutured together along a subduction zone, and this process is usually accompanied by mountain building in the vicinity of the suture (p. 208). Cratonic growth on a smaller scale, which is known as **continental accretion**, also entails mountain building, but this process occurs at the margin of a single large craton. Continental accretion can result from the suturing of an island arc or microcontinent to a large craton along a marginal subduction zone (p. 208). It can also result from the compression and metamorphism of sediments that have accumulated along a continental shelf. The latter process is sometimes referred to as **orogenic stabilization**, because it thickens the crust and hardens both unconsolidated sediments and soft sedimentary rocks (Figure 12-13). The Wopmay orogenic episode enlarged the Slave craton by both suturing of a small plate and orogenic stabilization (see Figure 12-1).

Stabilization is a cannibalistic process inasmuch as some of the sediment that is deposited and stabilized along a continental margin is derived from the interior of the continent by erosion. On the other hand, limestone that accumulates along a continental margin is precipitated from seawater or is secreted by organisms and thus represents an external contribution to the mass

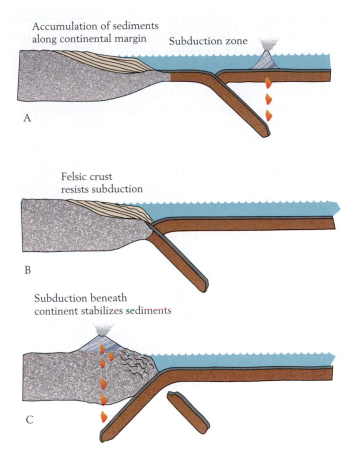

Figure 12-13 Stabilization of sediments that have accumulated along the margin of a continent. The continent comes to rest along a subduction zone (A). Because the continent has a low density and resists subduction (B), the direction of subduction is reversed. Igneous activity then adds rock to the continental margin and metamorphoses the sediments that have collected there (C).

of the continent—as do the igneous rocks and oceanic crust that become welded to a continental margin resting along a subduction zone.

Orogenic processes do not simply add material to continents; they also alter preexisting crust. Regional metamorphism, for example, often alters the character of preexisting rocks beyond recognition and resets their radiometric clocks so that the age of the crust can no longer be determined (p. 141). In reviewing the Precambrian history of individual Proterozoic cratons, we will encounter many examples of this process, called **remobilization**.

How do continents decrease in size? They can shrink by erosion, but this process operates so slowly that it has little overall significance. In addition, despite its low density, a small amount of continental crust is subducted into the mantle. Far more important is the process of continental rifting, which operates on many

Earth System Shift 12-2 | Was There a Snowball Earth?

During Neoproterozoic time, glaciers deposited tillites and icebergs released dropstones throughout the world, even near the equator. This extraordinary pattern has led some researchers to hypothesize that the entire planet was encrusted by ice—by glaciers on land and by sea ice on the ocean.

Many Neoproterozoic glacial deposits date to slightly before 700 million years ago and are assigned to the Sturtian ice age. Others date to slightly before 600 million years ago and are assigned to the Marinoan (or Varangian) ice age. Still others are of uncertain age, and some scientists believe that there was a third glacial interval before the Sturtian and a fourth after the Marinoan.

For glaciers to expand, some snow and ice formed in winter must survive summer temperatures. How might this have happened, even near the equator, more than half a billion years ago? In Neoproterozoic time the sun's output of radiation was about 6 percent lower than it is today (the sun has become progressively warmer since it formed as conversion of hydrogen to helium and heavier elements has made its core denser). In addition, Earth's continents were clustered into supercontinents at the time of the Sturtian and Marinoan ice ages, and large areas of these supercontinents were at high latitudes, where glaciers are likely to originate. Furthermore, the expansion of glaciers entails a positive feedback: when cool temperatures allow glaciers to expand, the high albedo of the glacial ice results in further cooling, which promotes further glacial expansion. With a relatively cooler sun and broad continental areas at high latitudes, it is not surprising that massive glaciers formed. The question is, could the entire planet have frozen over?

Adding to the puzzle are massive carbonate rocks, termed *cap carbonates*, that rest directly on top of both

Figure 1 **Neoproterozoic glacial deposits occur on all modern continents.** Dots depict the locations of these glacial deposits. The exact ages of many are uncertain. The geographic positions at which some of these deposits formed are not well established, but several tillites are known to have formed near the equator. (After D. A. D. Evans, *Amer. J. Sci.* 300:347–433, 2000.)

Sturtian and Marinoan tillites. These cap carbonates are formed by massive precipitation of calcium carbonate, which happens only in tropical seas. Thus, wherever they are found, climates must have shifted abruptly from frigid to tropical. Growth of large aragonite crystals right on the seafloor as part of the cap carbonates suggests that there was a high level of bicarbonate in the ocean to contribute to the precipitation of $CaCO_3$ (see Figure 10-11). This implies that carbon dioxide was abundant in the atmosphere, weathering rocks at a high rate to release large amounts of bicarbonate, which moved through rivers to the sea. Proponents of the snowball Earth hypothesis favor the idea that this carbon dioxide was released by submarine volcanoes, but remained trapped beneath a global blanket of sea ice during the glacial interval. Eventu-

Figure 2 **Light carbon isotopes characterize the cap carbonates above the Sturtian and Marinoan glacial intervals.** An interval above the Marinoan that displays similarly light isotopes may represent a younger ice age. Simple discs are the only structures below the Marinoan glacial interval that are generally considered to represent animal fossils.

Figure 3 Columnar crystals grew on the seafloor during deposition of the cap carbonates that rest on Marinoan glacial deposits in northwestern Canada. Originally aragonite, these crystals have been altered to calcite. Growth of stromatolites over the projecting surfaces of some crystals shows that those crystals grew upward to project above the surrounding seafloor. (Courtesy of N. P. James, Queens University.)

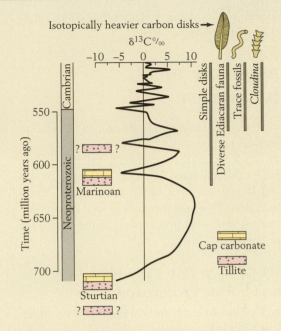

Figure 4 Yellowish cap carbonates overlie Marinoan tillites in southwestern Africa. (Paul F. Hofmann.)

ally, the argument goes, the carbon dioxide escaped to the atmosphere when cracks formed in the sea ice.

Another possibility is that the carbon dioxide was formed through oxidation of methane that was released by frozen methane hydrates that accumulated on continental slopes during the ice ages. The origin of the carbon dioxide from methane hydrates would explain another phenomenon: the shift of carbon isotopes to very light ratios in the cap carbonates. Because it is formed by multiple steps of bacterial metabolism, the methane of methane hydrates is isotopically light (p. 236).

Curiously, banded iron formations, which had ceased to form in the ocean more than a billion years before the Neoproterozoic glaciations, formed again at the time of these glaciations. Proponents of the snowball Earth hypothesis suggest that universal sea ice led to low-oxygen conditions in the waters below.

A challenge for the snowball Earth scenario is that it must account for the survival of life in an ocean covered by ice. One computer model of Neoproterozoic global climates left a band of ice-free ocean in the vicinity of the equator, where even animals could have survived. This model employed the relatively low level of solar radiation known to have been present in Neoproterozoic time as well as levels of atmospheric carbon dioxide similar to those of the present. In addition, forces produced by tides would have fractured bodies of ice on the surface of the ocean. The most direct challenge to the snowball Earth hypothesis comes from the Arabian Peninsula, where a series of glacial units document pulses of Marinoan glaciation into a marine basin. Multiple glacial expansions into such an area could not have occurred if the entire world was frozen over. In any event, Earth has never again experienced glaciation on a scale comparable to that of the Neoproterozoic ice age.

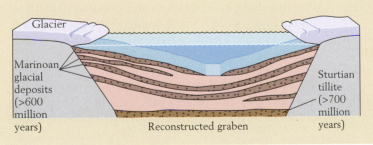

Figure 5 An ancient graben in the Arabian Peninsula contains a series of tillites. The oldest of these tillites represents the Sturtian ice age. A series of glacial deposits above this one, interbedded with marine strata, represent multiple glacial advances during the Marinoan ice age. Some of the Marinoan deposits are tillites near the basin margin that give way toward the basin center to marine strata with dropstones. (After J. Leather, P. A. Allen, M. D. Brasier, and A. Cozzi, *Geology* 30:891–894, 2002.)

scales. It can remove a small sliver of crust, or it can divide a large craton in half. Although continental rifting that took place more than half a billion years ago is difficult to document, evidence suggests that major rifting occurred late in Proterozoic time.

The Assembly of North America

Greenland today is a continent in its own right, but during Proterozoic and most of Phanerozoic time it was attached to North America. Laurentia is the name given to this combined landmass. The core of Laurentia was the crustal block that now forms most of the North American craton. This ancient block is well exposed today as the largest Precambrian shield in the world: the Canadian Shield (p. 250).

Continental accretion expanded Laurentia during Proterozoic time

The Canadian Shield constitutes a large portion of the North American craton, including a small part of the northern United States (Figure 12-14). Precambrian rocks also underlie the interior of the continent to the south, where they are overlain by a relatively thin veneer of Phanerozoic sedimentary rocks. Rocks obtained from wells that penetrate the Phanerozoic cover have provided a good picture of the general distribution of buried Precambrian rocks. These rocks, together with the exposed rocks of the Canadian Shield and Precambrian rocks that have been elevated by Phanerozoic mountain building in the American West, reveal that in the course of Proterozoic time, Laurentia gained territory by continental accretion.

Evidence that Laurentia was growing by accretion during Proterozoic time began to appear decades ago, when the recognition of structural trends and regional occurrences of rock units permitted geologists to recognize natural geologic provinces within the Canadian Shield. More recently, reliable radiometric dates for rocks of the Canadian Shield and for subsurface rocks bordering the shield have yielded a much more detailed picture. Uranium-lead techniques now yield dates with precision within a few million years for rocks that are about 2 billion years old. These dates are obtained from crystalline rocks, and thus they represent episodes of igneous and metamorphic activity.

Laurentia grew rapidly during Proterozoic time as it became sutured to other cratons. In fact, as we shall see shortly, near the end of the Proterozoic Eon it was united with nearly all of Earth's other landmasses to form a vast supercontinent only slightly smaller than Pangaea.

Figure 12-14 Major geologic features of North America. The Canadian Shield ends where sedimentary rocks of the interior lowlands lap over it on the south and west. The Cordilleran, Ouachita, and Appalachian orogens flank the North American craton on the west, south, and east.

The first stage in the formation of Laurentia, before it became part of a supercontinent, was the assembly of at least five microcontinents into a sizable craton (Figure 12-15). This amalgamation took place within only about 100 million years, between 1.95 billion and 1.85 billion years ago. Each of the microcontinents that were combined had formed during Archean time. Today these former microcontinents represent Archean terranes, which lie mostly within the Canadian Shield. The largest of these Archean terranes is the Superior Province, which crops out as far south as Minnesota. The Wyoming and Hearne provinces may actually have formed a single small plate; in the narrow zone of contact between the two, the geologic evidence is inconclusive. The Wyoming Province is exposed south of the Canadian Shield in mountainous uplifts in Wyoming (Figure 12-16) and also in the Black Hills, a blisterlike structure in South Dakota (see Figure 9-20).

Most of the Archean terranes were sutured directly together, but the Superior Province is separated from the Wyoming and Hearne provinces by a broad zone of rocks that formed at about the time of the suturing,

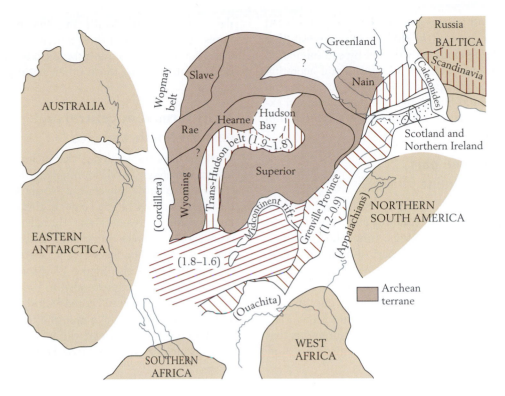

Figure 12-15 Geologic provinces of North America.
The diagram shows the composition of North America late in Proterozoic time, when this continent was attached to other landmasses (see Figure 12-19A). Numbers in parentheses represent times of origin in billions of years before the present. The provinces in the north are Archean terranes, representing Archean microcontinents that were amalgamated 1.95–1.85 billion years ago. The Wyoming and Hearne provinces may constitute a single terrane. The Trans-Hudson Belt consists of newly formed crust that was caught between the Superior terrane and an Archean terrane to the west. The origin of the broad province of the central United States 1.8–1.6 billion years ago expanded the craton toward the south. The Grenville Province formed when North America was sutured to Baltica and other landmasses that later became portions of Gondwanaland. Cross-hatching indicates orogenic belts. (After P. F. Hoffman, *Annu. Rev. Earth Planet. Sci.* 16:543–603, 1988.)

1.9 billion to 1.8 billion years ago (see Figure 12-15). This zone, known as the Trans-Hudson Belt, comprises both deep-sea sediments squeezed up between the converging cratons and crystalline rocks produced by an igneous arc.

South of the Archean terranes that were sutured together between 1.95 billion and 1.85 billion years ago is a broad zone of crust that formed shortly thereafter, between about 1.8 billion and 1.6 billion years ago. The rocks of this zone are exposed in uplifts from southern Wyoming to northern Mexico, and geologists have sampled them by drilling through the sedimentary rocks

Figure 12-16 A U-shaped glaciated valley in the **Beartooth Mountains of Wyoming.** Archean rocks form the core of the Beartooth uplift. The glacial scouring that made this valley U-shaped instead of V-shaped took place within the last 2 million years, during Earth's most recent Ice Age. (Martin G. Miller/Earth Lens.)

that blanket most of the central and western United States. The composition of these rocks suggests that they formed by igneous activity and by sedimentation along an island arc. The rate of continental accretion was very rapid by Phanerozoic standards. In the central and western United States, the continental margin expanded southward about 800 kilometers (500 miles) in 200 million years.

Laurentia became part of a larger continent

Thus far we have considered only regions of Laurentia that remain parts of North America and Greenland today. In fact, early in Proterozoic time, the combined landmass called Laurentia was part of a larger craton. Geologic similarities between the Wyoming Province and both eastern Antarctica and eastern Australia suggest that these regions were attached to one another well before a billion years ago (see Figure 12-15). Thus western North America was connected to cratons that today are positioned in the Southern Hemisphere. Similar evidence points to a connection between the terranes that now constitute Siberia and the northern Canadian region of Laurentia. Exactly when Laurentia became attached to these other landmasses is not yet known.

A rift formed in central and eastern North America

The growth of the landmass that included North America was threatened in Mesoproterozoic time by the greatest disturbance of the central North American craton during the last 1.4 billion years. In this episode of continental rifting, between about 1.2 billion and 1.0 billion years ago, lavas poured into elongate basins along a belt that extended from the Great Lakes region to Kansas (Figure 12-17; see also Figure 12-15). Had the crescent-shaped zone of rifting extended to the southern margin of the craton, the eastern United States would have drifted away as a separate small craton. But the rifting failed.

The rocks that formed within the failed midcontinent rift include hardened lavas known as Keweenawan

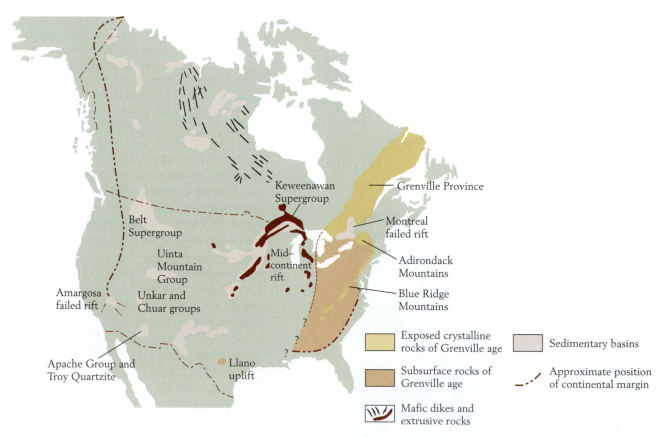

Figure 12-17 **The North American craton slightly more than a billion years ago.** The Keweenawan Supergroup, which includes mafic volcanics, accumulated in a midcontinent rift that ultimately failed. Some of the sedimentary basins of western North America were also failed rifts. (After J. H. Stewart, *Geology* 4:11–15, 1976.)

Figure 12-18
Columnlike features formed by joints in the Edwards Island flow, a body of volcanic rock in the Keweenawan Supergroup of northern Michigan. (Courtesy of William F. Cannon/U.S. Geological Survey.)

basalts, which are exposed near the southern border of the Canadian Shield (Figure 12-18). They contain ore deposits of native copper, a mineral consisting of elemental copper uncombined with other elements. Similar basin basalts lie to the southwest beneath the sedimentary cover of the Midwest, as has been revealed both by examination of rock cuttings taken from deep wells and by detection of strong magnetism from Earth's surface. Because these basalts are rich in iron and therefore are very dense, their presence is also associated with a feature known as the "midcontinental gravity high," which is a local increase in Earth's gravitational field as measured from the surface. While the basalts were forming, numerous mafic dikes were also emplaced across the Canadian Shield to the north (Figure 12-17 and p. 246)

The general configuration and composition of the Keweenawan Supergroup and its subsurface counterparts indicate the presence of a failed rift in the eastern United States (p. 202). The Keweenawan basalts, for example, are associated with red siliciclastic rocks and alluvial-fan conglomerates in what appear to be downfaulted grabens—configurations that tend to occur in newly forming continental rifts. Thus it would appear that about 1.3 billion years ago, a spreading center formed beneath the late Precambrian craton and began to rift it apart. This spreading center probably intercepted the eastern margin of the ancient craton. Whatever the reason, the Keweenawan rifting was obviously abortive. Rifting ceased before the continent was split, but left its mark in the enormous volumes of mantle-derived lavas that were disgorged along a belt more than 1500 kilometers (900 miles) long and 100 kilometers (60 miles) wide.

The Grenville Orogeny built mountains in eastern North America

While the midcontinent rifting was in progress, an episode of mountain building took place along the east coast of North America. The Grenville orogeny, which occurred slightly more than 1.1 billion years ago, was another step in the accretion of the North American continent, adding a belt that stretched from northern Canada to the southeastern United States (see Figures 12-15 and 12-17). The Grenville orogeny stabilized a large volume of sediments that had accumulated along the margin of eastern North America.

Crystalline rocks of Grenville age are best exposed in the Canadian portion of the Grenville Province. To the south, most crystalline rocks of Grenville age are buried, but some crop out here and there—for example, in the Adirondack uplift of New York State, the Blue Ridge Mountains of the central Appalachians, and an isolated prominence in Texas known as the Llano uplift (see Figure 12-17).

The relationship between the Grenville orogeny and the midcontinent rift remains a puzzle. What is clear is that the Grenville event entailed the collision of eastern North America with a landmass that would later become northern South America, where there are remnants of mountain systems that are the same age as the

Grenville orogenic belt (see Figures 12-15 and 12-19*A*). Forming another segment of the combined Neoproterozoic landmass was Baltica, which today forms northern Europe.

The Assembly and Breakup of Neoproterozoic Supercontinents

Between the time of the Grenville orogeny, slightly more than a billion years ago, and 500 million years ago, Earth underwent major episodes of continental suturing and fragmentation. At least one supercontinent—possibly two—formed and broke apart during this interval, which spanned the boundary between the Proterozoic and Phanerozoic eons.

The supercontinent Rodinia was assembled

The Grenville collision was actually part of a long zone of tectonic suturing that encircled much of the conti-

nent to which North America belonged. Similar orogenic belts in southern Africa, the Indian peninsula, and Australia illustrate that these regions, along with Laurentia, became attached to eastern Antarctica at this time. As a result of this extensive suturing, the landmasses that would later become Gondwanaland were wrapped around most of Laurentia (Figure 12-19*A*). The combined landmass thus formed, known as Rodinia, rivaled the Phanerozoic supercontinent Pangaea in total size. Rodinia was fully assembled by about 1 billion years ago.

Rodinia split apart to form the Pacific Ocean

Between about 800 million and 700 million years ago, Rodinia split in half. This was one of the most significant rifting events of all time, because it created the Pacific Ocean to the west of Laurentia. In western North America, tectonic episodes that preceded this rifting produced failed rifts that harbored large depositional basins in western Laurentia. From northern Canada to southern Arizona, these basins (shown in Figure 12-17)

RODINIA

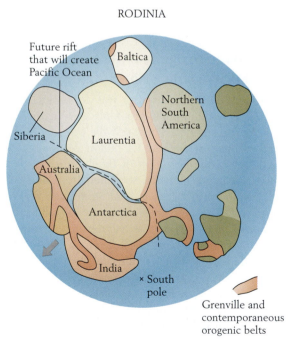

A   1 billion years ago

ANOTHER SUPERCONTINENT?

B   550 million years ago

Figure 12-19 **Changing paleogeographic patterns in Neoproterozoic and early Paleozoic time.** *A.* The likely configuration of the Neoproterozoic supercontinent about a billion years ago, at the end of the Grenville orogeny. Laurentia occupied the central position. *B.* The clustering of continents in a new configuration about 550 million years ago. By this time, other landmasses had broken away from the western margin of Laurentia, forming the Pacific Ocean, and had clustered on the eastern side of Laurentia; there, partly as a result of the Pan-African orogeny, they formed the landmass that would soon break away to become Gondwanaland. (After I. W. D. Dalziel, *GSA Today* 2:240–241, 1992; R. Unrug, *GSA Today* 7:1–6, 1997.

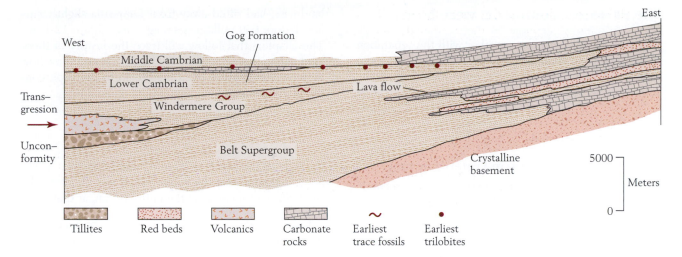

Figure 12-20 **The Belt Supergroup.** This stratigraphic sequence from Canada just north of Montana, ranging from the Mesoproterozoic through the Middle Cambrian, shows the relationship of the Belt Supergroup to its overlying strata. (After P. B. King, *The Evolution of North America*, Princeton University Press, Princeton, NJ, 1977.)

received large volumes of sediment. In the northern United States, the largest of these basin sequences is the Belt Supergroup (Figure 12-20), which ranges in age from about 1.4 billion to 1.5 billion years. In general, the Belt thickens toward the west, sometimes reaching a thickness of 16,000 meters (53,000 feet).

The Belt Supergroup formed in a northwesterly-trending failed rift (see Figure 12-17). Sandstones increase in abundance toward the western part of the sequence, while limestones increase toward the east, where sediments accumulated in shallower water. In general, however, mudstones predominate. The Belt apparently formed as a result of the accumulation of sediments in very shallow water during rapid subsidence. Salt crystals and mudcracks, both of which indicate a drying up of shallow bodies of water, are present in some Belt rocks, and shallow-water stromatolites are also abundant in the limestones. The Windermere Group, above the Belt, contains tillites of Neoproterozoic age (see Figure 12-20).

The new western border of Laurentia formed by rifting remained a passive margin for hundreds of millions of years. In fact, it has never since been sutured to another large craton. As we shall see, however, it has grown westward through suturing events that have added small bodies of crust. It has also, of course, been the scene of extensive mountain building and igneous activity that have added crust during the latter part of Phanerozoic time. Thus, at the start of the Phanerozoic Eon, the passive western margin of North America lay well to the east of its present position.

Did another Neoproterozoic supercontinent form?

The large block that separated from Laurentia as the Pacific Ocean formed was eventually to become the eastern segment of Gondwanaland. This block and Laurentia continued to separate, and the Pacific Ocean continued to grow, until their leading edges collided with opposites sides of the newly forming African craton. In this way a new supercontinent may have formed, with Africa in its center (Figure 12-19*B*). Whether such a supercontinent actually formed is uncertain, however, because some of the crustal blocks that would have been part of it may have broken away before others were sutured together.

We know for sure that between about 700 million and 500 million years ago, numerous small continental blocks were sutured together to form the large body of crust that today constitutes Africa. The orogenies that sutured these blocks together are collectively termed the Pan-African orogeny. This orogeny ended about 500 million years ago, about 40 million years after the start of the Paleozoic Era. By the start of the Paleozoic, however, both Laurentia and Baltica had broken away from the supercontinent (see Figure 12-19*B*). If they did so before the African crust was fully assembled, then a supercontinent never completely formed near the end of the Proterozoic Eon. We do know that nearly all elements of Earth's continental crust were clustered close together at this time.

The Paleozoic continents were born

The Pan-African suturing, together with the separation of Laurentia and Baltica as distinct continents, left more than half of Earth's continental crust locked up in the supercontinent Gondwanaland. Siberia, another large landmass, had rifted away from Laurentia slightly earlier, before 600 million years ago (see Figure 12-19B). The chapters that follow will trace the history of these landmasses throughout the Paleozoic Era and show how they eventually were combined to form the supercontinent Pangaea.

Chapter Summary

How did mountains form in Proterozoic time?

At least as early as 2 billion years ago, plate tectonic processes formed mountain belts similar to those of Phanerozoic age.

What global climatic and biological changes took place in the Proterozoic Eon before a billion years ago?

Continental glaciers spread over many regions more than 2 billion years ago, signaling global cooling. Stromatolites first became abundant about 2 billion years ago, and their success at this time may have resulted from the growth of continental shelves. Eukaryotes evolved through the ingestion of one kind of bacterial cell by another. Acritarchs are fossils of single-celled eukaryotic algae that lived as plankton. Fossils of multicellular eukaryotic algae first occur in rocks about 2.1 billion years old.

What evidence is there that oxygen began to build up in the atmosphere about 2.3 billion years ago?

The disappearance of banded iron formations about 2 billion years ago, together with the appearance of red beds, suggests that atmospheric oxygen had reached a moderate level early in Proterozoic time. Additional evidence comes from the rarity of the easily oxidized minerals pyrite and uraninite in rocks younger than 2 billion years. Finally, highly oxidized soils first appear in the geologic record and red beds become common about 2 billion years ago.

Why do some scientists use the label "snowball Earth" to describe our planet during part of Neoproterozoic time?

Slightly before 700 million years ago and again slightly before 600 million years ago, continental glaciers spread throughout the world, even to equatorial regions. It has been suggested that the entire planet was blanketed by ice, with sea ice covering the entire ocean, at that time.

What dramatic evolutionary changes occurred during the last 30 million years of Proterozoic time?

Animals may have originated, but certainly they diversified markedly and began to form a rich fossil record. Acritarchs also underwent major changes and began to experience high rates of origination and extinction.

What changes in the physical environment might have triggered the explosive diversification of animal life slightly after 600 million years ago?

Animal diversification might have been part of an evolutionary rebound from extinction caused by cooling associated with the great Marinoan ice age. Alternatively, a buildup of atmospheric oxygen might have triggered the initial adaptive radiation of major animal groups.

How did North America come into being?

The modern North American craton, with Greenland attached to form Laurentia, was assembled from smaller Archean cratons early in Proterozoic time.

What did continental suturing accomplish on a global scale during Neoproterozoic time?

About a billion years ago, it produced the supercontinent Rodinia, which contained nearly all of Earth's landmasses. After the partial breakup of Rodinia, a second supercontinent may have formed later in Proterozoic time. Near the end of Proterozoic time, rifting events formed Gondwanaland, Laurentia, and Baltica—the major continents of the Paleozoic Era.

Review Questions

   1. How are the basic features of the Wopmay orogenic belt typical of orogenic belts in general?

   2. What kinds of geologic evidence suggest that continental glaciers spread widely more than 2 billion years ago?

   3. List as many differences as you can between the Archean world and the world as it existed 1 billion years ago.

   4. What arguments favor the idea that little atmospheric oxygen existed on Earth until slightly before 2 billion years ago?

5. What evidence is there that eukaryotic organisms existed 2 billion years ago?

6. How does the history of North America illustrate continental accretion?

7. How does the Grenville orogeny relate to the supercontinent Rodinia?

8. How were the crustal elements of Gondwanaland assembled?

9. Using a world map, locate the modern positions of the various landmasses that make up the supercontinent shown in Figure 12-19*B*.

10. What major biological events occurred near the end of Proterozoic time?

11. What is curious about the Neoproterozoic rocks known as cap carbonates? (Hint: Refer to Earth System Shift 12-2.)

12. Life underwent extraordinary changes in the course of Proterozoic time. Using the Visual Overview on page 274 and what you have learned in this chapter, describe these changes and explain how some of them may have been related to changes in the chemistry of the atmosphere.

The boundary between the Ordovician and Silurian systems on Anticosti Island in eastern Canada. The light gray mounds along the beach are Upper Ordovician reefs; darker Lower Silurian strata stand above them. (Paul Copper, Laurentian University, Sudbury, Ontario.)

CHAPTER **13**

The Early Paleozoic World

The early Paleozoic interval of Earth's history tends to engage our interest because it was during that time that most animal phyla of the modern world appeared.

The establishment of the Cambrian and Ordovician systems in Britain more than a century ago illustrates how early geologists formally divided the stratigraphic record into useful intervals (p. 13). Early Paleozoic sedimentary rocks of marine origin were later found to be well displayed on the broad surfaces of many large cratons, reflecting the fact that, with brief interruptions, sea level rose in the course of the Cambrian Period and stood high during most of Ordovician time. As we saw in the previous chapter, life in the oceans diversified rapidly at the end of Proterozoic time. This diversification accelerated in what has been termed the Cambrian explosion of life; brief episodes of mass extinction also occurred during the Cambrian Period. Then, during the Ordovician Period, a greater variety of animal life evolved than had ever existed before. The Ordovician ended, however, in widespread global extinctions that were linked to a brief episode of glaciation near the south pole.

The Cambrian Explosion of Life

The story of early Paleozoic biotas is essentially a story of life in the sea. It is presumed that certain simple kinds of protists and fungi had made their way into freshwater habitats by this time, but no fossil record of early Paleozoic freshwater life is known. The terrestrial realm, too, was barren of all but the simplest living things. Neither insects nor vertebrate animals occupied the land before middle Paleozoic time.

The organisms that we will discuss in this chapter were the first biotas on Earth to leave a conspicuous fossil record—one that is plainly visible even to a casual observer—in many areas because they include many forms with skeletons composed of durable minerals. As we learned in Chapter 12, all but a few of the creatures that emerged during the initial radiation of animal life in the Neoproterozoic were soft-bodied. During the earliest segment of Cambrian time, the seas became populated by a different fauna, consisting of a great variety of animals with skeletons. Paleontologists have long debated the cause of this change, but now the prevailing view is that skeletons were widely deployed to thwart newly evolved predators, which readily victimized soft-bodied animals (see Earth System Shift 13-1).

Many Early Cambrian groups evolved skeletons

The Early Cambrian was the first time in Earth's history when many modern phyla of animals appeared, and many of them included taxa that evolved skeletons. This interval, which spanned about two-thirds of Cambrian time, was not marked by a uniform radiation of animal life, however. The Early Cambrian can be divided into three intervals, each characterized by a distinctive fauna.

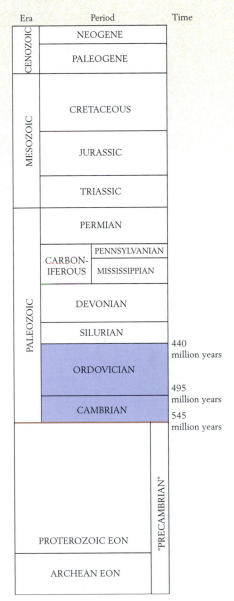

Era	Period	Time
CENOZOIC	NEOGENE	
	PALEOGENE	
MESOZOIC	CRETACEOUS	
	JURASSIC	
	TRIASSIC	
PALEOZOIC	PERMIAN	
	CARBON-IFEROUS: PENNSYLVANIAN / MISSISSIPPIAN	
	DEVONIAN	
	SILURIAN	
	ORDOVICIAN	440 million years
		495 million years
	CAMBRIAN	545 million years
"PRECAMBRIAN"	PROTEROZOIC EON	
	ARCHEAN EON	

Major Events of the Early Paleozoic

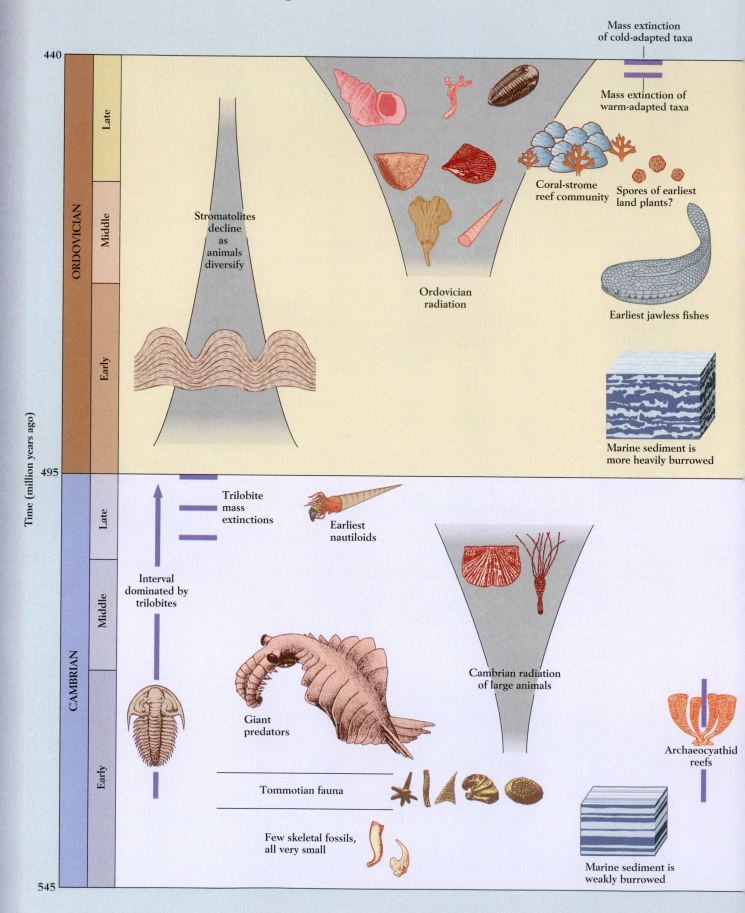

Time (million years ago)

440

ORDOVICIAN

Late

Middle

Early

495

Mass extinction of cold-adapted taxa

Mass extinction of warm-adapted taxa

Coral-strome reef community

Spores of earliest land plants?

Ordovician radiation

Earliest jawless fishes

Stromatolites decline as animals diversify

Marine sediment is more heavily burrowed

CAMBRIAN

Late

Middle

Early

545

Trilobite mass extinctions

Earliest nautiloids

Interval dominated by trilobites

Giant predators

Cambrian radiation of large animals

Archaeocyathid reefs

Tommotian fauna

Few skeletal fossils, all very small

Marine sediment is weakly burrowed

MIDDLE SILURIAN

Siberia

Euramerica

Gondwanaland

MIDDLE ORDOVICIAN

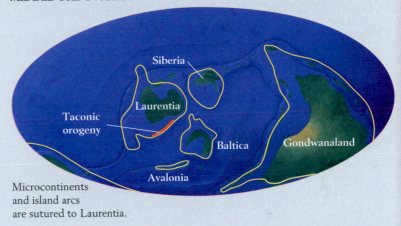

Siberia

Taconic orogeny

Laurentia

Baltica

Avalonia

Gondwanaland

Microcontinents and island arcs are sutured to Laurentia.

LATE CAMBRIAN

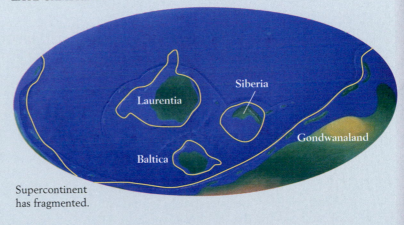

Laurentia

Siberia

Baltica

Gondwanaland

Supercontinent has fragmented.

Brief continental glaciation

$\delta^{18}O$ $\delta^{13}C$

Oxygen isotopes in the ocean shift toward heavier values as glaciers expand and the oceans cool.

Carbon isotopes in the ocean shift toward heavier values, perhaps because of an increase in the rate of carbon burial.

Taconic orogeny in eastern Laurentia

Global transgression

Shift from Aragonite to calcite seas

Sea level
Rising
Falling

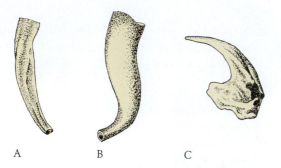

Figure 13-1 Earliest Cambrian fossils from Siberia and China. *Anabarites* (A) and *Cambrotubulus* (B) are tubular fossils of unknown relationships; they are about 3–5 millimeters long. *C*. A conodontomorph, the tooth of a predatory animal that may have been closely related to conodonts; this fossil is about 1 millimeter long. (After A. Y. Rozanov and A. Y. Zhuravlev in J. H. Lipps and P. W. Signor, eds., *Origin and Early Evolution of the Metazoa*, Plenum Press, New York, 1992.)

The lowermost Cambrian The lowermost portion of the Lower Cambrian record has yielded only simple skeletal fossils. Most of them are tube-shaped or vase-shaped, but one group, which appeared near the end of the interval, consists of teeth (Figure 13-1). This limited variety of forms represented only a slight advance over the modest variety of animals that existed in latest Proterozoic time.

The Tommotian fauna A much richer fauna appears abruptly in the central portion of the Lower Cambrian record. This so-called *Tommotian fauna*, named for the Tommotian Stage of Early Cambrian time, was first discovered in Siberia. It includes a host of small skeletal elements that cannot be assigned to any living phylum and that show no relation to any group of fossils found in post-Cambrian rocks (see Earth System Shift 13-1). The Tommotian fauna also contains the oldest known members of a few groups that survive to the present day—sponges, which are very simple animals (p. 67); mono-placophorans, which were ancestral to all present-day groups of mollusks (p. 70); and brachiopods (p. 71).

Large animals with skeletons The Tommotian fauna occupied the seas for perhaps only 3 or 4 million years. Then, for unknown reasons, most of its members disappeared, and a variety of new animals with skeletons emerged during the final few million years of Early Cambrian time. This new group of animals differed from the Tommotian fauna in two important ways: many were much larger, and most belonged to phyla that have survived to the present time.

Among the new animals were the trilobites, a group that survived to the end of the Paleozoic Era. Popular

UPPER CAMBRIAN

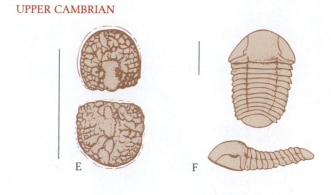

MIDDLE CAMBRIAN

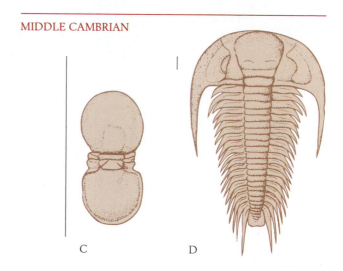

LOWER CAMBRIAN

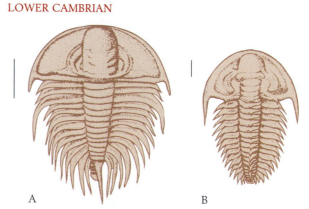

Figure 13-2 Typical Cambrian trilobites. *A. Olenellus*. *B. Holmia. C. Lejopyge. D. Paradoxides. E. Glyptagnostus. F. Illaenurus.* Trilobites were arthropods (invertebrate animals with segmented bodies and jointed legs). The soft body and the many legs were positioned beneath the flexible, jointed skeleton. Trilobites had mouthparts for chewing small pieces of food. Most species crawled over the seafloor, but some burrowed in sediment, and a few small species, including *Lejopyge* and *Glyptagnostus*, were planktonic. (Scale bars represent 1 centimeter [$\frac{3}{8}$ inch].) (After R. C. Moore, ed., *Treatise on Invertebrate Paleontology*, pt. O, Geological Society of America and University of Kansas Press, Lawrence, 1959.)

Figure 13-3 The shallow burrow of a Cambrian trilobite preserved as a cast on the underside of a bed of sediment. Scratches made by the appendages of the trilobite are clearly visible. The burrow is about 5 centimeters (2 inches) wide. (T. P. Crimes.)

with fossil collectors, trilobites were arthropods whose segmented skeletons were heavily calcified (p. 69). Lacking strong mouthparts for chewing, most trilobites were deposit feeders: they extracted small particles of organic matter from sediment. Some trilobites probably fed on very small animals, however. Most lived on the surface of the sediment, but some were shallow burrowers. A few kinds of trilobites were small planktonic (floating) forms that must have been suspension feeders (Figure 13-2C and E). Some tracks of benthic trilobites show scratch marks that their legs produced when they were digging in sediment (Figure 13-3). These trace fossils closely resemble some tracks of Neoproterozoic age. This similarity suggests that trilobite-like arthropods existed millions of years before trilobites, but failed to leave a skeletal fossil record because their skeletons were not mineralized with calcium carbonate.

Once present, trilobites diversified quickly, producing such a conspicuous fossil record throughout the remainder of the Cambrian that geologists commonly refer to the faunas that preceded them as the pretrilobite Cambrian faunas. The most abundant animal groups with skeletons that shared the late Early Cambrian seafloor with trilobites were monoplacophoran mollusks, inarticulate brachiopods (Figure 13-4), and a variety of echinoderms.

Of course, not all marine animals possessed skeletons in Early Cambrian time. A remarkably well-preserved fossil fauna in China has opened a unique window on the nature of the soft-bodied creatures that arose during this interval of primordial animal expansion. This Chengjiang fauna, though only about 30 million years younger than the soft-bodied Ediacaran fauna, contains a very different group of soft-bodied taxa. Among its members were jellyfishes and other cnidarians (p. 68), segmented worms (p. 69), arthropods that lacked mineralized skeletons, and a group of predatory worms known as priapulids, which survive today by way of only eight species that feed on other worms.

Arthropods are the most diverse animal group in the Chengjiang fauna. Some of the arthropod species had segmented bodies, but most had a flexible body covering that was folded along a central axis so that each half protected one side of the animal's body. The most spectacular Early Cambrian invertebrates were huge carnivores known as **anomalocarids**. These forms resembled arthropods, but only the appendages in front of the mouth were jointed. They were swimmers that propelled themselves with flaps positioned along their bodies and impaled prey on daggerlike spines along their frontal appendages. One of their species, which must have been a fearsome predator, reached a length of about 2 meters (6 feet) (Figure 13-5).

Figure 13-4 Cambrian brachiopods. The articulate genus *Eorthis* (left) lived on the surface of the sediment; the inarticulate genus *Lingula*, a burrowing form (right), survives in modern seas. The animals are shown at close to life size. (Chip Clark.)

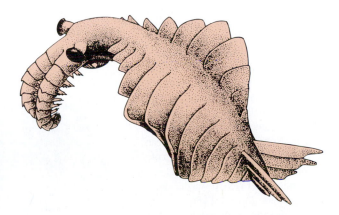

Figure 13-5 *Anomalocaris*, a giant Early Cambrian anomalocarid. This animal, which was about 2 meters (6 feet) long, swam by undulating the flaps along its body and captured prey with its frontal appendages, which bore sharp spines. (After D. Collins, *J. Paleontol.* 70:280–293, 1996.)

Earth System Shift 13-1 | Skeletons Evolve in Many Animal Groups

The origin of many kinds of skeletons, beginning with the evolution of Tommotian marine life, was a major development in the history of life. Why did so many kinds of animals evolve skeletons so abruptly? We know that skeletons support soft tissue and facilitate locomotion, but such adaptive functions cannot explain why so many different kinds of skeletons developed so suddenly. It has been suggested that a chemical change in the oceans was the cause: the concentration of calcium in the ocean rose during Neoproterozoic time and continued to rise throughout the Early Cambrian (see Figure 10-20). This trend would have favored the secretion of calcium carbonate and calcium phosphate—compounds most commonly secreted by Cambrian animals to form skeletons. The rapid evolution of a variety of external skeletons was probably a response to the evolution of advanced predators, however. The presence of several kinds of teeth in the Tommotian fossil record indicates that a variety of small predators were indeed present. It appears

that many kinds of small animals suddenly required the protection of a skeleton in order to survive.

Probably the appearance of new kinds of large predators promoted the evolution of skeletons by animals larger than the minute Tommotian creatures. Large predators are, in fact, well represented in the Chengjiang fauna. The largest Early Cambrian predator known was an anomalocarid that was about the size of a human (see Figure 13-5).

Predators not only feed on live animals, but also scavenge on dead animals. It appears that an absence of large scavenging animals accounts for the widespread preservation of the soft-bodied Ediacaran animals in late Proterozoic sediments (see Figure 12-10). Carcasses of these animals were preserved in clean, well-oxygenated sands of open marine settings, apparently because no scavengers were present to consume them quickly. The recent discovery of members of the Ediacaran fauna in Cambrian rocks at a single location in Australia provides evidence that this fauna did not become extinct at the end of Protero-

Figure 1 Fossils of the Tommotian fauna constitute the oldest diverse skeletonized fauna on Earth. All the specimens shown here are small; none exceeds a few millimeters in length. *A.* A fossil that appears to represent a primitive mollusk with a coiled shell. *B.* A spicule (small skeletal element) of a sponge. *C–E.* None of these specimens can be assigned to a familiar group of animals. (*A*, *B*, and *D* from S. C. Matthews and V. V. Missarzhevsky, *J. Geol. Soc. Lond.* 131:289–304, 1975; *C* and *E*, S. Bengston.)

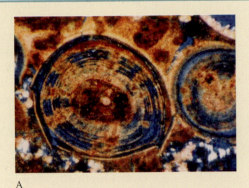
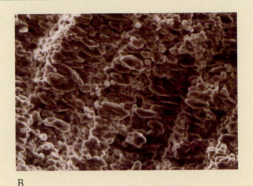

A   B

Figure 2 The internal structure of Ordovician oolites reveals that they were precipitated from seawater as calcite. Calcitic oolites grow by addition of concentric layers around a nucleus, which may be a piece of shell or a pellet of lime mud (A). Individual layers are formed of needle-like calcite crystals that radiate from the center (B). (Robert Goldhammer.)

zoic time, but simply failed to be preserved, except very rarely.

As predators became progressively more effective in attacking prey, not all skeletons that appeared in Cambrian time proved to be protective. Some of these skeletons, for example, were weak and flexible. A multitude of fossil plates from the skeletons of eocrinoids show that these stalked ancestors of modern sea lilies were abundant, yet their fossil skeletons are rarely intact; they normally fell to pieces soon after the animals died. Some of their relatives, however—cystoids, blastoids, and crinoids—had robust skeletons that are often found intact in the fossil record, and these animals flourished long after Cambrian time.

The history of trilobites also reflects the intensification of predation early in the Paleozoic Era. The shapes of the earliest trilobites reveal that they were unable to roll up to protect their soft underbellies with their external skeletons. During the Cambrian, these vulnerable forms gave way to ones that could roll up when threatened, for full-body protection.

Not surprisingly, the effects of seawater chemistry are evident in the early phases of skeletal evolution (see Figure 10-19). Reef-building taxa, which must secrete skeletal material rapidly to overcome the physical and biological forces that damage reefs, do not generally flourish unless seawater chemistry favors precipitation of the minerals they secrete (p. 240). Aragonite was precipitated on late Proterozoic seafloors, reflecting a high Mg^{2+}/Ca^{2+} ratio in seawater (see Earth System Shift 12-2). Aragonite seas persisted into Early Cambrian time, and archaeocyathids, which built the earliest skeletal reefs (see Figure 13-7), are believed to have secreted high-magnesium calcite, just as reef-building coralline algae do in the present-day aragonite sea.

Aragonite seas gave way to calcite seas at about the end the Cambrian Period. As would be expected, the coral-strome reef-building community, which arose in Ordovician time, consisted of taxa that secreted low-magnesium calcite: tabulate corals, colonial rugose corals, and stromatoporoid sponges. This reef-building fauna persisted into Silurian and Devonian time, as did calcite seas.

Figure 3 Colonial, reef-building corals of Ordovician age secreted skeletons of calcite. (Chip Clark.)

Early Cambrian animals had few modes of life

Although many animal taxa evolved during Early Cambrian time, they encompassed a narrower range of life habits than did later Paleozoic faunas. Most Early Cambrian seafloor dwellers, for example, lived close to the surface of the sediment. Most of those that burrowed lived only slightly below the surface of the sediment, and those that attached to the seafloor did not stand high above it.

Most free-living animals of the Early Cambrian were deposit feeders, extracting organic matter from sediment; these forms included trilobites and other arthropods as well as some echinoderms. Monoplacophorans, in contrast, grazed on algae growing on the seafloor. Suspension feeders, which collected organic matter and algal cells from the water, included brachiopods and attached echinoderms (p. 73). The most abundant of these echinoderms were the **eocrinoids**, whose name means "dawn crinoid." Recall that crinoids, or sea lilies, survive today (see Figure 3-33); eocrinoids were their evolutionary ancestors. Eocrinoids were obviously abundant in Cambrian seas because their skeletal plates are the principal components of many Cambrian limestones, but complete skeletons are rare because they tended to fall apart after death. Eocrinoids must have formed simple communities because all of them attached to the seafloor by very short stalks (Figure 13-6). Crinoids that lived later in the

Figure 13-6 An eocrinoid from the Cambrian of Utah. This animal, whose stem was only about 2.5 centimeters (1 inch) long, was buried suddenly by sediment that washed over it from the left, toppling the animal to the right. (Val Gunther, Brigham City, UT.)

Paleozoic Era, in contrast, formed complex communities of species that stood at various heights.

We have limited knowledge of life in the water above Early Cambrian seafloors, but we know that acritarchs persisted from the Proterozoic as important members of the phytoplankton and that the zooplankton included trilobites (see Figure 13-2C and E) as well as jellyfishes.

Stromatolites and animals Stromatolites are less abundant in Cambrian rocks than in Proterozoic rocks and more restricted in their occurrence. Whereas Proterozoic stromatolites commonly grew in subtidal settings, Cambrian forms were largely confined to the intertidal zone. This Cambrian restriction probably resulted from the effects of newly evolved grazing animals, including monoplacophoran mollusks. In modern seas, relatively few kinds of organisms can tolerate the unstable temperatures and salinities of the intertidal zone (p. 100), and presumably even fewer animals occupied this hostile environment during the Cambrian. Weak grazing pressure in the intertidal zone apparently allowed stromatolites to grow there during the Cambrian Period. Even so, Cambrian stromatolites are so riddled with holes made by animals that they display little layering and therefore have a very different appearance from Proterozoic stromatolites.

Reefs The oldest organic reefs with skeletal frameworks are low mounds that formed in Early Cambrian time, beginning in the Tommotian. The main builders of these reefs were **archaeocyathids**, which apparently were suspension feeders that pumped water through holes in their vase-shaped and bowl-shaped skeletons (Figure 13-7). Archaeocyathids were probably sponges, the simplest of which resemble them in general body plan (p. 67). Although archaeocyathids were the primary frame builders of Early Cambrian reefs, organisms of unknown taxonomic relationships actually contributed a larger volume of calcium carbonate to these reefs by encrusting archaeocyathid skeletons and binding them together.

At the end of Early Cambrian time, nearly all archaeocyathids became extinct. From then until mid-Ordovician time, all that remained were small, inconspicuous reeflike structures formed by the encrusting organisms that had previously lived with the archaeocyathids.

Evolutionary experimentation Evolution during the Cambrian Period produced many groups of animals that included only a few genera and species; indeed, some are classified as discrete classes or even phyla. This phenomenon is seen in the early Paleozoic history of the phylum Echinodermata. Today this phylum includes only a few large groups, such as starfishes and sea urchins, but quite a number of bizarre echin-

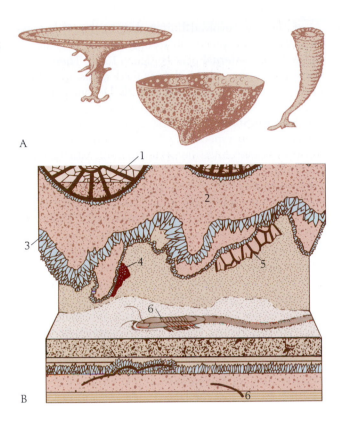

A

B

◀ **Figure 13-7** Archaeocyathids were the major reef builders of the Early Cambrian. *A.* Archaeocyathid species were vase-shaped or bowl-shaped. These animals were apparently sponges, pumping water through their porous walls. *B.* One of the world's oldest organic reefs, located in the Lower Cambrian Series of Labrador. The diagram shows the composition of the reef, which was constructed by several kinds of organisms, the most important of which were archaeocyathids (1) and calcareous algae (2). Cavities were encrusted with crystals of the mineral calcite (3), which were precipitated from seawater, and by organisms whose taxonomic relationships are uncertain (4, 5). Trilobites (6) left tracks on sediment flooring cavities in the reef as well as fossil remains within the sediment. (*A* from D. R. Kobluck and N. P. James, *Lethaia* 12:193–218, 1979; *B* after I. T. Zhuravleva, *Akad. Nauk. USSR, Geol. Geofiz. Novosibirsk* 2:42–46, 1960.)

many groups of invertebrates that appeared during Early Cambrian time, only a few—such as sponges, snails, brachiopods, and trilobites—flourished long afterward.

Later Cambrian diversification produced vertebrate animals

The Middle and Late Cambrian together spanned only about 15 million years. This interval was marked by the expansion of several preexisting animal groups, especially the trilobites. Remarkably, of some 140 families of trilobites recognized in Paleozoic rocks, more than 90 have been found in Cambrian strata. Echinoderms also continued to diversify, as did articulate brachiopods, which would become the most conspicuous fossils in younger Paleozoic rocks.

oderm classes evolved during Cambrian and Ordovician time (Figure 13-8). None included many species or genera, and most survived only a short time. Many body plans were "tried out" in this manner, but only a few succeeded. This pattern is sometimes referred to as evolutionary "experimentation," with the understanding that the evolution of new forms is not planned, but rather is caused blindly by nature. Of the

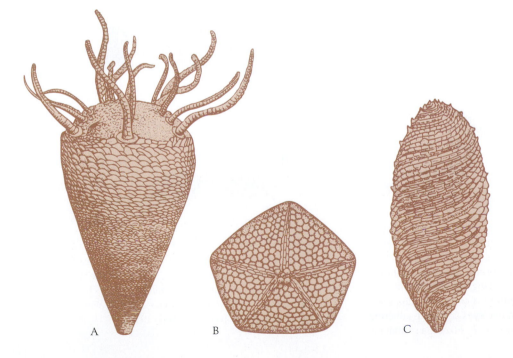

A   B   C

Figure 13-8 Strange Cambrian echinoderms that show no close relationship to any younger group. *A.* and *B.* Attached forms that apparently fed on organic matter suspended in the water. *C.* A flexible form that probably burrowed in sediment. Most living echinoderm groups, including starfishes and sea urchins, display a fivefold radial symmetry similar to that illustrated in *B.* (After R. C. Moore, ed., *Treatise on Invertebrate Paleontology*, pt. U, Geological Society of America and University of Kansas Press, Lawrence, 1966.)

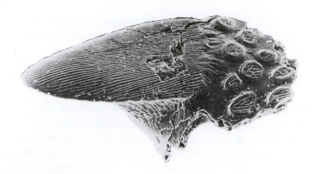

Figure 13-9 A tiny, bony plate of a small jawless fish from the Cambrian System of Wyoming. (U.S. Geological Survey.)

The earliest vertebrates

Conodonts also diversified in the course of the Cambrian Period. Their teeth, which are abundant in the fossil record, reveal nothing of their body form, but the recent discovery of fossils of their soft bodies has shown them to have been small swimming animals; the teeth themselves indicate that conodonts were the earliest known vertebrate animals (see Figure 3-35*B*). Similar small teeth in very early Cambrian faunas (see Figure 13-1*C*) may represent conodont ancestors.

Fishes also evolved during the Cambrian, but we know of their existence only through the preservation of isolated bony external plates (Figure 13-9). These early vertebrates were probably deposit feeders, like more fully preserved fishes of Silurian and Early Devonian age.

The Burgess Shale fauna

Strata in western North America have yielded a spectacular fauna of Middle Cambrian soft-bodied animals that invites comparison with the Early Cambrian Chengjiang fauna described earlier. The largest group of species in the North American soft-bodied fauna comes from the Burgess Shale, in the Rocky Mountains of British Columbia (Figure 13-10). Later in this chapter we will examine the environment in which the Burgess Shale formed, but for now we can simply note that it accumulated in a deep-water setting where soft-bodied animals were buried in the absence of oxygen and bacterial decay.

Among the Burgess Shale fossils is a species that represents the Chordata, the phylum to which vertebrate animals belong. *Pikaia*, the chordate genus of the Burgess Shale fauna, possessed a notochord—the structure that, in some Cambrian animal group that may never be singled out, evolved into a backbone (Figure 13-10*E*). Recall that the lancelet possesses only a notochord today (see Figure 3-34).

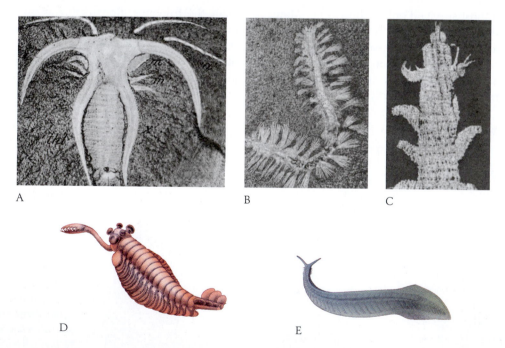

A

B

C

D

E

Figure 13-10 Animals without durable shells from the Burgess Shale of British Columbia. *A.* An arthropod related to the trilobites. *B.* A polychaete worm. *C.* An onychophoran, which had a wormlike body but walking legs that resembled those of the arthropods. *D.* *Opabinia*, a creature of unknown biological relationships that had five eyes and food-gathering pincers at the end of a strange nozzle. *E.* *Pikaia*, a swimming chordate; a notochord runs nearly the length of the animal. *A* and *B* are magnified 3 times; *C* is magnified 4 times; *D* was about 60 centimeters (2 feet) long; *E* is about life size. (*A–C*, courtesy Smithsonian Institution; *D* and *E*, drawings by Marianne Collins, reprinted by permission from S. J. Gould, *Wonderful Life*, W. W. Norton & Company, New York, 1989.)

Arthropods are the most abundant of the Burgess Shale fossils, and some of them resemble certain of the Chengjiang taxa (Figure 13-10A). Also present are anomalocarids smaller than the one depicted in Figure 13-5. In addition, both the Chinese and North American faunas include onychophorans (Figure 13-10C). Elongate animals with jointed legs, onychophorans are generally intermediate in form between segmented worms and arthropods. Today members of this group live as predators on moist forest floors, having somehow invaded the land (see Figure 3-28). Priapulid worms also occur in the Burgess Shale fauna, along with several types of segmented worms. An overall comparison of the Chinese fauna with the younger North American fauna indicates that evolutionary changes between Early and Middle Cambrian time were relatively minor for soft-bodied invertebrate animals.

Ordovician Life

The Ordovician Period was marked by a great evolutionary radiation of life in the seas. Many of the new animal groups remained successful for most of the Paleozoic Era. This radiation did not get under way until Middle Ordovician time, however. The Early Ordovician was a time of only modest evolutionary expansion.

Among Early Ordovician animals were floaters and swimmers

Trilobites suffered a major extinction at the end of Cambrian time, as we shall see later in this chapter. They recovered from this crisis, however, and remained the most abundant members of many marine communities throughout Early Ordovician time. Other groups, such as brachiopods and snails, that had originated on Cambrian seafloors became increasingly well represented early in Early Ordovician time.

In the waters above Ordovician seafloors, two groups expanded dramatically from modest evolutionary beginnings in the Cambrian: graptolites and nautiloids. Most graptolite species floated as zooplankton and were preserved in offshore settings, where their fragile colonies settled in muddy sediment after death (see Figure 5-24). Their widespread occurrence in the resulting black shales has made them useful for dating these rocks (see Figure 6-2). Nautiloids were swimmers that probably rested on the seafloor at times and also fed on bottom-dwelling animals. Like the living chambered nautilus (p. 48) and other cephalopod mollusks, they pursued their prey by means of jet propulsion and caught them with tentacles. Cambrian nautiloids are known only from China, where they arose shortly before the beginning of the Ordovician Period. Nautiloids diversified rapidly early in Ordovician time, however, and spread throughout the world.

A great radiation of life occurred later in the Ordovician

A spectacular evolutionary radiation during the Ordovician Period produced a threefold increase in the number of marine animal families. Occurring mainly in Middle and Late Ordovician time, this was the most dramatic evolutionary expansion of all time in the marine realm. Many new forms of life came to live in and on the sediment of the seafloor and in the overlying waters (Figures 13-11 and 13-12).

Life in sediment It is not easy for animals to evolve adaptations for living in sediment. The primary difficulty is in maintaining access to oxygen-bearing water. Some burrowers manage by pumping water down from above the surface of the sediment. Trace fossils reveal that an increasing variety of worms came to live in this way near the end of Proterozoic time (Chapter 12). Fabrics of sedimentary rocks reveal that animals burrowed through subtidal seafloors with increasing intensity between Cambrian and Late Ordovician time (Figure 13-13). Presumably this increase reflected a continued diversification of worms and other soft-bodied burrowers. Shelled burrowers diversified as well. As the Ordovician Period progressed, burrowing bivalve mollusks attained the position that they hold today: they became the most successful group of burrowing suspension feeders with shells. Some new kinds of trilobites were also smoothly contoured shallow burrowers (see Figure 13-11C).

Life on the seafloor A large variety of trilobites continued to scurry over the seafloor, and in the latter part of Ordovician time they were joined by an increasing array of other animals (see Figure 13-12). Jawless fishes continued to grub for food in the sediment. Snails grazed on algae that grew on the seafloor, and a few types were stationary forms that strained food from the water.

Articulate brachiopods diversified markedly; beginning in the Middle Ordovician, they began to produce the most conspicuous Paleozoic fossils. Some were attached to the seafloor by a fleshy stalk, but others rested freely on the sediment. Crinoids also expanded, forming complex communities in which species stood at various heights above the seafloor.

Rugose corals, also known as *horn corals* because of their typical shape, became well represented on the seafloor during the Ordovician Period. Some formed large colonies by budding, and certain colonial forms contributed to the growth of organic reefs. In fact, the most dramatic ecological development on Ordovician seafloors was the ascendancy of a new reef community, which produced large tropical reefs throughout the world. Early in the Ordovician, the most important reef builders were an extinct group of corals called **tabulate corals** and a group of sponges called **stromatoporoids**.

Figure 13-11
Ordovician invertebrate fossils. *A.* A straight-shelled nautiloid about 15 centimeters (6 inches) long. *B.* A spiny trilobite that lived on the sediment surface. *C.* A smooth-shelled burrowing trilobite. *D.* A snail (gastropod). *E* and *F.* Two kinds of articulate brachiopods. *G.* A bivalve mollusk that lived on the sediment surface. *H.* A branched bryozoan colony. *I.* A tabulate coral colony. *J.* A stromatoporoid colony. *K.* A rugose coral. (Courtesy Smithsonian Institution, photo by Chip Clark.)

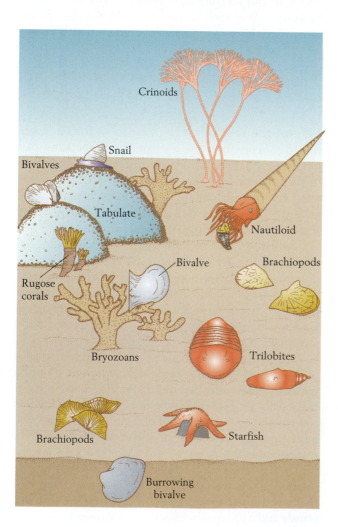

These groups, as well as the colonial rugose corals, formed massive reefs. As would be expected, these taxa formed calcite reefs in the calcite seas of the time (see Earth System Shift 13-1). We refer to these structures as **coral-strome reefs**. There had been no major reef-building community since the archaeocyathids all but disappeared at the end of Early Cambrian time. Many Ordovician coral-strome reefs were larger than those built by archaeocyathids, and the new reef community continued to thrive well beyond the Ordovician Period.

◀ **Figure 13-12** Life of a Late Ordovician seafloor in the area of Cincinnati, Ohio. Fossils of many of the groups of animals represented here can be seen in Figure 13-11. Note that at this early stage of Phanerozoic evolution, relatively few animals lived within the sediment. At the left a snail crawls over a large tabulate coral colony, and two bivalve mollusks are attached to another tabulate colony by strong threads. Another bivalve is similarly attached to a branch of a bryozoan colony. Two solitary rugose corals, lodged alongside the tabulate colonies, have their tentacles outstretched for food. Stalked crinoids are waving about at the top of the picture, feeding on suspended matter with their arms. To their right, a large nautiloid prepares to eat a trilobite that it has trapped in its tentacles; below the nautiloid's eye is a spoutlike siphon that the animal uses to expel water for jet propulsion. Two kinds of suspension-feeding brachiopods live on the seafloor. In the right foreground are trilobites of a type that left trace fossils, indicating a burrowing mode of life. In the central foreground a starfish prepares to devour a bivalve by prying apart the shell halves with its sucker-covered arms; then, by extruding its stomach, the starfish can digest the bivalve within its opened shell.

Upper Ordovician

Lower Ordovician

Pretrilobite Lower Cambrian

Figure 13-13 Fabrics produced by burrowing in lower Paleozoic sediments of the Great Basin of the western United States. The burrowing of animals disrupts layers of sediment. Each figure illustrates the degree of burrowing most commonly observed in shallow-water limestones for a particular interval. The intensity of burrowing increased markedly between earliest Cambrian and Early Ordovician time and between Early Ordovician and Late Ordovician time. (After M. L. Droser and D. J. Bottjer, *Geology* 16:233–236, 1988; 17:850–852, 1989.)

Predators The evolutionary expansion of nautiloids was the most conspicuous advance in predation during the Ordovician Period. Most Ordovician nautiloids had straight shells, some of which reached lengths of about 3 meters (10 feet). On the seafloor, starfishes appeared and assumed their role of feeding on other animals (see Figure 13-12).

Animals caused stromatolites to decline

Of all the Phanerozoic periods, only the Cambrian and Ordovician were characterized by abundant stromatolites. This abundance was carried over from late Precambrian time, but by the end of the Ordovician large stromatolites were rare. As we have seen, even in Cambrian time, grazing animals had largely restricted stromatolites to intertidal areas and destroyed their internal layering.

The few areas where stromatolites grow in the modern world offer clues to what happened to stromatolites during Ordovician time. The types of cyanobacteria that form stromatolites occur widely in modern seas, but cyanobacteria prosper well enough to form conspicuous stromatolites only in environments that are hostile to nearly all animals: fringes of land along the ocean that are flooded only occasionally, when tides are very high; subtidal channels in which very strong water movements exclude animals; and hypersaline lagoons (see Figure 5-32). In more normal marine environments, animals eat and burrow through algal mats; their ravages prevent the mats from forming stromatolites. Experiments have shown that when animals are excluded from small areas of seafloor in tropical climates, algal mats flourish, just as they did long ago. It seems evident that the great adaptive radiation of Ordovician life produced a variety of animals that prevented stromatolites from developing except in habitats where conditions excluded those animals. The severe restriction of stromatolites early in the Paleozoic Era permanently altered the nature of shallow seafloors.

Extinction set back marine diversification

Slightly more than 400 families of marine invertebrates have been recognized in Upper Ordovician rocks. As Figure 13-14 illustrates, the number of known Late Ordovician genera—about 1300—was close to the Paleozoic maximum. Some scientists have concluded that the leveling off of taxonomic diversity late in the Ordovician Period resulted from saturation of marine ecosystems: ecological crowding prevented further diversification. Figure 13-14 suggests otherwise, showing that large extinction events repeatedly set back diversification during the Paleozoic Era. These events, rather than

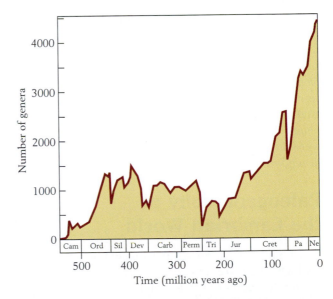

Figure 13-14 The number of genera of marine animals with skeletons that have existed at various times during the Phanerozoic Eon. (J. J. Sepkoski, *Geotimes*, March 1994.)

Figure 13-15 Late Ordovician fossils that may represent plants that lived on land. *A.* Spores that resemble those of modern land plants. *B.* A fossilized sheet of cells that resemble those covering the surfaces of some land plants. (From J. Gray et al., *Geology* 10:197–201, 1982.)

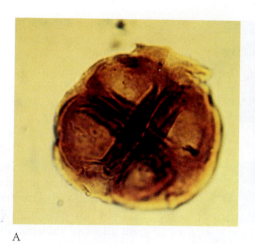

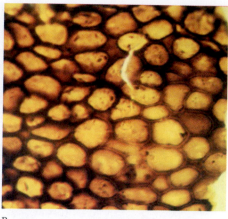

A

B

ecological crowding, apparently prevented marine diversity from rising appreciably above the Late Ordovician level before the end of Paleozoic time. Fewer large extinctions occurred during the Mesozoic and Cenozoic eras. The primary reason for the decline in rates of extinction as the Phanerozoic Eon progressed was that taxa that were particularly vulnerable to environmental change tended to disappear in major extinctions. Taxa that were more resilient survived and gave rise to similarly resilient taxa. Thus, through a kind of weeding-out process, animal life became increasingly resistant to extinction. The rarity of major extinction events during the Mesozoic and Cenozoic eras allowed animal life to become markedly more diverse.

Did plants invade the land?

The fossil record shows that small multicellular plants were well established in moist terrestrial environments during the Silurian Period. Although similar plants probably occupied Ordovician landscapes, evidence of their presence is not yet conclusive. It consists of fossilized sheets of cells similar to those that cover the surfaces of modern land plants as well as structures that resemble the spores released by primitive (nonseed) land plants of the modern world (Figure 13-15). If land plants did evolve during the Ordovician Period, they were probably restricted to moist habitats, as mosses are in modern times.

Paleogeography of the Cambrian World

As we saw in Chapter 12, paleomagnetism and other geologic evidence indicate that most cratons were clustered together near the end of Precambrian time and may have formed one giant supercontinent. The arrangement of continents late in Cambrian time was strikingly different. By that time, Gondwanaland and three smaller landmasses existed, and broad continental surfaces were positioned at low latitudes (Figure 13-16), where shallow-water limestones accumulated along their margins.

The Cambrian Period was notable for the progressive flooding of continents. The stage for this trend was set near the end of Precambrian time, when most of Earth's cratons stood largely exposed above sea level. As a result, only scattered local areas on modern continents yield a continuous record of shallow-water deposition across the Precambrian-Cambrian boundary. In Chapter 12 we discussed one of those areas: the depositional basin in western North America where the Belt Supergroup formed (see Figure 12-20).

As the Cambrian Period progressed, many parts of Gondwanaland remained above sea level, partly as a result of regional uplifts caused by orogenic activity between about 800 million and 500 million years ago. Other cratons, however, show evidence of continued encroachment of Cambrian seas until little of their total area remained exposed late in Cambrian time (see Figure 13-16). This flooding represented one of the largest and most persistent sea-level rises of the entire Phanerozoic Eon (see Figure 6-21). It was interrupted in North America only by a modest regression in Middle Cambrian and another during Late Cambrian time. As the seas began to encroach on broadly exposed continents slightly before the beginning of the Cambrian, siliciclastic sediments were eroded from the continents and accumulated around the continental margins. Figure 13-17 shows this pattern in North America. When the seas encroached farther over most continents during Middle and Late Cambrian times, a characteristic sedimentary pattern emerged. To understand the nature of this pattern, let us examine the geography of Laurentia, the landmass that included North America, Greenland, and Scotland (shown in Figure 13-16).

At all times during the Middle and Late Cambrian, some part of central Laurentia stood above sea level (see Figure 13-16). Around the margin of the continent,

LATE CAMBRIAN

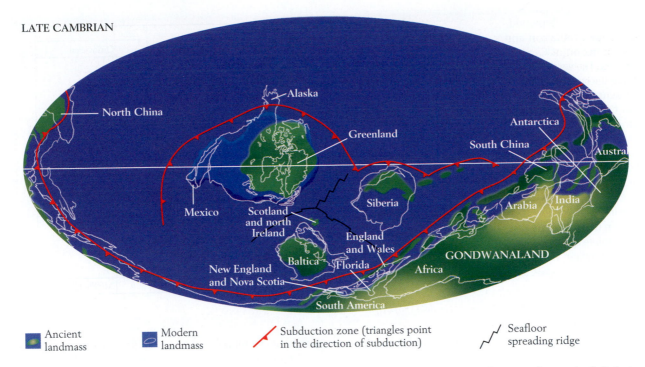

| Ancient landmass | Modern landmass | Subduction zone (triangles point in the direction of subduction) | Seafloor spreading ridge |

Figure 13-16 World paleogeography in Late Cambrian time. Laurentia was positioned at the equator and inundated by shallow seas. (Adapted from paleogeographic map by C. R. Scotese, PALEOMAP Project, University of Texas at Arlington, 1997.)

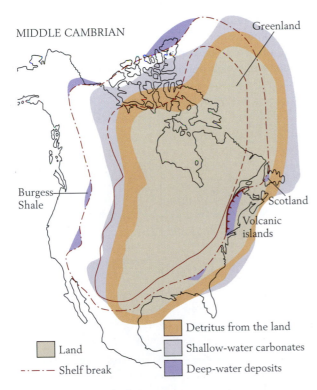

Land
— · — · — Shelf break
Detritus from the land
Shallow-water carbonates
Deep-water deposits

Figure 13-17 Concentric pattern of sediment deposition around the margin of Laurentia during Middle Cambrian time. A dashed line at the shoreline indicates uncertainty about its position. Note the location of the Burgess Shale, renowned for its fauna of soft-bodied invertebrates, at the base of the Middle Cambrian continental shelf in western Canada. (After A. R. Palmer, *Amer. Sci.* 62:216–224, 1974.)

belts of marine deposition were arranged in concentric patterns. Siliciclastic sediments derived from the craton were deposited in the innermost belt. This belt was essentially the same as the marginal siliciclastic belt that surrounded the continent during earliest Cambrian time, but it had shifted inland along with the shoreline. Seaward of this belt were broad carbonate platforms that were sometimes fringed by reefs. As the Cambrian progressed and seas shifted inland, limestones came to lie on top of nearshore sands. We viewed this pattern in the Grand Canyon region as an example of the concept of facies (see Figure 6-6). In this region, muds and breccias derived from the platform accumulated in deep water near the base of its steep slope, and the nearby subduction zone contributed volcanic sediments to the deep-water belt from offshore.

Trilobites, the dominant skeletonized animals of Middle and Late Cambrian oceans, were distributed around continents in a pattern corresponding to the arrangement of the sedimentary belts. Certain groups of these trilobites, including small forms that lived as plankton (see Figure 13-2C and E), are found primarily in deep-water deposits.

Episodic Mass Extinctions of Trilobites

Trilobite species that inhabited tropical seas—including small planktonic forms—were the ones that suffered in

Figure 13-18 Repeated evolutionary radiation and mass extinction of Cambrian trilobites. The vertical bars indicate the stratigraphic ranges of prominent species in Middle and Upper Cambrian deposits of North America. These ranges form clusters and thus delineate three successive radiations, each of which was terminated by a mass extinction (see arrows at left). A stratigraphic interval representing approximately 5 million years is shown on the right. Note that many trilobite species survived less than 1 million years. (After J. H. Stitt, *Oklahoma Geol. Surv. Bull.* 124:1–79, 1977.)

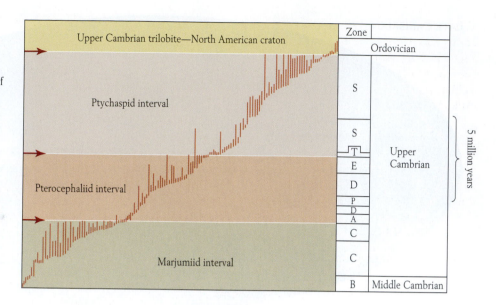

repeated mass extinctions of the Cambrian Period. The last of these crises marked the end of Cambrian time. Each mass extinction was followed by an evolutionary radiation that restored the diversity of shallow-water trilobites to a high level (Figure 13-18). These events are well documented in North America and thus far have been recorded elsewhere only in Australia.

Each evolutionary radiation of Cambrian trilobites proceeded for several million years, but each extinction was quite sudden. The fossil record reveals that each extinction took place during the deposition of a layer of sediment just a few centimeters thick and thus must have lasted no more than a few thousand years. The transition from one evolutionary radiation to another followed a characteristic pattern that is illustrated in Figure 13-19. Above the thin layer that records the extinction, in beds just a meter or so thick, a variety of new trilobite genera join species that survived the mass extinction. These beds record a brief time of biotic adjustment, in which opportunistic species flourished for

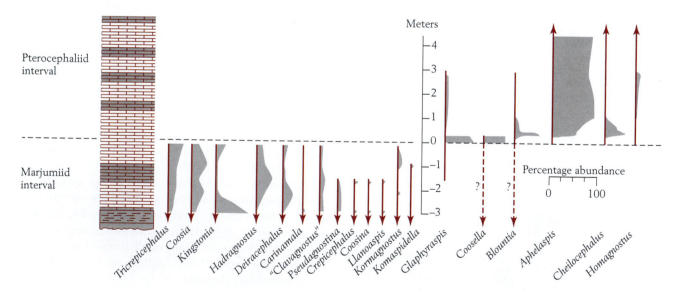

Figure 13-19 The pattern of trilobite mass extinction recorded in limestones at the top of the Cambrian Bonanza King Formation of Nevada. The gray band above each trilobite genus shows how the abundance of that genus relative to other genera changed through time. Many genera of the Marjumiid interval (see Figure 13-18) disappear at the boundary between the Marjumiid and Pterocephaliid intervals. Just above this level, only a few forms are found, and in the absence of other genera, these forms were very abundant. Arrows indicate that species are also found below or above the occurrences displayed in the diagram. (After A. R. Palmer, *Alcheringa* 3:33–41, 1979.)

a short time and then dwindled as new forms rose. The beds above contain a different fauna that included only about half as many species. From this fauna there issued a new evolutionary radiation that lasted several million years—until another mass extinction started the cycle again.

What led to the periodic mass extinction of trilobites? Elimination of shallow-water habitats by lowering of sea level can be ruled out because the extinctions did not all coincide with widespread regression of seas. In fact, a sizable regression occurred between two of the mass extinctions in Late Cambrian time. Because the adaptive radiation that preceded each mass extinction was associated with the deposition of tropical limestones, it has been suggested that a sudden, temporary cooling of the seas was the agent of the trilobites' periodic massive deaths. This idea gains support from evidence that the evolutionary radiation that followed each mass extinction issued from a group of trilobites that lived offshore in cool, deep waters marginal to the continent. These offshore trilobites did not suffer in the Cambrian mass extinctions.

Ordovician Paleogeography

Early in Ordovician time, Baltica was centered midway between the equator and the south pole (Figure 13-20A), but then it moved northward. By the end of the Ordovician it had moved into the tropics, and tropical limestones accumulated in the area of the present Baltic Sea. Some of these limestones, including oolite deposits, re-

sembled those that are forming today in the Bahama banks (p. 102). In contrast, Gondwanaland shifted so that its southern margin encroached on the south pole. Near the end of Ordovician time, a large glacier grew in this southern region, and associated changes in the world's oceans caused a profound global extinction event (Figure 13-20C).

Glaciation and sea-level lowering occurred at the end of the Ordovician

The ice cap that grew on Gondwanaland was centered near the south pole, in what is now northern Africa (see Figure 13-20C). Evidence of this glacial episode comes in many forms: tillites, scratches on bedrock, and dropstones in marine sediments (p. 108).

Near the end of the Ordovician Period, a global drop in sea level caused an unconformity to form on top of shallow-water strata throughout the world. Sea level fell because the ice cap that grew in Gondwanaland removed a significant amount of water from the global water cycle. In some areas of the central United States, rivers cut deep valleys near the end of the Ordovician, eroding rapidly downward to reach the new, suddenly lower level of the sea. The depth of their canyonlike valleys—about 50 meters (165 feet)—indicates that global sea level fell by at least this amount.

The Late Ordovician glacial interval lasted no longer than a million years—a much shorter interval than the other Phanerozoic ice ages—and probably resulted from a brief reduction of greenhouse warming (see Earth System Shift 13-2).

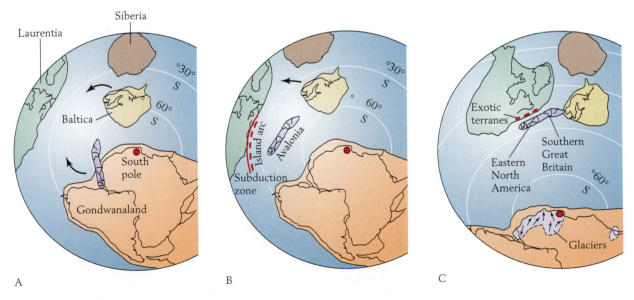

A   B   C

Figure 13-20 Movement of landmasses during Ordovician time. Avalonia was a fragment of Gondwanaland that moved toward Laurentia. Volcanic islands to its north collided with Laurentia, forming exotic terranes now located in Maine and eastern Canada. Near the end of the Ordovician, glaciers expanded over Gondwanaland near the south pole. (After B. A. van der Pluijm and R. Van der Voo, *Geol. Assn. Can. Spec. Paper* 41:127–136, 1995.)

Earth System Shift 13-2 | Glaciation Results in Mass Extinction

One of the largest mass extinctions of all time accompanied the ice age that occurred near the end of the Ordovician Period. It once seemed that the Late Ordovician glacial episode might have lasted several million years, but oxygen isotope ratios contradict this idea. Recall that when large glaciers grow on land, isotopically light H_2O accumulates in them. As a result, seawater becomes correspondingly enriched in oxygen 18, the heavier isotope. This condition, as well as cooling of seas during glacial expansion, produces heavier oxygen isotope values in the calcium carbonate secreted by marine organisms (p. 234). Fossil brachiopod shells reveal this kind of shift near the end of the Ordovician Period and show that it lasted only 0.5 to 1.0 million years. This, then, was the length of the glacial episode.

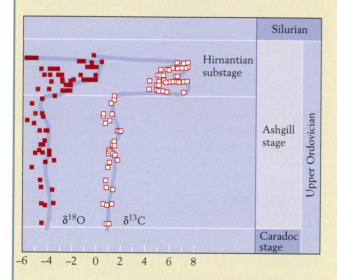

Figure 1 Oxygen and carbon isotope ratios shifted toward higher values late in the Ordovician Period. The data shown here, for brachiopod shells from the Baltic region, reveal a return to lower values near the end of the Ordovician. Data from other regions show the same pattern. (After P. J. Brenchley et al., *Geology* 22:295–298, 1994.)

The brevity of the drop in sea level explains why it does not appear in Figure 6-21, which displays sequences of deposition in North America that are bounded by unconformities. The sequences depicted there are separated by widespread unconformities that formed as sea level declined over millions of years. In general, such large but gradual changes in sea level have accompanied global changes in rates of seafloor spreading: sea level has risen or fallen along with the seafloor (p. 149). One such unconformity that formed earlier in the Ordovician Period marks the upper boundary of the first sequence of the Phanerozoic record (see Figure 6-21).

What caused glaciers to expand abruptly near the end of the Ordovician Period? Certainly the movement of Gondwanaland over the south pole was an essential element, but the supercontinent remained in this position for millions of years. Some additional factor must have come into play to trigger the glacial episode and then quickly bring it to an end. Carbon isotope ratios suggest that this factor was a reduction of Earth's greenhouse warming. Carbon isotope ratios in seawater, as recorded in brachiopod shells, shifted toward much heavier values during the glacial interval. Possibly an unusually large amount of organic matter was buried to produce this shift. Recall that organic carbon is isotopically light, so that its excess burial leaves isotopically heavy CO_2 in the ocean and atmosphere (p. 227). Perhaps, then, burial of carbon reduced the greenhouse effect in the Ordovician world, and the cooling that resulted caused glaciers to expand in the region of the south pole. If that was the case, something must have quickly ended the greenhouse cooling. In any event, more recent ice ages, including the one in which we live today, have lasted much longer than the Late Ordovician glacial episode.

The mass extinction associated with the Late Ordovician glaciation entailed two pulses. The first pulse preferentially eliminated tropical taxa. The coral-strome reef community, for example, was damaged by the loss of numerous species of reef-building corals and stromatoporoids. This is hardly surprising, because temperatures dropped throughout the world. Shifts in the distribution of marine life accompanied this climatic change: animals that had occupied seafloors in cool regions expanded into shallow seas nearer the equator. Apparently even shallow tropical seas became cooler, and because their inhabitants had no refuges, they died out.

Another distinctive pattern of the first pulse of extinction was the almost total disappearance of species that had been restricted to a single epicontinental sea on a single landmass. These species died out when sea level fell and eliminated the particular environments to which they had been adapted. Some species that lived in the open ocean also died out, apparently succumbing to changes in the temperature of the ocean.

The second pulse of extinction, which occurred when sea level rose at the end of the brief glacial interval, eliminated many of the cool-water species that had spread toward the equator when the glaciers expanded. Apparently the climatic warming that ended the glacial interval was responsible for their disappearance.

About half of all genera of marine animals died out as a result of environmental changes associated with the Late Ordovician ice age. Restoration of marine diversity to its previous level required some 50 million years.

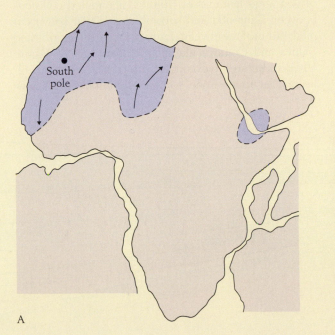

A

B

Figure 2 Glacial tills in northern Africa were deposited near the south pole in Late Ordovician time. *A.* The map shows the glaciated area (blue), directions of ice flow (arrows), and the position of the Late Ordovician south pole. *B.* Glacial sediments in the Arabian Peninsula; the arrow points to a cavity where an ice-transported boulder about 1 meter (about 1 yard) in diameter has weathered out of the tillite. (*B,* courtesy of H. A. McClure)

Late Ordovician climatic change caused a two-step mass extinction

During the Late Ordovician glacial event, marine life suffered one of the largest mass extinctions in Earth's history (see Figure 7-13). Many groups of brachiopods, trilobites, bryozoans, and corals died out on the seafloor. In the waters above, many species of acritarchs, graptolites, conodonts, and nautiloids disappeared.

There were actually two pulses of extinction in the Late Ordovician crisis. The first pulse coincided with the onset of glaciation and the second with the end of the glacial interval (see Earth System Shift 13-2). The first pulse is marked in the stratigraphic record by the disappearance of many fossil groups at the level of the unconformity that formed when sea level dropped as glaciers first expanded. Climatic cooling caused the extinction of many taxa adapted to warm conditions, including many members of the coral-strome reef community. The second pulse, which was not as severe, was marked by the disappearance of many fossil groups adapted to the cool conditions of the ice age. They died out as climates warmed and glaciers melted away.

Regional Events of Early Paleozoic Time

Some interesting regional events took place along the margins of Laurentia early in the Paleozoic Era. We will focus first on the active eastern margin of what is now North America, and then on its passive western margin.

The Taconic orogeny raised mountains in eastern Laurentia

Ordovician mountain-building events in eastern North America are collectively termed the *Taconic orogeny*. This event was the first of three orogenic episodes in what is now the Appalachian mountain belt. Chapters 14 and 15 will discuss the second and third Appalachian orogenies.

The Taconic orogeny did not result from collision of two large landmasses. Rather, it entailed collisions between Laurentia and several islands that had occupied the ocean between Laurentia, to the north, and Baltica and Gondwanaland, to the south (see Figure 13-20B). Before we examine the evidence for these collisions, let us examine some sedimentary rocks now

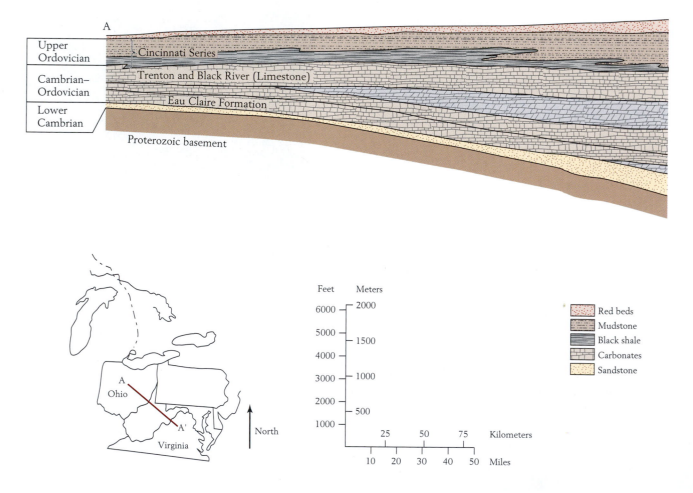

located in the central Appalachian Mountains. These rocks record a change in the pattern of deposition that offers evidence of the onset of orogenic activity.

The carbonate platform that bordered the eastern margin of Laurentia during the latter part of the Cambrian Period (see Figure 13-17) persisted through Early Ordovician time. Then, in mid-Ordovician time, the depositional framework changed drastically. Carbonate deposition ceased, and flysch deposits (see p. 211) were laid down in deep water (Figure 13-21). Deposition of black shale predominated at the outset of this activity; later, turbidites became prevalent. Sole marks (see Figure 5-33) in turbidites of the Martinsburg Formation indicate that these deposits were derived from upland regions to the east. This change represents the classic pattern for the onset of mountain building along a continental margin: the turbidites accumulated in a foreland basin (p. 211). Clearly, the shallow carbonate platform foundered, and dark sediments accumulated on top of it at considerable water depths.

Eventually sediment was being supplied by the eastern area of uplift faster than the foreland basin was subsiding. Near the end of the Ordovician Period, flysch gave way to molasse in the form of shallow marine and nonmarine clastics, some of which were red beds. Thus the Ordovician and Silurian Juniata and Tuscarora formations of the central Appalachians consist of coarse shallow marine and nonmarine deposits in the form of clastic wedges (p. 211), which taper out toward the northwest (see Figure 13-21). Cross-bedding and other features reveal that this was indeed the primary direction of transport.

Orogenies commonly result from the subduction of oceanic crust along a continental margin. The orogeny that is currently uplifting the Andes is an example (see Figure 9-15). In the Taconic orogeny, however, subduction was in the opposite direction: the margin of the carbonate platform that bordered Laurentia was wedged into a subduction zone (see Figure 13-20B). Several islands of an igneous arc collided with Laurentia during the Taconic orogeny. The rocks of these small landmasses form exotic terranes that are now embedded in the eastern margin of North America, forming parts of Newfoundland and New Brunswick in Canada and Maine in the United States. As Figure 13-22 indicates, both the rocks and the fossils of these terranes testify to their origin outside Laurentia. Many of the terranes are formed of rocks that differ from the cratonic rocks

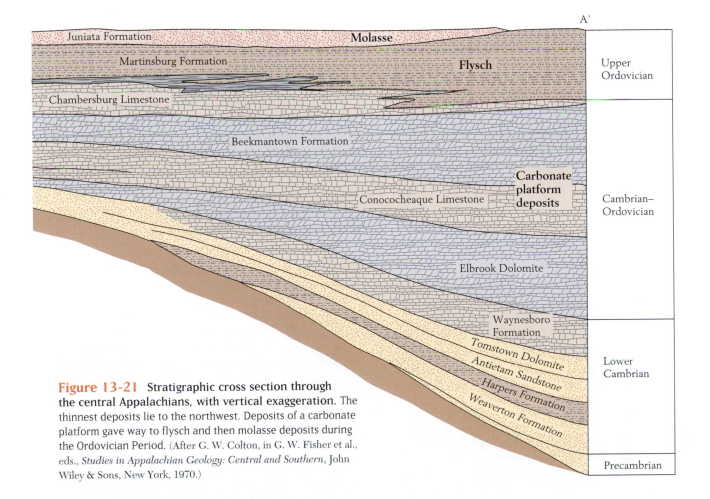

Figure 13-21 Stratigraphic cross section through the central Appalachians, with vertical exaggeration. The thinnest deposits lie to the northwest. Deposits of a carbonate platform gave way to flysch and then molasse deposits during the Ordovician Period. (After G. W. Colton, in G. W. Fisher et al., eds., *Studies in Appalachian Geology: Central and Southern*, John Wiley & Sons, New York, 1970.)

Figure 13-22 Evidence of exotic terranes in eastern North America. Both sedimentary rocks and fossils in exotic terranes contrast with those in adjacent Laurentian terranes.

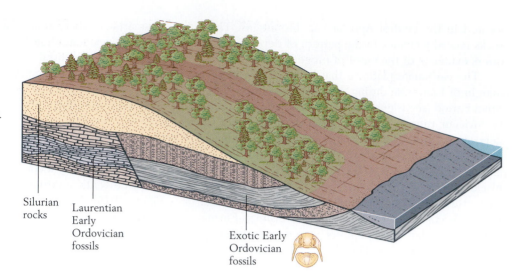

Silurian rocks

Laurentian Early Ordovician fossils

Exotic Early Ordovician fossils

of the same age that now lie immediately to their west. Some of these terranes also yield Cambrian or Early Ordovician fossils that differ from fossils of the same age in other parts of North America (Figure 13-23). Some of these fossils belong to taxa that occur in southern Great Britain. Others are unknown from any other

region and apparently represent taxa that in life were confined to islands.

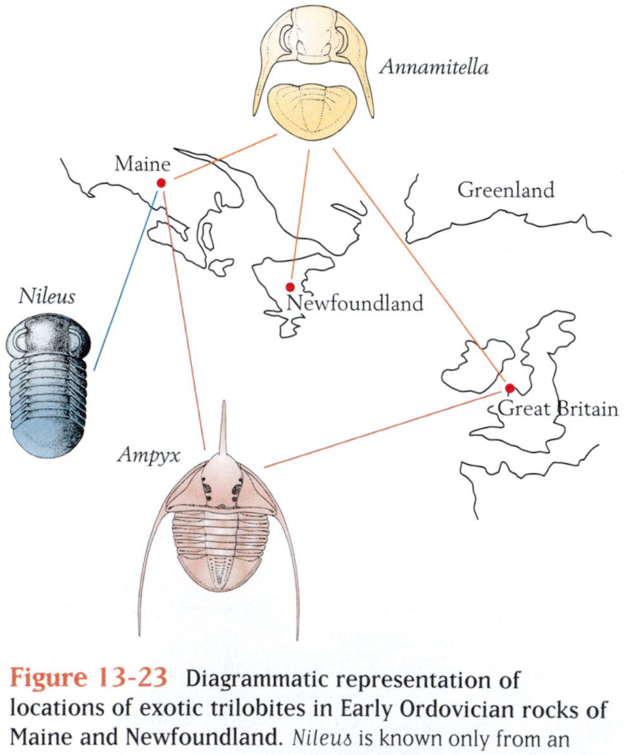

Before the advent of plate tectonics, the foreign-looking fossils in sections of eastern North America were quite puzzling. These fossils make sense only in the context of plate movements. Plate tectonic reconstructions reveal that southern Great Britain was part of a sliver of continental crust, called Avalonia, that rifted away from Gondwanaland during the Ordovician Period and encroached on Baltica. The exotic terranes that collided with Laurentia were islands of igneous arcs that had been positioned near Avalonia (see Figure 13-20C). We can no longer be surprised that they contain fossils resembling those of the British Isles.

Figure 13-24 shows how the margin of Laurentia, now positioned in eastern New York State, was deformed as small landmasses collided with it, and how the sedimentary record reflects those collisions. In this region, oceanic crust and the accretionary wedge adjacent to the neighboring island arc rode up over the carbonate platform. The depressed continental margin became a foreland basin, where dark sediment shed from the island arc accumulated as flysch in deep water.

As the collision proceeded, the foreland basin migrated inland, and large wedges of flysch that formed within it were thrust farther and farther westward, forming a thick pile of deep-water deposits on top of shallow-water carbonates. In places, ophiolites—slices of oceanic crust (p. 208)—were thrust up onto the craton along with turbidites. In advance of the foreland basin, the carbonate platform bulged up above sea level, creating unconformities in the record of carbonate deposition. The bulge formed because lithospheric material squeezed westward from beneath the foreland basin. Normal faults formed along the seaward margin of the bulge, and the foreland basin subsided in advance

Annamitella

Maine

Greenland

Nileus

Newfoundland

Great Britain

Ampyx

Figure 13-23 Diagrammatic representation of locations of exotic trilobites in Early Ordovician rocks of Maine and Newfoundland. *Nileus* is known only from an exotic terrane in Maine, apparently having occupied this terrane when it was an island far from Laurentia. *Ampyx* and *Annamitella* also occur in this terrane, as well as in southern Great Britain, which was part of Avalonia (see Figure 13-20C). *Annamitella* also occurs in Newfoundland.

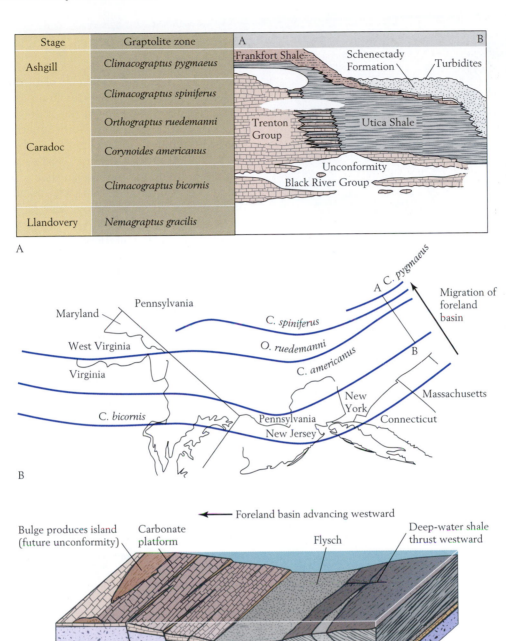

Stage	Graptolite zone
Ashgill	*Climacograptus pygmaeus*
Caradoc	*Climacograptus spiniferus*
	Orthograptus ruedemanni
	Corynoides americanus
	Climacograptus bicornis
Llandovery	*Nemagraptus gracilis*

Figure 13-24 Development of a foreland basin in what is now eastern New York during the Taconic orogeny. *A.* The deep-water Utica Shale was deposited above the shallow-water carbonate sediments of the Trenton Group; flysch, derived from an island arc to the east, then spread over the Utica Shale to form the Schenectady Formation and Frankfort Shale. *B.* A heavy line marks the western boundary of graptolite occurrences in each of the zones shown in *A.* Graptolites, which occupied deep-water environments, migrated progressively westward during Ordovician time, along with the foreland basin. *C.* The foreland basin subsided along normal faults as deep-water sediments were thrust westward. Upward bulging of the carbonate platform to the west elevated some areas above sea level. (After D. C. Bradley, *Tectonics* 8:1037–1049, 1989.)

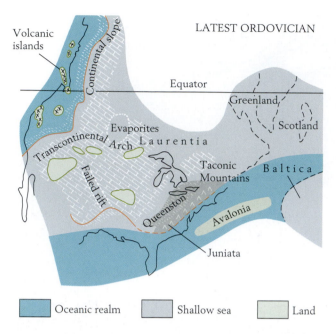

LATEST ORDOVICIAN

Volcanic islands

Continental slope

Equator

Greenland

Scotland

Evaporites

Laurentia

Transcontinental Arch

Failed rift

Taconic Mountains

Baltica

Queenston

Avalonia

Juniata

Oceanic realm	Shallow sea	Land

Figure 13-25 Patterns of deposition in North America during latest Ordovician time, when Baltica and Avalonia lay close to Laurentia. The Juniata and Queenston formations formed as clastic wedges of sediment shed inland from newly formed mountains to the east (see Figure 13-21). Carbonate sediments accumulated in the shallow seas that inundated almost all of the craton of North America. These deposits blanketed a failed rift in southern Oklahoma that had extended inland from the Gulf Coast in the early Paleozoic. Several low islands remained along the Transcontinental Arch. The continental margin bordering western North America remained stable, but volcanic islands lay offshore.

of the thrust sheets. Graptolite species that floated above deep water spread farther and farther west as the orogeny progressed, providing evidence of the basin's westward migration (see Figure 13-24*B*).

Radiometric dating of igneous rocks to the east of the zone occupied by the foreland basin provides absolute ages for igneous activity during the Taconic orogeny. The orogeny ended near the close of the Ordovician Period, apparently when the continental margin of Laurentia ultimately offered so much resistance that continental convergence could not continue.

Near the end of the Taconic orogeny, clastic wedges spread far to the west (Figure 13-25; see also Figure 13-21). Beyond the clastic belt, in the interior of the United States (southern Laurentia), limestones accumulated in shallow seas. Stratigraphic unconformities reveal that several low islands stood along the Transcontinental Arch in what is now the west-central United States. This arch was a gentle ridge that persisted throughout much of the Paleozoic Era; at various times during this era, one or more segments of it stood above shallow epicontinental seas.

A passive margin persisted in western Laurentia

Recall that a marine carbonate platform rimmed the craton of Laurentia during Cambrian time. Of course, the Taconic orogeny ended deposition along the carbonate platform in eastern North America during mid-Ordovician time, but the platform survived into middle Paleozoic time in western North America. Thus throughout Cambro-Ordovician time a passive margin bounded western North America. A stable continental shelf passed diagonally across what is now the southeastern corner of California (see Figure 13-25). Coarse, poorly sorted sediments derived from shallow-water environments accumulated at the base of the steep carbonate platform that formed the continental shelf. In many places there are great thicknesses of these limestones, composed in part of debris from shallow-water stromatolites and invertebrates (Figure 13-26). Deposits that accumulated on deep seafloors beyond the platform include black limey mudstones and limestones.

To the north, in British Columbia, the continental rise at the base of the carbonate bank was the depositional setting for the Burgess Shale, where a remarkable fauna of soft-bodied animals was preserved (see Figure 13-10). In 1909 Charles Walcott, the secretary of the Smithsonian Institution and an expert on Cambrian fossils, discovered a spectacularly well preserved fossil along a mountain trail. Careful examination of the strata above the trail turned up the layer, 2 meters

Figure 13-26 An outcrop of the Cambro-Ordovician Hales Limestone in Nevada, representing a deep-water setting along the western margin of Laurentia. The coarse fragments of shallow-water limestone in this graded bed were transported down the continental slope. (H. E. Cook and M. W. Taylor, U.S. Geological Survey, from Cook and Mullins, *AAPG Memoir* 33, 1983, p. 586, Fig. 80.)

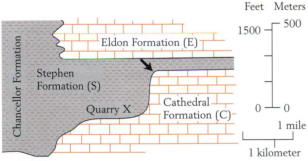

Figure 13-27 Location of the Burgess Shale quarry within the Stephen Formation in British Columbia. The formations labeled in the diagram are identified in the photograph by letters. The arrow in the diagram points to the edge of the ancient continental shelf. The Burgess Shale accumulated at the foot of the steep slope below this shelf edge. (Photograph from W. H. Fritz, *Proc. N. Amer. Paleontol. Conv.* J:1155–1170, Allen Press, Lawrence, KS, 1969.)

(7 feet) thick, from which the fossil-bearing block had fallen. Walcott organized a quarrying operation and removed nearly all the fossil-bearing material.

Extensive study of the Burgess assemblage since Walcott's day has revealed that the fauna is so well preserved because it was entombed in an oxygen-free environment from which destructive bacteria and scavenging animals were excluded. Stratigraphic evidence further indicates that the Burgess Shale was deposited at the foot of the steep carbonate shelf. In fact, the escarpment is still preserved in cross section in the mountainside some 200 meters (650 feet) above the beds that preserve the fossils (Figure 13-27). Presumably the carbonate bank stood close to sea level, so this figure of 200 meters approximates the depth at which the Burgess fauna was preserved. The Burgess fauna was collected from a series of turbidite beds. Within each

bed, calcareous siltstone grades upward into fine-grained mudstone. The beds apparently formed when turbidity currents descended the escarpment from one or more channels in the carbonate bank. Most of the animals found in the Burgess Shale probably lived along the continental margin and were swept farther down the steep continental slope by the turbidity currents. Possibly they were preserved in the absence of oxygen because they were buried very rapidly. Because several flows produced the same result, however, it seems more likely that the entire site was an oxygen-free basin near the foot of the continental slope—a depression filled with stagnant water from which oxygen had become depleted. The Santa Barbara Basin off the coast of California may be a modern analog. In any case, we must be grateful for this spectacular glimpse of the soft-bodied marine life of Middle Cambrian time.

Chapter Summary

What kinds of animal skeletons arose during the Cambrian Period?

The earliest Cambrian skeletal forms included a small variety of very small tubes and teeth. Following them was the more diverse Tommotian fauna of small animals, many of which are unknown from later intervals. Later Cambrian faunas were dominated by trilobites, but included the earliest vertebrates and large invertebrate predators.

How did Ordovician life differ from Cambrian life?

A great evolutionary radiation during the Ordovician produced a much more diverse invertebrate fauna that resembled that of later Paleozoic time. This fauna included many kinds of corals, bryozoans, brachiopods, mollusks, echinoderms, and graptolites.

Why did stromatolites decline during Cambrian and Ordovician time?

New groups of animals grazed on stromatolites and burrowed into them, making it difficult for them to grow successfully.

What kind of highly successful reef community developed during the Ordovician Period?

The coral-strome reef community, dominated by tabulate corals and stromatoporoid sponges, developed in the Ordovician and went on to thrive throughout almost all of middle Paleozoic time.

What was the configuration of land and sea during the Cambrian Period?

The continents were dispersed. Early in the Cambrian Period, many stood unusually high above sea level, but as the period progressed, the continents were increasingly flooded by a global sea level rise. Siliciclastic deposits fringed the land, and carbonate platforms developed along continental margins.

What pattern of mass extinction characterized Cambrian trilobites?

Trilobites suffered periodic mass extinctions during the latter part of the Cambrian. The last of these crises marked the close of the period.

What major continental movements took place late in the Ordovician Period?

While Baltica moved into the tropics, Gondwanaland encroached on the south pole.

Why did sea level drop suddenly near the end of the Ordovician Period?

Glaciers expanded in the polar region of Gondwanaland, locking up water. Oxygen isotopes in fossils indicate that this glacial event was relatively brief, and carbon isotopes suggest that it may have resulted from a weakening of greenhouse warming.

Why were there two pulses of marine extinction near the end of the Ordovician Period?

The first pulse, which represented one of the largest mass extinctions of all time, resulted from cooling and lowering of sea level when glaciers expanded. The second pulse, which removed fewer taxa, resulted from warming when the glacial episode came to an end.

How did the tectonic history of eastern Laurentia differ from that of western Laurentia during early Paleozoic time?

The Taconic orogeny occurred when the eastern margin of Laurentia was wedged into a subduction zone and islands of an igneous arc became attached to it, forming what are now exotic terranes in eastern North America. Whereas the carbonate platform bordering eastern North America was destroyed by Late Ordovician mountain building, the carbonate platform that bordered western North America remained intact into middle Paleozoic time.

Review Questions

1. Why do geologists know more about the life that colonized early Paleozoic seafloors than about the life that floated and swam above those seafloors?

2. What fossil evidence suggests that distinctive new kinds of predatory animals evolved during Cambrian time?

3. What evidence is there that the variety of animals that burrowed in marine sediments increased during early Paleozoic time?

4. What kinds of organisms formed reefs in Cambrian time? What kinds of organisms performed this role during the Ordovician?

5. What evidence is there that plants may have invaded the land before the end of the Ordovician Period?

6. What evidence is there that major extinctions of trilobites occurred very suddenly during the Cambrian?

7. On which landmasses is the climate likely to have been warmer in Late Cambrian time than it is today? (Hint: Compare Figure 13-16 with a map of the modern world.)

8. Review the history of sediment deposition along the eastern margin of North America and relate this history to plate movements.

9. What is the significance of the Burgess Shale? In what geographic region and environmental setting did it form?

10. Why do some Lower Ordovician rocks in Maine share some trilobite taxa with Great Britain but not with neighboring areas of the United States?

11. The Cambrian and Ordovician periods differed from one another in many ways. Using the Visual Overview on page 300 and what you have learned in this chapter, compare these two periods with respect to sea level, the distribution of landmasses on Earth, and the nature of life in the oceans.

The Old Red Sandstone forms cliffs along the coast of the Orkney Islands off northern Scotland. (John Forbes/PEP.)

CHAPTER 14

The Middle Paleozoic World

The Silurian and Devonian periods constitute middle Paleozoic time. The oceans of the world stood high during most of this interval, leaving a widespread sedimentary record on every continent. Marine deposition was interrupted in one region, however, by the most far-reaching plate tectonic event of middle Paleozoic time: the suturing of Baltica to Laurentia along a zone of mountain building.

In the northern British Isles, Silurian rocks were tilted by this collision of landmasses, and thus an angular unconformity separates them from the overlying Devonian sediments. It was farther south, in Wales, however, that in 1835 Roderick Murchison founded the Silurian System, along with the Cambrian. Five years later, Murchison and Adam Sedgwick formally recognized the Devonian System, naming it for the county of Devon, along the southern coast of England. They recognized that the fossils of this system were intermediate in character (we would now say intermediate in evolutionary position) between those of the Silurian System below and those of the Carboniferous System above. (The Carboniferous System, though younger than the Devonian, had been recognized earlier in the century.)

The broad, shallow epicontinental seas of Silurian and Devonian time teemed with life. In the tropical zone, a diverse community of organisms built reefs larger than any that had formed during early Paleozoic time. More advanced predators were also on the scene, including the first jawed fishes—a few of which were the size of large modern-day sharks. The Devonian Period was also distinguished by the progressive colonization of land habitats by new forms of life. Plants were restricted to marshy environments in Silurian time, but were forming large forests by Late Devonian time. The oldest known insects are also of Early Devonian age; and near the end of the Devonian Period, the first vertebrate animals crawled up onto land, the fins of their ancestors having been transformed into legs. Shortly before the end of the Devonian Period, however, a wave of mass extinction swept away large numbers of taxa. This great biotic crisis resulted from climatic cooling associated with glaciation near the south pole.

Renewed Diversification of Life

After the great mass extinction in the ocean at the close of the Ordovician Period, many of the decimated taxa diversified once again. Their recovery surpassed the Ordovician evolutionary radiation, yielding superior reef builders and swimming predators. Meanwhile, plants spread over the land, and near the end of the Devonian Period vertebrate animals invaded the terrestrial realm.

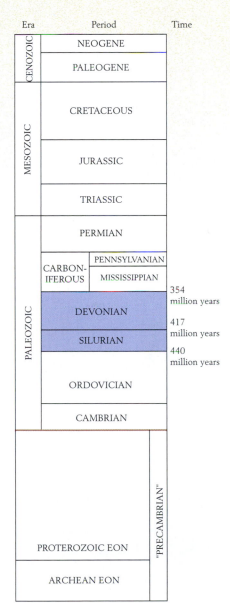

Era	Period	Time
CENOZOIC	NEOGENE	
	PALEOGENE	
MESOZOIC	CRETACEOUS	
	JURASSIC	
	TRIASSIC	
PALEOZOIC	PERMIAN	
	CARBON-IFEROUS PENNSYLVANIAN	
	CARBON-IFEROUS MISSISSIPPIAN	354 million years
	DEVONIAN	417 million years
	SILURIAN	440 million years
	ORDOVICIAN	
	CAMBRIAN	
"PRECAMBRIAN"	PROTEROZOIC EON	
	ARCHEAN EON	

Major Events of the Middle Paleozoic

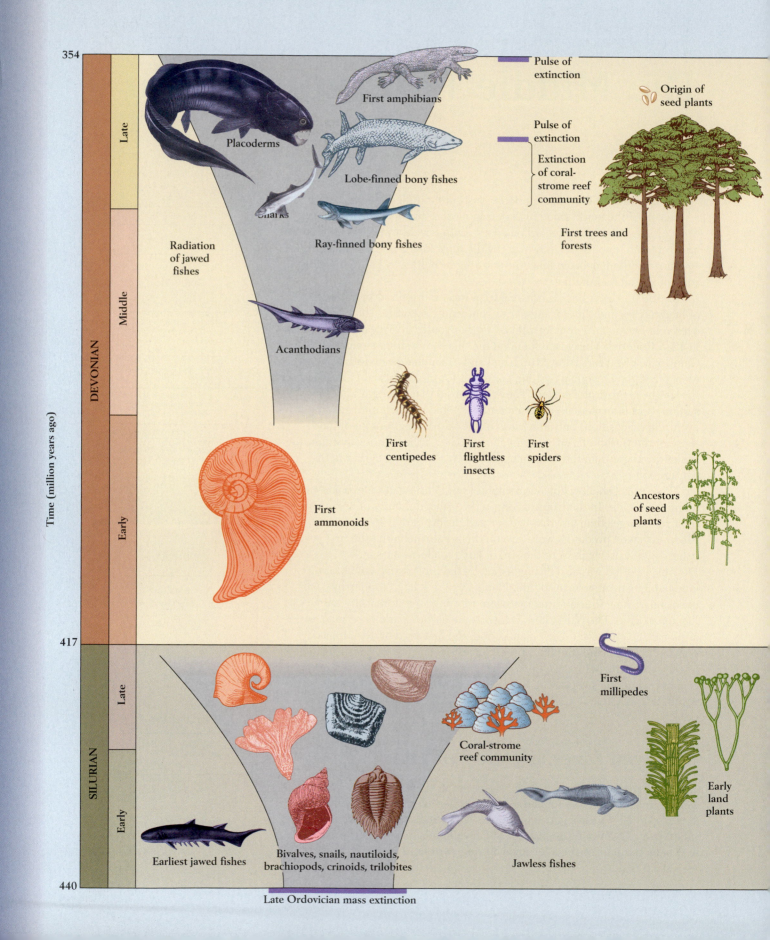

Time (million years ago)

354

DEVONIAN

Late

Middle

Early

417

SILURIAN

Late

Early

440

Placoderms

First amphibians

Lobe-finned bony fishes

Sharks

Ray-finned bony fishes

Radiation of jawed fishes

Acanthodians

First ammonoids

Pulse of extinction

Pulse of extinction

Extinction of coral-strome reef community

Origin of seed plants

First trees and forests

First centipedes

First flightless insects

First spiders

Ancestors of seed plants

First millipedes

Earliest jawed fishes

Bivalves, snails, nautiloids, brachiopods, crinoids, trilobites

Coral-strome reef community

Jawless fishes

Early land plants

Late Ordovician mass extinction

EARLY CARBONIFEROUS

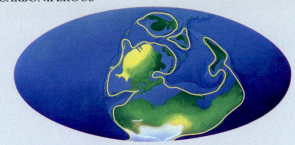

EARLY DEVONIAN

Euramerica

Acadian
orogeny

Gondwanaland

Laurentia and
Avalonia collide.

MID-SILURIAN

Acadian
orogeny

Euramerica

Avalonia

Gondwanaland

Laurentia and
Baltica collide.

MIDDLE ORDOVICIAN

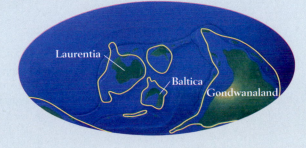

Laurentia

Baltica

Gondwanaland

Expansion of glaciers

Increased weathering
reduces CO_2 in
atmosphere.

$\delta^{18}O$

South

Antler
orogeny
in western
Laurentia

Acadian orogeny in
eastern Laurentia

North

Sea level

Rising

Falling

Life recovered and expanded in aquatic habitats

Most of the marine taxa that had flourished during the Ordovician Period rediversified after the mass extinction at the end of the Ordovician to become prominent members of the Silurian and Devonian marine biota. The trilobites failed to recover fully, however, and were less conspicuous in middle Paleozoic than in early Paleozoic seas. Brachiopods, on the other hand, attained higher diversities than ever before (Figure 14-1).

The bivalve mollusks expanded their ecological role by invading nonmarine habitats; some of the oldest known freshwater bivalves are found in the Upper Devonian strata of New York State. One of the most spectacular Early Silurian evolutionary radiations in the marine realm was that of the graptolites, which had nearly disappeared at the end of the Ordovician Period. During just the first 5 million years or so of Early Silurian time, the number of species of graptolites known in the British Isles increased from about 12 to nearly 60.

Luxuriant reefs Most of the Silurian radiations of marine life did not vastly alter marine ecosystems; instead, they refilled niches vacated by the mass extinction at the end of the Ordovician. Corals and stromatoporoids did diversify in new ways, however, and in some shallow seas they produced reefs much larger than any of Cambro-Ordovician age. The coral-strome reef community persisted for about 120 million years, until late in the Devonian Period. During this interval, colonial rugose corals came to outnumber tabulate corals as reef builders.

Coral-strome reefs occasionally attained substantial size during the Silurian Period, but in Devonian

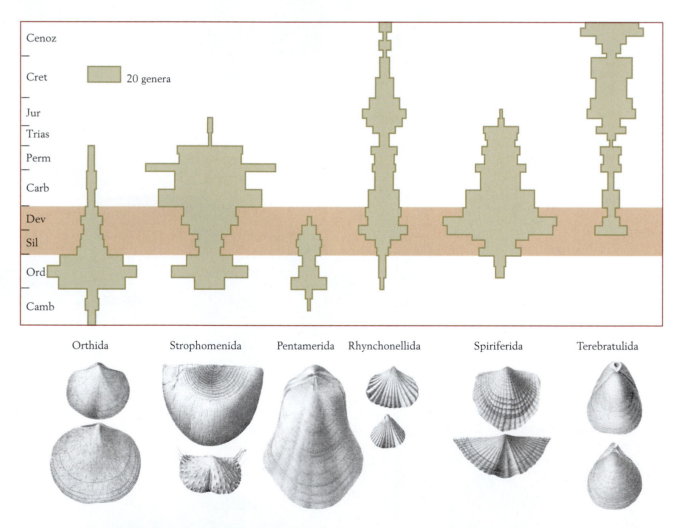

Figure 14-1 Middle Paleozoic articulate brachiopods.
Top: The diversity (numbers of genera) of the six major orders of Paleozoic brachiopods, all of which were well represented in middle Paleozoic (Silurian and Devonian) time. Bottom: Typical middle Paleozoic representatives of each brachiopod order. (Brachiopod illustrations from James Hall's volumes of the New York State Natural History Survey [1862–1894].)

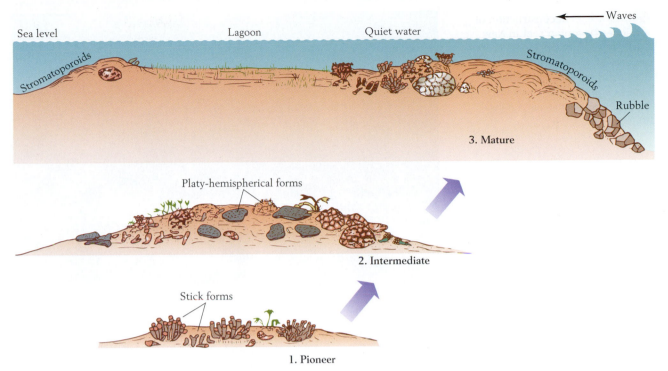

A

Figure 14-2 Coral-strome reefs flourished in the **Devonian period.** *A.* Ecological succession of a typical Devonian reef. (1) The pioneer community consisted of fragile, twiglike rugose corals and tabulates. (2) Broad and moundlike tabulate colonies were dominant during the intermediate stage of development. (3) In the mature stage, the reef grew close to sea level, and waves broke against a ridge of massive, encrusting stromatoporoids; behind them were species adapted to quieter water, and leeward of the reef was a lagoon populated by fragile, twiglike species. The leeward side of the lagoon was bounded by a small stromatoporoid ridge. *B.* A moundlike reef formed by tabulates and rugose corals. The fossils were collected in Michigan (see Figure 14-21) and were then reassembled at the Smithsonian Institution to recreate the reef. This moundlike reef represents the intermediate stage (2). (*A* after P. Copper, *Proc. 2nd Int. Coral Reef Symp.* 1:365–386, 1975; *B,* Smithsonian Institution.)

B

time they assumed enormous proportions. The largest one grew near the equator in Gondwanaland, in what is now Western Australia (see Figure 5-26). In areas subjected to strong wave action, the growth of coral-strome reefs followed a characteristic ecological succession (Figure 14-2). First, sticklike tabulates and rugose corals colonized an area of subtidal seafloor. A low mound was then formed when these fragile forms were encrusted by platy and hemispherical tabulates and colonial rugose corals. Finally, as the mound grew up toward the sea surface, stromatoporoids and algae encrusted the seaward side, forming a durable ridge.

Tabulates and colonial rugose corals occupied a zone of quieter water behind the ridge, and beyond them was a lagoon in which mud-sized sedimentary grains accumulated along with coarser skeletal debris from the reef. Pockets of fossils preserved in reef rock reveal that a wide variety of invertebrates inhabited coral-strome reefs: brachiopods and bivalve mollusks attached themselves to a typical reef, snails grazed over it, and crinoids and lacy bryozoans reached upward from its craggy surface. Although its fauna would look unusual to us today, a living, fully developed coral-strome reef (Figure 14-3) would certainly

Figure 14-3 Reconstruction of an Upper Devonian reef in New York State. Numerous kinds of corals are present. In the right foreground is a huge, spiny trilobite that measured about 45 centimeters (18 inches) in length. (Carnegie Museum of Natural History.)

seem as colorful and spectacularly beautiful as the coral reefs that flourish in the modern tropics (see Figure 4-28).

New swimming invertebrates Perhaps the greatest change in the nature of aquatic ecosystems during middle Paleozoic time resulted from the origin of new kinds of nektonic (swimming) animals, many of which were predators. The most prominent new invertebrate swimmers were the **ammonoids**. These coiled cephalopod mollusks evolved from a group of straight-shelled nautiloids during Early Devonian time (Figure 14-4). After giving rise to the ammonoids, the nautiloids persisted at low diversity. The ammonoids, in contrast, diversi-

fied rapidly. Because their species were distinctive, widespread, and relatively short-lived, they serve as guide fossils in rocks ranging in age from Devonian to latest Mesozoic. (Ammonoids died out along with the dinosaurs at the end of the Mesozoic Era.)

Another group of invertebrate predators that proliferated during middle Paleozoic time was the eurypterid arthropods. These distant relatives of scorpions were swimmers, and many had claws (Figure 14-5). Although the eurypterids appeared in the Ordovician Period and survived until Permian time, their most conspicuous fossil record is in middle Paleozoic rocks. Unlike ammonoids, eurypterids ranged into brackish and freshwater habitats.

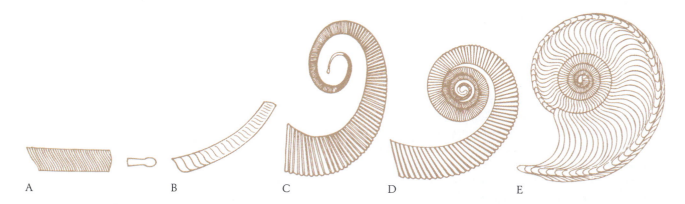

A   B   C   D   E

Figure 14-4 Shells of Lower Devonian cephalopod mollusks from the Hunsruck Shale of Germany reveal the apparent evolutionary sequence from nautiloids to early ammonoids. *A* and *B.* Fragments of nautiloids of the group that evolved into ammonoids. *C–E.* Early ammonoid species representing various degrees of coiling. The bulblike shape of

the first-formed part (tip) of each shell, together with other shared features, suggests that these species are closely related; the coiling sequence displayed here apparently represents the evolutionary sequence. (After H. K. Erben, *Biol. Rev.* 41:641–658, 1966.)

Figure 14-5 Reconstruction of a Late Silurian eurypterid. This animal was about 0.5 meters (20 inches) long. The appendages beneath its head bore sharp spikes for stabbing prey. (Chip Clark.)

Jawless fishes Other swimmers that were adapted to both marine and freshwater conditions were the fishes. The major groups are shown in Figure 14-6. Whereas only fragments of fish skeletons have been found in early Paleozoic sediments (see Figure 13-9), the Silurian and Devonian systems have yielded diverse, fully preserved fish skeletons, many of which are found in freshwater deposits.

We do not know when fishes first occupied freshwater habitats. The fact that all known Cambro-Ordovician fish remains have been found in marine deposits supports the idea that fishes evolved in the ocean. Most Silurian fish remains, unlike most Cambro-Ordovician fish fossils, come from freshwater deposits.

One of the most conspicuous new groups of fishes was the **ostracoderms**, whose name means "bony skin." Ostracoderms were small animals with paired eyes like those of higher vertebrates. Lacking jaws and covered by bony armor, ostracoderms differed from modern

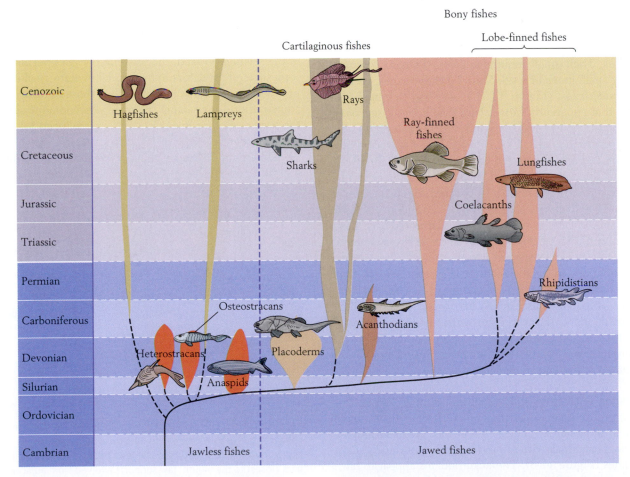

Figure 14-6 Geologic occurrence of various kinds of fishes. By Devonian time, all the major groups living today were in existence. The three major ostracoderm groups are colored orange. No ostracoderms and few placoderms survived beyond the Devonian Period. (After E. H. Colbert, *Evolution of the Vertebrates*, John Wiley & Sons, New York, 1980.)

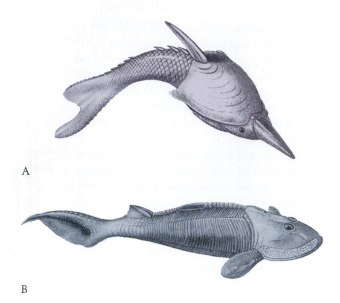

A

B

Figure 14-7 Reconstruction of Devonian ostracoderms (jawless fishes). *A. Pteraspis* was about 20 centimeters (8 inches) long. *B. Hemicyclaspis* was about 13 centimeters (5 inches) long. (After D. Dixon et al., *The Macmillan Illustrated Encyclopedia of Dinosaurs and Prehistoric Animals*, Macmillan Publishing Company, New York, 1988.)

Figure 14-8 Reconstruction of *Climatius*, a member of the most primitive group of jawed fishes, known as acanthodians. This animal was about 7.5 centimeters (3 inches) in length. (After D. Dixon et al., *The Macmillan Illustrated Encyclopedia of Dinosaurs and Prehistoric Animals*, Macmillan Publishing Company, New York, 1988.)

fishes. Their small mouths allowed them to consume only small items of food. Many of these fishes, such as *Hemicyclaspis* (Figure 14-7B), also had flattened bellies that apparently were adapted to a life of scurrying along the bottoms of lakes and rivers. The upper fin of the asymmetrical tail of *Hemicyclaspis* was elongate; when the animal wagged it back and forth for swimming, this structure would have pushed its head downward rather than upward. In contrast, the ostracoderm known as *Pteraspis* (Figure 14-7A) had a curved belly and an elongate lower fin in its tail that would have lifted it upward, suggesting a life of more active swimming well above lake floors and river bottoms. Ostracoderms lacked bony internal skeletons; it is assumed that they had cartilaginous internal skeletons, which were seldom preserved. They also lacked highly mobile fins, which provide more advanced fishes with stability and control of their movements. These animals continued to thrive throughout most of the Devonian Period but disappeared at the end of that interval.

Fishes with jaws Late in Silurian time a quite different group of small marine and freshwater fishes made their appearance. These were the **acanthodians**—elongate animals with numerous fins supported by sharp spines (Figure 14-8). The acanthodians appear to have been the first fishes to possess several features that were passed on to more advanced, modern fishes: their fins

were paired; scales rather than bony plates covered their bodies; and, most important, they had jaws. With the origin of jaws, a wide variety of new ecological possibilities opened up for vertebrate life—possibilities that related primarily to the ability to prey on other animals. Unlike ostracoderms, many acanthodians must have been predators that fed on small aquatic animals.

The evolution of jaws transformed the marine ecosystem. It also added a new dimension to life on land when jawed vertebrates emerged from aquatic environments near the end of the Devonian Period. Both in the sea and on land, the existence of jaws permitted sophisticated predators to evolve. Ultimately, evolution produced human jaws by remodeling those of fishes. Of course, there were many stages of development along the way, and other lines of evolution produced jaws of various other types.

Where did fishes get their jaws? Jaws evolved during the Devonian Period from bars that supported the

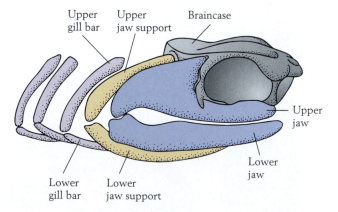

Upper gill bar   Upper jaw support   Braincase

Upper jaw

Lower jaw

Lower gill bar   Lower jaw support   Lower jaw

Figure 14-9 The origin of jaws. The braincase, jaws, and gill bars of a Carboniferous shark illustrate the primitive configuration of jaws in fishes. The jaws evolved from the gill bars closest to the mouth. Jaw supports evolved from adjacent gill bars. (After R. Zangrel and M. E. Williams, *Palaeontology* 18:333–341, 1975.)

gills of primitive fishes. Fossil sharks provide much of the evidence (Figure 14-9). In sharks, as in other primitive vertebrates, skeletal bars lie on either side of the throat between the gill slits—the openings that allow water to pass through the gills. Each bar has an upper and a lower part, which connect to form a backward-pointing V. The jaws of some primitive Devonian sharks resemble these bars in both shape and orientation. Unlike our jaws, those of primitive sharks were not fused to the skull: they were positioned directly in front of the gill bars and were aligned with them. In fact, the jaws of some fossil sharks give the appearance of being the first gill bars in the series, although they differ slightly in shape from the bars and bear teeth. It is difficult to avoid the conclusion that the jaws amount to modified gill bars. A strong similarity between the muscles that operate the jaws and those that close the gill slits reinforces this conclusion. Fossils of primitive sharks found in black shales in northeastern Ohio have been magnificently preserved under anoxic conditions similar to those that produced the Burgess Shale fauna of soft-bodied animals (p. 308). These fossils display traces of the jaw muscles of the sharks, along with many other anatomical features.

The similarity of gill bars and primitive jaws extends even to the teeth. Small, pointed structures called *denticles* toughen the skin of sharks, giving it the texture of sandpaper. The teeth that line the jaws of primitive sharks resemble the denticles of their skin. Thus it appears that evolution produced primitive teeth along the jaws simply by enlarging the denticles in the skin that overlay the ancestral gill bars.

Acanthodians declined near the end of Devonian time, but they left an evolutionary legacy of great ecological significance. We do not know precisely how more advanced groups of fishes were related to acanthodians, but during the Devonian Period, a great evolutionary radiation of descendant jawed fishes added new levels to the food webs of both freshwater and marine habitats. Soon very large fishes were feeding on smaller fishes, which in turn fed on still smaller fishes. At the top of this food web were the largest members of the group, the **placoderms**. A few of these heavily armored jawed fishes are known from uppermost Silurian and Lower Devonian freshwater deposits, and a wide variety of freshwater species existed by mid-Devonian time. Only secondarily did placoderms make their way into the oceans, and they were not highly diversified there until Late Devonian time. *Dunkleosteus*, a Late Devonian marine genus, attained a length of some 7 meters (23 feet). Like other placoderms, it had armorlike bone protecting the front half of its body (Figure 14-10), but its unarmored tail, which remained flexible for locomotion, was exposed to attack. *Cladoselache*, a small

Figure 14-10 The massive armored skull of *Dunkleosteus*, a placoderm fish of Late Devonian age. This skull is more than a meter (about 3 feet) long. Note the bony teeth and the armor protecting the eye. (Chip Clark.)

shark, is commonly found with the giant *Dunkleosteus* in black shales of northern Ohio (Figure 14-11).

Sharks were, in fact, among the most important groups of fishes in Devonian seas. Devonian sharks were primitive forms, and few grew much longer than 1 meter (3 feet).

Also arising in Devonian time were the **ray-finned fishes**. These jawed forms, which attained only modest success during the Devonian Period, went on to dominate Mesozoic and Cenozoic seas. They include most of the familiar modern marine and freshwater fishes, such as trout, bass, herring, and tuna. The term *ray-finned* refers to the fact that thin bones radiate from the body to support the fins of these fishes; these bones are visible through the transparent fins of living fishes. The oldest mid-Devonian ray-finned fishes, such as *Cheirolepis* (Figure 14-12), differed from modern representatives in having asymmetrical tails and diamond-shaped scales that did not overlap.

The origin of the ray-finned fishes was an event of great significance, but so was the origin of a related group of jawed fishes—one that included the lungfishes and lobe-finned fishes.

Fishes with lungs The Devonian Period was the time of greatest success for the **lungfishes**, only three genera of which survive today—one in South America, one in Africa, and one in Australia. (Presumably this fragmented distribution reflects the Mesozoic breakup of

Figure 14-11 The giant placoderm *Dunkleosteus* in pursuit of the Late Devonian shark *Cladoselache*. These creatures were, respectively, about 7 meters (23 feet) and 2 meters (6 feet) long. (After D. Dixon et al., *The Macmillan Illustrated Encyclopedia of Dinosaurs and Prehistoric Animals*, Macmillan Publishing Company, New York, 1988.)

Gondwanaland.) The Australian genus, *Neoceratodus*, so closely resembles the Triassic genus *Ceratodus* that it is commonly referred to as a "living fossil." The surviving lungfishes are named for the lungs that allow them to gulp air when they are trapped in stagnant pools during the dry season. Such lungs presumably served a similar function in Devonian time.

Lungfishes belong to a group known as **lobe-finned fishes** (Figure 14-13). These fishes derive their name from their paired fins, whose bones are not radially arranged, as in ray-finned fishes, but instead attach to their bodies by a single shaft. Most lobe-finned fishes have occupied freshwater habitats, but one unusual group, the coelacanths, invaded the oceans. A single coelacanth genus survives today in deep waters southeast of Africa and near Indonesia (see Figure 3-36). Lobe-finned fishes declined after the Devonian Period but left a rich evolutionary legacy. As we shall see, lobe-

finned fishes are the ancestors of all terrestrial vertebrates, including humans; their lungs were the predecessors of our own.

The effect of swimming predators The great diversification of jawed fishes and, to a lesser extent, the expansion of ammonoids and eurypterids must have had a profound effect on many relatively defenseless aquatic animals. These predators probably contributed to the decline in the trilobites' diversity in middle Paleozoic time. About 80 families of trilobites are known from the Ordovician, but only 23 families have been found in Silurian deposits. It seems likely that the weakly calcified external skeletons of these animals offered little resistance to the jaws of fishes, and certainly trilobites had no mechanism for rapid locomotion. The small, apparently defenseless ostracoderms, which died out late in the Devonian Period, must also have been easy prey for

Figure 14-12 *Cheirolepis*, a primitive ray-finned fish of mid-Devonian age. The tail of this animal was strongly asymmetrical, and its small, diamond-shaped scales did not overlap. It was about 55 centimeters (22 inches) long. (After D. Dixon et al., *The Macmillan Illustrated Encyclopedia of Dinosaurs and Prehistoric Animals*, Macmillan Publishing Company, New York, 1988.)

Figure 14-13 The lobe-finned fish *Eusthenopteron*. This large animal exceeded 50 centimeters (20 inches) in length. This unusually well preserved specimen is from Upper Devonian deposits at Scaumenac Bay, Canada.) (Swedish Museum of Natural History.)

jawed fishes. Ostracoderms even lacked the ability to burrow in sediment, which at least some trilobites could do (see Figures 13-3 and 13-11).

Plants invaded the land

It is difficult to imagine how the landscape looked in Precambrian and early Paleozoic times, before terrestrial plants became widespread. Moist terrestrial environments must have been populated by algae, cyanobacteria, and fungi, but forests and meadows were absent, and there must have been large areas of barren rock and soil with little or no humus (decayed organic matter). Thus one of the most important events revealed by the fossil record of Silurian and Devonian life was the invasion of terrestrial habitats by plants.

The basic requirements for the terrestrial existence of large multicellular plants are quite different from those for plants that live in water. Unlike water, air is much less dense than the tissues of a plant, so if a plant is to stand upright in air, it must have a rigid stalk or stem. A tall plant must also be anchored by a root system or a buried horizontal stem, either of which serves the further indispensable function of collecting water and nutrients from the soil.

The first upright plants to make their way onto land lacked the roots, leaves, and efficient means of transporting nutrients that made their descendants so successful. Essentially, these plants were simple rigid stems. Fragments of such early plants have been found in Silurian rocks. Silurian plants seem to have been pioneers that lived near bodies of water, and they may actually have been semiaquatic marsh dwellers rather than fully terrestrial plants.

Vascular plants Most large plants of the modern world are **vascular**; that is, their stems have one set of special tubes to carry water and nutrients upward from their roots and another to distribute the food that the plants manufacture for themselves. Most large modern plants also bear leaves, which serve to capture the sunlight necessary for photosynthesis.

A major adaptive breakthrough for life on land, before the evolution of roots and leaves, was the origin of vascular tissue. Figure 14-14 shows the tubes used for transporting water, nutrients, and food within a stem of the Early Devonian genus *Rhynia*.

A few kinds of vascular plants are found in nonmarine deposits of latest Silurian age. These plants had branched leaves as well as bulbous organs that shed spores. *Baragwanathia*, the largest such plant yet discovered, grew to a height of about 1 meter (3 feet) (Figure 14-15).

Spore plants Spores are reproductive structures that can grow into new adult plants when they are released into the environment. Ferns are familiar spore plants in

Figure 14-14 The Early Devonian vascular plant *Rhynia*. The reconstruction shows that a simple horizontal stem served the function of a root system, and that this primitive plant bore no leaves. The yellow structures are spore organs. In the cross section of the stem, the vascular tissue that transported water can be seen as a small dark area; around it, and visible as a narrow, light-colored ring, is the vascular tissue that transported food. (From M. E. White, *The Flowering of Gondwana*, Reed Books Pty. Ltd., 1990.)

the modern world. The fossil record of spores resembling modern ones extends well back into the Ordovician System (see Figure 13-15*A*), but while these older spores suggest that upright land plants existed much earlier than

Figure 14-15 One of the oldest known vascular plants. A specimen of the genus *Baragwanathia* from the Upper Silurian of Victoria, Australia. This plant was about 2.5 centimeters (1 inch) in diameter. (Francis M. Huber, Smithsonian Institution.)

the Late Silurian, they may in fact represent aquatic or semiaquatic species. In some Early Devonian forms, solitary spore organs stood atop upright stalks (see Figure 14-14), while other species displayed clusters of spore organs in similar positions, and in still other species, spore organs were arrayed along the upright stalks.

Whether or not land plants existed much before latest Silurian time, it was apparently near the end of the Silurian Period that vascular tissues evolved. As a result of this physiological breakthrough, a great evolutionary radiation took place in Early Devonian time. The plants that resulted were still relatively low, creeping forms that lacked well-developed roots and leaves, but during Early and Middle Devonian time, more complex plants evolved. The vascular tissues of early vascular plants such as *Rhynia* were confined to a narrow zone of the stem (see Figure 14-14) and so were mechanically weak and inefficient at conducting liquid. By Late Devonian time, however, some plants had developed vascular tissues that occupied a larger volume within the stem and were therefore mechanically stronger and also more efficient transporters of nutrients. Plant groups with these useful traits also evolved roots for support and effective

absorption of nutrients, as well as leaves for capturing sunlight. These plants seem to have competitively displaced such plants as *Rhynia*, which were less efficient at obtaining nutrients and synthesizing food.

Certain of the small plants that arose during Early and Middle Devonian time are classified as **lycopods**. This group includes the tiny club mosses of the modern world (Figure 14-16). In marked contrast, some late Paleozoic lycopods grew to the proportions of trees, and their petrified remains supply much of modern society's coal. Only tiny creeping lycopods resembling the primitive types of the Early Devonian have survived to the present. Trees did not appear until Late Devonian time. *Archaeopteris* was the first genus to assume tree proportions (see Earth System Shift 14-1).

All of the early spore plants, like those of the present day, had a complex reproductive cycle that would have confined them to habitats that were damp at least part of the year (see Figure 3-22). This reproductive cycle entails not only a conspicuous spore plant, such as a fern, but also a tiny, inconspicuous gamete-bearing plant, over whose surface a sperm must travel to fertilize an egg. The sperm requires moist conditions to make

A   B

C

Figure 14-16 **Small lycopods.** *A* and *B*. Reconstructions of *Protolepidodendron* (*A*) and *Asteroxylon* (*B*), primitive Devonian lycopods. *C*. The modern *Lycopodium* is a small form like the Devonian fossils, growing only a few centimeters tall.

In contrast to these oldest and youngest representatives, many lycopods of late Paleozoic time were large trees. (*C*, C. D. Klees/PEP.)

Figure 14-17 Reconstruction of an Early Devonian landscape. Many of the first land plants still bordered bodies of water. These plants were generally less than 1 meter (3 feet) tall. (Chase Studio, Inc.)

its journey. Because of the limitations imposed by their mode of reproduction, low-growing Early Devonian spore plants must have been restricted to marshes along bodies of water (Figure 14-17), and even *Archaeopteris* trees must have inhabited swamps.

Seed plants As the Devonian Period progressed, the appearance of a second adaptive innovation, the **seed**, liberated land plants from their dependence on moist conditions and allowed them to invade drier habitats (see Earth System Shift 14-1). Quite abruptly, large trees with strong, woody stems changed the face of Earth, forming the world's first forests in well-drained terrain, where only barren land had existed before. Because fertilization is an internal process in seed plants, environmental moisture is not necessary. The seed, which results from fertilization, is released as a durable structure that can sprout into a plant when conditions become favorable. Pollen, which represents a different part of the life cycle in seed plants, also tolerates a wide variety of environmental conditions. Pollen travels through the air to fertilize eggs so that seeds can form. Today most large land plants grow from seeds.

Animals moved ashore

Fragmentary remains of simple terrestrial animals are known from Late Silurian rocks, but few animals occupied terrestrial habitats before Devonian time. It is not surprising that animals colonized terrestrial habitats in large numbers only after vascular plants were well established, because a food web must be built upward from the base; herbivores require plant food. Assemblages of early terrestrial invertebrate animals are well preserved in the Lower Devonian Rhynie Chert of Scot-

land, in the Middle Devonian Gilboa Formation of northern New York State, and the Early Devonian Battery Point Formation of Quebec. The invertebrate fossils of these formations fall into two ecological categories. First, there are millipedes and flightless insects that fed on organic detritus. Second, there are scorpions, centipedes, and spiders—all of which were carnivores. Conspicuous in their absence are herbivores, such as leaf-eating or juice-sucking insects. Thus we can see that dead plant debris, not living plant tissue, provided the nutritional foundation for the earliest terrestrial animal communities.

Not until latest Devonian time did vertebrate animals make their transition onto land. Anatomical evidence indicates that the four-legged vertebrates most closely related to fishes are the **amphibians**—frogs, toads, salamanders, and their relatives (see p. 75). In fact, amphibians return to the water to lay their eggs and spend their juvenile period there. Then most kinds of amphibians metamorphose into air-breathing, land-dwelling adults. Living amphibians are small animals that differ substantially from the large fossil amphibians found in Paleozoic rocks.

The remains of some unusual vertebrate animals have been found in uppermost Devonian rocks of eastern Greenland. These fossils, some of which are assigned to the genus *Ichthyostega*, represent creatures that are strikingly intermediate in form between lobe-finned fishes and amphibians. The lobe fin itself is formed of an array of bones resembling that found in early amphibians; similarly, the complex teeth of lobe-finned fishes closely resemble the teeth of these amphibians (Figure 14-18). These features alone strongly suggest that amphibians evolved from lobe-finned fishes, and additional

Earth System Shift 14-1 | Plants Alter Landscapes and Open the Way for Vertebrates to Conquer the Land

Plants began to move onto land from the sea before Devonian time, but their effect on terrestrial environments was minimal (p. 312). During the Devonian Period, however, plants' more extensive invasion of the land had profound consequences.

One result of the global spread of terrestrial vegetation early in Devonian time was that, for the first time in Earth's history, plants carpeted the soil and gripped it with their roots, thereby stabilizing it against erosion. Precambrian and early Paleozoic rocks are characterized by braided-stream deposits, which reflect rapid erosion (p. 112). Only after vegetation stabilized the land and confined rivers to discrete channels could rivers meander and deposit sediment in orderly cycles (see Figure 5-16). Thus it is only in rocks of Early Devonian age or younger that we find meandering river deposits in the geologic record. Preserved root traces attest to the spread of vegetation over the floodplains of rivers during the Devonian Period.

The next step, the origin of plants that grew to tree proportions, took place in Late Devonian time. Probably the adaptive breakthrough responsible for the increase in the size of land plants was the evolution of broad leaves, which captured sunlight effectively and allowed for a higher overall rate of photosynthesis. In Late Devonian time, *Archaeopteris* formed Earth's first forests. These forests were restricted to marshy environments, however, because *Archaeopteris* was a spore plant that required moisture to reproduce (see p. 65). The origin of seed plants near the end of the Devonian Period was another evolutionary breakthrough that had enormous ecological consequences: it permitted forests to expand into well-drained habitats far from lakes and rivers. For the first time in Earth's history, forests spread widely over the landscape.

It is probably no accident that amphibian animals did not arise until shortly before the end of Devonian time. The earliest amphibians must have had

Figure 1 Braided stream deposits characterize pre-Devonian nonmarine rocks. This block of the Silurian Shawangunk Formation in eastern Pennsylvania includes cross-bedded sand and gravel deposited by braided streams that flowed from uplands formed by the Taconic Orogeny. The vertical dimension of this block is about 30 centimeters (one foot). (Steven M. Stanley.)

Figure 2 Traces of roots of early land plants. These traces were found in Devonian rocks along Gaspé Bay, Quebec, Canada. Branching traces are present above and to the right of the scale. (Courtesy of Jennifer M. Elick; J. M. Elick, S. G. Driese, and C. I. Mora, *Geology* 26:143–146, 1998.)

A

moist skin that was highly sensitive to dry air and sunshine, like the skin of the small newts and salamanders that today live among moist leaves. The earliest amphibians were much larger than modern newts and salamanders, however, and they could have found adequate shade to live their adult lives fully on land only after shrubs and trees formed a canopy of leaves.

The expansion of terrestrial vegetation over terrestrial landscapes did more than stabilize the landscape and provide shade for early amphibians. It also changed climates profoundly, with devastating consequences for life on Earth (see Earth System Shift 14-2).

B

Figure 3 *Archaeopteris*, which arose in Late Devonian time, was one of the first large trees. *A.* A segment of a branch with leaves. *B.* A reconstruction of the entire tree. (After C. R. Beck, *Biol. Rev.* 45:379–400, 1970.)

Figure 4 A reconstruction of a Late Devonian plant that bore some of the oldest known seeds at the ends of its branches. The individual seeds, four of which are shown, were about 1 centimeter (0.4 inches) in length. (J. M. Pettitt and C. B. Beck, *University of Michigan Paleontol. Contrib.* 22:139–154, 1968.)

Lobe-finned fish   Amphibian   Lobe-finned fish   Amphibian

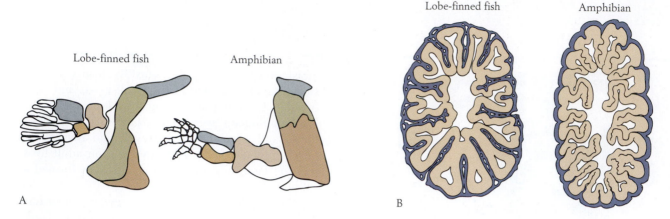

A   B

Figure 14-18 **Lobe-finned fishes and amphibians.** *A.* The shoulder and limb bones; shading identifies particular bones that the two groups have in common. *B.* Cross sections of teeth, showing the unusual, complex structure found in both groups.

features make the derivation a certainty. *Ichthyostega* had four legs, as do amphibians, but its skull structure was remarkably like that of a lobe-finned fish. The creature also had a fishlike tail—a feature of its ancestors that was probably not useful on land (Figure 14-19). Because of this intriguing combination of features, *Ichthyostega*, which was not discovered until the twentieth century, represents what is commonly termed a "missing link."

As we have seen, some early fishes possessed a lung long before amphibians evolved. They put this organ to use to breathe air occasionally, perhaps when a stream or lake dried up, but it was available for full-time exploitation by animals that moved onto land. This is yet another example of the "opportunism" of evolution. Unlike the gill supports that evolved into jaws earlier in vertebrate evolution, the lung required very lit-

Figure 14-19 **The primitive amphibian** *Ichthyostega.* The legs of *Ichthyostega* contrast with the fins of the approximately contemporary lobe-finned fish *Eusthenopteron* (left; see also Figure 14-13) and the lungfish *Rhynchodipterus* (right). The trunk and branches belong to the tree fern *Eospermatopteris.* (Drawing by Gregory S. Paul.)

tle evolutionary modification to open up an entirely new mode of life.

Why did vertebrates not invade terrestrial habitats until Late Devonian time? The likely answer is that the sun posed a problem for them. Early amphibians required protective shade that only large plants with abundant foliage could provide (see Earth System Shift 14-1).

The Paleogeography of the Middle Paleozoic World

Figure 14-20 displays the positions of continental landmasses during mid-Silurian time and during late Early Devonian time, about 65 million years later. During

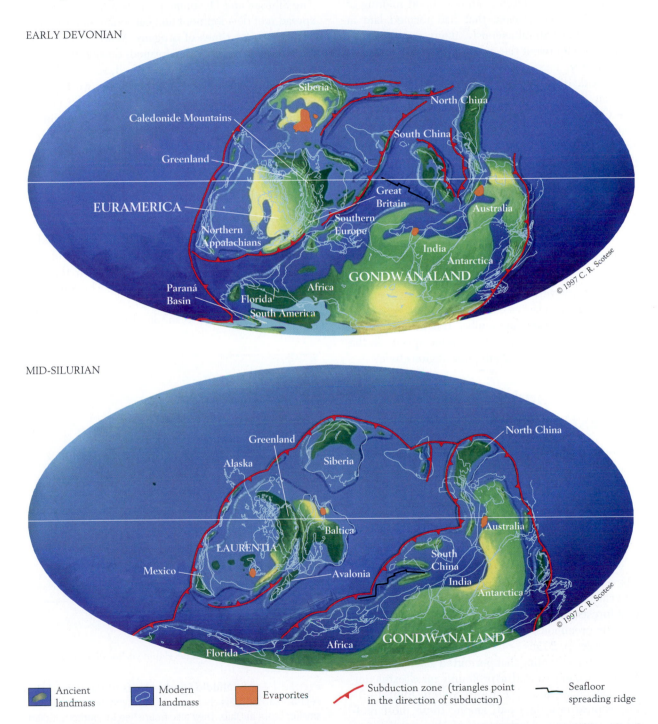

EARLY DEVONIAN

Siberia
Caledonide Mountains
North China
South China
Greenland
Great Britain
EURAMERICA
Australia
Northern Appalachians
Southern Europe
India
Antarctica
GONDWANALAND
Paraná Basin
Africa
Florida
South America
© 1997 C. R. Scotese

MID-SILURIAN

Greenland
North China
Alaska
Siberia
Baltica
Australia
LAURENTIA
South China
Mexico
Avalonia
India
Antarctica
© 1997 C. R. Scotese
Florida
Africa
GONDWANALAND

Ancient landmass   Modern landmass   Evaporites   Subduction zone (triangles point in the direction of subduction)   Seafloor spreading ridge

Figure 14-20 World geography during middle Paleozoic time. During this interval, Baltica and Avalonia collided with Laurentia to form Euramerica; this collision connected the landmasses that now form the British Isles.

During the Devonian Period, shallow seas near the south pole formed the Paraná Basin. (Adapted from paleogeographic maps by C. R. Scotese, PALEOMAP Project, University of Texas at Arlington, 1997.)

Devonian time, an important new geographic entity appeared, the continent of Euramerica, which was formed by the union of Laurentia, Baltica, and Avalonia. In general, the Silurian and Devonian were periods when sea level stood high in relation to the surfaces of major cratons. Early in Silurian time, sea level rose from its low position at the end of the Ordovician Period—a rise that is thought to have resulted from continued melting of the extensive polar glaciers that had formed late in Ordovician time. Simultaneously, many marine invertebrate taxa underwent the radiations described earlier in this chapter.

The wide distribution of organic reefs strongly suggests that middle Paleozoic climates were relatively warm. Climates were also relatively dry in many areas, as evidenced by accumulations of large volumes of evaporite deposits. Most evaporites of Silurian age formed within 30° or so of the ancient equator, but some Devonian evaporites accumulated farther north and south, apparently reflecting a broad distribution of warm climates during the Devonian Period (see Figure 14-20).

Early in Devonian time, a discrete marine province formed in the southern region of Gondwanaland (see Figure 14-20). This polar realm, called the Paraná Basin, was populated by a fauna adapted to cool water. Because the Paraná Basin lay within about 20° of the south pole, it is not surprising that it lacked coral-strome reefs. Also missing were bryozoans and ammonites. Burrowing bivalves formed a large percentage of the marine species in the Paraná Basin, just as they do in polar regions today.

Glaciation and a Great Mass Extinction

Earth experienced profound changes late in the Devonian Period. Tillites reveal that glaciers spread over a portion of southern Gondwanaland. The origin of these glaciers can be traced to biological causes. The ultimate cause was apparently the spread of forests over terrestrial landscapes for the first time in Earth's history (see Earth System Shift 14-2). Forests trap moisture, which is required for weathering, and the roots of plants also accelerate weathering (p. 234). Because weathering consumes carbon dioxide, the initial expansion of forests, in Late Devonian time, depleted this atmospheric greenhouse gas, causing climates to cool on a global scale (p. 233).

The climatic cooling that permitted glaciers to grow in Gondwanaland caused a profound mass extinction late in Devonian time (see Earth System Shift 14-2). Many tropical taxa died out, among them most reef-building organisms. Destruction of the coral-strome reef community and other prominent taxa transformed the marine ecosystem on a global scale. Paradoxically, terrestrial plants also experienced heavy extinction as

the result of the climatic change brought about by their formation of Earth's first forests.

Regional Events of Middle Paleozoic Time

The Silurian and Devonian periods were times of widespread reef development and carbonate deposition, but they were also times of orogeny. While eastern North America was being transformed during the Silurian from a highland to a carbonate shelf, reefs and evaporite deposits were forming farther to the west. Later, in the Devonian Period, Laurentia and Baltica united to form the continent of Euramerica, and mountains rose up in the Appalachian region. Mountains also formed in western North America, but reefs continued to grow there as well.

Eastern North America again became a passive margin

The Taconic orogeny in eastern Laurentia ended late in the Ordovician Period with the deposition of clastic wedges of sediment shed westward from the newly formed mountains (see Figure 13-25). The Silurian Period began with a continuation of this pattern. As the eastern mountains were subdued by erosion, however,

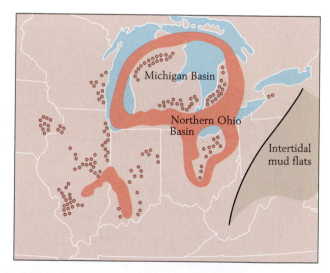

■ Barrier reefs

▒ Area of smaller reef development (locations generalized)

Figure 14-21 **Middle Silurian reefs of the Great Lakes region.** Barrier reefs encircled the Michigan Basin and a smaller basin in Ohio. They also flourished in southern Indiana and Illinois. Extensive mud flats now lay to the east in the Pennsylvania region, in contrast to the environments of coarse clastic deposition that occupied this area in Early Silurian time. (After K. J. Mesolella, *AAPG Bull.* 62:1607–1644, 1978.)

the site of clastic deposition became more broadly flooded by shallow seas, and late in Silurian time shallow-water carbonates accumulated along a new passive margin.

To the west, coral-strome reefs dotted shallow epicontinental seas (Figure 14-21). Here, however, the pattern of sedimentation and reef development changed drastically during Early Silurian time. First, two basins accumulated muddy carbonates. One was the Michigan Basin, in the area where the Great Lakes of North America are located today (Figure 14-21; see also Figure 9-22), and the other lay in what is now north-central Ohio. These basins were bounded by large barrier reefs and populated by scattered patch reefs. At this time siliciclastic muds were still accumulating on broad tidal flats to the east.

As the Silurian Period progressed, this pattern changed. To the east, in Pennsylvania and neighboring areas, the deposition of siliciclastic mud gave way to carbonate sedimentation. At the same time, the Michigan and central Ohio basins came to be only weakly supplied with seawater and thus turned into evaporite pans in which dolomite, anhydrite, and halite were precipi-

tated. The resulting deposits are a major source of rock salt today. It appears that barrier reefs in the Michigan Basin grew so high during Early Silurian time that they eventually restricted the flow of water into the basin. In time, evaporation and possibly a slight lowering of sea level led to the exposure and consequent death of the reefs. Although a weak flow of seawater into the basin replenished the water that had been lost by evaporation, the rate of evaporation was so high that evaporite minerals were precipitated around the margins of the basin and even at considerable depths within it. At first the center of the basin was moderately deep, but as evaporites accumulated, the water there grew progressively shallower until eventually the sea was excluded altogether.

Conditions within the evaporite basins were so inhospitable during the Late Silurian that reefs grew only to their southwest, in Indiana and Illinois. The most famous of these is the Thornton Reef of northern Illinois (Figure 14-22). The structure of this reef indicates the direction of the prevailing winds at the time it was growing: the stromatoporoid ridge obviously faced waves advancing from the southwest.

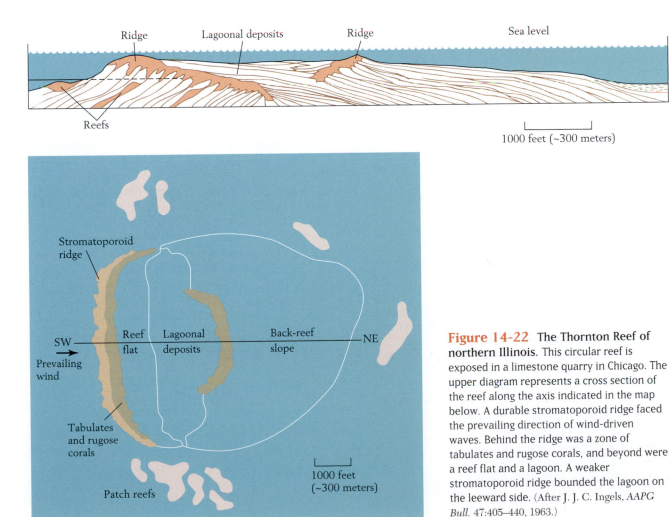

Figure 14-22 The Thornton Reef of northern Illinois. This circular reef is exposed in a limestone quarry in Chicago. The upper diagram represents a cross section of the reef along the axis indicated in the map below. A durable stromatoporoid ridge faced the prevailing direction of wind-driven waves. Behind the ridge was a zone of tabulates and rugose corals, and beyond were a reef flat and a lagoon. A weaker stromatoporoid ridge bounded the lagoon on the leeward side. (After J. J. C. Ingels, *AAPG Bull.* 47:405–440, 1963.)

Earth System Shift 14-2 | The Expansion of Plants over the Land Causes Global Climatic Change, Glaciation, and Mass Extinction

We have seen that the initial spread of land plants sta-bilized the landscape and provided shelter for early amphibians (see Earth System Shift 14-1). An increase in the intensity of weathering and climatic cooling were additional results of the spread of terrestrial veg-etation.

Recall that weathering of rocks in the modern world consumes atmospheric CO_2 (p. 233). Tree roots, as well as the fungi associated with them, release chemicals that weather rocky soil deeply and rapidly (p. 234). Fossils of root fungi of the kind associated with modern tree roots have, in fact, been identified in Devonian rocks. Thus the spread of moist forests—and then drier forests—during the Devonian Period must have accelerated weathering. This intensification of weathering must have reduced the concentration of CO_2 in the atmosphere, substantially weakening the greenhouse effect and cooling global climates. Given

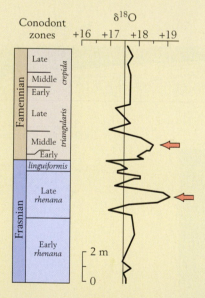

Figure 1 Oxygen isotope ratios in conodonts shifted toward higher values close to the Frasnian-Famennian boundary, when numerous extinctions occurred. The arrows point to two abrupt shifts, which are interpreted at representing episodes of climatic cooling. Conodont teeth are particularly useful for the study of isotopes in seawater because they consist of calcium phosphate, which is resistant to alteration by waters seeping through rocks. (After M. M. Joachimski and W. Buggisch, *Geology* 30:711–714, 2002.)

the reduction of greenhouse warming during the lat-ter part of Devonian time, it is not surprising that, as revealed by tillites, continental glaciers expanded at high latitudes in Gondwanaland. Thus began the sec-ond ice age of the Phanerozoic Eon.

The Late Devonian ice age, like that of Late Or-dovician time, was associated with one of the most devastating mass extinctions of all time. More than 40 percent of all marine genera disappeared (see Figure 7-13). The Devonian crisis was not instantaneous, however. The coral-strome reef community, which had prospered during Middle Devonian time, de-clined throughout Frasnian time (the Frasnian was the second-to-last age of Devonian time). This reef community was, of course, a tropical biota. Its de-struction exemplifies a general geographic pattern of the Late Devonian extinction: losses of tropical taxa were especially heavy. In contrast, inhabitants of the cold Paraná Basin (see Figure 14-20) were almost unscathed. In addition, corals that occupied deep, cool waters were relatively unaffected by the mass ex-tinction. These patterns suggest that cooling of seas played a major role in the extinction. Whereas non-tropical species could have migrated toward the equa-tor to find suitably warm waters as temperatures de-clined, no region would have remained warm enough to serve as a refuge for equatorial forms. The spread of silica-secreting sponges from deep habitats into shallow seas during Late Devonian time also suggests that climates cooled. Today these forms are adapted to cool waters in the ocean, and presumably they have been similarly adapted in the past. A change in the pattern of sedimentation also points to global cooling: the Late Devonian was marked by a striking reduc-tion in the accumulation of limestones (recall that most limestones are formed primarily of the skeletons of tropical organisms).

After the coral-strome reef community was al-ready decimated, a major pulse of marine extinction occurred at the boundary between the Frasnian and Famennian ages, the final two ages of the Devonian Period. Many previously diverse groups of major taxa, including some groups of brachiopods and am-monoids, abruptly disappeared. Oxygen isotopes in conodont teeth twice shifted toward heavier values close to the Frasnian-Famennian boundary. These shifts appear to have resulted primarily from climatic cooling rather than pulses of glaciation because there is no evidence of glacial expansion precisely at these times.

A final pulse of extinction took place at the end of the Famennian age (the very end of the Devonian Period). At this time, planktonic acritarchs and placoderm fishes—two groups that had flourished in Famennian oceans—suffered major extinctions. No marine placoderms survived into Carboniferous time, and acritarchs, although they survived, never recovered their previous diversity.

A smaller percentage of genera of marine animals died out during the Late Devonian mass extinction than during the Late Ordovician mass extinction, but this comparison is misleading. Although more marine genera died out at the end of the Ordovician, prominent communities of organisms survived. On the other hand, for some reason, the Devonian mass extinction, although weaker overall, had a more profound effect on marine communities. It destroyed the coral-strome reef community, for example. In addition, by eliminating most acritarch algae and all marine placoderm fishes, it devastated both the base and the top of the marine food web.

Fossil pollen reveals that terrestrial plants, which are especially vulnerable to climatic change, also suffered a major extinction at the end of Famennian time. This devastation of terrestrial floras is ironic, in that it was apparently the initial spread of these floras that, by reducing greenhouse warming, was the ultimate agent of mass extinction.

A

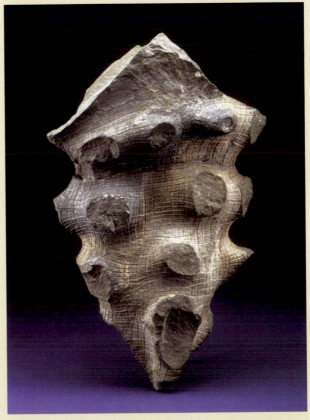

B

Figure 2 The Late Devonian mass extinction eliminated some taxa and caused others to shift to new habitats. The atrypid brachiopods (A) were a diverse tropical group that died out altogether. Siliceous sponges (B), which were adapted to cool waters, expanded from deep water into shallow seas. The gridlike markings are molds of the sponge's skeleton. (Courtesy Smithsonian Institution, photo by Chip Clark.)

Euramerica formed during the second Appalachian orogeny

The most famous angular unconformity in the world occurs in Scotland between Devonian beds of the nonmarine Old Red Sandstone and the nearly vertical Silurian marine beds on which they rest. It was at the locality shown in Figure 14-23 that James Hutton recognized the meaning of stratigraphic unconformities in 1788. Thus it came to be understood that the Old Red Sandstone was deposited after a Silurian episode of mountain building. The Old Red crops out over large areas of Scotland. Chunks of its distinctive red sandstones (see p. 326) can be found in parts of Hadrian's Wall, built by the Roman emperor Hadrian across northern England in the second century.

The Old Red includes not only rocks of Early, Middle, and Late Devonian age, but also rocks representing Late Silurian and earliest Carboniferous times. For a long time geologists found it puzzling that such a large volume of sediment could have been shed from highlands in the British Isles when most of the islands' area

Figure 14-23 Angular unconformity between the Old Red Sandstone (top left) and Silurian rocks (bottom left) at Siccar Point, Berwickshire, Scotland. The Silurian rocks were tilted and folded when Baltica collided with Laurentia to form Euramerica (see Figure 14-20). The Old Red Sandstone was subsequently deposited near the margin of this continent. (The British Institute of Geological Sciences.)

formed a depositional basin. Now the puzzle has been solved within the framework of plate tectonics by the reassembly of the landmass that formed when Laurentia and Baltica were united during mid-Paleozoic time to form the continent aptly named Euramerica.

Continental suturing and the Acadian orogeny

The second Phanerozoic interval of mountain building in the Appalachian region is known as the *Acadian orogeny*. Like the Taconic orogeny before it, the Acadian entailed a collision of landmasses, but this time the landmasses were larger. In fact, the Acadian event resulted from a double collision along the eastern margin of Laurentia. In the north, the collision was with Baltica; in the south, it was with the microcontinent Avalonia (see Figure 14-20).

The Acadian orogeny began in the north, in mid-Silurian time. Here the suturing of Laurentia to Baltica produced an orogenic belt now located in northeastern North America and Greenland. This suturing also created the Caledonide Mountains, which now lie along the Atlantic coast of Norway.

Orogenic activity then progressed southward as Avalonia, which was evidently part of the same lithospheric plate as Baltica, collided with eastern Canada and the northeastern United States. Recall that Avalonia was an elongate microcontinent that broke away from Gondwanaland and moved close to Laurentia during the Ordovician Period. Small islands moving ahead of it became attached to Laurentia during the Taconic orogeny (see Figure 13-20). Seaward of these exotic terranes are others, known as *Avalon terranes*, that represent segments of Avalonia attached during the Acadian orogeny, along with elements of oceanic crust (Figure 14-24). The Avalon terranes share fossil taxa that are unknown from North American rocks immediately to the west. These foreign taxa, which also occur in larger fragments of Gondwanaland, include the distinctive Early Cambrian trilobite *Paradoxides*.

Farther south, another elongate microcontinent became sutured to the eastern United States to form the Carolina terrane (see Figure 14-24). Metamorphism has obscured the original nature of this terrane, but a few surviving fossil remains, including those of *Paradoxides*, point to an origin in or close to Gondwanaland.

Paradoxides also occurs in southern England and southern Ireland, which together formed the northern segment of Avalonia and became attached to Scotland and northern Ireland during the Acadian orogeny. Thus the British Isles were assembled. It is ironic that Scotland and northern Ireland, which historically have been in political conflict with England and southern Ireland, respectively, are geologically distinct from these southern regions. Originally they were attached to Greenland (see Figure 13-16).

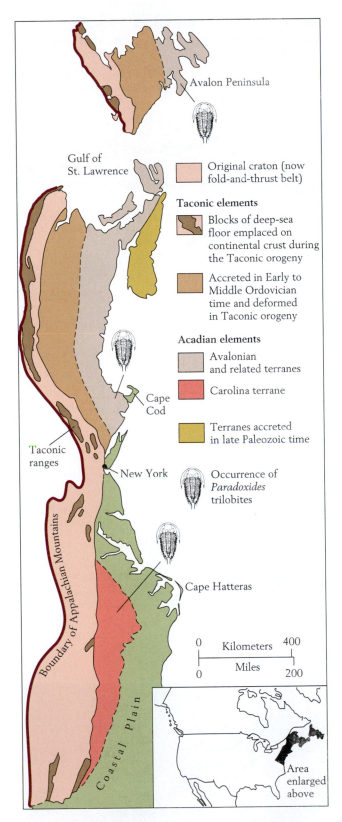

Figure 14-24 Exotic terranes of eastern North America. Distinctive fossils, including those of the Early Cambrian trilobite *Paradoxides*, characterize terranes accreted during the Acadian orogeny. (After H. Williams and R. D. Hatcher, *Geology* 10:530–536, 1982.)

The assembly of continental Europe would not come until later in the Paleozoic Era. During the Devonian Period a microcontinent destined to become southern Europe was encroaching on the newly forming Euramerica (see Figure 14-20). This microcontinent had rifted away from Gondwanaland after the departure of Avalonia. It would attach to the former landmass of Baltica (northern Europe) to create what we know as continental Europe late in the Paleozoic Era, when Gondwanaland also was sutured to Euramerica. At this time the British Isles and continental Europe became locked within the supercontinent Pangaea (see Figure 8-5). The Atlantic and Mediterranean coastlines of Europe did not form until Pangaea began to fragment early in the Mesozoic Era.

The second great depositional cycle in the Appalachian region Let us now return to the eastern margin of Laurentia to examine the rocks that provide a record of Avalonian mountain building. During the Silurian Period, erosion subdued the mountains that formed during the Taconic orogeny, and a new passive margin came into being along Laurentia's east coast; an immense carbonate platform soon extended along much of the continent's length. Reflecting the global prosperity of the coral-strome community, it included more extensive reefs than the Cambro-Ordovician carbonate platform of the same region (see Figure 13-17). Like the earlier platform, however, this one suddenly subsided, to be replaced by a foreland basin as mountains began to rise up to the east. Before this subsidence, the Oriskany Sandstone, a sandy beach deposit, spread over the uppermost carbonate rocks of the Helderberg Group (Figures 14-25*A* and 14-26). Then shallow-water sedimentation suddenly gave way to deposition of deep-water flysch. In New York State, black muds of the Marcellus Formation came first, and they were followed by turbidites and shales of the Hamilton Group. A similar transition took place farther south, in Maryland. In accordance with the typical pattern of foreland basin deposition (p. 210), an increased supply of sediment from the adjacent mountain belt then pushed back the waters of the basin, and deep-water deposits gave way to shallow marine and nonmarine molasse.

The enormous Catskill clastic wedge is a regressive body of molasse that records a westward progradation of sedimentary environments during the Acadian orogeny (Figure 14-25*B*). Nonmarine red beds of the Catskill wedge include sandstones and conglomerates that accumulated in braided streams near highlands; at lower elevations, meandering rivers formed depositional cycles with coarse sediments at the base and muds at the top. In some areas upward-coarsening deltaic cycles formed along the coastline.

A

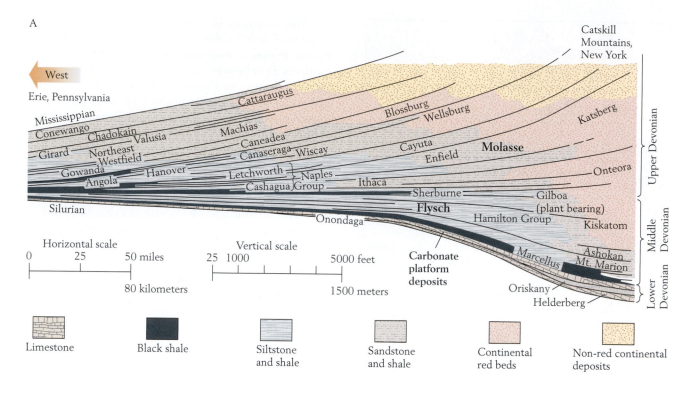

Figure 14-25 Devonian rocks of New York State that record a regression caused by the Acadian orogeny. *A.* Shallow-water carbonates (the Helderberg Group and Oriskany Sandstone) gave way to deep-water deposits (the Marcellus Shale and Hamilton Group). Next, coarse molasse deposits shed from the mountains formed a clastic wedge that spread westward. *B.* Environments of deposition of the Catskill clastic wedge and associated deposits. Braided streams meander seaward from the feet of the mountains to the east. Eventually they empty into lagoons behind barrier islands. Muds are deposited offshore. (*A* after P. B. King, *The Geological Evolution of North America*, Princeton University Press, Princeton, NJ, 1977; *B* after J. R. L. Allen and P. F. Friend, *Geol. Soc. Amer. Spec. Paper* 106:21–74, 1968.)

The cycle of deposition associated with the Acadian orogeny was much like the one associated with the Taconic orogeny. In both cycles passive-margin deposition was followed by accumulation of flysch and then molasse (compare Figures 13-21 and 14-25).

▶ **Figure 14-26** The Oriskany Sandstone in western Maryland. This rock unit was tilted late in the Paleozoic and is now weathering to loose sand. It formed along a beach, but it is overlain by deep-water shales that mark the onset of the Acadian orogeny. (S. Stanley.)

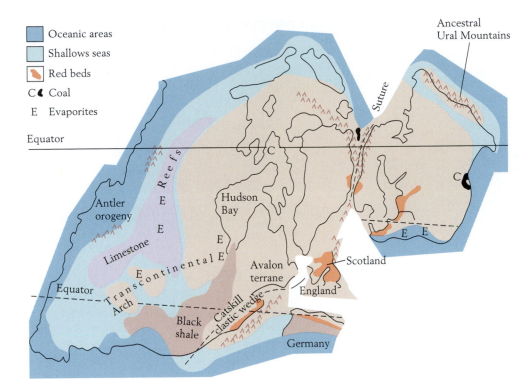

Oceanic areas

Shallows seas

Red beds

C C Coal

E Evaporites

Figure 14-27
Euramerica during Late Devonian time. The central mountain belt running from north to south formed as Baltica converged with Laurentia (see Figure 14-20). Red beds were concentrated in the south, near the equator. Deep-water deposits accumulated in what is now central Germany. Shales accumulated in North America west of the Catskill clastic wedge, and limestones, reefs, and evaporites formed farther west. The Antler orogeny affected the western margin of the continent.

Interior and eastern North America Figure 14-27 depicts general environments of Euramerica late in the Devonian Period. During much of Devonian time, an arm of land may have extended southwestward across what is now the western interior of North America, where the Transcontinental Arch persisted from early Paleozoic time (see Figure 13-25) and little or no sediment accumulated. The Devonian equator passed through Euramerica, so that prevailing winds, blowing from the east, must have supplied abundant moisture to the continent's eastern margin. It is therefore not surprising that coal deposits, formed from early land plants, are found here. Evaporites accumulated in the rain shadow to the west, in the area of North America where the Rocky Mountains now stand. Along the east and west coasts of Euramerica, the climate was at least intermittently hot and dry enough for caliche nodules (p. 106) to form in abundance in low-lying areas that are now represented by ancient soils.

Late in the Devonian Period a mud-floored seaway lay to the west of the coastal mountains and the area of molasse deposition, extending northward to the Hudson Bay area (see Figure 14-27). The giant placoderm *Dunkleosteus* flourished in this sea, together with other fishes now preserved in the black shales of northern Ohio (see Figures 14-10 and 14-11). Near the end of Devonian time, the deposition of black muds extended farther to the west, leaving a vast area of eastern and central North America blanketed with these sediments.

Reef building and orogeny in western North America Along the continental shelf west of Euramerica, in what is now western Canada, coral-strome reef complexes developed during the latter half of the Devonian Period (Figure 14-28). Some of the reefs here took the form of elongate barriers or atolls with central lagoons. Many of these reefs now lie deeply buried, and their porous textures have created traps for petroleum. South of the belt of reef growth, carbonates were deposited in shallow tropical seas (see Figure 14-27).

Through Silurian and Devonian time, the western margin of North America remained approximately where it had been during the Ordovician Period. In middle Paleozoic time, however, an island arc stood offshore. In the Klamath Mountains and in the Sierra Nevada of present-day northern California, ophiolite sequences, which include graywackes, shales, cherts, and volcanics, record the presence of this Klamath Arc (Figure 14-29A). Rocks in Nevada show that this simple geographic picture became more complex between Middle Devonian and early Mississippian time; they reveal closure of the basin between the Klamath Arc and the western margin of Euramerica. In central Nevada, deep-sea deposits like those of northern California can be seen to have been thrust as far as 160 kilometers (100 miles) onto the craton (Figure 14-29B). The principal thrust fault along which this movement occurred is called the Roberts Mountains Thrust.

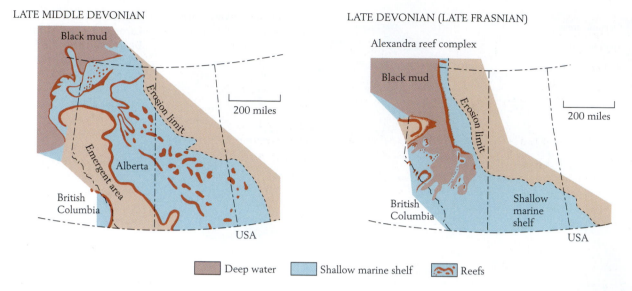

| Deep water | Shallow marine shelf | Reefs |

Figure 14-28 The distribution of reefs in western Canada during the latter part of Devonian time. Late in Frasnian time, reefs ceased to grow, and black mud spread onto the continental shelf. Northeast of the erosion limit, Middle and Late Devonian rocks have been eroded away. (After E. R. Jamieson, *Proc. N. Amer. Paleontol. Conv.* J:1300–1340, 1969.)

The collision of the island arc and continental margin that produced the Roberts Mountains Thrust is known as the *Antler orogeny* (see Figure 14-29). This was the first sizable episode of mountain building in the Cordilleran region of North America during Phanerozoic time. The remainder of the Cordilleran story, which is told in the following chapters, has mountain building as its dominant theme.

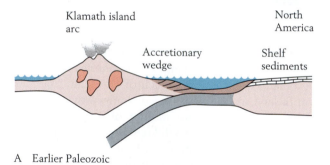

▶ **Figure 14-29** The likely mechanism by which the Klamath Arc was added to the North American continent by the Antler orogeny. The basin between the craton and the Klamath Arc (A) closed. As the continental crust was thrust beneath the volcanic crust of the Klamath Arc, deep-sea sediments slid onto shallow-water carbonates along the Roberts Mountains Thrust (B).

Chapter Summary

Why was there a great expansion of marine life at the start of the Silurian Period?

Many groups of marine life depleted during the great mass extinction at the end of the Ordovician Period recovered their high diversity during the Silurian Period. Other important forms, such as the ammonoids and the jawed fishes, were new.

What groups of animals are known to have invaded fresh water during middle Paleozoic time?

Jawed fishes and mollusks expanded into freshwater habitats during middle Paleozoic time.

What was the nature of the initial conquest of land by vascular plants and animals?

The Silurian Period witnessed the invasion of land by vascular plants, followed in Devonian time by the invasion of arthropods (scorpions, spiders, and insects) and vertebrate animals (amphibians). During the Devonian Period, spore plants were joined by seed plants, which did not require moist habitats for reproduction and thus were able to invade drier terrain that had previously been barren.

What reef community occupied Silurian and Devonian seas?

Coral-strome reefs flourished throughout middle Paleozoic time, as they had during Ordovician time. They were especially well developed in the Great Lakes region and near the western continental margin of North America.

In what way was Late Devonian time an interval of biotic crisis?

Late in the Devonian Period, a great mass extinction eliminated many forms of marine life, including nearly all members of the coral-strome reef community and nearly all placoderm fishes.

What evidence is there that climatic cooling caused the Late Devonian mass extinction?

This mass extinction coincided with the spread of glaciers in Gondwanaland. Shifts of oxygen isotopes in conodonts toward heavier values reflect climatic cooling. In addition, species that occupied tropical regions, including members of the coral-strome reef community, died out in especially large numbers, apparently because the climatic conditions to which they were adapted disappeared. Siliceous sponges, which are adapted to cool conditions, migrated from deep water to shallow water in Late Devonian time. The rate of deposition of limestones, most of which are deposited in warm water, also declined markedly.

What may have caused the Late Devonian climatic cooling?

The evolution of trees in Late Devonian time led to the initial spread of forests over terrestrial landscapes. The deep roots of trees increased rates of weathering. Because weathering consumes atmospheric carbon dioxide, the spread of forests probably reduced greenhouse warming of Earth.

What caused mountain building in the Appalachian region during Devonian time?

The Acadian orogeny occurred when Laurentia, Baltica, and the microcontinent Avalonia united to form Euramerica. As a result, exotic terranes were attached to eastern North America.

How does the sedimentary record in eastern North America reflect the occurrence of the Acadian Orogeny?

This record reveals that shallow-water deposition gave way to the accumulation of deep-water flysch and then molasse deposits in a foreland basin.

Review Questions

1. In what important ways did invertebrate life change between Ordovician time and Devonian time?

2. What animals have the oldest extensive fossil record in freshwater sediments?

3. In what way did terrestrial environments of the Late Devonian Period look different from those of Early Devonian time?

4. What evidence do fossil bones and teeth provide that amphibians evolved from fishes?

5. Where was the landmass that now forms southern Europe toward the end of the Devonian Period?

6. How did the landmass that now forms the British Isles come into being?

7. How do reefs of middle Paleozoic age illustrate ecological succession?

8. Reefs are commonly porous structures that serve as traps for petroleum. If you wanted to drill for oil in Devonian reefs, what geographic regions would seem most promising?

9. What caused large quantities of sediment to accumulate in the south-central part of Euramerica during the Devonian Period?

10. What evidence is there of a decline in the concentration of atmospheric carbon dioxide during Devonian time? (Hint: Refer to Figure 10-9A and Earth System Shift 14-2.)

11. Land areas changed dramatically in the course of middle Paleozoic time. Using the Visual Overview on page 328 and what you have learned in this chapter, describe how continental surfaces changed during Silurian and Devonian time with respect to their distribution on Earth, their topography, and their colonization by terrestrial life.

During Late Carboniferous time, logs of spore trees accumulated in broad swamps to form coal deposits that are now exploited by modern societies. In this museum reconstruction of a Late Carboniferous forest, ferns and seed ferns form the undergrowth beneath lycopod trees. On the fallen log on the right is a cockroach, one of the many kinds of insects that evolved during Late Carboniferous time. (Field Museum of Natural History, Chicago, Neg. GEO75400c.)

CHAPTER 15

The Late Paleozoic World

The late Paleozoic interval of geologic time includes the Carboniferous Period, when coal formed from the remains of new kinds of plants that grew in swamps, and the subsequent Permian Period, when many organisms died out in two biotic crises.

The late Paleozoic world was marked by major climatic changes. Glaciers spread over the south polar region of Gondwanaland between Early and Late Carboniferous time and then receded during the Permian Period. At the time of their expansion, a major global extinction occurred, and during the glacial interval the ocean was inhabited by hardy taxa that experienced unusually low rates of extinction. Subsequently, a general drying of climates during the Permian Period led to a contraction of coal swamps and a widespread accumulation of evaporites. Increased aridity also led to the extinction of many kinds of spore plants and amphibians, both of which required moist conditions. Seed plants and mammal-like therapsids inherited Earth. Two additional mass extinctions occurred in late Paleozoic time, one a few million years before the end of the Permian Period and the other at the end of the Permian. Each of these crises may have resulted from a massive outpouring of lava in Asia. The second one was the greatest mass extinction of all time, decimating life both in the ocean and on land. This devastating event brought the Paleozoic Era to an end.

A major plate tectonic event took place in Late Carboniferous time: the attachment of Gondwanaland to Euramerica, which was accompanied by mountain building in Europe and in eastern North America. By the time this major suturing event was completed, almost all of the supercontinent of Pangaea was in place.

The Carboniferous System was formally recognized in Britain in 1822, early in the history of modern geology. Its name was chosen to reflect the system's vast coal deposits, which had long been mined for fuel. Actually, only the upper part of the Carboniferous System harbors enormous volumes of coal; the lower part contains an unusually large proportion of limestone. Recognizing this distinction, American geologists, late in the nineteenth century, began to refer to the lower, limestone-rich Carboniferous interval as the Mississippian because of its excellent exposure along the upper Mississippi Valley and to the upper, coal-rich interval as the Pennsylvanian because of its widespread occurrence in the state of Pennsylvania. Soon the Mississippian and the Pennsylvanian were informally recognized as separate systems in North America, and in 1953 the U.S. Geological Survey officially granted them this status. A dramatic expansion of glaciers in Gondwanaland occurred close to the end of Mississippian time, causing sea level to drop. As a result, an unconformity separates Mississippian and Pennsylvanian strata throughout the world.

Although this difference between the upper and lower parts of the Carboniferous System is evident elsewhere in the world, most geologists in

Era	Period		Time
CENOZOIC	NEOGENE		
CENOZOIC	PALEOGENE		
MESOZOIC	CRETACEOUS		
MESOZOIC	JURASSIC		
MESOZOIC	TRIASSIC		
PALEOZOIC	PERMIAN		251 million years
PALEOZOIC	PERMIAN		292 million years
PALEOZOIC	CARBON-IFEROUS	PENNSYLVANIAN	320 million years
PALEOZOIC	CARBON-IFEROUS	MISSISSIPPIAN	354 million years
PALEOZOIC	DEVONIAN		
PALEOZOIC	SILURIAN		
PALEOZOIC	ORDOVICIAN		
PALEOZOIC	CAMBRIAN		
"PRECAMBRIAN"	PROTEROZOIC EON		
"PRECAMBRIAN"	ARCHEAN EON		

Visual Overview
Major Events of the Late Paleozoic

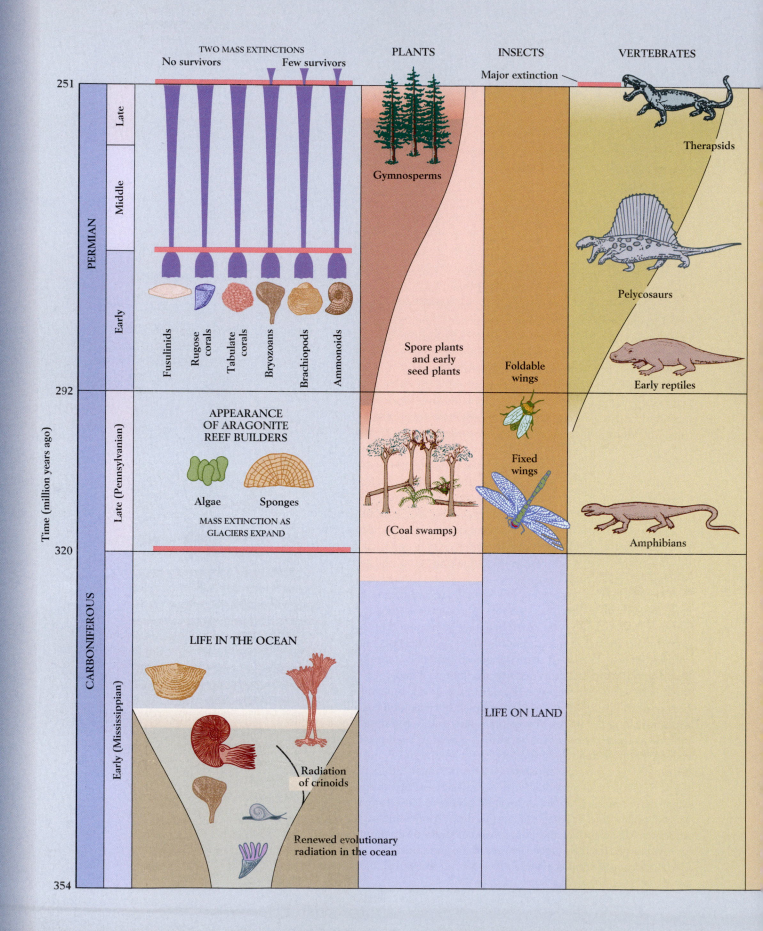

TWO MASS EXTINCTIONS

No survivors   Few survivors

PLANTS   INSECTS   VERTEBRATES

Major extinction

251

Time (million years ago)

292

320

354

PERMIAN — Late / Middle / Early

CARBONIFEROUS — Late (Pennsylvanian) / Early (Mississippian)

Fusulinids
Rugose corals
Tabulate corals
Bryozoans
Brachiopods
Ammonoids

Gymnosperms

Spore plants and early seed plants

Foldable wings

Fixed wings

Therapsids

Pelycosaurs

Early reptiles

APPEARANCE OF ARAGONITE REEF BUILDERS

Algae   Sponges

MASS EXTINCTION AS GLACIERS EXPAND

(Coal swamps)

Amphibians

LIFE IN THE OCEAN

LIFE ON LAND

Radiation of crinoids

Renewed evolutionary radiation in the ocean

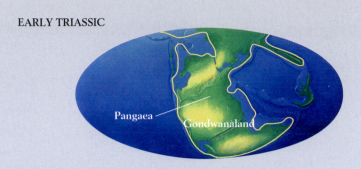

EARLY TRIASSIC

Pangaea

Gondwanaland

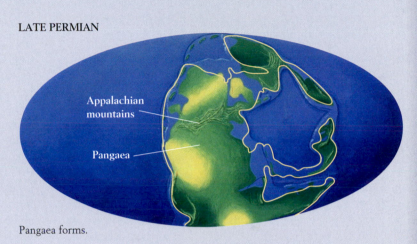

LATE PERMIAN

Appalachian
mountains

Pangaea

Pangaea forms.

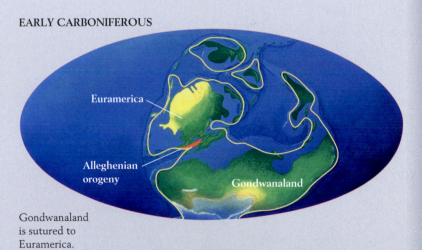

EARLY CARBONIFEROUS

Euramerica

Alleghenian
orogeny

Gondwanaland

Gondwanaland
is sutured to
Euramerica.

EARLY DEVONIAN

Euramerica

Gondwanaland

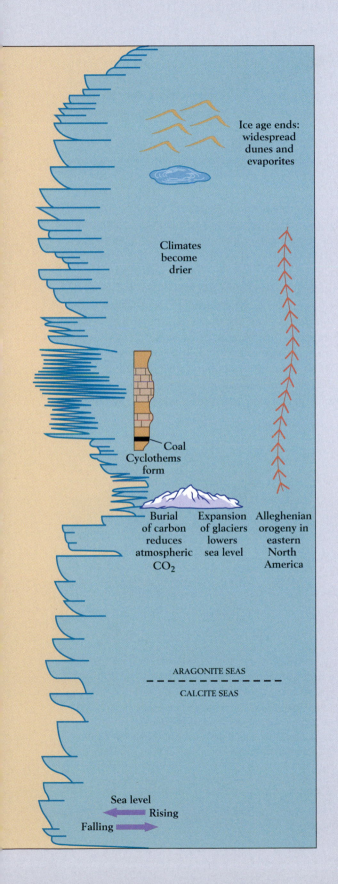

Ice age ends:
widespread
dunes and
evaporites

Climates
become
drier

Coal

Cyclothems
form

Burial
of carbon
reduces
atmospheric
CO_2

Expansion
of glaciers
lowers
sea level

Alleghenian
orogeny in
eastern
North
America

ARAGONITE SEAS

- - - - - - - - - - - - - - - -

CALCITE SEAS

Sea level

Rising

Falling

Europe continue to recognize just one system: the Carboniferous. Because this book covers world geology, we will follow the European practice, but occasionally it will be helpful to reiterate that the Lower and Upper Carboniferous systems are equivalent to the Mississippian and Pennsylvanian, respectively.

Roderick Murchison, who established the Silurian System and coestablished the Devonian, recognized the Permian System in 1841 and named it for rocks in the town of Perm in Russia. Permian rocks were later identified throughout the world.

Late Paleozoic Life

Marine life of the late Paleozoic interval did not differ markedly from that of Late Devonian time except for the appearance of new reef builders. Changes on land were far more profound. Numerous insects of remarkably modern appearance evolved during this time, and many new kinds of spore trees colonized broad swamps, where their remains eventually formed coal deposits. During Permian time, these plants were replaced by seed trees. In parallel fashion, amphibians, which like spore plants were tied to water for reproduction, initially dominated terrestrial habitats. Later they were replaced by more fully terrestrial reptile groups. By the end of Permian time, terrestrial vertebrates displayed a variety of new adaptations for feeding and locomotion, many of which resembled those of mammals.

New forms of life emerged in Paleozoic seas

Some forms of marine life never recovered from the mass extinction of Late Devonian time. Tabulate corals and stromatoporoids, for example, never again played a major ecological role. The ammonoids, on the other hand, rediversified quickly and once again assumed a prominent ecological position; indeed, ammonoid fossils are widely employed to date late Paleozoic rocks (see Figure 14-4). Also persisting from Devonian time as mobile predators were diverse groups of sharks and ray-finned bony fishes. Gone shortly after the start of the Carboniferous Period, however, were the armored placoderms that had ruled Devonian seas. Their absence reflected a general trend: although the late Paleozoic is not known for dramatic changes in the composition of marine life, heavily armored taxa of nektonic (swimming) animals tended to give way to more mobile forms. Apparently, as the Paleozoic Era progressed, the ability to swim rapidly became a near necessity, probably because of the presence of increasingly effective predators in the oceans. After Devonian time, armored fishes never again dominated marine habitats, and heavy-shelled nautiloids also declined in number. In contrast to these heavy, awkward forms, the swimmers

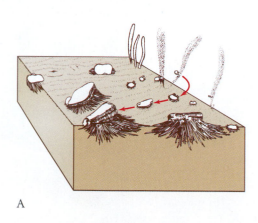

A

B

Figure 15-1 **Modes of life of late Paleozoic spiny brachiopods of the productid group.** A. The life habits of this mud-dwelling species changed as it matured (arrows). The juvenile brachiopods appear to have been attached to stalks of algae by curved spines. When the algae died, the small brachiopods came to rest on fine-grained sediment. As the brachiopods grew, their long spines served as "snowshoes," preventing the animals from sinking into the sediment. Thus the brachiopods could pump water in and out between the two halves of their shells to obtain food and oxygen without being clogged by mud. B. A group of Permian brachiopods of the genus *Prorichthofenia*. The lower halves of the shells of these coral-like animals were cone-shaped rather than cup-shaped, and throughout their lives their spines were attached to hard objects—in this case, the shells of neighboring brachiopods. The upper halves of the shells were flattened lids. (A after R. E. Grant, *J. Paleontol.* 40:1063–1069, 1966; B, courtesy Smithsonian Institution.)

Figure 15-2 Reconstruction of an Early Carboniferous (Mississippian) meadow of crinoids. Sharks, which also were well represented in Early Carboniferous seas, cruise above the crinoids. (Chase Studios, Inc.)

that thrived in late Paleozoic time—the ammonoids and especially the sharks and ray-finned fishes—were highly mobile.

Brachiopods rebounded from the Late Devonian mass extinction to resume a prominent ecological role. A group of spiny brachiopods known as *productids* enjoyed particular success. Some immobile productids employed their spines to anchor or support themselves in sediment, and a group of Permian productids developed cone-shaped shells that were attached by spines to the frameworks of solid reefs (Figure 15-1). Like the brachiopods, burrowing and surface-dwelling bivalves continued to thrive in late Paleozoic time, as did gastropods.

Crinoids (sea lilies)—animals that were attached to the seafloor and captured floating food that came within reach of their waving arms—expanded to their greatest

diversity early in the Carboniferous Period, forming meadows on many areas of the shallow seafloor (Figure 15-2). During this time, these organisms contributed vast quantities of carbonate debris to the rock record (Figure 15-3), leading to widespread deposition of limestone during the Early Carboniferous (Mississippian) Period. Other animal groups also contributed to the formation of Carboniferous limestones. Lacy bryozoans were sheetlike colonial animals that stood above the seafloor and fed on suspended organic matter (Figure 15-4). These organisms not only contributed skeletal debris to limestones but also trapped sediment to form mound-shaped structures.

The **fusulinids**, a group of large foraminifera that lived on shallow seafloors, included only a few genera in Early Carboniferous time, but underwent an enormous

Figure 15-3 Early Carboniferous limestone composed largely of skeletal debris from crinoids. The cylindrical fossils are segments of crinoid stems, which are corrugated, so that in life they stacked one on top of another like poker chips. (S. Stanley.)

Figure 15-4 A lacy, fanlike bryozoan of late Paleozoic age. In life this colony stood upright and was the size of a small fern. (Courtesy Smithsonian Institution, photo by Chip Clark.)

A

B

Spiral
growth

Figure 15-5 **Fusulinid foraminifera.** *A.* These unusually large single-celled creatures secreted skeletons that were commonly shaped like grains of wheat, which many resembled in size. *B.* A cross section shows their spiral mode of growth. (*A,* courtesy Smithsonian Institution; *B,* R. C. Douglass, U.S. Geological Survey.)

evolutionary radiation during the Late Carboniferous and Permian (Figure 15-5). Some 5000 species have been found in Permian rocks alone. Although they were single-celled, amoeba-like creatures with shells, some fusulinid species exceeded 10 centimeters (4 inches) in length. Their abundance and rapid evolution make fusulinids useful guide fossils for Upper Carboniferous and Permian strata. They also became major constituents of limestone.

Aragonitic reef builders flourished in aragonite seas

Organic reefs were poorly developed after the Late Devonian collapse of the coral-strome community. Coral-strome reefs had been built by organisms that secreted calcite skeletons in calcite seas—seas with a low magnesium-calcium (Mg^{2+}/Ca^{2+}) ratio (p. 241). The Mg^{2+}/Ca^{2+} ratio rose early in the Carboniferous Period, and calcite seas gave way to aragonite seas (see Figure 10-20). It appears that few kinds of organisms can build massive reefs if the Mg^{2+}/Ca^{2+} ratio of seawater does not favor their skeletal mineralogy. Thus aragonitic algae built Late Carboniferous reefs, and aragonitic sponges assumed a large role in the growth of Permian reefs (Figure 15-6).

Trees grew in swamps

Plants gave the Carboniferous Period its name, and in no other geologic interval are plant fossils more conspicuous. Soft coal (p. 46) from this period typically contains recognizable stems and leaves. Coal deposits developed chiefly in lowland swamps, where fallen tree trunks accumulated in large numbers. Because it takes several cubic meters of wood to make one cubic meter of coal, the vast coal beds of Late Carboniferous age

Figure 15-6 A Permian calcareous sponge of the genus *Girtyocoelia.* This reef builder had pea-sized chambers that were interconnected. Water that passed through the chambers was expelled through holes after food particles were filtered from it. (Chip Clark.)

must represent an enormous biomass of original plant material. Wetlands were far more extensive than they are today.

A small number of genera, each represented by many species, emerged as the dominant late Paleozoic flora of the coal swamps and adjacent habitats. The most important coal-swamp genera were *Lepidodendron* and *Sigillaria,* two types of lycopod trees that contributed many of the logs that were buried and compressed to form coal (Figure 15-7). As we have seen, the lycopods present during Early and Middle Devonian time had all been small plants (see Figure 14-16). Like smaller lycopods, *Lepidodendron* and *Sigillaria* were spore plants that were confined to swampy areas by their mode of reproduction (p. 65). *Lepidodendron* was the more successful of the two genera; some of its species grew more than 30 meters (100 feet) tall and measured 1 meter (3 feet) across at the base.

At the feet of the treelike plants in Late Carboniferous time was an undergrowth that consisted primarily of ferns and fernlike plants. Although some of them were spore plants like the modern ferns, many others were so-called **seed ferns**, which, as their name suggests, reproduced by means of seeds (Figure 15-8). Seed

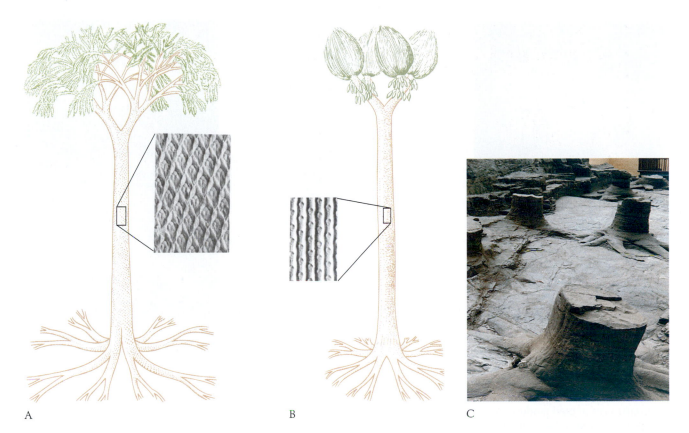

A   B   C

Figure 15-7 The dominant trees of Carboniferous coal swamps belonged to the lycopod group. *A.* The genus *Lepidodendron* has a trunk with a spiral pattern of scars where leaves were formerly attached. Note the similar spiral arrangement of branches in the early vascular plant *Baragwanathia* (see Figure 14-15); this small, simple plant may have been an ancestor of the treelike lycopods. *B.* The trunk of *Sigillaria* has vertical columns of leaf scars. *C.* The roots of lycopod trees are often preserved as fossils. Those shown here are in Scotland. (*A* and *B*, photographs of leaf scars, Field Museum of Natural History, Chicago, Negs. 75444 and 75410; *C*, The British Institute of Geological Sciences.)

ferns are difficult to distinguish from spore ferns on the basis of their foliage alone, and they were not recognized as a separate group until 1904. Many seed ferns were small, bushy plants, but others were large and tree-like. *Glossopteris*, the famous plant so abundant in Gondwanaland, was a seed fern. Its species were trees with tonguelike leaves (see Figure 8-3).

Upland floras expanded

Not all Late Carboniferous vegetation occupied coal swamps. In fact, seed ferns and spore plants called **sphenopsids** were more abundant on higher ground (Figure 15-9). Fossils of these groups are more common in sands and muds that accumulated along levees and floodplains of rivers than in coals that accumulated in permanently wet coal swamps. Late Carboniferous sphenopsids, though often tree-sized, were similar in general form to modern sphenopsids called horsetails, which grow to a height of about a meter (3 feet). They were characterized by branches that radiated from discrete nodes along the vertical stem and by horizontal underground stems that bore roots.

Another important group of Late Carboniferous plants that occupied high ground was the **cordaites**, tall trees that often reached 30 meters (100 feet) in height

Figure 15-8 A frond and seeds of a Late Carboniferous seed fern. (Theodore Clutter/Photo Researchers.)

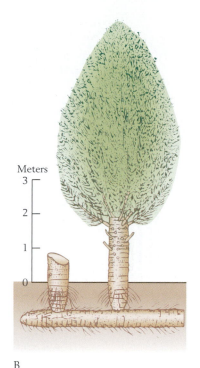

A   B   C

Figure 15-9 The Late Carboniferous sphenopsid plant
Calamites resembled modern horsetails. Branches such as
the one preserved in the fossil (*A*) were positioned at intervals
along the segmented tree trunk (*B*), just as they are on the
segmented stalk of living horsetails (*C*), which grow to only
about a meter (3 feet) in height. (*A*, James L. Amos/Photo
Researchers; *C*, Jeff Foott/Survival Anglia.)

(Figure 15-10). As seed plants, cordaites were liberated
from moist habitats (p. 66). They formed large woodlands resembling modern pine forests. In fact, cordaites
belonged to the group known as **gymnosperms** ("naked-
seed plants"), which include the living **conifers**, or
cone-bearing plants (pines, spruces, redwoods, and
their relatives). The seeds of these plants are lodged in
exposed positions on cones or other reproductive organs. Gymnosperm seeds thus differ from the covered
seeds of flowering plants, a group that did not emerge
until the Mesozoic Era.

During the Permian Period, gymnosperms, including conifers, came to dominate terrestrial environments.
Figure 15-11 shows the foliage of one of these conifers,
which resembled the needled branches of certain living
conifers. Paleozoic conifers, like modern ones, bore
naked seeds on cones. Gymnosperm floras, after expanding in Late Permian time, prevailed throughout
Mesozoic time (Figure 15-12) and thus are often thought
of as representing Mesozoic vegetation, despite their
rise to dominance in latest Paleozoic time.

Figure 15-10
Reconstruction of a tall
cordaite tree of Late
Carboniferous time. Cordaites
were seed plants that formed large
forests on dry ground.

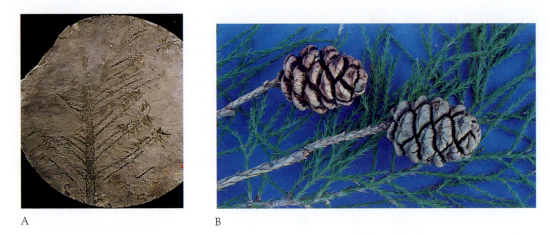

Figure 15-11 An early conifer of Late Carboniferous age. *A.* Like living conifers, *Walchia* had needled branches and reproduced by means of cones. *B.* A living conifer species that is related to the redwoods is shown for comparison. (*A*, Museum für Naturkunde, Humboldt-University, Berlin; *B*, Phil A. Dotson/Photo Researchers.)

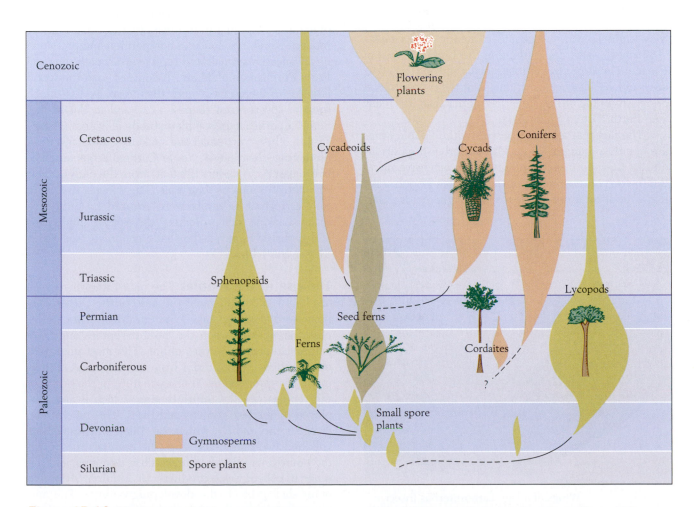

Figure 15-12 The history of major groups of swamp- and land-dwelling plants. Spore plants dominated Silurian, Devonian, and Carboniferous floras. Seed ferns are the oldest known seed plants. Gymnosperms (naked-seed plants) dominated Mesozoic floras, but the early conifers, which belonged to this group, diversified greatly during the Permian Period, while spore plants declined. (After A. H. Knoll and G. W. Rothwell, *Paleobiology* 7:7–35, 1981.)

Animals diversified on land and invaded freshwater habitats

In late Paleozoic freshwater habitats, aquatic ray-finned fishes continued to diversify and were joined by freshwater sharks that have no close modern relatives. For the first time, mollusks also became conspicuous in freshwater environments; the shells of many species of clams are found in freshwater and brackish sediments associated with coal deposits.

Insects On land, one group of invertebrate animals, the insects, assumed an important ecological role—one that they have never relinquished. The oldest known insects are of Early Devonian age, but they were wingless forms. By Late Carboniferous time, however, many kinds of insects had wings (Figure 15-13). The earliest flying insects differed from most modern species in that they could not fold their wings back over their bodies; the only two living orders of insects that lack this ability are the dragonflies and the mayflies, both of which occur in Upper Carboniferous deposits. Giant dragonflies—animals with wingspans of nearly half a meter (18 inches)—have been found in the Carboniferous of France. The large sizes of these forms and of a few other types of insects provide evidence that the concentration of oxygen in the atmosphere was higher than it is today (see Earth System Shift 15-1).

The fact that advanced insects with foldable wings are found in younger Upper Carboniferous deposits indicates that the insects underwent an extensive radiation before the beginning of the Permian Period. Like many modern insect species, some of these Late Car-

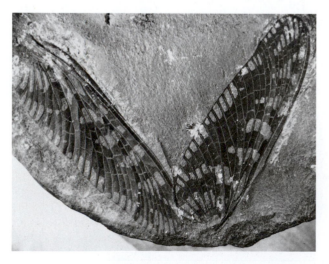

Figure 15-13 Wings of a Late Carboniferous insect from the Mazon Creek Formation of Illinois. The color pattern is beautifully preserved in this remarkable specimen. (F. M. Carpenter, *Proc. N. Amer. Paleontol. Conv.* I:1236–1251, 1969.)

boniferous forms had eggs that hatched into caterpillar-like larvae. In addition, some of these ancient insects possessed specialized egg-laying organs or mouthparts adapted to sucking juices from plants. The legs of some species were highly modified for grasping prey, leaping, or running. Indeed, many of these insects appear to have been as highly adapted for particular modes of life as are insects of the modern world.

Amphibians It can be difficult to distinguish between aquatic and terrestrial vertebrates of late Paleozoic time because many four-legged animals lived along the shores of lakes, rivers, shallow seas, or swamps and divided their time between land and water. The only vertebrates to populate the Early Carboniferous landscape were amphibians, many of which retained aquatic or semiaquatic habits throughout their lives. Carboniferous amphibians, however, did not closely resemble their modern relatives. The frogs, toads, and salamanders that comprise most living species of Amphibia are small, inconspicuous creatures; they seem to be the only kinds of animals belonging to this class that can thrive in the modern world in the face of competition and predation by advanced mammals and reptiles. Carboniferous and Early Permian amphibians, in contrast, had the world largely to themselves, and thus displayed a much broader spectrum of shapes, sizes, and modes of life. Some superficially resembled stubby alligators (Figure 15-14), others were small and snakelike, and a few were lumbering plant eaters. Some Carboniferous amphibians measured 6 meters (20 feet) from the ends of their snouts to the tips of their tails. The species that were fully terrestrial as adults were covered by protective scales. Lacking such protection, the earliest amphibians had probably required the shade of forests (see Earth System Shift 14-1).

The rise of the reptiles The oldest known reptiles are found in deposits near the base of the Upper Carboniferous (Pennsylvanian) System. Most of the skeletal differences between the earliest reptiles and their amphibian ancestors were minor. The most important way in which reptiles differ from amphibians is in their mode of reproduction. The key feature in the origin of the reptiles was the **amniote egg**, which is also employed by modern reptiles and birds. This type of egg provides the embryo with a sac filled with nutritious yolk and two other sacs: one (the *amnion*) to contain the embryo and the fluid in which it develops, and the other to collect waste products. In addition, a durable outer shell protects the developing embryo. The amniote egg apparently originated in Carboniferous time, when reptiles evolved.

Because the amniote egg is in essence a self-contained pond, it allowed vertebrates for the first time

Figure 15-14 An Early Permian scene beside a body of water. *Dimetrodon* (see Figure 15-15) threatens *Eryops*, an amphibian that was about 2 meters (6 feet) long. Early insects are visible in the foreground, and the small reptile *Araeoscelis* is climbing a tree of the genus *Cordaites*. The vine is *Gigantopteris*, and the small plants are *Lobatannularia*. (Drawing by Gregory S. Paul.)

to live and reproduce away from bodies of water. There is an interesting parallel here with the evolution of the seed in plants. As we have seen, spore plants, like amphibians, require environmental moisture during part of their life cycle (p. 75). The origin of the more advanced groups—the seed plants and reptiles—represented a transition to a fully terrestrial existence.

Later reptiles developed yet another feature of great importance: an advanced jaw that could apply heavy pressure upon closing so as to slice food by means of bladelike teeth, which they also evolved. Carboniferous amphibians and early reptiles had jaws that could snap closed quickly but could apply little pressure. Moreover, they had pointed teeth that could kill prey by puncturing them but could not slice or tear food apart. As a result, these animals were forced to swallow their victims whole.

Despite the origin of reptiles in Late Carboniferous time, amphibians continued to prosper into Early Permian time. During the Permian Period, however, reptiles diversified and apparently began to replace amphibians in various ecological roles, probably because the reptiles not only possessed more advanced jaws and teeth but were also faster and more agile on their feet.

Permian rocks of Texas have yielded large faunas of amphibians and reptiles that reveal this pattern. By Early Permian time, the **pelycosaurs**, fin-backed reptiles and their relatives, had become the top carnivores of widespread ecosystems (Figure 15-15). Their stratigraphic occurrence suggests that many lived in swamps and that some may have been semiaquatic. *Dimetrodon*, one such carnivore (see Figures 15-14 and Figure 15-15), was about the size of a jaguar and had sharp, serrated teeth. Whereas even the Permian carnivorous amphibians, such as the alligator-like *Eryops*, were forced to swallow small prey whole, *Dimetrodon* could tear large animals to pieces.

A new level of metabolism *Dimetrodon* and other pelycosaurs had a skull structure that in some ways resembled that of the mammals, which evolved from them. Their descendants, the **therapsids**, were even more similar to mammals (Figure 15-16). Therapsids' legs were positioned more vertically beneath their bodies than were the sprawling legs of primitive reptiles, or even pelycosaurs. In addition, the jaws of therapsids were complex and powerful, and the teeth of many species were differentiated into frontal incisors for

Earth System Shift 15-1   Weakened Greenhouse Warming Results in a Great Ice Age

The expansion of glaciers in the Southern Hemisphere during mid-Carboniferous time locked up water on land as ice and caused sea level to drop dramatically, as evidenced by a disconformity in shallow marine deposits throughout the world. Climates at middle and high latitudes became colder and more seasonal, resembling those of the present world. Fossil trees that grew at high latitudes in Late Carboniferous time document the new climatic pattern: they are known for their distinctive seasonal growth rings. Strongly seasonal growth of wood produces conspicuous rings in the cross sections of tree trunks. In contrast, fossil trees that grew near the Late Carboniferous equator were of the tropical type, lacking seasonal growth rings.

During the late Paleozoic ice age, continental glaciers expanded to their largest volume in all of Phanerozoic time. Climatic cooling associated with this glacial expansion caused a major extinction of marine life and created a new state of the marine ecosystem that persisted until the ice age ended during the Permian Period. All major groups of marine animals suffered substantial losses in this biotic crisis. The recovery of marine diversity during the ice age came slowly. The species that persisted into the new world were broadly adapted forms that were able to deal with seasonal climatic changes, and many were also broadly distributed. Not only were they resistant to extinction,

Figure 1 Rocky surface in South Africa smoothed and scratched by glacier. Gravelly sediments of the Permian Dwyka Tillite rest against this surface. The point on the green 15-centimeter (about 6-inch) scale shows the direction of ice flow. (J. C. Crowell and L. A. Frakes, *GSA Bull.* 83:2887–2912, 1972.)

Figure 2 Marine strata around the world display a disconformity in mid-Carboniferous time. This disconformity reflects the drop in sea level that resulted from the growth of enormous glaciers at low latitudes in the Southern Hemisphere. Letters with subscripted numbers denote fossil horizons that permit correlation around the world. (After W. B. Saunders and W. H. C. Ramsbottom, *Geology* 14:208–212, 1986.)

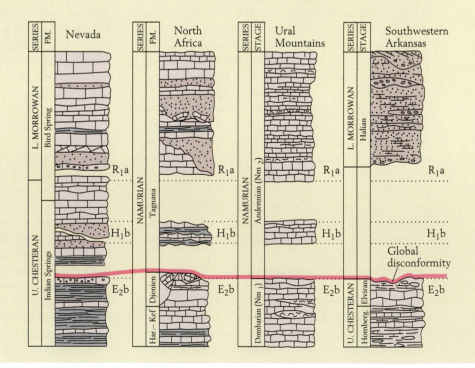

but because they dispersed widely and did not readily give rise to isolated populations, their speciation rates were also low. In fact, rates of both speciation and extinction were lower during this great ice age than at any other substantial interval of Phanerozoic time. Rates of speciation and extinction did not rise again until the ice age ended in the Permian Period, about 35 million years later.

It appears likely that changes in greenhouse warming initiated the massive Carboniferous glaciation. Extensive burial of reduced carbon in Carboniferous coal swamps must have greatly lowered the concentration of carbon dioxide in the atmosphere, weakening greenhouse warming of Earth's surface (see Figure 10-5). Thus it appears to be no accident that the glaciation more or less coincided with the interval of widespread coal swamps. Recall that the spread of forests in Late Devonian time had already reduced the atmospheric concentration of CO_2 by intensifying the weathering of continental surfaces (p. 436).

Recall also that bacterial decomposition of organic matter entails respiration and therefore consumes oxygen (p. 226). The increased burial of organic carbon during the coal-swamp era, by protecting organic matter from bacterial consumption, must have caused oxygen to accumulate in the atmosphere. The buildup of atmospheric oxygen to levels above that of the present world probably explains the appearance of gigantic insects in Late Carboniferous time (see p. 364), because it is insects' limited ability to assimilate oxygen that governs their maximum body size.

Just as a weakening of greenhouse warming triggered the late Paleozoic ice age, an intensification of greenhouse warming brought this glacial interval to an end. When climates became drier in Late Carboniferous and Permian time, levels of atmospheric carbon dioxide must have risen, for two reasons. First, as coal swamps dried up, rates of carbon burial declined, and as a result, more plant debris was consumed at Earth's surface by bacteria, whose respiration released carbon dioxide into the atmosphere. Second, dry conditions reduced rates of weathering, which consumes carbon dioxide, so that more of this gas was left in the atmosphere.

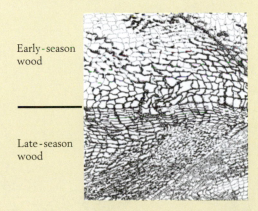

Figure 3 A section of fossil wood of late Paleozoic age from Antarctica shows the details of the boundary between two seasonal growth rings. Each space in the woody tissue was occupied by a single cell. When the late-season wood of the earlier (lower) layer formed, the cells that formed were small, and the tree grew slowly. Growth was interrupted during winter, but spring then stimulated the growth of early-season wood, resulting in large cells and rapid growth. The number of growth rings indicates the age of a tree. Growth rings are not well developed in tropical climates. In late Paleozoic time, high-latitude climates were strongly seasonal, producing growth rings in trees not only in Antarctica, but also in the far north, in Siberia. (After a photograph by J. M. Schopf.)

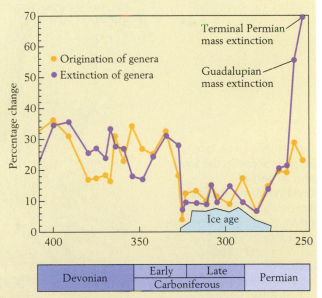

Figure 4 Rates of origination and extinction of genera in the ocean dropped to the lowest levels in all of Phanerozoic time at the start of the late Paleozoic ice age. These rates returned to normal levels precisely when the ice age ended, partway through the Permian Period. (After S. M. Stanley and M. G. Powell, *Geology* 31:877–880, 2003.)

Figure 15-15 Skeletons of pelycosaurs from the Lower Permian Series of Texas. The back fins of the herbivore *Edaphosaurus* (left) and the carnivore *Dimetrodon* (right), which were supported by long vertebral spines, served an uncertain function. Some paleontologists believe that skin stretched between the spines was used to catch the sun's rays, allowing the animal to raise its temperature to a level above that of its surroundings. From snout to tail, *Dimetrodon* exceeded 2 meters (6 feet) in length. (Field Museum of Natural History, Chicago, Neg. GEO85820.)

nipping, large lateral fangs for puncturing and tearing, and bladelike molars for shearing and chopping food.

Many experts believe that the therapsids were **endothermic**, or warm-blooded: by virtue of a high metabolic rate, they maintained their body tempera-tures at relatively constant levels that were usually above those of their surroundings. Hair similar to that of mod-ern mammals may have insulated therapsids' bodies (see Figure 15-16). Even if they were endothermic, how-ever, therapsids may not have kept their body temper-

Figure 15-16 A Late Permian scene in the South African part of Gondwanaland. Therapsids move along an ice-covered stream in a snowy environment, where they may have been able to live by virtue of being endothermic. In this reconstruction, they are shown as having hair, which is associated with endothermy in mammals. The largest animal is *Jonkeria*; in the background is *Trochosaurus*. The animals at lower right are *Dicynodon*, and the very small form is *Blattoidealestes*. (Drawing by Gregory S. Paul.)

atures at levels as constant as those of mammals. In any case, the upright postures and complex chewing apparatuses of advanced Permian therapsids show that these active animals approached the mammalian level of evolution in anatomy and behavior.

The endothermic condition allows animals to maintain a sustained level of activity—to hunt prey or to flee from predators with considerable endurance. **Ectothermic** (or cold-blooded) reptiles, in contrast, must rest frequently in order to soak up heat from their environment. Endothermic metabolism, along with advanced jaws, teeth, and limbs, may account not only for the success of the therapsids during Permian time but also for the decline of the pelycosaurs, which were probably ectothermic. In fact, while pelycosaurs were declining to extinction during Late Permian time, therapsids were undergoing a spectacular adaptive radiation. More than 20 families of these advanced animals seem to have evolved in just 5 or 10 million years, and they were the dominant groups of large animals in Late Permian terrestrial habitats. Therapsids seem to have represented an entirely new kind of animal—one so advanced that it was able to diversify very quickly.

Paleogeography of the Late Paleozoic World

During late Paleozoic time the major continents moved closer and closer together, until by early Mesozoic time nearly all of them were fused together as the supercontinent Pangaea (p. 181). Early in the Carboniferous Period the continents were already tightly clustered (Figure 15-17). As the period progressed, continental glaciers grew at high latitudes in the Southern Hemisphere and persisted into the Permian Period. Meanwhile, warm conditions prevailed in equatorial regions. Thus there was a strong temperature gradient between the equator and the poles. Coal deposits formed extensively during Late Carboniferous time, accumulating at both low and high latitudes. Global climates changed dramatically early in the Permian Period, when glaciers melted in Gondwanaland and the spread of arid conditions caused coal swamps to contract throughout the world.

Warm, moist conditions were widespread in Early Carboniferous time

Sea level, which had declined near the end of the Devonian Period, rose at the start of Early Carboniferous time, so that warm, shallow seas spread across broad continental surfaces at low latitudes. As a result, limestones accumulated over large areas, often with crinoid debris as their major component.

Warm, moist conditions prevailed in broad continental areas near the equator. Thus coal-swamp floras, which first became established early in Carboniferous time, flourished along the northeastern margin of Euramerica (see Figure 15-17). Trade winds must have continued to bring this continent moisture from the oceans to the northeast. In contrast, the western part of the continent was in the rain shadow of the Appalachian Mountains. Here, across what is now central and western North America, evaporites and limestones accumulated in broad, shallow seas.

In mid-Carboniferous time, continents collided and a great ice age began

During the middle portion of Carboniferous time, the northward movement of Gondwanaland caused that continent to collide with Euramerica. This collision formed a supercontinent that stretched from pole to pole, constituting most of Pangaea. The mountains thus formed in southern Europe are known collectively as the Hercynides, and the orogeny as a whole is known as the *Hercynian* (or Variscan). Hercynian mountains also formed in northwestern Africa. In North America the Hercynian orogeny, known there as the *Alleghenian*, in effect continued where the Acadian orogeny left off, extending the Appalachian mountain chain southwestward and forming the adjacent Ouachita Belt in Oklahoma and Texas.

Global climatic cooling transformed the world in mid-Carboniferous time. This cooling was apparently brought on by a weakening of greenhouse warming, which resulted from the burial of vast quantities of reduced carbon in coal swamps (see Earth System Shift 15-1). The greatest ice age in all of Phanerozoic time got under way, and a major extinction decimated life in the sea. The expansion of glaciers caused sea level to drop substantially. The resulting global unconformity in marine strata, along with attendant changes in environments of deposition around the world, is partly what led American geologists to separate what Europeans call the Early and Late Carboniferous into two distinct periods: the Mississippian Period and Pennsylvanian Period. The ice age ushered in a new state of the marine ecosystem, a world in which winters were much colder than they had been before. Hardy taxa adapted to the harsh conditions persisted for long intervals of time during the ice age.

On land, latitudinal temperature gradients steepened during Late Carboniferous time—that is, there were extreme differences in temperature between the equator and the poles. Continental glaciers pushed northward to within nearly 30° of the equator, a latitude where subtropical conditions have prevailed during most of Phanerozoic time. It seems amazing that

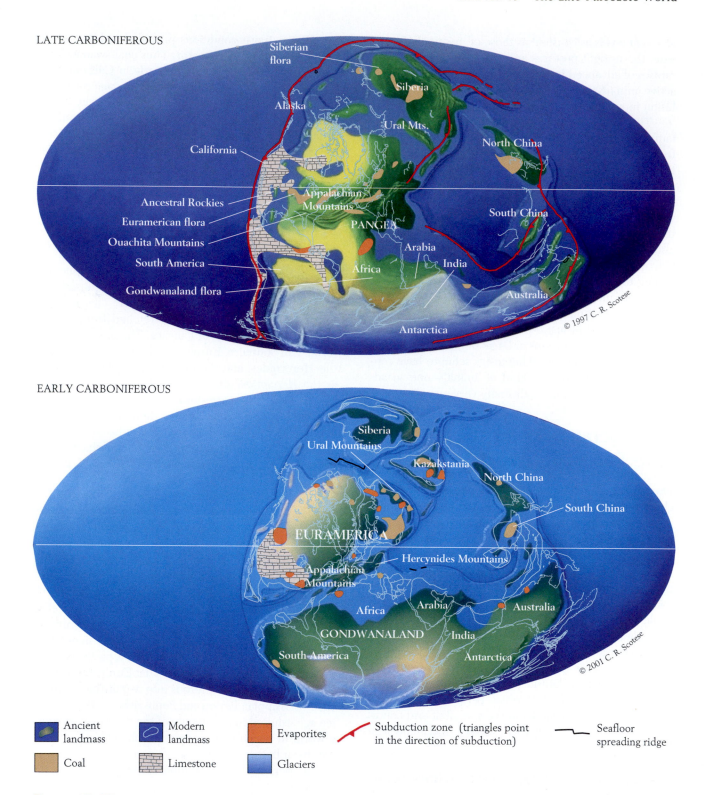

LATE CARBONIFEROUS

Siberian flora
Siberia
Alaska
Ural Mts.
California
North China
Ancestral Rockies
Euramerican flora
Appalachian Mountains
South China
Ouachita Mountains
PANGEA
South America
Arabia
India
Gondwanaland flora
Africa
Australia
Antarctica
© 1997 C. R. Scotese

EARLY CARBONIFEROUS

Siberia
Ural Mountains
Kazakstania
North China
South China
EURAMERICA
Hercynides Mountains
Appalachian Mountains
Africa
Arabia
Australia
GONDWANALAND
India
South America
Antarctica
© 2001 C. R. Scotese

Ancient landmass	Modern landmass	Evaporites	Subduction zone (triangles point in the direction of subduction)	Seafloor spreading ridge
Coal	Limestone	Glaciers		

Figure 15-17 World geography in Carboniferous time. The major continents were tightly clustered on one side of the globe in Early Carboniferous (Mississippian) time. Coal deposits formed near the sea that bordered eastern Euramerica, and limestones and evaporites accumulated in the epicontinental sea that flooded western Euramerica. Glaciers spread over Gondwanaland near the south pole. Late in Early Carboniferous time Gondwanaland and Euramerica collided. Coal deposits formed over a larger total area at this time than at any other period in Earth's history. In Gondwanaland, continental glaciers spread to remarkably low latitudes and were separated from tropical coal swamps (formed by the Euramerican flora) by steep temperature gradients. The Gondwanaland and Siberian floras flourished under cooler conditions. (Adapted from paleogeographic maps by C. R. Scotese, PALEOMAP Project, University of Texas at Arlington, 1997 (top), 2001 (bottom).)

tropical coal swamps flourished in North America and western Europe not much farther north than the northernmost Carboniferous glaciers (see Figure 15-17).

Coal deposits formed north of the glaciers in Gondwanaland as well (see Figure 8-7). Nonetheless, the flora that produced the coal deposits in this southern region differed substantially from the equatorial Euramerican flora. *Lepidodendron* and *Sigillaria*, the dominant Euramerican plants, were present in the Southern Hemisphere, but many of the southern plants are unknown from northern continents. The Gondwanaland flora was adapted to the cool climates of the glacial regime in the south (see Figures 8-3 and 15-17). Siberia, which lay near Earth's other pole, also had a distinctive flora adapted to cold conditions.

Dry habitats expanded in Permian time

In Permian time, the suturing of Siberia to eastern Europe along the Ural Mountains resulted in the nearly complete assembly of Pangaea (Figure 15-18). Southeast Asia remained the only separate landmass of large size; it would become attached to Pangaea early in the Mesozoic Era. The mountain building resulting from the continental collisions that produced Pangaea resulted in a landscape in which many mountain ranges stood high above lowlands. Pangaea was so large that much of it lay far from moisture-providing oceans. As a result, numerous dune deposits and evaporites accumulated, especially in the dry trade wind belt (Figure 15-19). The Permian, in fact, has a greater volume of salt deposits than any other geologic system. In addition, the polar regions of Pangaea became quite cold in winter because of the low heat capacity of continents relative to that of oceans (p. 92).

As a result of complex topographic conditions and steep climatic gradients, the floras of the Permian Period were more provincial than those of any other Phanerozoic period—with the possible exception of the most recent interval, when continents have been widely dispersed and the waxing and waning of continental glaciers have produced much geographic differentiation.

Throughout most of the world, floras changed during Permian time as climates became more arid. Plants

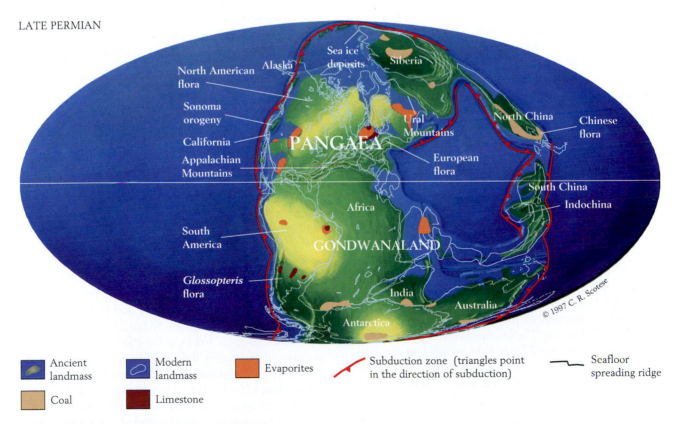

LATE PERMIAN

Figure 15-18 World geography in Late Permian time. At this time, the ocean separating Europe from Asia was closing along the Ural Mountains to form the supercontinent Pangaea. The Permian landmasses had a complex topography, in which many mountain ranges stood high above lowlands. Four distinctive floras are labeled on the map. Floras produced coal deposits only at high latitudes, while equatorial regions were dry. Small glaciers formed in Siberia, which now encroached on the north pole. (Adapted from paleogeographic maps by C. R. Scotese, PALEOMAP Project, University of Texas at Arlington, 1997.)

Figure 15-19 Cross-bedded dune deposits of the Permian Coconino Sandstone, which crops out in the vicinity of the Grand Canyon in Arizona. (David M. Dennis/ Tom Stack & Associates.)

adapted to moist conditions gave way to ones favored by drier habitats. Thus, in the north, conifers and other gymnosperms replaced coal-swamp floras (see Figure 15-11). Many late Paleozoic conifers apparently resembled conifers of the modern world in their ability to thrive under relatively dry conditions.

An important consequence of the drying of continents was that less reduced carbon came to be buried in coal swamps. The result was that carbon dioxide built up in the atmosphere and the resulting intensification of greenhouse warming brought the great late Paleozoic ice age to an end.

Coal nonetheless continued to form in two regions during the Permian. While climates became drier in most regions, lycopods persisted and continued to form coal in a portion of China, which was a warm, moist island surrounded by tropical seas (see Figure 15-18). Because it was not far from the ocean, the southern portion of the Permian supercontinent also remained moist. There, the flora named for the tree *Glossopteris* (see Figure 8-3), which had come to dominate moist landscapes at the beginning of the Permian, persisted and formed coal until the end of the period.

Mass extinctions ended the Paleozoic Era

Two mass extinctions occurred in the Permian, the second of which was the most devastating mass extinction

of all time for animal life. This was the first great mass extinction to strike vertebrate animals on land, eliminating the vast majority of therapsid genera. It was the destruction of the marine fauna, however, that led nineteenth-century geologists to designate the end of the Permian Period as the end of the Paleozoic Era— literally, the era of ancient life. The terminal Permian crisis struck all major animal taxa in the ocean, entirely sweeping away several prominent Paleozoic groups, including the rugose and tabulate corals (see Earth System Shift 15-2).

Regional Events of Late Paleozoic Time

The Paleozoic Era witnessed two orogenies in what is now the United States: the final episode of mountain building in the Appalachian region and a related orogeny in the southwestern region. Both orogenies resulted from the collision of Euramerica with Gondwanaland. Coal deposits formed in western Europe as well as in North America, and tectonic events produced mountains in the Rocky Mountain region. An enormous reef complex (now the site of large petroleum reservoirs) formed and came to an end in western Texas, and an orogenic episode visited the Pacific margin of North America.

The Alleghenian orogeny formed the Appalachian Mountains

The mid-Carboniferous collision of Euramerica with Gondwanaland resulted in the *Alleghenian orogeny*. This collision shifted the region of mountain building southward along the eastern margin of North America. The conspicuous fold-and-thrust belt of the central and southern Appalachians formed at this time. This still-mountainous region is a zone of low-temperature deformation called the Valley and Ridge Province (Figure 15-20). In this zone, large slices of crust slid westward along thrust faults and were crumpled by pressure from the east. To the east of the Valley and Ridge Province, separated from it by the Blue Ridge Mountains, lies the Piedmont Province. Rocks of the Piedmont were metamorphosed and highly deformed because of their proximity to the zone of suturing. The Blue Ridge is a band of Proterozoic rocks that were metamorphosed about a billion years ago, during the Grenville orogeny (Chapter 12). These rocks were elevated along a large thrust fault during the Alleghenian orogeny.

The Valley and Ridge Province includes rocks representing the passive margin that preceded the Taconic

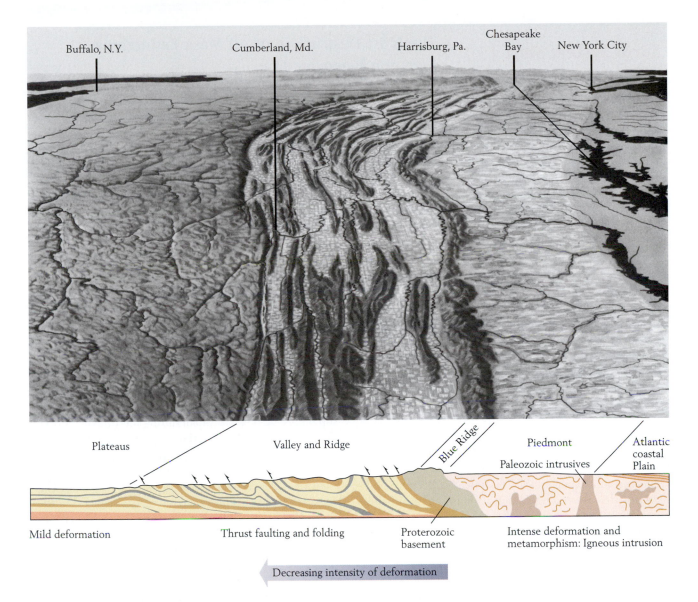

Buffalo, N.Y.　　Cumberland, Md.　　Harrisburg, Pa.　　Chesapeake Bay　　New York City

Plateaus　　Valley and Ridge　　Blue Ridge　　Piedmont　　Atlantic coastal Plain

Paleozoic intrusives

Mild deformation　　Thrust faulting and folding　　Proterozoic basement　　Intense deformation and metamorphism: Igneous intrusion

Decreasing intensity of deformation

Figure 15-20 **Aerial view and idealized cross section of the Appalachian region.** The aerial view is toward the northeast. Sediments of the Atlantic Coastal Plain lap up on the worn-down eastern portion of the Appalachian system. To the west of the Atlantic Coastal Plain, the low-lying Piedmont Province is separated from the Valley and Ridge Province by the conspicuous Blue Ridge Province. (After J. S. Shelton, *Geology Illustrated*, W. H. Freeman and Company, New York, 1966.)

orogeny and the flysch-molasse cycle that formed during that mountain-building event (see Figure 13-21). This province also displays the similar sequence of rocks that formed before and during the Acadian orogeny (see Figure 14-25). The Alleghenian Orogeny followed quickly after the Acadian orogeny, at a time when the continental margin was still an upland from which molasse was spreading westward. The Alleghenian orogeny simply built new mountains and perpetuated the accumulation of molasse (Figure 15-21). Some of the molasse deposits, such as the Pottsville Formation of Pennsylvania, harbor commercially valuable coal beds that formed in swamps bordering rivers. Most of these coal beds are positioned in the upper portions of meandering-river cycles (see Figure 5-16).

The Alleghenian orogeny culminated in the folding and thrust faulting of Paleozoic rocks of all ages, including those that had accumulated as flysch and molasse during the Taconic, Acadian, and Alleghenian orogenies. The ridges of the Valley and Ridge Province still stand as the Appalachian Mountains because they are underlain by ancient roots of lower density than the mantle. Recent isostatic uplift has created uplands in this old orogenic belt.

Between 80 and 85 percent of the species alive near the end of the Permian died out in the terminal Permian mass extinction. This was the largest percentage of species eliminated by any mass extinction in all of Phanerozoic time. The total number of taxa that died out was smaller than it might otherwise have been, however, because there had been only a partial evolutionary recovery from the mass extinction that occurred only 7 or 8 million years earlier, at the end of the Guadalupian Age, in Middle Permian time. In fact, fewer species were living in the ocean at the time of the terminal Permian mass extinction than at any time since early in the Ordovician Period, when marine animals were still undergoing their initial evolutionary radiation. The first Permian crisis eliminated perhaps 70 percent of marine species. Although this event was less devastating than the terminal Permian crisis, it removed more marine genera than the mass extinction that wiped out the dinosaurs and brought the Mesozoic Era to an end.

A distinctive aspect of the first (Guadalupian) mass extinction was its destruction of the organic reef community. A second was its elimination of about three-quarters of all genera of fusulinid foraminifera. The fusulinid species that disappeared included all species that were longer than 6 millimeters ($\frac{1}{4}$ inch) and all species whose skeletons had complex wall structures. Symbiotic algae probably lived within these fusulinids, as they do in large living foraminifera. Perhaps, then, the large fusulinid species became extinct because their symbiotic algae died out.

The even more severe crisis at the end of the Permian damaged all marine taxa, but it totally eliminated the rugose and tabulate corals and the trilobites—although the latter two groups had been on the decline for some time. The ammonoids, crinoids, and bryozoans hung on by a thread: only a handful of species of each group made its way into the Mesozoic Era.

The impact of the terminal Permian mass extinction was also evident in the terrestrial realm. In fact, whereas the Guadalupian crisis had no apparent effect on land animals, the terminal Permian crisis was so severe that almost 20 families of Permian therapsids failed to survive into Triassic time. Only in the Karroo Basin of South Africa are there highly fossiliferous terrestrial strata that span the Permo-Triassic boundary. *Lystrosaurus*, one of the few therapsid genera to survive the crisis in this region, became very abundant, dominating therapsid faunas of earliest Triassic time. After the crisis, this piglike herbivore (see Figure 8-10) became an opportunistic creature, and its populations swelled enormously. Probably the disappearance of predators in the mass extinction accounts for *Lystrosaurus'* population explosion.

Terrestrial vegetation was also transformed at the end of the Permian. In most regions woody conifers and other gymnosperms died out, to be replaced by small, nonwoody lycopods. In other words, forests disappeared. The devastation of terrestrial vegetation is reflected in a remarkable phenomenon: the largest concentration of fungal remains in the entire geologic record occurs in shallow marine deposits of latest Permian age. Apparently, at the end of Permian time, a huge volume of decaying, fungus-infested wood was washed into shallow seas throughout the world.

Recent studies of deep-sea deposits have shed new light on the cause of the two Permian crises. Both crises were associated with an interval of anoxia in the deep sea. The Permo-Triassic strata that reveal this anoxia accumulated in the Pacific Ocean far from land and were transported westward on the Pacific plate, to be thrust up onto southwestern Japan during the Jurassic Period. These strata indicate that the deep sea was well oxidized until the very end of Guadalupian time. Guadalupian cherts in the uplifted rocks consist of the remains of radiolarians, which secrete siliceous skeletons (p. 64). The great abundance of radiolarians that descended to the Permian deep-sea floor after death indicates that upwelling must have supplied abundant nutrients to the shallow waters of the Pacific Ocean where the radiolarians lived, nourishing phytoplankton upon which the radiolarians fed. Surface water in other regions of the ocean must have moved downward to replace the upwelling water. This downwelling water must have carried oxygen-rich waters to the deep sea. In fact, the cherts formed from radiolarian skeletons are stained red by hematite, a highly oxidized iron mineral (Chapter 12).

Gray sediments suddenly replaced red cherts at the end of Guadalupian time in the strata uplifted onto Japan, signaling the onset of low-oxygen conditions in the deep sea. Significantly, this change coincided with the Guadalupian mass extinction, indicating that this extinction was associated with a restructuring of the oceans: very little oxygen-rich water was descending from the surface to the deep sea.

Higher up in the Japanese sequence, the deep-sea sediments become black, rather than gray, reflecting total stagnation of the deep ocean. When this severe stagnation began, anoxic sediments began to accumulate even on continental shelves around the margins of the Pacific Ocean. Apparently, for several million years, only the uppermost waters of the ocean were oxygenated. In

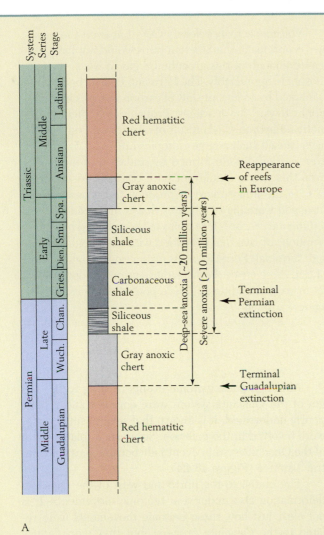

A

B

Figure 1 Uplifted rocks in Japan document an episode of deep-water anoxia in Late Permian time. *A.* When anoxia began at the end of Guadalupian (Middle Permian) time, gray chert replaced hematitic (highly oxidized) red chert. An interval of severe anoxia, marked by even darker sediments, began at the time of the terminal Permian extinction. Deposition of hematitic chert resumed in Middle Triassic time, when reefs began to grow again in shallow water. *B.* Rocks representing this Middle Triassic recovery interval. (*A,* After Y. Isosaki, *Science* 276:235–238, 1997; *B,* courtesy of Y. Isosaki, University of Tokyo.)

other words, the ocean at this time was highly stratified, with little vertical movement of water masses.

Early Triassic sediments show a trend opposite to that of the Permian: first they become gray, and then they become red. Thus they record the return of fully oxygenated conditions to the deep sea over several million years, which means that oxygen-laden water was again descending from the surface to great depths. It is striking that organic reefs, which were eliminated from shallow seas by the terminal Permian mass extinction, returned exactly when the deep sea became fully oxygenated once again, near the start of Middle Triassic time.

Cool conditions had persisted in polar regions into Late Permian time even though continental glaciers had shrunk. It is in these regions that cold, dense waters would have descended to oxygenate the deep sea, just as they do today, when polar regions are also frigid (see Figure 4-23). Thus stagnation of the deep sea in Late

Permian time must have resulted from a warming of polar regions, which brought an end to the downwelling of oxygen-rich waters. In other words, there must have been an episode of global warming. Terrestrial deposits also point to a dramatic warming and drying of climates. The Permian *Glossopteris* flora of the Southern Hemisphere had formed coal under cool, moist conditions. *Glossopteris* and coal do not occur above the Permo-Triassic boundary, however. In their place are red beds that formed in a warmer, drier climate. River deposits in South Africa also indicate that climates became more arid: at the Permo-Triassic boundary, meandering river deposits give way to gravelly braided-stream deposits that represent habitats too dry to support enough plant cover to stabilize stream banks.

Presumably, a sudden warming and drying of climates accounts for many of the terrestrial extinctions that occurred in the terminal Permian crisis. Similarly, warming of the seas may have been responsible for many of the marine extinctions of both Permian crises. Nonetheless, other environmental changes may also have contributed to the greatest mass extinction of all time, depending in part upon the ultimate cause of the climatic changes. This ultimate cause is still debated.

(continued)

Earth System Shift 15-2 | The Most Destructive of All Mass Extinctions Ends the Paleozoic Era (continued)

There is a good chance that the terminal Permian crisis was in some way associated with the volcanic event that produced the so-called Siberian traps—a massive body of rock in northeastern Asia that represents the greatest outpouring of basalt onto a continent in all of Phanerozoic time. Radiometric dating gives an age about 251 million years for the Siberian traps, which is also the age obtained by uranium-lead dating of a volcanic ash bed just a few centimeters from the Permo-Triassic boundary in China. In addition, new dating has shown that a smaller, but still enormous, body of volcanic rock formed suddenly in Asia sometime between 259 and 256 million years ago. The timing of this earlier eruption makes it a possible cause of the Guadalupian extinction event. It occurred in southwestern China, not Siberia, but is believed to have issued from the same plume that rose from deep within the mantle to form the Siberian traps a few million years later.

Volcanic activity releases CO_2, but even the massive eruptions of the Siberian traps would not have released enough of this greenhouse gas to warm the atmosphere appreciably. The Permian-Triassic boundary is marked by an abrupt shift of carbon isotopes in limestones toward lower values. Probably only melting of methane hydrates, as a result of ocean warming, could have caused such a large isotopic shift (p. 235). However, carbon dioxide formed by oxidation of methane from this source would not have contributed enough CO_2 to the atmosphere to produce the pronounced, sustained warming of climates that is indicated by the sedimentary record. It remains to be explained how the eruptions of lava that coincided with the Guadalupian and terminal Permian mass extinctions might have caused these crises. Perhaps these eruptions somehow set in motion additional, more powerful agents of climatic warming.

Orogenies also occurred in the southwestern United States

The late Paleozoic was also a time of mountain building along a zone extending from Mississippi across Oklahoma and Texas to Utah (Figure 15-22). Here the Ouachita Mountains formed as a westward continuation of the Appalachians when Gondwanaland and Euramerica collided (see Figure 15-17). Today traces of the two mountain chains meet at right angles beneath flat-lying younger deposits. Although the zone of contact is not well understood, the exposed segment of the Ouachitas is a fold-and-thrust belt resembling the Appalachian Valley and Ridge Province (Figure 15-23). One difference is that the folded rocks of the Ouachitas, which range in age from Ordovician to middle Pennsylvanian, consist of deep-water black shale and flysch deposits that have been thrust northward against shelf-edge carbonates of similar age. In other words, deformation took place offshore from the continental margin (see Figure 15-22), and after it began, the rate of de-position in the adjacent basin increased. In fact, the deformed region seems to have behaved as an unusually deep foreland basin, in which enormous volumes of deep-water Carboniferous deposits continued to accumulate on top of already-deformed older deposits. These younger, thicker deposits were, in turn, folded and thrust northward. Most of this deformation was completed before the start of the Permian Period.

Although plate tectonic events relating to the origin of the Ouachita system were complex and remain poorly understood, it is known that several small plates were involved in the collision. Some small plates south of the Ouachita system eventually became parts of Central America (Figure 15-24).

The craton to the north and west of the Ouachita deformation also underwent tectonic movements. It is not clear just how these cratonic movements were related to the Ouachita orogeny, but they were largely

▶ **Figure 15-21** **Stratigraphic cross section through the central Appalachians west of the Blue Ridge Province, with vertical exaggeration.** Folds and faults are not shown. The thickest deposits lie to the southeast, in the Valley and Ridge Province. The thinnest deposits lie to the northwest, in the plateau region west of the Appalachians. This package of Paleozoic strata represents three cycles of deposition related to orogenic activity to the east. The Taconic and Acadian orogenies produced complete tectonic cycles of deposition, including passive margin deposition, flysch, and molasse. The Alleghenian orogeny, however, occurred when mountains produced during the Acadian orogeny were still shedding sediment to the east. Therefore, the Alleghenian orogeny simply piled Carboniferous molasse on top of the Devonian molasse produced by the Acadian event. (After G. W. Colton, in G. W. Fisher et al., eds., *Studies of Appalachian Geology: Central and Southern*, John Wiley & Sons, New York, 1970.)

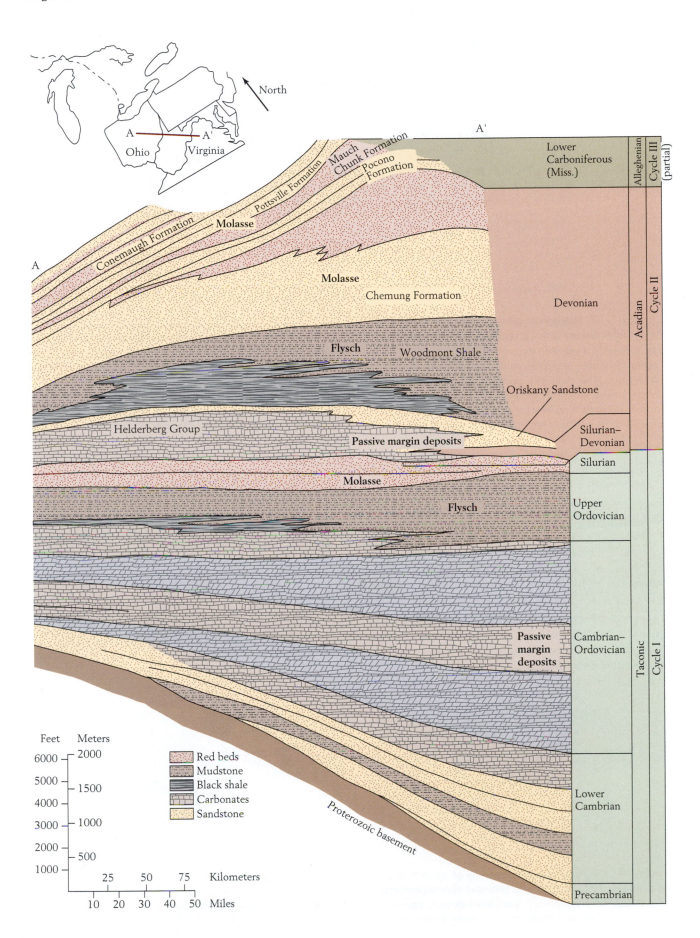

North

Ohio   Virginia

A   A'

Mauch
Chunk Formation
Pottsville Formation
Pocono
Formation
Conemaugh Formation
Molasse

Lower
Carboniferous
(Miss.)

Alleghenian   Cycle III
(partial)

Molasse

Chemung Formation

Devonian

Flysch   Woodmont Shale

Oriskany Sandstone

Acadian   Cycle II

Helderberg Group

Passive margin deposits

Silurian–
Devonian

Silurian

Molasse

Flysch

Upper
Ordovician

Passive
margin
deposits

Cambrian–
Ordovician

Taconic   Cycle I

Proterozoic basement

Lower
Cambrian

Precambrian

Feet   Meters

6000   2000

5000   1500

4000   1000

3000

2000   500

1000

Red beds
Mudstone
Black shale
Carbonates
Sandstone

25   50   75   Kilometers

10   20   30   40   50   Miles

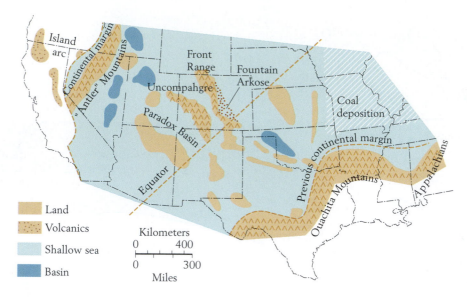

Figure 15-22 Paleogeography of the southwestern United States during Late Carboniferous (Pennsylvanian) time. Partway through the Late Carboniferous interval, the Ouachita Mountains began to form at the southern end of the Appalachians. Most of this deformation occurred offshore, uplifting oceanic sediments and welding them onto the previous continental margin. Coal-bearing cyclothems formed in marginal marine environments in the midcontinent region. To the west, shallow seas covered most of the craton, but many uplifts and basins developed from Texas to eastern Nevada, apparently in association with the Ouachita orogeny. The highest uplifts, the Front Range and the Uncompahgre, are together known as the Ancestral Rocky Mountains. Farther west, mountains produced by the earlier Antler orogeny still bordered the continent, and an island arc lay along a subduction zone offshore in what is now California and Oregon.

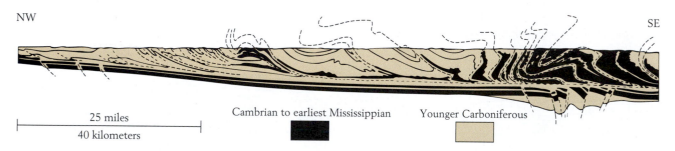

Figure 15-23 Cross section through the Ouachita Mountains as they exist today in the southeastern corner of Oklahoma. Here large volumes of deep-water sediment have been thrust northwestward. Note the general similarity between the style of deformation here and in the Appalachian fold-and-thrust belt (see Figure 15-20). (After J. Wickham et al., *Geology* 4:173–180, 1976.)

Figure 15-24 Approximate positions of small plates south and east of the Ouachita fold-and-thrust belt late in Paleozoic time and today. Since late Paleozoic time, some of these plates have shifted southward, leaving the Gulf of Mexico in their place. (After A. G. Smith and J. C. Briden, *Mesozoic and Cenozoic Paleocontinental Maps*, Cambridge University Press, Cambridge, 1977.)

Figure 15-25 Northward view along the Front Range of the Rocky Mountains at Boulder, Colorado. The core of the Rockies lies to the right. Tilted upward along its margin are the so-called Flatirons. The Flatirons consist of the Fountain Arkose, formed of sediment shed from the Ancestral Rocky Mountains, which lay slightly to the west in Late Carboniferous time. The Fountain Arkose was later tilted upward when the modern Rockies formed. (Ann Duncan/Tom Stack & Associates.)

vertical: enormous areas in what is now the southwestern United States became transformed into a series of uplifts and basins (see Figure 15-22). Many of these structural features are bounded by steeply dipping faults. The basins accumulated Late Carboniferous (Pennsylvanian) and, in some cases, Permian deposits. Clastic debris shed from the uplifts was deposited rapidly in nearby basins as coarse arkose. Recall that arkose is a sedimentary rock that consists largely of feldspar, a mineral that weathers to clay if it is not buried rapidly (p. 41).

Two of the uplifts, the Front Range and Uncompahgre uplifts, are commonly referred to as the Ancestral Rocky Mountains. The Ancestral Rockies developed during Late Carboniferous time; they were elevated and then subdued by erosion in an area where portions of the Rocky Mountains stand today. It is estimated that the Uncompahgre uplift rose to between 1.5 and 3.0 kilometers (1 or 2 miles) above the surrounding seas, which flooded much of western North America. This elevation is comparable to that of the modern Rockies above the Great Plains to the east. The Front Range uplift is named for the Front Range of the modern Rockies, which now extends slightly farther east than the late Paleozoic uplift. Growth of the Ancestral Rockies elevated underlying Precambrian rocks, which later were leveled by erosion. The Precambrian roots of the Ancestral Rockies can still be observed where more recent secondary uplift has caused rivers to cut deep gorges.

At places in the basin between the Uncompahgre and Front Range uplifts, arkosic sands and conglomerates accumulated to thicknesses exceeding 3 kilometers (2 miles). The Fountain Arkose, which formed along the eastern flank of the Front Range uplift, was later upturned along the front of the modern Rockies when they were uplifted. Here, through differential erosion, the Fountain stands out in central Colorado as a series of spectacular ridges (Figure 15-25).

The Ancestral Rockies lay close to the Late Carboniferous equator, where easterly equatorial winds must have prevailed. It is therefore understandable that the ancient mountains produced a rain shadow to their west. Here, in the Paradox Basin (see Figure 15-22), great thicknesses of evaporites—primarily halite (p. 30)—accumulated.

Coal deposits formed within cyclothems

In Late Carboniferous (Pennsylvanian) time, as rivers flowing from the Appalachians continued to form molasse deposits in eastern North America, coal swamps spread over the floodplains of those rivers and over the margins of the epicontinental seas to the west. Some coal deposits that are now separate were probably connected at the time they were formed. The Michigan Basin, however, formed in isolation (see Figures 9-21 and 9-22), and the basins in New England and eastern Canada may have done so as well (Figure 15-26).

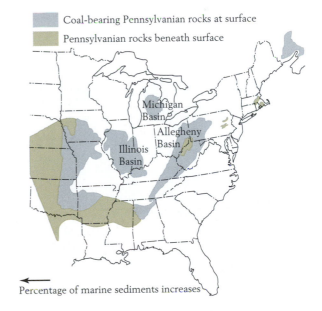

Figure 15-26 Distribution of coal-bearing cyclothems of Pennsylvanian (Upper Carboniferous) age in eastern North America. Before undergoing erosion, the cyclothems in the areas east and south of the Illinois Basin were even more extensive. The main area of coal formation extended from the margin of the early Appalachian Mountains in the east to an area of fully marine deposition to the west.

In the nearly flat midcontinent region of the United States, Late Carboniferous (Pennsylvanian) depositional cycles of a special kind were formed. The coal beds of these cycles formed from peat that accumulated in swamps bordering shallow seas. Here, as in similar settings on other continents, coal beds are thin but widespread, occurring within cycles that include marine deposits. Dozens of similar cycles are commonly found superimposed on one another. Such cycles in coal beds are known as **cyclothems** in North America and as *coal measures* in Britain.

Shifting environments and the origin of cyclothems

The fact that many marine and nonmarine depositional environments are often represented in just a few vertical meters of stratigraphic section in a cyclothem indicates that the depositional gradient was very gentle. Clearly, only a slight vertical movement of the sea or of Earth's crust accounted for a substantial advance or retreat of water, with related shifting of environments.

The coal now found within cyclothems began its formation as peat within coal swamps that occupied lowland areas neighboring the sea. They were fed by the rivers whose deposits lie beneath them (Figure 15-27). It is possible that a broad swamp was, in effect, a broad river into which inland streams emptied—one that flowed so slowly that its movement could not have been observed with the naked eye. The Everglades "river" of Florida, which also flows over a very flat region, takes this form today. The water of the Everglades remains fresh except near the edge of the sea.

Cyclothems were formed by alternating transgressions and regressions of shallow seas. A transgression resulted in the deposition of marginal marine peat (future coal) on top of nonmarine deposits and then of marine sediments on top of the peat (see Figure 15-27). Regression reversed the sequence, burying marine deposits beneath peat and then nonmarine sediments (Figure 15-28).

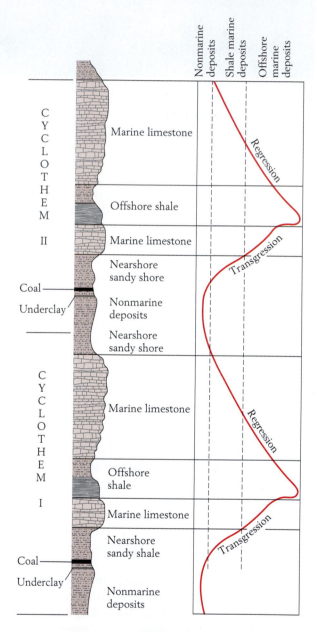

Figure 15-28 Two idealized cyclothems. In each cyclothem, coal swamps migrated over nonmarine deposits as a transgression proceeded. Above the coal are deltaic and other marginal marine sediments, which were deposited above the coal as the sea spread inland. The marginal marine deposits are succeeded, in turn, by marine limestones that represent fully marine conditions and then by black shales that represent deep environments that were present in the region at the time of maximum transgression. In time, sea level began to fall again, and the depositional sequence was reversed.

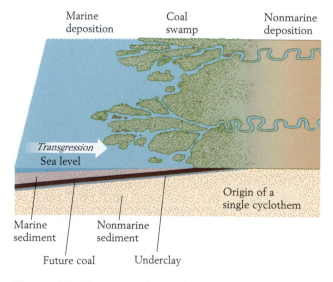

Figure 15-27 The development of a transgressive sequence as sea level rises and the shoreline shifts inland. A transgression produces the lower part of a cyclothem, where marine sediments are superimposed on nonmarine sediments. Underclay is the nonmarine material on which coal swamp plants grew and then died to produce the peat that ultimately would become coal.

Glaciers and sea level The oscillations in sea level responsible for the cyclothems were so rapid that they must have resulted from the repeated expansion and contraction of glaciers in Gondwanaland. Why, then, do we not find similar cycles representing the modern Ice Age, during which continental glaciers have expanded and retreated many times? The explanation seems to be that in recent times—even during interglacial periods of relatively high sea level, such as the one in which we live—the continents have remained relatively emergent. The seas are rising and falling over steeply sloping continental margins; they are not invading and receding from vast, almost flat interior lowlands as they did when they formed the cyclothems of Kansas, Illinois, and neighboring regions.

Reefs formed in the Delaware Basin of western Texas

The Delaware Basin of Texas and New Mexico is one of the most famous geologic features in the world, both because of its economic importance in harboring petroleum and because of its spectacular scenery. Although it has not been occupied by the sea for more than 200 million years, it remains a topographic basin. A person can

stand in its center today and view ancient carbonates that formed as banks or reefs around its margin during the Permian Period (Figure 15-29).

During the latter part of Late Carboniferous (Pennsylvanian) time, the shallow seas that had shifted back and forth over the coal basins of the central United States withdrew westward, never to return. In earliest Permian time they remained only in Texas and neighboring areas, where they were connected to the seas that still flooded the western margin of North America (Figure 15-30). The Ancestral Rockies still stood high, as a mountainous island, and the young Ouachita Mountains bordered the southern margin of North America as what must have been a rugged and imposing mountain range.

During early and middle Paleozoic time, marine deposits had accumulated in the area that now forms western Texas, which was a broad, shallow basin on the continental shelf. During the uplift of the Ouachita range and other Carboniferous uplands, a small fault block formed within the western Texas basin, dividing it into the Delaware Basin and the Midland Basin (Figure 15-31). Both of these basins subsequently received large thicknesses of sediment that have yielded enormous quantities of petroleum.

A

B

Figure 15-29 **The Delaware Basin.** *A.* The Capitan Reef Limestone of the Guadalupe Mountains rims the basin. *B.* A cross section, from northwest to southeast, shows the configuration of the basin. (Figure 15-31 shows the location of the cross section.) (*A*, National Park Service.)

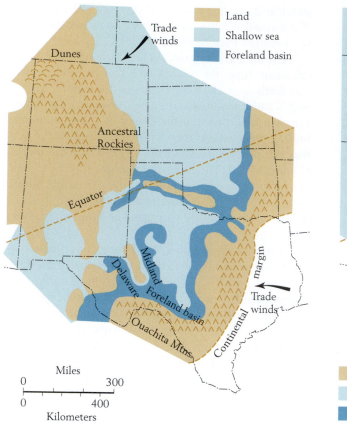

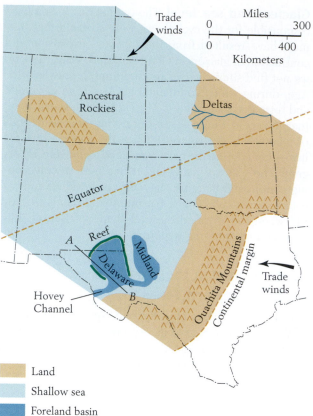

Figure 15-30 The paleogeography of Texas and neighboring regions in earliest Permian time. Shallow seas flooded only parts of western North America, including this region between the Ouachita Mountains and the Ancestral Rockies. Mild deformation of the craton northwest of the Ouachitas continued from Late Carboniferous time, and the Delaware and Midland basins formed in western Texas as part of the Ouachita foreland basin. (After J. M. Hills, *AAPG Bull.* 56:2303–2322, 1972.)

Figure 15-31 The paleogeography of Texas and neighboring regions when reefs encircled the Delaware Basin in Late Permian time. During this interval a narrow passageway (Hovey Channel) connected the Delaware Basin with the open ocean to the west, but the Midland Basin was eventually filled with sediment. Line *A–B* shows the location of the cross section in Figure 15-29.

Reef growth While reefs grew upward around the Delaware Basin during Permian time, the Midland Basin to the east became filled with sediment (see Figure 15-31). Shallow seas then flooded the Delaware Basin, along with surrounding areas. Although by this time the Ancestral Rockies had been lowered by erosion, they still formed a large island to the northwest. The Delaware Basin lay very close to the Permian equator, and the Ouachita chain must have left the basin in the rain shadow of trade winds blowing from the east. The shallow seas that surrounded the basin were sites of carbonate and evaporite deposition in what was obviously an arid climate.

As time passed and sea level rose, the reef grew upward more rapidly than the Delaware Basin filled in, and eventually the reef stood high above a basin that was some 600 meters (2000 feet) deep (Figure 15-32*B*).

Although the waters have long since withdrawn, the structural basin remains today (see Figure 15-29). Early in its history, when the Delaware Basin was relatively shallow (Figure 15-32A), its floor was inhabited by snails, deposit-feeding bivalves, sponges, and brachiopods. Later, when the basin deepened, these animals decreased in variety; the only abundant marine fossil remains in the younger basin deposits are conodonts, radiolarians, and ammonoids, all of which lived high in the water column and sank to the bottom when they died. Also present are plant spores that were blown into the basin from the land. We can conclude that late in the basin's history, when the Capitan Limestone was forming (see Figure 15-32B), the floor of the deep basin was poorly oxygenated; oxygen used up in the decay of organic matter was not replenished, and few bottom-dwelling animals could survive.

The reeflike structure that rims the Delaware Basin was built during the Guadalupian Age, the

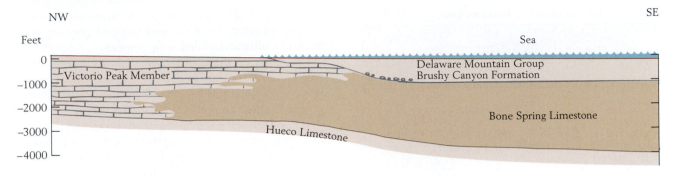

A   At close of lower Guadalupian time

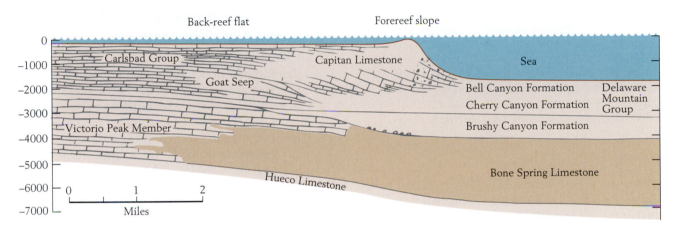

B   At close of upper Guadalupian time

Figure 15-32 Profiles of the reef that now forms the Guadalupe Mountains. *A.* A cross section early in Guadalupian time. *B.* A cross section later in Guadalupian time, when the reef had grown rapidly upward while the basin deepened. Note that the reef advanced toward the basin center as it grew upward. (After N. D. Newell et al., *The Permian Reef Complex of the Guadalupe Mountains Region, Texas and New Mexico*, W. H. Freeman and Company, New York, 1953.)

second-to-last age of the Permian. The reef was formed primarily by aragonitic sponges (see Figure 15-6), algae, and lacy bryozoans (see Figure 15-4). The crest of the reef was covered by shallow water that also bathed an extensive back-reef flat (see Figure 15-32B). Rubble from the reef periodically tumbled down the forereef slope into the basin. The ancient talus slope is present today in bedding that dips at angles as high as 40°. In the talus are many beautifully preserved fossils whose originally calcareous hard parts have been replaced by durable silica; among them are skeletons of sponges and fusulinid foraminifera that lived in shallow habitats, but were periodically washed down the slope to lodge in the talus. Some were swept into the basin by turbidity currents that left conspicuous graded beds in the rocks of the basin; these beds constitute the Delaware Mountain Group. Most

sediments of this unit are dark sands and silts that periodically washed into the basin, apparently when sea level was low and the reef surface was exposed to erosion.

When the older bedding surfaces of the carbonates that ring the Delaware Basin are traced laterally, a different configuration becomes apparent. These older bedding surfaces show that the early reefs, now constituting the Goat Seep Formation, stood in much lower relief above the basin (see Figure 15-32). From its earliest days until late in the Permian, the Delaware Basin was connected with the open sea to the southwest through what is called Hovey Channel (see Figure 15-31). Early in the evolution of the basin, when the reefs were low, this connection, and the resulting pattern of water circulation, permitted oxygen-rich waters to reach the basin floor so that animals could live there

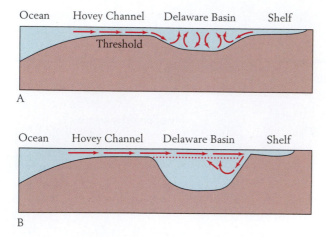

Figure 15-33 Patterns of water circulation in the Delaware Basin. *A.* Early in its history, the basin was shallow enough that well-oxygenated surface water reached its floor. *B.* Later, when the basin had deepened, good circulation was restricted to the upper waters, and the bottom waters became stagnant. (After N. D. Newell et al., *The Permian Reef Complex of the Guadalupe Mountains Region, Texas and New Mexico*, W. H. Freeman and Company, New York, 1953.)

(Figure 15-33*A*). Later the basin deepened, but Hovey Channel remained shallow, a conformation that caused the bottom waters of the basin to become stagnant and poor in oxygen, excluding almost all forms of life (Figure 15-33*B*).

Death of the reef Eventually, near the end of Permian time, the Delaware Basin filled with evaporites. As we have seen, the climate of Texas and neighboring areas became arid toward the end of the Permian, and the rate at which waters evaporated from the Delaware Basin may have increased. Hovey Channel must also have become so constricted that the rate of evaporation

there occasionally exceeded the rate at which new water was supplied. Thus the reef stopped growing, and the basin ultimately filled with evaporites; distinct layers in some of these evaporite deposits extend over many hundreds of square kilometers (see Figure 6-15). It is possible that this layering reflects seasonal changes similar to those responsible for glacial varves.

The Delaware Basin evaporites remained in place for a long period, protecting the magnificent geologic record along the walls and floor of the basin. Fresh water later dissolved the evaporites in many areas, exposing the ancient reef-encircled structure (see Figure 15-29).

The Sonoma orogeny expanded the North American continent

What was happening to the west of the Delaware Basin and the Ancestral Rocky Mountains? During late Paleozoic time the western margin of the North American craton passed through what is now northwestern Nevada (see Figure 15-22). Hundreds of kilometers from the margin of the craton, volcanoes were active again in the area that is now California and slightly to the north. Here coarse clastic deposits were shed from the volcanic island arc into the surrounding seas. An orogenic episode in Nevada in latest Permian and Early Triassic times, called the *Sonoma orogeny*, was remarkably similar to the Antler orogeny (see Figure 14-29). In the Sonoma orogeny, as in the Antler, marine deposits were thrust upward over the continental margin. The Sonoma orogeny was of great significance in that it entailed the complete closure of the basin between the volcanic island arc and the North American continent. While some of the deep-sea deposits of the basin were thrust onto the continent, others were welded onto the continental margin along with the volcanic terrane of the arc. The result was a considerable westward growth of the North American crust.

Chapter Summary

How did marine life of late Paleozoic time differ from that of middle Paleozoic time?

Marine life of late Paleozoic time in many ways resembled life of the middle Paleozoic, but the coral-strome reef community was gone. In addition, four groups that expanded enormously contributed large volumes of skeletal debris to limestones: first crinoids and lacy bryozoans, and later algae and fusulinid foraminifera.

How did terrestrial floras change on a global scale in late Paleozoic time?

In Carboniferous time the coal-swamp floras, which were dominated by trees of the genera *Lepidodendron*

and *Sigillaria*, played a major ecological role, as did seed ferns and, on drier land, sphenopsids and cordaites. During the Permian Period, however, climates in the Northern Hemisphere became warmer and drier, and conifers and other gymnosperms came to dominate landscapes.

How did coal beds that formed in Carboniferous swamps come to form part of depositional cycles?

Some of the swamps bordered meandering rivers, which produced depositional cycles. Others bordered shallow seas and migrated with transgressions and regressions of these seas.

How did continental glaciers come to blanket the south polar region of Gondwanaland in mid-Carboniferous time?

A weakening of greenhouse warming that resulted from burial of carbon in coastal swamps apparently initiated this glacial episode.

How did the late Paleozoic ice age affect life in the ocean?

The onset of this glacial interval caused a major extinction, and rates of extinction and origination in the ocean remained low until the glacial interval ended in Permian time.

What changes occurred in terrestrial faunas during late Paleozoic time?

Insects underwent a great evolutionary radiation during Carboniferous time. During the Permian Period, amphibians were displaced from terrestrial habitats by early pelycosaurs, but these soon gave way to the more advanced, mammal-like therapsids.

How was Pangaea formed late in Paleozoic time?

In mid-Carboniferous time Gondwanaland became sutured to Euramerica, causing widespread orogenies, including the Alleghenian in the Appalachian region. In Permian time, Siberia was sutured to eastern Europe, forming the Ural Mountains. Only what is now Southeast Asia then remained separate from the supercontinent.

Why was the Permian Period a time of hot, dry conditions and widespread evaporite deposition?

Because of its great size, Pangaea developed widespread aridity, and coal swamps shrank.

What major biotic changes occurred in the latter part of Permian time?

Two mass extinctions, separated by just a few million years, occurred at this time. The second of these events, at the very end of the Permian, was the largest mass extinction of animals that has ever occurred.

What uplifts and basins formed in North America late in Paleozoic time?

The collision of Euramerica with Gondwanaland not only elevated mountains in the Appalachian region but also created the Ouachita Mountains in Oklahoma and Texas. Uplifts and basins formed north and west of the Ouachitas. The Delaware Basin in western Texas became encircled by a reef complex in Permian time, and near the end of the Permian this basin was filled by evaporite deposits. Beginning in the latest Permian, the Sonoma orogeny resulted in continental accretion along the western margin of North America.

Review Questions

1. What caused shallow seas to expand and contract over the midcontinental United States many times during Late Carboniferous time?

2. If you could examine Late Carboniferous (Pennsylvanian) coal-bearing cycles in the field, how would you determine whether the coal deposits formed along a river or a shallow sea? (Hint: Refer to p. 380 and Figures 5-16 and 15-28.)

3. What groups of terrestrial plants that existed in late Paleozoic time survive today?

4. What justification is there for dividing the Carboniferous Period into the Mississippian and Pennsylvanian intervals?

5. How did the history of glacial activity in Late Carboniferous time relate to the deposition of coal?

6. In eastern North America, mountain building progressed from New England to Texas during Paleozoic time. How does this pattern relate to continental movements? (Review the relevant parts of Chapters 13 and 14, as well as the relevant part of this chapter.)

7. In what ways were therapsids adaptively superior to the amphibians and reptiles that preceded them?

8. Why were evaporite deposits widespread in Europe and North America during Late Permian time?

9. What apparently caused greenhouse warming of Earth to weaken during the Carboniferous and then strengthen again during the Permian?

10. What important changes occurred in the deep sea during Permian time, and how may these changes have related to major extinctions?

11. The formation of Pangaea during the Carboniferous Period was a major geologic event. Using the Visual Overview on page 356 and what you have learned in this chapter, describe how sea level, climates, and life changed late in the Paleozoic Era, after the formation of Pangaea. How did some of these changes relate to one another and to the existence of Pangaea?

Dinosaur tracks of Jurassic age in the Painted Desert in Arizona. (Tom Bean.)

CHAPTER 16

The Early Mesozoic Era

The Mesozoic Era, or "interval of middle life," began with the Triassic Period. The Triassic and the subsequent Jurassic Period together constitute slightly more than half of the era. Rocks representing these periods are especially well exposed and well studied in Europe.

Near the transition from the Paleozoic Era to the Mesozoic Era, the great supercontinent Pangaea took its final form, encompassing all the major segments of Earth's continental crust. Pangaea was so large that much of its terrain lay far from any ocean and, as a result, became arid. During Jurassic time, however, sea level rose and marine waters spread rapidly over the land, leaving an extensive record of shallow marine deposition. Later in early Mesozoic time, Pangaea began to fragment, and before the end of the Jurassic Period, Gondwanaland was once again separate from the northern landmasses.

Life of early Mesozoic time differed substantially from that of the Paleozoic Era. For many groups of animals, recovery from the Late Permian mass extinction was sluggish, but by the end of the Triassic Period, mollusks had reexpanded to become more diverse than they had been during the Paleozoic Era. Their success has continued to the present time. The marine ecosystem was also transformed during Triassic and Jurassic time by the addition of modern reef-building corals as well as large reptiles, which joined fishes as swimming predators. On land, gymnosperm plants, which had conquered much of the land during the Permian Period, continued to flourish, and flying reptiles and birds appeared as well. The most dramatic event in the terrestrial ecosystem was the emergence and diversification of the dinosaurs. Mammals arose slightly after dinosaurs in the Triassic Period, but they remained small and relatively inconspicuous throughout the Mesozoic Era.

The Triassic System is bounded by the terminal Permian extinction below and by another extinction above. It was the unique fauna of this system that led Friedrich August von Alberti to distinguish the Triassic in 1834. Alberti originally named the system the Trias for its natural division in Germany into three distinctive stratigraphic units.

At first the Jurassic System also had a shorter name, Jura, a label that was borrowed from a portion of the Alps in which the system is especially well exposed. The Jurassic was not formally established by a published proposal; instead, it gradually came to be accepted as a valid system during the first half of the nineteenth century, when its many distinctive marine fossils were widely investigated.

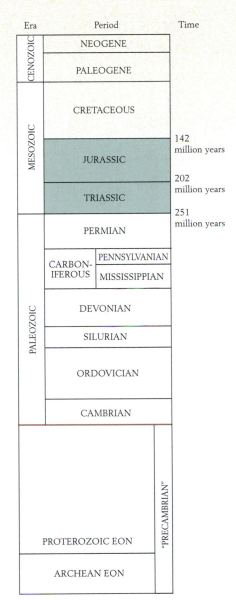

Era	Period	Time
CENOZOIC	NEOGENE	
CENOZOIC	PALEOGENE	
MESOZOIC	CRETACEOUS	
MESOZOIC	JURASSIC	142 million years
MESOZOIC	TRIASSIC	202 million years
PALEOZOIC	PERMIAN	251 million years
PALEOZOIC	CARBONIFEROUS — PENNSYLVANIAN	
PALEOZOIC	CARBONIFEROUS — MISSISSIPPIAN	
PALEOZOIC	DEVONIAN	
PALEOZOIC	SILURIAN	
PALEOZOIC	ORDOVICIAN	
PALEOZOIC	CAMBRIAN	
PROTEROZOIC EON	"PRECAMBRIAN"	
ARCHEAN EON	"PRECAMBRIAN"	

Life in the Oceans: A New Biota

By the end of the great extinction that brought the Paleozoic Era to a close, several previously diverse groups of marine life had vanished and others had

387

Major Events of the Early Mesozoic

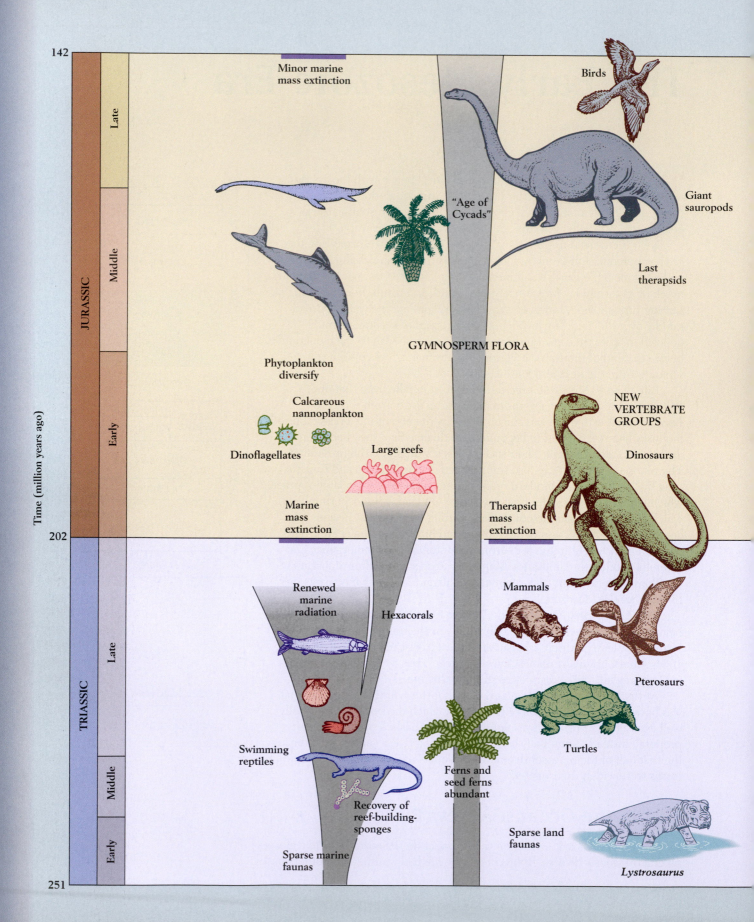

Time (million years ago)

142

JURASSIC

Late

Middle

Early

202

TRIASSIC

Late

Middle

Early

251

Minor marine mass extinction

Birds

Giant sauropods

"Age of Cycads"

Last therapsids

GYMNOSPERM FLORA

Phytoplankton diversify

Calcareous nannoplankton

Dinoflagellates

Large reefs

Marine mass extinction

Therapsid mass extinction

NEW VERTEBRATE GROUPS

Dinosaurs

Renewed marine radiation

Hexacorals

Mammals

Swimming reptiles

Pterosaurs

Turtles

Ferns and seed ferns abundant

Recovery of reef-building sponges

Sparse land faunas

Sparse marine faunas

Lystrosaurus

LATE CRETACEOUS

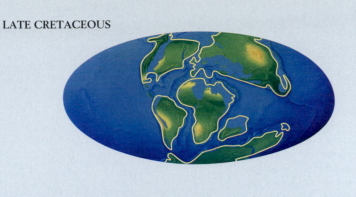

LATE JURASSIC

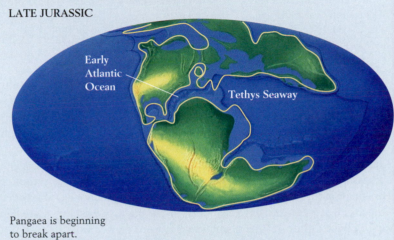

Early Atlantic Ocean

Tethys Seaway

Pangaea is beginning to break apart.

EARLY TRIASSIC

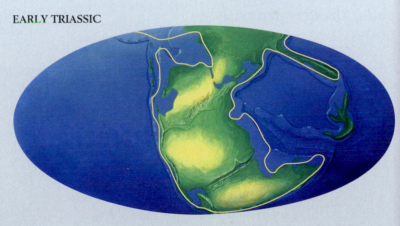

EARLY CARBONIFEROUS

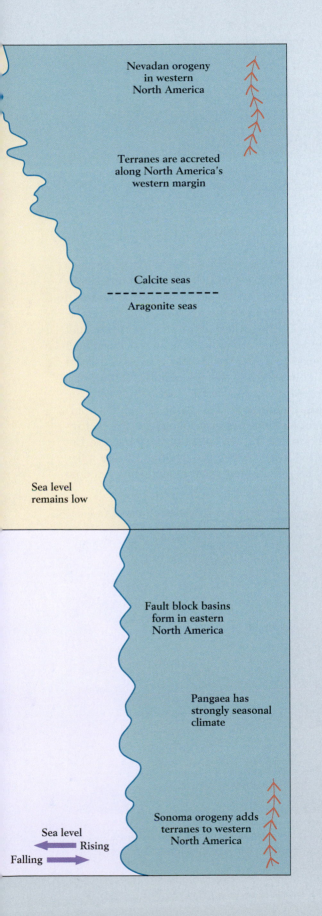

Nevadan orogeny in western North America

Terranes are accreted along North America's western margin

Calcite seas

- - - - - - - - - - - - - - -

Aragonite seas

Sea level remains low

Fault block basins form in eastern North America

Pangaea has strongly seasonal climate

Sonoma orogeny adds terranes to western North America

Sea level

← Rising

← Falling

Figure 16-1 Life of a Late Jurassic seafloor. As in Paleozoic time, many animals lay exposed on the seafloor, but some of them were of new types, such as the irregularly shaped oyster *Lopha*, which cemented itself to other shells, and the coiled oyster *Gryphaea*. Other new animals, including sea urchins such as *Nucleolites*, lived within the sediment. (After F. T. Fursich, *Palaeontology* 20:337–385, 1977.)

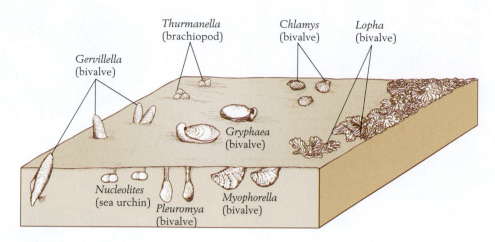

Gervillella (bivalve)
Thurmanella (brachiopod)
Chlamys (bivalve)
Lopha (bivalve)
Gryphaea (bivalve)
Nucleolites (sea urchin)
Pleuromya (bivalve)
Myophorella (bivalve)

become rare. Gone were fusulinid foraminifera, lacy bryozoans, rugose corals, and trilobites. The most common fossils in Lower Triassic rocks are mollusks. The ammonoids made a dramatic recovery after almost total annihilation: although only two ammonoid genera are known to have survived the Permian crisis, Lower Triassic rocks have yielded more than a hundred genera of ammonoids. The adaptive radiation that produced these genera seems to have issued from the single genus *Ophiceras*. Other groups of marine life were slower to recover, but by Late Triassic time the seas once again teemed with a variety of animals.

Benthic life recovered after the Permian crisis

One remarkable consequence of the loss of marine life at the end of the Permian was a return of stromatolites to shallow subtidal environments in many parts of the world. These cyanobacterial structures flourished briefly in the Early Triassic, just as they had done for a longer interval before the evolution of animals that ate and burrowed through them (p. 306). Bivalve mollusks, like ammonoids, are frequently found in Lower Triassic rocks, although their diversity is somewhat limited. Both bivalves and gastropods expanded in number and in variety, however, to become prominent groups of early Mesozoic marine animals. As in the Paleozoic Era, some of the bivalves burrowed in the seafloor, while others rested on the sediment surface (Figure 16-1).

In addition to ammonoids and bivalves, only brachiopods make a modest showing in Lower Triassic rocks; all other marine invertebrates are rare. The brachiopods also diversified during the Triassic and Jurassic periods, but they subsequently declined, and today they are relatively rare.

Sea urchins, which had existed in limited variety during the Paleozoic Era, diversified greatly during the first half of the Mesozoic Era. Some of the new forms that emerged at this time were surface dwellers, like most of the Paleozoic sea urchins, but others lived within the sediment as actively burrowing deposit feeders (see Figure 16-1).

Reefs did not recover from the terminal Permian extinction until Middle Triassic time, when vertical mixing in the oceans increased, as shown by the return of oxygenated conditions to the deep sea (see Earth System Shift 15-2). The earliest of the new reefs were built by sponges and algae resembling the ones that had built Late Permian reefs. As the Triassic progressed, however, the dominant role in reef building shifted to a newly evolved group that is still successful today: the **hexacorals** (Figure 16-2). This group includes not only colonial reef builders, but also solitary species that resemble the solitary rugose corals of the Paleozoic Era (see Figure 13-12). During the Middle Triassic a few species of hexacorals built small mounds that stood no

Figure 16-2 Triassic hexacorals. The fragments of colonies shown here are 3 to 4 centimeters (1.5 inches) across. (George Stanley.)

more than 3 meters (10 feet) above the seafloor. By latest Triassic time, reefs were much larger, some having been constructed by more than 20 species.

Some of the early coral mounds grew in relatively deep waters, so it appears that the earliest hexacorals, unlike the corals that form large tropical reefs today, did not live in association with symbiotic algae, which require strong sunlight (see p. 99). Perhaps it was not until latest Triassic or Early Jurassic time, when hexacorals began to form large reefs, that this symbiotic relationship was established.

Because of the success of bivalve and gastropod mollusks, sea urchins, and reef-building hexacorals, by Late Jurassic time seafloor life looked more like it does today than it had in the Paleozoic Era (see Figure 16-1). Still missing were many kinds of modern arthropods, but the group that includes crabs and lobsters got off to a modest evolutionary start during the Jurassic Period.

Pelagic life included new groups of phytoplankton and numerous swimming predators

Presumably many kinds of planktonic organisms in Triassic and Jurassic seas left no fossil record. The **dinoflagellates**, however, produced many fossils in the form of *cysts*, which were durable resting stages that these algae formed when the environment became inhospitable. Dinoflagellates underwent extensive diversification during mid-Jurassic time and remain an important group of producers in modern seas (see Figure 3-16A). The **calcareous nannoplankton**, another important group of living algae (see Figure 3-16C), made their first appearance in earliest Jurassic time. Today these tiny, spherical floating forms are most diverse in warm seas (see Figure 4-27), and their armor plates rain

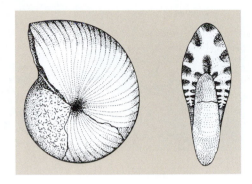

Figure 16-3 The Jurassic ammonoid *Phylloceras*. The suture pattern of this genus is shown below the shell. (The suture is the juncture between the convoluted internal partitions, or septa, and the coiled outer shell.)

down on the seafloor to become prominent constituents of deep-sea sediments (see p. 125).

Higher in the food web, the ammonoids and belemnoids played major roles as swimming predators. The ammonoids' evolutionary recovery after the Permian crisis led to great success throughout the Mesozoic Era. Individual ammonoid species, however, survived for relatively brief intervals—often a million years or less—so they are extremely useful as index fossils for Mesozoic rocks (Figure 16-3). The **belemnoids**, which were squidlike relatives of the ammonoids, also pursued prey by jet propulsion (Figure 16-4). They evolved in late Paleozoic time but remained inconspicuous until the

A

B

Figure 16-4 Belemnoids were squidlike cephalopod mollusks that were related to ammonoids but lacked external shells. *A.* Belemnoids as depicted in a museum reconstruction. Like ammonoids, they were predators that swam by jet propulsion. *B.* The most commonly preserved part

of a belemnoid is the cigar-shaped counterweight. This heavy structure, shown here on a rock surface displaying numerous ammonoids, acted to offset the buoyant effect of gas within the shell, thereby maintaining balance. (*A*, Field Museum of Natural History, Chicago, Neg. GEO80826; *B*, Chip Clark.)

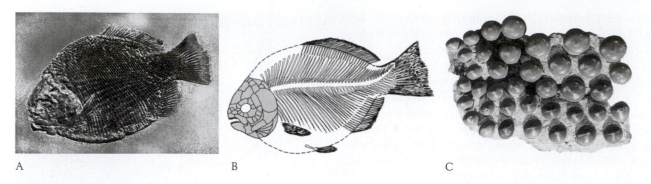

A   B   C

Figure 16-5 The Jurassic fish *Dapedius*. *A.* Unlike modern fishes, this early form had scales that barely overlapped one another. *B.* It also had asymmetrical skeletal supports within its tail (note the upturned spinal column). *C.* This particular genus had knobby teeth for crushing shellfish. (*A* and *C*, Museum Hauff, Holzmaden, Germany.

Mesozoic Era, when many types evolved. Conodonts, the toothlike structures that are now known to have belonged to small fishlike vertebrates (see Figure 3-35), have also proved useful in the correlation of Triassic rocks, but by Jurassic time, conodont-bearing animals no longer existed.

Paleozoic ray-finned bony fishes gave rise to forms that were successful in early Mesozoic time but were still more primitive than most of their modern-day descendants. The scales that covered the bodies of these fishes were diamond-shaped structures that overlapped slightly or not at all (Figure 16-5), in sharp contrast to the circular, strongly overlapping scales of nearly all modern bony fishes. Presumably the primitive diamond-shaped scales were less protective against infections and parasites than the modern kind. Other features that distinguished early Mesozoic bony fishes from their modern counterparts were skeletons that consisted partly of cartilage rather than entirely of bone; relatively simple, primitive jaws; and highly asymmetrical tails that resembled those of Paleozoic bony fishes. Some early Mesozoic bony fishes had teeth shaped like rounded pegs that served to crush durable items of food—probably small shellfish (see Figure 16-5C). Bony fishes underwent many changes during the Mesozoic Era, and few species with these primitive traits survived. One especially useful feature that developed during this time was the *swim bladder*, a sac of gas that allows advanced fishes to regulate their buoyancy. The swim bladder evolved from the lung, which was present in some primitive fishes.

Sharks were also well represented in early Mesozoic seas. Some had teeth adapted for crushing shellfish, like those of the bony fish shown in Figure 16-5C. Some modern groups of sharks appeared during the Jurassic Period, among them the family that includes the modern tiger shark.

Many reptiles that resembled the popular conception of sea monsters emerged in early Mesozoic seas.

Among them were the **placodonts**, which, like many early Mesozoic fishes, were blunt-toothed shell crushers (Figure 16-6). The placodonts' broad, armored bodies gave them the appearance of enormous turtles. Cousins of the placodonts were the **nothosaurs** (Figure 16-7), which have been found in Early Triassic deposits and seem to have been the first reptiles to invade the marine realm. Nothosaurs had paddlelike limbs resembling those of modern seals. It seems likely that, like

A

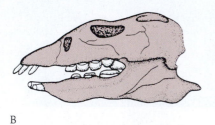

B

Figure 16-6 Reconstruction of a Triassic placodont. This aquatic reptile (*A*) was about 1.5 meters (5 feet) long. It used its large, rounded teeth (*B*) to crush shelled marine invertebrates of the seafloor. (*A*, A. Dan Varner.)

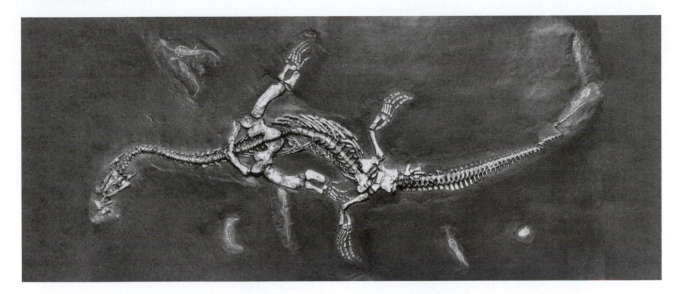

Figure 16-7 The nothosaur *Ceresiosaurus*. This specimen, which was about 2.2 meters (7 feet) long, was preserved with small nothosaurs of a different family in the Middle Triassic Muschelkalk of Germany. Like modern seals, these animals probably fished along the shore. (Museum für Geologie und Paläontologie, Tübingen, Germany.

seals, they were not fully marine, but lived along the seashore and periodically plunged into the water to feed on fish.

Placodonts and nothosaurs did not survive the Triassic Period. The more fully aquatic **plesiosaurs** evolved from the nothosaurs in mid-Triassic time, however, and played an important ecological role for the remainder of the Mesozoic Era. Plesiosaurs apparently fed on fish, and in Cretaceous time, they attained the proportions of modern predatory whales, reaching some 12 meters (40 feet) in length. The limbs of plesiosaurs were winglike paddles that propelled them through the water in much the same way that birds fly through the air (Figure 16-8).

By far the most fishlike reptiles of Mesozoic seas were the **ichthyosaurs,** or "fish-lizards," many of which must have been top predators in marine food webs (Figure 16-9). Superficially, ichthyosaurs bear a closer resemblance to modern dolphins, which are marine mammals, than to fishes; outlines of skin preserved in black shales under low-oxygen conditions show the dolphinlike profiles of some ichthyosaurs. The ichthyosaurs had upright tail fins, however, in contrast to the horizontal pair of rear flukes that propels dolphins through

Figure 16-8 Late Jurassic plesiosaurs from England mounted in swimming position. Note the paddlelike limbs. These two animals illustrate the two body types of plesiosaurs. *Cryptoclidus*, above, has a long neck and a short head, whereas *Peloneustes*, below, has a short neck and a long head. *Cryptoclidus* is about 3 meters (10 feet) long. (Museum für Geologie und Paläontologie, Tübingen, Germany.)

Figure 16-9 An ichthyosaur that died in the act of giving birth. The infant's head apparently stuck in the mother's birth canal, and both animals died. These individuals were preserved in a Jurassic deposit in Germany. The mother was about 2 meters (6 feet) long. (Staatliches Museum für Naturkunde.)

the water. The extension of the backbone into the ichthyosaur tail bent downward, in contrast to the upward curve that characterized early Mesozoic bony fishes (see Figure 16-5). Large eyes supplemented other adaptations for rapid swimming in pursuit of prey. Ichthyosaurs were fully marine and thus could not easily have laid eggs; instead, they bore live young. In fact, skeletons of ichthyosaur embryos have been found within the skeletons of adult females or in the process of being born (see Figure 16-9).

Surprising as it may seem, the last important group of early Mesozoic marine reptiles to evolve were the early **crocodiles**, which, as we shall see, were related to the dinosaurs. Although crocodiles evolved in Triassic time as terrestrial animals, some were adapted to the marine environment by earliest Jurassic time. In fact, some crocodiles became formidable marine predators whose finlike tails were well adapted for rapid swimming.

Life on Land

The presence of dinosaurs during the Mesozoic Era gave the biotas of large continents an entirely new character, but Mesozoic land plants were also distinctive. Because these plants were positioned at the base of the terrestrial food web, we will review them first.

Gymnosperms dominated the Mesozoic flora

Major groups of land plants survived the great mass extinction that brought the Paleozoic Era to an end. As we learned in Chapter 15, the late Paleozoic floras began to decline long before the end of the Permian Pe-

riod. In effect, the transition from the late Paleozoic kind of flora to the Mesozoic kind of flora began before the start of the Mesozoic Era.

Among the groups that decreased in diversity long before the end of Permian time were the lycopod trees, which formed coal swamps, and the sphenopsid and cordaite trees, which inhabited higher ground. Persisting into the Mesozoic Era in greater numbers were ferns and seed ferns. Seed ferns, however, were reduced in abundance and apparently failed to survive into Jurassic time. Ferns, of course, survived and, in fact, were much more diverse and abundant during the Triassic than they are today.

Most of the trees that stood above Triassic ferns belonged to three groups of gymnosperms that had originated during the Permian Period, all of which have surviving members today. The most diverse of these three groups was the one comprising the **cycads** and **cycadeoids**. Cycads are tropical trees that superficially resemble palms but are relatively rare in the modern world (see Figures 3-4 and 4-19). Cycadeoids, which are similar in form and closely related to cycads, are extinct; their trunks are well known as early Mesozoic fossils. Second to the cycads and cycadeoids in diversity were the conifers. With the possible exception of the pine family, all of the modern conifer families were present in early Mesozoic time. Ginkgos constituted the third group of tree-forming gymnosperms in Triassic time (Figure 16-10). The single living species of ginkgo looks more like a hardwood tree (that is, like an oak or a maple) than a conifer, and, like hardwoods, it sheds its leaves seasonally. This surviving species of ginkgo is a true living fossil whose record extends back some 60 million years to the Paleocene Epoch (see Figure 16-10A), early in the Cenozoic Era.

A

B

Figure 16-10 The ginkgo is a living fossil. *A.* Leaves of the living ginkgo, *Ginkgo biloba.* *B.* Similar leaves of Jurassic age. (*A*, Dave Watts/NHPA; *B*, Martin Land/Science Photo Library/ Photo Researchers.)

These three tree-forming groups are united as gymnosperms because they were all characterized by exposed seeds. The seeds of pines and other modern

Figure 16-11 Reconstruction of a Mesozoic landscape. Cycads, cycadeoids, and ferns appear in the foreground, and conifers are visible along the horizon. (Drawing by Z. Burian under the supervision of Professor J. Augusta.)

conifers, for example, rest on the projecting scales of their cones. There is a reason for this configuration: whereas flowering plants, which did not evolve until Cretaceous time, can attract insect and bird pollinators, most gymnosperms rely primarily on wind to carry their pollen from tree to tree.

Cycads, cycadeoids, conifers, and ginkgos formed the forests of the Jurassic Period, but the cycads were so dominant that the Jurassic interval has been called the Age of Cycads. Both Triassic and Jurassic landscapes, however, would have looked more familiar to us than Paleozoic landscapes, largely because of the presence of conifers that closely resembled modern evergreens (see Figure 15-11). Even so, the absence of flowering plants such as grasses and hardwood trees would have made Mesozoic floras appear monotonous to a modern observer (Figure 16-11).

The Age of Dinosaurs began

The terminal Permian mass extinction devastated therapsid faunas on land. Remaining at the start of Triassic time were only a few predatory genera and the bulky herbivore *Lystrosaurus* (see Figure 8-10).

Early mammals Although therapsids rediversified during the Triassic Period to play a dominant ecological role once again, they barely survived into the Jurassic Period. Nonetheless, they left an important legacy in the form of the mammals, which evolved from them near the end of Triassic time. Mammals remained small and peripheral throughout the Mesozoic Era; apparently no species grew larger than a house cat (Figure 16-12). Their problem was that the **dinosaurs** evolved slightly earlier in Triassic time and thus got off to an evolutionary head start.

Figure 16-12 Early dinosauromorphs of the Triassic genus *Lagosuchus* intimidating a smaller mammal. *Lagosuchus,* which was about 30 centimeters (1 foot) tall, closely resembled the earliest dinosaurs. Dinosauromorphs were the ancestors of dinosaurs. (Drawing by Gregory S. Paul.)

Dinosaur origins The **dinosaurs** (formally termed Dinosauria) were members of the Dinosauromorpha, a group that includes the birds, which evolved from dinosaurs (see Figure 3-10*B*). The dinosaurs inherited their advanced locomotory ability from early dinosauromorphs, which evolved early in the Triassic Period (see Figure 16-12). Early dinosauromorphs probably spent much of their time standing or walking on all fours, but some were adapted to rise up and run on two legs in the fashion of ostriches and other flightless birds. The upper portion of the legs of many early dinosauromorphs extended straight downward beneath their bodies rather than sprawling slightly out to the side, as they did in therapsids. This feature, which facilitated running, was passed on to the dinosaurs and seems to have been a key to their success.

The first dinosaurs resembled early dinosauromorphs that sometimes traveled on their two hind legs, but the dinosaurs' skulls were differently formed and their teeth were more highly developed. Dinosaurs did not become gigantic until Jurassic time, but some Triassic species did reach lengths of more than 6 meters (20 feet). Figure 16-13 displays a small species along with early dinosauromorphs that stood on all fours and

Figure 16-13 Terrestrial life of Late Triassic time in Argentina. The plants are of the widespread seed fern genus *Dicroidium*. The largest animals depicted here are dinosauromorphs of the genus *Saurosuchus*, which were about 7 meters (25 feet) long. Confronting them are three small, primitive dinosaurs of the genus *Herrerasaurus*. The dead animal is the rynchosaurian reptile *Scaphonyx*. Two small dinosauromorphs of the genus *Ornithosuchus* are scampering off in the foreground. In the left foreground is the long-legged primitive crocodile *Trialestes*. (Drawing by Gregory S. Paul.)

a primitive long-legged crocodile. The crocodiles, like the dinosaurs, evolved from early dinosauromorphs late in the Triassic Period.

The dinosaurs' rise to dominance By Late Triassic time, therapsids lived alongside increasing numbers of dinosaurs, other still-diverse dinosauromorphs, and smaller amphibians and reptiles. A few kinds of large amphibians persisted as well. The dinosaurs did not rise to dominance, however, until a mass extinction at the end of the Triassic Period all but eliminated the therapsids. The known fossil record of Early Jurassic time is too poor to permit us to piece together the details of the dinosaurs' dramatic rise to dominance. Nonetheless, fossil remains of huge dinosaurs even in Lower Jurassic rocks indicate that these animals evolved rapidly; the oldest dinosaur giants are found in Australia.

Dinosaurs fall into two groups, which are characterized by different pelvic structures (Figure 16-14). All of the "bird-hipped" (ornithischian) dinosaurs were herbivores, whereas the "lizard-hipped" (saurischian) group included both herbivores and carnivores. Some species in each group traveled on two legs and others on four. The largest of all dinosaurs were the **sauropods**, lizard-hipped herbivores that moved about on all fours. Important aspects of dinosaur biology are outlined in Earth System Shift 16-1.

By Late Jurassic time, both bird-hipped and lizard-hipped dinosaurs were quite diverse. The most spectacular Jurassic assemblage of fossil dinosaurs in the world is found in the Upper Jurassic Morrison Formation, which extends from Montana to New Mexico. At Como Bluff, Wyoming, dinosaur bones were so common in the nineteenth century that a local sheepherder constructed a cabin of them because they were the most readily available building material. The Morrison dinosaurs, which include more than a dozen genera, are representative of the kinds of dinosaurs that lived

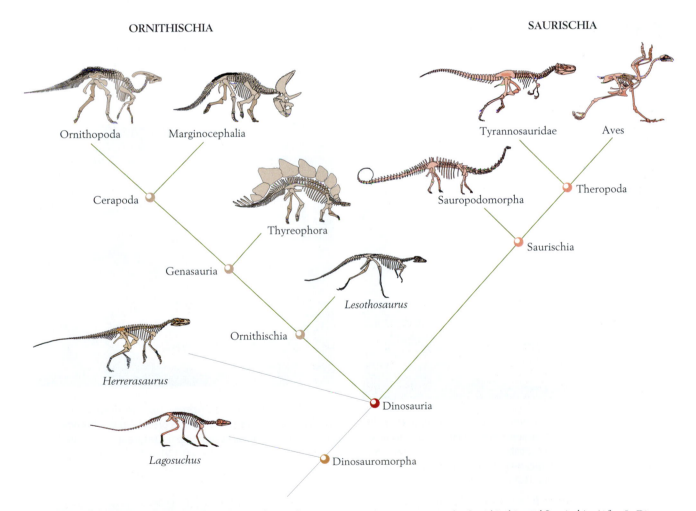

ORNITHISCHIA

Ornithopoda   Marginocephalia

Cerapoda

Thyreophora

Genasauria

Lesothosaurus

Ornithischia

Herrerasaurus

Lagosuchus

Dinosauria

Dinosauromorpha

SAURISCHIA

Tyrannosauridae   Aves

Sauropodomorpha   Theropoda

Saurischia

Figure 16-14 **The phylogenetic relationships of dinosaurs.** *Lagosuchus* (bottom) was an early dinosauromorph. The early dinosaur genus *Herrerasaurus* (reconstructed in Figure 16-13) was distinct from the two more advanced dinosaur groups, the Ornithischia and Saurischia. (After L. Dingus and T. Rowe, *The Mistaken Extinction: Dinosaur Evolution and the Origin of Birds*, W. H. Freeman and Company, New York, 1997.)

Earth System Shift 16-1 | The Rise of the Dinosaurs

After the terminal Triassic mass extinction, which all but eliminated the therapsids, dinosaurs diversified to rule the land for about 140 million years. There are widespread misconceptions about dinosaurs. They have often been portrayed as hulking creatures that lumbered about on widely separated feet and legs that bent outward from their bodies. The truth is that not all dinosaurs were of massive proportions; many were less than 1 meter (3 feet) long. The orientations of dinosaur limbs in their sockets indicate that the legs were positioned almost vertically beneath the body, and fossil dinosaur tracks confirm that this posture was typical. The left and right tracks are nearly in line, signifying that both feet were positioned beneath the body. In fact, some dinosaurs were as agile as ostriches, which are famous for their great speed.

Another misconception is that dinosaurs were reptiles. Although they were descendants of reptiles, they were actually quite different from reptiles, and in fact, were in many ways more similar to birds, which evolved from them.

Dinosaurs laid eggs and, like many large bird species, clustered them in nests on the ground. Many dinosaur species laid eggs that were more pointed at one end than the other, like those of a chicken but typically more elongate. The pointed end was sometimes thrust downward into the soil. Other species laid eggs that were spherical or cylindrical with rounded ends. Some species laid their eggs in disarray, but many arranged them in rows, circles, or spiral patterns.

Figure 1 A remarkably well preserved specimen of the small dinosaur *Protarchaeopteryx*. Filaments that form a fringe along the back are probably precursors of feathers. This animal was about 60 centimeters (2 feet) long. (Courtesy of the Peabody Museum of Natural History, Yale University 2004.)

Figure 2 Bones of *Oviraptor*. This specimen, whose name ironically means "egg stealer," was preserved in a brooding posture on top of her eggs in Upper Cretaceous strata of China. The skeleton is about 0.7 meters (28 inches) across. (Mick Ellison, Department of Vertebrate Paleontology, American Museum of Natural History.)

Some species that positioned their eggs in an orderly way deposited them in two or more layers beneath the surface of the ground. Extensive fieldwork has uncovered numerous egg-bearing dinosaur nests in Late Cretaceous strata of northwestern Montana. Some of these nests are bunched on two hills, which retain the topography that made them islands in a large lake when dinosaurs occupied them. These dinosaurs were isolating their offspring from predators, like populations of large water birds that nest on islands today.

Many dinosaurs were social animals, traveling in herds. Testifying to this behavior are fossil trackways produced by many individuals of a single species traveling together. In addition, members of some species communicated with one another. Duck-billed dinosaurs, for example, had a tall, crested skull, which functioned like the resonating chambers of a trumpet. The unique shape of this structure in each species allowed its members to call out to one another over long distances. A Late Cretaceous death assemblage in Montana provides additional evidence of social behavior. Sampling of this assemblage, which is spread over nearly a square kilometer (more than a quarter of a square mile), indicates that it includes the remains of about 10,000 dinosaurs be-

Figure 3 The skull crest of *Parasauralophus*, which was about 1 meter (3 feet) long, probably served as a resonating chamber for trumpeting. (Royal Tyrell Museum of Paleontology/Alberta Culture and Multiculturalism.)

longing to a single plant-eating species. Lying above the bone bed is a layer of volcanic ash, which was apparently spewed from a volcano that first emitted toxic gas and dust that killed an enormous herd of dinosaurs. This scene of carnage paints a more uplifting picture when we bring the victims back to life and imagine how they and other dinosaur populations must have swarmed over western North America in vast herds during Cretaceous time, much as the American bison did before human hunting devastated its populations.

For many years, it was debated whether dinosaurs were endothermic (warm-blooded, like birds and mammals). A large variety of evidence indicates that this was indeed the case:

- Ectothermic ("cold-blooded") animals have little endurance because their low rate of metabolism provides relatively little energy. A mammal has a high rate of metabolism that not only elevates the temperature of its body above that of its environment, but also allows it to move about at a fast pace for an extended period of time. In contrast, a reptile can run only a short distance before running out of energy and having to soak up energy from the sun. Although dinosaurs had an evolutionary head start on mammals, it is unlikely that they could have suppressed mammals ecologically for 140 million years if they had not matched them in rate of metabolism—if they had not possessed the endurance to run them down.

- Additional evidence derives from the fact that in communities of dinosaurs, carnivores usually made up less than 10 percent of the total biomass, as is the case in living and fossil mammal communities. The reason that carnivores are relatively rare in mammal communities is that mammals need a great deal of food to maintain their endothermic metabolism; one might say that mammalian carnivores need to live among large numbers of herbivores in order to capture enough prey to stoke their metabolic furnaces. In contrast, predators commonly represent 40 percent or so of the volume of living tissue in communities of ectothermic animals. Having low rates of metabolism, they need relatively little food, and many can sustain themselves on small populations of prey animals. The low percentage of predators in many communities

(continued)

of dinosaurs argues that these creatures, like mammals, were endothermic.

- Fossil trackways indicate that dinosaurs were too active in everyday life to have been ectothermic. Unless old or infirm, dogs and other relatively large mammals normally trot, rather than walking, when on the move. They can maintain this level of activity only because of their endothermic metabolism. Reptiles, in contrast, occasionally make short dashes, but ordinarily move around at a walk. They lack the energy to maintain a trot as their standard gait. Fossil trackways provide a sample of the average pace of extinct animals. The stride length of a moving animal increases with its speed (a sprinter takes long strides, for example, whereas a jogger takes short steps). A mammal, trotting around as it normally does, leaves tracks that are relatively far apart for the length of its legs. In contrast, a reptile, walking around as it normally does, leaves tracks that are relatively close together. The foot length of any kind of animal, which can be measured from a track, tells us how long the animal's legs were. Measurement of numerous fossil tracks and trackways indicates that dinosaurs, like mammals, habitually took long strides for their foot and leg length. Thus, like mammals, they moved about rapidly in everyday life. This high level of activity would have been impossible had dinosaurs not been endothermic.

- In Upper Cretaceous rocks of Mongolia, paleontologists have found a parent dinosaur preserved in the act of brooding her eggs, like a bird. This behavior of keeping eggs warm is strong evidence of endothermy.

- Nests of baby dinosaurs found in Upper Cretaceous strata of Montana—clusters of juvenile skeletons with broken eggshells in a depression—also show

Figure 4 This *Maiosaura* hatchling, about 50 centimeters (20 inches) long, was found in a nest. (Museum of the Rockies, Montana State University, Bozeman.)

Figure 5 Fossilized tail feathers of *Protarchaeopteryx* from Chinese rocks close to the Jurassic-Cretaceous boundary. (Mick Ellison, Department of Vertebrate Paleontology, American Museum of Natural History.)

They almost certainly served to insulate an endothermic animal that would otherwise have lost too much body heat to survive because its small size endowed it with a relatively large surface area for its body volume.

- Oxygen isotopes indicate a relatively uniform body temperature for dinosaurs. Isotope ratios in bone reflect the temperature at which the bone was secreted. The extremities of a reptile are warmer in summer than in winter—a condition reflected in variable oxygen isotope ratios in leg and foot bones. For a mammal, these ratios are similar throughout the body. Dinosaur skeletons generally resemble mammals in this regard, although a small percentage of species display enough variability to suggest that they were not fully endothermic.

One might ask how the largest dinosaurs, which had relatively small heads and jaws, were able to chew up enough food to maintain their high rates of metabolism. The answer is that these giant herbivores used their mouths and jaws only for gathering and swallowing plant food. In the animals' intestinal tracts were "gizzard stones," like those of birds but much larger, that ground up coarse food after it was swallowed.

The word "dinosaur" has come to serve as a label for anything that is badly outmoded, but this designation is unfair. Dinosaurs were quite advanced animals, even by modern, mammalian standards. For all we know, dinosaurs would fare well in the modern world if they could somehow be resurrected and released into it to confront mammals. As we shall see in the next chapter, the sudden extinction of the dinosaurs at the end of the Cretaceous Period was an accident, not a result of biological inferiority, and it was only through their disappearance that the mammals were able to diversify markedly and evolve large body sizes.

that dinosaurs cared for their young. In addition, these nests show that dinosaurs grew rapidly. Having hatched at a length of just a few inches, the young dinosaurs reached about 1.5 meters (5 feet) in length before leaving the nest at the end of the warm season, perhaps only 3 or 4 months after hatching. The implied rate of growth is much higher than that of sizable ectothermic animals such as crocodiles or large tortoises.

- Feathers have been preserved with small dinosaur skeletons in Mongolia. Some of these feathers were small structures that would not have aided in flight.

Figure 16-15 The dinosaur fauna of the Morrison Formation. *Camarasaurus* walks toward the right in the background, and *Diplodocus* is drinking water (both are sauropods). At the far right is *Stegosaurus* (a stegosaurian). Reclining on the left is *Allosaurus*, a large carnivorous dinosaur that reached about 12 meters (40 feet) in length. In the foreground are dragonflies, pterosaurs, turtles, and a crocodile. The trees are conifers. (Drawing by Gregory S. Paul.)

throughout the world during Late Jurassic time. Several of the common Morrison species are shown in Figure 16-15. A skeleton of *Allosaurus*, the large carnivore in this reconstructed scene, is seen in Figure 16-16.

Frogs and turtles Two groups of small vertebrates that remain successful in the modern world also became established in Triassic time. One was the frogs, which were then as now amphibians of small body size. The oldest known fossil displaying the form of a modern frog is of earliest Jurassic age, but froglike skeletons have also been found in Triassic rocks. The other modern group was the turtles, although the earliest of these reptiles lacked the ability to pull their heads and tails fully into their protective shells.

Creatures that took to the air Late in the Triassic Period, vertebrate animals invaded the air for the first time as the **pterosaurs** came into being (Figure 16-17). These animals had long wings and hollow bones that

▶ **Figure 16-16** *Allosaurus*, a huge carnivorous dinosaur of Jurassic age that roamed the American West. This specimen was about 12 meters (40 feet) long, and its skull was nearly a meter (3 feet) in length. (Chip Clark.)

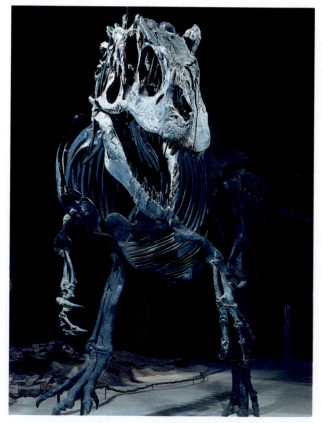

served to facilitate flight. Some species had long tails as well. The structure of pterosaur skeletons reveals their capacity for flight, but the great length of their wings suggests that most species flapped their wings primarily when taking off and then, once airborne, soared on air currents without much flapping. The behavior of pterosaurs when not in flight has been widely debated. Most species appear to have been able to walk and also to climb adeptly with the aid of hooklike claws.

The oldest known fossil birds are of Late Jurassic age. The first clue to their existence was a feather discovered in 1861 in the fine-grained Solnhofen Limestone of Germany, followed a few months later by the discovery of an entire skeleton of the species to which the feather belonged (Figure 16-18). This feathered animal was given the name *Archaeopteryx*, which means "ancient wing." *Archaeopteryx* had a skeleton so much like that of a dinosaur that it would be regarded as one

▶ **Figure 16-17** The pterosaur, or "flying lizard," *Pterodactylus*. This skeleton was preserved intact in the Upper Jurassic Solnhofen Limestone of Germany; it measures about 60 centimeters (2 feet) long. Many of the bones are hollow, like those of a bird. (Franz Hoeck, Bayerische Staatssammlung für Paläontologie und Geologie Munchen.)

A   B

Figure 16-18 Fossil remains of *Archaeopteryx lithographica*, the oldest known bird, from the Solnhofen Limestone. *A.* The existence of a bird during Solnhofen deposition was first indicated by the discovery of a feather. The asymmetry of the feather suggests that it aided in flight; flightless living birds have feathers that are symmetrical about the central shaft. *B.* The bird itself was soon found, and impressions of long feathers are clearly visible around it in the fine-grained limestone. This animal was about the size of a crow. Despite the feathers, *Archaeopteryx* had a skeleton and teeth similar to those of dinosaurs. (*A*, J. H. Ostrom, Peabody Museum of Natural History, Yale; *B*, Walther-Arndt-Fonds, Fordererkreis der naturwissenschaftlichen Museen Berlins e.V.)

were it not for its birdlike plumage. This animal is a classic missing link—in this case, the link between birds and their flightless ancestors. The teeth, large tail, and clawed forelimbs of *Archaeopteryx*, which are absent from advanced birds, reflect its dinosaur ancestry. *Archaeopteryx* lacks a breastbone and so is assumed to have possessed weak flying muscles. It was probably a clumsy flier by the standards of modern birds. Unfortunately, the hollow bones of birds are fragile, and no other bird bones have been found within the Jurassic System.

The Paleogeography of the Early Mesozoic Era

At the start of the Mesozoic Era, all of the major landmasses of the world were united as the supercontinent Pangaea (Figure 16-19). Near the end of Triassic time, Pangaea began to break apart, but continental movement is so slow that even by the end of Jurassic time, the newly forming continental fragments were barely separated. Thus, throughout the early Mesozoic Era, Earth's continental crust was concentrated on one side of the globe. Sea level rose slightly at the start of Triassic time (see p. 389). As in Late Permian time, however, most landmasses stood above sea level, forming one vast continent. At the start of the Triassic, the Tethys Seaway was an embayment of the deep sea projecting into the portion of equatorial Pangaea that today constitutes the Mediterranean. Later in Triassic and Jurassic time, rifting extended the Tethys Seaway between Eurasia and Africa and all the way westward between North and South America to the Pacific.

The size of Pangaea affected climates and distributions of organisms

Although the dominant land plants of the Triassic Period differed from those of the Permian, the distributional pattern of floras on Pangaea remained much the same: a Gondwanaland flora existed in the south and a Siberian flora in the north (see Figure 16-19). The Euramerican flora grew under warmer, drier conditions at low latitudes; in fact, unusually extensive deposition of evaporites attests to the presence of arid climates far from the equator. This condition resulted in part from the sheer size of Pangaea, which left many land areas far from oceans.

Pangaea also had strongly seasonal climates: because of its great size, the ocean only weakly affected its temperature, and the low heat capacity of land therefore caused it to become very hot in summer and cold in winter (p. 92). When summer came to the northern half of Pangaea, hot air rising from the land must have drawn strong monsoonal winds from the southern half (p. 93); similarly, powerful winds must have flowed southward across the equator when it was summer in the Southern Hemisphere.

The survival of only a small percentage of Permian species into Triassic time produced some striking biogeographic distributions. The scalloplike bivalve genus *Claraia*, for example, occupied an enormous area of the ocean. Similarly, on land, the large herbivorous therapsid

EARLY TRIASSIC

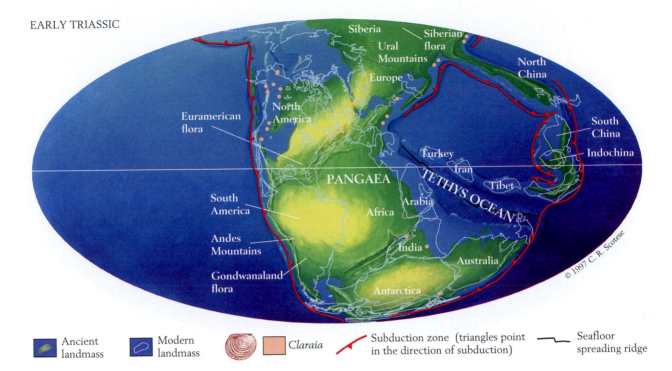

Figure 16-19 **World geography of Early Triassic time.** The Euramerican flora occupied a broad, warm belt across the middle of Pangaea, and the Siberian and Gondwanaland floras occupied regions to the north and south. The bivalve mollusk genus *Claraia* was broadly distributed along both the eastern and western borders of Pangaea. The widespread evaporites illustrated here represent the entire Triassic Period, which was relatively warm and dry even at high latitudes. (Adapted from paleogeographic maps by C. R. Scotese, PALEOMAP Project, University of Texas at Arlington, 1997.)

Lystrosaurus ranged over large areas of the globe; its fossil remains have been found on several continents that formed part of Gondwanaland (see Figure 8-10). It appears that *Lystrosaurus* was abundant at the start of the Triassic because few therapsid carnivores were present to prey on it. As other vertebrate groups radiated to high diversities in the course of Triassic time, most of their species were wide-ranging. Many vertebrate families occupied territory that is now divided among several modern continents. In fact, the Triassic is the only period for which the distributions of fossil land vertebrates clearly indicate that all of Earth's continents were connected.

Pangaea began to fragment

The most spectacular geographic development of the Mesozoic Era was the fragmentation of Pangaea, an event that began in the region of the Tethys Seaway. As the Triassic Period progressed, the seaway spread farther and farther inland, and eventually the craton began to rift apart. The Tethys subsequently became a deep, narrow arm of the ocean separating what is now southern Europe from Africa. During the Jurassic Period, this rifting spread westward, ultimately separating North and South America (Figure 16-20).

The rifting of Pangaea progressed southward through time. South America and Africa did not separate to form the South Atlantic until the Cretaceous Period. In fact, all of the Gondwanaland continents remained attached to one another until Cretaceous time.

North America began to break away from Africa in mid-Jurassic time. Interestingly, this rifting generally followed the old Hercynian suture. Rifting occurred as some of the arms of a series of triple junctions joined, tearing Pangaea in two (see Figure 9-3).

The rifting that formed the Atlantic Ocean had another important consequence. When continental fragmentation begins in an arid region near the ocean, evaporite deposits often form (p. 41). As rifting began in Pangaea, extension produced normal faults between Africa and the northern continents. Zones bounded by such faults sank to form troughs, into which water from the Tethys to the east periodically spilled and then evaporated. Evaporites that were formed in this way now lie on opposite sides of the Atlantic, both in Morocco and offshore from Nova Scotia and Newfoundland (Figure 16-21).

During Middle and Late Jurassic time, one arm of rifting passed westward between North and South America, giving rise to the Gulf of Mexico. Intermittent influxes of seawater into this rift, apparently from the Pacific Ocean, produced great thicknesses of evaporites. Today these evaporites, known as the Louann Salt, lie beneath the Gulf of Mexico and in the subsurface of Texas. Because its density is low, the Louann Salt has in some places pushed up through younger sediments

LATE JURASSIC

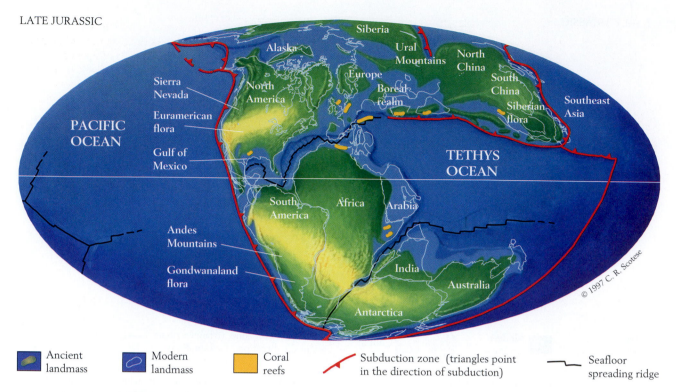

Ancient landmass	Modern landmass	Coral reefs	Subduction zone (triangles point in the direction of subduction)	Seafloor spreading ridge

Figure 16-20 Geography of the Late Jurassic world. The three floras persisted from Triassic time. The Tethyan marine realm, which was characterized by tropical life, including reef corals, extended from the eastern Pacific across the newly forming Mediterranean and Gulf of Mexico to the western Pacific. The cooler Boreal realm lay to the north. (Adapted from paleogeographic maps by C. R. Scotese, PALEOMAP Project, University of Texas at Arlington, 1997.)

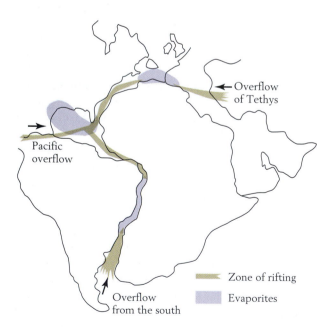

Figure 16-21 Early Mesozoic evaporites. Evaporites accumulated during the early stages of the rifting that formed the Atlantic Ocean as seawater overflowed intermittently into newly forming fault block basins. Those in southern South America and Africa did not form until Early Cretaceous time. (After K. Burke, *Geology* 3:613–616, 1975.)

to form *salt domes* (Figure 16-22), many of which are associated with reservoirs of petroleum and sulfur. The rifting that formed the South Atlantic did not begin until Early Cretaceous time, when salt deposits formed as seawater spilled inland from the south (see Figure 16-21).

Tropical and nontropical zones were evident, even in Europe

Although sea level underwent only minor changes during Late Triassic and Early Jurassic time, it subsequently rose, with minor oscillations, until Late Jurassic time. Then, very late in the Jurassic Period, while remaining relatively high, sea level underwent more rapid oscillations that caused epicontinental seas to flood western North America and Europe. Paleontologists have long recognized that two biogeographic provinces of marine life existed in Europe during the Jurassic Period: a southern province, which was centered in the region of the Tethys Seaway and designated the Tethyan realm, and a northern province, which is labeled the Boreal realm.

Coral reefs were largely restricted to the Tethyan realm, as were limestones and certain groups of warm-adapted mollusks. Thus it is apparent that the Tethyan realm was essentially a tropical province. The transition

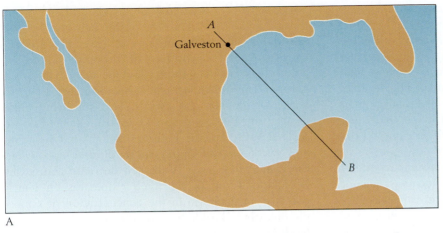

Figure 16-22 Position of the
Louann Salt within the sediments
beneath the Gulf of Mexico. These
Jurassic salt deposits, which rise into
domes in some places because of their
low density, accumulated when the
Gulf of Mexico began to form by
continental rifting. (After O. Wilhelm
and M. Dwing, *Geol. Soc. Amer. Mem.*
83, 1972.)

A

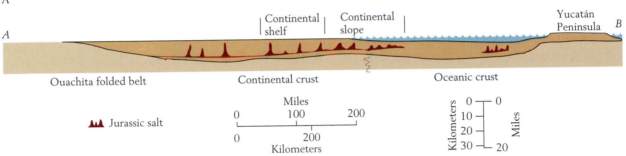

B

from the Tethyan to the Boreal realm resembled what we see today between tropical southern Florida, with its carbonates and coral reefs, and subtropical northern Florida, where siliciclastic sediments prevail and reefs are absent.

There is no doubt that temperature gradients from equator to poles were gentle throughout the Jurassic Period. Plants that appear to have required warmth (the Euramerican flora of Figure 16-20) occupied a broad belt extending to about 60° north latitude. Even the Gondwanaland flora to the south and the Siberian flora to the north included groups of ferns whose modern relatives cannot tolerate frost. These high-latitude floras do not seem to have been tropical, however; they contained few members of the cycad group, which has always been restricted to warm regions (see Figure 4-19).

Mass Extinctions

A mass extinction marked the end of the Triassic Period. It took a heavy toll on marine life and utterly transformed the terrestrial ecosystem. Heavy extinction also occurred near the end of the Jurassic Period.

Dinosaurs inherited the land after the terminal Triassic mass extinction

The mass extinction that brought the Triassic Period to a close was one of the largest of all time. This event struck both on land and in the sea. In the marine realm, about half of all genera of animals suffered extinction,

a slightly larger percentage than died out with the dinosaurs at the end of the Mesozoic Era. Conodonts and placodont reptiles died out altogether. Most species of bivalves, ammonoids, plesiosaurs, and ichthyosaurs were also lost, although all of these groups recovered in Jurassic time. On land, very few therapsid or large amphibian species survived into Jurassic time. Dinosaur taxa survived in larger numbers.

The best fossil record of the Triassic-Jurassic transition for terrestrial biotas is in strata deposited in several fault block basins of eastern North America, which will be described in the following section. Here, the fossil record of pollen and spores (collectively known as plant *microfossils*) reveals that most species of seed plants disappeared abruptly and that ferns spread rapidly over the land. Fern spores dominate the microfossil flora of a stratigraphic interval less than a meter thick. Strata in Morocco reveal a similar pattern for land plants. Fossilized bones are rare in rocks of the fault block basins of eastern North America, but footprints of vertebrate animals are common, and they reveal the sudden disappearance of nearly all therapsids close to the time of the ferns' abrupt expansion. Following this crisis there was a brief interval when only a small variety of vertebrate animals, most of them dinosaurs, occupied the land. Clearly, terrestrial biotas were devastated by a sudden mass extinction.

Dinosaurs, which for some reason survived the terminal Triassic crisis preferentially, quickly diversified. The first large carnivorous dinosaurs appeared within perhaps 30,000 years of the mass extinction, and within about 100,000 years of the crisis dinosaurs had attained a high level of diversity, which they maintained for the remainder of the Jurassic Period. It is now apparent that the dinosaurs' success resulted from the almost total elimination of the therapsids at the end of Triassic time. The therapsids, benefiting from an evolutionary head start, had suppressed the dinosaurs, just as both of these groups suppressed the mammals, which came on the scene even later than the dinosaurs.

It is striking that the terminal Triassic mass extinction occurred during a brief interval when a vast amount of magma emerged from Earth's mantle as Pangaea began to break apart. This magma produced dikes, sills, and flood basalts along a rift zone now extending from Newfoundland to southern Africa. These igneous bodies all formed between 191 and 205 million years ago. Many of them occur within fault block basins or bound them. Taken together, these igneous bodies constitute the Central Atlantic Magmatic Province, which may be the largest province of continental basalts ever to form on Earth. Sills in several areas just a few meters above the stratigraphic level of the mass extinction all yield radiometric ages of about 202 million years, making this the approximate date of the mass extinction.

Volcanic activity releases carbon dioxide, an important greenhouse gas. Might the massive igneous outpourings of latest Triassic time have caused a pulse of greenhouse warming that triggered the mass extinction? Fossil plants suggest that this is what happened (Figure 16-23). The fraction of cells on the surface of a leaf that are *stomates*—pore cells that admit carbon dioxide used in photosynthesis—decreases as the concentration of carbon dioxide in the atmosphere

increases (fewer stomates are needed when carbon dioxide is plentiful). The relationship between the proportion of cells on the surfaces of leaves that are stomates and the concentration of carbon dioxide in the atmosphere has been established for many living plant groups. This relationship has been applied to fossil leaves of ginkgos and cycads (two still-living plant groups; see Figures 4-19 and 16-10) collected above and below the Triassic-Jurassic boundary in Greenland and Sweden to determine carbon dioxide levels when the plants were alive. A marked decrease across this boundary in the proportion of stomates on leaves points to an increase in the concentration of atmospheric carbon dioxide from slightly more than twice its modern level to seven or eight times its modern level (see Figure 16-23). These values imply that Earth's average global temperature, which had stood about 3°C warmer than today, rose another 3°–4°C during the Triassic-Jurassic transition. This sudden warming may explain the extinction of many species. Land plants are especially sensitive to climatic changes. Many terrestrial animals may have died out because of the disappearance of plant species; the brief interval when ferns opportunistically took the place of larger plants must have been an interval of starvation for many herbivores. With the disappearance of herbivores would have come starvation for carnivores.

Many extinctions occurred near the end of Jurassic time

The primary beneficiaries of the terminal Triassic extinction on land were the dinosaurs, which radiated rapidly during the Jurassic and then continued to dominate

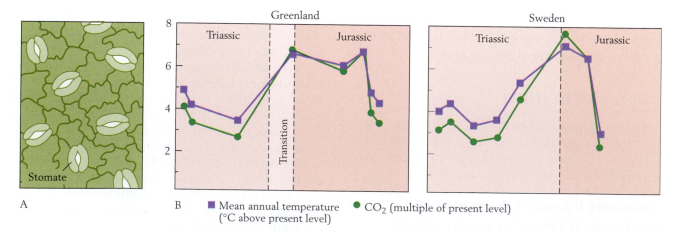

Figure 16-23 Increases in the concentration of carbon dioxide in the atmosphere and in mean annual temperature on Earth during the Triassic-Jurassic transition. *A.* The fraction of the cells in a leaf that are stomates indicates the level of CO$_2$ in the atmosphere.

B. Increases in the proportion of stomates in fossil ginkgo and cycad leaves indicate a rise in atmospheric CO$_2$ levels, and therefore a rise in global temperatures, over the Triassic-Jurassic transition. (After J. C. McElwain, D. J. Beerling, and F. I. Woodward, *Science* 285:1386–1390, 1999.)

terrestrial habitats throughout the remainder of the Mesozoic Era (see Earth System Shift 16-1). At the end of the Jurassic Period there was moderately heavy extinction of life, both in the ocean and on land. Many dinosaurs died out late in Jurassic time, but the fossil record is not complete enough to indicate whether the extinction was sudden. In any event, with the dawning of the Cretaceous Period, animal life on land had a new aspect. The stegosaurian dinosaurs failed to make the transition, as did the larger sauropods (see Figure 16-15). New kinds of dinosaurs populated the Cretaceous world.

Tectonic Events in North America

Early in the Mesozoic Era, tectonic events in eastern and western North America were in sharp contrast. Eastern North America began rifting away from Europe and Africa to form a new continental margin. Western North America, on the other hand, expanded westward as numerous island arcs and small exotic terranes were sutured to its Pacific margin.

Fault block basins formed in the east

During Early and Middle Triassic time, erosion subdued the Appalachian Mountains, which were centrally located in Pangaea. In Late Triassic time, long, narrow depositional basins bounded by faults developed on the gentle Appalachian terrain (Figure 16-24). These basins formed when Pangaea was splintered by normal faults on either side of the great rift that began to divide the continent and form the Atlantic Ocean (see Figure 9-3). One string of aligned basins extended from New York City to northern Virginia and received sediments known as the Newark Supergroup. Here, during a Late

Triassic and Early Jurassic interval of subsidence, nonmarine sediments accumulated to a thickness of nearly 6 kilometers (4 miles); it is these sediments that provide details of the terminal Triassic mass extinction. Early Mesozoic basins resembling those of eastern North America are also found in Africa and South America, but these basins contain thick evaporite deposits that formed when water from the Tethys Seaway periodically spilled into them (see Figure 16-21).

The most revealing basin sediments are in eastern North America. One particularly well-studied basin, which occupies present-day Connecticut and Massachusetts, was bounded on the east by a large normal fault along which the basin subsided continually while sediments accumulated from an eastern source area (Figure 16-25A). Several types of depositional environments existed within this basin. Coarse conglomerates that wedge out to the west accumulated as alluvial fans that spread from the eastern fault margin. Many sand-sized sediments of the basin are stream deposits. The fact that most of these deposits are composed of red arkose (p. 41) suggests that deposition in this area was rapid, because apparently there was little time for feldspars to disintegrate into clay.

Lakes within many of the eastern North American basins were floored by well-laminated muddy sediment. Cycles now visible in these sediments reflect expansion and contraction of the lakes, which for the most part must have been quite shallow. During some dry intervals, evaporite minerals were precipitated from the shrinking waters, but abundant fossil fish remains indicate that at other times the waters were hospitable to life. In fact, freshwater fishes underwent spectacular adaptive radiations in some of the larger lakes. These radiations resembled those that have occurred very recently in the African Great Lakes (see Figure 7-10). It

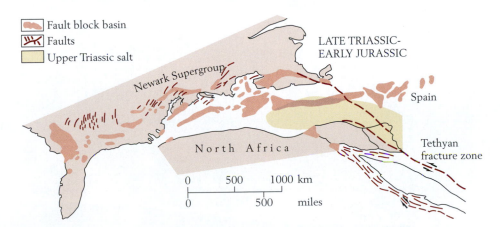

Figure 16-24 Geologic features of eastern North America and nearby regions during Late Triassic and Early Jurassic time. In eastern North America, block faulting produced elongate depositional basins, most of which paralleled the enormous rift that eventually formed the Atlantic Ocean. Salt deposits accumulated from the sporadic westward spilling of seawater from the Tethys Seaway, where the Mediterranean was forming as a result of Africa's movement in relation to Europe. (After W. Manspeizer et al., *Geol. Soc. Amer. Bull.* 89:901–920, 1978.)

West   East

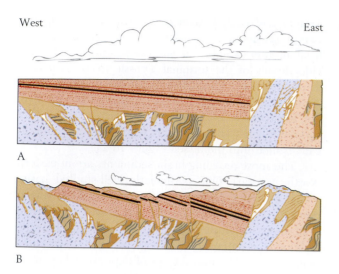

A

B

Figure 16-25 Cross sections of the early Mesozoic fault block basin that passes through central Connecticut. Part of the Newark Supergroup was deposited in this basin (see Figure 16-24) *A.* The basin late in its depositional history, when great thicknesses of sediment had accumulated. As the basin subsided, lavas welled up periodically, forming dikes and sills (heavy black lines). Gravels from the uplands to the east spread into the basin as alluvial fans. *B.* The basin was eventually destroyed by extensive faulting.

is interesting to note that these modern African lakes occupy rift valleys much like those in which the Newark Supergroup was deposited (see Figures 9-2 and 9-3).

Although dinosaur tracks are common in rocks representing lake margins, conditions in the basins seldom favored the preservation of dinosaur skeletons, except in eastern Canada, where a number of bones have been found. Some of the ancient soils in the basin that extends through Connecticut and Massachusetts contain caliche nodules, indicating that the climate here was warm and seasonally arid (see Figure 5-1). Apparently bones decayed rapidly under these conditions, so that relatively few were preserved as fossils.

Periodically, mafic magmas welled up through faults, forming dikes and widespread sills within fault block basins. One of the largest of these sills forms the Palisades along the Hudson River near New York City (Figure 16-26). This is one of the mafic bodies of the Central Atlantic Magmatic Province that have yielded radiometric dates for the Triassic-Jurassic boundary, which occurs just a few meters below it.

At least some of the North American basins continued to subside until Early Jurassic time, when deposition ended with a final episode of faulting. After this time, the basins had apparently moved so far westward along with the North American plate that they were no longer affected by the mid-Atlantic rifting. The fact that some of the basins are located several hundred kilometers from the present margin of North America (see Figure 16-24) indicates how extensive the fracturing of a large continent can be; many small breaks and ruptures occur, rather than a clean parting of the crust.

North America grew westward

Throughout the Triassic Period, much of the American West was the site of nonmarine deposition. Shallow seas expanded and contracted along the margin of the craton, but for the most part remained west of present-day Colorado. Through marginal orogenic activity and ac-

Figure 16-26 The Palisades sill, exposed along the Hudson River across from New York City. (Breck P. Kent.)

cretion of exotic terranes, the North American craton expanded westward during Triassic and Jurassic time.

Terrestrial and marine environments As in the Permian Period, the climate of western North America remained largely arid. At times, however, there was sufficient moisture to permit the growth of large trees belonging to the Euramerican flora. The river and lake sediments in Utah and Arizona that are collectively known as the Chinle Formation, for example, erode spectacularly in some places to reveal the well-known Petrified Forest of Arizona (Figure 16-27). In southwestern Utah the Chinle is overlain by the Wingate Sandstone, a desert dune deposit. Above the Wingate lies a river deposit called the Kayenta Formation, on top of which rests the Navajo Sandstone. The Navajo, also a desert dune deposit, ranges upward in the stratigraphic sequence from approximately the position of the Triassic-Jurassic boundary. The Navajo is famous for its large-scale cross-bedding in the neighborhood of Zion National Park (see Figure 5-10*C*).

During Middle and Late Jurassic time, as sea level rose throughout the world (see p. 389), the waters of the Pacific Ocean spread farther inland in a series of four transgressions, each more extensive than the last. The first

such transgression went no farther than British Columbia and northern Montana, but the last one spread eastward to the Dakotas and southward to New Mexico, forming the so-called Sundance Sea (Figure 16-28). Eventually, as mountain building progressed along the Pacific coast in Late Jurassic time, the Sundance Sea retreated.

Subduction and the accretion of new terranes
During the Mesozoic Era, the western margin of North America expanded by the addition of numerous island arc terranes and other small plates (Figure 16-29), in a manner analogous to the addition of exotic terranes to eastern North America in Paleozoic time, during episodes of mountain building in the Appalachian region (p. 352). This mode of continental accretion actually began earlier. The Antler orogeny of Devonian and Early Carboniferous time entailed the collision of the Klamath Arc with the western margin of North America (see Figure 14-29). This event added a sliver of exotic terrane, called the Roberts Mountains terrane, to the western margin of North America.

In late Paleozoic time, after the Antler episode of accretion, the Golconda Arc approached the Pacific margin of North America. Early in the Triassic Period this island arc collided with the North American conti-

Figure 16-27 Silicified logs that have weathered out of the Triassic Chinle Formation in the Petrified Forest of Arizona. (Tom Bean.)

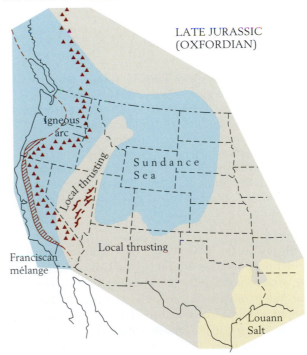

LATE JURASSIC (OXFORDIAN)

Igneous arc

Sundance Sea

Local thrusting

Local thrusting

Franciscan mélange

Louann Salt

Figure 16-28 Geologic features of western North America during Late Jurassic (Oxfordian) time. In mid-Triassic time, the western margin of the continent had ridden up against a subduction zone. As a result, during the Jurassic Period a belt of igneous activity extended for hundreds of kilometers parallel to the Pacific coast. At this time, thrust faulting was limited largely to the state of Nevada. The Sundance Sea flooded a large interior region from southern Canada to northern Arizona and New Mexico.

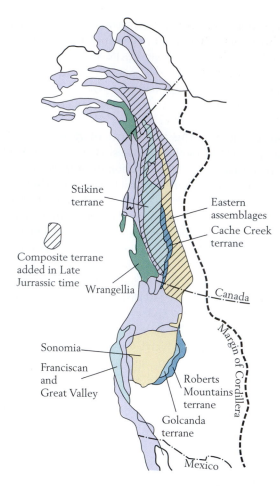

Figure 16-29 Exotic terranes in western North America. Exotic terranes are indicated by colored regions. Early in the Triassic Period, the Sonomia and the Golconda terranes were sutured to the Roberts Mountains terrane. In Late Jurassic time a large composite terrane was sutured to Canada. It consisted of the Stikine terrane, the Cache Creek terrane, and the so-called Eastern assemblages, all of which had been united during the Triassic Period before colliding with North America. (After J. B. Saleeby, *Annu. Rev. Earth Planet. Sci.* 15:45–73, 1983.)

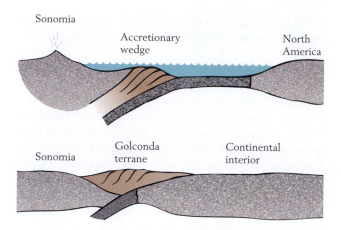

Figure 16-30 The accretion of Sonomia and the Golconda terrane to the western margin of North America. The eastern portion of the microcontinent of Sonomia was formed by the eruptions of an island arc. The Golconda terrane formed from an accretionary wedge that was squeezed between Sonomia and North America as the western margin of North America became wedged against the island arc that bordered Sonomia. (Figure 16-29 shows the location of Sonomia and the Golconda terrane in North America today.)

nent, just as the Klamath arc had done in the earlier Antler orogeny. The suturing of the Golconda Arc, during what is known as the *Sonoma orogeny*, differed from the Antler orogeny in one important way: rather than simply adding a narrow slice of island arc terrane to North America, it also attached a broad microcontinent, known as Sonomia (Figure 16-30). Today Sonomia comprises southeastern Oregon and northern California and Nevada (see Figure 16-29). Squeezed in between Sonomia and the Roberts Mountains terrane was the Golconda terrane, formed largely of the accretionary wedge associated with the Golconda Arc (see Figure 16-30).

After the Sonoma orogeny ended, early in the Triassic Period, there was a brief interlude of tectonic quiescence along the west coast of North America. Then, in mid-Triassic time, the continental margin once again

came to rest against a subduction zone and experienced an orogenic episode that extended from Alaska all the way to Chile. Mountain building along the Pacific coast of North America during the Mesozoic Era resembled the growth of the Andes to the south, which has continued to the present day (p. 212).

Subduction of the oceanic plate beneath the margin of North America thickened the continental crust by leading to the production of intrusive and extrusive igneous rocks. The oldest intrusives of the Sierra Nevada were emplaced during Jurassic time (Figure 16-31); larger volumes were added during the Cretaceous.

The Mesozoic history of the Pacific coast of North America is highly complex. At times more than one subduction zone lay offshore, and exotic slivers of crust were added to the continental margin. Near the end of the Jurassic Period the continent accreted westward when the Franciscan sequence of deep-water sediments and volcanics was forced against the craton along a subduction zone. The Franciscan sediments include graywackes and dark mudstones, together with smaller amounts of chert and limestone. Before becoming attached to North America, the Franciscan sequence constituted an accretionary wedge, whose sediments were deformed and metamorphosed along the subduction zone at high pressures and relatively low temperatures; they represent a mélange (see Figure 8-23). When the continental margin eventually collided with the accretionary wedge, the Franciscan rocks were piled up against the continent, along with the Great Valley sequence of deep-sea turbidites, which accumulated in the forearc basin (Figure 16-32). This Late Jurassic

Figure 16-31 The Sierra Nevada at Yosemite National Park, California. Many of the granitic rocks that formed the Sierra Nevada were emplaced during the Jurassic Period. (Peter Kresan.)

event coincided approximately with folding and thrusting farther east. These tectonic events of Jurassic age are collectively known as the *Nevadan orogeny*. Orogenic activity related to the Nevadan orogeny continued well into the Cretaceous Period, although it is not generally assigned that name in that period.

Farther north, from northern Washington State to southern Alaska, a large exotic terrane collided with the margin of North America, resulting in substantial westward accretion. This exotic terrane was actually a composite block, formed of several smaller terranes (see Figure 16-29). The presence of diverse suites of Paleozoic rocks and fossils in these small terranes indicates that they were once separate entities. They do, however, share rock units of Triassic age, an indication that they were a single unit during the Triassic Period. The entire composite terrane was then accreted to North America late in Jurassic time, along the subduction zone that bordered the continent.

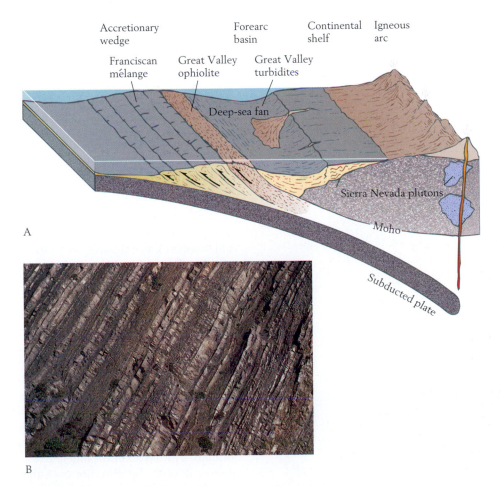

Figure 16-32 The Pacific margin of northern California in Late Jurassic time. *A.* The Franciscan mélange formed an accretionary wedge along the marginal subduction zone. The Great Valley ophiolite was a zone of seafloor that was squeezed up along the eastern margin of the accretionary wedge, and the Great Valley turbidites formed along the continental margin. Today the Great Valley ophiolite and turbidites still occupy a low region, the Central Valley of California. West of the Central Valley, portions of the Franciscan mélange have been elevated as part of the Coast Ranges. *B.* Turbidites that now lie along the western margin of the Sacramento Valley in California, where they have been tilted to a high angle by tectonic activity. (*A,* after R. K. Suchecki, *J. Sediment. Petrol.* 54:170–191, 1984; *B,* Martin G. Miller/Earth Lens.)

Deposition in a foreland basin To the south, in the western United States, the eastward thrusting and folding of Late Jurassic time greatly altered patterns of deposition as far east as Colorado and Wyoming. The Sundance Sea spread over a broad foreland basin east of the belt of folding and thrusting. This was the most extensive marine incursion since late Paleozoic time (see Figure 16-28). In latest Jurassic time, however, the folding and thrust faulting that extended over Nevada, Utah, and Idaho produced a large mountain chain.

The shedding of large volumes of clastics eastward from these mountains eventually drove back the waters of the Sundance Sea, leaving only a small inland sea to the north (Figure 16-33). What remained in Colorado, Wyoming, and adjacent regions was a nonmarine foreland basin in which molasse deposits accumulated. Apparently, on the gentle profile of the foreland basin, even the lowest depositional environments were above sea level, because there was no initial deposition of marine flysch. The molasse of the foreland basin was deposited in rivers, lakes, and swamps, creating the famous Morrison Formation, which has yielded the world's most spectacular dinosaur faunas (Figure 16-34;

Figure 16-34 Excavation of dinosaur fossils in the Morrison Formation. These partly intact skeletons remain embedded in the rock at Dinosaur National Monument, Utah. (Dinosaur Nature Association.)

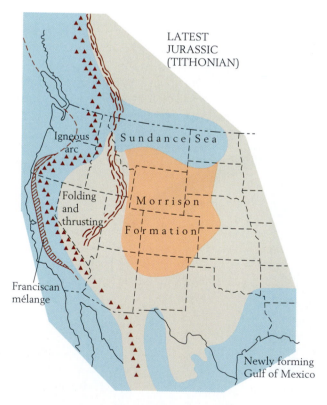

LATEST JURASSIC (TITHONIAN)

Figure 16-33 Geologic features of western North America during latest Jurassic (Tithonian) time. A fold-and-thrust belt now extended for hundreds of kilometers roughly parallel to the coastline, but far inland. Tectonic activity had driven the Sundance Sea from the western interior, leaving a nonmarine basin where the nonmarine Morrison Formation accumulated.

see also Figures 16-15 and 16-16). Some of the Morrison dinosaur skeletons are partly intact, and the remains of as many as 50 or 60 individuals may occur within a small area. These patterns of preservation suggest that the Morrison dinosaurs were buried during floods.

The Morrison Formation consists of sandstones and multicolored mudstones deposited over an area of about 1 million square kilometers. Caliche soil deposits indicate that the climate was seasonally dry during at least part of the Morrison depositional interval, while the scarcity of crocodiles, turtles, and fishes in lake deposits of the Morrison suggests that many lakes of the region were saline. Dinosaurs are found in deposits representing all of the Morrison environments—rivers, lakes, and swamps. This broad environmental distribution suggests that none of the species, not even the huge sauropods (see Figure 16-15), were adapted specifically for a life of wading in large bodies of water. The Morrison Formation spans the last 10 million years or so of the Jurassic Period and is overlain by the nonmarine Cloverly Formation of Early Cretaceous age, which contains a completely different fauna of dinosaurs, apparently because of major extinctions at the end of the Jurassic Period.

Chapter Summary

What groups of animals were conspicuous in Triassic and Jurassic seas, and what groups that had been prominent in late Paleozoic time were absent?

Important groups of marine life during the early Mesozoic included bivalve, gastropod, and ammonoid mollusks, brachiopods, sea urchins, hexacorals, bony fishes, sharks, and swimming reptiles. Conspicuously absent were tabulate and rugose corals, trilobites, and fusulinid foraminifera.

What reptile groups were important carnivores in Triassic and Jurassic seas?

Placoderms crushed shells to feed on marine invertebrates early in Triassic time, and nothosaurs fed on fishes near the shore. Later to evolve were ocean-going predators: ichthyosaurs, plesiosaurs, and marine crocodiles.

What kinds of plants played major roles on land in early Mesozoic time?

Ferns were abundant during the Triassic Period, but gymnosperm plants dominated Jurassic landscapes.

Why did dinosaurs replace therapsids as the dominant vertebrate animals on land?

Nearly all therapsid species died out in the terminal Triassic mass extinction. A larger percentage of dinosaur species survived, and thus the dinosaurs assumed numerical dominance during the Jurassic Period.

What groups of vertebrate animals evolved the ability to fly during early Mesozoic time?

Flying reptiles evolved near the end of the Triassic Period, and birds evolved from small dinosaurs in Late Jurassic time.

What may have caused a major mass extinction during early Mesozoic time?

One of the most severe mass extinctions of all time marked the end of the Triassic Period. Massive volcanism at this time associated with the breakup of Pangaea may have caused lethal greenhouse warming through release of carbon dioxide.

What was the configuration of Earth's landmasses near the beginning of the Mesozoic Era, and how did this configuration change early in the era?

Nearly all of Earth's continental crust was consolidated into the supercontinent Pangaea at the start of the Mesozoic Era. Even at the end of the Jurassic Period, all of the continents remained close together. Evaporites mark zones where Pangaea began to rift apart early in the Mesozoic Era.

Why did large bodies of nonmarine sediment accumulate in eastern North America during early Mesozoic time?

Fault block basins formed during the rifting episode that eventually formed the Atlantic Ocean between North America and Africa, and these basins received thick deposits of sediment.

What evidence is there for the existence of a foreland basin during Jurassic time in western North America?

The Sundance Sea formed inland from mountains that rose up during the Nevadan orogeny in western North America during the Jurassic Period, but marine waters were eventually expelled by the influx of sediments. Dinosaur fossils are abundantly preserved in the molasse sediments that were then deposited in the vicinity of Utah and western Colorado.

Review Questions

1. What important groups of Paleozoic marine animals were absent from Triassic seas?

2. Why should we not be surprised that stromatolites spread over Early Triassic seafloors?

3. What two kinds of flying vertebrates evolved during early Mesozoic time?

4. How did reefs formed by hexacorals during the Jurassic Period differ from those formed during Triassic time?

5. What accounts for the dinosaurs' rise to dominance at the start of the Jurassic Period?

6. What evidence is there that dinosaurs brooded their eggs and tended their young long after they hatched?

7. What may be the largest province of continental volcanic rocks ever to form came into being early in the Mesozoic Era. Where and why did it form?

8. What was the geographic setting in which the most spectacular known assemblage of Jurassic dinosaurs was preserved?

9. In what areas is there evidence that new ocean basins started to form in Triassic and Jurassic time? What is that evidence?

10. By what mechanism did western North America expand westward during early Mesozoic time?

11. Vertebrate life underwent spectacular evolutionary changes early in the Mesozoic Era. Using the Visual Overview on page 388 and what you have learned in this chapter, review the ways in which vertebrate animals expanded their ecological role in the ocean, on land, and in the air during the Triassic and Jurassic periods.

Chalk, the soft, powdery rock that is unusually abundant in the Upper Cretaceous Series in many areas. The chalk deposits shown here stand above the coastline of southeastern England, where they form the famous White Cliffs of Dover. (David Woodfall/NHPA.)

The Cretaceous World

The Cretaceous was the last period of the Mesozoic Era, and it ended with a mass extinction that wiped out the dinosaurs and many other forms of life that flourished in the Mesozoic. The Cretaceous Period was in many ways an interval of transition. Some Cretaceous sediments are lithified, like nearly all those of older systems; many others, however, consist of soft muds and sands, like most deposits of the Cenozoic Era. Fossil biotas of the Cretaceous Period also display a mixture of archaic and modern features. They include members of diverse taxa that failed to survive the Cretaceous—among them the dinosaurs and ammonoids— as well as diverse modern taxa, such as flowering plants and the largest subclass of fishes in the world today. It was during the Cretaceous Period that continents moved toward their modern configuration. At the start of the Cretaceous, continents were tightly clustered, but by its end, the Atlantic Ocean had widened and the southern portion of Pangaea—the former Gondwanaland—had fragmented into most of its daughter continents.

The Cretaceous System was first described formally in 1822. Its name derives from *creta*, the Latin word for chalk, which is a soft, fine-grained kind of limestone that accumulated over broad areas of the Late Cretaceous seafloor.

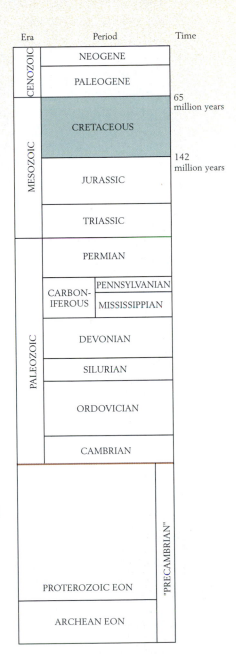

Cretaceous Life

Life of the Cretaceous Period was a curious assortment of forms, some of which resembled taxa of the present world and others that seem strikingly prehistoric. In the marine realm, remarkably modern types of bivalve and gastropod mollusks populated Late Cretaceous seas along with enormous coiled oysters and other now-extinct bivalves. Diverse fishes of the modern kind occupied the same waters as a variety of ammonoids, belemnoids, and reptilian sea monsters—none of which have any close living relatives. Dinosaurs, however, continued to rule the land, while mammals remained quite small by modern standards.

Pelagic life was modernized

The appearance of new groups of single-celled organisms gave marine plankton a thoroughly modern character by the end of Cretaceous time. The primary change among the phytoplankton was the evolutionary expansion of the **diatoms** (see Figure 3-16*B*). Diatoms may have existed during the Jurassic Period, but they did not radiate extensively until mid-Cretaceous time. Together with dinoflagellates and, in warm seas, calcareous nannoplankton, diatoms must have accounted for much of the marine photosynthesis in Cretaceous time, as these three groups do in modern seas (p. 97). Today

Major Events of the Cretaceous

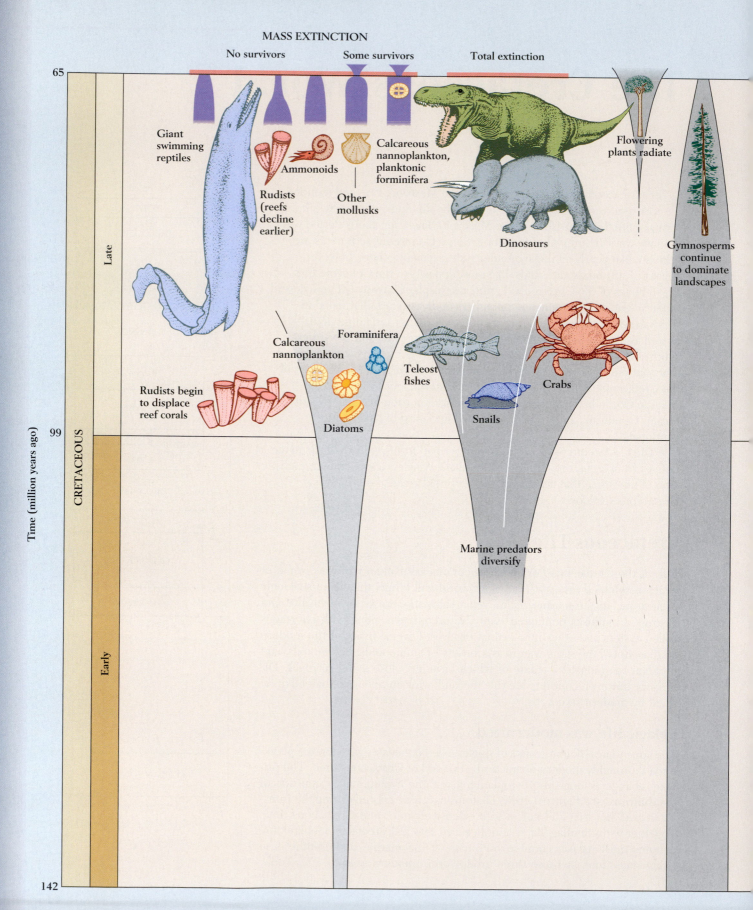

MASS EXTINCTION

No survivors   Some survivors   Total extinction

65

Giant
swimming
reptiles

Ammonoids

Rudists
(reefs
decline
earlier)

Other
mollusks

Calcareous
nannoplankton,
planktonic
forminifera

Dinosaurs

Flowering
plants radiate

Late

Gymnosperms
continue
to dominate
landscapes

Foraminifera

Calcareous
nannoplankton

Teleost
fishes

Crabs

Rudists begin
to displace
reef corals

Snails

Diatoms

99

Time (million years ago)

CRETACEOUS

Marine predators
diversify

Early

142

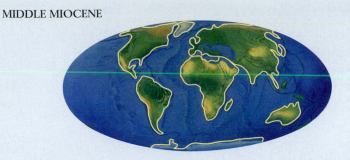

MIDDLE MIOCENE

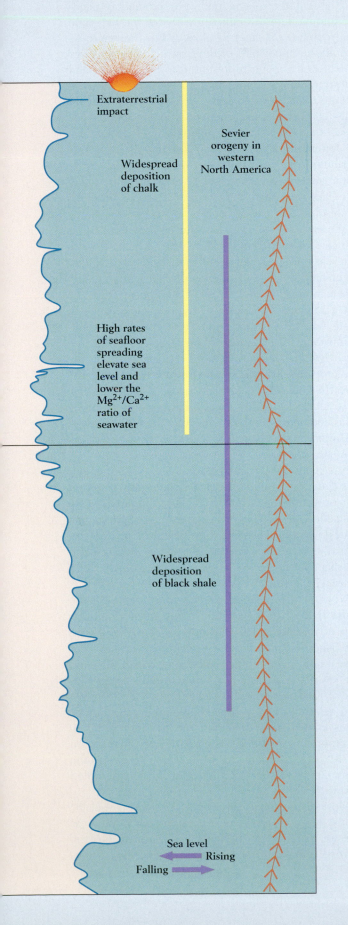

Extraterrestrial
impact

Sevier
orogeny in
western
North America

Widespread
deposition
of chalk

High rates
of seafloor
spreading
elevate sea
level and
lower the
Mg^{2+}/Ca^{2+}
ratio of
seawater

Widespread
deposition
of black shale

Sea level

Rising

Falling

LATEST CRETACEOUS

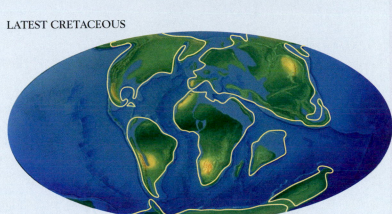

LATE CRETACEOUS

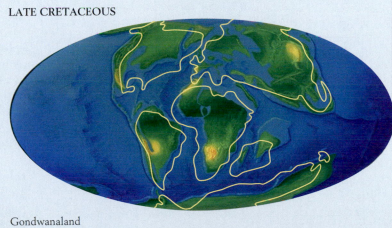

Gondwanaland
breaks apart.

LATE JURASSIC

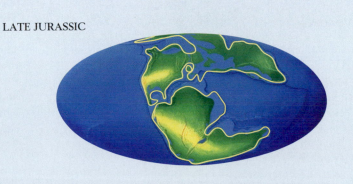

diatoms are the dominant contributors to the siliceous oozes of the deep sea, and their abundant accumulation in deep-sea sediment began during the Cretaceous Period (p. 125).

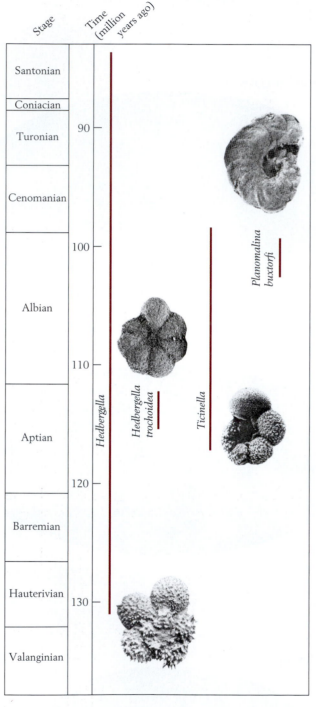

Figure 17-1 Early planktonic foraminifera (Globigerinacea). Cretaceous stages are shown at the left. These species average about $\frac{1}{2}$ millimeter ($\frac{1}{50}$ of an inch) in diameter. (After A. Boersma.)

Higher in the pelagic food web, the modern planktonic foraminifera diversified greatly for the first time. This group has a meager fossil record in Jurassic rocks; not until the upper part of the Lower Cretaceous System is it well enough represented to be of great value in biostratigraphy (Figure 17-1).

Late Cretaceous adaptive radiations of two of the single-celled planktonic groups altered depositional patterns in the pelagic realm: since mid-Cretaceous time, both foraminifera and calcareous nannoplankton have contributed vast quantities of calcareous sediment to oceanic areas (see Figure 5-35), whereas before about 100 million years ago, little or no calcareous ooze was present on the deep-sea floor.

We saw in Chapter 10 that during Late Cretaceous time, calcareous nannoplankton were so abundant in warm seas that the small plates that armored their cells accumulated in huge volumes as the fine-grained limestone commonly known as chalk. It appears that the great abundance of calcareous nannoplankton, which secrete calcite, resulted from the low magnesium-calcium (Mg^{2+}/Ca^{2+}) ratio of Late Cretaceous seas (see Figure 10-21). The most famous chalk deposits in the world crop out along the southeastern coast of England, where they are known formally as the Chalk (p. 416). Similar chalks formed in many other regions, including Kansas and nearby regions and along the Gulf Coast of the United States, as well as southern Europe, Africa, and Australia.

Still higher in the pelagic food web of Late Cretaceous time, the ammonoids and belemnoids persisted as major swimming carnivores. The ammonoids serve as valuable index fossils for the Cretaceous System, just as they do for the Triassic and Jurassic. Among the Cretaceous ammonoids were many species with straight, cone-shaped shells and others with coiled shells (Figure 17-2E).

New on the scene in Cretaceous time were the **teleost fishes**, a subclass that today is the dominant group of marine and freshwater fishes. Teleosts are bony fishes characterized by such features as symmetrical tails, overlapping scales, specialized fins, and short jaws that are often adapted to take particular kinds of food. By Late Cretaceous time, a wide variety of teleosts already existed, including the largest species known from the fossil record (Figure 17-3). This group also included close relatives of the modern sunfish, carp, and eel, as well as members of the salmon family. Similarly, Cretaceous sharks resembled present-day forms.

Most of the top carnivores of Cretaceous pelagic habitats, however, were not at all modern. Whereas whales of one kind or another have occupied the "top carnivore" role during most of the Cenozoic Era, reptiles

Figure 17-2 Cretaceous invertebrate fossils. *A.* A burrowing bivalve mollusk. *B.* A coiled oyster, the size of a small grapefruit. *C.* A crab. *D.* A carnivorous gastropod (snail). *E.* An ammonoid. (Courtesy Smithsonian Institution, photo by Chip Clark.)

were the largest marine carnivores until the end of Cretaceous time. Ichthyosaurs and marine crocodiles were rare by this time, but plesiosaurs still thrived, some exceeding 10 meters (35 feet) in length. Some members of the Late Cretaceous pelagic community of the western interior of the United States are depicted in Figure 17-4. Huge marine lizards known as **mosasaurs** were probably the most formidable marauders of Cretaceous seas; some grew to be longer than 15 meters (45 to 50 feet). Unlike the plesiosaurs, whose shapes are those of long-distance swimmers, mosasaurs probably lurked inconspicuously and ambushed their prey. Figure 17-4 also portrays the flightless diving bird *Hesperornis* and a species of marine turtle that grew to a length of nearly 4 meters (13 feet).

Benthic life was also modernized

Life on the seafloor began to take on a modern appearance during the Cretaceous Period. One noteworthy feature was the decline of the brachiopods, which had suffered greatly in the mass extinction at the end of the Paleozoic Era but had experienced a moderate expansion again early in Mesozoic time. Sea urchins and the hexacorals diversified but underwent no startling adaptive changes. Other major groups produced distinctive new representatives that have survived to the present. Some of them are described below.

Foraminifera A large percentage of the families of benthic foraminifera in existence today appeared dur-

Figure 17-3
Xiphactinus, a Cretaceous fish. At about 5 meters (16 feet) in length, this fish is the largest known teleost. A careful look reveals that the animal shown here died with a good-sized fish in its belly. (Courtesy Smithsonian Institution.)

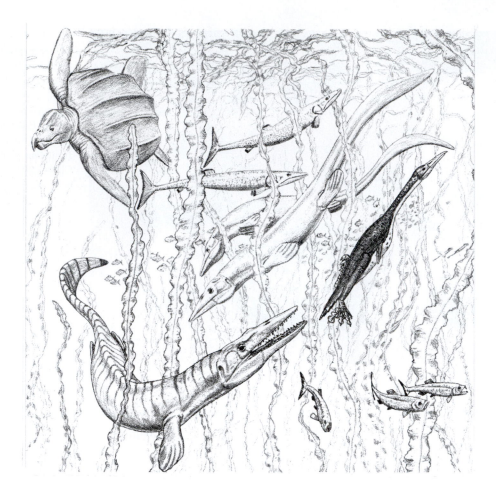

Figure 17-4 Reconstruction of marine life preserved in the Upper Cretaceous Pierre Shale of the western interior of the United States. The animals are shown swimming in a bed of kelp, which are algae of large proportions. The giant turtle at the upper left is *Archelon*, which reached a length of almost 4 meters (13 feet). The striped animal at the lower left is the mosasaur *Clidastes*, and beyond it is a pair of mosasaurs of the genus *Platecarpus*. *Clidastes* is in pursuit of the diving bird *Hesperornis*. The teleost fishes are *Cimolichthyes* (the pikelike pair near the turtle) and *Enchodus* (the small fishes on the lower right). (Drawing by Gregory S. Paul.)

ing the Cretaceous Period, so this group had a modern aspect at that time (Figure 17-5).

Bryozoans The most abundant modern bryozoans are the *cheilostomes*, which commonly encrust marine surfaces, including the hulls of boats, in the form of low-growing mats (see Figure 3-31). Cheilostomes originated in Jurassic time but did not enjoy success until the Late Cretaceous, when they expanded to include more than a hundred genera.

Burrowing bivalve mollusks Early Cretaceous burrowing bivalves resembled those of the Jurassic, but by the end of the period new genera had appeared as well, including many that were rapid burrowers or burrowed deeply into the sediment (see Figure 17-2A).

Gastropod mollusks During the Cretaceous Period the aptly named Neogastropoda, or "new snails," produced

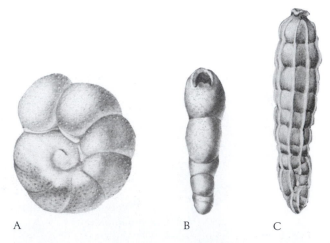

A   B   C

Figure 17-5 Genera of benthic foraminifera that arose during the Cretaceous Period. A. *Anomalinoides* (×68). B. *Pleurostomella* (×48). C. *Siphogenerinoides* (×46). (H. Tappan.)

many modern families and genera (see Figure 17-2*D*). Unlike most earlier snails, these animals are generally carnivorous, feeding on such prey as worms, bivalves, and other snails. Some live in the sediment, others on the sediment surface. Many modern seashells popular with collectors belong to neogastropod species.

Crabs A more or less modern type of crab had evolved during the Jurassic Period, but a much greater variety of crabs appeared during the Cretaceous Period (see Figure 17-2*C*).

Surface-dwelling bivalve mollusks Among bivalve mollusks living on the surface of the seafloor, coiled oysters and other groups that had existed during Jurassic time evolved species of enormous size (see Figure 17-2*B*). Among these large forms, the **rudists** were of special significance because they lived like corals, forming large tropical reefs (Figure 17-6; see also Figure 10-16*A*). Rudists grew a cone-shaped lower shell and a lidlike upper shell. These curious animals attached to hard objects (often other rudists) and grew upward, some reaching heights of more than 1 meter (3 feet).

Before their demise at the end of the Cretaceous Period, the rudists apparently flourished at the expense of reef-building corals. Shallow-water reefs built in Early Cretaceous time, like those built during the Jurassic Period and in the modern world, were formed primarily by

hexacorals. In mid-Cretaceous time, however, rudist bivalves replaced corals as the primary builders of tropical reefs. Rudists, like reef-building corals, apparently grew rapidly by feeding on symbiotic algae that lived and multiplied in their tissues. Only after the rudists died out with the dinosaurs did corals prevail on reefs once more. Rudists also formed broad, low banks, many of which were dominated by a single rudist species.

The corals declined at a time when the Mg^{2+}/Ca^{2+} ratio of seawater reached its lowest Phanerozoic level (see Figure 10-19). This chemical condition may have hampered corals in their precipitation of aragonitic skeletons (p. 241). In contrast, the dominant Late Cretaceous rudists consisted primarily of calcite, so the calcite seas of the time may have favored them.

Modern marine predators proliferated

Many of the general changes that occurred in benthic marine life during Jurassic and Cretaceous time seem to have been related to the great expansion of modern types of marine predators. Among the new predators were the advanced teleost fishes, modern crabs, and carnivorous gastropods. Many of these new predators were efficient at penetrating shells: fish by biting, crabs by crushing or peeling with their claws, and some of the gastropods by drilling holes. The contrast between Paleozoic and Mesozoic predation on the seafloor is exemplified by the absence during Paleozoic time of large arthropods with crushing claws and by the rarity of holes drilled by predators in fossilized Paleozoic brachiopod and bivalve shells.

The decline of brachiopods and stalked crinoids, both of which were moderately well represented in early Mesozoic seas, probably resulted from the diversification of modern predators. The few species of stalked crinoids that survive today live in deep water; in shallow waters, predation by fish is probably too severe to permit their existence. Today, shallow-water crinoids are unattached forms that hide in crevices in coral reefs during the day and come out at night; they can also move away from predators by swimming. As for modern brachiopods, more species live in temperate seas than in tropical seas, where predation by crabs, fish, and snails is severe. By the end of the Mesozoic Era, relatively few sedentary species of animals lived on the surface of the seafloor in the mode typical of many groups of Paleozoic brachiopods (see Figure 13-12). The ability to swim or to burrow actively appears to have been the best defense against predation, except for species that had defensive spines or unusually heavy protective shells.

Figure 17-6 A reef formed by a population of rudists of the genus *Durania* preserved in the Upper Cretaceous of Egypt. The cap-shaped upper valves that fitted on the cone-shaped lower valves have been eroded away. (T. Steuber, Institut für Geologie und Mineralogie, Universität Erlangen, Nürnberg.)

Flowering plants expanded on land

The greatest change in terrestrial ecosystems during the Cretaceous Period was the ascendancy of the flowering plants (angiosperms), although gymnosperm floras resembling those of Triassic and Jurassic age continued to dominate the land during Cretaceous time. The most conspicuous change during Early Cretaceous time was in the types of gymnosperms that predominated: conifers became the most numerous species of trees, and the Age of Cycads came to a close. The angiosperms that now made their appearance included not only plants with conspicuous flowers, but also hardwood trees, such as maples and oaks, and grasses. The key reproductive feature that distinguishes angiosperms from gymnosperms (naked-seed plants) is the enclosure of the seed (p. 66).

The earliest angiosperm floras Fossils of the Atlantic Coastal Plain in Maryland document the early phase of the evolutionary radiation of flowering plants. Here, within a sedimentary interval representing only about 10 million years of mid-Cretaceous time, both fossil leaves and fossil pollen increase in variety and in complexity of form (Figure 17-7). The early leaves have simple, smooth outlines, and their supporting veins branch in irregular patterns. Later leaves include varieties with marginal lobes and with veins that fol-

low more regular geometric patterns. The more regular patterns probably transported fluids more efficiently and gave the leaves greater strength to withstand tearing.

Secrets of angiosperm success One special feature of the flowering plants is their ability to provide a food supply for their seeds by a process known as *double fertilization*. One fertilization event produces a seed, and a second fertilization event produces a supply of stored food for that seed, such as the nutritional part of a kernel of corn or a grain of wheat. The rapid manufacture of this food supply allows for the quick release of a well-fortified seed. Because gymnosperms lack this double-fertilization mechanism, it takes much longer for the parent plants to supply their seeds with enough food to enable the progeny to survive on their own. As a result, most gymnosperms have reproductive cycles of 18 months or longer. In contrast, thousands of flowering plant species can grow from a seed and then release seeds of their own in just a few weeks.

A second reproductive mechanism of flowering plants that has contributed enormously to their success

Figure 17-7 The pattern of initial adaptive radiation of flowering plants. These fossil leaves and pollen are found in strata (Patuxent through Raritan) of the Cretaceous Potomac Group of Maryland representing about 10 million years of mid-Cretaceous time. Both pollen (left) and leaves exhibit an increase in complexity and variety of form through time. (After J. A. Doyle and L. J. Hickey, in C. B. Beck, ed., *Origin and Early Evolution of the Angiosperms*, Columbia University Press, New York, 1976.)

is the ability of flowers to attract insects. Insects benefit from the nutritious nectar that the flowers provide, and the flowers benefit because the insects unknowingly carry pollen from one flower to another, fertilizing the plants on which they feed. This attraction is often specialized—that is, a particular kind of insect often feeds on a particular kind of plant. This specialization provides a unique mechanism for speciation. If a flower of a new shape, color, or scent develops within a small population of plants, the flower may attract a different kind of insect than the one that visited its ancestors. The plants with the new kind of flower will thus be reproductively isolated from their ancestral species; in other words, the new forms will become a new species. In general, new kinds of insects create opportunities for the development of new species of plants (with new kinds of flowers); similarly, new kinds of plants create feeding opportunities for new species of insects. This reciprocity has apparently accelerated rates of speciation in both flowering plants and insects. These high rates of speciation have permitted the frequent development of new adaptations. Thus the symbiotic relationship between flowering plants and insects has played a major role in the great success of both groups since mid-Cretaceous time. The fossil record of flowers, though meager, shows a diversification of these showy organs in Late Cretaceous time—a diversification that clearly reflects the association of particular species of angiosperms with particular kinds of insects. A few other kinds of animals, including hummingbirds, also visit flowers and fertilize angiosperms today.

Flowering plants diversified late in Cretaceous time, but for unknown reasons they were unable to dislodge gymnosperms and ferns from many terrestrial habitats. Plants that were buried catastrophically where they stood by a volcanic eruption in Wyoming provide a snapshot of the composition of a flora near the end of Cretaceous time. This flora was dominated by ferns, gymnosperms, and a single species of low-growing palm (a flowering plant). Many other species of flowering plants were present, but in low abundance; furthermore, all the angiosperm species were nonwoody or were small trees or shrubs. This and other Cretaceous floras indicate that angiosperms were quite diverse in Late Cretaceous time, but remained ecologically marginal.

The fossil record indicates that angiosperms evolved in the tropics and spread poleward during the Cretaceous. They flourished primarily in unstable habitats, especially along riverbanks, while gymnosperms and ferns continued to dominate most terrestrial environments. Sycamores were one of the most abundant groups of Cretaceous flowering plants; interestingly, even today sycamores tend to grow in unstable environments along streams.

Dinosaurs dwarfed early mammals

Owing to a patchy fossil record, Early Cretaceous vertebrate faunas are poorly known, but Late Cretaceous faunas are well known from collecting sites in Wyoming, Montana, Alberta, and Asia.

In the American West, Late Cretaceous dinosaurs formed a community that has been compared with the modern mammalian fauna of the African plains. Instead of antelopes, zebras, and wildebeests, there were many species of duck-billed dinosaurs (Figure 17-8). These fast-running herbivores probably traveled in herds and may have trumpeted signals to one another by passing air through complex chambers in their skulls (see Earth System Shift 16-1). They also tended their young after birth (Figure 17-9). In place of rhinoceroses, there were horned dinosaurs with beaks and teeth for cutting harsh vegetation. Sharing the Late Cretaceous plains with these herbivores were fearsome predators, including the largest carnivorous land animals of all time, such as *Tyrannosaurus*. Here, too, were terrestrial crocodiles (Figure 17-10) that grew to the remarkable length of 15 meters (45 to 50 feet).

The skies above the plains where the dinosaurs roamed were populated—perhaps sparsely—by the two groups of flying vertebrates that had evolved earlier in the Mesozoic: flying reptiles and birds (see Figures 16-17 and 16-18). Most Cretaceous birds were large wading birds and shorebirds that lived like modern herons and cranes; there were no songbirds of the kind that surround us today. Flying reptiles were among the most spectacular of all Cretaceous animals. While they may have relied heavily on passive soaring on the wind, it appears that at times they flapped their wings in flight. The largest known species, represented by fossils from the uppermost Cretaceous of Texas, is estimated to have had a wingspan of at least 11 meters (35 feet) (see Figure 17-8). Members of this species, like modern vultures, may have soared through the sky in search of carrion. Given its size, it may have fed primarily on dinosaur carcasses.

Throughout the Mesozoic Era, mammals evolved new traits, some of which are shown in Figure 17-11. Nonetheless, mammals remained quite small until the end of the Cretaceous because dinosaurs had originated before them and had been better poised to take over terrestrial habitats when the terminal Triassic mass extinction decimated the therapsids (p. 407). It was probably through predation that the dinosaurs prevented mammals from evolving body sizes larger than that of a house cat.

The oldest known mammals are of Late Triassic age. The bones that form the joint between the jaw and skull of mammals establish their identity, as they differ from the bones that formed this joint in mammal-like reptiles.

Figure 17-8
Reconstruction of a Late Cretaceous fauna of Alberta. On the left is the armored herbivorous dinosaur *Edmontonia* in front of the duck-billed herbivore *Kritosaurus*. The duck-billed herbivores to their right belong to the genus *Corythosaurus*. The ferocious carnivore to the right of center is *Tyrannosaurus*; it confronts horned dinosaurs of the genus *Chasmosaurus*, with the large head shields, and *Monoclonius*, with the long horn. Passing overhead in the foreground are pterosaurs (flying reptiles) of the genus *Quetzalcoatlus*. The water birds flying in the distance have feathered wings, in contrast to the naked wings of the pterosaurs. (Drawing by Gregory S. Paul.)

Figure 17-9 **Painting of a maiasaur dinosaur tending her young.** (Museum of the Rockies, Montana State University, Bozeman.)

With bodies only about 15 centimeters (6 inches) long and pointed snouts, the earliest mammals resembled modern shrews. Their fossil remains reveal a remarkable amount about their mode of life. Their pointed, cutting teeth show that they were carnivorous, and their small size would have restricted them to a diet made up largely of insects. Their mouth structure indicates that they were endothermic: they had a secondary palate, the bony structure that in all mammals separates the nasal air passages from the mouth so that they can breathe while they eat. Endothermy entails such a high metabolic rate that breathing cannot be interrupted for long. Reptiles, in contrast, can suspend breathing temporarily during their meals. Fossil skulls reveal that the earliest mammals had brains that were large for their body size, and that extensive regions of the brain were associated with hearing and smell. The fact that these senses are particularly useful after dark suggests that these small creatures were nocturnal, avoiding the much larger dinosaurs, which presumably were active in daylight. Early mammals appear also to have suckled their young, as modern mammals do. The pattern of tooth development provides the evidence for this conclusion. Lower vertebrates have

Figure 17-10 Skull of a huge terrestrial crocodile, *Phobosuchus*, which probably fed on Late Cretaceous dinosaurs of small and medium size. The length of the head equaled the height of a large man. (British Museum of Natural History.)

functional teeth early in life, and many species replace worn or lost teeth more than once. Mammals, in contrast, do not have functional teeth for quite some time, because early in life their only food is their mother's milk. Because mammals do not need teeth until long after birth, they generally have only two sets of teeth, the baby teeth that appear in infancy and the adult teeth that replace them. Finally, the earliest mammals had rear feet that were adapted for grasping; this characteristic points to a life of tree climbing.

Although mammals remained small and relatively inconspicuous until the end of the Cretaceous, they did diversify to a degree. The first herbivorous forms, which evolved during the Jurassic Period, had gnawing teeth like those of modern rodents. By Late Cretaceous time, the two large modern groups of mammals were present: the placental forms, which include most modern species, and the marsupials, which carry their young in pouches and are the dominant group in Australia today (p. 75).

Among the small Cretaceous mammals was the genus *Purgatorius* (Figure 17-12). This animal would have had no special interest for a human observer during its lifetime, but from our modern perspective it has

Figure 17-11 Stages in the evolution of mammals from mammal-like reptiles. Among the most important changes were the evolution of more highly differentiated, specialized teeth; enlargement of the brain; modification of jawbones into ear bones; and reduction of the number of bones forming the jaw to one, the dentary (shown in color). (After R. E. Sloan.)

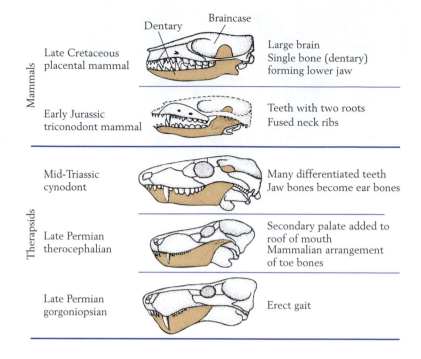

Mammals

Late Cretaceous placental mammal — Large brain / Single bone (dentary) forming lower jaw

Early Jurassic triconodont mammal — Teeth with two roots / Fused neck ribs

Therapsids

Mid-Triassic cynodont — Many differentiated teeth / Jaw bones become ear bones

Late Permian therocephalian — Secondary palate added to roof of mouth / Mammalian arrangement of toe bones

Late Permian gorgoniopsian — Erect gait

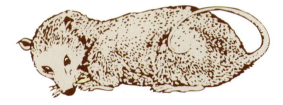

Figure 17-12 *Purgatorius*, a member of the group of early mammals that was ancestral to the primates, the order that includes humans. *Purgatorius* was the size of a rat.

great significance. *Purgatorius* belonged to a group of animals that was ancestral to modern primates, including humans, and some taxonomists even assign this genus to the primate order. When a mass extinction at the end of the Cretaceous Period swept away the dinosaurs, many small mammals were fortunate enough to survive. Among them was *Purgatorius*, whose fossil remains are found in very early Cenozoic deposits. With the oppressive dinosaurs gone, the surviving mammals proliferated in a spectacular evolutionary radiation, which produced the great diversity of mammals in the modern world. Had the group of ratlike animals to which *Purgatorius* belonged died out with the dinosaurs, we humans would never have evolved.

Paleogeography of the Cretaceous World

Because the Cretaceous System has undergone less metamorphism and erosion than older geologic systems,

it is represented on modern continents by an extensive record of shallow marine and nonmarine sediments and fossils. Cretaceous sediments and fossils are also widespread in the deep sea, in contrast to the sparse deep-sea records for the Triassic and Jurassic periods. This difference reflects the fact that movements of plates across Earth's surface are rapid enough so that by now a large percentage of deep-sea sediments older than the Cretaceous System have been swallowed up along subduction zones. The relative abundance of Cretaceous sediments in the ocean basins and on the continents helps us to interpret the geographic, oceanographic, and climatic patterns of the period. Additional information is drawn from the Upper Cretaceous fossil record of flowering plants. As we saw in Chapter 4, these organisms are particularly sensitive to climatic conditions.

Continents fragmented and narrow oceans expanded

Although Pangaea had begun to break apart early in the Mesozoic Era, the smaller continents that had formed from the supercontinent remained tightly clustered at the beginning of the Cretaceous Period. The continued fragmentation of Pangaea and the dispersion of its daughter continents were among the most important developments in Cretaceous global geography. At the start of the Cretaceous Period, Gondwanaland was still intact. By Late Cretaceous time, however, South America, Africa, and peninsular India had all become discrete entities; of the present-day continents that represent fragments of Gondwanaland, only Antarctica and Australia remained attached to each other (Figure 17-13).

LATEST CRETACEOUS

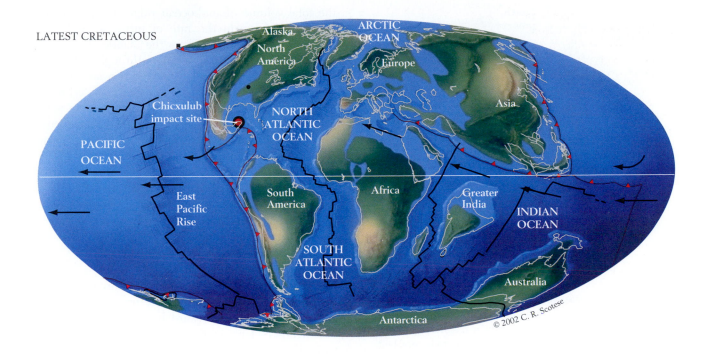

LATE CRETACEOUS

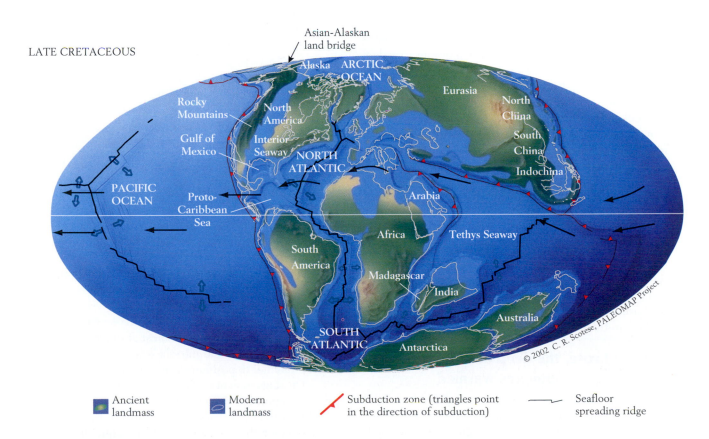

■ Ancient landmass	▨ Modern landmass	╱ Subduction zone (triangles point in the direction of subduction)	— Seafloor spreading ridge

Figure 17-13 **Changes in global geography during Late Cretaceous time.** Fragments of Pangaea moved farther apart. Greenland rifted away from North America, forming the Labrador Sea. Early in Late Cretaceous time, the Interior Seaway of North America extended from the Gulf of Mexico to the Arctic Ocean, and a strong current (black arrows) flowed westward through the Tethys Seaway. In latest Cretaceous time, the Interior Seaway retreated from the Arctic Ocean. Northward movement of Africa and lowering of sea level constricted the flow of water through the Tethys Seaway. (Adapted from paleogeographic maps by C. R. Scotese, PALEOMAP Project, University of Texas at Arlington, 1997.)

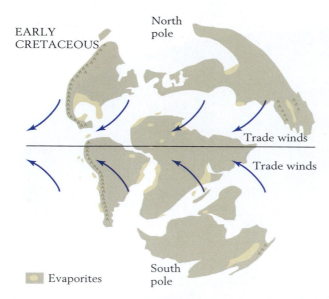

Figure 17-14 Sites of Early Cretaceous evaporite deposition. Most evaporites accumulated along margins of restricted shallow seas recently formed by the breakup of Pangaea. Evaporites are best developed in the trade wind belt.

The fragmentation and separation of continents during Cretaceous time caused narrow oceans to widen. Greenland finally broke away from North America, but remained attached to Scandinavia, with Great Britain wedged in between them (see Figure 17-13). A small, triangular piece of lithosphere that had been part of Gondwanaland remained attached to North America. It forms a portion of the southeastern United States, including Florida (see Figures 14-20 and 17-13).

Also notable were the Early Cretaceous openings of the South Atlantic Ocean, Gulf of Mexico, and Caribbean Sea. As we have seen, evaporites had formed during the Jurassic Period when marine waters spilled into the rifts that later widened to form the Gulf of Mexico and the South Atlantic (see Figure 16-21). Early in Cretaceous time, these basins remained narrowly connected to the rest of the world's oceans. Evaporites accumulated along the basin margins in these restricted bodies of water, especially in the trade wind belt (Figure 17-14).

Sea level rose, the deep ocean stagnated, and climates warmed

Throughout Early Cretaceous time global sea level rose, with only minor interruptions (see p. 419). As a result, sea level stood perhaps as high throughout mid-Cretaceous time as at any other time in the Phanerozoic, and extensive marine deposits blanketed most continents. On the North American craton these deposits constitute the Zuni sequence (see Figure 6-21). The rise in sea level apparently resulted from an expansion of

the total volume of mid-ocean ridges in combination with an increase in the rate of intrusion of Earth's oceanic crust by plumes from the mantle. The resulting increase in the flow of seawater through hot oceanic crust lowered the Mg^{2+}/Ca^{2+} ratio in seawater (p. 241). In particular, broad areas of seafloor originated in the eastern Pacific Ocean during the interval from about 125 million to 80 million years ago. This interval coincided with a remarkably long interval when Earth's magnetic field failed to reverse its polarity (Figure 17-15). Recall that the magnetic field originates through movements in Earth's core (p. 135). It has therefore been suggested that movement of a large plume of magma upward from near the core-mantle boundary may have altered movements within the core while simultaneously building extensive oceanic crust.

The Tethys Seaway A dominant feature of the Cretaceous world was the tropical Tethys Seaway, whose waters were driven westward by trade winds without obstruction by large landmasses. These waters warmed while flowing across the broad equatorial Pacific. The channel between Eurasia and Africa deflected the Tethyan current so that it carried heat from the Pacific to a latitude about 40° north of the equator, across continental crust that now forms southern Europe (see Figure 17-13). From there, the Tethyan current crossed the Atlantic, passed through the Gulf of Mexico, and flowed back to the Pacific, between North and South America, which were not yet connected. Tropical rudist reefs and low banks flourished throughout the Tethys Seaway; oxygen isotope ratios for Tethyan rudists from Greece and Turkey are displayed in Figure 10-16.

Ocean stagnation in mid-Cretaceous time The middle part of the Cretaceous Period was marked by intervals when marine muds rich in organic matter accumulated on continents, eventually forming black shales (see Figure 10-7). Some of the abundant organic matter within these shales has been transformed into large volumes of petroleum. The dark muds accumulated because of unusually poor circulation within the ocean and stagnation of much of the water column. As Figure 17-16 indicates, these anoxic waters must at times have spilled over from oceanic areas into shallow seas, leading to the epicontinental deposition of dark muds.

The stagnation of mid-Cretaceous seas contrasts with the strong mixing of the oceans today. In modern seas, cold waters in polar regions sink to the deep sea and spread along the seafloor toward the equator, carrying with them oxygen from the atmosphere (p. 95). Only a thin zone of the ocean is characterized by a low concentration of oxygen (see Figure 17-16A); this zone lies beneath the photic zone, where photosynthesis produces oxygen. The light color of the sediments on the present

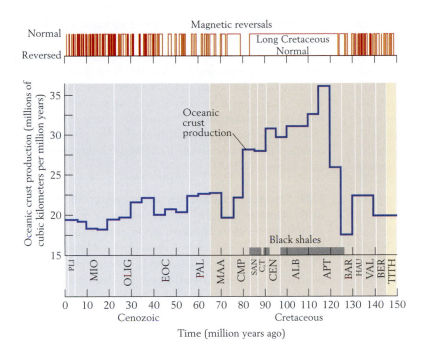

Figure 17-15 Correspondence in mid-Cretaceous time of a high rate of oceanic crust production with accumulation of abundant black shales and absence of magnetic reversals. (After R. L. Larson, *Geology* 19:963–986, 1991.)

seafloor reflects an abundance of oxygen in the deep sea. In mid-Cretaceous time, polar seas were apparently too warm for their surface waters to descend and spread oxygen throughout the deep waters of the ocean. As a result, the low-oxygen zone was greatly expanded, so that it extended upward into the photic zone (see Figure 17-16*B*).

The Tethys Seaway may have generated the waters of the deep sea during mid-Cretaceous time. At any time in Earth's history, the densest waters of the ocean have sunk to form the water mass of the deep sea. During the mid-Cretaceous interval, when polar regions were relatively warm, the densest shallow waters were probably bodies of hypersaline water formed by high rates of evaporation along continental margins in the dry trade wind belt.

The global climate in mid-Cretaceous time The mid-Cretaceous world has sometimes been described as a greenhouse world, in the sense that the average surface temperature was high as a result of a high concentration of carbon dioxide in the atmosphere. In fact, there is no evidence that Earth was warmer overall than during Triassic or Jurassic time. Furthermore, there are indications that atmospheric carbon dioxide was on the decline during the latter part of the Cretaceous (see Figure 10-13). Nonetheless, studies of oxygen isotopes in fossil rudists (see Figure 10-16) and in planktonic foraminifera indicate that tropical seas were slightly warmer in mid-Cretaceous time than they are today.

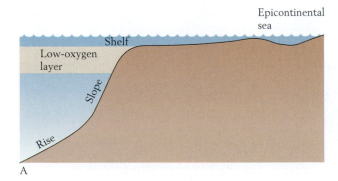

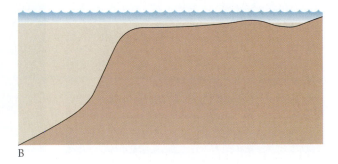

Figure 17-16 Expansion of the low-oxygen layer in the ocean at a time when the deep sea is warm. *A.* At times like the present, when cold, dense water at the poles is sinking to the deep sea, it supplies deep waters with oxygen, so that the low-oxygen layer is relatively thin. *B.* When polar waters are warmer and less dense, they do not sink, so they fail to supply oxygen to the deep sea. Under these conditions, the warm waters below the depth of wave activity are relatively stagnant, and the low-oxygen layer thickens, extending even into some epicontinental seas. (After A. G. Fischer and M. A. Arthur, *Soc. Econ. Paleontol. Mineral. Spec. Publ.* 25:19–50, 1977.)

The fossil record of life on land provides abundant evidence that warm temperatures spread to high latitudes during mid-Cretaceous time. Fossil leaves of warm-adapted plant species occur in Cretaceous deposits of northern Alaska, Greenland, and Antarctica (Figure 17-17). Dinosaurs also lived within about 15° of the Cretaceous south pole. Their fossil remains have been discovered near Melbourne, Australia, along with fossil plants that suggest an average annual temperature of about 10°C (50°F). These dinosaurs had to endure several weeks of darkness in winter, but the warm climate must have provided them with food throughout the year.

Some scientists favor the idea that hypersaline waters warmed polar regions during the Cretaceous. While strongly hypersaline waters descended to the deep sea, less hypersaline waters may have descended to intermediate depths at low latitudes and flowed poleward. Upward mixing of these intermediate waters may have delivered much of the heat that warmed climates at high latitudes.

The presence of warm climates at high latitudes had important consequences for wind-driven circulation in the oceans. At times like the present, when steep temperature gradients extend from the equator to the poles, large regional temperature contrasts produce strong winds (as, for example, cold, dense air masses push under warmer, lighter ones). Upwelling caused by strong winds stirs the ocean (p. 99). During mid-Cretaceous time, when temperature gradients were gentle, winds blowing over the ocean would have been weaker, on average, than they are today. Upwelling would therefore have been weaker as well. Thus the absence of strong wind-driven circulation contributed to the general stagnation of the lower portion of the mid-Cretaceous ocean (see Figure 17-16B).

Late Cretaceous changes in oceanic circulation and climate Oxygen isotope ratios in foraminifera shifted significantly between mid-Cretaceous time and the final Cretaceous age, named the Maastrichtian. These isotopic shifts, measured in specimens extracted from deep-sea cores, signal a general change in oceanic circulation. A shift toward lighter oxygen in foraminifera that inhabited the deep-sea floor indicates that the temperature of the deep sea rose from about 14°C to 20°C in mid-Cretaceous time and then fell, reaching about 9°C during the Maastrichtian. Isotope ratios in planktonic foraminifera indicate that surface waters at high latitudes also cooled between mid-Cretaceous and Maastrichtian time. It appears, therefore, that the stratified ocean of mid-Cretaceous time gave way to a new

A   B

Figure 17-17 Warm-adapted plants spread to high latitudes in the mid-Cretaceous. This fossil leaf from the Cretaceous System of Greenland (A) resembles the leaves of a modern breadfruit (B), a tropical plant. (A, Swedish Museum of Natural History, photo by Yvonne Arremo, Stockholm; B, Gerry Ellis/Wildlife Collection.)

kind of ocean in which waters that cooled at high latitudes sank to the deep sea, carrying oxygen with them.

A shift in the composition of terrestrial floras at high latitudes also points to polar cooling in Late Cretaceous time. This polar cooling may have resulted from changes in oceanic circulation, or it may have resulted from a reduction of atmospheric carbon dioxide and weakened greenhouse warming (see Figure 10-13).

Cooling of tropical seas may have been responsible for an extinction event that decimated the rudist reef fauna early in Maastrichtian time. As a result of this decline, the rudists were already an impoverished group when they finally died out altogether in the mass extinction that also killed off the dinosaurs and brought the Mesozoic Era to an end.

The Terminal Cretaceous Extinction

The Mesozoic Era came to a dramatic end with a mass extinction that was quite sudden on a geologic scale of time. Many forms of life that had played major ecological roles for tens of millions of years disappeared. The most prominent in the minds of modern humans were the dinosaurs, but many other groups of animals and plants died out as well. Both the gymnosperms and the angiosperms suffered heavy losses. In the ocean, ammonoids disappeared, as did reptilian "sea monsters," including mosasaurs, plesiosaurs, and giant turtles (see Figure 17-4). About 90 percent of all species of calcareous nannoplankton and planktonic foraminifera died out. On the seafloor, many groups of mollusks disappeared, including the remaining groups of reef-building rudists. The extinction of the rudists, like earlier extinctions of the Phanerozoic Eon, exemplified the fragility of reef ecosystems in general.

Discoveries during the past few years have established beyond a reasonable doubt that the collision of an asteroid with Earth caused the terminal Cretaceous mass extinction. The first of these discoveries, reported in 1981, was a high concentration of the rare heavy metal iridium—an *iridium anomaly*—at the stratigraphic level of the extinction. Iridium is rare in Earth's crust but more highly concentrated in meteorites that are asteroids or fragments of asteroids. Other discoveries followed quickly (see Earth System Shift 17-1).

The terminal Cretaceous impact issued a warning

In contemplating the consequences of the asteroid impact that ended the Mesozoic Era, scientists have gained new insights into global disturbances in general. For ex-

ample, the idea that atmospheric dust would refrigerate the world after such an impact led to the conclusion that a large-scale nuclear war would produce a "nuclear winter." The dust and soot from explosions and fires caused by a nuclear war would darken and cool the planet, causing massive deaths of humans and other forms of life far from areas of nuclear attack.

Fossils disguised the timing of the extinction

When it was first suggested that an extraterrestrial impact caused the terminal Cretaceous extinction, some aspects of the fossil record seemed to conflict with this idea. Several fossil groups appeared to have died out over several million years. Few species of dinosaurs, for example, were found in the uppermost few meters of Cretaceous sediment in Montana and nearby regions of Canada, where the fossil record of latest Cretaceous dinosaurs is the best in the world. Thus it appeared that the dinosaurs died out gradually near the end of the Cretaceous Period. The impact hypothesis, however, stimulated paleontologists to scour the uppermost Cretaceous fossil record more thoroughly than they had done before. The result was the discovery that many species of dinosaurs once thought to have died out long before the end of the Cretaceous may well have survived to the very end: their fossils have now been found within a meter of the iridium anomaly. The lesson here is that the imperfection of the fossil record—or our imperfect knowledge of that record—can fool us.

Opportunistic species flourished in the aftermath of the extinction

Most of the groups of animals and plants that survived the terminal Cretaceous extinction at reduced diversity expanded again during the Cenozoic Era. For a time, however, life on Earth was impoverished in many ways. Among the survivors, some species of calcareous nannoplankton underwent an especially interesting ecological change. Immediately after many species in this group died out or became rare at the very end of the Cretaceous, a few species blossomed to great abundance in the ocean. Perhaps these ecological opportunists (p. 84) were especially tolerant of abnormal conditions. After new species evolved during the Cenozoic Era, the opportunists declined in abundance. One species, however, survived for more than 150 million years: *Braarudisphaera bigelowi* (Figure 17-18) exists even today, but is confined to marginal marine lagoons. The fossil record shows that this form has occasionally spread to the open ocean and undergone population explosions since the Cretaceous, apparently because unusual conditions have briefly favored it.

Earth System Shift 17-1 | Death from Outer Space

In 1981 a team of researchers led by the physicist Luis Alvarez and his son Walter, a geologist, reported an abnormally high concentration of the element iridium precisely at the level of the Cretaceous-Paleogene boundary in a stratigraphic section at Gubbio, Italy. Soon a comparable iridium anomaly was found at the same stratigraphic level in many other parts of the world, in rocks of both marine and terrestrial origins. Because iridium is very rare on Earth but more abundant in meteorites, the Alvarez team advanced the hypothesis that a large asteroid had struck Earth at the end of the Cretaceous Period, producing a great explosion that dispersed iridium-rich dust high into the atmosphere. The dust from such an explosion would have spread around the globe and then settled to produce a thin layer of iridium-rich sediment in nearly all depositional environments. On a geologic scale of time, these events would have been instantaneous. An asteroid of average meteoritic composition would have had to be about 10 kilometers (6 miles) in diameter to produce the total amount of iridium that forms the worldwide anomaly.

Three types of sedimentary grains point to an extraterrestrial source for the widespread iridium anomaly at the terminal Cretaceous boundary. All three types can form only under very intense heat or pressure of the kind that develops when a large extraterrestrial body collides with Earth:

1. One kind of grain displays sets of parallel welded fractures that formed under enormous pressure. Grains of this type occur at sites where meteorites are known to have formed craters on Earth. These "shocked" grains have turned up in many parts of the world at the level of the iridium anomaly.

2. A second type of grain found at the level of the iridium anomaly is a "microspherule," a nearly spherical grain that resembles window glass in its molecular structure. In other words, it is largely uncrystallized: it cooled so rapidly after having been liquefied that its chemical elements failed to assemble into a consistent geometric pattern. Microspherules, like shocked grains, occur where meteorites are known to have struck Earth. They formed when droplets of rock, liquefied by the enormous heat generated during the impact, were thrown into the atmosphere, where they quickly cooled.

3. Grains of the third type are microscopic diamonds, which have been discovered at the level of the iridium anomaly at several sites in North America. Diamonds form only at extremely high pressures (p. 31).

Completing one of the greatest triumphs of modern geology, researchers have actually succeeded in locating the crater created by the deadly asteroid impact. This so-called Chicxulub crater is a ringlike structure, about 200 kilometers (120 miles) in diameter, straddling the shoreline of Mexico's Yucatán Peninsula. The impact caused an explosion that opened a cavity about 100 kilometers (60 miles) in diameter in the center of the ringlike structure. To produce such a large crater, the asteroid would have had to be about 10 kilometers (6 miles) in diameter—the dimension also estimated from the magnitude of the iridium anomaly. A core obtained by drilling the crater from a ship yielded rocks formed by cooling of magma that was produced by the heat of the impact. Argon-argon radiometric dating of these rocks showed that they formed 65 ± 0.4 million years ago. This time is within the narrow range of possible dates established for the Upper Cretaceous boundary elsewhere in the world.

The huge Chicxulub crater represents one of the largest impact structures produced by an extraterrestrial body in the last 4 billion years—after the interval

Figure 1 The terminal Cretaceous iridium anomaly can be detected in terrestrial deposits near Drumheller, Alberta, Canada. The anomaly is located in the thin clay bed at the lower end of the white pen that is perched on the outcrop. The last dinosaur bones occur just below the iridium anomaly. (Francois Gohier, San Diego.)

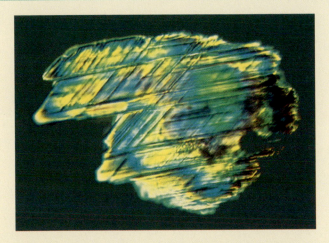

Figure 2 Shocked quartz grains, such as this one from Montana, are found in uppermost Cretaceous deposits throughout the world. These grains display sets of parallel welded fractures that result from exposure to extremely high pressures. (Glen A. Izett, U.S. Geological Survey.)

in Earth's history when large impacts occurred frequently (see p. 257). Given the unusually large size of the Cretaceous impactor, it is no surprise that its effects were so devastating.

Some of the effects of the impact were strongest in the region immediately surrounding the impact site. Terrestrial floras of western North America suffered especially high extinction rates, for example; about 75 percent of all species died out. In contrast, plants in Australia and New Zealand were virtually unscathed. In addition, the abundance of microspherules at the top of the Cretaceous System decreases with distance from the Chicxulub site. At localities in Mexico relatively close to the impact crater, the microspherules are concentrated in a layer about 1 meter (3 feet) thick; the equivalent layer in Texas is only about 10 centimeters (4 inches) thick, and still farther away, in New Jersey, the layer is a mere 5 centimeters (2 inches) thick. Much smaller concentrations of microspherules are found in regions of the world even more distant from the Chicxulub site. Pieces of fractured rock, apparently blasted from the impact crater, are found in latest Cretaceous strata in areas not far from the Chicxulub site.

Opinions have varied as to how the impact of an asteroid 10 kilometers in diameter would disturb environments on Earth. Here are some consequences that many scientists have attributed to the Chicxulub impact:

1. *Perpetual night.* Dust particles and tiny droplets of liquid called aerosols would have been blown high into the atmosphere and spread around the world, screening out nearly all sunlight. Many of these particles would have remained aloft for many months, perhaps preventing plants from conducting photosynthesis.

2. *Global refrigeration.* The darkening of the skies may have plunged the entire planet into cold, wintry weather for several months.

3. *Heat from the settling dust.* As dust returned to Earth, friction caused by its descent through the atmosphere would have produced a vast amount of heat. As a result, after experiencing a severe cold snap, life might have been subjected to abnormal warmth.

4. *Delayed greenhouse warming from aerosols.* The aerosols, being less dense than the dust particles, would have remained in the atmosphere long after the dust had settled, and by trapping solar radiation they would have prolonged the greenhouse warming that followed the impact.

5. *Sudden greenhouse warming from release of carbon dioxide.* More pronounced global warming—possibly the most profound climatic effect of the impact—would have resulted from the release of carbon dioxide from carbonate rocks that the asteroid penetrated. When the carbonates were heated and sheared at the time of impact, they must have liberated this gas, as happens less abruptly when limestone or dolomite is subjected to metamorphism in

Figure 3 Microspherules are found in uppermost Cretaceous deposits throughout the world. These grains come from a thin clay layer at the terminal Cretaceous boundary in Wyoming. Although they have undergone chemical alteration, they were once glassy structures that cooled rapidly from droplets of molten rock. (Bruce F. Bohor, U.S. Geological Survey.)

(continued)

Earth System Shift 17-1 | Death from Outer Space (continued)

Figure 4 The Chicxulub crater marks the site of the asteroid impact. *A.* Location of the crater. *B.* The crater as graphically portrayed by variations in the strength of Earth's gravitational field. Its outer ring has a diameter of about 200 kilometers. The crater flares out to the northwest, indicating that the impactor arrived from the southeast with a low-angle trajectory. *C.* Chunks of striated dolomite blasted from the impact crater are present at the level of the Cretaceous-Paleocene boundary in Belize, on the Yucatán Peninsula to the southeast of the impact site. (*B*, V. L. Sharpton, Lunar Planetary Institute, Houston; Courtesy of Kevin O. Pope, from *C*, A. C. Ocampo, K. O. Pope, and A. G. Fischer, *Geol. Soc. Amer. Spec. Paper* 307:75–88, 1996.)

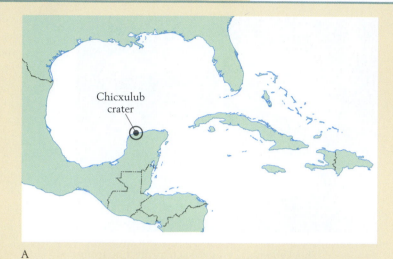

A

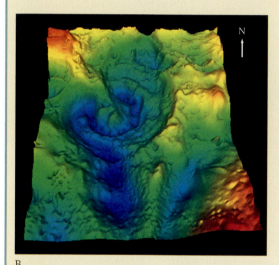

B

C

A similar pattern of opportunism is evident among plants. A phenomenon called a *fern spike* is present at the level of the anomaly in terrestrial sediments of west-

Figure 17-18 The calcareous nannoplankton species *Braarudisphaera bigelowi* (×3135). (From B. U. Haq and A. Boersma, eds., *Introduction to Micropaleontology*, Elsevier, New York, 1978.)

ern North America. Pollen becomes rare in the sediment close to the anomaly, and then, within an interval of just a few centimeters, the spores of ferns heavily dominate the assemblage of plant microfossils. This pattern suggests that communities of flowering plants died out suddenly and were replaced in the landscape by a heavy growth of ferns. Often when a fire or volcanic eruption wipes out a forest today, ferns readily invade the cleared land. Thus ferns are ecological opportunists (p. 84). Above the fern spike and iridium anomaly are fossil pollen and leaves of flowering plants that represent a flora quite different from the kind that existed during Late Cretaceous time. Recall that a similar fern spike marks the terminal Triassic mass extinction in terrestrial strata (p. 407).

On land, both angiosperms and mammals were beneficiaries of the terminal Cretaceous extinction.

a mountain belt (p. 232). It is estimated that the volume of the carbonate rocks thus affected by the impact was so large that the carbon dioxide released would have elevated the average global temperature of Earth's lower atmosphere by at least 4.5°C and perhaps by as much as 13.5°C. In fact, fossil plant leaves provide evidence of a brief but dramatic rise in the level of atmospheric carbon dioxide precisely at the Cretaceous-Paleocene boundary. This evidence comes from the proportion of stomate cells on the surfaces of fossil leaves (see Figure 16-23). The proportions of stomate cells on fossil leaf surfaces indicate that levels of carbon dioxide in Earth's atmosphere 1–2 million years before and after the mass extinction were similar to or slightly higher than that of the present. In contrast, stomates are very sparse in fossil fern fronds found in New Mexico within a thin layer positioned just a few centimeters above the iridium anomaly, representing the brief interval when ferns monopolized terrestrial landscapes. Experiments with modern ferns closely related to those fossil ferns have revealed that atmospheric carbon dioxide levels at least seven times the present level are required for ferns to grow fronds with as few stomates per square centimeter as are found in the fossils. Here, then, is independent evidence that the greenhouse effect was dramatically intensified. The warming must have subsided rapidly, however, as increased rates of weathering removed carbon dioxide from the atmosphere; temperatures probably approached normal levels again within a few hundred thousand years.

6. *Acid rain.* Some of the rocks that the asteroid penetrated in forming the Chicxulub crater were sulfate evaporites. The impact must have released oxides of sulfur from these evaporites, and chemical reaction of these compounds with water in the atmosphere would have produced sulfurous and sulfuric acid. The result would have been acid rain, which may have harmed many forms of life.

7. *Fires.* Wildfires would have raged across the land, triggered by the fiery cloud that burst from the impact site. The most severe fire damage to life should have been relatively close to the Chicxulub site.

The lethal consequences of the impact were global. Dinosaurs and ammonoids vanished from both hemispheres. Extinction rates differed among different types of organisms, however. Interestingly, freshwater vertebrates in west-central North America—fishes, amphibians, turtles, and crocodiles—suffered relatively little extinction at the end of the Cretaceous, despite their proximity to the impact. Perhaps these animals benefited from immersion in water, which because of its high heat capacity would have undergone less cooling or heating than the air above. In the oceans, virtually all dinoflagellate species survived the terminal Cretaceous crisis. Presumably their ability to form protective cysts under unfavorable circumstances allowed them to survive (p. 62). The mammals obviously fared better than the dinosaurs. Whatever the reason for the mammals' relatively low incidence of extinction, we humans must be grateful for it because some of the mammalian survivors were our ancestors.

Angiosperms quickly rose to dominance over gymnosperms, and in the absence of dinosaurs, mammals underwent a spectacular diversification that led to the mammal-dominated world to which we now belong.

North America in the Cretaceous World

Mountain building continued in western North America during the Cretaceous Period, and it produced an enormous foreland basin that became flooded by a seaway that extended from the Gulf Coast to the Arctic Ocean. The Gulf Coast itself was fringed by rudist reefs and banks. A rudist-rimmed carbonate bank also stretched along a large segment of the adjacent Atlantic coast until midway through the Cretaceous Period, when it gave way to the deposition of mud and sand that continues today.

Cordilleran mountain building continued

During Cretaceous time, an important change took place in the pattern of igneous activity in western North America. Subduction of the Franciscan mélange along the western margin of the continent continued, as did the associated igneous activity (Figure 17-19; see also Figure 16-32). Nonetheless by Late Cretaceous time, although volcanic and plutonic activity persisted in the Sierra Nevada region, the northern igneous activity had shifted eastward to Nevada and Idaho. This pattern contrasted with that of the Late Jurassic Epoch, when igneous activity in the north had been centered near the coast, in northern California and Oregon (see Figure 16-33). The eastward migration of igneous activity in the northern United States apparently resulted from a decrease in the angle of subduction. This change would have resulted from an increased rate of westward movement of the North American plate, which would have caused faster rollback of the subducted Pacific plate (see Figure 9-16). The subducted crust would therefore have failed to sink deep enough to melt until it had extended

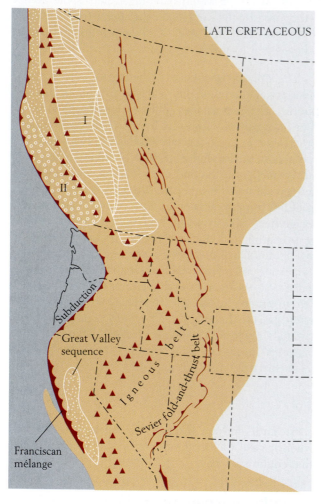

Figure 17-19 Late Cretaceous geologic features of western North America. Subduction produced the Franciscan mélange in California. North of California, igneous activity resulting from subduction was located far to the east of the continental margin; this activity, together with the folding and thrusting to the east, represented the latter part of the Sevier orogeny. In Canada, the margin of the continent consisted of two blocks of exotic terranes (I and II) that had been sutured to North America earlier in the Mesozoic Era; each of these blocks consisted of two or more slivers of crust that were welded together to form the block before it was attached to North America.

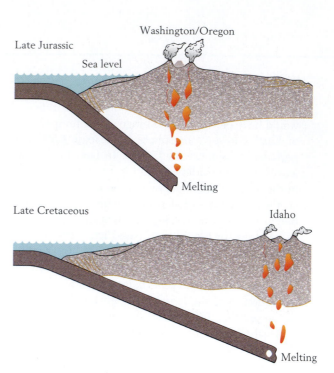

Figure 17-20 A likely explanation for the eastward migration of igneous activity in the Cordilleran region during Cretaceous time. The subducted plate began to move downward at a reduced angle, so that it reached the depth of melting only after passing far to the east.

far inland (Figure 17-20). The fold-and-thrust belt in front of the mountainous igneous region also shifted inland in the northern United States. By Late Cretaceous time, folding and thrusting extended eastward as far as the Idaho-Wyoming border.

A major episode of igneous activity and eastward folding and thrusting coincided approximately with the Cretaceous Period; although this episode was not entirely divorced from earlier and later tectonic activity, it is separately identified as the *Sevier orogeny*. East of this orogenic belt lay a vast foreland basin, which in Late Cretaceous time was occupied by a narrow seaway stretching from the Gulf of Mexico to the Arctic Ocean.

The orogenic belt that occupied western North America during the latter half of Cretaceous time was unusually broad, apparently because of low-angle subduction. In its development of a foreland basin and certain other features, however, this orogenic belt was typical. For example, it was symmetrical (see Figure 9-11): the Sevier folding and faulting east of the belt of igneous activity mirrored on a larger scale the Franciscan deformation at the continental margin (see Figure 17-19).

The Mesozoic history of western Canada is far more complicated. Recall that during the Jurassic Period, a sizable microcontinent was sutured to this region of

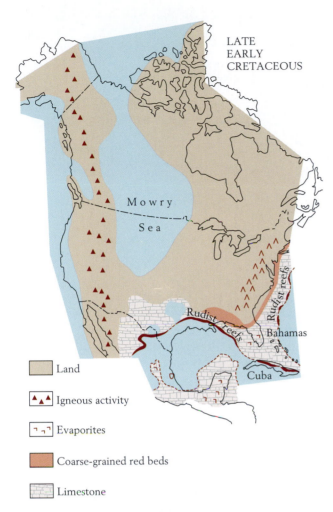

LATE
EARLY
CRETACEOUS

Mowry

Sea

Rudist reefs

Rudist reefs

Bahamas

Cuba

Land

Igneous activity

Evaporites

Coarse-grained red beds

Limestone

Figure 17-21 Geography of North America late in Early Cretaceous time. The Mowry Sea, where black muds were deposited, spread southward from the Arctic Ocean. A carbonate platform bordered by rudist reefs encircled the Gulf of Mexico, and carbonate deposition extended far to the north along the East Coast. (After G. D. Williams and C. R. Stelck, *Geol. Assn. Can. Spec. Paper* 13:1–20, 1975.)

North America (see Figure 16-29). This theme of continental accretion continued into the Cretaceous Period, when a small microcontinent was attached along the western margin of the first. This new landmass, like the one accreted during the Jurassic, was a composite of two or more terranes. They had become amalgamated during the Jurassic Period and were attached to North America during Cretaceous time.

A seaway connected the Gulf of Mexico and Arctic Ocean

Shortly before the end of Early Cretaceous time, during the Albian Age, Arctic waters spread southward, flooding a large area of western North America to form the Mowry Sea (Figure 17-21), named for the Mowry Formation that accumulated within it. The Mowry consists mostly of *oil shale*, a well-laminated shale in which dark layers rich in fish bones and scales alternate with thicker, lighter layers. The Mowry Sea formed as a part of the great mid-Cretaceous marine transgression that resulted in the deposition of black shales on many continents (see Figure 10-7). To the south, the Gulf of Mexico was part of the tropical Tethyan realm, and rudist reefs flourished around its margin.

The Mowry Sea made brief and intermittent contact with the Gulf of Mexico before the end of Early Cretaceous time, but an enduring connection was established at the start of the Late Cretaceous. The result of this contact was the enormous Cretaceous Interior Seaway, which occupied the foreland basin to the east of the Sevier orogenic belt. Until just before the end of the Cretaceous Period, this seaway extended from the Gulf of Mexico to the Arctic Ocean (see Figure 17-13). Most of the sediments deposited here were shed from the Cordilleran mountains that formed to the west. The history of the seaway is especially well understood because of excellent stratigraphic correlations based on abundant fossil ammonoids; in addition, ash falls from volcanic eruptions to the west provided numerous marker beds, many of which can be dated radiometrically (see Figure 6-14).

Barrier islands bounded much of the seaway. Behind them stood lagoons bordered by broad swamps. On the western margin of the seaway the swamps gave way to plains, which were succeeded near the mountains by alluvial fans.

The western shoreline of the seaway shifted back and forth, primarily in response to the rate of sediment supply. At all times conglomeratic sediments were shed eastward from the neighboring mountains as clastic wedges, but at times of particularly active thrusting or uplift these wedges prograded especially far to the east (Figure 17-22). Because of the great weight of sediments on the western side of the seaway, subsidence was more rapid there than farther east. Along the western margin, in nonmarine environments, Late Cretaceous dinosaurs left a rich fossil record.

The Upper Cretaceous strata of the Interior Seaway represent large depositional cycles, one of which is illustrated in Figure 17-23. Each cycle consists of an interval of transgression followed by an interval of regression. In addition to changing rates of sediment supply, global changes in sea level and the changing rate of subsidence of the seaway floor must have influenced these patterns of transgression and regression. At times of low sediment supply and maximum lateral expansion of the seaway, chalks were laid down in the center of the basin. The most famous of these deposits is the Niobrara Chalk, which occupies the

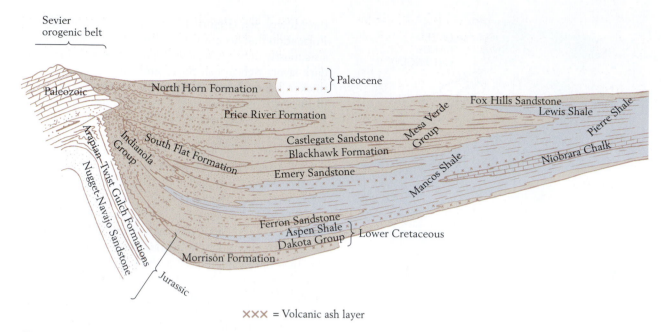

Figure 17-22 **Cross section of Upper Cretaceous sediments in central Utah.** These sediments were deposited in the foreland basin east of the Sevier orogenic belt. Clastic wedges in the west grade eastward into finer-grained marine sediments. (After R. L. Armstrong, *Geol. Soc. Amer. Bull.* 79:429–458, 1968.)

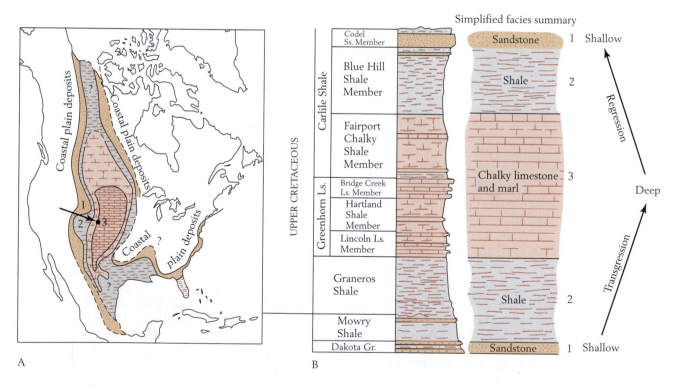

Figure 17-23 **The early Late Cretaceous "Greenhorn" depositional cycle of the North American Interior Seaway.** *A.* The three facies in the simplified facies summary are shown in map view. *B.* The stratigraphic section represents the vicinity of eastern Colorado (arrow on map), where the cycle developed by an oscillation of the shoreline (transgression and regression). During the transgression, the area of chalky limestone and marl (limey clay) deposition in the center of the basin (facies 3) expanded; first facies 2 and then facies 3 spread into eastern Colorado. During the regression, the area of deposition of facies 3 contracted, and facies 2 and then facies 1 shifted into eastern Colorado. (After E. G. Kauffman, *Mountain Geologist* 6:227–245, 1969.)

middle of a transgressive-regressive cycle. The Niobrara has yielded beautifully preserved fossil vertebrates (see Figures 17-3 and 17-4).

Just before the end of the Cretaceous Period, the seas retreated southward from the Interior Seaway, and a new pulse of mountain building began along its western margin. This *Laramide orogeny* continued well into the Cenozoic Era. Except for a brief and less extensive incursion just after the beginning of the Cenozoic Era, the seas have never returned to the western interior of North America.

The modern continental shelf formed in eastern North America

Seismic studies reveal a great thickness of sediments beneath the continental shelf bordering eastern North

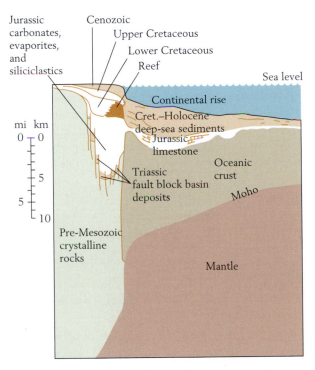

Figure 17-24 **Cross section of the continental shelf and deep sea off the coast of New Jersey.** Here the opening of the Atlantic is recorded by early Mesozoic sediments deposited within fault block basins. During Early Cretaceous time, a reef-rimmed carbonate platform extended this far north under the influence of warm Tethyan ocean currents. During Late Cretaceous time, carbonate deposition gave way to siliciclastic deposition, which has predominated to the present day. (After R. E. Sheridan et al., *The Geology of Continental Margins*, Springer-Verlag, New York, 1974.)

America (Figure 17-24). These sediments consist of deposits laid down during the early Mesozoic episodes of rifting that formed the modern Atlantic Ocean. At the base are fault block basin deposits like those of the Newark Supergroup that are exposed on the continent to the west (see Figure 16-24). Next come large thicknesses of Jurassic carbonates that accumulated in the narrow, young Atlantic Ocean as passive margin deposition commenced (see Figure 17-24).

Above the Jurassic carbonates are more carbonates from the Early Cretaceous interval, when, under much warmer climatic conditions than exist today, reef-rimmed carbonate banks bordered the ocean from Florida to New Jersey (see Figure 17-21). Before the end of Early Cretaceous time, reef growth gave way to deposition of predominantly siliciclastic sediments. This change marked the beginning of the growth of the large clastic wedge that forms the modern continental shelf. The clastic wedge consists largely of sands and muds from the Appalachian region laid down in nonmarine and shallow marine settings. These sediments were apparently supplied by uplands produced by renewed uplift of the Appalachian mountain belt to the west, after it had been largely leveled during Jurassic time.

Off the coast of New Jersey, the total thickness of Cretaceous and Cenozoic sediments is approximately 3 kilometers (2 miles) (see Figure 17-24). Only to the south, in southern Florida, did carbonate deposition persist to the present. Here, about 3 kilometers of sediments, consisting mainly of carbonates, accumulated during the Cretaceous Period, and another 2 kilometers or so were added during the Cenozoic Era.

The Chalk Seas of Europe

To many geologists, the term *chalk* means specifically the soft, fine-grained limestones of western Europe, although, as we have seen, chalky rocks are found elsewhere, including the Cretaceous System of North America. The Cretaceous chalks of Europe are spectacularly displayed as the White Cliffs of Dover and the coastal cliffs of Denmark (see p. 416 and Figure 10-21). They accumulated throughout almost all of Late Cretaceous time, when high seas led to extensive flooding of western Europe.

Except in the newly forming Alps to the south, tectonic activity was largely absent from western Europe during Late Cretaceous time, when the chalk accumulated. Here and there, relatively stable blocks stood relatively high as islands or, during transgressions, as shallow seafloors (Figure 17-25). Chalk accumulated almost continuously in the basins that surrounded these massifs.

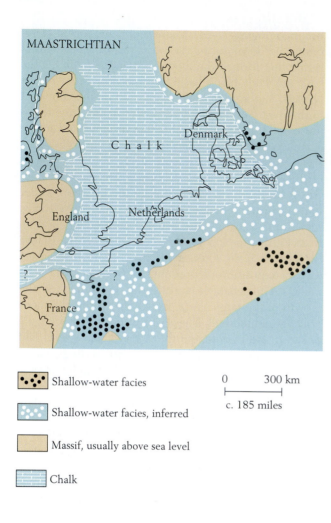

Shallow-water facies

Shallow-water facies, inferred

Massif, usually above sea level

Chalk

0   300 km

c. 185 miles

◄ **Figure 17-25** Paleogeography of northwestern Europe during Maastrichtian time. Chalk deposition was centered in the North Sea basin. Several stable blocks (massifs) formed islands around which marginal facies developed; these facies consisted mainly of coarse limestones but included siliciclastics as well. (After E. Hakansson et al., *Spec. Publ. Int. Assoc. Sedimentol.* 1:211–233, 1979.)

The lower portions of the chalk often contain clay, but the remainder is relatively pure calcium carbonate, consisting of minute skeletal debris that is about 75 percent planktonic in origin.

The general presence of oxygenated conditions on the floor of the chalk seas is demonstrated by the widespread occurrence of a fauna of bottom-dwelling species in the chalk—among them bryozoans, arthropods, foraminifera, brachiopods, bivalves, sea urchins, and soft-bodied burrowers. Knowing the typical thickness of the European chalk and the total time elapsed during its deposition, we can estimate that it accumulated at a rate perhaps as great as 15 centimeters (6 inches) per 1000 years. Because the very low Mg^{2+}/Ca^{2+} ratio of Late Cretaceous seawater favored the precipitation of calcite, the productivity of calcareous nannoplankton was greater in both the chalk seas of Europe and the Interior Seaway of North America than it has ever been anywhere in the ocean since Cretaceous time (see Figure 10-20).

Chapter Summary

How did plankton in the ocean become more modern in their general aspect during the Cretaceous Period?

With the evolutionary expansion of the dinoflagellates, diatoms, and calcareous nannoplankton during the Cretaceous Period, the phytoplankton assumed a modern character. Similarly, the diversification of the planktonic foraminifera contributed to the modernization of the zooplankton.

How does seawater chemistry account for the name "Cretaceous"?

Because calcareous nannoplankton benefited from the very low Mg^{2+}/Ca^{2+} ratio in seawater, their shieldlike plates rained down on the seafloor to produce thicker deposits of chalk than are known from other geologic intervals. *Creta* is the Latin name for "chalk."

How did the evolutionary expansion of predatory taxa transform the marine ecosystem during Cretaceous time?

Crabs, teleost fishes, and carnivorous snails diversified markedly during Cretaceous time and caused the decline of vulnerable sedentary groups such as brachiopods and stalked crinoids.

How did the reef ecosystem change during the Cretaceous Period?

In mid-Cretaceous time, rudist bivalves temporarily displaced corals as the primary builders of organic reefs.

How did terrestrial floras change during the Cretaceous Period?

On land, angiosperms diversified, but gymnosperms and ferns remained more abundant in most environments.

What happened to Gondwanaland during the Cretaceous Period?

Gondwanaland broke apart during the Cretaceous Period, forming the South Atlantic and other oceans.

What was the Tethys Seaway?

The Tethys Seaway was a tropical seaway that carried warm waters from the Pacific Ocean through the Mediterranean region and the Gulf of Mexico and back to the Pacific Ocean again.

What was the structure of the ocean during mid-Cretaceous time?

Latitudinal temperature gradients were gentle and oceanic circulation was sluggish. The ocean was stratified, and deep and mid-depth waters were depleted of oxygen. High rates of production of oceanic crust elevated sea level, and organic-rich black muds accumulated in epicontinental seas.

What was the tectonic and geographic configuration of western North America during Late Cretaceous time?

Orogenic activity shifted eastward from its Jurassic position. To the east of the orogenic belt, the Interior Seaway stretched from the Gulf of Mexico to the Arctic Ocean.

What evidence is there that the collision of an asteroid with Earth caused the mass extinction at the end of the Cretaceous Period?

At the Cretaceous-Paleogene boundary, on a global scale, there is a high concentration of the heavy metal iridium, as well as shocked mineral grains, microspherules formed from droplets of melted rock, and tiny diamonds.

Review Questions

1. What accounts for the abundance of chalk in the Cretaceous System?

2. What were the most prominent groups of swimming predators in Cretaceous seas?

3. What general climatic and oceanographic conditions characterized the mid-Cretaceous world?

4. How did climatic and oceanographic conditions change a few million years before the end of the Cretaceous Period? What were the consequences for the reef-building rudists?

5. What conditions may account for the formation of widespread black shales in seas that spread over continental surfaces at certain times during the Cretaceous Period?

6. What modern continents that were once part of Gondwanaland remained attached to each other at the end of the Cretaceous Period?

7. Why did thick siliciclastic deposits accumulate in the western interior of North America during Cretaceous time?

8. How did Greenland become separated from North America?

9. What happened along the passive margin of the eastern United States during the Cretaceous Period?

10. What evidence is there that an asteroid struck Earth at the end of the Cretaceous Period?

11. Major physical and chemical events altered life during and at the very end of the Cretaceous Period. Using the Visual Overview on page 418 and what you have learned in this chapter, describe these events and their biological consequences.

Pinkish terrestrial Eocene sediments rest atop tan marine Cretaceous sediments in Badlands National Park, South Dakota. (Tom Bean.)

CHAPTER 18

The Paleogene World

The transition between the Cretaceous and Paleogene periods marked a major shift in Earth's history. Scarcely any belemnoids survived, and ammonoids, rudists, and marine reptiles disappeared from the seas. What remained at the opening of the Cenozoic Era were marine taxa that persist as familiar inhabitants of modern oceans, among them bottom-dwelling mollusks and teleost fishes. On land, the flowering plants of the Paleogene Period resembled those of latest Cretaceous time in many ways, but animal life changed dramatically. Taking the place of the dinosaurs were the mammals, which were universally small and inconspicuous at the start of the Paleogene but in many ways resembled modern mammals by the period's end.

The most profound geographic change during Paleogene time was a refrigeration of Earth's polar regions, which resulted in the formation of the Antarctic ice cap and chilling of the deep sea. Paleogene mountain-building events in western North America foreshadowed Neogene uplifts of such ranges as the Sierra Nevada and the Rocky Mountains. For the most part, the siliciclastic sediments that record these and other Paleogene events are unconsolidated, or soft, whereas most carbonates are lithified.

Early in the history of modern geology, the marked difference between Mesozoic and Cenozoic biotas was readily apparent. Subdivision of the Cenozoic Era itself is less clear-cut. Today many geologists divide the era into two periods: the Paleogene Period, which includes the Paleocene, Eocene, and Oligocene epochs, and the Neogene Period, which includes the Miocene, Pliocene, Pleistocene, and Holocene epochs. This Paleogene-Neogene classification has become increasingly popular during the past two decades because it yields two periods of similar duration. Traditionally, however, the Cenozoic Era has been divided into two periods of quite different lengths: the Tertiary, which encompasses the interval from the Paleocene through the Pliocene, and the Quaternary, which includes only the Pleistocene and Holocene epochs (an interval of less than 2 million years).

The first formally recognized Paleogene epoch was the Eocene, which Charles Lyell established in 1833 on the basis of deposits found in the Paris and London basins (see Figure 6-1). Lyell named and described the Eocene Series in his great book *Principles of Geology*, the work that popularized the uniformitarian view of geology (p. 6). It was not until 1854 that Heinrich Ernst von Beyrich distinguished the Oligocene Series from the Eocene in Germany and Belgium on the basis of fossils. Still later, in 1874, W. P. Schimper established the Paleocene Series on the basis of distinctive fossil assemblages of terrestrial plants in the Paris Basin.

Era	Period	Epoch	Time
Cenozoic	NEOGENE	Holocene	
		Pleistocene	
	PALEOGENE	Pliocene	
Mesozoic	CRETACEOUS	Miocene	
			24 million years
	JURASSIC	Oligocene	
	TRIASSIC	Eocene	
Paleozoic	PERMIAN	Paleocene	
	CARBON-IFEROUS (PENNSYLVANIAN / MISSISSIPPIAN)		65 million years
	DEVONIAN		
	SILURIAN		
	ORDOVICIAN		
	CAMBRIAN		
	"PRECAMBRIAN" — PROTEROZOIC EON / ARCHEAN EON		

445

Major Events of the Paleogene

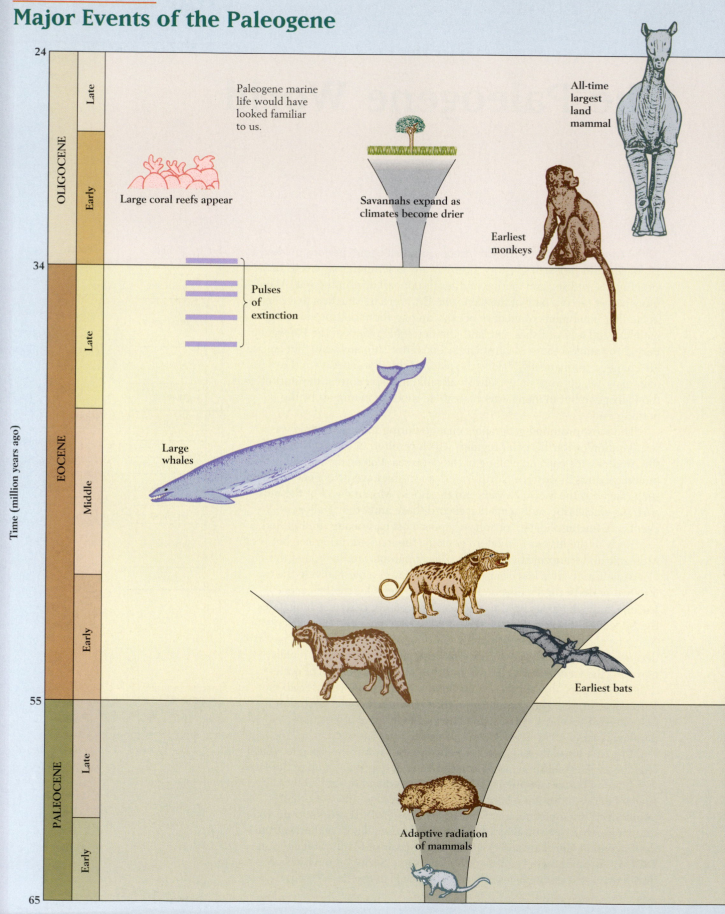

Time (million years ago)

24

OLIGOCENE

Late

Early

Paleogene marine life would have looked familiar to us.

Large coral reefs appear

Savannahs expand as climates become drier

All-time largest land mammal

Earliest monkeys

34

EOCENE

Late

Pulses of extinction

Large whales

Middle

Early

Earliest bats

55

PALEOCENE

Late

Early

Adaptive radiation of mammals

65

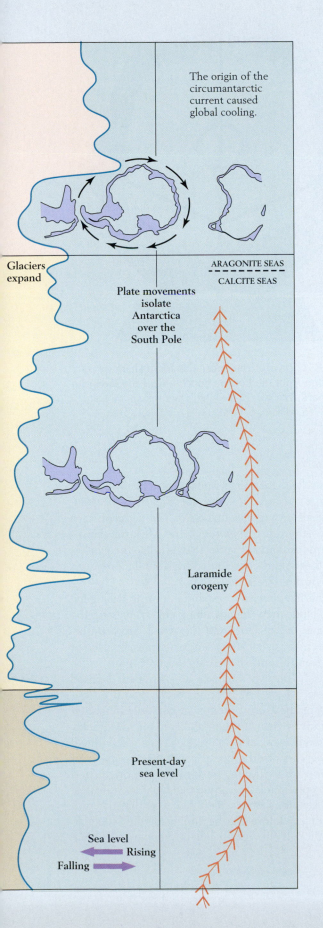

The origin of the circumantarctic current caused global cooling.

Glaciers expand

ARAGONITE SEAS

- - - - - - -

CALCITE SEAS

Plate movements isolate Antarctica over the South Pole

Laramide orogeny

Present-day sea level

Sea level

← Rising

Falling →

MIDDLE MIOCENE

MIDDLE EOCENE

LATE CRETACEOUS

Life of the Paleogene

Paleogene life is so familiar to us that it requires no special introduction; its most interesting features are the major expansions and contractions of certain taxonomic groups in all parts of the globe. Nor do we need a special section on Paleogene paleogeography, because the Paleogene lasted only 41 million years—a span in which paleogeographic changes occurred on such a small scale that they are best considered as regional events.

Marine life recovered

The present marine ecosystem is for the most part populated by groups of animals, plants, and single-celled organisms that survived the extinction at the end of the Mesozoic Era to expand during the Cenozoic. Many groups of benthic foraminifera, sea urchins, cheilostome bryozoans, crabs, snails, bivalves, and teleost fishes survived in sufficiently large numbers to assume prominent ecological positions in Paleogene seas.

Only three species of planktonic foraminifera appear to have survived the terminal Cretaceous extinction, but they gave rise to a remarkably rapid evolutionary radiation—one that yielded 17 new species, assigned to 8 new genera, within the first 100,000 years of Paleocene time.

Calcareous nannoplankton, which had suffered severe losses at the end of the Cretaceous Period, also rediversified rapidly during the Paleogene. These forms, as well as the diatoms and dinoflagellates, which were not so adversely affected, have accounted for much of the ocean's productivity throughout the Cenozoic Era, just as they did in Cretaceous time. Nonetheless, after the Cretaceous, the calcareous nannoplankton formed conspicuous chalk deposits again only early in Paleogene time (see Figure 10-21). Presumably the sharp rise in the Mg^{2+}/Ca^{2+} ratio of seawater during the Cenozoic Era reduced their productivity (see Figure 10-20).

Interestingly, the corals, having been displaced as dominant reef builders by the rudists during the Cretaceous, failed to take rapid advantage of the rudists' demise: they built few massive reefs during Paleocene and Eocene time. The influence of the Mg^{2+}/Ca^{2+} ratio of seawater on these aragonitic forms seems to have been the opposite of its effect on the calcareous nannoplankton, which secrete calcite. Only during the Oligocene, when the Mg^{2+}/Ca^{2+} ratio reached a high level, did Cenozoic corals begin to form massive reefs throughout tropical seas.

Although many elements of Paleogene marine life closely resembled those of Late Cretaceous age, some forms were dramatically new. Perhaps the most distinctive marine organisms of this period were the whales, which evolved during the Eocene Epoch from

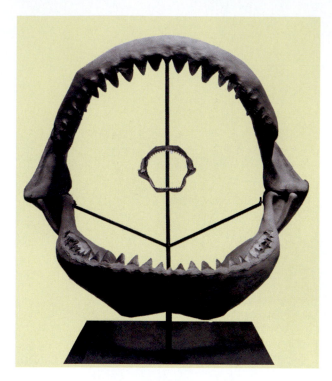

Figure 18-1 Jaws of the enormous fossil shark *Carcharodon* from the Eocene Epoch engulf the jaws of a modern shark. The jaws of the fossil shark were more than 2 meters (6.5 feet) across. (Field Museum of Natural History.)

carnivorous land mammals and quickly achieved success as large marine predators (see Figure 7-16). Joining the whales as replacements for the reptilian "sea monsters"—the top carnivores of the Mesozoic Era—were enormous sharks (Figure 18-1). Unlike the whales, however, these sharks descended from similar creatures that lived during Cretaceous time.

The marine ecosystem expanded during Paleogene time to include new niches along the fringes of the oceans. Sand dollars, for example, which are the only sea urchins able to live along sandy beaches, evolved at this time from biscuit-shaped ancestors (Figure 18-2). New kinds of bivalve mollusks also invaded exposed sandy coasts, where few benthic species had lived before. Both of these new groups successfully inhabited shifting sands by virtue of their ability to burrow quickly into the sand again after they have been dislodged. Other newcomers to the ocean margins were the penguins, a group of swimming birds of Eocene origin, and possibly the pinnipeds, the group that includes walruses, seals, and sea lions. It is widely believed that the pinnipeds evolved before the beginning of the Neogene Period, although this group left no known Paleogene fossil record.

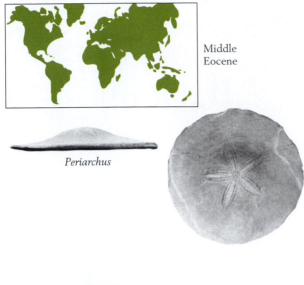

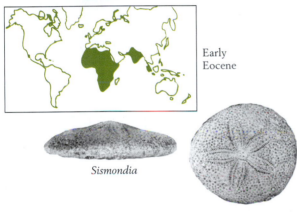

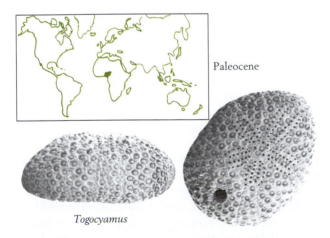

Figure 18-2 The evolution of sand dollars. The biscuit-shaped Paleocene ancestors of sand dollars (*Togocyamus*) are known only from Africa. *Sismondia* is an intermediate genus that has been found in Africa and India. These animals were the size of small cookies. True sand dollars (*Periarchus*) were present throughout the world by Middle Eocene time. (Photos, P. M. Kier, Smithsonian Institution.)

Flowering plants rose to dominance

Land plants did not experience major evolutionary changes early in Paleogene time. Instead, the greatest change in terrestrial vegetation was ecological: flowering plants assumed a much larger role after the terminal Cretaceous extinction, while gymnosperms and ferns played a lesser role. In the process, modern families of flowering plants evolved. By the beginning of Oligocene time, some 34 million years ago, about half of all genera of flowering plants were ones that are alive today, and although many modern plant genera had not yet evolved, forests had taken on a distinctly modern appearance.

One major event in plant evolution that did take place during the Paleogene interval was the origin of the grasses, although these usually low-growing flowering plants did not reach their full ecological potential until Late Oligocene and Miocene times. Early grasses may have been confined to wooded or swampy areas. The mode of growth of early grasses, like that of the modern sedges that form marshlands along continental coastlines, did not allow their leaves to grow continuously and thus to recover from heavy grazing by the kinds of animals that abound in open country. It was only an adaptive breakthrough—the origin of the continuous growth process, which forces us to cut our lawns every week or two—that ultimately enabled grasses to invade open country with great success. Once they were able to survive the effects of heavy grazing, grasses quickly spread to form vast expanses of grasslands.

Mammals radiated dramatically in the Paleocene and Eocene

The mammals, having inherited the world from the dinosaurs, underwent a remarkably rapid adaptive radiation during the early part of the Cenozoic Era (Figure 18-3). It is this radiation that gave the era its informal name: the Age of Mammals. It was probably primarily through predation rather than competition that the dinosaurs had prevented the mammals from undergoing any great evolutionary expansion during Mesozoic time.

Early in the Paleocene Epoch, when they first had the world to themselves, mammals were small creatures, most of which resembled modern rodents; no mammal seems to have been substantially larger than a good-sized dog. Furthermore, most mammal species tended to remain generalized in both feeding and locomotory adaptations; ground-dwelling species generally retained a primitive limb structure that caused the heels of the hind feet and the "palms" of the front feet to touch the ground as the animals moved about. Perhaps 12 million years later, however, by the end of Early Eocene time, mammals had diversified to the extent that most of their

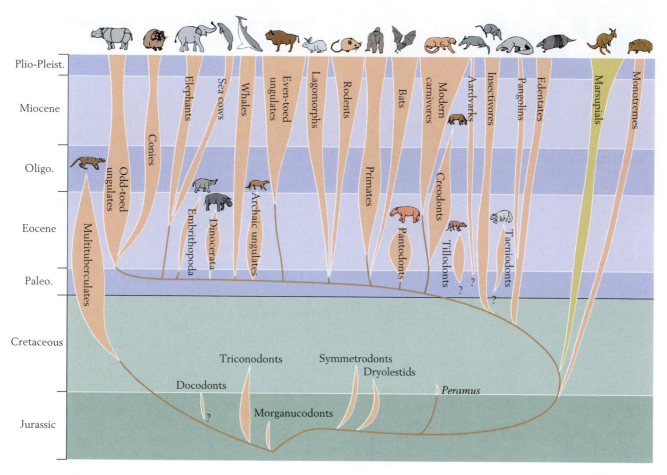

Figure 18-3 The pattern of adaptive radiation of mammals in the Cenozoic Era. It is thought that the two large groups of modern mammals, the placentals and the marsupials, had a common ancestry in the Cretaceous Period. The multituberculates, which failed to survive into Neogene time, apparently evolved separately, as did the monotremes, primitive survivors of which include the Australian platypus, which lays eggs instead of giving birth to live young. For the placental mammals, each vertical bar represents an order. Note that most of the new Cenozoic orders had already evolved by the beginning of the Eocene Epoch, about 10 million years after the dinosaurs disappeared. (D. R. Prothero.)

modern orders were in existence (see Figure 18-3). Bats already fluttered through the night air (Figure 18-4), for example, and, as we have seen, large whales swam the oceans.

Also included among the Paleocene mammals were groups that had survived from Cretaceous time, such as marsupials, rodentlike forms called multituberculates, and the placental mammals called insectivores. Primates, the order to which humans belong, evolved before the end of the Paleocene. Although these small animals were quite different from monkeys, apes, or humans in many respects, by Early Eocene time they were climbing with grasping hind limbs and forelimbs that foreshadowed our own hands and feet (Figure 18-5). Primitive doglike mammalian carnivores known as mesonychids were common early in the Paleocene (Figure 18-6). The relatively primitive carnivorous mammals known as creodonts diversified during the Paleocene; some of them

were doglike or catlike in their adaptations. The earliest members of the horse family had evolved by the end of Paleocene time; the first such animals were no larger than small dogs (Figure 18-7), but by the end of the epoch, larger herbivorous mammals, some the size of cows, had appeared.

The diversity of mammals continued to increase in the Eocene Epoch. The number of mammalian families doubled to nearly a hundred, approximating that of the world today. In addition, modern varieties of hoofed herbivores appeared. Most animals of this kind are known as **ungulates**, and they are divided into **odd-toed ungulates** (living horses, tapirs, and rhinos) and **even-toed ungulates**, or cloven-hoofed animals (cattle, antelopes, sheep, goats, pigs, bison, camels, and their relatives). The odd-toed ungulates expanded before the even-toed group did, but primitive even-toed ungulates were also present early in the Eocene (Figure 18-8); of the modern

Figure 18-4 A fossilized Eocene bat from the Green River Formation of Wyoming. (Senckenberg Museum.)

even-toed types, camels and relatives of the present-day chevrotain (the Oriental "mouse deer") evolved before the end of the epoch. In addition, the earliest members of the elephant order appeared during Early Eocene time; *Moeritherium*, the earliest genus well known from the fossil record, was a bulky animal slightly larger than a pig, with rudimentary tusks and a short snout rather than a fully developed trunk (Figure 18-9A). The **rodents**, which had originated in the Paleocene, continued to diversify as well, but they may have attained their success at the expense of the archaic multituberculates, which were also specialized for gnawing seeds and nuts. As the rodents expanded during the Eocene, the multituberculates declined; they finally became extinct early in the Oligocene Epoch.

Figure 18-5 Reconstruction of the early primate *Cantius*, a small genus of Early Eocene time. This arboreal animal had large toes on its hind feet and nails much like our own, and it apparently jumped from limb to limb. (After R. T. Bakker.)

Figure 18-6 Reconstruction of a mid-Paleocene biota of New Mexico. The large trees represent the sycamore genus *Platanus*, which has survived to the present time. A small insectivore (*Deltatherium*) rests on a small branch. On the ground, mesonychid carnivores of the genus *Ankalagon* feed on the small crocodile *Allognathosuchus*. Ferns and sable palms (or fan palms) constitute the undergrowth in the background. (Drawing by Gregory S. Paul.)

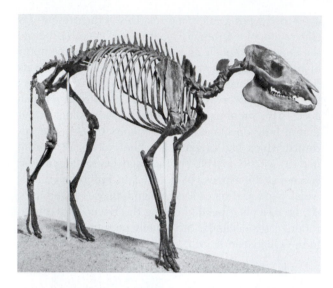

Figure 18-7 *Hyracotherium* (also known as *Eohippus*), the earliest genus of the horse family. This animal, which was present in Late Paleocene and Eocene time, was no larger than a small dog. It had four toes on each front foot (see Figure 7-20) and three on each hind foot. (Field Museum of Natural History.)

Figure 18-8 *Diacodexis*, an early even-toed ungulate, or cloven-hoofed herbivore. *Hyracotherium* (Figure 18-7), in contrast, was an early odd-toed ungulate. The limb structure of *Diacodexis* shows that it was an unusually adept runner and leaper for Early Eocene time. (After K. D. Rose, *Science* 216:621–623, 1984.)

Figure 18-9 Early members of the elephant group. *A. Moeritherium*, which evolved in Late Eocene time, was about a meter (3 feet) tall. It probably wallowed in shallow waters and grubbed for roots or other low-growing vegetation. *B. Palaeomastodon*, which was about twice as tall, possessed a rudimentary trunk and tusks.

A   B

Figure 18-10 Large terrestrial predators that had evolved by Early Eocene time. The animals that superficially resemble dogs are giant mesonychids of the genus *Pachyhyaena*, which were the size of small bears. The flightless birds guarding their chicks are members of the genus *Diatryma*, which stood about 2.4 meters (8 feet) tall. (Drawing by Gregory S. Paul.)

A   B

Figure 18-11 The long-legged Eocene duck *Presbyornis*. A. The large numbers of fossils of this wading bird found in Green River sediments of Wyoming and Utah suggest that it lived in enormous colonies. B. *Presbyornis* left tracks revealing that webbed feet supported it when it walked in mud. Along with the tracks are lines of probing marks made by the beak while the bird searched for food. Its legs were too long to allow it to swim, but modern ducks have inherited its webbed feet and employ them as paddles to propel them through the water. (A after a drawing by J. P. O'Neill; B, S. Stanley.)

Early Paleogene birds were large

New terrestrial predators of the Paleocene Epoch included not only mammals, but also the diatrymas, which were huge flightless birds with powerful clawed feet and enormous slicing beaks (Figure 18-10). The diatrymas disappeared toward the end of the Eocene Epoch.

The monstrous diatrymas were not the only birds of the Eocene, but flying birds were much less diverse then than they are today. Most species were large shorebirds that waded in shallow water when not in flight (Figure 18-11). Not yet present were many other modern kinds of birds, including the songbirds that are so numerous today.

Modern groups of hoofed animals, carnivores, and primates expanded in the Oligocene

Mammals became increasingly modern during the Oligocene Epoch. As many Eocene families died out, many of the groups that are living today expanded. The horse family had disappeared from Eurasia during the Eocene, but a few horse species survived in North America. Other odd-toed ungulates that enjoyed greater success during the Oligocene Epoch than in previous intervals were the rhinoceroses, which included *Paraceratherium*, the largest land mammal ever known to have existed on Earth (Figure 18-12), and the rhinolike brontotheres (Figure 18-13). Many more large mammals populated the land during the Oligocene Epoch than during the Eocene.

Figure 18-12 *Paraceratherium*, the largest land mammal of all time. This Oligocene giant from Asia belonged to the rhinoceros family. It stood about 5.5 meters (18 feet) at the shoulder—the height of the top of the head of a good-sized modern giraffe. (Drawing by Gregory S. Paul.)

Figure 18-13 An Early Oligocene mammalian fauna from Nebraska and South Dakota. The huge animal at the top of the painting is the brontothere *Brontotherium*. Below it is the early tapir *Protapirus*, and to its right is *Subhyracodon*, an early rhinoceros. Below *Subhyracodon*, in succession, are *Merycodon*, a sheeplike animal; *Hyaenodon*, an archaic hyena-like animal eating a lizard; and the saber-toothed cat *Hoplophoneus*. At the upper left is the small rhinoceros *Hyracodon*, and to its right are the large, piglike *Archaeotherium* and the ancestral camel *Proebrotherium*, in front of which is the horned herbivore *Protoceras*. (Courtesy Smithsonian Institution, painting by Jay H. Matternes.)

As the Oligocene Epoch progressed, odd-toed ungulates were outnumbered for the first time by even-toed ungulates, including deerlike animals and pigs, which became especially diverse. This trend has continued to the present day, when there are many more species belonging to the deer and antelope families than to the horse and rhino families. Elephants became larger early in the Oligocene and developed rudimentary trunks and tusks (Figure 18-9*B*).

Among the carnivorous mammals, mesonychids and creodonts were rare in most regions of the world after Eocene time. They were replaced by members of the Carnivora, the order to which most modern mammalian carnivores belong. For example, the dog, cat, and weasel families, which had their origins in Eocene time, radiated during the Oligocene Epoch to produce such advanced forms as large saber-toothed cats (Figure 18-14), bearlike dogs, and animals that resembled modern wolves.

An especially important aspect of the modernization of mammals during the Oligocene Epoch was the appearance of monkeys and apelike primates. The genus *Aegyptopithecus*, an arboreal (tree-climbing) animal the size of a cat, had teeth resembling those of an ape but a head and a tail resembling those of a monkey (Figure 18-15). In Neogene time, before the appearance of humans, apes attained considerable diversity in Africa and Eurasia.

Frogs and insects were modernized in Paleogene time

As for other forms of vertebrate life, reptiles and amphibians were relatively inconspicuous during Paleogene time. The first record of the Ranidae, the largest family of living frogs, is in the Eocene Series, but the fossil record of this group of fragile animals is poor, so we do not know precisely when the Ranidae originated or attained high diversity.

There is no question that the insects took on a modern appearance with the origin of several modern families in Paleogene time. A few Oligocene forms, which have been preserved with remarkable precision in amber, closely resemble living species.

Figure 18-14 The Oligocene cat *Dinictis*, which was approximately the size of a modern lynx. This animal possessed advanced adaptations for running and springing upon prey. Its canine teeth were elongated for stabbing, but *Dinictis* was not a member of the true saber-toothed cat group. (Drawing by Gregory S. Paul.)

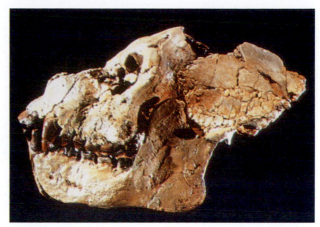

Figure 18-15   The Oligocene primate *Aegyptopithecus*, whose name reflects its discovery in Egypt. The skull resembles that of a monkey, but the teeth are apelike. The brain of *Aegyptopithecus* was unusually large for the size of the animal, presumably reflecting a high level of intelligence for the Oligocene world. (Drawing by S. F. Kimbrough; photo by E. Simons.)

Paleogene Climates

Early in the Paleogene Period, continents were arranged much as they are today, but were bunched more closely together (Figure 18-16). The average global temperature rose very high early in the period, but the onset of cooler conditions—as well as drier conditions on land—produced widespread extinction late in Eocene time.

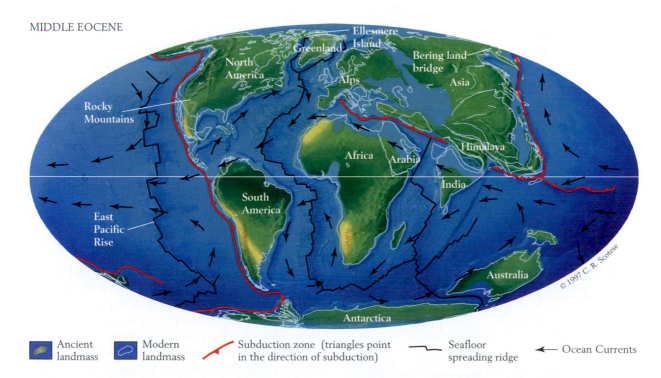

MIDDLE EOCENE

Figure 18-16   **World geography during Middle Eocene time.** Substantial areas of continental crust were inundated during this interval. Relatively warm climates prevailed even as far north as Ellesmere Island, where a diverse fauna of terrestrial plants and vertebrate animals flourished. The Bering land bridge, a neck of continental crust, stood above sea level throughout the Eocene interval, permitting animals to migrate between North America and Eurasia. (Adapted from paleogeographic map by C. R. Scotese, PALEOMAP Project, University of Texas at Arlington, 1997.)

The Eocene began with a pulse of warming

The very warm interval of the Paleogene Period began during the transition between the Paleocene and Eocene epochs. Marking the boundary between these two epochs in cores from the deep sea is an abrupt shift of oxygen isotope ratios in foraminifera toward lighter values (Figure 18-17). This shift is seen in both planktonic species and species that lived on the floor of the deep sea. The total volume of glacial ice on Earth was quite small at that time, so the isotopic shift does not reflect melting of glaciers and return to the ocean of waters enriched in oxygen 16 (see Figure 10-19). Instead, the shift signals a temperature change. It indicates that even near Antarctica, both surface and deep-sea waters warmed by several degrees to a temperature of about 18°C (64°F) within less than 3000 years. More than 70 percent of the foraminiferal species of the deep sea died out, probably because of the warming and also because the deep sea became depleted of oxygen: cool polar waters no longer transported oxygen to the deep sea. The deep sea did not become totally anoxic, however, because deep-sea sediments did not turn dark with reduced organic matter.

The carbon isotope ratios of deep-sea foraminifera also shifted abruptly to very low levels (see Figure 18-17). Similar pronounced shifts toward isotopically light carbon are seen in organic matter produced by marine phytoplankton and in plant fossils on land. Thus it seems apparent that carbon dioxide used in photosynthesis throughout the world suddenly became highly enriched in carbon 12. The terminal Paleocene shift toward iso-

topically light carbon was so pronounced that melting of frozen methane hydrates along continental slopes seems to be the only reasonable cause. Recall that the carbon in this frozen material is isotopically very light (p. 236).

What caused this release of methane at the end of Paleocene time is still debated. A slight warming of water along continental slopes could have triggered the release, but so could a massive slumping of slope sediments. In any event, a positive feedback would have resulted. As a greenhouse gas, methane is perhaps 30 times more powerful than carbon dioxide (p. 235). Therefore, a small amount of greenhouse warming from an initial release of methane would have led to further melting of methane hydrates—and further warming. Greenhouse warming would have weakened in a few years, as the methane added to Earth's atmosphere oxidized to carbon dioxide.

Climatic warming at the end of Paleocene time is reflected in a major change in the composition of terrestrial floras of the Green River Basin (see Figure 18-23 below). Habitat changes at this time, presumably causing changes in vegetation, appear to have permitted certain mammalian taxa to undertake a great migration. At the end of the Paleocene, primates, odd- and even-toed ungulates, and members of the opossum family all made their first appearance in North America. These groups, which had arisen in Eurasia slightly earlier, suddenly migrated from Siberia to Alaska, via a land bridge that today is flooded by the Bering Sea, and spread southward. Thus a brief climatic event profoundly reorganized mammalian faunas on a global scale.

Warmth extended to high latitudes

The global climate remained warm throughout Early Eocene time. As exemplified by fossil floras in England, warm conditions spread to very high latitudes. Today England has a stable but relatively cool climate. It lies farther north than any U.S. state but Alaska. It was similarly positioned during Early Eocene time, yet it was cloaked by a virtually tropical jungle that shared many taxa with the modern tropical region of Malaysia. Needless to say, England has never since experienced such warmth.

Other northerly regions were also bathed in balmy climates during the Eocene. Fossil palm leaves are known from Wyoming (Figure 18-18), and even from as far north as southern Alaska. These palms were part of a diverse subtropical flora that spread from Asia around the northern rim of the Pacific to North America.

An especially interesting Eocene fossil assemblage of plants and animals comes from Ellesmere Island, which lies well within the Arctic Circle, north of eastern Canada. Even this biota is subtropical, sharing many Eocene taxa with western North America and Eurasia. During the interval of Eocene warmth, more than 40

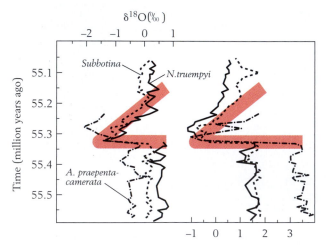

Figure 18-17 Abrupt shifts of oxygen and carbon isotope ratios in deep-sea foraminifera toward lighter values at the end of the Paleocene Epoch. The specimens for which data are plotted here are from the vicinity of Antarctica. (After J. P. Kennett and L. D. Stott, *Nature* 353:225–229, 1991.)

Figure 18-18 Fossil palm frond from the Eocene Green River Formation of Wyoming, far north of where palms live today. (Chip Clark.)

it is today. Computer modeling of levels of atmospheric carbon dioxide yields a similar result (see Figure 10-13). Greenhouse warming may nonetheless have been stronger during the Eocene than it is today because water vapor, like carbon dioxide, is a greenhouse gas. Forests were much more widespread in the Eocene than they are today, and because trees emit water through the process of transpiration (see p. 19), forests send large amounts of water into the atmosphere. Much of this water, especially in moist tropical regions, is returned to the forests in the form of rain (p. 87). In effect, water is retained in forested regions, moving from soil through trees to the atmosphere and back again. Thus the widespread tropical and subtropical forests of the Eocene Epoch must have maintained a high concentration of water vapor in the atmosphere over broad areas of Earth. This water vapor must have resulted in substantial greenhouse warming.

Isotope ratios for planktonic foraminifera show that, despite the warmth at high latitudes, tropical seas were only about as warm as they are today. In other words, the temperature gradients from the equator to the poles were relatively gentle. This condition would require that an enormous amount of heat be transported from lower latitudes toward the poles. There is evidence that some of this heat was transported northward by air movements. As clouds move across continents, the moisture within them becomes isotopically lighter as water molecules containing the heavier oxygen isotope (^{18}O) rain out preferentially. The oxygen in remarkably well-preserved fossil wood found on Axel Heiberg Island, within about 10° of the north pole, is exceptionally light in its isotopic composition. Thus it appears that the trees that formed this wood were receiving moisture from air masses that had traveled great distances across North America after forming at relatively low latitudes. Air containing a large amount of water vapor can carry much more heat than dry air. As air masses carrying clouds moved northward over heavily forested North American terrain, they must have brought much heat to the north polar region.

Climatic change, glaciation, and a mass extinction marked the Eocene-Oligocene transition

We often think of the modern Ice Age as having begun when glaciers expanded over Greenland, Europe, and northern North America slightly more than 3 million years ago. In fact, the modern glacial age of Earth history began more than 30 million years ago, when a massive glacier came to occupy Antarctica (Earth System Shift 18-1). The origin of this south polar glacier, which has never melted away since that time, was part of a global climatic change that produced widespread

percent of the plant species of the Ellesmere flora had leaves with smooth margins—an indication that the average temperature resembled that of southern California today. Also present were large tortoises and alligators, which require warm winter temperatures. On Ellesmere Island today, temperatures hover around −18°C (about 0°F) throughout the year. In Late Eocene time, however, the island remained above freezing even in winter.

The warm Eocene temperatures cannot be explained by greenhouse warming caused by an elevated concentration of carbon dioxide in Earth's atmosphere. The proportion of stomate cells on the surfaces of fossil leaves of ginkgo trees, which are living fossils (see Figure 16-10), has been applied to this question, just as it has been used to estimate atmospheric carbon dioxide levels during the Triassic-Jurassic transition (see Figure 16-23). Comparison of the stomate densities of fossil ginkgos with those of modern ginkgos grown under a variety of carbon dioxide concentrations indicates that the concentration of carbon dioxide in the atmosphere in Early Eocene time was not much higher than

Earth System Shift 18-1 | Global Cooling and Drying Begins

Fossil plants attest to dramatic changes in Earth's climate after the Early Eocene interval of global warmth. Climates became cooler and drier in many regions of the world, and a mass extinction resulted. Remarkably, these profound global changes can be traced to a regional plate tectonic event near the south pole.

Flowering plants (angiosperms) are commonly viewed as the thermometers of the past 100 million years. As we have seen, their value derives in part from the strong correlation between mean annual temperature and the percentage of species within a flora that have leaves with smooth margins (ones lacking teeth or lobes; see Figure 4-20). Leaf margin analyses for fossil floras of North America reveal that three pulses of cooling, each more severe than the one before, occurred in Middle and Late Eocene time.

Events that took place in the deep sea point to the cause of the cooling episodes. Study of deep-sea cores has revealed shifts toward isotopically heavier oxygen in foraminifera that inhabited the deep-sea floor. This shift reflects cooling in the deep sea. The most pronounced cooling occurred near the end of Eocene time, when the cold bottom layer of the ocean came into being. Cold, dense polar water began to sink to the deep sea, as it does today (see Figure 4-23), and

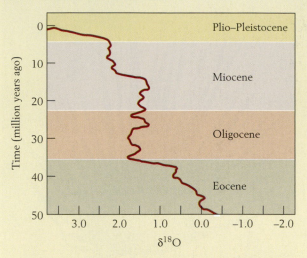

Figure 2 Oxygen isotopes in deep-sea foraminifera changed markedly during Eocene time. The strong shift toward high values of oxygen 18 near the end of the Eocene signals both cooling of deep-sea waters and the buildup of ice sheets. (After K. S. Miller, R. G. Fairbanks, and G. S. Mountain, *Paleoceanography* 2:201–219, 1987.)

deep-sea temperatures dropped by perhaps 4°C to 5°C (7°F to 9°F). The cooling was quite rapid: fossils from deep-sea cores indicate that changes in the bottom-dwelling fauna occurred in less than 100,000 years. Upwelling of these cold waters lowered temperatures far from the poles.

The large size of the isotopic shift in deep-sea foraminifera also reflects growth of glaciers on Antarctica, with oxygen 16 accumulating preferentially in the glacial ice (see Figure 10-18B). The appearance of ice-rafted sand in deep-sea sediments offshore from Antarctica provides additional evidence of this glaciation.

The refrigeration of the deep sea and the expansion of glaciers in Antarctica were related phenomena. Antarctica had extended over the south pole since the Cretaceous Period, but it had remained warm because its shores were bathed in relatively warm waters from lower latitudes. Movement of Australia away from Antarctica automatically formed the circumpolar current that has persisted to the present day (see Figure 4-22). Waters from the South Atlantic, Indian, and South Pacific oceans become trapped in this current, becoming progressively colder while cycling round and round Antarctica. Initially, when the passageway around Antarctica was constricted and cooling was relatively weak, glaciers covered only part of Antarctica. As Australia moved farther away from Antarctica, the circumpolar current must have strengthened, causing further cooling and glacial expansion. In fact, a sudden mid-Oligocene decline of global sea level must

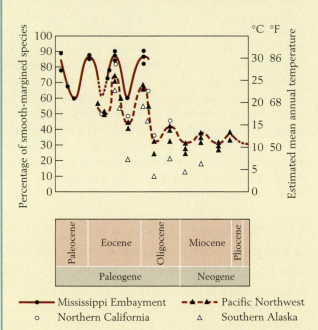

Figure 1 Changes in percentages of smooth-margined leaves in fossil floras reveal changes in temperature during Eocene and Early Oligocene time. The fact that the curves for different regions of North America follow parallel paths where they overlap in time suggests that the trends hold for large areas of Earth's surface. (After J. A. Wolfe, *Amer. Sci.* 66:694–703, 1978.)

have resulted from substantial expansion of the Antarctic glacial ice cap (see p. 447).

Marine life experienced widespread extinction during the Eocene-Oligocene transition. The history of planktonic foraminifera recorded in deep-sea cores reveals five successive pulses of change between Middle Eocene and Early Oligocene time. The final pulse of extinction appears to have coincided with the final episode of cooling in North America, as indicated by changes in terrestrial floras. The species that disappeared in these pulses were mostly spiny forms adapted to warm conditions. The pattern of extinction of calcareous nannoplankton has not yet been documented in great detail, but during the final 7 million years of Eocene time the total number of nannoplankton species in the world declined markedly. Pulses of extinction also occurred on shallow seafloors. Many of the species that died out—especially mollusks—were adapted primarily to warm conditions.

Numerous species of mammals also disappeared late in Eocene time. It seems evident that climatically induced changes in the terrestrial flora played a major role in the mammalian extinctions. Many of these floral changes resulted from increased aridity rather than cooling. Before the Eocene climatic change, moist tropical and subtropical forests had cloaked much of North America and Eurasia. In contrast, during the Oligocene Epoch, dry woodlands with large, grassy clearings occupied large areas of these continents.

The shrinking of forests must have created a positive feedback for these general climatic changes. Forests not only have a low albedo, but also retain moisture. As forests shrank during the Eocene-Oligocene transition, they not only would have absorbed less heat from the sun, but also would have retained less moisture in the atmosphere to serve as a greenhouse gas and to transport heat from low to high latitudes. Thus loss of forest would have accentuated the cooling and drying of climates, leading to further contraction of forests—and further cooling and drying. In addition, the cooler oceans provided less moisture to the atmosphere through evaporation.

Inevitably, the regional changes in terrestrial vegetation led to the extinction of many species of mammals. This extinction occurred earlier in North America than in Eurasia, being concentrated at the Middle-Late Eocene boundary. Mammalian species that died out in North America and Eurasia included many tree climbers as well as many herbivores whose teeth required a diet of soft leaves. Among the latter were the massive brontotheres (see Figure 18-13). In contrast, many of the new Oligocene species possessed molar teeth well suited to grinding the coarse vegeta-

tion of open terrain. An excellent fossil record in Mongolia reveals a shift from forest ecosystems to more arid, open ecosystems. Here, medium-sized herbivores disappeared. Oligocene faunas featured large herbivores, which would have had considerable stamina for outrunning predators in open country, as well as many new kinds of rodents and lagomorphs (rabbits and their relatives), which could have burrowed in the ground or found cover in low vegetation.

Since global cooling began more than 30 million years ago, forests have never been as widespread on Earth as they were before this pivotal climatic change. On the other hand, grasslands and other relatively arid habitats have occupied much broader regions of the world than they did before. One might say that during the Eocene-Oligocene transition Earth shifted into its modern environmental regime.

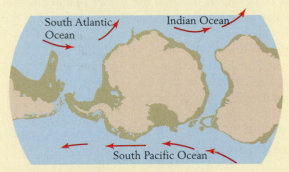

A MIDDLE EOCENE

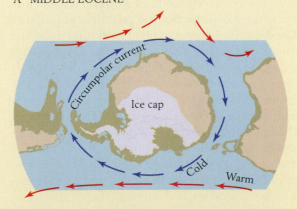

B EARLY OLIGOCENE

Figure 3 The circumpolar current originated during the Eocene-Oligocene transition. *A.* In Middle Eocene time, warm currents flowed from the southern oceans to warm Antarctica (red arrows). *B.* By Early Oligocene time, South America and Australia had moved far enough from Antarctica to allow waters from the southern oceans to become trapped in a current around Antarctica, where they became very cold. As a result, glaciers grew on Antarctica. (From L. A. Lawyer, L. M. Gahagan, and M. F. Coffin, *Antarctic Res. Ser., Amer. Geophys. Union* 56:7–30, 1992.)

extinction and transformed ecosystems throughout the world. This climatic change issued from the region where the glacier grew: Australia rifted away from Antarctica, permitting a large ocean current to encircle the polar continent. Because of its isolation, this current became very cold, and as a result, so did Antarctica and the deep sea.

Regional Events of Paleogene Time

In examining important regional events of the Paleogene Period, we will travel to the ends of Earth—first to the south pole, where Antarctica became separated from Australia and developed its icy cover, and then toward the north pole, where land areas of North America and Eurasia were more closely connected than they are today. Then we will look at the history of the Cordilleran region of North America, where the Laramide orogeny yielded many structures that remain conspicuous today in the Rocky Mountains. We will also examine the nature of deposition along the Gulf Coast of North America, where large volumes of petroleum are trapped in soft Paleogene sediments. Finally, we will see how the impact of a meteorite near what is now the District of Columbia established the site of the Chesapeake Bay.

Positions of land and sea changed near the poles

As we have seen, the cooling of Antarctica and expansion of glaciers there during the Eocene Epoch resulted from the movement of plates that had previously formed parts of Gondwanaland. Australia rifted away from Antarctica, leaving it isolated in a polar position. Major geologic and geographic changes also took place during the Paleogene Period within 30° latitude of the north pole. Many of these changes can be read from the ages of various segments of the deep-sea floor. The Arctic deep-sea basin existed during the Cretaceous Period, but until nearly the end of Cretaceous time it remained separated from the Atlantic, except by way of shallow seas, because North America, Greenland, and Eurasia were still united as a single landmass. Recall that mid-Atlantic rifting proceeded northward during Cretaceous time, splitting Greenland from North America (see Figure 17-13). Early in the Paleogene, this rifting shifted to a younger spreading zone between Greenland and Scandinavia, eventually establishing a broad connection between the Arctic Ocean and the Atlantic Ocean (see Figure 18-16).

In contrast, continental crust has continued to separate the Arctic and Pacific deep-sea basins; today, Alaska and Siberia remain connected by a stretch of

continental crust, despite the fact that this connecting segment now lies submerged beneath the Bering Sea. During much of the Cenozoic Era, this segment stood above sea level and served as a land corridor, known as the *Bering land bridge*, between North America and Eurasia. This corridor allowed mammals and land plants to migrate between Asia and North America.

Mountain building continued in western North America

Mountain-building activity in the Cordilleran region of North America continued into the Paleogene Period, but with a number of changes. Figure 18-19, which summarizes the orogenic history of the eastern Cordilleran region, shows that the Sevier orogeny spanned almost the entire Cretaceous Period. In latest Cretaceous time, however, a different style of tectonic activity was initiated in the American West. This activity persisted through the Paleocene and well into Eocene time.

The Laramide orogeny The episode characterized by the new tectonic style is known as the *Laramide orogeny*. The northern and southern segments of the Laramide were typical of orogenies in general. In the

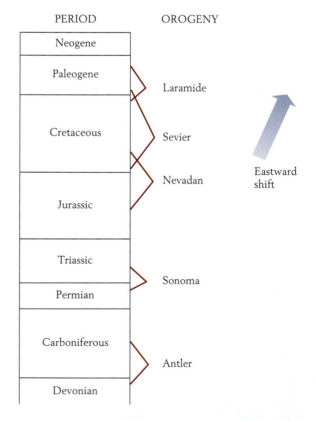

Figure 18-19 **Summary of major orogenic events in the eastern Cordilleran region.** Between Jurassic and Paleogene time, orogenic activity migrated eastward.

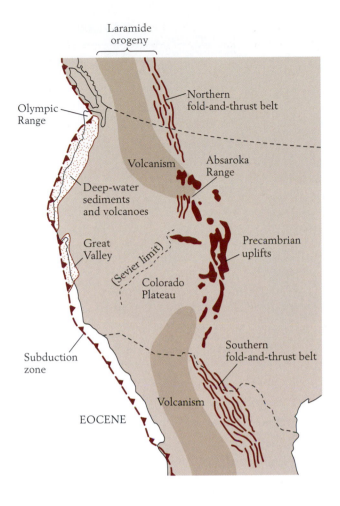

EOCENE

◀ **Figure 18-20** Geologic features of western North America during Eocene time. Subduction continued along the west coast. Marine sediments were deposited in the Great Valley of California, and deep-water sediments and volcanics accumulated in a forearc basin to the north, in Washington and Oregon. Farther inland in the north and south, the Laramide orogeny produced a band of volcanism and, still farther inland, a belt of folding and thrusting. In Colorado and adjacent regions, however, the orogeny was expressed as a series of Precambrian uplifts that extended far to the east of the Cretaceous Sevier orogenic belt. They may have been formed by a slight clockwise rotation of the Colorado Plateau in relation to the continental interior.

18-20). Thrust sheets of enormous proportions are spectacularly exposed in the Canadian Rockies (Figure 18-21). A similar pattern of tectonic activity persisted in the southern United States and Mexico.

The unusual features of the Laramide orogeny were in the central part of the western United States, where a broad area of tectonic quiescence extended from the Great Valley of California to the Colorado Plateau (see Figure 18-20). East of this inactive region, in a curious pattern of tectonic activity, large blocks of underlying crystalline Precambrian rock were uplifted in a belt extending from Montana to Mexico. The largest of these blocks were centered in Colorado, where the Ancestral Rocky Mountains had been uplifted more than 200 million years earlier, late in the Paleozoic Era (p. 379).

The pattern of subduction What created the unusual tectonic pattern that characterized the central part

north, extending from the United States into Canada, there remained an active belt of igneous activity and, inland from it, an active fold-and-thrust belt (Figure

Figure 18-21 The Lewis Thrust Fault in the fold-and-thrust belt of the northern Rocky Mountains. In this view the fault is exposed on the side of a mountain in Glacier National Park, Montana. The tree line follows the fault, separating bare Proterozoic rocks from the Cretaceous rocks over which they have been thrust. (Martin G. Miller/Visuals Unlimited.)

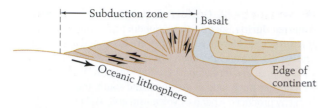

Figure 18-22 Formation of the Olympic Range in an embayment of the Pacific coast of Washington State. Subduction piled sediments against basaltic rocks. (After D. E. Kari and G. F. Sharman, *Geol. Soc. Amer. Bull.* 86:377–389, 1975.)

of the Cordilleran region during the Laramide orogeny? Note that the Paleogene uplifts were, for the most part, positioned well to the east of Sevier orogenic activity (see Figure 18-20), and recall in addition that an eastward migration of orogenic activity had taken place during the Mesozoic Era, culminating in the Sevier orogeny (p. 438). The widely favored explanation for the earlier eastward shift applies to the Laramide shift as well: a central segment of the subducted plate that passed beneath North America assumed a still lower angle, extending far to the east before becoming deep enough to melt and send magma to the overlying crust (see Figure 17-20). As in Cretaceous time, the lower angle of subduction must have resulted from an increase in the rate of westward movement of North America, which accelerated the rollback of the subducted plate (see Figure 9-16).

Farther west, along the coast, the Great Valley of California continued to receive marine sediment, while northern California and the Sierra Nevada region remained as highlands. A separate basin in Washington and Oregon received deep-water sediments and layers of pillow lava. Here the Olympic Range began to form along a sharp inward bend of the subduction zone (Figure 18-22; see also Figure 18-20).

The eastern uplifts and basins A belt of Laramide uplifts stretches from southern Montana to New Mexico. Because this is the region in which the central and southern Rocky Mountains would develop during Neogene time, Paleogene events that preceded the uplift of this segment warrant special attention. Deformation here began in latest Cretaceous time with the origin of uplifts and basins trending from north to south (Figure 18-23); the uplifts are typically anticlines bounded on one or both sides by relatively steep thrust faults. Elevated Precambrian rocks are commonly exposed in their interior regions. In Utah and Wyoming these structures lie along the eastern margin of the northern fold-and-thrust belt. The uplift farthest to the east formed the Black Hills of South Dakota (Figure 18-24; see also Figure 9-20). In Colorado, many ranges were formed

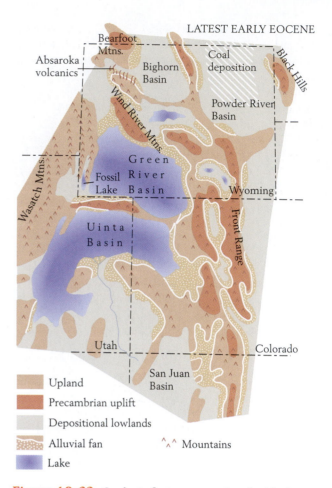

LATEST EARLY EOCENE

Upland	
Precambrian uplift	
Depositional lowlands	
Alluvial fan	^∧^ Mountains
Lake	

Figure 18-23 Geologic features associated with the Laramide uplifts of Colorado and adjacent regions at the end of Early Eocene time. The cores of the major uplifts consist of crystalline Precambrian rock. The Green River Formation, which is well known for its oil shales and splendid fossils, was accumulating in the Green River and Uinta basins. To the north, volcanism formed the Absaroka Mountains, where Yellowstone Park is now located. The Black Hills of South Dakota represent the easternmost uplift.

Figure 18-24 The craggy peaks of the Black Hills that stand high above the Great Plains. (Paul Horestead.)

by the elevation of large bodies of rock along steeply dipping thrust faults. It has been suggested that these uplifts were produced by a slight clockwise rotation of the Colorado Plateau, which behaved as a rigid crustal block, as a result of forces generated along the subduction zone to the west (see Figure 18-20).

Today, of course, many peaks and ridges of the central and southern Rocky Mountains stand at very high elevations; the Front Range uplift, for example, rises far above the high plains of eastern Colorado (see Figure 15-25). It is inappropriate, however, to evaluate the effects of the Laramide orogeny simply by viewing the Rocky Mountains today because, as we shall see, the high elevations of the modern Rockies reflect post-Laramide uplift. During the Laramide orogeny, erosion nearly kept pace with uplift in most areas of the United States, so that the regional topography remained less rugged than it is today. Basins in front of elevated areas were receiving large volumes of rapidly eroded material.

Before the end of Early Eocene time the regional north–south pattern had weakened, and individual basins were experiencing independent histories. Most of these basins accumulated alluvial and swamp deposits with abundant fossil mammal remains, and at times some were occupied by lakes. By late in Early Eocene time, lakes came to occupy large areas within several of the basins, and these lakes survived, sometimes at reduced size, throughout much of the Eocene Epoch (see Figure 18-23). The famous Green River deposits accumulated in and around the lake that occupied the basin for which they are named. Plant remains in these sediments, including fossil palms, reveal that climates in this region were much warmer in Eocene time than they are today, even subtropical at times (see Figure 18-18). These lake deposits, which are very finely layered, have commonly been termed *oil shales* because algal material within them has broken down to yield vast quantities of petroleum (see Figure 5-3). Unfortunately, this petroleum is disseminated throughout the rock and thus has proved difficult to extract. Nonetheless, the Green River deposits—the largest body of ancient lake sediments known—may eventually serve as a valuable source of fuel. The fine undisturbed layers of these deposits account for the remarkable preservation of a host of animal and plant fossils, including delicate creatures such as bats (see Figure 18-4).

Most of the depositional basins that lay between Montana and New Mexico were filled with sediment by the end of Eocene time. The Laramide orogeny came to an end, and the uplands that it had produced were largely leveled; thus, as the Oligocene Epoch dawned, a monotonous erosion surface stretched across western and central North America, interrupted by only a few isolated hills.

The present-day Rockies The modern Rocky Mountains are the product of a renewed uplift during Neogene time. Today, high up in many areas of the Rockies, a person can look into the distance and view a nearly flat surface formed by the tops of mountains (Figure 18-25). This is what remains of the broad erosional surface that existed at the end of the Eocene Epoch, but this so-called subsummit surface now stands high above the Great Plains as a result of Neogene uplift.

During the Oligocene Epoch, most of the Laramide uplifts had been leveled, and a thin veneer of deposits spread as far east as South Dakota. The Badlands of South Dakota consist of rugged terrain carved mostly from Eocene and Oligocene deposits that formed part of this veneer (see Figure 18-13). These deposits have yielded rich faunas of fossil mammals, and changes in the nature of their fossil soils reveal that aridity increased here, as it did in many parts of the world. Forests of Late Eocene age gave way to open woodlands and finally, in Late Oligocene time, to still drier savannahs.

Figure 18-25 The subsummit surface formed by the flat-topped Rocky Mountains. This surface, seen here near Derby Peak, Colorado, formed near the end of the Eocene Epoch and has since been uplifted and dissected by erosion. (Michael Collier.)

The Yellowstone hot spot appeared

Another interesting group of rocks found in the region of Paleogene basins and uplifts are the volcanics that form the Absaroka Range in western Wyoming and Montana. A large portion of Yellowstone National Park lies within the Absarokas; here the still-active geysers and hot springs serve as evidence that igneous activity has not ceased completely. Recall that Yellowstone represents a hot spot that is apparently the result of a plume of magma rising from Earth's mantle (p. 194). Volcanism here was episodic, and all of the Eocene volcanic episodes were catastrophic, destroying entire forests and, we must assume, the animal life within them. Fossilized leaves, needles, cones, and seeds reveal the presence of lowlands with subtropical vegetation. Today, remnants of the Eocene forests can be seen at high elevations in Yellowstone National Park, where trees are preserved upright as stumps buried by lavas, mud flows, and flood deposits of volcanic debris (Figure 18-26).

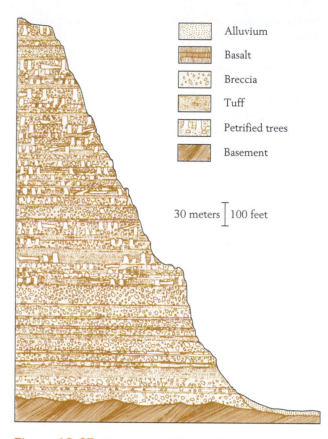

Alluvium	
Basalt	
Breccia	
Tuff	
Petrified trees	
Basement	

30 meters | 100 feet

Figure 18-27 Succession of 27 petrified forests now exposed along the Lamar Valley in Yellowstone National Park. Each of these Eocene forests was destroyed by a volcanic eruption. (After E. Dorf, *Sci. Amer.*, April 1964. ©1964 by Scientific American, Inc. All rights reserved.)

Figure 18-26 Petrified stumps standing upright in the Absaroka volcanics at Specimen Ridge, Yellowstone National Park. Some logs here have diameters of about 1 meter (3 feet). (Joe Case/Visuals Unlimited.)

More than 20 successive forests, all of them killed in this way, have been recognized (Figure 18-27).

Deposition continued along the Gulf Coast

Unlike the Cordilleran region, the Gulf Coast of North America has remained an area of tectonic quiescence throughout the Cenozoic Era. The sea retreated southward in latest Cretaceous time, draining the Interior Seaway (see Figure 17-13). During Paleogene time, marine waters still occupied the Mississippi Embayment, an inland extension of the Gulf of Mexico, where a thick sequence of Eocene marine sediments accumulated (Figure 18-28). During the Oligocene Epoch, the seas withdrew to the approximate position of the present shoreline of the Gulf of Mexico and then spread inland again, but not as far as they had in Eocene time. The total thickness of Paleogene sediments near the present coastline of the Gulf of Mexico exceeds 5 kilometers (3 miles), largely because of the enormous quantities of sediment carried to the region by the Mississippi

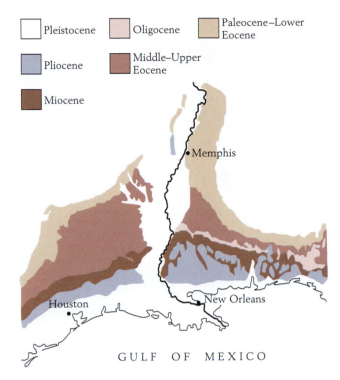

Pleistocene   Oligocene   Paleocene–Lower Eocene

Pliocene   Middle–Upper Eocene

Miocene

• Memphis

Houston •

New Orleans

GULF OF MEXICO

Figure 18-28 The segment of the Gulf coastal plain known as the Mississippi Embayment. Here marine sediments of Paleogene age were deposited farther inland than those of Neogene age.

River system. The Cenozoic clastic wedge in this region, though largely buried, has been studied in great detail during the successful search for petroleum there.

A meteorite created the site of the Chesapeake Bay

The Chesapeake Bay, along the middle Atlantic coast of the United States, is the largest estuary in the world (an *estuary* is a broad valley drowned by the ocean). It is notable that the bay is not the drowned mouth of a single river. Instead, several rivers converge in the Chesapeake Bay to empty their waters into the Atlantic Ocean. One of these rivers is the Potomac, which flows through the District of Columbia close to the bay.

In 1986, geologists of the U.S. Geological Survey drilled through sediments beneath a coastal marsh bordering the Chesapeake Bay in Virginia. In the first 300 meters (1000 feet) or so of drilling, they encountered ordinary sands and muds. Then, to their surprise, they pulled up about 60 meters (200 feet) of drill core containing rubbly deposits: breccias (p. 40) containing angular clasts of many sizes and compositions, some of which were measured in meters (Figure 18-29*A*). Fossils just above the rubbly deposits showed that they were laid down about 36 million years ago.

Seismic studies and further drilling showed that the rubbly deposits occupied a subsurface circular depression (Figure 18-29*C*). They also showed that the depression had a complex concentric configuration, with a peak in the center—a configuration similar to that of some meteorite craters on the moon. Together with the rubbly infilling, these observations indicated that the Chesapeake depression is a crater formed by a meteorite impact (a rebound after an impact can produce a central peak). Confirming this conclusion are abundant shocked mineral grains in the rubble resembling those found at the Cretaceous-Paleogene boundary (see Earth System Shift 17-1).

Seismic profiling also revealed a smaller, more triangular crater near the head of Toms Canyon, a submarine canyon about 150 kilometers (90 miles) east of Atlantic City, New Jersey (Figure 18-29*B*). The elongate shape of this structure suggests that a cluster of small meteorites formed it, and dating of associated deposits shows this crater to be the same age as the Chesapeake crater. Apparently the meteorites that formed the Chesapeake Bay and Toms Canyon craters were fragments of a larger meteorite that broke apart in Earth's atmosphere.

Because sea level was higher in Late Eocene time than it is today, the Toms Canyon impact must have occurred in water between 500 and 700 meters (about 1500–2000 feet) deep. The result of the impact would have been a tidal wave that, as it spread westward and touched bottom in shallower water, would have crested at a height of tens, or even hundreds, of meters. This tidal wave deposited a layer of sediment that extends from New Jersey to North Carolina. Above this unit is a thin layer of sediment containing abundant microspherules, or spherical molds where these grains once occurred. Like the similar microspherules found at the Cretaceous-Paleogene boundary, these grains are beads of glass formed when the heat of the impact melted sediment or rock that later cooled and settled to Earth. Microspherules occur at the same stratigraphic level as the Chesapeake and Toms Canyon craters in shallow-water sediments of the Gulf Coast and in deep-sea sediments of the Gulf of Mexico and Caribbean.

Calculations based on nuclear weapons tests indicate that the Chesapeake Bay meteorite was between 3 and 5 kilometers (about 2–3 miles) in diameter. The meteorite that caused the mass extinction at the end of the Cretaceous Period was about 10 kilometers (6 miles) in diameter. We can therefore conclude that the minimum size of a impacting meteorite that will cause a global biotic crisis must be somewhere between 3 and 10 kilometers in diameter. Fortunately, impacts of meteorites as large as this seldom occur.

Today large rivers flowing from the Valley and Ridge and Piedmont provinces of the Appalachians

A

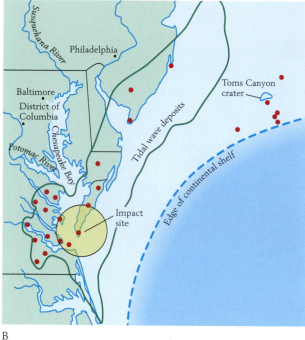

B

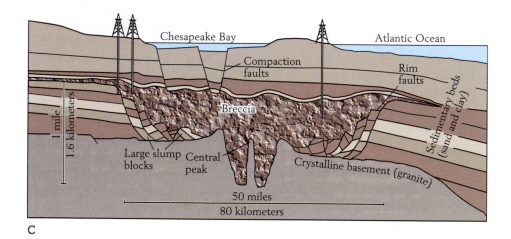

C

Figure 18-29 Features associated with the impact of a meteorite at what is now the mouth of the Chesapeake Bay. The red dots are sites where sediment cores have been obtained by drilling. *A.* Cores about 0.6 meters (2 feet) long from the impact site consisting of impact breccia. *B.* Map showing the Chesapeake Bay impact site, the smaller Toms Canyon crater, formed by the impact of several small fragments, and the distribution of sedimentary debris deposited by the tidal wave from the Toms Canyon impact. *C.* Idealized east–west cross section of the crater, with considerable vertical exaggeration. (Courtesy of Wylie Poag, U.S. Geological Survey.)

converge at the Chesapeake Bay. The largest of these rivers are the Susquehanna and Potomac (see Figure 18-29*B*). There can be little doubt that this confluence of rivers developed because the thick pile of sediments in and above the Eocene impact crater settled over time, creating a slight depression that attracted drainage from the uplands to the north and west. Nor-

mal faults in the sediments above the breccia indicate uneven compaction of the breccia (see Figure 18-29*C*). These compaction faults constitute a geologic hazard for the Chesapeake region in that future movements along them will cause earthquakes—jolting reminders of the regional catastrophe that took place 36 million years ago.

Chapter Summary

Would life in Paleogene seas have looked familiar to modern humans?

Yes, marine life of the Paleogene Period bore a strong resemblance to that of the modern world.

Would terrestrial vegetation of the Paleogene Period also have looked familiar to us?

Yes, plants on land also resembled those of the present time, except that grasses were not widespread until late in Paleogene time.

How similar were terrestrial vertebrate animals of the Paleocene Epoch to those of the present world?

Terrestrial vertebrates were more primitive than those of the present. Although most of the mammalian orders alive today existed in Paleogene time, many Paleogene families and genera are now extinct. Songbirds had not yet evolved; most species of birds were large wading animals that resembled storks or herons.

What climatic conditions characterized the Early Eocene world?

Fossil floras and oxygen isotope ratios in planktonic foraminifera show that very warm climates extended to high latitudes early in the Eocene Epoch. Much of Earth was cloaked in tropical or subtropical forests.

What global climatic changes occurred late in the Eocene Epoch, and what caused them?

Australia rifted away from Antarctica late in Eocene time, and the circumpolar current came into being, trapping water in the south polar region. Waters trapped in this current cooled as they circled Antarctica, and cold, dense waters began to descend to form the cold zone of the deep sea that has persisted to the present time. Cold, deep waters mixed with shallower waters and cooled them so that they released less moisture to the atmosphere through evaporation. These changes resulted in cooler and drier climates worldwide.

How did climatic changes in Late Eocene time affect life on Earth?

Dry woodlands and open terrain replaced dense forests in many parts of the world. Many species became extinct both on land and in the sea.

What mountain-building event occurred in western North America early in the Cenozoic Era?

The Laramide orogeny, continuing from latest Cretaceous time, produced northern and southern fold-and-thrust belts that were separated by an unusual zone of Precambrian uplifts. By the end of the Eocene Epoch, the Laramide orogeny had ended, and the mountains it produced had been subdued by erosion.

What was unusual about the pattern of Paleocene and Eocene mountain building between southern Montana and New Mexico?

It occurred far inland, owing to the low angle at which the subducted plate descended into the asthenosphere. In addition, most uplifts were anticlines bounded by steeply dipping thrust faults, in contrast to the conventional fold-and-thrust belts to the north and south.

Review Questions

1. What groups of animals that played important roles in Late Cretaceous ecosystems were absent from the Paleocene world?

2. Which group changed more from the beginning of the Paleogene Epoch to the end—marine or land animals? Explain your answer.

3. What seems to have prevented corals from forming massive reefs before Oligocene time?

4. Which animals took the place of large Mesozoic reptiles in Paleogene oceans?

5. How does the fossil record of flowering plants reveal climatic change during Paleogene time?

6. What happened to the deep sea at the end of the Paleocene Epoch?

7. How did changes in patterns of oceanic circulation bring about glaciation in Antarctica?

8. How did the location of the Laramide orogeny differ from that of the Cretaceous Sevier orogeny? What might explain this change?

9. What did the region of the modern Rocky Mountains look like at the end of the Eocene Epoch?

10. What is the origin of the geologic features that intrigue tourists at Yellowstone National Park?

11. What produced the basin occupied by the Chesapeake Bay, where rivers converge and their waters flow into the Atlantic Ocean?

12. Climatic changes occurred throughout the world during the latter part of the Paleogene Period. Using the Visual Overview on page 446 and what you have learned in this chapter, explain what caused these changes and describe their effects on life in the sea and on land.

Mt. Ranier, a volcanic peak in the state of Washington that formed during the Neogene Period. (Mark J. Bilak/Visuals Unlimited.)

CHAPTER 19

The Neogene World

Because it leads nearly to the present, the Neogene Period holds special interest for us. It was during the Neogene that the modern world took shape—that is, when global ecosystems acquired their present state and prominent topographic features assumed the configurations we observe today.

No mass extinction marked the transition from the Paleogene to the Neogene. During the scant 24 million years of the Neogene, however, Earth's physical features and life have changed significantly. The most far-reaching biotic changes were the spread of grasses and weedy plants and the modernization of vertebrate life. Snakes, songbirds, frogs, rats, and mice expanded dramatically, and apes—and then humans—evolved. The Rocky Mountains and the less rugged Appalachians took shape during Neogene time, as did the imposing Himalaya. The Mediterranean Sea almost disappeared and then rapidly formed again.

The most widespread physical changes on Earth, however, were climatic. Glaciers expanded across large areas of North America and Eurasia late in Neogene time. Although this modern Ice Age is commonly thought of as corresponding to the Pleistocene Epoch, it actually began during the Pliocene and presumably will continue long into the future. We still live in the Ice Age. The spreading of glaciers has been episodic, and today we live during the latest of many intervals between pulses of extensive glaciation. Ice caps in the far north remain poised to spread southward, as they have done repeatedly during the past 3 million years or so.

The three Neogene epochs—the Miocene, Pliocene, and Pleistocene— were founded by Charles Lyell in 1833 in his *Principles of Geology*. Lyell distinguished the epochs of the Neogene Period on the basis of his observations of marine strata and fossils in France and Italy, noting that about 90 percent of the molluscan species found in Pleistocene strata are still alive in modern oceans, but that Pliocene strata contain fewer surviving species and Miocene strata, fewer still. It was not until later in the nineteenth century, however, that glacial deposits were recognized on land and found to correlate with the marine record.

Era	Period		Epoch	Time
CENOZOIC	NEOGENE		Holocene	
			Pleistocene	1.8 million years
	PALEOGENE		Pliocene	5.3 million years
			Miocene	
MESOZOIC	CRETACEOUS			24 million years
			Oligocene	
	JURASSIC		Eocene	
	TRIASSIC			
			Paleocene	
PALEOZOIC	PERMIAN			
	CARBONIFEROUS	PENNSYLVANIAN		
		MISSISSIPPIAN		
	DEVONIAN			
	SILURIAN			
	ORDOVICIAN			
	CAMBRIAN			
	PROTEROZOIC EON	"PRECAMBRIAN"		
	ARCHEAN EON			

Worldwide Events of the Neogene Period

In general, the animals and plants that inhabit Earth today are representative of Neogene life, so once again we are dealing with a period whose general life forms require no special introduction. Furthermore, continents have moved very little during the brief Neogene interval of geologic time. On the other hand, shifts in climate have led to major changes in environments and life.

469

Visual Overview
Major Events of the Neogene

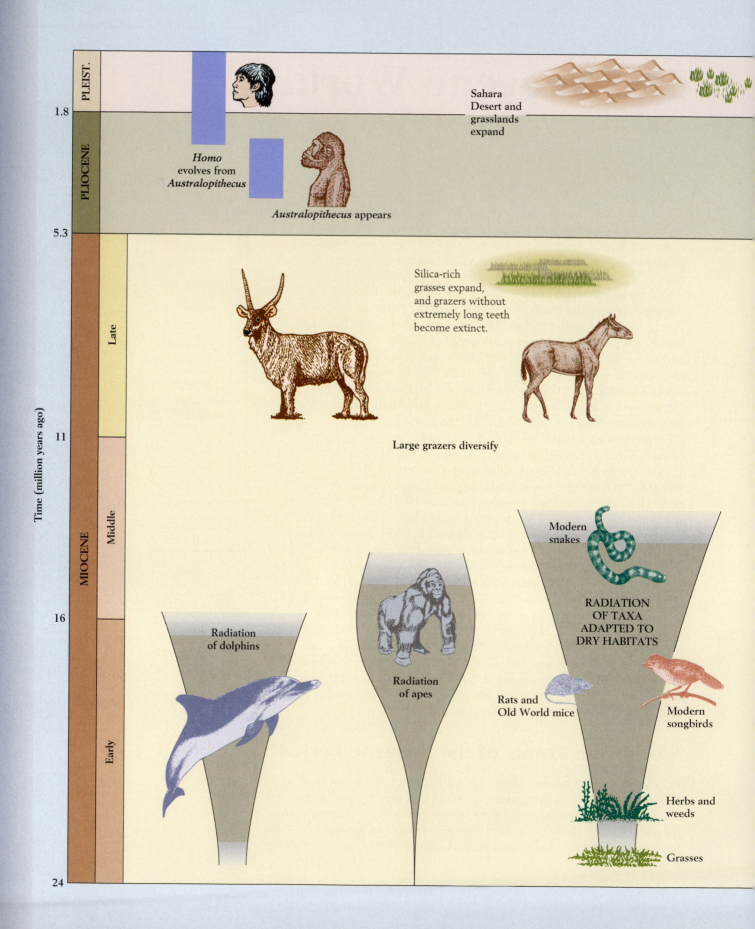

Time (million years ago)

PLEIST.

1.8

PLIOCENE

5.3

MIOCENE

Late

11

Middle

16

Early

24

Homo evolves from *Australopithecus*

Australopithecus appears

Sahara Desert and grasslands expand

Silica-rich grasses expand, and grazers without extremely long teeth become extinct.

Large grazers diversify

Radiation of dolphins

Radiation of apes

Modern snakes

RADIATION OF TAXA ADAPTED TO DRY HABITATS

Rats and Old World mice

Modern songbirds

Herbs and weeds

Grasses

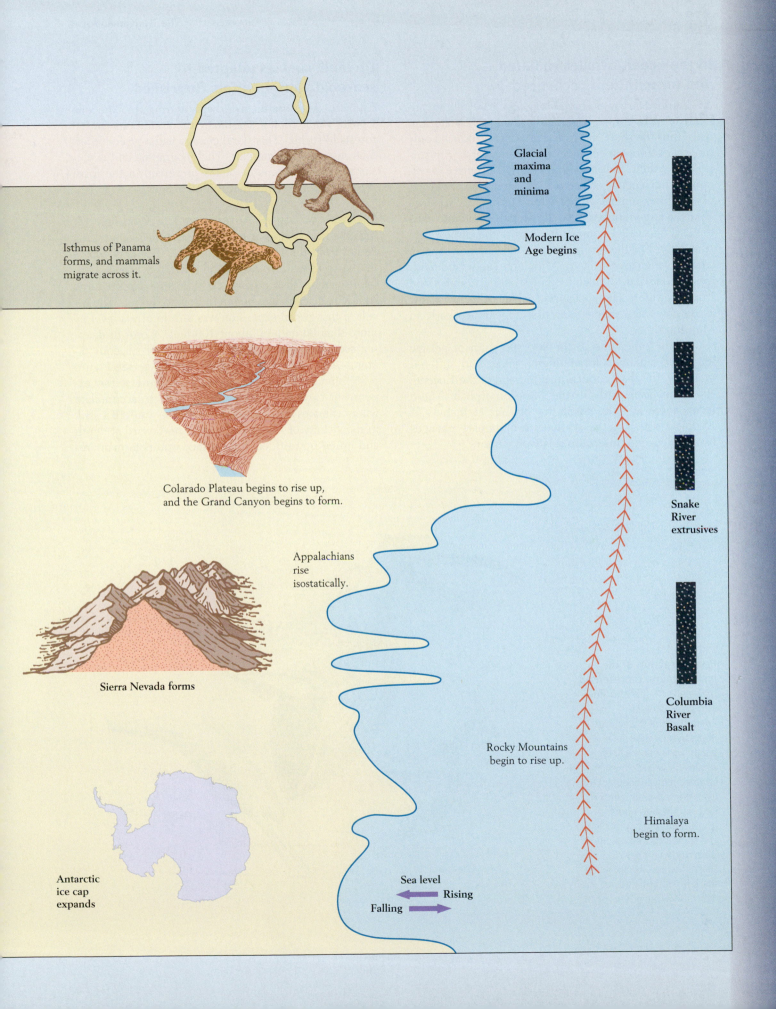

Isthmus of Panama forms, and mammals migrate across it.

Colarado Plateau begins to rise up, and the Grand Canyon begins to form.

Appalachians rise isostatically.

Sierra Nevada forms

Antarctic ice cap expands

Glacial maxima and minima

Modern Ice Age begins

Snake River extrusives

Columbia River Basalt

Rocky Mountains begin to rise up.

Himalaya begin to form.

Sea level

Rising

Falling

In the ocean, whales radiated and foraminifera recovered

As Charles Darwin observed long ago, invertebrate animals tend to evolve less rapidly than vertebrate animals; thus the short Neogene Period has produced only modest evolutionary changes for invertebrates. Not surprisingly, then, the most dramatic evolutionary development in Neogene oceans was the expansion of a group of vertebrate animals: the whales. During the Miocene Epoch a large number of whale species came into existence (Figure 19-1); among them were the earliest representatives of the groups that include modern sperm whales, which are carnivores with large teeth, and modern baleen whales, which feed by straining zooplankton from seawater. The specialized whales that we know as dolphins also made their first appearance early in Miocene time.

At the other end of the size spectrum of pelagic life, the planktonic foraminifera, which had suffered greatly in the mass extinction at the end of the Eocene, expanded again early in the Miocene Epoch. These forms serve as index fossils for oceanic sediments of Neogene age. On the seafloor, evolutionary changes from Paleocene time were relatively minor.

On land, species adapted to seasonally dry habitats flourished

Climates deteriorated worldwide during the Miocene Epoch, when cooler, seasonally dry conditions caused forests to shrink. This floral change had a profound effect on terrestrial animals. The geographic and evolutionary modifications of biotas preserved in the fossil record help us reconstruct these changes.

Further cooling of Antarctica The Miocene cooling, like the cooling of Late Eocene time (see p. 457), issued from the region of Antarctica. The appearance in the southeastern Pacific of ice-rafted coarse sediments shows that during the Miocene Epoch large Antarctic glaciers had begun to flow to the sea. The result of this glacial expansion was a drop of sea level by more than 30 meters (about 100 feet). Cores of deep-sea sediment also reveal that the belt of siliceous diatomaceous ooze that encircled Antarctica (see Figure 5-35) simultaneously expanded northward at the expense of carbonate ooze, which tends to accumulate where climates are warmer. Cold, dense water that sank in the vicinity of Antarctica spread northward at depth and mixed upward, cooling surface waters in many re-

Figure 19-1
Reconstruction of the marine fauna represented by fossils of the Middle Miocene Calvert Formation of Maryland. Representing the whale family were early baleen whales (*Pelocetus*), which strained minute zooplankton from the seawater (center); long-snouted dolphins (*Eurhinodelphis*, lower left); and short-snouted dolphins (*Kentriodon*, upper right). Sharks include the six-gilled shark *Hexanchus* (lower right). (Drawing by Gregory S. Paul.)

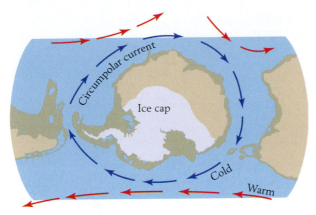

A Early Oligocene

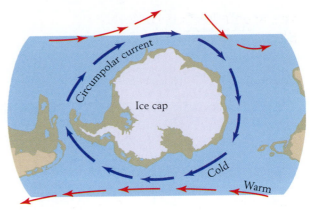

B Late Miocene

Figure 19-2 **Strengthening of the circumpolar current during Miocene time.** *A.* In Early Oligocene time Australia had moved far enough from Antarctica to allow waters from the southern oceans to become trapped in a current around Antarctica, where they became very cold. As a result, glaciers grew on Antarctica. *B.* By the beginning of Late Miocene time, about 11 million years ago, Australia had moved farther from Antarctica, and South America had also rifted away from it; the broader, deeper passageway thus formed around Antarctica accommodated a stronger circumpolar current. As a result, Antarctica was colder and its ice cap was larger. (From L. A. Lawyer, L. M. Gahagan, and M. F. Coffin, *Antarctic Res. Ser., Amer. Geophys. Union* 56:7–30, 1992.)

gions. Ever since Early Miocene time, assemblages of plankton at high latitudes have differed greatly from those at low latitudes.

Detailed studies of plate movements in the vicinity of Antarctica reveal the ultimate cause of Miocene climatic change. Recall that the rifting of Australia away from Antarctica in Late Eocene time trapped waters in a circumpolar current, where they cooled and, by spreading equatorward and mixing upward, changed climates throughout the world (see Earth System Shift 18-1). The passage for the current between Antarctica and Australia was initially narrow, however, and the passage between Antarctica and South America was not only narrow but also shallow. During Miocene time, Australia moved farther away from Antarctica, and South America also began rifting away from Antarctica, allowing a deep current to flow between the two continents. The result was a stronger circumpolar current and further cooling of Antarctica—as well as further cooling of climates throughout the world (Figure 19-2).

The cooler seas of the Miocene sent less water to the atmosphere through evaporation, so that many continental regions became not only cooler but also drier. As a result, during Miocene time, grasslands adapted to seasonal dryness expanded into areas that had once supported open woodlands and dense forests.

An explosion of grasses and herbs As in Late Cretaceous and Paleogene times, flowering plant fossils represent our best gauge of these climatic shifts. In the world of plants, the Neogene Period might be described as the Age of Grasses or the Age of Herbs. Whereas grasses are widespread on Earth today, they occupied relatively little land early in the Cenozoic Era. The diversity of grasses has exploded during the latter part of the era, however; today there are some 10,000 species of grasses (Figure 19-3). *Herbs,* or *herbaceous plants,* are small, nonwoody flowering plants that die back to the ground after releasing their seeds. (Defined in this way, herbs include many more plants than the few that we use to season our food.) The Compositae, a family of herbs that includes such seemingly diverse members as daisies, asters, sunflowers, and lettuces, appeared near the beginning of the Neogene Period, only 20 million or 25 million years ago. Today this family contains some 13,000 species, including nearly all of the plants that ecologists refer to as *weeds* (see Figure 19-3). As any gardener knows, weeds are exceptionally good invaders of bare ground. They may not compete successfully against other plants to retain the space they invade, but they soon disperse their seeds to other bare areas that have been cleared by fires, floods, or droughts and spring up anew.

Grasses and herbs flourish in seasonally arid climates—grasses because the lack of summer moisture excludes shade-generating trees, and weeds, for the same reason and also because seasonal dryness causes other plants to die back, providing bare ground that weeds can invade. The recent success of grasses and herbs is primarily the result of the worldwide deterioration of climates during Oligocene and Miocene times.

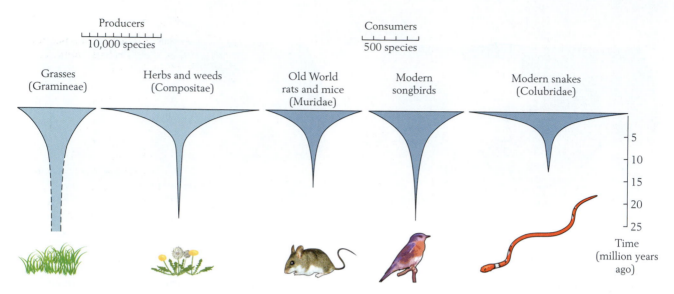

Figure 19-3 Cascading evolutionary radiations of terrestrial taxa during the past 25 million years. Grasses and weeds diversified dramatically, and their enormous production of seeds was partly responsible for a great expansion of rats, mice, and songbirds. Colubrid snakes, which feed on rats and mice and the eggs and chicks of songbirds, then radiated as well. (After S. M. Stanley, in P. D. Taylor and G. P. Larwood, *Major Evolutionary Radiations*, Clarendon Press, Oxford, 1990.)

Cascading radiations in open terrain Most of us are so interested in the origins of large mammals that we tend to ignore the great success of smaller creatures. In fact, the Neogene Period might well be called the Age of Rats, the Age of Mice, the Age of Snakes, or the Age of Songbirds—because all four of these groups have undergone tremendous evolutionary radiations over the past few million years (see Figure 19-3). Radiations cascaded upward through food webs: the radiations of grasses and weeds at the base of food webs stimulated the radiation of small herbivores, which, in turn, stimulated the diversification of snakes. Many species of rats and mice eat the seeds of grasses and herbs. To a large extent, therefore, the success of these small rodents during Neogene time has resulted from the success of both the grasses and the Compositae, the family to which most herbs belong. Also poorly represented before Neogene time were the *passerine birds*, or songbirds and their relatives, which are highly conspicuous today. As the contents of our birdfeeders attest, songbirds also benefited from the seed production of new species of herbs—although some songbirds owe their success to their ability to capture insects.

Snakes have obviously benefited from the proliferation of rats, mice, and songbirds; few other predators can pursue small rodents down their burrows without digging or make their way along tree branches to consume eggs and chicks in songbirds' nests. Before the start of Neogene time, there were few snakes except for members of the primitive boa constrictor group. The snake family Colubridae, which includes most common snake species—for example, garter snakes, corn snakes, rattlesnakes, and copperheads—arose only about 15 million years ago, yet today it contains about 1400 species.

In summary, the increased aridity of Miocene time stimulated radiations that cascaded up the food web, from grasses and herbs to rats, mice, and songbirds, and then to snakes. With songbirds representing a striking exception, we might succinctly label the modern world the Age of Weeds and Vermin!

Diversification of large mammals Of course, groups of large animals also developed their modern characteristics during the Neogene Period. Among the herbivores, the horse and rhinoceros families dwindled after Middle Miocene time in a continuation of the general decline of the odd-toed ungulates. Meanwhile, the even-toed, or cloven-hoofed, ungulates expanded, especially through the adaptive radiation of both the deer family and the family called the Bovidae, which includes cattle, antelopes, sheep, and goats. The giraffe family and the pig family also radiated during the Miocene Epoch, but the number of species in these families has since declined. Similarly, many types of elephants, including those with long trunks, experienced great success during Miocene and Pliocene time, but later declined. Today only three elephant species survive: two large-eared African elephant species, one adapted to woodlands and the other to savannahs, and the smaller, more docile Indian elephant—the species that is commonly trained to perform in circuses.

Figure 19-4
Reconstruction of a portion of the so-called *Hipparion* fauna of Asia. This diverse fauna occupied open country in Asia about 10 million years ago, in Late Miocene time. *Hipparion* is the galloping horse (center). The elephant on the left, with downward-directed tusks, is *Dinotherium*. In the foreground, short-legged hyenas of the genus *Percrocuta* look on from their den.
(Drawing by Gregory S. Paul.)

Carnivorous mammals also assumed their modern character in the course of the Neogene Period; this group included the dog and cat families, both of which had appeared during Paleogene time. The bear and hyena families were important Miocene additions to the carnivore group.

Many Neogene mammal groups expanded successfully because of the spread of grassy woodlands (Figure 19-4). Several herbivore groups, such as antelopes, cattle, and horses, evolved many species that were well adapted for long-distance running over open terrain and for grazing on harsh grasses with the aid of tall molar teeth. Grasses contain tiny fragments of silica that wear down a grazer's teeth throughout its lifetime, so only tall teeth can tolerate a diet of grasses. Also on the increase were the groups of rodents, such as prairie dogs, that are adapted for burrowing in prairies. As we might expect, the diversification of herbivores in savannahs and woodlands fostered the success of fast-running carnivores that were well adapted for attacking herbivores in open country—groups such as hyenas, lions, cheetahs, and long-legged dogs.

The spread of C_4 grasses and the extinction of large herbivores Between 7 million and 6 million years ago, near the end of the Miocene Epoch, North American mammals suffered the largest extinction event that has struck them since the crisis at the end of the Eocene (p. 459). Species of large herbivores that lacked extremely tall molar teeth formed the largest group of victims. A similar extinction event occurred in Asia. It was not the spread of grasslands that eliminated these species, because they were predominantly forms whose tall molars were adapted for grazing on grasses. Only grazing species whose molars were extremely tall survived, however. What changes in the environment could have caused this kind of selective extinction?

As it turns out, the extinction event coincided with a major change in the kinds of grasses that formed grasslands. The evidence of this change comes from stable isotopes (carbon 12 and carbon 13). When plant material decays, it releases carbon into the soil. The isotope ratio of this carbon reflects the ratio in the plants that produced it. Similarly, the carbon isotope ratio in herbivores reflects that of the plants they eat. Analyses of ancient grassland soils and of the teeth of the herbivores that grazed on those grasslands has revealed that a pronounced shift toward heavier carbon isotope ratios occurred on many continents between 7 million and 6 million years ago (Figure 19-5). This shift reflects the partial replacement of the group of grasses known as C_3 grasses by the group known as C_4 grasses. These two groups of grasses have different physiologies: when C_4 grasses extract carbon dioxide from the atmosphere, they assimilate a larger fraction of carbon 13 than C_3 grasses do.

Recall that silica in grasses is what makes them highly abrasive to herbivore teeth. It happens that C_4 grasses have about five times as many silica bodies in their tissues as C_3 grasses. Clearly, the expansion of C_4 grasses near the end of the Miocene caused problems for many grazing herbivores that lacked extremely tall teeth. Animals whose teeth are worn down by grasses end up suffering from malnutrition, which leads to

death. That is why looking a gift horse in the mouth is actually a good idea: a horse with heavily worn teeth is an old animal with a short life expectancy. Apparently when C_4 grasses first spread into many regions of the world near the end of the Miocene, they so rapidly abraded the teeth of grazers that lacked extremely tall molars that those species could not produce enough offspring before death to sustain their populations. The result was extinction.

What caused the sudden spread of C_4 grasses near the end of Miocene time? These grasses grow better than C_3 grasses under low carbon dioxide conditions. It was therefore hypothesized that a drop in the concentration of carbon dioxide in Earth's atmosphere resulted in the spread of C_4 grasses. As it turns out, carbon isotope ratios for residues of marine phytoplankton in sediments refute this hypothesis. During photosynthesis, phytoplankton incorporate a smaller proportion of carbon 13, relative to carbon 12, as the concentration of carbon dioxide in their environment declines. Once phytoplankton die, their organic matter is normally degraded by bacteria that alter its isotopic composition. As a result, it is impossible to know the original carbon isotopic composition of organic carbon in sediments. Calcareous nannoplankton, however, produce a group of organic compounds known as alkenones that are not isotopically altered by bacteria. As a result, fossil alkenones provide faithful records of atmospheric oxygen levels for ancient times. Studies of alkenones indicate that carbon dioxide levels in the upper ocean, which reflect levels in the atmosphere, actually increased, rather than decreasing, between 15 million and 5 million years ago (Figure 19-6).

It is now clear that global climatic change caused the spread of C_4 grasses. To flourish, C_3 grasses require a cool, moist growing season. Thus they prevail at relatively high latitudes and in regions such as California that have Mediterranean climates. In contrast, C_4 grasses predominate where the moist season is warm—in tropical savannahs, for example. Today in North America, C_4 grasses form grasslands in Mexico, whereas C_3 grasses form grasslands in central Canada; grasslands have equal biomasses of C_3 and C_4 grasses at the latitude of South Dakota. In contrast, C_3 grasses, along with taller C_3 plants, were dominant in central North America during Miocene time almost until the end of the epoch. Then, about 6 million years ago, savannahs dominated by C_4 grasses spread across central North America, signaling the disappearance of a cool growing season. To survive, grasses had to be able to grow when the growing season became warmer, taking advantage of occasional summer rains. Thus, as grasslands expanded on a global scale, C_4 grasses replaced C_3 grasses in many regions.

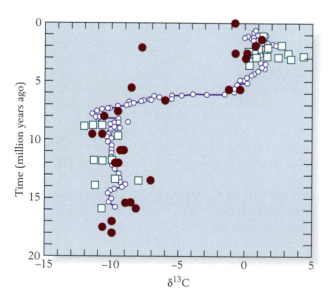

Figure 19-5 Major shifts in carbon isotopes indicating the spread of C_4 grasses between 7 million and 6 million years ago. The plotted values are carbon isotope ratios from ancient soils and mammal teeth from Pakistan and North America. (After T. E. Cerling, Y. Wang, and J. Quade, *Nature* 361:344–346, 1993.)

Atmospheric CO_2 (parts per million)

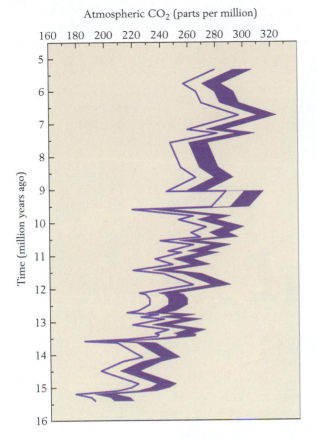

Figure 19-6 Estimate of the concentration of carbon dioxide in Earth's atmosphere between 15 million and 5 million years ago. The values plotted here are based on carbon isotope ratios in alkenones. The dashed line and the boundaries of the shaded band represent results of three slightly different methods of calculating the values. (After M. Pagani, K. H. Freeman, and M. A. Arthur, *Science* 285:876–879, 1999.)

The radiation of primates What about primates, the group to which we humans belong? In general, primates favor forests over savannahs; in fact, most live in trees. As we have seen, monkeys were present by Oligocene time; the oldest group includes the so-called Old World monkeys, which now live in Africa and Eurasia. Before the end of the Oligocene interval, however, a distinct group of monkeys was present in South America. How their ancestors got there remains uncertain. These New World monkeys, which differ from their Old World counterparts in that most possess prehensile (grasping) tails, probably had a separate evolutionary origin. In any event, monkeys on both sides of the Atlantic underwent evolutionary radiations during Neogene time.

Apes, which evolved in the Old World, diversified markedly during the Miocene Epoch, but declined in number of species as forests shrank late in Neogene time. We will discuss apes and apelike animals when we examine the origins of humans, which belong to the

same superfamily, the Hominoidea (see Figure 3-7). The most recent phases of human evolution have taken place within the climatic context of the modern Ice Age; therefore, before we discuss the Hominoidea, it is appropriate to examine the major global events of this fascinating interval of geologic time.

The Modern Ice Age of the Northern Hemisphere

Early in the Pliocene Epoch, relatively warm climates spread to high latitudes. Partway through the Pliocene, however, the Northern Hemisphere plunged into the modern Ice Age. Climates in many regions have been cooler and drier ever since, even at times when ice sheets have shrunk back.

Early Pliocene climates were relatively warm

As the Pliocene Epoch got under way, about 5 million years ago, sea level rose, leaving marine deposits inland of modern coastlines in such areas as California, eastern North America, the Gulf Coast of North America (see Figure 18-28), and countries bordering the North Sea and the Mediterranean. Fossil faunas and floras also reveal that in many parts of the globe, climates were more equable at this time than they are today because winters were warmer. Average annual temperatures were also higher in many regions. Pollen analyses, for example, indicate that southeastern England was subtropical, or nearly so. Especially in the Northern Hemisphere, however, this warm interval came to a sudden close with the start of the modern Ice Age, about 3.2 million years ago.

Continental glaciers formed in the Northern Hemisphere

Figure 19-7A depicts Earth at a time of full glacial expansion, or what is termed a **glacial maximum**. Throughout the Ice Age, glacial maxima have alternated with times of glacial recession, such as the present. During these warmer intervals, or **glacial minima**, the Greenland ice cap has been the only continental glacier to survive in the Northern Hemisphere (see Figure 4-13).

A wide variety of evidence documents the modern Ice Age, revealing details of the timing and geographic distribution of continental glaciation:

1. *Erratic boulders.* Certain large rocks that sit on Earth's surface far from exposures of the bedrock from which they have broken—called *erratic boulders*—are too large to have been transported by

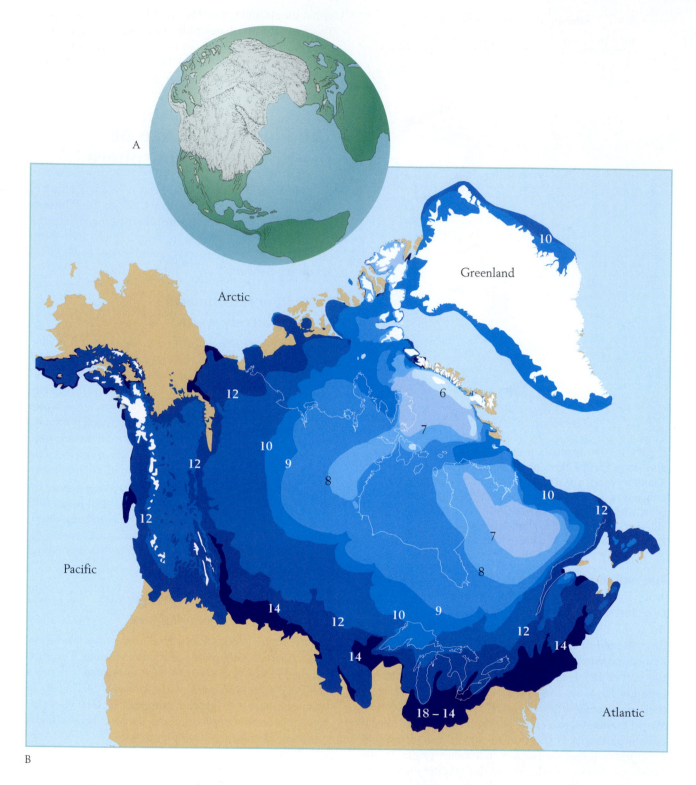

Figure 19-7 Reconstruction of glaciation patterns during the most recent glacial episode. *A.* At the most recent glacial maximum, about 20,000 years ago, large continental glaciers were centered in North America, Greenland, and Scandinavia. *B.* Numbers show borders of glacial ice on the North American continent at various times, in thousands of years ago, as indicated by radiocarbon dating. Continental glaciers disappeared from North America slightly after 6000 years ago. (*A* after a drawing by A. Sotiropoulos in J. Imbrie and K. P. Imbrie, *Ice Ages*, Enslow, Short Hills, NJ, 1979; *B* courtesy of Arthur Dyke, Geological Survey of Canada.)

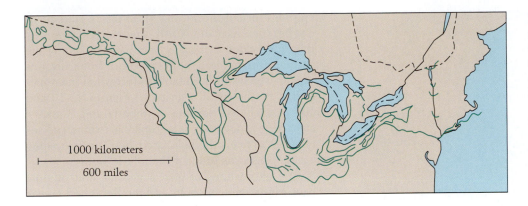

Figure 19-8 Positions of moraines that mark the southern limits of ice sheets in eastern North America during the most recent glacial maximum. (After D. M. Mickelson et al., in S. C. Porter, ed., *Quaternary Environments of the United States*, vol. 1, Longman, London, 1983.)

rivers, and it is difficult to imagine that any agent other than continental glaciers might have transported them.

2. *Glacial till.* A mixture of boulders, pebbles, sand, and mud that has been plowed up, transported, and then deposited by glaciers, till is difficult to confuse with sediments deposited by other mechanisms, especially where it rests at Earth's surface and forms moraines or is associated with outwash deposits (p. 108). Glacial moraines form much of Cape Cod, Massachusetts, where they extend into the marine realm. Retreating glaciers commonly left terminal moraines behind them, and as these glaciers melted back, they often left shallow basins in which water accumulated behind the moraines. The Great Lakes of North America occupy such basins; they did not exist before the modern Ice Age (Figure 19-8).

3. *Depression of the land.* Earth's crust remains depressed in regions that lay beneath large glaciers a few thousand years ago. Hudson Bay, the only epicontinental sea that exists today in North America, occupies such a depressed region in eastern Canada.

4. *Glacial scouring.* Glaciers smoothed the sides of mountains that they scraped past. Mount Monadnock, in New Hampshire, stood partially above surrounding ice sheets, as some mountains of Antarctica do today (Figure 19-9). The lower part of Mount Monadnock, which was smoothed by flowing glaciers, stands in sharp contrast to the upper part, which remains rugged. Small glaciers, known as alpine or mountain glaciers, left their marks along valleys within the Rockies and other mountain chains, where they flowed during the modern Ice Age. Especially spectacular are U-shaped valleys that glaciers sculpted from valleys that were once shaped like a V (see Figure 12-16).

5. *Lowering of sea level.* One important effect of each major expansion of ice sheets during the modern Ice Age was a profound lowering of sea level as great quantities of water were locked up on land. During major glacial expansions, most of the surfaces that now form continental shelves stood above sea level. Rivers cut rapidly downward through the soft sediments of continental shelves to form valleys that exist today as submarine canyons, having been excavated further by submarine turbidity currents. During some glacial episodes, sea level dropped slightly more than 100 meters (330 feet) below its present position.

Today there remain only two ice caps of the sort that expanded to cover broad areas many times during the modern Ice Age. One of these modern ice caps covers much of Greenland (see Figure 4-13) and the other covers nearly all of Antarctica (see Figure 19-9). Today about three-quarters of the world's fresh water is locked up in glacial ice, and about 90 percent of it belongs to the Antarctic ice cap. It may seem impressive that glaciers now contain about 25 million cubic kilometers of ice, but it has been estimated that the volume of ice was nearly three times as great during glacial maxima of the Pleistocene

Figure 19-9 Dark rocks of the Prince Charles Mountains projecting above the surface of the modern Antarctic ice cap. Many North American mountains were similarly buried in ice during the Pleistocene Epoch. (D. Parer and E. Parer-Cook/Auscape International.)

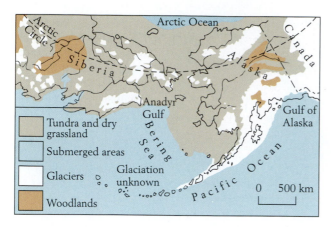

Figure 19-10 The geography of the Bering land bridge during the most recent glacial maximum. Though positioned at a high latitude, the Bering land bridge was dry and free of large glaciers. (After D. M. Hopkins, in D. M. Hopkins, ed., *The Bering Land Bridge*, Stanford University Press, Stanford, CA, 1967.)

Epoch, with the largest ice sheets averaging about 2 kilometers (1.2 miles) in thickness. The total volume of ice has been calculated from the volume of water that would have to be removed from the ocean to lower sea level slightly more than 100 meters. Shelves of ice projected into the sea, and these shelves, together with the icebergs and pack ice that broke loose from them, spread over half the world's oceans.

6. *Migration of species.* As ice sheets have expanded repeatedly and sea level has dropped, the resulting geographic changes have allowed species to migrate to new regions. Regression of the seas during glacial

episodes turned the Bering Strait into a land corridor between Asia and North America, and it was by this land bridge that many mammals, including the first humans, entered the New World. Ironically, this region, which was hospitable to terrestrial mammals during the height of glaciation (Figures 19-10 and 19-11), included portions of Siberia and Alaska—areas that we now view as inhospitable to most species but that remained unglaciated because prevailing weather patterns brought them little snow.

The alternations of glacial maxima and minima have caused climatic belts and the floras and faunas that occupy them to shift over distances measured in hundreds of kilometers. Fossils of mammals such as the muskrat, which today does not range south of Georgia, reveal that climates in Florida were cool when glaciers pushed southward into the northern United States. Other fossil occurrences, such as those of hippopotamuses in Britain, show that during at least some glacial minima, climates were warmer than they are today.

One of the most useful fossil indicators of Pleistocene climates is the pollen of terrestrial plants (Figure 19-12). Pollen assemblages reveal climatic change by indicating the shifting of floras to the north or south. Figure 19-13 shows the southward movement of floras in Europe by about 20° latitude during the most recent glacial maximum there.

The chronology of glaciation can be read in isotope ratios

The Pleistocene Epoch is often thought of as the modern Ice Age, but the Ice Age actually began long before the end of Pliocene time. The most detailed chronol-

Figure 19-11
Reconstruction of the mammalian fauna that occupied the dry grassland in the Alaskan portion of the Bering land bridge, about 12,000 years ago, during the most recent glacial interval. Of the 61 species depicted here, 11 are extinct; among them are the woolly mammoth, the American mastodon, the long-horned bison, a lion, and a saber-toothed cat. (Courtesy Smithsonian Institution, painting by Jay H. Matternes.)

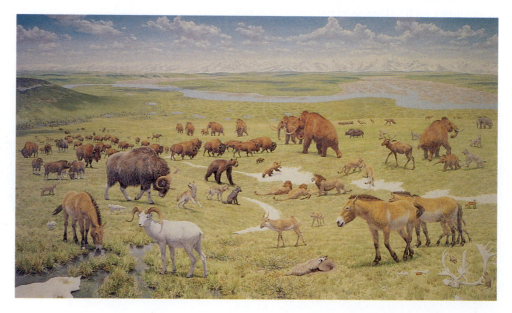

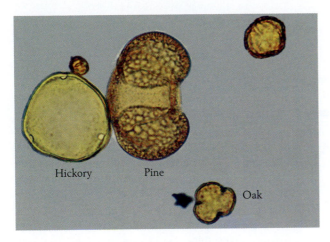

Figure 19-12 Three common forms of pollen found in Pleistocene sediments of eastern North America. (Paige Newby, Brown University.)

ogy of Ice Age glaciation comes from oxygen isotope ratios of foraminiferal skeletons preserved in deep-sea sediments. Recall that this ratio shifts toward heavier values during glacial episodes for two reasons. First, a disproportionate amount of the lighter isotope, oxygen 16, accumulates in glaciers, leaving the ocean enriched in oxygen 18. Second, as temperatures decline, foraminifera take up a larger percentage of oxygen 18 from the seawater in which they live.

Slightly before 3 million years ago there were marked increases in the ratio of oxygen 18 to oxygen 16 in the skeletons of foraminifera in many oceanic areas (Figure 19-14). This change resulted from a brief

episode of widespread cooling, which terrestrial floras also document. In northwestern Europe, for example, several subtropical species of land plants, including palms, disappeared.

Continental glaciers began to expand slightly thereafter, and large oscillations of oxygen isotope ratios in planktonic foraminifera indicate that these glaciers expanded and contracted every few tens of thousands of years (see Figure 19-14). The isotope ratios indicate that by about 2.5 million years ago the Northern Hemisphere had moved fully into the Ice Age. In addition, deep-sea deposits of this age in the North Atlantic record the first occurrence of numerous sand grains released by melting icebergs. In other words, continental glaciers were now flowing to the North Atlantic, releasing sediment-laden blocks of ice.

Climatic changes altered floras

Terrestrial floras underwent dramatic changes as well. About 2.5 million years ago, northwestern Europe lost the last of many subtropical plant taxa that it had shared with Malaysia earlier in Pliocene time. It was not only cooling of climates that affected floras, but also increased aridity. Many regions became drier as climates cooled because cooler sea surfaces released less water to the atmosphere through evaporation. Increased aridity in Africa led to a great expansion of the Sahara Desert. Pollen from deep-sea cores off the west coast of Africa reveals that before about 3 million years ago tropical forest frequently extended to a latitude close to the present southern limit of the Sahara. The Sahara itself was at that time a small desert far from the ocean.

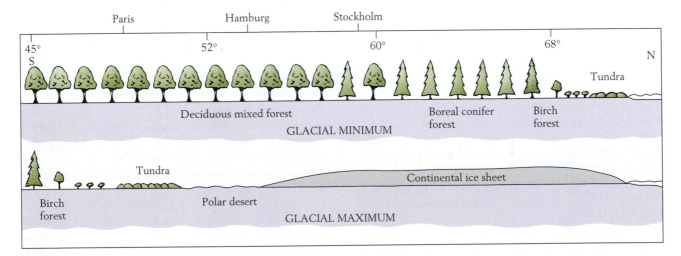

Figure 19-13 North–south migration of vegetation in Europe during the Pleistocene Epoch. During a glacial maximum (lower diagram), continental glaciers spread southward to the vicinity of Hamburg and tundra shifted to the latitude of Paris. (After T. Van Der Hammen, in K. K. Turekian, ed., *The Late Cenozoic Glacial Ages*, Yale University Press, New Haven, CT, 1971.)

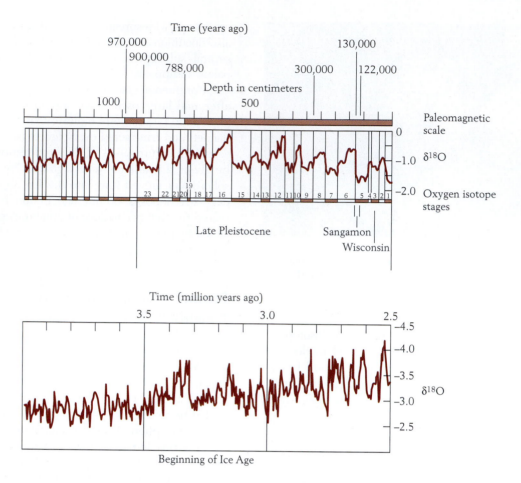

Figure 19-14 Oxygen isotope fluctuations in foraminifera from deep-sea cores. Large peaks in the graphs, representing increases in the relative abundance of oxygen 18, represent glacial maxima. The ratios in the lower graph show the initial buildup of ice sheets between 3.5 million and 3.0 million years ago. Oxygen isotope stages (1–23) for the late Pleistocene, representing glacial maxima and minima, are shown below the upper graph. The paleomagnetic time scale above the graph provides dates for several levels in the core. (Upper graph after N. J. Shackleton and N. D. Opdyke, *Geol. Soc. Amer. Mem.* 145:449–464, 1976; lower graph after R. Tiedmann, M. Sarnthein, and N. J. Shackleton, *Paleoceanography* 9:619–638, 1994.)

Glaciers expanded and contracted many times

Three large glacial centers developed in the Northern Hemisphere—one in North America, one in Greenland, and one in Scandinavia (see Figure 19-7A). Because the northern Atlantic Ocean was adjacent to these three glacial centers, it was more profoundly affected by the Pleistocene glaciers than any of the world's other major oceans except the Arctic. Land areas adjacent to the Atlantic also suffered marked climatic change. When glaciers grew to their maximum extent, pack ice choked large areas of the North Atlantic (Figure 19-15), just as it now occupies bays adjacent to northern Canada in winter. Farther south, along the east coast of North America, glaciers flowed southward to New Jersey, and tundra occupied what is now Washington, D.C.

Glacial maxima and glacial minima are represented by peaks and valleys in oxygen isotope curves (see Figure 19-14). Today we live during the glacial minimum established when the most recent glaciers melted back between about 15,000 and 12,000 years ago. The most recent glacial maximum extended from about 35,000 to about 12,000 years ago and constituted the Wisconsin Stage. By the time this glacial maximum peaked about 20,000 years ago, sea level had dropped to at least 100 meters (330 feet) below its present level. The Wisconsin glacial interval was preceded by the Sangamon glacial minimum, about 125,000 years ago, when sea level stood slightly higher than it does today. Because of its relative recency, the last interval of glacial expansion (the Wisconsin) has left the best record of glacial deposits. Furthermore, the time of maximum expansion of Wisconsin glaciers places their organic deposits, including fossilized wood, well within the range of radiocarbon dating.

PRESENT

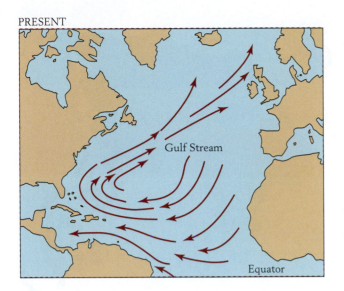

GLACIAL MAXIMUM

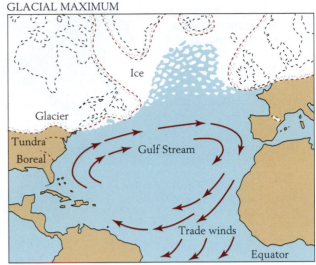

Figure 19-15 The Atlantic Ocean and neighboring regions today and during the most recent glacial maximum.

Detailed stratigraphic studies in the region of the Great Lakes have revealed that the late Wisconsin glacial advance was a complex event, consisting of pulses of glacial expansion separated by partial retreats (see Figure 19-8). In addition, individual lobes of the North American ice sheet did not always expand and contract at the same rate. Ultimately, however, as the ice sheet receded (see Figure 19-7), glacial lobes retreated into large basins that became the Great Lakes (Figure 19-16).

Mountain glaciers

Principal lakes of Wisconsin age

Southern limit of Wisconsin continental ice

Figure 19-16 Locations of glaciers and lakes in the United States during the most recent glacial interval. Lake Agassiz and the ancestral Great Lakes formed to the south of the continental glacier as it retreated northward near the end of the glacial interval. (After C. B. Hunt, *Natural Regions of the United States and Canada*, W. H. Freeman and Company, New York, 1974.)

Vegetation patterns changed during glacial maxima

Figure 19-17 compares world vegetation patterns at the most recent glacial maximum with those of the present. Having expanded when the Ice Age was getting under way, the Sahara Desert grew still larger during glacial maxima because of reduced evaporation from cool seas. Figure 19-18 shows the southern expansion of dune activity in the Sahara Desert during the most recent glacial maximum. Deserts expanded in a similar way on other continents.

Rain forests shrank at the start of the Ice Age, then shrank even more drastically during glacial maxima. The African rain forest was repeatedly restricted to three small areas. Populations of rain forest species were therefore fragmented, and some, such as the gorilla, have failed to recolonize the entire rain forest up to the present day, remaining in separate areas that served as refuges when the rain forest contracted. In their isolation, the two existing gorilla populations have evolved into separate subspecies (see Figure 19-18). Other taxa appear to have evolved many new species as a result of geographic isolation of small populations at times when the rain forest contracted.

During the Wisconsin glacial interval, north–south temperature gradients steepened in the Northern Hemisphere, both in shallow seas and on land. Winter temperatures fell by a few degrees in most tropical areas, but plummeted at latitudes north of 30° in the Northern Hemisphere.

There were exceptions to the general pattern of increased aridity during glacial maxima. Among the exceptions was the Great Basin in the American West, where water accumulated to form numerous lakes in areas that are now arid (see Figure 19-16). Apparently

Figure 19-17 Vegetation patterns reconstructed for the most recent glacial maximum, about 20,000 years ago, compared with those of today. (After J. M. Adams et al., *Nature* 348:711–714, 1990.)

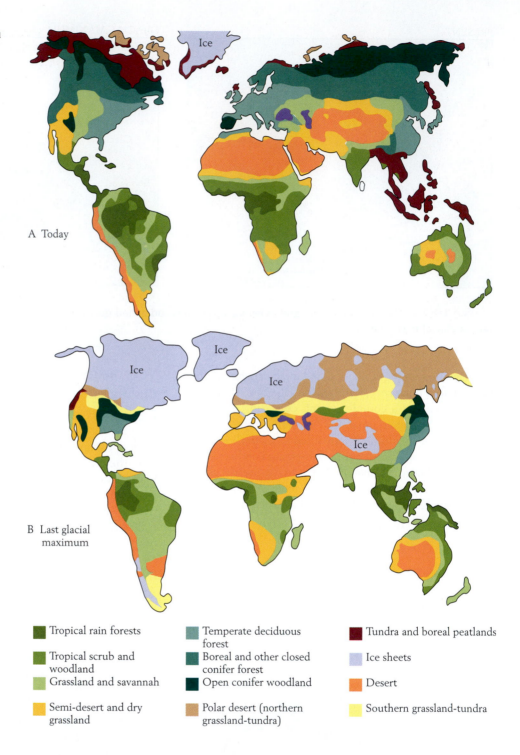

A Today

B Last glacial maximum

Tropical rain forests	Temperate deciduous forest	Tundra and boreal peatlands
Tropical scrub and woodland	Boreal and other closed conifer forest	Ice sheets
Grassland and savannah	Open conifer woodland	Desert
Semi-desert and dry grassland	Polar desert (northern grassland-tundra)	Southern grassland-tundra

the great mountain of ice to the north deflected winds from the Pacific Ocean, causing them to follow a more southerly course and bring moisture to the Great Basin. The Great Salt Lake in Utah is a remnant of the largest western Ice Age lake, known as Lake Bonneville.

Annual layers in cores from the Greenland ice cap indicate that the climate of Greenland warmed suddenly many times during the last glacial interval. This record alerts us to the basic instability of Earth's climate and

serves warning as to what may happen in coming years (Earth System Shift 19-1).

Changes in oceanic circulation may have triggered the Ice Age

The cause of the late Neogene glaciation in the Northern Hemisphere has been a matter of debate. Although a relatively low level of carbon dioxide in Earth's

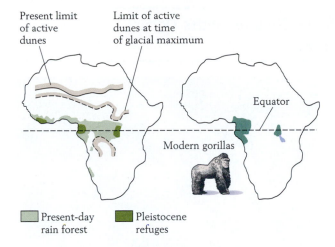

Present limit of active dunes

Limit of active dunes at time of glacial maximum

Equator

Modern gorillas

Present-day rain forest

Pleistocene refuges

◀ **Figure 19-18** **Environmental changes in Africa during the most recent glacial maximum.** Dune activity shifted toward the equator, marking the southward expansion of the Sahara Desert. The rain forest shrank into three small areas. Repeated contractions of this kind have left gorillas divided into two populations that have evolved to become separate subspecies.

atmosphere may have set the stage for this glacial episode, carbon isotopes in alkenones indicate that atmospheric carbon dioxide had already declined approximately to its modern level by more than 5 million years ago, or about 2 million years before the Ice Age began (see Figure 19-6).

One possibility is that changes in oceanic circulation brought about the late Neogene Ice Age. The key event here would have been the origin of the Isthmus of Panama when an eastward-moving island arc lodged between North and South America. The presence of the isthmus today makes the North Atlantic Ocean much saltier than the North Pacific (Figure 19-19). The Atlantic is saltier because the dry trade winds blowing westward from the Sahara Desert evaporate large amounts

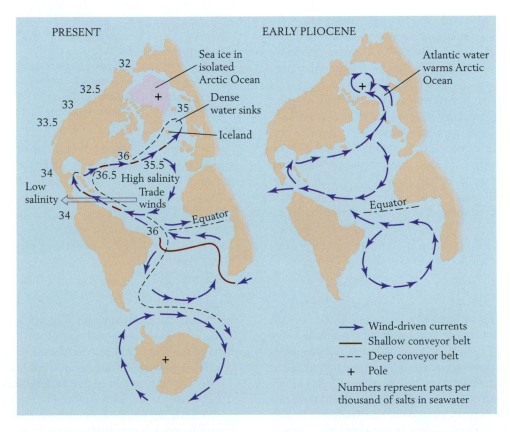

PRESENT

EARLY PLIOCENE

32

32.5

33

33.5

34

Low salinity

34

36

35

Sea ice in isolated Arctic Ocean

Dense water sinks

Iceland

36

35.5

36.5 High salinity

Trade winds

Equator

Atlantic water warms Arctic Ocean

Equator

→ Wind-driven currents
─ Shallow conveyor belt
--- Deep conveyor belt
+ Pole

Numbers represent parts per thousand of salts in seawater

Figure 19-19 **The oceanic conveyor belt may have influenced the temperature of the Arctic Ocean.** Today the trade winds carry to the Pacific water that they evaporate from the surface of the Atlantic, making the Atlantic more saline. (Numbers show salinities for the two oceans.) The dense, saline Atlantic water sinks just north of Iceland, driving the oceanic conveyor belt. During Early Pliocene time, before the Isthmus of Panama was in place, mixing with Pacific waters should have lowered the salinity of the Atlantic. The more buoyant Atlantic waters may then have flowed northward into the Arctic, keeping the polar region warm. The uplift of the Isthmus of Panama may have triggered the Ice Age by elevating the salinity of Atlantic waters and causing them to sink north of Iceland, as they do today; this change would have deprived the Arctic Ocean of heat from the Atlantic. (After S. M. Stanley, *J. Paleontol.* 69:999–1007, 1995.)

Earth System Shift 19-1　Shockingly Rapid Climatic Shifts Occur during the Ice Age

Detailed studies of the last glacial maximum, between 80,000 and 20,000 years ago, have revealed astoundingly sudden climatic changes separated by longer cooling trends. Part of the evidence comes from a particular species of planktonic foraminifera that survives today and is known to flourish in very cold water. Additional evidence comes from oxygen isotope ratios in annual layers of ice that can be observed in cores of the Greenland ice cap. A thick layer of ice accumulates every year in summer, whereas a thin layer accumulates in winter, when the colder surface waters of the ocean supply less moisture for precipitation. Oxygen isotopes in the ice reflect temperature changes from year to year in the Greenland region.

The foraminiferal and isotope records are in accord, showing that climatic oscillations were grouped into long-term cooling cycles that lasted an average of 10,000 to 15,000 years. The average temperature declined within a cycle, but each cycle ended with an abrupt warming event, during which the temperature jumped by several degrees Celsius within only about 10 years. Interestingly, not long before each pulse of warming, a Heinrich event occurred. A *Heinrich event* is a massive discharge of icebergs that release sedimentary debris to the sea as they melt. Heinrich layers are conspicuous in North Atlantic deep-sea cores. Heinrich events occurred when climates were very cold and glaciers surged to the sea. (Mountain glaciers occasionally surge in the present world, though on a much smaller scale, and sometimes they launch numerous icebergs into the sea; see Figure 4-14.) Why sudden warming followed Heinrich events is unclear, as is the reason for the lengthening of the cooling cycles.

Figure 1 Annual layers of ice are visible in this glacier in the Andes of South America. Melting of the glacier has exposed these bands. (Lonnie G. Thompson, Byrd Polar Research Institute, Ohio State University.)

Whatever may have caused the abrupt warming episodes, such sudden changes serve warning that the global climate can change dramatically as a result of natural causes within the space of a single decade. How rapidly, we must ask, may future global warming occur as we humans release the greenhouse gas carbon dioxide into Earth's atmosphere through our burning of fossil fuels?

of water from its surface. This evaporation not only increases the salinity of the Atlantic, but also results in lower salinity for the Pacific west of the Isthmus of Panama because much of the evaporated water is transported across Central America and enters the eastern Pacific as rainfall.

The clockwise circulation of the North Atlantic carries the saline water formed in the trade wind belt northward. This dense saline water cools as it moves northward in the Gulf Stream and thus becomes even denser. Finally it sinks just north of Iceland. This sinking is the primary driving force for a huge loop of moving water known as the *oceanic conveyor belt*. The water that sinks north of Iceland flows back to the south and ultimately bends eastward into the Pacific, where it surfaces and then returns to the Atlantic, completing the loop of the conveyor belt. The waters of the North Pacific are less saline and therefore less dense than those of the North Atlantic, so they do not sink to form a conveyor belt when they became cold in the far north.

Because the waters of the Atlantic sink at the brink of the Arctic Ocean and loop back to the south, the Arctic Ocean is deprived of Atlantic warmth. Without inflow of warm Atlantic surface waters, the upper Arctic Ocean assumes the cold temperature that is typical of a relatively isolated body of water at a very high latitude. The key role of the isthmus is as a barrier that prevents the saline Atlantic waters from mixing with the waters of the Pacific. Thus it keeps the Atlantic waters dense enough to sink just north of Iceland, before they reach the Arctic Ocean.

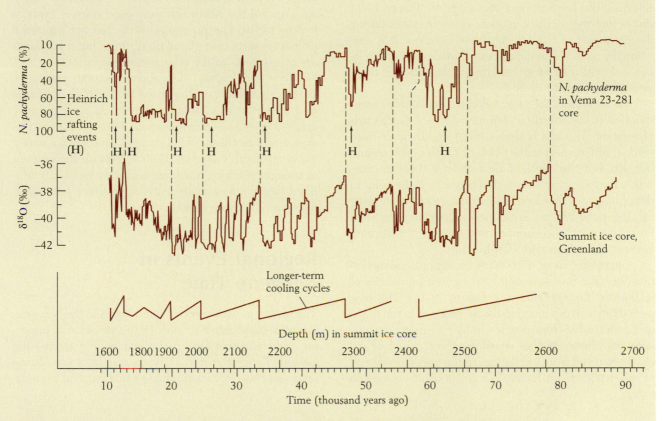

Figure 2 Evidence of rapid climatic changes during the most recent interval of glacial expansion. Increases in the abundance of *Neogloboquadrina pachyderma* in relation to other species of planktonic foraminifera in the North Atlantic (upper graph) indicate cooling. Oxygen isotope ratios in annual layers in a Greenland ice core (lower graph) show the same pattern. Intervals of net cooling in the North Atlantic, averaging 10,000 to 15,000 years in duration and punctuated by temperature oscillations, ended with sudden warming by several degrees within only about 10 years. (After G. Bond et al., *Nature* 365:143–147, 1993.)

A variety of evidence, including the presence of nearshore mollusks of mid-Pliocene age, indicates that the Isthmus of Panama was emplaced by plate movements between 3.5 million and 3 million years ago, about the time the Ice Age of the Northern Hemisphere got under way. Before the isthmus formed, Atlantic waters would have flowed freely into the Pacific through the gap between North and South America (see Figure 19-19). Mixing of the waters of the Atlantic and Pacific should have maintained the two oceans at similar levels of salinity. Being less dense than they are today, North Atlantic waters may have flowed into the Arctic Ocean before cooling sufficiently to become dense enough to sink. By flowing into the Arctic Ocean, the Atlantic waters would have kept it warmer than it is today. The uplift of the Isthmus of Panama would

have changed that pattern. If it suddenly caused waters of the North Atlantic to descend north of Iceland by increasing their salinity, then the sudden isolation and cooling of the Arctic Ocean may have cooled the entire Arctic region, and that cooling may have brought on the Ice Age.

There is, in fact, evidence that when the isthmus formed, the Atlantic water to its east became more saline. Oxygen isotope ratios for planktonic foraminifera of this region became heavier between 4 million and 3 million years ago, although temperatures underwent little change and continental glaciers had not yet expanded. The increase in oxygen 18 to the east of the isthmus apparently resulted from an increase in the salinity of the waters there, with evaporation preferentially removing oxygen 16, the lighter isotope (p. 239).

Changes in Earth's rotational movement may have affected glacial cycles

Just as interesting as the cause of the modern Ice Age is the source of the glacial oscillations that have characterized the Ice Age since its inception. Figure 19-14 shows that the oscillations were more frequent during the early part of the Ice Age than later on. It is now generally agreed that changes in Earth's rotation on its axis and in its rotation around the sun have caused these oscillations (Figure 19-20).

Early in the Ice Age, glacial oscillations corresponded to the so-called *obliquity cycle* of Earth's axis of rotation. The axis is always tilted slightly away from vertical with respect to the plane of Earth's orbit around the sun, but the angle of tilt oscillates through time, with a periodicity of about 41,000 years. At the point in the obliquity cycle when the axis is farthest from vertical, the polar regions are aimed most directly toward the sun during the summer and receive a maximum amount of sunlight and solar heating.

Beginning about 850,000 years ago, glacial oscillations became less frequent, shifting to a periodicity of 90,000 to 100,000 years (see Figure 19-14). This new periodicity corresponded to that of changes in the shape of Earth's orbit from nearly circular to more strongly elliptical. These changes constitute the *eccentricity cycle*. They result from oscillations in the gravitational pull of other planets on Earth. When the orbit changes so as to bring Earth closer to the sun, it receives more solar heat than it does when it is farther away.

Also affecting Earth's climate is the *precession cycle*, for which the periodicity averages about 20,000 years. Precession is rotation of Earth's tilted axis, which aims toward the North Star only once in every cycle as it does today. The precession cycle has not exerted a primary control over glacial oscillations, but it has modified the effects of the other rotational variables.

It is not known why, about 850,000 years ago, Earth's oscillations in the shape of Earth's orbit came to govern the expansion and contraction of glaciers, overshadowing the obliquity cycle. It also remains to be explained why these orbital changes exert such a strong effect on Earth's climate. Their effects on the amount of sunlight reaching Earth are relatively small. Unidentified factors, presumably entailing positive feedbacks, must amplify these effects.

Regional Events of Neogene Time

The history of the western United States in the Neogene Period is highlighted not only by the elevation of imposing mountains that form part of our scenery today—the Cascade Range, the Sierra Nevada, and the Rocky Mountains—but also by climatic changes that resulted from the

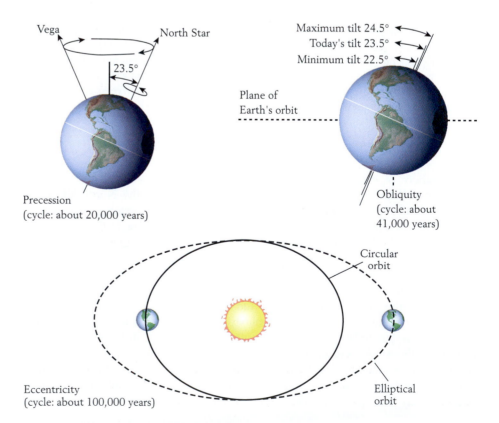

Figure 19-20 Periodic changes in Earth's rotational movements influence climate. (Courtesy of Denis Tasa, Tasa Graphic Arts, Inc.)

uplifting of those mountains. Tectonic movements were milder in and around the western Atlantic Ocean, but the passive margin of eastern North America accumulated sediments that contain a rich fossil record.

Mountains rose up throughout the American West

The pre-Neogene history of mountain building in the Cordilleran region, described in earlier chapters, is summarized in Figure 18-19. By late Paleogene time, uplifts resulting from the final mountain-building episode of the western interior, the Laramide orogeny, had been largely subdued by erosion, which set the stage for the Neogene events that produced the Rocky Mountains. In the broad region west of the Rockies, the Neogene Period was a time of widespread tectonic and igneous activity, which built most of the mountains standing there today.

Provinces of the American West Lying between the Great Plains and the Pacific Ocean are several distinct physiographic provinces that have taken shape largely in Neogene time, primarily as a result of uplift and igneous activity (Figure 19-21). Let us briefly review the present characteristics of these provinces before considering how they have come into being.

The lofty, rugged peaks of the Rocky Mountains, some of which stand more than 4.5 kilometers (14,000 feet) above sea level, could only be of geologically recent origin. We have seen that the widespread sub-summit surface of the Rockies was all that remained of the Laramide uplifts by the end of the Eocene Epoch, about 40 million years ago (see Figure 18-25). One question we must answer, then, is how the Rocky Mountain region became mountainous again during the Neogene Period.

Adjacent to the Rockies in the "Four Corners" area where Colorado, Utah, New Mexico, and Arizona meet is the oval-shaped Colorado Plateau, much of which stands about 1.5 kilometers (1 mile) above sea level. The Phanerozoic sedimentary units here are not intensively deformed. Some, however, are gently folded in a step-like pattern, and others, especially to the west, are offset by block faults (Figure 19-22). Cutting through the plateau is the spectacular Grand Canyon of the Colorado River (see Figure 1-5). About 10 million years ago, however, neither the Colorado Plateau nor the Grand Canyon existed. The origin of these features forms another part of our story.

West of the Rockies and the Colorado Plateau, within the belt of Mesozoic orogeny, lies the Basin and Range Province. This is an area of north–south-trend-

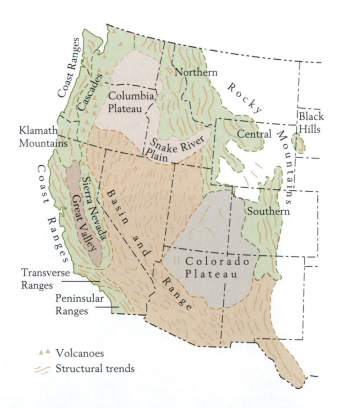

Figure 19-21 Major geologic provinces of the western United States. The topographic map at the left shows the relations of the provinces to topographic features. (Topographic map after U.S. Geological Survey, *National Atlas of the United States of America*.)

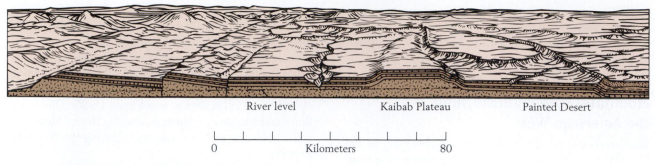

Figure 19-22 The western part of the Colorado Plateau north of the Grand Canyon. This high-standing region is characterized by block faulting (left) and gentle, steplike folds (right). (After P. B. King, *The Evolution of North America*, Princeton University Press, Princeton, NJ, 1977.)

Figure 19-23 The possible pattern of block faulting in the Basin and Range Province that might have been responsible for lateral extension of the crust.

ing fault block basins and intervening ridges (Figure 19-23)—features of Neogene origin. A large area of this province forms the Great Basin, an arid region of interior drainage (p. 110). Volcanism has been associated with some faulting episodes. The thickness of Earth's crust in the Basin and Range Province ranges from about 20 to 30 kilometers, in contrast to thicknesses of 35 to 50 kilometers in the Colorado Plateau. The thinning and block faulting in the Basin and Range Province point to considerable lateral extension of the crust.

Farther north, centered in Oregon, is a broad area covered by the volcanic rocks of the Columbia Plateau and Snake River Plain. Today the climate here is cool and semiarid; only about one-quarter of the plateau area

is cloaked in forest and woodland, and sagebrush and drier conditions characterize about half of the terrain. In Oligocene time, however, lavas had not yet blanketed the region, and, as remains of fossil plants reveal, a large forest of redwood trees grew there.

Along the western margin of the Columbia Plateau stand the lofty peaks of the Cascade Range (p. 468). These cone-shaped volcanoes represent the igneous arc associated with subduction of the Pacific plate along the western margin of the continent. Volcanism began here in Oligocene time and continues to the present.

The Cascade volcanic belt passes southward into the Sierra Nevada, a mountainous fault block of granitic rocks. The plutons forming the Sierra Nevada were emplaced in east-central California during Mesozoic time, before igneous activity at this latitude shifted inland. As we shall see, however, the present topography of the Sierra Nevada is of Neogene origin. This mountain range is unusual in that, throughout its length of some 600 kilometers (350 miles), it is not breached by a single river. That is why it represented such a formidable obstacle to early pioneers attempting to reach the Pacific (Figure 19-24).

Figure 19-24 The eastern face of the Sierra Nevada. This face is formed of rocks uplifted on one side of a fault and is partly dissected by youthful valleys. The view is from the Owens Valley, Inyo County, California. (Greg Vaughn/Tom Stack & Associates.)

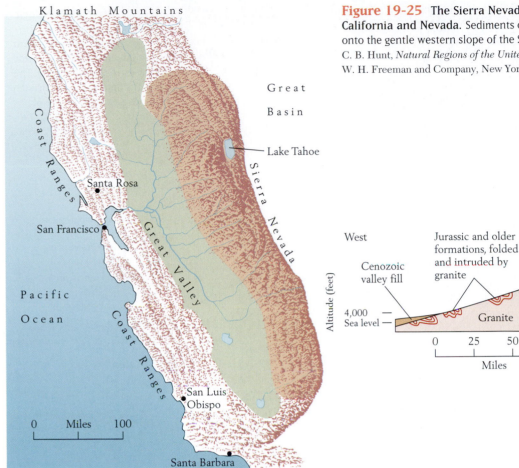

Figure 19-25 The Sierra Nevada fault block of California and Nevada. Sediments of the Great Valley lap onto the gentle western slope of the Sierra Nevada. (After C. B. Hunt, *Natural Regions of the United States and Canada*, W. H. Freeman and Company, New York, 1974.)

The Sierra Nevada bounds the Basin and Range Province to the east and stands above the Great Valley of California to the west. The Great Valley is an elongate basin containing large volumes of Mesozoic sediment (the Great Valley Sequence; see Figure 16-32). This sediment was eroded from the plutons of the Sierra Nevada region long before the modern Sierra Nevada formed by block faulting (Figure 19-25). Resting on top of these sediments are Cenozoic deposits, some of which accumulated during marine invasions of the Great Valley and others during times of nonmarine sedimentation.

West of the Great Valley are the California Coast Ranges (Figure 19-26). These uplifts consist of slices of crust that include crystalline rocks representing Mesozoic orogenic activity, Franciscan mélange of deep-water origin (see Figure 16-32), and Cenozoic rocks. To the south, the Transverse and Peninsular ranges are formed of similarly faulted and deformed rocks, but these ranges lie inland of the main belt of Franciscan rocks in the region of intensive Mesozoic igneous activity. Striking features of all of these mountainous terrains are the great faults that divide the crust into sliver-shaped blocks. The longest and most famous of these faults is the San Andreas, which extends for about 1600 kilometers (1000 miles). Until the great San Francisco earthquake of 1906, it was not widely recognized that the San Andreas was still active. The earthquake of 1906 was produced by a sudden horizontal movement of up to 5 meters (16 feet) along the fault. Geologic features cut by the San Andreas fault show that its total movement during the past 15 million years has amounted to about 315 kilometers (190 miles). Continued movement at this rate for the next 30 million years or so would bring Los Angeles northward to the latitude of San Francisco, through which the fault passes. As we shall see, the faulting and uplifting of the Coastal Ranges of California are probably related not only to the Neogene uplift of the Sierra Nevada but also to the origins of the Basin and Range topography to the east.

The Cascade Ranges of Washington have quite a different history. These relatively low mountains, which lie to the west of the Cascade volcanics, consist of oceanic sediments and volcanics that were deformed primarily during Eocene time in association

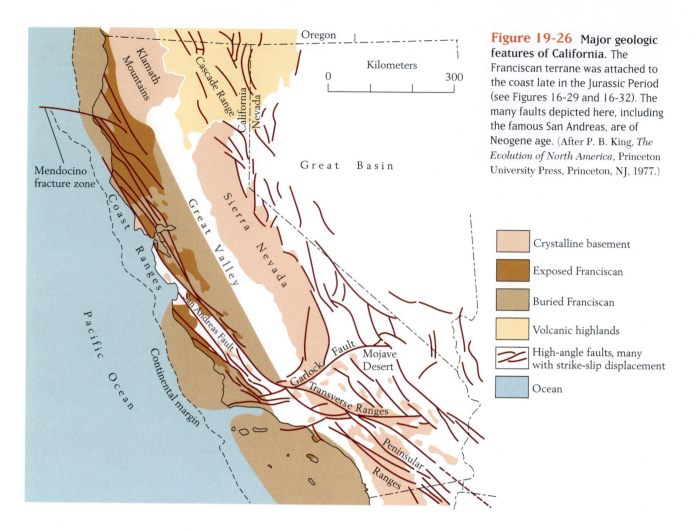

Figure 19-26 Major geologic features of California. The Franciscan terrane was attached to the coast late in the Jurassic Period (see Figures 16-29 and 16-32). The many faults depicted here, including the famous San Andreas, are of Neogene age. (After P. B. King, *The Evolution of North America*, Princeton University Press, Princeton, NJ, 1977.)

with subduction along the continental margin (see Figures 18-20 and 18-22).

Tectonic and volcanic events in western North America Geologic features of western North America in Miocene time are shown in Figure 19-27. Subduction continued beneath the continental margin in the northwestern United States, and the resulting igneous arc produced peaks in the Cascade Range, where volcanism continues today. To the south, in California, the Middle Miocene interval was a time of faulting and mountain building; elements of the modern Coast Ranges and other nearby mountains were raised, and the seas were driven westward. Meanwhile, as in Paleogene time, the Great Valley remained a large marginal marine embayment, and during Miocene time it received great thicknesses of siliciclastic sediments, most of which were shed from the Sierra Nevada, which rose to at least its present height during Miocene time.

During Miocene time, the Basin and Range Province began forming to the east of the Sierra Nevada. Volcanism began during Paleogene time, and the Basin and Range topography began to form near the beginning of the

Miocene. To the north of the Basin and Range Province, great volumes of basalt spread from fissures at the site of the Yellowstone hot spot, which had shifted eastward from its Paleogene location. Most of the great Columbia Plateau was formed by outpourings of lava between about 16 million and 13 million years ago (see Figure 2-13); individual basalt flows of this plateau range in thickness from 30 to 150 meters (100 to 500 feet), and in places the total accumulation reaches about 5 kilometers (3 miles).

Geologists have reconstructed the histories of the Colorado Plateau and Rocky Mountains by studying the times at which rivers have cut through well-dated volcanic rocks. Many of the rivers of these regions existed before uplift began in the Miocene Epoch, and they cut downward as the land rose, producing deep gorges. Part of the Grand Canyon, for example, was incised during the elevation of the Colorado Plateau between about 10 million and 8 million years ago. Uplift of the Rockies began slightly earlier, in Early Miocene time, and terrain that now forms the southern Rockies has since risen between 1.5 and 3.0 kilometers (1 to 2 miles). Uplift of both the Colorado Plateau and the Rockies accelerated about 5 million years ago; much of the present eleva-

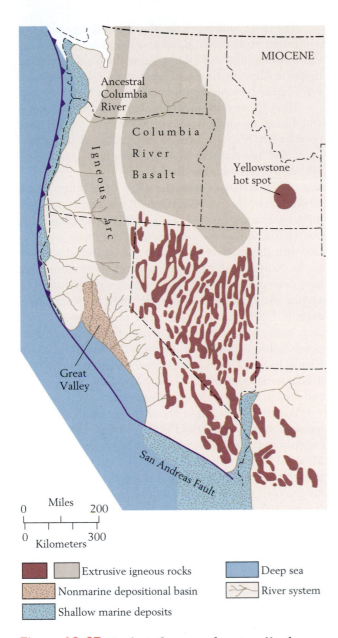

MIOCENE

Ancestral
Columbia
River

Columbia
River
Basalt

Yellowstone
hot spot

Igneous arc

Great
Valley

San Andreas Fault

Miles
0   200

0   300
Kilometers

Extrusive igneous rocks

Nonmarine depositional basin

Shallow marine deposits

Deep sea

River system

Figure 19-27 Geologic features of western North America in Miocene time. West of the San Andreas fault, coastal southern California lay farther south than it does today. The Great Valley of California was, for the most part, a deep-water basin from which a nonmarine depositional basin extended to the north. Volcanoes of the early Cascade Range formed along an igneous arc inland from the subduction zone along the continental margin. Igneous rocks were extruded along north–south-trending faults in the Great Basin, and farther north the Columbia River Basalt spread over a large area. (After J. M. Armentrout and M. R. Cole, *Soc. Econ. Paleontol. Mineral. Pacific Coast Paleogeogr. Symp.* 3:297–323, 1979.)

Ogallala Formation. Caliche nodules are abundant in many parts of the Ogallala, indicating the presence of seasonally arid climates (p. 106). The Ogallala is a thin, largely sandy unit, most of which lies buried under the Great Plains from Wyoming to Texas, and it serves as a major reservoir of groundwater. Unfortunately, this ancient water is not being replenished as rapidly as it is being drawn from underground. As a result, severe water shortages may one day strike many areas of the central United States.

During the Pliocene and Pleistocene epochs, igneous activity continued in the volcanic provinces of Oregon, Washington, and Idaho (Figure 19-28). Many of the scenic volcanic peaks of the Cascades have formed within the past 2 million years or so (see p. 468). Beginning in Late Miocene time and continuing sporadically to the present, the flow of basalt from fissures has produced the Snake River Plain, which amounts to an eastward extension of the Columbia Plateau (see Figure 19-21) and has been formed by the same Yellowstone hot spot.

Faulting and deformation continued in California during the Pliocene and Pleistocene epochs. Since the beginning of the Pliocene Epoch, about 5 million years ago, the sliver of coastal California that includes Los Angeles has moved northward on the order of 100 kilometers (60 miles). The Great Valley has, of course, remained a lowland to the present day, but during Pliocene and Pleistocene time it became transformed from a marine basin into a terrestrial one. Early in the Pliocene Epoch, seas flooded the basin from both the north and the south, but as the epoch progressed, uplift associated with movement along the San Andreas fault eliminated the southern connection. Eventually nonmarine deposition prevailed throughout the Great Valley, which is now one of the world's richest agricultural areas as well as a site of large reservoirs of petroleum.

Major climatic changes occurred in western North America in latest Miocene and Pliocene time. The Basin and Range Province, which had been covered by forests throughout most of Miocene time, became carpeted by savannah and eventually turned largely into a desert. This shift toward more arid conditions was a rain-shadow effect of the rising Sierra Nevada to the west (see Figure 4-17). Streams cut downward rapidly during the tectonic uplift, and radiometric dates for volcanic rocks incised by streams draining the Sierra Nevada indicate this great fault block rose up rapidly during the past 5 million years. During Miocene time, the Sierra Nevada had never risen to more than about half its present height.

Mechanisms of uplift and igneous activity What led to the many tectonic and igneous events of Neogene time in the American West? It might seem likely

tion of these uplands was attained during this relatively recent interval.

Sediments derived from the Rocky Mountains spread eastward late in Miocene time, creating the

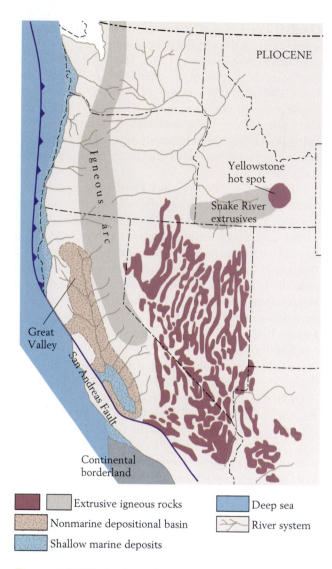

PLIOCENE

Yellowstone
hot spot

Snake River
extrusives

Igneous arc

Great
Valley

San Andreas Fault

Continental
borderland

		Extrusive igneous rocks		Deep sea
		Nonmarine depositional basin		River system
		Shallow marine deposits		

Figure 19-28 Geologic features of western North America in Pliocene time. The Great Valley of California was a shallow basin that received nonmarine sediments except where a shallow sea flooded its southern portion. The igneous arc continued to form volcanoes of the Cascade Range, and igneous rocks continued to be extruded along faults in the Great Basin. The Snake River extrusives spread over a large area of southern Idaho above the Yellowstone hot spot, which had shifted westward. (After J. M. Armentrout and M. R. Cole, *Soc. Econ. Paleontol. Mineral. Pacific Coast Paleogeogr. Symp.* 3:297–323, 1979.)

that the secondary uplift of the Colorado Plateau and Rocky Mountains, which took place long after the Laramide orogeny, resulted from simple isostatic adjustment (see Figure 1-14). Geophysical studies, however, show that these uplifts do not have deep roots that might have caused them to bob up. Instead, for some reason, swelling of Earth's mantle below seems to have elevated broad areas of the American West.

The block faulting of the Basin and Range requires a different explanation. Basin and Range events and the extensive faulting and folding along the California coast began in Miocene time and seem to be related in some way to plate tectonic movements along the Pacific coast.

How these movements have produced the Basin and Range Province is a controversial issue. The most popular hypothesis relates to the famous San Andreas fault. The San Andreas is a transform fault (see Figure 9-1) associated with the East Pacific Rise, a large oceanic rift that passes into the Gulf of California and is offset along the San Andreas fault where it encounters thick continental crust (Figure 19-29). Spreading along the rise should have ceased when the rise came into contact with the subduction zone at the western boundary of the North American plate; instead, movement must have been propagated along one or more transform faults such as the San Andreas, passing along the continental margin. Crustal shearing adjacent to a strike-slip fault such as the San Andreas will automatically cause extensional faulting similar to that seen in the Great Basin (see Figure 19-29). This hypothesis is deficient in one regard: it fails to account for the broad elevation of the Basin and Range Province during Neogene time.

We are more certain about the general pattern of tectonic activity along the Pacific coast. North America encountered the Pacific plate near the beginning of the Miocene Epoch. Movements along the San Andreas and other faults that have formed since that time account for the complex slivering and deformation in the Coast Ranges and neighboring areas (see Figure 19-26).

Glaciation in the American West With the onset of the Ice Age in the Northern Hemisphere, frigid conditions brought glaciation to mountainous regions of the western United States, just as they foster glaciation in Alaskan mountains today (see Figure 4-14). The Sierra Nevada, for example, was heavily glaciated, as were portions of the Rocky Mountains (see Figure 19-16). Today broad U-shaped valleys in both mountain systems testify to the scouring activity of Pleistocene glaciers (see Figure 12-16).

One of the greatest controversies of modern geology erupted in 1923, when J. Harlan Bretz, a professor at the University of Chicago, advanced what he modestly called the "outrageous hypothesis" that catastrophic floods had swept across a broad region of the northwestern United States as Earth emerged from the last glacial maximum. Far from being outrageous, Bretz's bold idea turned out to be correct.

Bretz based his idea of catastrophic flooding on his studies of the so-called channeled scablands in the eastern part of Washington State—a landscape of bare rock that has obviously been scoured by water. Its topogra-

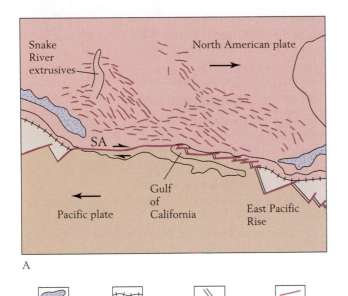

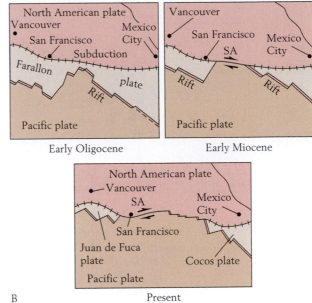

Early Oligocene   Early Miocene

Present

A

Arc volcanism   Trench   Spreading ridge   Transform fault

B

Figure 19-29 Plate tectonic features that may account for the structure of the Basin and Range Province in western North America. *A.* At present, the spreading ridge known as the East Pacific Rise passes into the Gulf of California. The East Pacific Rise abuts the North American continent and is offset westward along the San Andreas fault (SA). It appears that shearing forces resulting from relative movement of terrain on either side of the San Andreas fault (heavy arrows) has pulled the crust apart, producing the north–south-trending faults of the Basin and Range Province. *B.* Near the beginning of Miocene time, the rift zone between the Pacific and Farallon plates encountered the thick crust of North America along the subduction zone that bordered the continent. Unable to pass inland, the rift was divided along a strike-slip fault (the San Andreas). (Based on J. H. Stewart, *Geol. Soc. Amer. Mem.* 152:1–31, 1975. After Atwater.

phy includes water-carved channels, some of which are offset by steps that are sites of ancient waterfalls (Figure 19-30). The channels of the scablands also display remarkable depositional features, including giant ripples of gravelly sediment that are typically about 8 meters (27 feet) tall and spaced about 100 meters (330 feet) apart (Figure 19-31). So little soil carpets the scablands that Bretz concluded that they must have formed quite recently, at the time when Wisconsin glaciers were melting back. Radiometric dating has since shown that many features of the scablands were indeed formed between about 20,000 and 11,000 years ago.

Figure 19-30 Scabland topography in eastern Washington State. Here, floodwaters have carved a channel 100–150 meters (330–500 feet) deep in the Columbia River Basalt. (Victor R. Baker, The University of Arizona, Tucson.)

Figure 19-31 Giant ripples, averaging 8 meters (about 27 feet) in height, that were deposited during catastrophic flooding in eastern Washington State. (Victor R. Baker, The University of Arizona, Tucson.)

When Bretz proposed that this catastrophic flooding had occurred, critics argued that there was no source for the voluminous floodwaters that his hypothesis required. Soon, however, Bretz and others recognized that the likely source was Lake Missoula, a body of water that had formed in front of the glaciers that capped the Rocky Mountains (see Figure 19-16). The configuration of the lake was well known from well-layered sediments that display annual varves (see Figure 5-6). Lake Missoula was dammed by a lobe of glacial ice. The volume of the lake is estimated to have been about 2000 cubic kilometers (500 cubic miles).

Years after Bretz presented his argument, calculations showed that catastrophic collapse of Lake Missoula's ice dam would have produced currents deep enough and swift enough to have formed the enormous ripples of the scablands. Unconformities in the rippled sediments and in the sediments of Lake Missoula indicate that as many as 40 catastrophic flows may have occurred. According to one proposal, the ice that dammed the lake stretched across a valley, and a flood occurred when the level of the lake rose to a point at which the water pressure at the bottom separated the ice from the rocks on which it rested. The ice dam then collapsed and the waters of the lake burst through, rushing westward to the Pacific Ocean, scouring the landscape, and depositing gravelly sediment in the form of giant ripples. Floating ice then began to pile up again in the constricted valley, forming a new dam, and the sequence of events repeated itself. After about 40 episodes of this kind, the glaciers receded far enough that they no longer supplied enough meltwater and floating ice to re-form Lake Missoula. All that remained of the lake was a sequence of well-layered sediments, punctuated by many unconformities.

Today J. Harlan Bretz is widely viewed as a hero for advancing a novel idea and defending it rationally

on the basis of sound observations. The floods that he brought to light stand as the largest ever identified on Earth.

The Appalachians bobbed up and shed sediment eastward

Although the margins of the Atlantic Ocean were relatively quiescent during the Neogene Period, they did experience mild vertical tectonic movements—and these movements, together with more profound changes in sea level, had major effects on shoreline positions.

Global sea level has never stood as high during the Neogene Period as it did during much of Cretaceous or Paleogene time. For this reason, Neogene marine sediments along the margin of the Atlantic Ocean stand above sea level in only a few low-lying areas. Among the most impressive of the Miocene deposits found here are those of the Chesapeake Group, which form cliffs along the Chesapeake Bay in Maryland. The Chesapeake Group accumulated in the Salisbury Embayment during a worldwide sea-level rise between about 16 million and 14 million years ago. The Salisbury Embayment is one of several depressions of the American continental margin (Figure 19-32). Inhabiting the waters of the Salisbury Embayment was a rich fauna that included

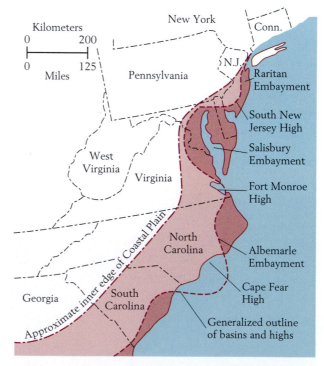

Figure 19-32 Elevated regions and depositional embayments along the mid-Atlantic coast during the Miocene Epoch. (After J. P. Menard et al., *Geological Society of America Northeast-Southeast Sections, Field Trip Guidebook 7a,* 1976.)

many large vertebrates, especially whales, dolphins, and sharks (see Figure 19-1). The fact that most of the fossils of baleen whales in this fauna represent juvenile animals suggests that the embayment was a calving ground. Probably sharks were numerous because the young whales were especially vulnerable prey. Land mammal bones are also found here and there in the Chesapeake Group, indicating that the waters of the embayment were shallow. Pollen from nearby land plants settled in the Salisbury Embayment, leaving a fossil record that shows a warm temperate flora near the base of the Chesapeake Group slowly giving way upward in the sedimentary sequence to a flora adapted to slightly cooler conditions.

The Chesapeake Group and older buried deposits to the south consist primarily of siliciclastic sediments shed from the Appalachians to the west. Erosional features associated with the Appalachians reveal that these ancient mountains have a complex history. The existing topographic mountains that we call the Appalachians are the products of secondary isostatic uplift. The Ap-

palachian orogenic belt was largely leveled by erosion long before the end of the Mesozoic Era. In many regions there have been three or more additional intervals of uplift and erosion during the Cenozoic Era. The erosion that followed intervals of uplift left ridges of resistant folded rock standing above elongate valleys, but when erosion was especially intense, preexisting rivers cut through the ridges as they and their tributaries carved out the valleys (Figure 19-33).

The Caribbean Sea was born

Although the Caribbean Sea is now an embayment of the Atlantic Ocean, it was once connected to the Pacific (Figure 19-34). During the Cretaceous Period, the floor of the Caribbean, which consists of oceanic rocks, was a small segment of the Pacific plate that was pushing toward the Atlantic. During the Cenozoic Era, however, the Caribbean seafloor has lain along the north coast of South America while the Atlantic plate has been subducted beneath it. The Caribbean region became a discrete plate late in Cenozoic time, when a

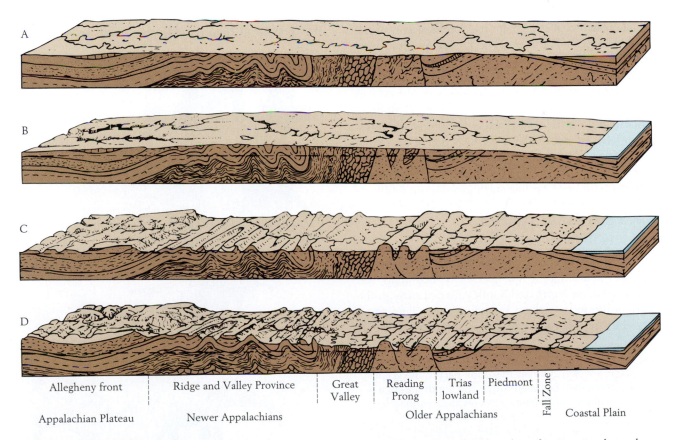

Allegheny front	Ridge and Valley Province		Great Valley	Reading Prong	Trias lowland	Piedmont	Fall Zone	
Appalachian Plateau	Newer Appalachians				Older Appalachians			Coastal Plain

Figure 19-33 Episodic uplift has rejuvenated topography in the Ridge and Valley Province of the Appalachians. *A.* A modest amount of relief characterized some regions early in the Cretaceous Period. *B–D.* Later intervals of uplift and erosion produced the modern topography, characterized by ridges of resistant rocks, such as sandstone, and valleys of easily eroded rock, including shale. (After D. W. Johnson, *Stream Sculpture on the Atlantic Slope*, Columbia University Press, New York, 1931.)

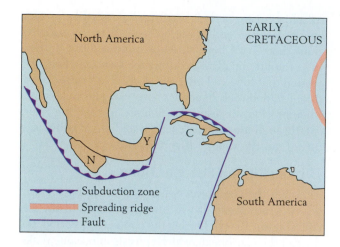
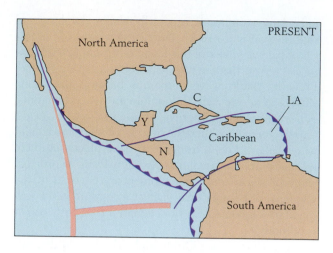

Figure 19-34 Tectonic development of the Caribbean Sea. A segment of Pacific oceanic crust that has overridden Atlantic crust along a subduction zone now extends northward from South America east of the Lesser Antilles. N, Nicaragua; Y, Yucatán; C, Cuba; LA, Lesser Antilles. (After G. W. Moore and L. Del Castillo, *Geol. Soc. Amer. Bull.* 85:607–618, 1974.)

new subduction zone came to connect the subduction zone bordering North America with the one bordering South America. The Greater Antilles—Cuba, Puerto Rico, Jamaica, and Hispaniola—represent an ancient mountain belt that is actually the southern end of the North American Cordillera. The Lesser Antilles represent an igneous island arc west of the subduction zone, together with islands formed by deformation associated with subduction. The Yucatán Peninsula is a broad carbonate platform that lies to the west of the Caribbean. At the margin of this peninsula is the site of the asteroid impact that caused the mass extinction at the end of the Cretaceous Period (see Earth System Shift 17-1). As we have seen, the Bahamas are an ancient carbonate platform positioned farther north (see Figure 5-30).

North and South America exchanged mammals

Before the Isthmus of Panama formed, slightly before 3 million years ago, a few species of mammals had passed between North and South America early in the Neogene Period—perhaps by swimming or floating on logs—but the terrestrial faunas of the two continents

had remained largely separate. South America had been a great island continent and, like Australia, was populated by many marsupial mammals. The marsupials of Australia and South America have common ancestors that populated these continents as well as Antarctica

▶ **Figure 19-35** Animals that took part in the great faunal interchange between North and South America when the Isthmus of Panama was elevated, connecting the continents. The animals shown in North and Central America are immigrants from the south; they include armadillos, sloths, porcupines, and opossums. The animals shown in South America are immigrants from the north; among them are rabbits, elephants, deer, camels, and members of the bear, dog, and cat families. More animals migrated southward than northward. (Drawing by M. Hill Werner, courtesy of L. G. Marshall.)

when all three were part of a single Mesozoic landmass; in fact, Eocene mammalian faunas resembling those of South America occur in Antarctica. By Pliocene time, however, when the isthmus developed, South American marsupials had diverged greatly from Australian marsupials, and the South American fauna included several groups of placental mammals whose ancestors had reached the continent from the north early in Cenozoic time, or even earlier. Among the South American marsupials present when the isthmus formed were members of the opossum family, and among the placentals were sloths and armadillos that dwarf their relatives in the modern world (see Figure 7-5).

Once the isthmus had formed, more North American species invaded South America than vice versa (Figure 19-35). Among those that reached South America were members of the camel, pig, deer, horse, ele-

phant, tapir, rhino, rat, skunk, squirrel, rabbit, bear, dog, raccoon, and cat families. Migrating in the opposite direction were monkeys, anteaters, armadillos, porcupines, opossums, and some less familiar animals.

The Himalaya rose to become Earth's highest mountain range

The Himalaya, having formed during Neogene time, is a relatively young mountain system. Partly because of its youth, the Himalaya is the tallest mountain range on Earth (p. 198). The Himalayan front rises abruptly from the flat Ganges Plain; not far from the front, Mount Everest, the tallest mountain on Earth, towers to 8848 meters (5.5 miles) above sea level. Even the broad Tibetan plateau, which lies to the north of the Himalayan front (Figure 19-36), stands at an average

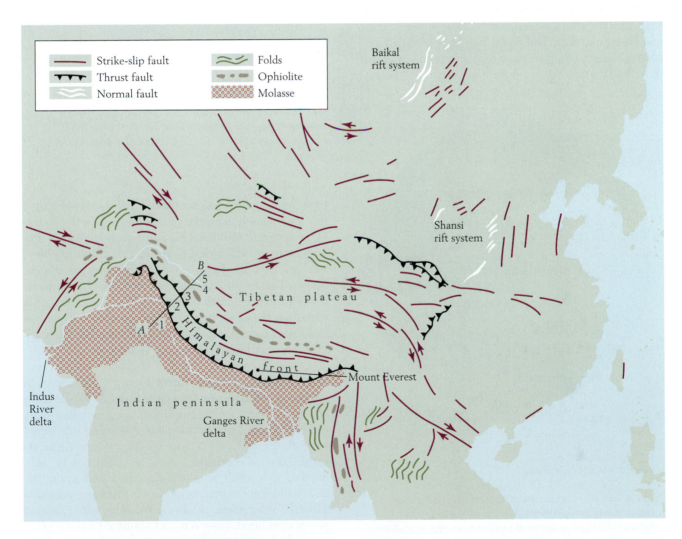

Figure 19-36 Geologic features of the Himalayan region. The high-standing Tibetan plateau is bounded by thrust faults, especially in the south, and molasse is being shed southward from the plateau. Numerous strike-slip faults throughout the region seem to have permitted the Asian crust

to squeeze eastward as the Indian peninsula has pushed northward. (The numbers refer to the cross section in Figure 19-39C.) (After P. Molnar and P. Tapponier, *Sci. Amer.*, April 1977. © 1977 by Scientific American, Inc. All rights reserved.)

elevation of about 5 kilometers (3 miles) above sea level—higher than any mountain peak in the 48 contiguous United States.

Plate movements The Himalaya is part of a great series of mountain chains of Cenozoic origin that stretch from Spain and North Africa to Southeast Asia (Figure 19-37). All of these chains formed as a result of the northward movement of fragments of Gondwanaland. The Alps and other Cenozoic mountains of the Mediterranean region formed as the African plate, one of these fragments, moved northward against the Eurasian plate. The Indian peninsula, which projects southward from the Himalaya, was another fragment of Gondwanaland. By late in the Mesozoic Era, this fragment was moving northward as an island continent within the large Australian plate (Figure 19-38). The collision of this Indian craton with Eurasia created the Himalaya.

When did the Indian craton arrive? During Eocene time, shallow seas covered much of the Indian craton, and limestones were laid down over large areas. Coarse sediment derived from mountainous terrain was first deposited on top of the limestones in Late Miocene

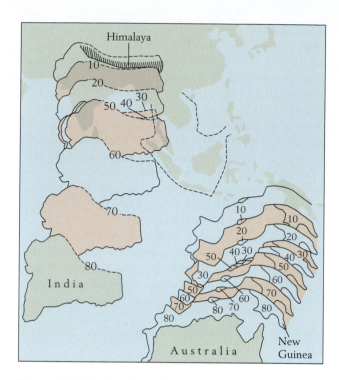

Figure 19-38 Northward movement of the Indian craton between 80 million and 10 million years ago. Numbers represent the times (million years ago) when geographic boundaries reached various positions. (After C. McA. Powell and B. D. Johnson, *Tectonophysics* 63:91–109, 1980.)

time. Apparently it was not until shortly before this time that mountain building began. Sediments in the Indian Ocean provide additional evidence of the timing of orogenic activity. The oldest deep-sea turbidites deposited offshore from the Indus and Ganges rivers (see Figure 19-36) date to the Middle Miocene. The rivers themselves cannot be much older, and they came into being when the Himalaya began to form, perhaps 20 million years ago. Indeed, much of the Himalaya has been uplifted during the last 15 million years.

The pattern of orogenesis Figure 19-39 shows in greater detail how the Himalaya formed. When India was approaching Eurasia, riding on the Australian plate, the northern margin of this plate was being subducted beneath Eurasia (see Figure 19-36). When India arrived, being a continental mass, it could not be subducted. As a result, subduction ceased, and so did the associated igneous activity along the southern margin of Tibet. Convergence of the Australian and Eurasian plates continued, however, and about 20 million years ago India began to wedge beneath the southern margin of Tibet without descending into the asthenosphere (Figure 19-39*B*).

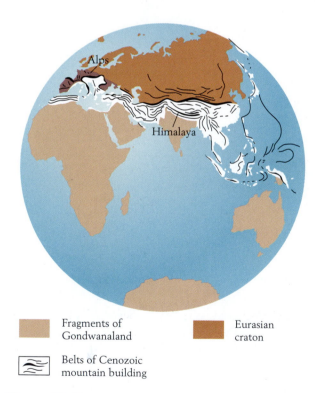

| Fragments of Gondwanaland | Eurasian craton |

Belts of Cenozoic mountain building

Figure 19-37 The series of mountain chains that formed along the southern margin of Eurasia when fragments of Gondwanaland moved northward against the large northern continent during the Cenozoic Era.
(After H. Cloos, *Einführung an die Geologie*, Verlag von Gebrüder Borntraeger, Berlin, 1936.)

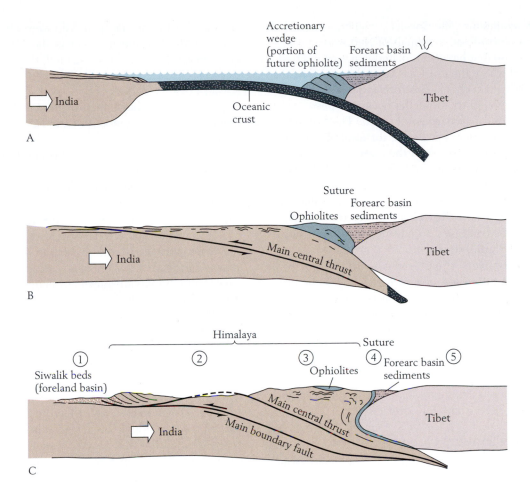

Figure 19-39 Cross section showing how the Himalaya formed when the Indian craton wedged beneath the margin of Eurasia. *A.* Slightly before 20 million years ago the Indian craton was rafted toward Eurasia as part of a plate being subducted beneath Tibet. *B.* About 20 million years ago the Indian craton began to wedge beneath Tibet and fractured along the main central thrust. Movement along this fault thickened the crust, forming mountainous terrain. Compression in the suture zone uplifted and deformed the accretionary wedge that had bordered Tibet, producing ophiolites. *C.* Today movement has shifted to a new thrust fault, the main boundary fault, and the crust has further thickened. Molasse, including the Siwalik beds, has been accumulating to the south, and older molasse deposits are being deformed in the vicinity of the main boundary fault. (Numbers refer to zones identified in Figure 19-36 along line AB.) (After P. Molnar, *Amer. Sci.* 74:144–154, 1986.)

Sediments of the forearc basin that had bordered Tibet were squeezed up along the suture, along with material of the accretionary wedge and solid oceanic crust, to form ophiolites. At some unknown time another dramatic event took place: the northern margin of India, consisting of sediments and underlying continental crust, broke away from the rest of the Indian craton. The remaining Indian craton then slid beneath the margin of Eurasia for at least 100 kilometers (about 60 miles) along a huge thrust fault, now known as the *main central thrust*. This fault can be seen today in many areas of the Himalaya, where valleys have cut deep into the mountains. Movement along the main central thrust ceased sometime before 10 million years ago, and a new fault, the *main boundary fault*, developed below it (Figure 19-39C). Movement along this fault has continued to the present day. The result of movement along the two faults has been a great thickening of the Indian crust as the margin of India has underthrust the slices of crust that have broken from it. This underthrusting has made the Himalaya the tallest mountain chain of the modern world.

A fold-and-thrust belt has formed above the main central thrust and main boundary fault where these faults approach the surface at the southern margin of the Himalaya (see Figure 19-39C). In the foreland basin to the south of this belt, a huge body of sediment, the Siwalik beds, has formed from material that has eroded from the mountains. The Siwaliks, which have yielded large numbers of fossil mammals of Neogene age, constitute molasse that has accumulated in a foreland basin that formed where the crust has been depressed by the adjacent mountain chain. This foreland basin has never been deep enough to admit the ocean, however, so it has received only nonmarine sediments. The famous Siwalik beds of Pakistan and India, for example, provide a nearly continuous record for the interval from 11 million to 1 million years ago (see Figure 19-39C). They document the composition of the rich faunas that occupied the spreading savannahs of Late Miocene and Pleistocene age (see Figure 19-4). As the fold-and-thrust belt has continued to advance, it has deformed some of the Siwalik strata.

The great rivers of eastern Asia, which flow from the Himalaya to the sea, also formed in Miocene time during the uplift of the Himalayan region. Because of the high relief and abundant rainfall of the region, the Indus and Ganges of India and the several large rivers of Southeast Asia contribute huge volumes of sediment to the ocean each year. Earthquakes still rumble through the Himalayan region as a result of movement along the faults, and there is every reason to believe that mountain building here is far from over.

The Tethys Seaway came to an end

The collision of the African plate and India with Eurasia during the Cenozoic Era destroyed what remained of the Tethys Seaway. Today only vestiges of the seaway remain in the form of the isolated Mediterranean, Black, Caspian, and Aral seas. As Africa moved northward, movements of small plates in the Mediterranean region uplifted the Alps.

At the end of the Miocene Epoch the Mediterranean Sea underwent spectacular changes. The first strong hint that geologists had of these changes was the discovery in 1961 of pillar-shaped structures in seismic profiles of the Mediterranean seafloor. These structures looked very much like the salt domes of Jurassic age in the Gulf of Mexico (see Figure 16-22), but if the strange features were indeed salt domes, the salt could have formed only by evaporation of restricted Mediterranean waters. In 1970 the presence of evaporites was confirmed by drilling that brought up anhydrite in cores of latest Miocene age. The idea that the Mediterranean had somehow turned into a shallow hypersaline basin was confirmed by the discovery of halite (rock salt) near the center of the eastern Mediterranean basin.

Further evidence that the Mediterranean shrank by evaporation at the end of Miocene time was the discovery of deep valleys filled with Pliocene sediments lying beneath the present beds of such rivers as the Rhône in France, the Po in Italy, and the Nile in Egypt. Rivers

such as the Rhône, Po, and Nile were already flowing into the Mediterranean earlier in the Miocene Epoch, and when the waters of the sea fell, the rivers cut deep canyons. In attempting to find solid footing for the Aswan Dam, geologists of the Soviet Union discovered a canyon buried beneath the present Nile delta and judged it to rival the modern Grand Canyon of Arizona in size.

Clearly, at the end of the Miocene Epoch, the single narrow connection between the Mediterranean Sea and the Atlantic Ocean nearly closed, probably as a result of the lowering of global sea level at the time (p. 471). Rates of evaporation similar to those of the Mediterranean region today would dry up an isolated sea as deep as the Mediterranean in a mere thousand years. During the crisis enough water must have flowed weakly into the Mediterranean from the Atlantic to keep it from drying up altogether.

All of this happened between about 6 million years ago, when the eastern passage to the Atlantic closed, and 5 million years ago, when the Mediterranean basin refilled with deep water. Five-million-year-old deep-water microfossils in sediments on top of evaporites attest to the refilling. Apparently the connection with the Atlantic was enlarged again when the natural barrier at Gibraltar was suddenly breached. It has therefore been suggested that the first Atlantic waters must have been carried into the deep basin by a waterfall that would have dwarfed Niagara Falls.

Human Evolution

In addition to the single species that now constitutes the human family, the superfamily Hominoidea at present includes just four species of the ape family Pongidae—the common chimpanzee, the pygmy chimpanzee, the gorilla, and the orangutan—together with six species of the gibbon family Hylobatidae (see Figure 3-7). The human family, Hominidae, did not evolve from modern apes. Instead, modern humans—as well as modern apes—evolved from other apes that died out several million years ago.

Early apes radiated in Africa and Asia

Although an extensive Plio-Pleistocene fossil record has been uncovered for the Hominidae, very few known fossil remains of any age represent the Pongidae. Furthermore, the fossil record of the superfamily Hominoidea in latest Miocene and earliest Pliocene time (8 million to 5 million years ago) is unfortunately very poor. Farther back in the Miocene Series, fossils reveal that two extinct hominoid groups we can loosely call early apes underwent such a great evolutionary expansion that we might refer to the Miocene as the Age of

Apes. Among these early forms must be ancestors of both modern apes and modern humans, but the evolutionary connections are not yet understood.

The oldest fossils of these early apes come from African sediments about 20 million years old. These extinct groups of apes first spread from Africa to Eurasia about 15 million to 16 million years ago—just a few million years after Africa, having moved thousands of kilometers northward after the breakup of Gondwanaland, finally collided with Eurasia and allowed the exchange of mammals between the two landmasses. Hominoids made their way northward to Eurasia during Middle Miocene time, as did many other previously isolated groups of African mammals, including elephants and giraffes (giraffes have since become extinct in Eurasia). The early apes underwent large evolutionary radiations in both Africa and Eurasia. Some species were so large that they must have spent most of their time on the ground, but others were probably arboreal animals.

Throughout most of Miocene time the Old World was much more heavily populated with apes than Africa is today. By the end of the Miocene, however, only a single genus of early apes seems to have survived. This was the aptly named *Gigantopithecus*, a gorilla-sized creature that lived on into the Pleistocene. After this decline of apes, close to the boundary between the Miocene and Pliocene epochs, there was an evolutionary event of great significance: the emergence of the earliest hominids from some unknown group of apes. These hominids constitute a distinct subfamily within the Hominidae and are informally termed the *australopithecines*. They are of special interest to modern humans because we are their only living descendants.

The earliest hominid lived between 6 million and 7 million years ago

The molecular clock, applied to humans and chimpanzees, indicates that the human family, Hominidae, branched from early apes between 5 million and 8 million years ago. Thus it was particularly gratifying when, in 2002, anthropologists discovered a 6- to 7-million-year-old fossil skull in the northern African country of Chad that was intermediate in form between the skull of an ape and that of a human. This new form was christened *Sahelanthropus*. Its skull is apelike, with a massive, ridgelike brow, and it housed a brain about the size of an ape's (Figure 19-40). Its face, on the other hand, was short and relatively flat, like that of a human. Its teeth were also more human than apelike in form. Additional fossil hunting will someday reveal more about this half-ape–half-human creature. Remains of another hominid genus, *Ardipithecus*, are known from Ethiopian strata ranging from 4 to 6 million years in age (Figure 19-41), but they are fragmentary.

Figure 19-40 The skull of *Sahelanthropus* from Chad. *Sahelanthropus* was an early genus of the human family that retained a braincase and brow ridges resembling those of an ape. (© Michael Brunet.)

The australopithecines resembled both apes and humans

The most diverse australopithecine genera are *Australopithecus* and the closely related genus *Paranthropus*. *Australopithecus* was the immediate ancestor of our genus, *Homo*. It evolved before 4 million years ago (see Figure 19-41). In many respects, *Australopithecus* was intermediate in form between apes and humans.

Australopithecus was much shorter than a modern human. Females averaged slightly more than a meter (3.5 feet) in height, and males, about 1.3 meters (4.5 feet). Members of *Australopithecus* weighed roughly the same as chimpanzees—and, as in chimpanzees, males were much larger than females. The brain of *Australopithecus*, in relation to its body size, was also only slightly larger than that of a chimpanzee. Thus it is not surprising that the skull of *Australopithecus* was more apelike than human in form, featuring heavy bony ridges above the eye sockets and a large, projecting jaw (Figure 19-42). In fact, the bony ridges strengthened the skull for the attachment of heavy muscles that operated the large jaw to chew on coarse plant food. *Australopithecus* may never have fashioned stone tools, perhaps because it lacked the intelligence to do so.

The body of *Australopithecus*, unlike the brain, displayed features of both apes and humans. Whereas the pelvis of an ape is narrow and elongate, that of *Australopithecus* was broad, resembling the human pelvis, which

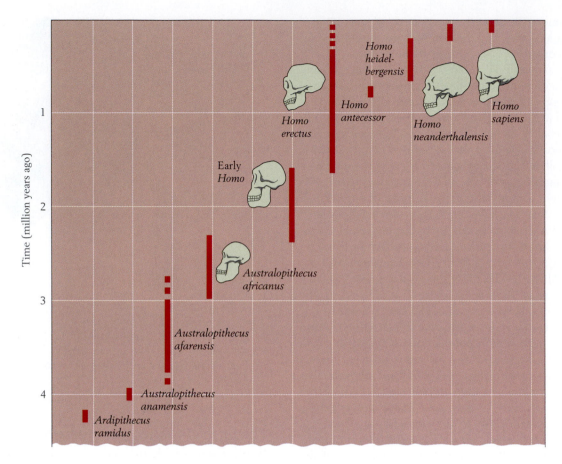

Figure 19-41 Stratigraphic ranges of species of *Ardipithecus, Australopithecus,* and *Homo,* as recognized from fossil data. Early *Homo* includes more than one species.

supports the body on two legs (Figure 19-43). Tracks beautifully preserved in volcanic ash in the geographic region occupied by *Australopithecus afarensis* provide direct evidence of upright, two-legged walking; in fact, they are remarkably similar to the footprints of modern humans (Figure 19-44). Thus it is clear that, when on the ground, *Australopithecus* walked upright like humans, rather than on all fours in the manner of apes.

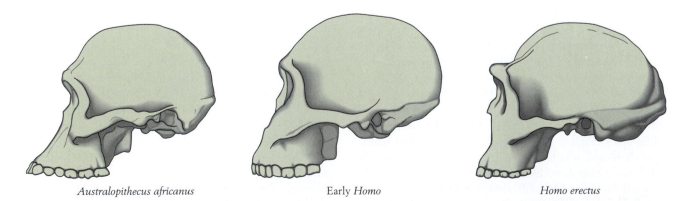

Australopithecus africanus   Early *Homo*   *Homo erectus*

Figure 19-42 Comparison of the skulls of *Australopithecus africanus,* early *Homo,* and *Homo erectus,* illustrating an evolutionary increase in brain size and flattening of the face. (After F. C. Howell, in V. J. Maglio and H. B. S. Cooke, eds., *Evolution of African Mammals,* Harvard University Press, Cambridge, MA, 1978.)

Figure 19-43 Reconstruction of the skeleton of *Australopithecus afarensis* from partial skeletal remains of the individual named Lucy. This skeleton, from Hadaf, Ethiopia, is about 3.2 million years old. Although Lucy walked upright, many of her other anatomical features, including her long arms, suggest that she frequently climbed trees. (Denver Museum of Natural History, photo by Rick Wicker.)

Although *Australopithecus* walked on two legs when on the ground, it also possessed adaptations for climbing trees adeptly. Among these were long arms relative to the length of its legs, long, curved fingers that resembled those of chimpanzees, and long big toes that were capable of grasping (Figure 19-45). It also had an upward-directed shoulder socket of the kind that improves the climbing ability of apes. In addition, the inner ear bones of *Australopithecus* were like those of an ape, not like those that provide humans with the balance required for running on two legs. Probably the fastest gait for *Australopithecus* was a lope or jog.

In summary, the locomotory adaptations of *Australopithecus* represented an adaptive compromise between moving about on the ground and climbing trees. Although relatively long, its arms were shorter for its body size than those of an ape. Thus it would have been

Figure 19-44 Tracks made by *Australopithecus* in volcanic ash at Laetolil, Tanzania, more than 3 million years ago. (© J. Reader/Photo Researchers Inc.)

less adept than a chimpanzee at moving about in the trees while, at the same time, being less mobile than a modern human when moving about on the ground.

The human genus made a sudden appearance

The human genus, *Homo*, evolved from *Australopithecus*. The oldest bones thus far assigned to *Homo* are about 2.4 million years old. It appears that by about 2 million years ago, two or more species of *Homo* were in existence. Because the taxonomy of these early forms is not yet well established, it is convenient to group them all under the informal label "early *Homo*."

Some fossil skulls of early *Homo* reveal a large cranial capacity, one of *Homo*'s trademarks. Whereas the average volume of space for the brain is only about 450 cubic centimeters in an *Australopithecus* skull, in the early *Homo* skull shown in Figure 19-42 it is about 760 cubic centimeters, and fragments of less well-preserved skulls indicate brain capacities well above

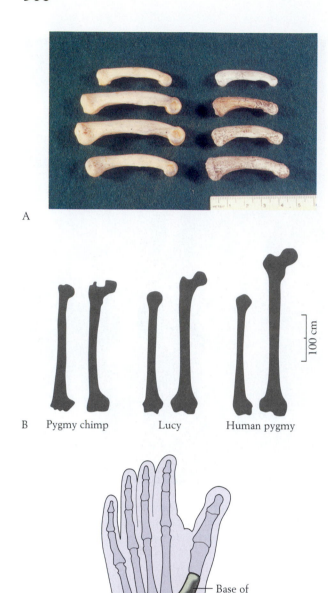

A

B   Pygmy chimp   Lucy   Human pygmy

100 cm

Base of big toe

1 cm

C

Figure 19-45 Adaptations of *Australopithecus* for climbing. *A.* The strong, curved fingers of the individual named Lucy (right) resemble those of a chimpanzee (left), which are adapted to climbing. *B.* Lucy also resembled a pygmy chimpanzee more than a pygmy human in having a short upper arm bone (left member of pair) relative to the length of her upper leg bone (right member of pair); these three hominoids have been chosen for comparison because they have the same body weight. *C.* Foot bones of *Australopithecus africanus*, showing the base of the long, divergent big toe, which would have served well for climbing. (*A*, courtesy of Steven Stanley; *B*, after S. M. Stanley, *Paleobiology* 18:237–257, 1992; *C*, after L. Oliwenstein, *Science* 269:476–477.)

800. The skull of early *Homo* exhibits additional features that make it more human in form than the skull of *Australopithecus*. The teeth, for example, are smaller.

Fossil remains of the pelvis and thigh bones of early *Homo* do not differ greatly from those of modern humans. These bones appear to have belonged to large individuals that spent almost all of their time on the ground.

Early *Homo* appears to have put its large brain to good use in the manufacture of stone tools. In fact, some types of fossils included within this group have been assigned to a species named *Homo habilis*, or "handy man," because stone tools have been found in some of the deposits from which such fossils have been collected. These tools—the oldest known in the geologic record—include sharp flakes of stone and many-sided "core" stones that were left after the flakes were broken away (Figure 19-46). Such tools are referred to as *Oldowan* because they were first found at Olduvai Gorge in Tanzania, the site of many fossil hominid discoveries. The oldest known jawbone of *Homo* and the oldest known Oldowan tools, which come from farther north, in Ethiopia, are both about 2.4 million years old.

Australopithecines, like modern chimpanzees, may have supplemented their largely vegetarian diet by devouring small animals, but meat apparently formed a much larger part of early *Homo*'s diet. Scratched bones of other mammals found in association with Oldowan tools indicate that early *Homo* used its stone tools to sever meat from bones. It is uncertain whether most of

Figure 19-46 Simple stone choppers from Olduvai Gorge, Tanzania, representing the Oldowan culture, which arose about 2.5 million years ago. Flakes broken from these "core" implements were used for cutting. (Courtesy of Ken Mobray/American Museum of Natural History.)

this meat was the product of active hunting or of scavenging on carcasses killed by other animals.

In any event, the rather abrupt appearance of *Homo* requires an explanation. The origin of early *Homo* and the extinction of *Australopithecus* probably resulted from changes that occurred in the climate and vegetation of Africa between about 2.6 and 2.4 million years ago (Earth System Shift 19-2).

Homo erectus resembled us

About 2 million years ago, *Homo erectus* evolved from early *Homo* in Africa (see Figures 19-41 and 19-42). This new species seems not to have differed greatly from early *Homo*, but it was slightly more similar to modern humans. In addition, *Homo erectus* was the first hominid species to have migrated beyond Africa. By 1.9 million years ago it had reached China, where it has been referred to as "Peking man," and Java, where it has been called "Java man." *Homo erectus* produced impressive hand axes and cleavers as part of its distinctive stone tool culture, known as *Acheulian* (Figure 19-47).

It had long been thought that *Homo erectus* died out about 300,000 years ago, but a site in Java that has yielded 12 skulls of this species has recently been estimated to be between 30,000 and 50,000 years old. If this date is valid, *Homo erectus* may have survived only on islands of Indonesia after more modern humans had occupied Africa and Eurasia.

The recent discovery in Africa of the 1.6-million-year-old skeleton of an 11- or 12-year-old boy has revealed the remarkable resemblance between *Homo erectus* and modern humans (Figure 19-48). This species resembled us in body size but had a smaller brain. Its

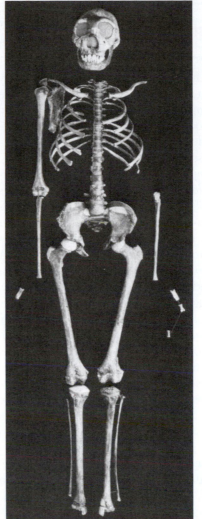
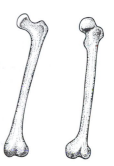

H. erectus   H. sapiens

Figure 19-48 Nearly complete skeleton of an 11- or 12-year-old *Homo erectus* boy from strata 1.6 million years old. The boy's height was about 5 feet 3 inches (1.6 meters), and had he lived to adulthood, he would have grown to a height of about 6 feet (1.8 meters). On the right, the neck of the boy's femur is seen to be much longer than that of a femur belonging to a typical modern human, who has a wider pelvis. (David L. Brill, Fairburn, Georgia.)

Figure 19-47 Tools representing the widespread Acheulian culture of *Homo erectus*. These hand axes, cleavers, and other stone implements are scattered profusely over the ground at an 800,000-year-old collecting site in Kenya. (Photograph by Willard Whitson; courtesy of Ian Tattersall.)

cranial capacity averaged about 1000 cubic centimeters, which was greater than that of early *Homo* (see Figure 19-42), but smaller than that of modern humans (typically 1200–1500 cubic centimeters). The pelvis of *Homo erectus* was narrower than ours, and here we see why this extinct species necessarily had a smaller brain. The size of the pelvis limits the size of the brain in hominids because a baby's head must pass through its mother's pelvis during birth. The narrow pelvis of *Homo erectus* would not have permitted the birth of a modern baby of average size.

Not only did *Homo erectus* have a smaller brain than ours, but its skull differed from ours in having a

Earth System Shift 19-2	The Human Genus Arises at a Time of Sudden Climatic Change

Australopithecus had existed with little evolutionary change for at least 1.5 million years before it gave rise to the modern human genus, *Homo*. Why, we must ask, did *Australopithecus* give rise to *Homo* about 2.5 million years ago and then disappear?

The answer seems to relate to the tree-climbing habit of *Australopithecus*. It is not difficult to understand why *Australopithecus* would have climbed trees frequently. First of all, many of the fruits and seedpods that it ate must have grown on trees. Second, in order to avoid predators, it must have needed to sleep in trees and occasionally to flee into them during the day. A band of *Australopithecus* individuals probably slept in a grove of trees and fed in and around it during the day, staying close enough to the grove to flee into the trees if predators approached. When the local food supply dwindled, they would have been forced to migrate across dangerous open country to a new grove. Today this behavior characterizes both baboons, which are very large monkeys, and chimpanzees. *Australopithecus* would have been as defenseless as are these other large primates against lions, hyenas, and other ferocious predators. Like them, it lacked advanced weapons and was slower on the ground than large four-legged herbivores such as antelopes and zebras. Furthermore, more species of large predators occupied Africa then than today.

The large brain of early *Homo* could not evolve until its ancestors abandoned their tree-climbing habit. The problem was that infants endowed with this large brain were physically helpless, and a mother who needed both arms free to climb trees could not have carried such a helpless infant about with her. The origin of this large brain can be attributed mainly to a change in the pattern of infant development. The brains of all newborn primates (including monkeys, apes, and humans) account for about 10 percent of their total body weight—a very large proportion. The brains of all these species grow rapidly before birth. In monkeys and apes, however, this high rate of brain growth slows dramatically shortly after birth, so that the brains of adults are only moderately larger than those of newborns. In *Homo*, however, the brain continues to grow rapidly for about a year after birth. This is the main reason why adult humans have such large brains. The size of the human brain increases only moderately after the age of 1, but an average year-old human infant already has a very large brain capacity— more than twice that of an adult chimpanzee.

Figure 1 *Dinofelis* was one of the formidable array of large, swift Pliocene predators that would have menaced *Australopithecus. Dinofelis* was larger than a leopard and possessed a pair of daggerlike teeth. Contemporaneous predators included a long-legged hyena species, now extinct, as well as the modern brown and spotted hyenas and the modern lion, leopard, and cheetah. (After Dixon et al., *The Macmillan Illustrated Encyclopedia of Dinosaurs and Prehistoric Animals*, Macmillan Publishing Company, New York, 1988.)

This extension of the high rate of brain growth through the first year of life was achieved through a general evolutionary delay of maturation. Our permanent teeth do not replace our baby teeth, for example, until we are much older than apes are when they undergo these changes. A key feature of the delayed development of humans is that it produces great intelligence at a very early age, so that small children can engage in relatively advanced learning.

We humans mature so slowly that our infants are physically helpless much longer than the infants of any other species of mammals. The disadvantage of the need to provide years of care to our offspring is more than offset, however, by the enormous advantages that result from the expansion of the brain: humans are much more intelligent than any other species of mammals, and for this reason we are able to cope with a host of environmental challenges. We are neither physically powerful nor fleet of foot, yet we have come to dominate Earth.

Nonetheless, it is quite evident why australopithecines did not evolve a large brain for more than 1.5 million years of existence. As we have seen, they were required to spend a significant portion of their lives in trees. Newborn australopithecines, like newborn chimpanzees, had to be mature enough to cling to mothers whose arms were occupied with tree climbing. Under these circumstances an evolutionary change that produced helpless infants and a large brain was impossible. Why did australopithecines finally abandon their habitual tree climbing?

Recall that climates became cooler and drier over broad regions of the world about 2.5 million years ago, when continental glaciers expanded in the Northern Hemisphere. Fossil pollen reveals that terrestrial floras in Africa underwent a dramatic change at this time. Forests shrank and grasslands expanded. This is exactly the kind of change that would be expected to force animals such as australopithecines to abandon dependence on tree climbing. Presumably the australopithecines faced an ecological crisis: when they no longer had numerous trees to climb, they lost important sources of food and were suddenly more exposed to predators. Many populations of *Australopithecus* died out during this crisis—the genus disappeared— but at least one population evolved into *Homo*. A large brain was so valuable for avoiding predators and developing advanced hunting techniques that its value overshadowed the problems that resulted from the prolonged helplessness of human infants.

Skeletal remains of a new species recently found in Ethiopia may provide a glimpse of the transition to *Homo*. These fossils have been assigned to the species *Australopithecus garhi*. They are about 2.5 million years old, meaning that they represent a creature that existed at about the time of the great climatic change in Africa. This animal resembled other species of *Australopithecus* in the size of its brain and the proportions of its arms, but had much longer legs. Perhaps its long legs represented an evolutionary response to the need to move effectively on the ground as forests shrank. Oldowan stone tools (see Figure 19-46) found nearby may be artifacts of this species. If they are, we might infer that *Australopithecus* had possessed the brainpower to create stone tools for more than a million years before doing so, but never took advantage of this capacity until it was forced to survive largely on the ground.

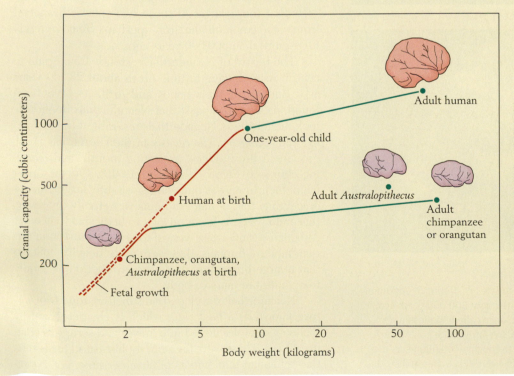

Figure 2 Brain growth patterns in primates. Primates experience a high rate of brain growth before birth (dashed red lines). Apes and monkeys shift to a slower rate of brain growth soon after birth (green line). In humans, however, the high fetal rate of brain growth persists through the first year after birth (solid red line). It is during this early interval of life outside the womb that the human brain attains most of its adult size. (After S. M. Stanley, *Paleobiology* 18:237–257, 1996.)

low, sloping forehead inherited from *Australopithecus*. Other primitive features of the skull were prominent brow ridges, a projecting mouth, and a heavy lower jaw. The brain of *Homo erectus* was nonetheless larger than that of its predecessors. A high level of intelligence and the ability to move about effectively on the ground were probably traits that endowed it with the ability to make its way out of Africa and spread throughout Asia.

Homo heidelbergensis resembled us even more

Apparently a descendant of *Homo erectus*, *Homo heidelbergensis* was named for the place where it was first found, Heidelberg, Germany. Though not all well dated, skulls assigned to *Homo heidelbergensis* generally fall within the 200,000- to 700,000-year age range,

A

B

Figure 19-49 **Advanced species of *Homo*.** *A.* A skull of *Homo heidelbergensis*, dating to about 600,000 years ago, from Kabwe, Zambia; this form, once called Rhodesian man, had a sloping forehead and massive brow ridges. *B.* An early skull of *Homo sapiens*, dating to roughly 100,000 years ago, from Qafzeh, Israel. (Pascal Goelgheluck/Photo Researchers, Inc.)

so that this species overlapped in time with *Homo erectus*. It coexisted with this more ancient species in Africa and ranged farther north in Europe. *Homo heidelbergensis* had a larger brain than *Homo erectus* (Figure 19-49*A*). Although resembling modern humans in cranial capacity, *Homo heidelbergensis* failed to advance stone tool technology, continuing to manufacture tools of the Acheulian type (see Figure 19-47).

Preceding *Homo heidelbergensis* in 780,000-year-old strata of southern Spain are fossils assigned to the species *Homo antecessor*. This older species, which presumably evolved from *Homo erectus*, shared some features of the skull with *Homo heidelbergensis* and is thought to have given rise to this younger species. Curiously, however, *Homo antecessor* in some ways resembles *Homo sapiens* more than *Homo heidelbergensis*. Possibly, then, it also gave rise to this modern species through some other line of descent.

The Neanderthals emerged in Eurasia

The creatures known as Neanderthals appeared at least 150,000—and probably 200,000—years ago (see Figure 19-41). Although they were regarded for a time as representing a variety of our species, Neanderthals are now known to represent a separate species that goes by the name *Homo neanderthalensis*. A recent comparison of DNA from a limb bone of a Neanderthal with the DNA of modern humans uncovered so many differences that the two forms clearly do not represent a single species. Furthermore, the number of genetic differences between *Homo sapiens* and *Homo neanderthalensis* suggests that the lineages of these two species branched from some common ancestor—perhaps *Homo antecessor*—more than 500,000 years ago.

The record of Neanderthals extends from Spain to central Asia, ranging in time up to about 28,000 years ago. Enough of their bones and artifacts have been found in caves to suggest that Neanderthals frequently took shelter there. Neanderthals resembled *Homo erectus* and *Homo heidelbergensis* in their long, low skull, prominent brow ridges, projecting mouth, and receding chin (see Figure 19-41 and 19-49). On the other hand, their brain was quite large—slightly larger, on average, than that of modern humans. Actually, however, this brain was smaller in relation to body size than ours. Although Neanderthals were of somewhat shorter stature than modern humans, they were more heavily built, so that a large proportion of the brain functioned simply to operate the massive body.

Neanderthals developed a distinctive stone tool culture known as *Mousterian*, which is characterized by flakes of stone fashioned into knives and scrapers (Figure 19-50). These were far more sophisticated tools than the Acheulian implements of *Homo erectus*, yet Nean-

Figure 19-50 Mousterian flint tools from western France. These tools were made by Neanderthals. (Courtesy of Ian Tattersall.)

derthals never attached stone points to spears. Instead, they hunted animals with wooden spears sharpened at one end. Neanderthals obviously had a difficult life under the Ice Age conditions of Europe. Their arms were more powerful than those of modern humans, reflecting the physical effort they put forth. In addition, their skeletons display many wounds, including broken bones that healed. They nurtured infirm members of their society, however, as indicated by skeletons of individuals who had survived for many years in a crippled state and must have lived under the care of others. Neanderthals also appear to have had religion. Burial sites reveal that Neanderthals sometimes prepared their dead for a future life by interring them with flint tools and cooked meat (Figure 19-51). In the Zagros Mountains of Iraq, a Neanderthal man who died after a skull injury was buried in a bed of boughs and flowers that can be identified from the pollen they left beneath the skeleton. It is fortunate for us that Neanderthals buried their dead. In effect, they intentionally produced much of their own fossil record, preserving many complete skeletons to the benefit of modern science.

Homo sapiens evolved in Africa and spread north

Both application of the molecular clock (p. 164) to modern human populations and occurrences of fossil skulls indicate that *Homo sapiens* evolved in Africa somewhat earlier than 150,000 years ago. Sophisticated tools, made of bone and stone, that are characteristic of *Homo sapiens* also appear in the fossil record of East Africa in sediments ranging in age from about 100,000 to 200,000 years. The oldest fossils assigned to our species are three skulls from Ethiopia that were described in 2003. Radiometric dating of associated volcanic rocks established the age of these fossils as 154,000–160,000 years. With the

human fossils were advanced tools, including knives, and bones of hippopotamuses that the humans butchered.

Perhaps our species then required several tens of thousands of years to make its way northward to the icy climates of Europe. The oldest remains of modern humans in Europe date to about 40,000 years ago. It is a striking fact that the Neanderthals then disappeared only about 12,000 years later. Could the timing have been a matter of chance? Certainly it is possible that our species' aggressive behavior caused the Neanderthals' extinction. Perhaps *Homo sapiens* was the victor in competition for occupancy of caves or control of territory in general.

In any event, our species brought to Europe far more advanced technology than that of Neanderthals. What is known as the *Cro-Magnon culture* of *Homo sapiens* quickly emerged in Europe. Cro-Magnon culture initially incorporated Mousterian elements, but

Figure 19-51 Burial of a young Neanderthal, as reconstructed from a fossil skeleton. The position of the skeleton shows that the head was cradled on one arm. Also present were the charred bones of animals that apparently were left as food for the dead individual in an afterlife. (Painting by Z. Burian under the supervision of Professor J. Augusta, Professor J. Filipa, and Dr. J. Moleho.)

soon developed into the more sophisticated *Late Neolithic culture*, in which a variety of specialized tools were invented.

Even more innovative was the artwork of the Cro-Magnon people, which reflected new use of the powers of imagination. Their magnificent cave paintings, primarily of animals, can still be admired in France and Spain (Figure 19-52). Other artifacts show that their artistic efforts included modeling in clay, carving friezes, decorating bones, and fabricating jewelry from teeth and shells. All of this reflected one unique aspect of the history of *Homo sapiens*: rapid cultural evolution. Since early in the history of *Homo sapiens*, its culture has been changing rapidly. In contrast, cultures of earlier species of the human family, once in place, remained almost stagnant.

A second unique aspect of our culture that has strengthened through time is an ability to manipulate Earth's environment. As we shall see in Chapter 20, this expanding power now threatens our well-being.

Figure 19-52 A cave painting of the Cro-Magnon people. (Ferrero/Labat/Auscape International.)

Chapter Summary

How did marine life of Neogene time differ from that of Paleogene time?

Invertebrate life in the oceans underwent only minor changes during Neogene time, but whales radiated rapidly.

What happened to grasses and grasslands early in Neogene time?

On land, grasses and herbs, benefiting from a cooling and drying of climates, diversified and occupied more territory during Miocene time.

How did climatic changes during Neogene time cause changes in vertebrate faunas on land?

As climates became drier and grasslands expanded, rats, mice, and songbirds, many of which eat the seeds of grasses and herbs, underwent major evolutionary radiations; snakes, which prey on rodents and the eggs and chicks of songbirds, also diversified markedly.

Why might we label the Miocene Epoch the Age of Apes?

Apes radiated for the first time during the Miocene Epoch; their diversity has since dwindled.

Why did global climates change during the Pliocene Epoch?

The modern Ice Age began about 3.2 million years ago with the expansion of continental glaciers in the Northern Hemisphere. As a result, climates in many regions became cooler and drier, and grasslands expanded while forests shrank.

Why is it reasonable to say that we still live in an Ice Age in the Northern Hemisphere?

The modern Ice Age has been characterized by glacial maxima and minima dictated by periodic changes in Earth's rotational motions. During the past 850,000 years, in synchrony with changes in the shape of Earth's orbit (the eccentricity cycle), continental glaciers have expanded and contracted nine times in the Northern Hemisphere. Expansions have lowered sea level by as much as 100 meters. Glaciers are now in a contracted state, but they will probably expand again.

What tectonic events elevated mountains in the American West in Neogene time?

Block faulting has elevated ridges in the Great Basin; the Sierra Nevada is the westernmost of these uplifts. The Rocky Mountains, which had been subdued by erosion, rose again to great heights, apparently because of swelling of the underlying mantle.

What kinds of volcanism affected western North America in Neogene time?

Lavas have welled up along faults in the Basin and Range Province; subduction has produced the Cascade Range and other volcanic peaks near the continental margin; and the Yellowstone hot spot shifted to produce the basalts of the Columbia Plateau and the Snake River Plain.

What plate tectonic changes occurred in Neogene time in the area between North and South America?

In the western Atlantic region, the Caribbean Sea, which is bounded on the east by an island arc, developed its modern configuration during Neogene time, and the Isthmus of Panama was uplifted, permitting extensive biotic interchange between North and South America.

Why did the Himalaya begin to rise up during the Miocene?

The craton that now constitutes the Indian peninsula collided with the southern margin of Asia and was wedged beneath it.

How did a drop in sea level at the end of Miocene time affect the Mediterranean Sea?

The Mediterranean Sea became weakly connected to the Atlantic Ocean about 6 million years ago and briefly became hypersaline. Soon thereafter, reenlargement of the connection allowed the Mediterranean to fill with normal marine waters again.

How does the modern human genus, *Homo*, differ from its ancestor, *Australopithecus*?

Australopithecus possessed a small body, a relatively small brain, and spent much of its life in trees. *Homo*, the modern human genus, has a much larger brain. It evolved in Africa about 2.5 million years ago, and soon spent all of its time on the ground. The origin of *Homo* may have been a consequence of the shrinkage of forests, which made it difficult to sleep and find food in trees. By about 150,000 years ago, humans of the modern type were present.

Review Questions

1. In what ways did mammals become modernized during the Neogene Period?

2. What factors influenced the isotopic composition of oxygen in skeletons of marine organisms during the modern Ice Age?

3. What changes in the geographic distribution of land animals did the uplift of the Isthmus of Panama produce?

4. How did the Rocky Mountains develop their present configuration in the course of the Neogene Period?

5. How and when did the Sierra Nevada form?

6. What kinds of volcanic activity occurred in the American West during Neogene time?

7. How has climatic change altered the general distribution of terrestrial vegetation since early in Pliocene time?

8. How did the Appalachian Mountains develop their present configuration in the course of Neogene time?

9. What evidence is there that about 7 million years ago a major change occurred in the kinds of grasses that populate the world? How did this change affect animals?

10. How did members of the genus *Australopithecus* differ from modern humans?

11. Two phases of global climatic change occurred during the Neogene Period. The first one was gradual, the second one sudden. Using the Visual Overview on page 470 and what you have learned in this chapter, review each of these phases of climatic change and describe how each affected life on Earth.

The upper limit of trees and the lower limit of a mountain glacier in Olympic National Park in Washington State. (Breck P. Kent.)

CHAPTER 20

The Holocene

The Holocene, sometimes called the Recent interval, is the time that extends back from the present to the most recent retreat of continental glaciers in the Northern Hemisphere. The Holocene was originally established as the final epoch of the Cenozoic Era. Whereas the Miocene, Pliocene, and Pleistocene epochs, which preceded the Holocene, are assigned to the Neogene Period, however, the Holocene currently belongs to no geologic period. Technically, then, it cannot be formally recognized as an epoch, which must be a subdivision of a period. On the other hand, the Holocene could itself be elevated to the status of a geologic period, although it would be a very short one. It stands out as a unique geologic interval—the interval during which humans have altered Earth's environment, first by hunting, later by cutting down trees and planting crops, and eventually by building cities, burning fossil fuels, and creating vast networks of communication and transportation.

The boundary between the Pleistocene and the Holocene remains loosely defined. We will begin this chapter with events that took place about 15,000 years ago because those events marked the beginning of the end of the most recent glacial interval in the Northern Hemisphere and were quickly followed by the first major impacts of humans on Earth's ecosystems.

Holocene sediments and fossils are fully within the range of radiocarbon dating, so they can be dated with great precision. In addition, because the vast majority of Holocene species of plants, animals, and protists survive today, paleontologists can use the fossil records of these species to characterize habitats with great accuracy.

Holocene history holds special importance for us today because it reveals how the Earth system has approached its present state. Future changes in the global environment will build on trends of the recent past. The geologic record of the Holocene reveals the speed with which environments can change and the ways in which the changes affect the species that still populate our planet. This record, together with other patterns of change seen in the geologic record of the more distant past, provides lessons that will help us confront environmental changes in the future.

Era	Period	Epoch	Time
CENOZOIC	NEOGENE	Holocene	15,000 years ago
CENOZOIC	NEOGENE	Pleistocene	1.8 million years
CENOZOIC	PALEOGENE	Pliocene	5.3 million years
CENOZOIC	PALEOGENE	Miocene	
MESOZOIC	CRETACEOUS	Oligocene	24 million years
MESOZOIC	JURASSIC	Eocene	
MESOZOIC	TRIASSIC	Paleocene	
PALEOZOIC	PERMIAN		
PALEOZOIC	CARBONIFEROUS — PENNSYLVANIAN		
PALEOZOIC	CARBONIFEROUS — MISSISSIPPIAN		
PALEOZOIC	DEVONIAN		
PALEOZOIC	SILURIAN		
PALEOZOIC	ORDOVICIAN		
PALEOZOIC	CAMBRIAN		
"PRECAMBRIAN"	PROTEROZOIC EON		
"PRECAMBRIAN"	ARCHEAN EON		

The Retreat of Glaciers

Soon after the last glacial maximum, about 20,000 years ago, continental glaciers began to melt away. Fossil insects, dated by the radiocarbon method, show that temperatures in the Rocky Mountain region began to rise about 15,000 years ago; the insect species that provide this information survive today, and their temperature tolerances are well known. The meltwaters of

Major Events of the Holocene

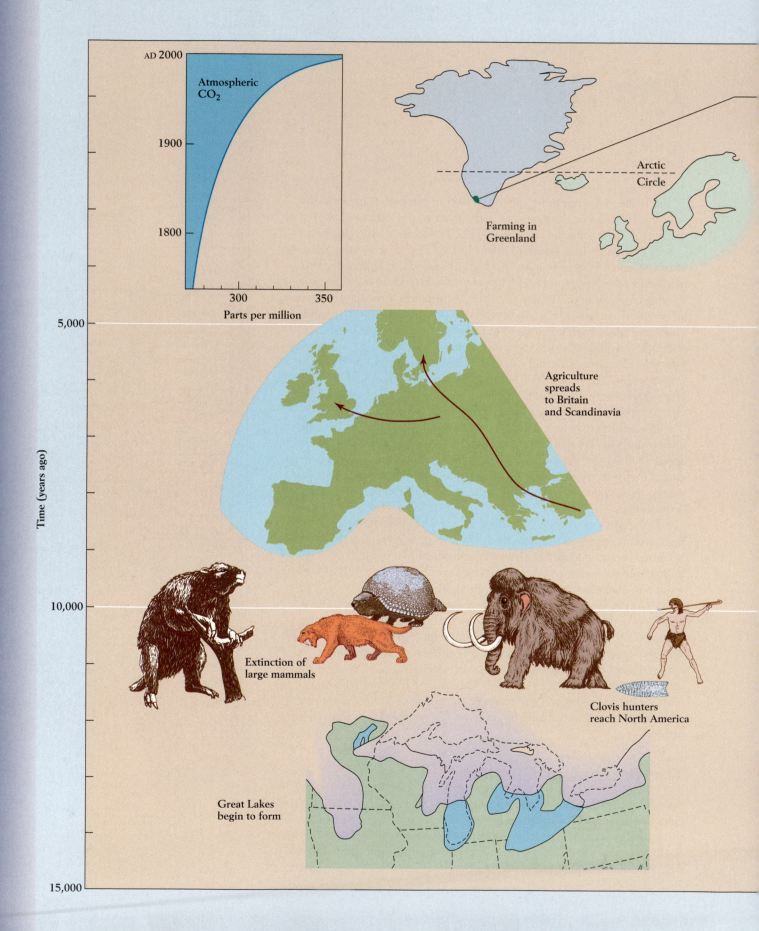

AD 2000

1900

1800

Atmospheric CO_2

300   350

Parts per million

Time (years ago)

5,000

10,000

15,000

Arctic
Circle

Farming in
Greenland

Agriculture
spreads
to Britain
and Scandinavia

Extinction of
large mammals

Clovis hunters
reach North America

Great Lakes
begin to form

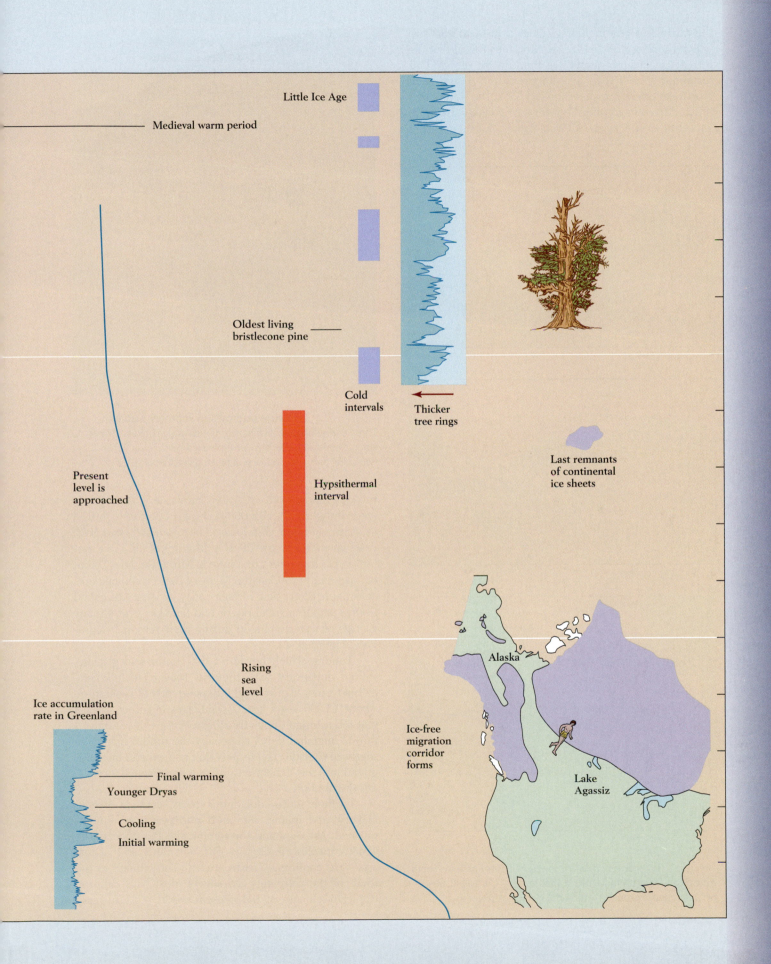

Little Ice Age

Medieval warm period

Oldest living
bristlecone pine

Cold
intervals

Thicker
tree rings

Hypsithermal
interval

Present
level is
approached

Last remnants
of continental
ice sheets

Rising
sea
level

Ice accumulation
rate in Greenland

Ice-free
migration
corridor
forms

Alaska

Lake
Agassiz

Final warming

Younger Dryas

Cooling

Initial warming

the shrinking continental glaciers flowed to the sea, which therefore began to rise. Moraines reveal that the retreat of glaciers was slow at first, but accelerated after about 15,000 years ago (Figure 20-1). As the great North American ice sheet melted back, the waters that drained from its southern border formed lakes that were ancestral to the modern Great Lakes. Farther west, a

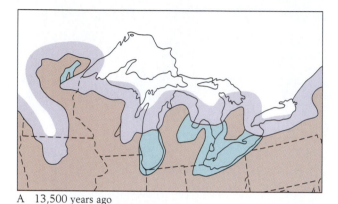

A   13,500 years ago

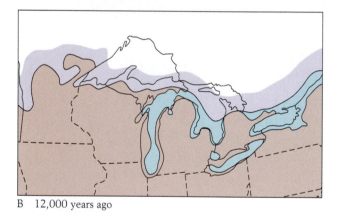

B   12,000 years ago

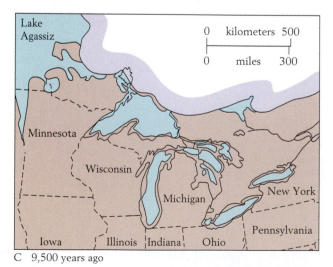

```
0        kilometers   500
0          miles      300
```

C   9,500 years ago

Figure 20-1 **The retreat of glaciers from eastern North America after the last glacial maximum.** The Great Lakes occupied depressions left when the glaciers retreated.

Figure 20-2 Prairie potholes in southern Canada. These depressions, which harbor ponds and marshes, formed when chunks of ice left by retreating glaciers melted after becoming partly buried in sediment. (Bates Littlehales.)

broad, shallow body of water known as Lake Agassiz formed slightly after 12,000 years ago, constantly shifted its shape, and then disappeared between 4000 and 5000 years later (see Figure 20-1C). Mounds of ice a few meters across remained for a time in Minnesota, the Dakotas, and southern Canada. Many of these remnants created depressions that remain today as what are called *prairie potholes*: small ponds and marshes where many species of waterfowl stop during their seasonal migration (Figure 20-2).

Tundra, which bordered the continental glaciers, shifted northward with them as they retreated. Surprisingly, insects migrated independently of vegetation. About 13,000 years ago, assemblages of insect species that today inhabit taiga (northern evergreen coniferous forest) migrated north to occupy the tundra habitat in southern Canada. The insects that now inhabit taiga migrated into this habitat only after it approached its present distribution.

Farther south, deciduous trees, such as beech, hickory, and maple, migrated northward as climates warmed. As happened throughout the Pleistocene, the various species migrated at different rates, so that forests were continually restructured. The climate to the south of the retreating ice sheets must have differed from any climate of the present world because fossil

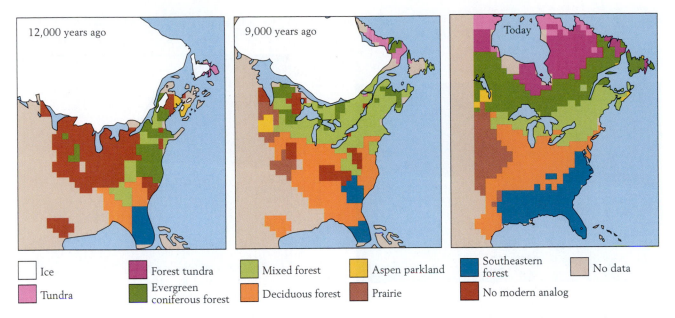

12,000 years ago   9,000 years ago   Today

☐ Ice	Forest tundra	Mixed forest	Aspen parkland	Southeastern forest	No data
Tundra	Evergreen coniferous forest	Deciduous forest	Prairie	No modern analog	

Figure 20-3 Recombination of tree species in forest communities during migration of forests as the North American glacier melted back. Communities of trees having no analog in the modern world occupied broad areas between 12,000 and 9000 years ago, even though the species that formed these communities are all alive today. (After J. T. Overpeck, R. S. Webb, and T. Webb, *Geology* 20:1071–1074, 1992.)

pollen reveals that 12,000 years ago an evergreen forest unlike any modern flora existed there (Figure 20-3). Spruce trees were common, but pine trees, an abundant component of modern northern evergreen coniferous forests, were absent. On the other hand, such hardwood trees as oak, elm, and ash abounded.

Because individual tree species have shifted their ranges independently of one another during the climatic shifts of the modern Ice Age, there has been constant reshuffling of species within communities (see Figure 20-3). The discovery of this phenomenon has overturned the long-popular idea that each major plant community on Earth today, such as the temperate deciduous forest of the eastern United States or the Douglas fir forest of the Pacific Northwest, is an ancient entity whose component species have evolved in association with one another. Instead, such communities are transient assemblages that have formed since the last glacial maximum.

Abrupt Global Events of the Early Holocene

Earth's emergence from the last glacial maximum was not smooth. The transition to the present climatic conditions was marked by temporary reversals to cold conditions and, more generally, by sudden climatic changes. Between 15,000 and 7000 years ago three abrupt sea-level rises corresponded to sudden global shifts in climate.

Fossil corals of the island of Barbados in the Caribbean Sea provide evidence of the three Holocene episodes of rapid elevation of sea level. Here scientists have made use of the reef-building "moosehorn" coral, *Acropora palmata*, which always grows close to sea level. Fossil specimens of this species, which can be dated by the uranium-thorium method, indicate the level of the sea in relation to the island for various past intervals. Barbados has been rising tectonically since late Pleistocene time. Colonies of *Acropora palmata* that grew during previous glacial minima, when sea level was close to its present position, now stand many meters above sea level. Their position indicates that Barbados has been rising tectonically at an average rate of more that 30 centimeters per year for the past 125,000 years. Cores obtained by drilling through the limestones of Barbados have yielded fossil *Acropora palmata* from many depths below the surface (Figure 20-4). Radiometric dating by the radiocarbon and uranium-thorium methods (p. 142) has established the ages of these corals. Their depths below the surface, corrected for tectonic uplift of the island, then indicate the position of sea level back to the glacial maximum, about 18,000 years ago. They show that sea level rose rapidly slightly after 15,000, 12,000, and 8000 years ago.

It is possible that sea level also rose rapidly by a few meters at other times during the general eustatic (worldwide) rise of the past 18,000 years. Certainly there have been abrupt pulses of warming in the far north. Sometimes the Greenland glacier has responded

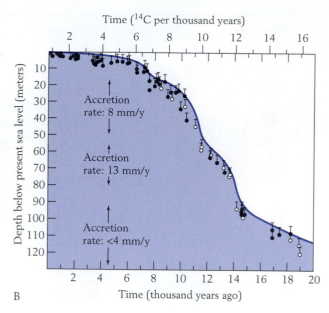

A

B

Figure 20-4 Holocene rises in sea level, indicated by occurrences of a reef-building coral species in Barbados. *A. Acropora palmata* is a reef-building coral species that lives only close to the sea surface. *B.* Dots in the graph show fossil occurrences of this species at various depths below modern sea level, dated by the radiocarbon method (black) and uranium-thorium method (white). (*A*, Marty Snyderman/Visuals Unlimited; *B* after P. Blanchon and J. Shaw, *Geology* 23:4–8, 1995.)

differently than other ice sheets to global warming: when climatic warming has caused ice sheets to shrink in North America and Europe, annual layers of glacial ice in Greenland have thickened because of the increased supply of moisture from the warmer North Atlantic Ocean (p. 486). Changes in the thickness of these layers point to a remarkable sequence of events: as Earth began to emerge from the last glacial maximum, there was a sudden reversal of warming, and Ice Age conditions returned for more than a thousand years. After this period of cooling, known as the Younger Dryas, the planet finally moved into the glacial minimum in which we now live (Earth System Shift 20-1).

The First Americans

About the time of the Younger Dryas, humans colonized North America on a grand scale. The timing of this migration was no accident. After moving northward from Africa to Eurasia, modern humans had spread rapidly across that vast continent. By 30,000 years ago they had reached Siberia, where climates were very cold but a weak supply of moisture prevented large glaciers from forming. Soon thereafter they reached the Bering land bridge, where many large animals roamed (see Figures 19-10 and 19-11), but glaciers blocked easy passage to Alaska. A small number of archeological sites farther south suggest that a few bands of humans made their way to the Americas about 30,000 years ago, perhaps by sea. Only after glaciers began melting rapidly, however,

did a large corridor open through Alaska and western Canada to unglaciated portions of North America. This ice-free avenue passed between the glacier that still occupied the northern Rocky Mountains and the retreating continental ice sheet to the east. Along it trod the Clovis people, who developed the first widespread culture in the New World and became the ancestors of nearly all Native Americans. They reached North America slightly before 11,000 years ago.

Archeological studies have revealed little about the Clovis people, except that they hunted adeptly in groups to take down elephants and bison with spears. Each of three species of elephants occupied a particular North American habitat when the Clovis hunters arrived (Figure 20-5). The woolly mammoth occupied tundra and grassy terrain close to the great ice sheets in both Eurasia and North America. Frozen carcasses in Siberia and Alaska retain the long hair that insulated the woolly mammoth's body. These carcasses also display small ears and a short trunk, features that would have reduced heat loss. Stomach contents have also been preserved, revealing a diet of grasses and tundra plants. Populations of the mastodon, a heavily built elephant, were concentrated in eastern forests, where twigs and conifer needles formed much of its diet. Vast herds of the great southern mammoth roamed prairies of the Midwest and Southwest. These animals stood as tall as 3.4 meters (more than 11 feet) at the shoulder and had strongly curved molars that served well for grinding up harsh prairie grasses.

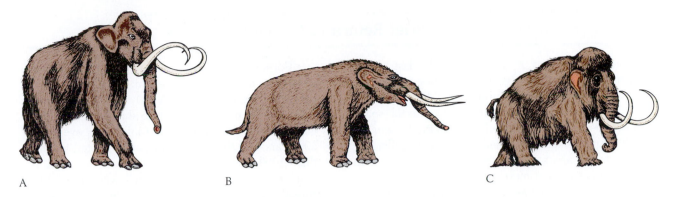

A

B

C

Figure 20-5 The last three species of elephants to inhabit North America. *A.* The southern mammoth (*Mammuthus columbi*) formed great herds on the prairies. *B.* The mastodon (*Mammut americanum*) occupied eastern forests. *C.* The woolly mammoth (*Mammuthus primigenius*) lived in cold northern regions.

Clovis hunters used a distinctive spear point (Figure 20-6*A*). It was fluted; that is, a channel was chipped into each side opposite the point to accommodate a slot in a spear shaft. The spears of Clovis hunters were cut short so that they could be hurled by a spear thrower—a hand-held device that generated greater velocity than a hunter could produce by throwing a spear he held in his hand. The archeological record offers abundant evidence of Clovis hunters' ability to take down large game (Figure 20-6*B*). In the 1920s a cowboy stumbled upon a bison skeleton near Folsom, New Mexico, with a spear point lodged between its ribs. Since then, anthropologists have discovered Clovis sites displaying spear points lodged in mammoth skulls or vertebrae, or otherwise associated with mammoth bones.

A Sudden Extinction of Large Mammals

Between about 12,000 and 10,000 years ago, many species of large mammals disappeared from North and South America. The extinction of these creatures, which transformed the terrestrial ecosystem, has been the subject of great controversy. Were these species killed off by Clovis hunters or by widespread climatic change? Before considering the arguments on each side, let us look at the victims. The bones of many of them have been found in the La Brea tar pits in Los Angeles, along with remains of mammals that survive to the present day (Figure 20-7). Here large herbivores became mired

A

B

Figure 20-6 Clovis hunting. *A.* A spear point found at Folsom, New Mexico, between the ribs of a giant bison. *B.* A reconstruction of Clovis hunters attacking a giant bison. (Denver Museum of Natural History.)

Earth System Shift 20-1 | Brief Return to Glacial Conditions

Although Earth was at the peak of the last glacial maximum about 20,000 years ago, it began to move into the present glacial minimum slightly more than 5000 years later. About 14,680 years ago, in the space of only about 10 years, annual layers in the Greenland glacier approximately doubled in thickness, reflecting a warming of nearby oceans that led to an increase in evaporation that sent more water into the atmosphere to fall on the glacier as snow. The sudden warming that produced the change remains unexplained, but about 500 years later a massive movement of ice into the North Atlantic created Heinrich layers (p. 486).

For unknown reasons, a long cooling trend began after the sudden warming event. This trend was accentuated slightly before 13,000 years ago, when the Northern Hemisphere began to shift back toward full glacial conditions. The shift took place over about 200 years, and the cold conditions persisted for more than 1000 years. This cold interval is called the Younger Dryas, after a genus of flowers (*Dryas*) that spread southward with the frigid conditions and left pollen in sediments to reveal the climatic reversal. Sea ice also spread southward, and planktonic foraminifera from deep-sea cores show that the relative amount of

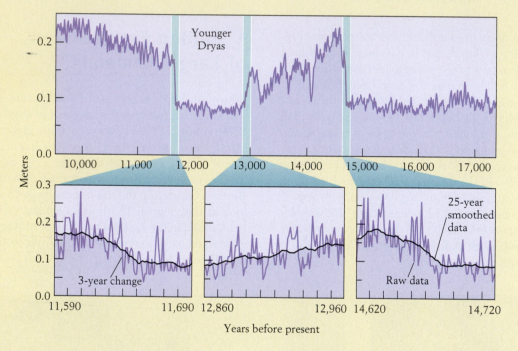

Figure 1 **Cores of the Greenland ice sheet display annual layers that vary in thickness and can be counted.** Snow accumulated more rapidly here when the climate warmed and evaporation from warmer seas supplied more moisture to the atmosphere. Measurement of annual layers shows that about 14,680 years ago, climates warmed abruptly within a single decade. A gradual cooling trend followed, and then an episode of more rapid cooling began about 13,000 years ago and lasted about 200 years, producing the Younger Dryas cold interval. The Younger Dryas ended about 11,600 years ago, in the space of only about 3 years. (After R. B. Alley et al., *Nature* 312:527–529.)

in sticky pools of tar that oozed up from sediments below. The trapped creatures attracted numerous carnivores, many of whom also became entrapped and sank into the soft tar.

The most striking feature of the mammal species that disappeared early in the Holocene is their large average body size. Among the large species of North American herbivores to disappear were all three species

of American elephants (see Figure 20-5); a beaver as large as a black bear; five species of horses; three members of the deer family; two species of wild oxen; three species of musk oxen; the only surviving North American camel, which stood about 20 percent taller than the modern Middle Eastern camel; a large bison, whose horns spread to more than 2 meters (about 7 feet) (see Figure 20-6B); two species of giant armadillos, each of

Figure 2 A mountain glacier formed a terminal moraine in southern New Zealand during the Younger Dryas cooling event. After forming the moraine, which is now covered by trees, the glacier melted back as Earth finally emerged from the last glacial maximum. (George Denton, University of Maine.)

Terminal moraine

oxygen 18 increased markedly in the North Atlantic. Warm, saline water from the south apparently failed to reach the cold North Atlantic. As a result, the downwelling that had driven the oceanic conveyor belt (see Figure 19-19) shifted southward.

Younger Dryas cooling was global in scale, causing mountain glaciers to expand well below their present limits. The Younger Dryas suddenly ended about 11,600 years ago, however. Amazingly, oxygen isotope measurements in ice cores reveal that the climatic warming that ended this event took place within just 3 years, with most of it occurring during a single year. Heinrich layers also formed, pointing to a massive launching of icebergs. Isotope studies suggest that the climate of Greenland warmed by about 7°C. Presumably, gradual warming suddenly triggered some kind of positive feedback, leading to an abrupt accentuation of the warming trend. The abrupt end of the

Younger Dryas moved Earth fully into the Holocene interval of glacial retreat in which we now live.

The final sudden rise in sea level deduced from the Barbados reef rock took place about 7600 years ago (see Figure 20-4). No Heinrich layers mark this event in the Northern Hemisphere; in fact, too little of the Laurentide ice sheet remained to release icebergs. It seems most likely that a catastrophic loss of ice from the Antarctic ice cap elevated sea level. At this time climates were restructured in ways that are not fully understood. Many parts of Africa, for example, became drier.

Global environmental changes of the early Holocene occurred with frightening speed. Knowledge of the instability of the natural system must heighten our concern about the future effects of human activities, including the destruction of rain forests and the release of greenhouse gases into the atmosphere.

which weighed about a ton; and several species of large ground sloths (see Figure 7-5 and p. 152), one of which weighed about 3 tons.

Several species of large North American carnivores also died out: the so-called short-faced bear, which stood nearly 2 meters (about 6 feet) tall at the shoulder and was probably capable of taking down horses and bison; the dire wolf, a fearsome creature that was probably

30 percent heavier than the living wolf species; a lion that was perhaps a subspecies of the modern African lion but was about 50 percent heavier; a species of cheetah; and three species of large saber-toothed cats.

It may be that the big carnivores died out because their prey disappeared. This is especially likely to have happened to the saber-toothed cats, which probably used their bladelike teeth to pierce the thick hides and

Figure 20-7 Mammals of the La Brea tar pits, now located in the city of Los Angeles. In the foreground a saber-toothed cat defends a zebra carcass against huge dire wolves. Beyond them a lion pounces on a bison calf. Southern mammoths roam in the distance. On the right are tall camels, and in front of them are a giant ground sloth and a scavenging condor. (Painting by Mark Hallett. © 1999 Mark Hallett Illustrations, Salem, Oregon.)

fatty tissues of elephants. These teeth were fragile and would have broken easily against the shoulder blade or rib of a horse or deer. The Freisenbahn cave in Texas served as a lair for *Homotherium*, one of the American saber-toothed cats. From this single cave paleontologists have recovered the skeletal remains of more than 30 of these large cats—many representing juveniles—along with the bones of about 70 juvenile mammoths. Clearly the saber-tooths favored vulnerable young mammoths as prey and dragged them into the cave. Probably all three species of saber-toothed cats died out because their elephant prey disappeared.

Interestingly, several species of large birds, including eagles, vultures, and a condor, disappeared from North America along with the large mammals. These birds probably scavenged on carcasses of large herbivores after carnivores were through with them (see Figure 20-7); when many species of large herbivores died out, the scavenging birds probably began to starve along with the carnivores.

Did human hunting eliminate many species of large mammals?

The question remains: Why did the large herbivores die out? The *overkill hypothesis* focuses on human hunting

as the cause. According to this hypothesis, large bands of proficient Clovis hunters invaded regions that few, if any, humans had entered before. The prey they encountered would not have been fearful of humans, having never seen them before. Thus, they would have been easy marks, and when a band of hunters exhausted the supply of large prey in one area, they would have moved on to another. Accordingly, some scientists envision that a wave of Clovis hunters swept across North America and into South America, devastating populations of elephants, bison, oxen, horses, camels, and ground sloths.

Researchers who favor the overkill hypothesis note that the Clovis people possessed an unprecedented level of hunting skill and expanded rapidly throughout the New World. These scientists cite several additional patterns of supporting evidence:

1. Hunters favor large animals as prey because a single kill provides abundant meat.

2. Large animals are especially conspicuous to hunters.

3. Extinctions of large animals were especially numerous in North and South America, where humans had recently arrived and animals may not have feared them. Few extinctions occurred in Eurasia and Africa, where large animals had long experience of human hunters.

4. Those species of large North American herbivores that did not become extinct had recently migrated from Eurasia, where they had learned to fear human hunters. These species included familiar species of the present: mountain lion, moose, elk, caribou, deer, musk ox, bighorn sheep, and mountain goat.

Opponents of the overkill hypothesis argue that early humans could not have overwhelmed entire species of large animals. Besides, the extinctions did not begin in the north and spread southward, as would be expected if a wave of hunters had swept from Alaska to South America. Finally, there is no archeological evidence that Clovis people killed any large mammals other than elephants and bison.

Did climatic change cause the extinction of large mammals?

Some researchers have proposed that changes in climate and vegetation caused the extinction of large mammals. They point to particular changes as potential agents of extinction:

1. The Younger Dryas interval began and ended suddenly, and temperatures shifted abruptly.

2. Certain habitats that had supported many kinds of large herbivores disappeared at the beginning of the Holocene. In particular, the northern dry grassland region that supported a variety of nutritious herbs and shrubs (and many species of large mammals; see Figure 19-11) gave way to less nutritious prairies.

Opponents of the idea that climatic change caused the extinctions argue that species that died out had weathered comparable changes when glaciers had waxed and waned throughout the Pleistocene.

Perhaps hunting and climate together caused the extinctions

The overkill and climatic change scenarios may both be partly correct. Most species of large animals are inherently vulnerable to extinction because their populations are small. Perhaps the extinctions between 12,000 and 10,000 years ago resulted from the combined impact of sudden shifts of vegetation during the Younger Dryas and a simultaneous intensification of human hunting.

Many kinds of large marsupial mammals died out in Australia late in Pleistocene time. Among them were two species of kangaroos that stood over 2.5 meters (about 8 feet) tall and a lionlike marsupial with no present-day counterpart. Unfortunately, the timing of the relevant events in Australia is not yet pinned down. Humans apparently became well established there about 35,000 years ago, and the extinctions occurred

soon thereafter, between about 30,000 and 25,000 years ago. On the other hand, Australia was also becoming cooler and dryer when the huge mammals disappeared.

Climatic Fluctuations of the Last 10,000 Years

Even after the Younger Dryas event, climates did not shift along a smooth path toward their present state. An early warm interval was followed by a net decline in global temperatures, and along the way there were several intervals of marked cooling. We began to emerge from the most recent cold spell just a few decades ago and are now in a warming phase.

The hypsithermal interval was a brief period of global warmth

Continental glaciers all but disappeared between about 9000 and 6000 years ago, and climates became warmer than they have ever been since. During this *hypsithermal* ("high heat") *interval*, eastern North America warmed enough that dwarf birch shrubs replaced tundra in some areas. Fossil needles of hemlock reveal that during the hypsithermal interval this conifer species was able to invade mountainous elevations of New England where the mean annual temperature today is 2°C too low for hemlock to survive. From this information and additional evidence from fossil pollen, researchers have estimated that mean annual temperatures in North America and Europe during the hypsithermal interval were about 2°C warmer than they are today. Temperatures were similarly elevated in the Southern Hemisphere as far south as Antarctica.

Because huge masses of ice melt slowly, remnants of continental glaciers persisted into the hypsithermal interval in North America and Europe. Glaciers have existed only in mountainous regions of these large northern continents since the last remnants of continental ice sheets disappeared about 6500 years ago.

Humans invented agriculture

One of the most significant transitions in human history began when some populations abandoned a mobile life of hunting and gathering and settled down to domesticate plants and animals for food. These activities spread throughout Europe between about 9000 and 6000 years ago, during the hypsithermal interval.

Agriculture first emerged in the Zagros Mountains, near the borders between modern Iraq, Iran, and Turkey. Archeological sites reveal that populations here lived in caves before 12,000 years ago and hunted wild animals, especially sheep and goats. Pollen in lake sediments reveals that climates of the region became moister after

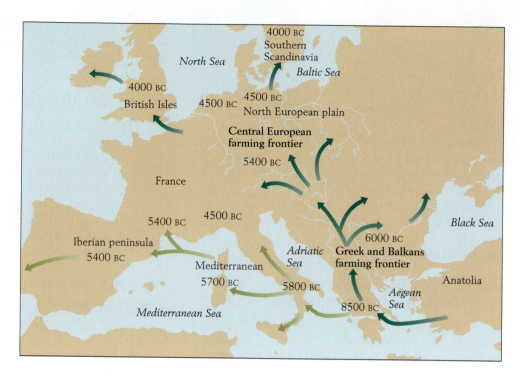

12,000 years ago and trees invaded what had been a dry, grassy habitat. Wild wheat and barley probably migrated into the region about the same time. These events may have coincided with the end of the Younger Dryas cold interval (see Earth System Shift 20-1). In any event, grinding tools found in caves indicate that local human populations were producing flour between 10,000 and 9000 years ago. Remains of houses and domestic animals in Mesopotamia, near the Zagros Mountains, reveal that people were living in the open, rather than in caves, and raising sheep, goats, pigs, and dogs by 9000 years ago. Farmers reached Greece from Turkey before 8000 years ago by colonizing one island after another in the Aegean Sea (Figure 20-8). From Greece, agriculture spread across Europe by fits and starts, reaching the British Isles and Scandinavia by 6000 years ago. Animal bones preserved at archeological sites reveal that cattle replaced sheep, goats, and pigs as the main large domesticated animals in Europe before 7000 years ago, perhaps reflecting an increase in the use of dairy products.

There is no evidence that regional climatic changes guided the initial expansion of agriculture. On the other hand, agriculture could not have spread throughout Europe until glacial and near-glacial conditions gave way to temperate climates. Certainly the abrupt onset of the hypsithermal interval provided favorable conditions for farming.

Glaciers, tree lines, and tree rings record climatic change

Mountain glaciers behave like thermometers, expanding down valleys when the climate cools and retreating

when the climate warms. Radiocarbon dating of spruce stumps provides a chronology for the upward and downward movements of tree lines in mountainous regions. Similarly, radiocarbon dating of fossil wood collected from glacial moraines reveals the vertical movements of glaciers since the hypsithermal interval. As it turns out, the major movements of tree lines and glaciers in mountains of Alaska and Europe have generally paralleled one another, indicating that these movements reflect climatic changes throughout the Northern Hemisphere.

Studies of annual growth rings in ancient trees provide an independent record of climatic conditions that generally supports the record of glaciers and tree lines (Figure 20-9A). The trunks of trees expand between the bark and preexisting wood. In nontropical areas, which have a distinct warm growing season, tree growth is rapid early in this season, when moisture is abundant; at this time large, thin-walled cells form. Growth slows in the late summer and fall, when moisture is less abundant; at this time small, thick-walled cells form. The result is one wide and one narrow ring of growth each year (see Earth System Shift 15-1). When climates become warmer, if they do not also become very dry, the growing season lengthens and growth accelerates; wider growth rings result. Thus trees provide a record of climatic warming and cooling. The bristlecone pine trees of western North America have exceptionally long life spans and therefore provide exceptionally lengthy records of climatic change (Figure 20-9B). Rings counted in small cores extracted from the trunk of the oldest known bristlecone, aptly named Methuselah, show it to be about 4600 years old. This tree is probably the oldest living plant on Earth.

Figure 20-9 Annual growth rings reveal the climates in which very old trees have grown.
A. Annual growth rings in a cross section of the trunk of a Ponderosa pine. B. A very old bristlecone pine
in the White Mountains of California. (A, Tom Bean; B, Anna E. Zuckerman/Tom Stack & Associates.)

Rings measured in cores from bristlecone pines growing near the tree line in the White Mountains of California have yielded estimates of temperature changes for an interval extending back more than 3400 years (Figure 20-10). Although regional climatic variations are to be expected, the overall pattern is in remarkable agreement with that derived from elevations of tree lines and glaciers in both Alaska and Europe. These three independent kinds of evidence reveal four intervals of marked global cooling during the past 6000 years.

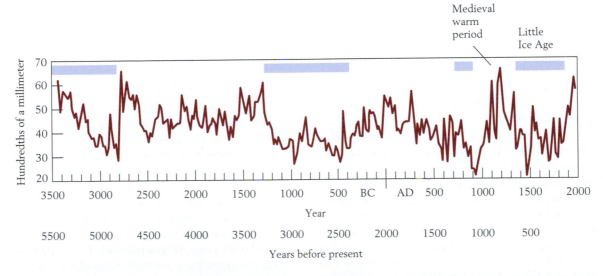

Figure 20-10 Cold intervals of the past 5500 years recorded by widths of annual growth rings in bristlecone pines near the upper tree line of the White Mountains of California. (Data from V. C. La Marche, in H. H. Lamb, *Climate History and the Modern World*, Routledge, London, 1995.)

Figure 20-11 *Hunters in the Snow.* Pieter Brueghel painted this scene in 1575, late in the first severe winter that Holland experienced during the Little Ice Age. (Brueghel, Pieter the Elder, from the series of paintings of "The Seasons," oil on oakwood, 1565. Erich Lessing/Art Resource.)

Temperatures have fluctuated since the hypsithermal interval

About 5800 years ago the hypsithermal interval gave way to the first major cold interval since the retreat of continental glaciers. Fossil pollen reveals that the white pine migrated northward in eastern North America at this time. Similarly, the lodgepole pine spread northward in the Pacific Northwest. This cold interval ended about 4900 years ago. A second such interval began about 3300 years ago and ended almost 2400 years ago (see Figure 20-10). A third cold interval extended from about AD 700 to AD 900. After this third pulse of cooling came a warm interval. This so-called *Medieval warm period* permitted the Norse Vikings to flourish in the northern Atlantic region, plundering other countries by sea between about AD 700 and AD 1200. In AD 985 they established an outpost in western Greenland and raised cattle and herded sheep there, where frigid conditions prevent such activities today. The climate began to cool in the thirteenth century, however, and sometime close to AD 1500, after sea ice isolated Greenland from Europe, the Norse colony died out.

The climatic deterioration that destroyed the Norse colony in Greenland marked the beginning of a long, cold interval that lasted until about 1850. During this so-called *Little Ice Age*, glaciers expanded along mountain valleys not only in North America and Europe but also in New Zealand. New England and northern Europe suffered bitter winters and short summers that produced major crop failures. Potatoes replaced wheat on many European farms because they require a shorter

growing season. Figure 20-11 is a scene painted in February 1575, during the first of many severe winters that descended on Holland over a 200-year period. George Washington and his shivering troops at Valley Forge in 1777 and 1778 were actually fortunate to have experienced a winter that was relatively mild for the times.

A global warming trend occurred during the first half of the nineteenth century. A warm, moist summer in 1846 favored the spread of the potato fungus in Ireland and led to the famine that caused many deaths and the emigration of many Irish people to North America. The Little Ice Age came to an end about this time, and Earth's average annual temperature remained relatively stable for the rest of the nineteenth century.

Severe droughts have occurred during Holocene time

Moisture conditions, like temperatures, have varied greatly in many regions during the Holocene. Fossils of lake-dwelling life, along with pollen blown into lakes, reveal four intervals when African lake levels were low, as a result of some combination of reduced rainfall and increased evaporation.

Pictures taken from satellites reveal that numerous dunes, now stabilized by vegetation, were active near the Colorado-Nebraska border at four times within the past 10,000 years (Figure 20-12). Exactly when these dunes were active is not well established, but the most recent activity was less than 1000 years ago. At that time the climate must have been too dry to support the short prairie grass that grows in the area today.

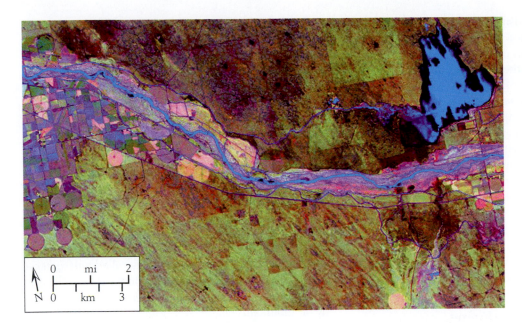

Figure 20-12 Satellite image showing ancient dunes that are now stabilized, adjacent to the Platte River near Greeley, Colorado. These dunes, visible in the lower portion of the image, were active earlier in the Holocene, when climates in this region were drier than they are today. (Image courtesy of Roberta H. Yuhas, CSES/CIRES, Univ. of Colorado at Boulder, © 1997, all rights reserved.)

Analyses of annual growth rings of pines on the western slope of the Sierra Nevada indicate that during the past 1000 years California has seldom received as much precipitation as it does today. Narrow rings during the growing season indicate dry conditions. Mono Lake, which occupies this region of the Sierra, was lowered in 1940, when the city of Los Angeles began to divert streams that flow into it. The lake's decline exposed ancient stumps that are the remnants of trees that grew on dry land when the climate was drier than it is today and the lake level was very low. Radiocarbon dating has shown that some of those trees died about AD 1112 and others about AD 1350. Both groups of trees had grown at the site for more than 50 years, while the climate remained very dry.

The recent history of climatic changes in both California and the Great Plains serves warning to present-day inhabitants of these regions. California is already suffering from a water shortage and will face severe problems if its climate returns to its more customary Holocene state. Similarly, it is possible that the Great Plains will again experience climates like those that a few centuries ago made them even drier than they were in the 1930s, when drought turned them into the so-called Dust Bowl.

Sea Level

Although the melting of glaciers after the last glacial maximum caused sea level to rise dramatically (see Figure 20-4), the pattern of changes in sea level in relation to land has varied from place to place. Local shorelines have shifted not only because of the eustatic rise, but also because of regional elevation and subsidence of the land. Thus, even at a time of eustatic rise, sea level has declined in relation to the land in some regions: in other words, the shoreline has regressed (see Figure 6-20).

Sea level rose rapidly in the early Holocene

The initial Holocene rise of sea level by more than 100 meters as continental glaciers melted had profound effects along continental margins. Between about 8500 and 7600 years ago shallow-water reefs became established off Florida and many Caribbean islands at depths of 30 to 15 meters below present sea level. These reefs died abruptly about 7600 years ago, when sea level rose further, flooding broad areas behind them (see Figure 20-4). Settling of suspended sediment from the newly formed shelves behind the reefs may have killed them. In any event, the existing reef tracts in Florida and the Caribbean became established when sea level approached its present level, slightly after 7600 years ago.

Rising seas also flooded broad river valleys that had incised continental shelves when sea level dropped more than 100 meters at times of glacial expansion. Sedimentation has not totally filled these drowned river valleys, so that today they form broad estuaries such as Chesapeake, Delaware, and Mobile bays.

In the west, small, steep rivers that drain the Coast Ranges of California cut downward into the narrow continental shelf to the west during the last glacial maximum. The Holocene eustatic rise turned these valleys into estuaries by flooding them and causing them to partially fill with sediment. Tectonic subsidence has accentuated the flooding of some of these estuaries, including San Francisco Bay.

In northern regions, rising seas invaded broad valleys that had been widened by glaciers. The resulting fingerlike inland extensions of the sea, known as fjords, are best developed along the coast of Norway.

The sea level curve for Barbados (see Figure 20-4) was produced under the assumption that the island has been undergoing tectonic uplift at a constant rate. To the degree that this rate has varied, the curve is distorted. Attempts to produce eustatic curves for other regions suffer from similar uncertainties. In fact, many experts believe that sea level rose to virtually its present level by about 7000 years ago, when nearly all of the meltwater of continental glaciers had been released. These researchers believe that no significant eustatic change has occurred since that time, only regional changes in relative sea level that have resulted from elevation or subsidence of coastal zones.

Coastlines have shifted during the past 7000 years

The distribution of glaciers during the last glacial maximum has contributed to regional differences in relative sea level change during the last several thousand years. Where large glaciers stood 20,000 years ago, their enormous weight depressed the continental lithosphere, and since their retreat this lithosphere has been rebounding (Figure 20-13). As a result, some coastlines of the North Atlantic Ocean, including those of Scotland and much of Scandinavia, are still rising, and the seas are regressing from the land.

Ironically, in other areas, Earth's surface is subsiding because ice sheets have melted away. These areas were elevated when large glaciers grew nearby. The elevation occurred because as Earth's surface subsided beneath the weight of a large ice sheet, material of the lithosphere squeezed out in all directions. This material accumulated around the margins of the ice sheet, where it elevated the crust to form what is called a *peripheral bulge*. (Note that such a bulge is comparable to the one that can form in front of the foreland basin produced by a mountain chain; see Figure 9-11.) Peripheral bulges that formed during the last glacial maximum are still subsiding as a result of the disappearance of the ice sheets. One of these bulges is now subsiding in southern Great Britain (see Figure 20-13B). Another passes

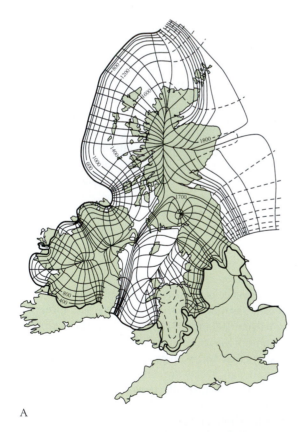

A

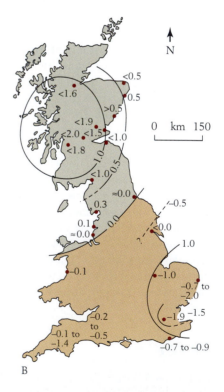

B

Figure 20-13 Effects of the weight of glaciers on elevations in Great Britain. *A.* Reconstruction of the surface topography of glaciers, in meters above present sea level, during the most recent glacial maximum. *B.* Estimated current rates of vertical crustal movement, in millimeters per year. The northern region, having been depressed by glacial ice, is rebounding; the southern region, having been elevated by the glacier to the north as a peripheral bulge, is sinking. (*A* after G. S. Boulton et al., in F. W. Shotton, ed., *British Quaternary Studies: Recent Advances*, Clarendon Press, Oxford, 1971; *B* after I. Shennan, *J. Quaternary Sci.* 4:77–89, 1989.)

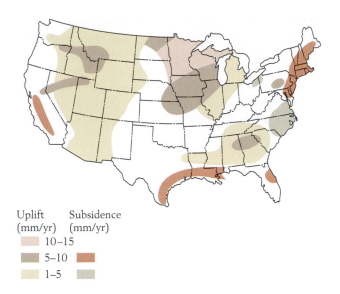

Uplift   Subsidence
(mm/yr)   (mm/yr)
       10–15
       5–10
       1–5

Figure 20-14 Rates of uplift and subsidence of Earth's crust in the United States today. (After S. P. Hand, National Oceanic and Atmospheric Administration.)

along the northeastern margin of North America, having formed southeast of the Laurentide glacier. Because of the subsidence of this bulge, the sea is transgressing over the land from Maine to Delaware (Figure 20-14). As the coast of New England has subsided, intertidal marshes that became established there about 7600 years ago have transgressed inland. The rate of transgression has been decreasing, however, as the subsidence of the land has slowed.

Forces unrelated to glaciers also cause vertical movements of coastal terrain. For example, tectonic uplift has probably augmented glacial rebound in elevating land in the vicinity of Seattle, Washington. Here gla-

cial sediments deposited in shallow seas during the final retreat of glaciers from Puget Sound now stand about 50 meters above sea level. In contrast, subsidence of the Gulf Coast in the region of the Mississippi delta is producing a transgression. This region has been subsiding for millions of years under the weight of the sediment accumulating there (p. 116).

We have seen that humans have accelerated the rate at which sea level is rising in relation to the land in the vicinity of the Mississippi delta (see p. 116). First, by extracting water from wells, humans are accelerating the rate at which the coastal plain subsides. Second, by damming tributaries of the Mississippi, they have reduced the influx of river-borne sediment to marginal marine environments. A stronger influx of sediments would at least partly offset the effect of subsidence.

Along the coast of New Jersey, Holocene subsidence of the peripheral bulge that formed south of the Laurentide ice sheet has caused barrier island–lagoon complexes to transgress over older sediments of the Coastal Plain Province (Figure 20-15A). Rivers here deposit too little sand to build a gentle shore face in front of the barrier islands. Instead, storm waves scour the seafloor, creating a steep shore face and destroying the transgressive record of lagoon deposits. In other words, processes of destruction are winning the battle over processes of accumulation.

In contrast to the transgressive pattern along the New Jersey coast, a regressive pattern is evident along Galveston Island in Texas (Figure 20-15B). Here radiocarbon dates from mollusk shells obtained from boreholes show that a regression has occurred during the Holocene. A high rate of sand supply from the Mississippi Delta region has caused this barrier island to

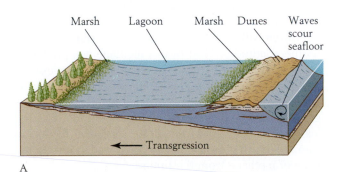

A

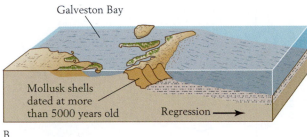

B

Figure 20-15 Movement of barrier island-lagoon complexes during the Holocene after sea level approached its present position. *A.* In New Jersey the complex is shifting inland because the land is subsiding, having been elevated as part of the peripheral bulge during the most recent advance of the Laurentide glacier; with little sediment being supplied, the sea is transgressing over the land. *B.* Galveston Island, Texas, is positioned in a region that has

not been subsiding and has been receiving an abundant supply of sediment; it has been prograding seaward for more than 5000 years, as indicated by radiocarbon dating of mollusk shells. (*A* after A. G. Fischer, *AAPG Bull.* 45:1656–1666, 1961; *B* after H. A. Bernard and R. J. LaBlanch, in H. E. Wright and D. G. Frey, eds., *Quaternary Geology of the United States*, Princeton University Press, Princeton, NJ, 1965.)

expand seaward by about 4 kilometers (2.5 miles) during the past 5300 years, while relative sea level has changed very little.

The Twentieth and Twenty-First Centuries: The Impact of Humans

Earth's average temperature has risen slightly in recent years, and most experts believe that at least part of this increase is a result of human burning of fossil fuels, which has oxidized carbon and released the greenhouse gas CO_2 into Earth's atmosphere. Further human-induced greenhouse warming may have profound consequences for habitats and life around the globe.

Human activities cause greenhouse warming

Figure 20-16 shows the buildup of CO_2 in the atmosphere since the acceleration of the Industrial Revolution in the nineteenth century. The data for the interval before 1957 are measurements of CO_2 trapped in Antarctic ice and extracted from cores, and those for the interval since 1957 are measurements of CO_2 taken directly from air samples. At present, the rampant destruction of forests, especially in the tropics, may be contributing about half as much CO_2 to the atmosphere as the burning of fossil fuels. Trees are often burned during deforestation, or if they are cut and removed, plant debris that remains is burned. Even the rotting of abandoned plant debris releases CO_2 through respiration (p. 226).

Only about half of the CO_2 produced by human activities ends up in the atmosphere. Roughly a quarter

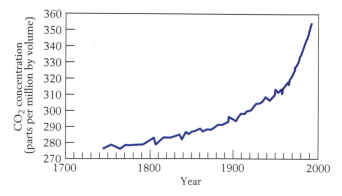

Figure 20-16 The concentration of carbon dioxide in the atmosphere since 1700. (After R. T. Watson et al., in J. T. Houghton, G. J. Jenkins, and J. J. Ephraums, eds., *Climate Change: The IPCC Scientific Assessment*, Cambridge University Press, 1990.)

of it ends up in the oceans, where part of it remains as dissolved CO_2 and part is converted to biomass through photosynthesis by phytoplankton. The remaining quarter of the CO_2 generated by humans is probably turned into plant biomass through the expansion of forests. These forests may actually benefit from the increased atmospheric concentration of CO_2, a raw material for their production of sugars.

Has the increase of atmospheric CO_2 elevated global temperatures significantly during the past century? Since 1920 the average temperature at Earth's surface has increased by about 0.7°C (Figure 20-17A). Most experts believe that much of the sharp increase since 1980 is the result of humans' addition of CO_2 to Earth's atmosphere. In some places, such as the upper reaches of Mount Kilimanjaro, in equatorial Africa, recent global warming is evident in the shrinkage of glaciers (Figure 20-18).

Figure 20-17*B* displays the results of a computer model devised to estimate the temperature effects of carbon dioxide emissions resulting from human activities. Predicted future temperatures shown in this figure are based on the premise that civilization will exercise no controls over future CO_2 emissions. Figure 20-17*B* displays several curves, which are based on slightly differing values for parameters used in the model. The estimates for the increase in mean global temperature over the average for 1961–1990 by the end of the twenty-first century range from about 2°C to 5°C.

Tundra can release methane, a powerful greenhouse gas

Carbon dioxide in not the only greenhouse gas in Earth's atmosphere. In fact, as a greenhouse gas, methane is about 20 times as powerful as carbon dioxide. The low concentration of methane in the present atmosphere makes it an unimportant greenhouse gas, but its role could increase in the future. Huge quantities of methane are frozen within soils deep beneath the tundra at high northern latitudes. Most of this methane was produced by bacteria in marshy environments during Arctic summers. If future global warming releases enough methane gas from the tundra, this gas may add significantly to the greenhouse warming caused by carbon dioxide. This release of methane could constitute a positive feedback, warming Earth's climate, which could lead to further release of methane from the tundra.

Future climatic change will have serious consequences

For people now living in regions with long, cold winters, the prospect of a warmer climate may seem welcome. Unfortunately, such a simplistic view ignores many likely consequences of future global warming.

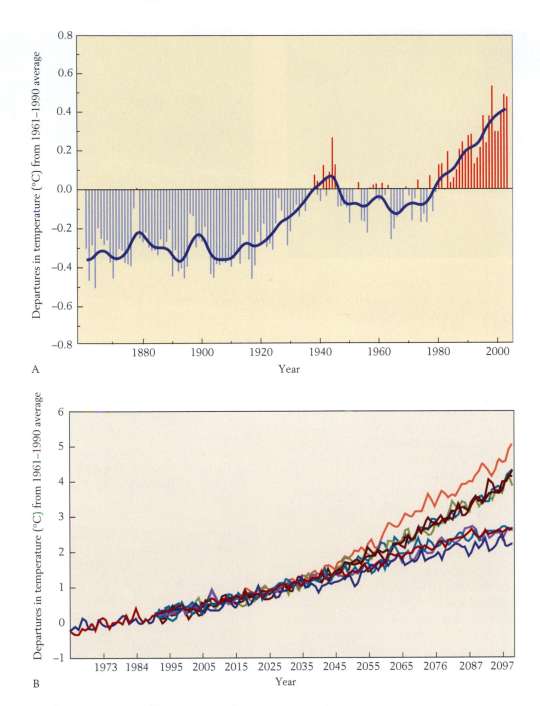

Figure 20-17 Increase in global average temperature. *A.* Increase in average global temperature since 1920, illustrated by using the average value for the 1961–1990 interval as a standard for comparison. *B.* Estimates of the global warming that will occur during the twenty-first century if humans make little effort to reduce carbon dioxide emissions. The different curves were produced from global circulation models for the ocean and atmosphere for which inputs were varied slightly. (Hadley Centre.)

Migration As we have seen, the fossil record of pollen shows that when climates change rapidly, communities of plants and animals are reshuffled (p. 518). Thus future global warming will yield new communities of plants and animals. Some populations will suffer from

a diminished supply of resources or from exposure to new competitors, predators, or diseases. Even simple dislocations of plant and animal populations will be problematic. Some people may be pleased if optimal conditions for certain crops shift to their region, but

A

B

Figure 20-18 Reduction of the ice cap on Mount Kilimanjaro between February 1993 (A) and February 2000 (B). This volcanic mountain stands close to the equator in Africa. (NASA.)

what of the farmers in other regions who can no longer depend on those crops for their livelihood?

Temperature changes alone will damage some ecosystems. Although coral reefs require warm seas, they cannot tolerate extreme warmth. If tropical waters heat up by 2°C or 3°C, the warmest regions of the Pacific will become inhospitable to reef-building corals. Reefs may expand to other regions that were previously too cool to support them, but they will require centuries to become well established.

Water supply in terrestrial environments On land, global warming will create severe problems of water supply for both natural and human communities. Scientists have offered widely varying predictions of the geographic patterns of increased and decreased wetness that will result from global warming. These predictions are based in part on patterns of wetness inferred for the hypsithermal interval of the early Holocene (p. 525). Measures of wetness (and dryness) take into account not only the amount of precipitation but also the rate of water loss through evaporation and through transpiration by plants (p. 19). Despite many uncertainties, we can make three generalizations about wetness and global warming:

1. Many areas that derive most of their precipitation from nearby oceans will become wetter because warm seas supply more water through evaporation than cold seas. Thus the West Coast and Gulf Coast of North America may become wetter.

2. Monsoonal rains will increase because seasonal warming of the land, which draws them inland, will

intensify (p. 476). As a result, southern India will experience even heavier torrential rains during summer months than it does today.

3. Many areas in the middle of large continents will become drier, primarily because higher temperatures will cause rates of evaporation to increase. Thus farmers in the American Midwest may face hard times.

Increased evaporation will make some semiarid (dry grassland) regions even drier. *Desertification* of this kind is accelerated by a positive feedback: when plants begin to disappear, erosion increases, and moisture is lost through reduction of plant cover and soil. Thus the system becomes even drier, and still more vegetation disappears. Figure 20-19 identifies the semiarid regions of the present world that will be susceptible to desertification. One such region is southern Africa, which has been exceptionally warm since 1980 and has experienced two major droughts in the 1990s. This association suggests that dry conditions will accompany global warming in southern Africa. Increased aridity would place human populations there in jeopardy because they already suffer from a shortage of water.

Increased CO_2 will affect plants

Future increases in atmospheric CO_2 will actually "fertilize" plants because they use this compound to manufacture sugars. Recall, however, that C_3 plants benefit more than C_4 plants from increased CO_2 levels (p. 476). Among domestic plants, wheat and rice are C_3 plants, whereas corn is a C_4 plant. Wheat and rice might therefore be

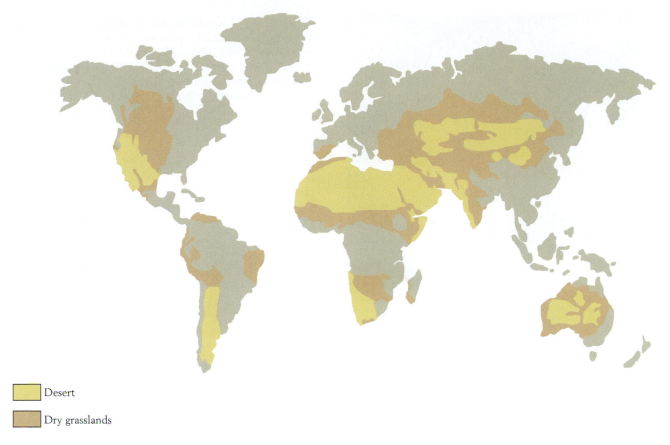

Desert

Dry grasslands

Figure 20-19 **Dry regions of the modern world.** The dry grasslands are in danger of future desertification.

expected to prosper as atmospheric CO_2 rises, but the situation is not so simple. As C_3 plants, wheat and rice will not tolerate the hot, dry conditions that will result from global warming in many areas as well as corn will. Obviously, predictions of the biotic effects of greenhouse warming are fraught with uncertainties.

Sea level will change

The amount of warming depicted in Figure 20-17 would cause a eustatic rise of about 50 centimeters (20 inches) by the year 2100. This figure compares to an estimated rise of 4–10 centimeters during the past 100 years. As we have seen, the relative rise in sea level would vary from place to place (p. 530). Melting of mountain glaciers would contribute a significant portion of the increase in the volume of the ocean, but an even larger amount would result from the expansion of the ocean as it heats up slightly. Surprisingly, partial melting of ice caps may contribute little to the eustatic rise. The Greenland ice cap may shrink only slightly because, while more snow may melt during summer, higher rates of evaporation from warmer seas may produce more snow in winter.

A predicted increase in snowfall for the south polar region may actually expand the Antarctic ice cap (Figure 20-20A). On the other hand, increased ice flow to the ocean (Figure 20-20B) may diminish the Antarctic ice cap. The ice cap that covers 99 percent of Antarctica today contains 90 percent of the world's ice. If all of this frozen water were to melt, global sea level would rise by more than 60 meters (197 feet). Were future global warming to melt away even a small fraction of the Antarctic ice, many cities and towns would be flooded.

The Antarctic ice cap, which is in places more than 4 kilometers (about 2.5 miles) thick, is divided into the east and west Antarctic ice sheets. Each of these sheets is dome-shaped, and ice flows outward to the sea from the elevated center along ice streams at rates as high as a few kilometers per year. The flowing ice forms shelves that float in bays along nearly half the Antarctic coastline.

The ice shelves are frozen to the rocky sides of the bays in which they reside, and their undersurfaces are also attached to bedrock. Because the ice shelves are attached to the land, they resist the flow of the ice streams that feed them. Ice nonetheless flows through the central regions of the ice shelves, and ice-

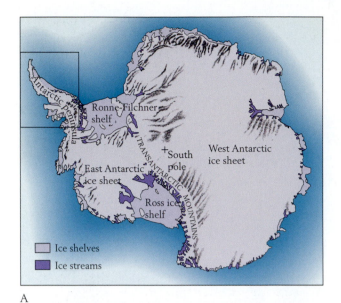

A

B

Figure 20-20 Ice sheets and ice shelves of Antarctica. *A.* The distribution of ice shelves and the ice streams that feed them. *B.* An ice shelf releasing icebergs. *C.* Loss of ice shelves since 1989. The dates indicate when particular ice shelves completely disintegrated. (*B*, Tom Brakefield/The Stock Market; *C*, J. Kaiser, *Science* 297:1494–1495, 2002.)

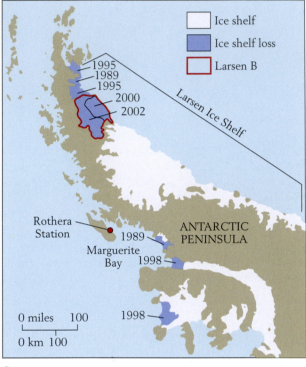

C

bergs sporadically break from them and float away. Thus Antarctica is continually losing glacial ice. On the other side of the ledger, it gains ice through the accumulation of snow. Antarctica amounts to a frigid desert, however, receiving an average of only about 1 meter (3 feet) of snow per year. One giant iceberg can carry away the equivalent of more than a year's precipitation for the entire continent! The question is: What is happening, and what will happen in the next few decades, to rates of addition and subtraction of Antarctic ice?

During the past 50 years, five small ice shelves have retreated dramatically along the Antarctic peninsula, which extends toward South America. Increased temperatures have recently caused the disintegration of these small Antarctic ice shelves through massive discharge of icebergs. Since 1957, the average temperature in Antarctica has risen by nearly one-quarter of 1°C per decade. During this time, the areas where ice shelves have disintegrated, being relatively far from the

south pole, have become too warm to sustain ice shelves. The Larsen ice shelf is especially vulnerable to global warming because it lies exposed far from the south pole along the narrow peninsula (Figure 20-20C). This ice shelf has shrunk back from the tip of the peninsula in steps since 1989. In 2002 the Larsen B ice shelf, which was the size of Rhode Island, fragmented and disintegrated within just 3 weeks.

Warming not only melts the surfaces of ice shelves, but also weakens them and increases the rate at which they release icebergs. Might future warming from an enhanced greenhouse effect greatly diminish the volume of Antarctic glacial ice and elevate sea level? We know that ice shelves have been lost in the past: fossil marine diatoms beneath the Ross ice shelf reveal that it collapsed sometime during late Pleistocene time, and that the place where this shelf now rests became covered by a shallow bay. The loss of ice shelves alone, however, would have little effect on sea level because

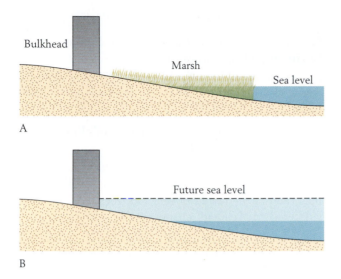

Bulkhead

Marsh

Sea level

A

Future sea level

B

◀ **Figure 20-21** A bulkhead built slightly landward of a coastal marsh at a time when sea level is rising. After sea level rises from its initial position (A), the presence of the bulkhead eliminates the marginal marine environment where the marsh must grow (B).

floating ice is, in effect, part of the ocean even before it melts. The key question is, might warming diminish the size of the Antarctic ice sheets, releasing stored water to the ocean?

The East Antarctic ice sheet, being smaller and centered close to the ocean, appears to be the more vulnerable to melting. The Ronne-Filcher and Ross ice shelves hold back the East Antarctic ice sheet, however. They are each the size of France and lie relatively close to the south pole, so that a very large amount of warming would be required to destabilize them. Furthermore, while global warming will increase the rate at which ice shelves shed icebergs, it may compensate by increasing the supply of snowfall to glacial ice. Warmer temperatures will increase the rate of evaporation from the sea surface adjacent to Antarctica, and the result-

ing increase in atmospheric water vapor will result in more snow.

The truth is, we do not know whether global warming will shrink the Antarctic ice cap and elevate sea level substantially or whether it will instead expand the ice cap by producing a large increase in regional snowfall. Our uncertainty about the future effects of Antarctic ice on sea level is more than a little unsettling.

Even a 1-meter rise of sea level would endanger the homes of about a billion people, as well as one-third of the world's crop-growing areas. Many of the world's wetlands would be lost because obstructions that humans have emplaced, including landfills and bulkheads, would prevent coastal marshes from migrating inland with shorelines (Figure 20-21). Rising seas during storms would pose the greatest threat. Some low-lying coastal cities, including the architecturally rich Italian city of Venice, would be destroyed (Figure 20-22). The country of Bangladesh, about half of which is already within 5 meters (about 16 feet) of sea level, would experience even more frequent disastrous floods than it has suffered in recent years.

Recall that an ice core from the Greenland glacier has shown that when Earth's climate is warming, it can shift toward much warmer conditions in a single year—for reasons we do not understand. Clearly, humankind must make every effort to curb our generation of greenhouse gases that enter Earth's atmosphere.

Figure 20-22 Flooding in Venice, Italy, during a very high tide in 1990.
(Luigi Tazzari/Gamma Liaison.)

Chapter Summary

Although we still live in the modern Ice Age, Earth has emerged from the most recent glacial maximum; when did this happen?

Earth emerged from the last glacial maximum slightly after 15,000 years ago.

Did Earth move directly from the last glacial maximum to the present glacial minimum?

No, annual layers in the Greenland ice sheet show that Earth began to emerge from the last glacial maximum, but then continental glaciers expanded again for more than a thousand years, during what is called the Younger Dryas interval. Only after this return of glacial conditions did Earth move fully into the present glacial minimum.

Did species that form modern plant communities evolve together to form well-integrated associations?

During the most recent shrinkage of continental glaciers, plant species migrated independently of one another, as happened throughout the modern Ice Age. Thus modern plant communities are assemblages of species that did not evolve in association with one another but have assembled in response to environmental changes during the past 20,000 years.

Have climatic changes been gradual during Holocene time?

No, some dramatic global climatic changes have occurred within just a few years, and many of them have been rapidly reversed.

When did humans migrate from Eurasia to North America?

Humans may have reached North America as early as 30,000 years ago, but large populations, characterized by advanced hunters, did not arrive until slightly before 11,000 years ago.

What may have caused numerous species of large mammals to disappear from North America between 12,000 and 11,000 years ago?

Human hunters may have contributed to their extinction, but so may climatic changes associated with the Younger Dryas cold interval.

How did climatic conditions permit humans to begin engaging in agriculture?

Agriculture spread rapidly throughout Europe when climates warmed at the end of the Younger Dryas cold interval.

How do tree rings provide records of climatic change?

Trees generally grow more rapidly when the climate warms, lengthening the growing season. Widths of annual rings in the trunks of very old trees therefore provide a record of climatic changes over the last 3400 years.

What evidence is there of major climatic change between the fourteenth and nineteenth centuries?

The Medieval warm period was followed by the Little Ice Age, which began after AD 1300 and lasted until about AD 1850. Tree rings record this transition.

Why is there reason to worry that, through natural climatic changes, the Great Plains and western North America may become much drier sometime soon?

At many times during the past 10,000 years, climates in these regions have been more arid than they are today.

Why may shorelines shift, even without human influence, in the near future?

Global sea level approached its present position about 7000 years ago. Since that time, positions of coastlines have shifted substantially because of regional elevation or subsidence of land resulting from glacial rebound, subsidence of peripheral bulges that formed around glaciers, or subsidence due to the accumulation of sediments.

How will human-induced global warming affect soil moisture in the interior regions of continents?

By increasing evaporation rates, future global warming will create drier soil conditions in continental regions far from the ocean.

How will changes of sea level cause problems for humans if global temperatures continue to rise?

Melting of glaciers will elevate sea level, posing problems for coastal populations.

Review Questions

1. How did the Great Lakes form?

2. How does fossil pollen indicate that modern plant communities are temporary associations of species?

3. How do fossil corals record Holocene sea level changes?

4. What kinds of large mammals disappeared from North America between 12,000 and 11,000 years ago?

5. What evidence favors the idea that human hunters contributed to the extinction of the large mammals?

6. What evidence favors the idea that climatic changes contributed to the extinction of the large mammals?

7. How does glacial ice in Greenland that is thousands of years old provide a record of climatic change?

8. Why is it not surprising that agriculture spread from the Middle East to Europe, rather than in the opposite direction?

9. What evidence is there that at times during the last 10,000 years climates in many areas have been drier than they are today?

10. What conditions have affected relative sea level in particular regions during the past 7000 years, since global sea level approached its present position?

11. Global environmental change is of great concern in the world today. Using the Visual Overview on page 516 and what you have learned in this chapter, review the causes and effects of changes in global climate and sea level that have occurred during the past 14,000 years or that may occur during the next few centuries.

APPENDIX

Stratigraphic Stages

In many parts of the world the geologic record has been divided into stages. As discussed in Chapter 6, stages are time-stratigraphic units. For the most part, the stages recognized in Europe have become the standard stages with which stages defined elsewhere are correlated. Correlations remain imperfect, however, as do estimates of the absolute ages of stage boundaries. This appendix is a reference for students who encounter unfamiliar stage names in their studies. Figure 1 lists major Paleozoic and Mesozoic stages that were first defined in Europe and shows how a number of North American stages are currently believed to correlate with them. Figure 2 presents the same kind of information for Cenozoic stages, showing how European stages are thought to relate to American stages that are based on biostratigraphic zones for fossil land mammals.

TIME (million years ago)		SYSTEM		SERIES	STAGE (European)	STAGE (North American)
CENOZOIC			QUATERNARY	PLEISTOCENE		
		NEOGENE		PLIOCENE		
				MIOCENE		
	24		TERTIARY	OLIGOCENE	(See Figure 2)	
		PALEO-GENE		EOCENE		
				PALEOCENE		
MESOZOIC	65	CRETACEOUS		UPPER	Maastrichtian	
					Campanian	
					Santonian	
					Coniacian	
					Turonian	
					Cenomanian	
				LOWER	Albian	
					Aptian	
					Barremian	
					Hauterivian	
					Valanginian	
	142				Berriasian	
		JURASSIC		UPPER	Tithonian	
					Kimmeridgian	
					Oxfordian	
				MIDDLE	Callovian	
					Bathonian	
					Bajocian	
					Aalenian	
				LOWER (LIAS)	Toarcian	
					Pliensbachian	
					Sinemurian	
	202				Hettangian	
		TRIASSIC		UPPER	Rhaetian	
					Norian	
					Carnian	
				MIDDLE	Ladinian	
				LOWER	Anisian	
	250				Scythian	

TIME (million years ago)	SYSTEM	SERIES	STAGE (European)	STAGE (North American)	(SERIES)
	PERMIAN	UPPER	Tatarian	Ochoan	
			Ufimian/Kazanian	Guadalupian	
		LOWER	Kungurian	Leonardian	
			Artinskian		
			Sakmarian	Wolfcampian	
			Asselian		
292	CARBON-IFEROUS	PENNSYLVANIAN — UPPER	Stephanian	Virgilian	
			Westphalian	Missourian	
				Desmoinesian	
				Atokan	
				Morrowan	
320		MISSISSIPPIAN — LOWER	Namurian	Springerian	
			Visean	Chesterian	
				Meramecian	
				Osagean	
			Tournaisian	Kinderhookian	
354	DEVONIAN	UPPER	Famennian	Chautauquan	
			Frasnian	Senecan	
		MIDDLE	Givetian		
			Eifelian	Erian	
		LOWER	Emsian		
			Siegenian	Ulsterian	
			Gedinnian		
417	SILURIAN	UPPER	Ludlovian	Cayugan	
		LOWER	Wenlockian	Niagaran	
			Llandoverian	Medinan	(SERIES)
440	ORDOVICIAN	UPPER	Ashgillian		Champlanian
			Caradocian		
		LOWER	Llandeilian	Chazyan	
			Llanvirnian	Whiterockian	
			Arenigian		Canadian
			Tremadocian		
495	CAMBRIAN	UPPER	Dolgellian	Trempealeauan	Croixan
			Maentwrogian	Franconian	
				Dresbachian	
		MIDDLE	Menevian		Albertan
			Solvan		
		LOWER	Lenian		
			Aldabanian		Waucoban
			Tommotian		
			Nemakitian-Daldynian		
545					

PALEOZOIC

STAGE (North American)	(SERIES)
Richmondian	Cincinnatian
Maysvillian	
Edenian	
Trentonian	Mohawkian
Black River	
Ashbyan	

Detail of shaded area

Figure 1 Major Paleozoic and Mesozoic stages of Europe and North America. Through correlation and absolute dating, efforts are being made to extend the European stages to all parts of the world.

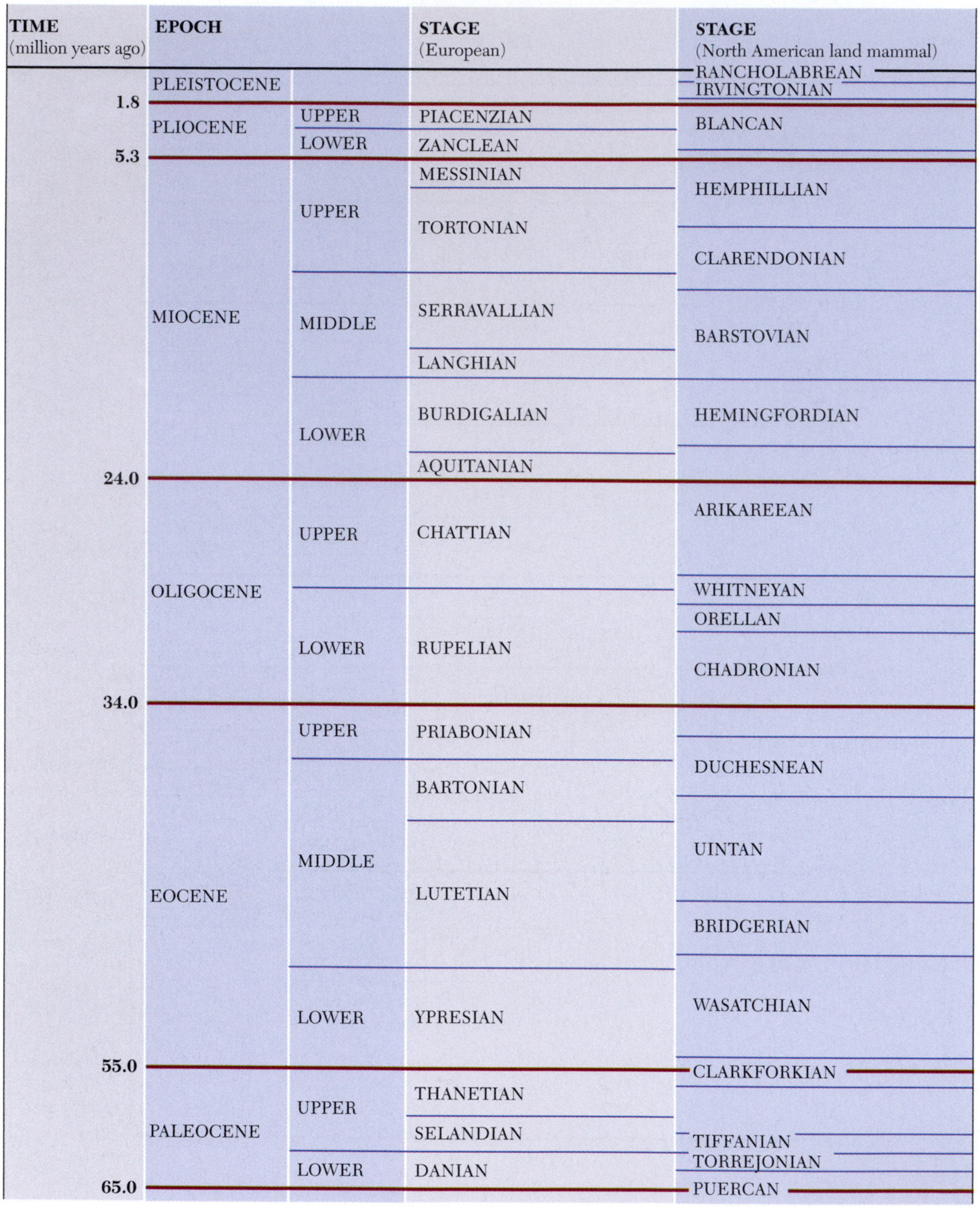

TIME (million years ago)	EPOCH	STAGE (European)		STAGE (North American land mammal)
	PLEISTOCENE			RANCHOLABREAN
				IRVINGTONIAN
1.8	PLIOCENE	UPPER	PIACENZIAN	BLANCAN
		LOWER	ZANCLEAN	
5.3	MIOCENE	UPPER	MESSINIAN	HEMPHILLIAN
			TORTONIAN	CLARENDONIAN
		MIDDLE	SERRAVALLIAN	BARSTOVIAN
			LANGHIAN	
		LOWER	BURDIGALIAN	HEMINGFORDIAN
			AQUITANIAN	
24.0	OLIGOCENE	UPPER	CHATTIAN	ARIKAREEAN
		LOWER	RUPELIAN	WHITNEYAN
				ORELLAN
				CHADRONIAN
34.0	EOCENE	UPPER	PRIABONIAN	DUCHESNEAN
		MIDDLE	BARTONIAN	
			LUTETIAN	UINTAN
				BRIDGERIAN
		LOWER	YPRESIAN	WASATCHIAN
55.0	PALEOCENE	UPPER	THANETIAN	CLARKFORKIAN
			SELANDIAN	TIFFANIAN
				TORREJONIAN
		LOWER	DANIAN	PUERCAN
65.0				

Figure 2 **Major Cenozoic stages of Europe and North America.** The North American stages, which are currently only crudely correlated with the European stages, are based largely on fossil occurrences of land mammals in the Midwest and West. The most recent epoch, not listed in this diagram, is known as the Holocene or Recent. This brief epoch began at the end of the Pleistocene, roughly 12,000 years ago.

Glossary

Terms used in these definitions that are also defined in this glossary are in many instances *italicized* for the reader's convenience.

Abyssal plain The broad expanse of seafloor lying between about 3 and 6 kilometers (about 2 to 4 miles) below sea level.

Acanthodians Small, elongate marine and freshwater fishes of middle Paleozoic age with jaws, numerous paired fins supported by sharp spines, and scales rather than bony plates.

Accretionary wedge A body of *rocks* that have accumulated above an oceanic *plate* undergoing *subduction*. Slices of *mélange* pile up along *thrust faults* to form the wedge.

Acritarchs An extinct group of apparently *eukaryotic phytoplankton* whose earliest representatives are in Proterozoic *rocks*.

Active lobe (of a delta) The site on a *delta* where functioning *distributary channels* cause the delta to grow seaward.

Active margin The border of a continent along which *subduction* occurs, producing *igneous* activity and deformation.

Actualism The interpretation of ancient *rocks* by applying the results of analyses of modern-day geologic processes in accordance with the principle of *uniformitarianism*.

Adaptation A feature of an organism that serves one or more functions useful to the organism.

Age The division of geologic time smaller than an *epoch*.

Albedo The percentage of solar radiation reflected from Earth's surface. That percentage is higher for ice than for land or water, and usually higher for land than for water.

Alluvial fan A low, cone-shaped structure that forms where an abrupt reduction in slope—for example, the transition from a highland area to a broad valley—causes a stream to slow down.

Ammonoids Cephalopod mollusks related to the modern chambered nautilus that were highly successful swimming predators in late Paleozoic and Mesozoic seas.

Amniote egg The type of egg laid by *reptiles* and *birds*, having a nutritious yolk and a hard outer shell to protect the embryo from the dry environment. The amniote egg is named for the amnion, the sac that contains the embryo.

Amphibians Vertebrate animals that hatch and spend their juvenile period in water, then usually metamorphose into air-breathing, land-dwelling adults, but return to water to lay their eggs. Among modern amphibians are frogs, toads, salamanders, and their relatives.

Angiosperm A plant that produces flowers or flowerlike structures and that produces covered seeds.

Angular unconformity An *unconformity* separating horizontal *strata* above from older strata that have been tilted and eroded.

Anhydrite The *mineral* that consists of calcium sulfate ($CaSO_4$), or the *rock* composed of that mineral.

Animalia A kingdom of multicellular *eukaryotes* whose members digest food outside their cells and then absorb the products.

Anomalocarids A group of Cambrian *arthropods* that propelled themselves through the water with flaps positioned along the body and impaled prey on daggerlike spines along the frontal appendages.

Anoxia The condition in which oxygen (O_2) is virtually absent from an environment.

Anoxic In a state of anoxia.

Anticline A fold that is concave in a downward direction—that is, the vertex is the highest point.

Aragonite A form of calcium carbonate that precipitates from watery solutions in nature and is secreted by some organisms to form a skeleton. Aragonite is an important *mineral* in many *limestones*.

Aragonite needles Slender crystals of the *mineral aragonite* that constitute most *carbonate muds* in the modern ocean. Some of the needles form by direct precipitation from *seawater* and some by the collapse of the skeletons of organisms, principally algae.

Archaebacteria "Old bacteria"; a group of *prokaryotes* that forms one of the six kingdoms of organisms.

Archaeocyathids Primary frame builders of Early Cambrian *organic reefs*, probably *suspension-feeding* sponges that pumped water through holes in their vase-shaped and bowl-shaped skeletons.

Arkose A *rock* consisting primarily of *sand*-sized particles of feldspar. Most arkose accumulates close to the source area of the feldspar because feldspar weathers quickly to *clay* and seldom travels far.

Arthropod Any of a variety of *invertebrates* (*insects*, spiders, *crustaceans*) with a segmented body and an external skeleton.

Asteroids Small, planetlike bodies of the solar system.

Asthenosphere The *ultramafic* layer of Earth lying below the *lithosphere*. The asthenosphere is marked by low seismic velocities, an indication that it is partly molten.

Atmosphere The envelope of gases that surrounds Earth.

Atoll A circular or horseshoe-shaped *organic reef* growing on a submerged volcano.

Axial plane An imaginary plane that cuts through a fold, dividing it as symmetrically as possible.

Axis of a fold The line of intersection between the *axial plane* of a fold and the *beds* of folded *rock*.

Backswamp A broad vegetated area that lies adjacent to a *meandering river* and becomes covered with water when the river overflows its banks.

Banded iron formation Complex *rock* that consists of *oxides*, sulfides, or carbonates of iron interlayered with thin beds of *chert*. Most rocks of this type are older than about 2 billion years.

Barrier island An elongate island composed of *sand* heaped up by ocean waves that lies approximately parallel to the shoreline.

Barrier island–lagoon complex The set of marginal marine environments that consists of a *barrier island*, the *lagoon* behind it, and (usually) *tidal flats*, *marshes*, and sandy beaches.

Barrier reef An elongate *organic reef* that parallels a coastline and is large enough to dissipate ocean waves, leaving a quiet-water *lagoon* on its landward side.

Basalt A fine-grained, *extrusive, mafic igneous rock*; the dominant rock of oceanic *crust*.

Bed A distinct layer (*stratum*) of *sediment* thicker than 1 centimeter.

Bedding The arrangement of a *sedimentary rock* into discrete layers (*strata*) thicker than 1 centimeter (*beds*).

Belemnoids Squidlike mollusks that were common in Mesozoic seas. These swimming predators balanced the buoyant effect of gas in the shell with a cigar-shaped counterweight.

Benthic life *See* **Benthos**.

Benthos The bottom-dwelling life of an ocean or freshwater environment.

Big bang The enormous explosion that created the expanding universe.

Biogenic sediment *Sediment* consisting of *mineral* grains that were once parts of organisms.

Biogeography The study of the distribution and abundance of organisms on a broad geographic scale.

Biomarker A distinctive organic compound preserved in the *rock* record that indicates the existence of a particular kind of organism.

Biostratigraphic unit A body of *rock*, such as a *zone*, defined on the basis of its *fossil* content and having approximately time-parallel upper and lower boundaries.

Biota The *flora* and *fauna* of an *ecosystem*.

Biozone *See* **Zone**.

Birds A class of *endothermic vertebrates* that evolved from *dinosaurs* and are characterized by feather-covered bodies and by forelimbs that have evolved into wings.

Bivalves A group of aquatic *mollusks* with shells divided into two halves (valves). Among this group are clams, mussels, oysters, and scallops.

Boulder A piece of *gravel* larger than 256 millimeters (about 10 inches).

Boundary stratotype An *outcrop* where the boundary between two *time-rock units* is formally recognized.

Brachiopods A group of marine *invertebrates* that have shells divided into two halves and a frilly loop-shaped structure that pumps water and sieves food particles from it.

Brackish Having a *salinity* lower than that of normal *seawater* and higher than that of *fresh water*, ranging from 30 to 0.5 parts salt per 1000 parts water.

Braided stream A stream that has many intertwining channels separated by bars of coarse *sediment*. Braided streams develop where sediment is supplied to the stream system at a very high rate—on an *alluvial fan*, for example, or in front of a melting *glacier*.

Breccia A *rock* that contains large amounts of angular *gravel* or *cobbles*.

Bryozoans A group of aquatic *invertebrates* that reproduce by budding and form colonies.

Burial metamorphism *Metamorphism* produced when *rocks* are buried so deep that they are exposed to temperatures and pressures high enough to change their mineralogy.

Calcareous nannoplankton Small, nearly spherical unicellular algae that secrete minute, overlapping, shieldlike plates of calcium carbonate that serve as armor against attackers.

Calcareous ooze Fine-grained sediment of the deep sea formed largely of skeletons of single-celled planktonic organisms consisting of calcium carbonate.

Calcite A form of calcium carbonate that precipitates from watery solutions in nature and is secreted by some organisms to form a skeleton. Calcite is an important *mineral* in many *limestones*.

Caliche Nodular calcium carbonate that accumulates in the layer of *soil* below the *topsoil* in warm climates that are relatively dry.

Carbonate mineral A *mineral* in which the basic building block is a carbon atom linked to three oxygen atoms. *Calcite*, *aragonite*, and *dolomite* are the most abundant carbonate minerals found in *sediments* and *sedimentary rocks*.

Carbonate mud An accumulation of fine-grained calcium carbonate formed when calcareous skeletons collapse or when calcium carbonate precipitates directly from shallow tropical seas; a major component of *sediments* that harden to form *limestone*.

Carbonate platform A marine structure that is composed largely of calcium carbonate and that stands above the neighboring seafloor on at least one of its sides.

Carbonate rock A *sedimentary rock* that consists primarily of *carbonate minerals*. The dominant mineral is nearly always either *calcite*, in which case the rock is *limestone*, or *dolomite*, in which case the rock is *dolostone*.

Carbonate sediment Unconsolidated *sediment* that consists primarily of *carbonate minerals*, usually *aragonite* or *calcite*.

Carbonization The process by which a residue of carbon is left on the surface of an *impression* of an organism after liquids and gases have escaped during fossilization.

Carnivore An animal that feeds on other animals or animal-like organisms; a *consumer* that feeds on other *consumers*.

Catastrophism The outmoded doctrine that sudden, violent, and widespread events caused by supernatural forces formed most of the *rocks* that are visible at Earth's surface.

Cell The smallest unit of life; a module that contains a genetic code and can replicate itself.

Cephalopods A group of marine *mollusks* (including squids, octopuses, and cham-

bered nautiluses) that pursue prey by jet propulsion, squirting water out through a small opening in the body.

Chalk Fine-grained, well-lithified limestone that consists largely of the plates of calcareous nannoplankton.

Chemical sediment *Sediment* created by precipitation of one or more inorganic materials from natural waters, sometimes as a result of evaporation.

Chemosynthesis The breakdown of simple chemical compounds within a cell for the production of energy.

Chert (flint) An impure *sedimentary rock*, often gray, that consists primarily of extremely small quartz crystals precipitated from watery solutions.

Chlorite A green, micalike *mineral* that occurs in *metamorphic rocks*, primarily in *schist*; it is a prominent constituent of Archean *greenstone belts*.

Chloroplast A body within a cell of a plant or plantlike protist that serves as the site of *photosynthesis* within the cell. Chloroplasts are apparently evolutionary descendants of *cyanobacteria* that became trapped in other single-celled organisms.

Chordate Any of a group of animals characterized by a notochord, a flexible rodlike structure that may develop into a vertebral column. *Vertebrates* are chordates.

Chromosome One of several elongate bodies in which *DNA* is concentrated within the nucleus of a cell.

Chron A formally recognized *polarity time-rock unit*.

Chronostratigraphic unit *See* **Time-rock unit**.

Circumpolar current The circular flow of water around Antarctica, resulting from the juncture of the southern segments of the three gyres in the oceans of the Southern Hemisphere.

Cladogram A diagram showing the evolutionary relationships of *taxa* as reconstructed from the distribution of derived traits.

Clast A fragment of rock produced by destructive forces.

Clastic wedge A wedge-shaped body of *molasse*.

Clay *Siliciclastic sedimentary* particles that are smaller than $\frac{1}{256}$ millimeter; also, a member of the clay *mineral* family, which includes *silicates* that resemble micas.

Cnidarians Aquatic *invertebrates* whose bodies consist of two layers of *tissue* separated by a jellylike inner layer, and which capture food with tentacles armed with stinging cells. Examples are jellyfishes, sea pens, and modern *corals*. Cnidarians reproduce sexually and also asexually, by budding.

Coal *Rock* formed by the low-grade *metamorphism* of stratified plant debris. Coal burns readily because organic carbon compounds account for more than 50 percent of its composition.

Cobble A piece of *gravel* measuring between 8 and 256 millimeters (about 10 inches).

Comet An icy body of the solar system smaller than a planet.

Community, ecological *Populations* of several *species* living together in a *habitat*.

Competition, ecological The condition in which two *species* vie for an environmental resource, such as food or space, that is in limited supply.

Components, principle of The principle that a body of *rock* is younger than any other body of rock from which any of its components is derived.

Conglomerate A *rock* that contains large amounts of rounded *gravel*.

Conifers Plants (pines, spruces, redwoods, and their relatives) whose seeds are exposed on cones rather than covered, like the seeds of flowering plants.

Conodonts Extinct fishlike vertebrates that left an extensive fossil record of their teeth.

Consumer An animal or animal-like organism that obtains nutrition by consuming the organic material of other forms of life.

Contact metamorphism Local *metamorphism* caused by an *igneous intrusion* that bakes nearby *rocks*.

Continental accretion The marginal growth of a continent along a *subduction zone* by mountain building or by addition of a *microplate*.

Continental drift The movement of continents with respect to one another over Earth's surface.

Continental rise A slightly elevated region along the base of the *continental slope*. The continental rise is formed of *sediment* transported down the slope, often by *turbidity currents*.

Continental shelf An extension of a continental landmass beneath the sea.

Continental slope The sloping submarine portion of a continent, extending from the *continental shelf* to the *continental rise* or the *abyssal plain*.

Convection Rotational flow of a fluid resulting from an imbalance in its density. Convection often occurs because the fluid below is heated and becomes less dense than the fluid above, or because the fluid above is cooled and becomes more dense than the fluid below.

Convective cell One of a number of rotational units believed to operate within Earth's *mantle* as a result of *convection*.

Cope's rule The generalization that body size tends to increase during the *evolution* of a group of animals.

Corals *Cnidarians* that secrete skeletons of calcium carbonate. Some form large colonies of interconnected individuals by budding.

Coral-strome reefs *Organic reefs* produced between Ordovician and Devonian time by tabulate and rugose corals and stromatoporoid sponges.

Cordaites A group of late Paleozoic *gymnosperms* that often reached 30 meters (100 feet) in height and formed large woodlands resembling modern pine forests.

Core (of Earth) The central part of Earth below a depth of 2900 kilometers. It is composed largely of iron and is molten on the outside, with a solid central region.

Coriolis effect The tendency of a current of air or water flowing over Earth's surface to bend to the right in the Northern Hemisphere and to the left in the Southern Hemisphere.

Correlation The procedure of demonstrating correspondence between geographically separated parts of a *stratigraphic unit*.

Covalent bond A chemical bond in which electrons are shared rather than exchanged.

Craton The portion of a continent that has not undergone *tectonic* deformation since Precambrian or early Paleozoic time.

Crinoids Sea lilies; a group of *echinoderms* that sieve food from the water with featherlike arms and pass it to a centrally positioned mouth with tube feet. Some species swim by waving their arms; others attach to the seafloor by a long, flexible stalk.

Crocodiles Marine *reptiles* that evolved in Triassic time as terrestrial animals but became adapted to aquatic environments.

Cross-bedding (cross-stratification) A *sedimentary structure* in which groups of *strata* lie at angles to the horizontal.

Cross-stratification See **Cross-bedding**.

Crust The outermost layer of the *lithosphere*, consisting of *felsic* and *mafic rocks* less dense than the rocks of the *mantle* below.

Crustaceans A group of *arthropods* distinguished by a head formed of five fused segments, behind which are a thorax and an abdomen formed of additional segments. Among the crustaceans are lobsters, shrimps, and crabs.

Crystalline rocks *Igneous* or *metamorphic rocks*.

Cyanobacteria Photosynthetic *prokaryotes* that originated in Archean time and that form *stromatolites*.

Cycadeoids A group of *gymnosperms* that were prominent in the Mesozoic Era.

Cycads A diverse group of *gymnosperms* whose few modern descendants superficially resemble palm trees.

Cyclothems *Sedimentary* depositional cycles that include *coal beds*. Most cyclothems are of Late Carboniferous (Pennsylvanian) age.

Declination The angle that a compass needle makes with the line running to the geographic north pole, reflecting the fact that the magnetic pole (to which the compass needle points) does not coincide with the geographic pole.

Deep-focus earthquake An earthquake produced by movements along or within a *subducted slab* of *lithosphere* more than 300 kilometers (about 190 miles) below Earth's surface.

Deep-sea floor The continental slope and abyssal plain.

Deep-sea trench An elongate depression of the *deep-sea floor* formed where one *lithospheric plate* is *subducted* beneath another.

Delta A depositional body of *sand*, *silt*, and *clay* formed when a river discharges into a body of standing water so that its current dissipates and drops its load of *sediments*. This structure takes its name from the Greek letter Δ, which it resembles in shape.

Delta front The submarine slope of a *delta* extending downward from the *delta plain*. The delta front is usually the site of accumulation of *silt* and *clay*.

Delta plain The upper surface of a *delta*, characterized by *distributary channels* and their *natural levees* and intervening swamps.

Deposit feeders Seafloor dwellers that extract organic matter from *sediment*.

Desert A terrestrial environment that receives less than about 25 centimeters (10 inches) of rain per year and consequently supports a low *diversity* of plants.

Diatoms Unicellular algae that secrete skeletons of opal in two parts, which fit together like the top and bottom of a petri dish.

Dike A sheetlike or tabular body of *intrusive igneous rock* that cuts upward through *sedimentary strata* or *crystalline rocks*.

Dinoflagellates Unicellular algae that employ two whiplike structures (flagella) for limited locomotion, but are transported chiefly by movements of the water in which they drift.

Dinosaurs A group of extinct terrestrial *vertebrates* that evolved from *reptiles* and were confined to the Mesozoic Era. Dinosaurs are defined by their distinctive pelvic structure.

Dip The angle that a tilted *bed* or *fault* forms with the horizontal.

Disconformity An *unconformity* above *rocks* that underwent *erosion* before the *beds* above the unconformity were deposited. The *strata* above and below a disconformity are horizontal.

Distributary channels Channels on a *delta plain* that radiate out from the mainland, carrying river water to the ocean in several directions.

Diversity The number of *species* that live together in a *community* or belong to a *taxon*.

DNA Deoxyribonucleic acid, the molecule that carries chemically coded genetic information and is passed from generation to generation.

Dollo's law The principle that any substantial evolutionary change is virtually irreversible because genetic changes are not likely to be reversed in an order exactly opposite to the order in which they originally occurred.

Dolomite A *mineral* that consists of calcium magnesium carbonate, with calcium and magnesium present in nearly equal proportions.

Dolostone A *sedimentary rock* consisting largely of the mineral dolomite.

Dropstone A stone dropped to the bottom of a lake or ocean from a melting body of ice afloat on the surface.

Dune A body of sand piled up by wind.

Easterlies Winds that form near Earth's poles, where cold, dense air descends and flows toward the west under the influence of the *Coriolis effect*.

Echinoderms A group of marine *invertebrates* characterized by fivefold radial symmetry, radial rows of tube feet that terminate in suction cups, and an internal skeleton or skeletal elements formed of *calcite* plates.

Ecological niche The ecological position of a species in its environment, including its requirements for certain kinds of food and physical and chemical conditions and its interactions with other species.

Ecology The study of the factors that govern the distribution and abundance of organisms in natural environments.

Ecosystem An environment together with the group of organisms that live within it.

Ectothermic Cold-blooded: characterized by a body temperature that is not internally regulated, but controlled by the environment.

Endothermic Warm-blooded: characterized by a body temperature that is internally regulated.

Eocrinoids Evolutionary ancestors of *crinoids*, abundant in Cambrian seas, that formed simple *communities*.

Eon The largest unit of geologic time. There are three eons: the Archean, Proterozoic, and Phanerozoic.

Epicontinental sea A shallow sea formed when ocean waters flood an area of a continent far from the continental margin.

Epoch A division of geologic time shorter than a *period*.

Equatorial countercurrent The eastward-flowing global ocean current that carries the water that has been piled up by the *equatorial current*.

Equatorial current The global ocean current pushed westward along the equator by the *trade winds*.

Era A division of geologic time shorter than an *eon* but including two or more *periods*.

Erathem A *time-rock unit* consisting of all the *rocks* that represent a geologic *era*.

Erosion The group of processes that loosen *rock* and transport the resulting products downhill.

Eubacteria True bacteria; a group of *prokaryotes* that forms one of the six kingdoms of organisms.

Eukaryote An organism whose cells are characterized by a nucleus with *chromosomes*, *mitochondria*, and other complex internal structures. All organisms except *archaebacteria* and *eubacteria* are of this type.

Eustatic event A change in sea level throughout the world.

Evaporite A *mineral* or *rock* formed by precipitation of crystals from evaporating water.

Even-toed ungulates A group of cloven-hoofed herbivores that includes cattle, antelopes, sheep, goats, pigs, bison, camels, and their relatives.

Event stratigraphy The use of geologic records of sudden events, such as the relocation of shorelines or the deposition of volcanic ash, for the *correlation* of *rocks* of widely separated regions.

Evergreen coniferous forest A high-latitude forest, often adjacent to *tundra* and always dominated by *conifers*, such as spruces, pines, and firs.

Evolution The process by which particular forms of life give rise to other forms by way of genetic changes.

Evolutionary breakthrough An evolutionary innovation that affords a group of organisms a special ecological opportunity and often leads to the *evolutionary radiation* of that group.

Evolutionary convergence The evolution of similar features in two or more different biological groups, or *taxa*.

Evolutionary radiation The rapid origin of many new *species* or higher *taxa* from a single ancestral group.

Exotic terrane A block of *lithosphere* that has been *sutured* to a much larger continent.

Exposure See **Outcrop**.

Exterior drainage A drainage pattern in which lakes and rivers carry runoff from a region beyond the borders of that region.

Extinction The total disappearance of a *species* or higher *taxon*.

Extrusive (volcanic) igneous rock An *igneous rock* formed from *lava* that has flowed onto Earth's surface from a *vent*.

Facies The set of characteristics of a *rock* that represents a particular depositional environment.

Failed rift A *rift* that projects inland from a continental margin but fails to divide the continent into two separate landmasses.

Fault A surface along which *rocks* have broken and moved.

Fault block basin A valley formed by the subsidence of a block of *crust* between *normal faults*.

Fauna The animals and *protozoans* of an *ecosystem*.

Felsic rock A silicon-rich *igneous rock* that contains only a small percentage of iron and magnesium. Granite is the most abundant example. Felsic rocks dominate the *crust* of continents.

Fern A seedless *vascular plant* that reproduces by means of *spores*.

Fissile Having a tendency to break along *bedding* surfaces; a property of some *sedimentary rocks*, especially *shales*.

Fission-track dating The dating of a *rock* according to the number of fission tracks produced by the decay of uranium 238. In the process of decaying, uranium 238 atoms eject subatomic particles that leave microscopic tracks in the surrounding rock.

Fissure A crack in a body of *rock*, often filled with *minerals* or *intrusive igneous rock*.

Flint See **Chert**.

Flood basalt *Extrusive igneous rock* of *mafic* composition formed from lava that has flowed widely over Earth's surface.

Floodplain A broad lowland flooded by a stream or river.

Flora The plants and plantlike *protists* of an *ecosystem*.

Flux The rate at which a *reservoir* gains or loses its contents, and thus expands or contracts.

Flysch *Shales* and *turbidites* that accumulate in deep water within a *foreland basin* bordering an active mountain system.

Fold-and-thrust belt The tectonic zone of a mountain chain characterized by folds and *thrust faults* and positioned adjacent to the *metamorphic belt*, but farther away than the metamorphic belt from the *igneous* core of the mountain chain.

Folding Tectonic bending of *rocks* into *anticlines* and *synclines* or other contorted configurations.

Foliation The alignment of platy *minerals* in *metamorphic rocks*, caused by the high pressure applied during *metamorphism*.

Food chain The sequence of nutritional steps in an *ecosystem*, with *producers* at the bottom and *consumers* at the top.

Food web The nutritional structure of an *ecosystem* in which more than one *species* occupies each level. Thus there are usually several *producer* species and several *consumer* species in a food web.

Foraminifera Marine protozoans that form a chambered skeleton by secreting *calcite* or *aragonite* or cementing grains of *sand* together. Long filaments of their protoplasm extend through pores in the skeleton and interconnect to form a sticky net in which they catch food.

Forearc basin The basin between an *igneous arc* and the associated *subduction zone*.

Foreland basin An elongate depositional basin that lies between a mountain chain and the continental interior.

Formation The fundamental *rock unit*; a body of *rock* characterized by a particular set of *lithologic* features and given a formal name.

Fossil The remains or tangible traces of an organism preserved in *sediment* or *rock*.

Fossil succession The vertical ordering of *fossil taxa* in the geologic record, reflecting the operation of *evolution* and *extinction*.

Fresh water Natural water that contains less than 0.5 parts per 10,000 salt by weight.

Fringing reef An elongate *organic reef* that fringes a coastline and has no *lagoon* on its landward side.

Fungi A kingdom of unicellular and multicellular *eukaryotes* whose members absorb food materials into their cells and digest them there.

Fusulinids A group of large *foraminifera* that lived on shallow Paleozoic seafloors.

Gabbro A *mafic igneous rock*; the coarse-grained, *intrusive* equivalent of *basalt*.

Gamete A sex cell (egg or sperm) that carries half the normal complement of *chromosomes* and combines with another sex cell to produce a new individual possessing the normal complement.

Gastropods Snails, the largest and most varied class of *mollusks*.

Gene A unit of inheritance consisting of a segment of *DNA* that performs a particular function.

Gene pool The sum total of the genetic components of a population.

Geochronologic unit *See* **Time unit**.

Geologic system A *time-rock unit* consisting of all the *rocks* that represent a geologic *period*.

Glacial maximum An interval during the recent Ice Age when glaciers expanded close to their maximum extent.

Glacial minimum An interval during the recent Ice Age when glaciers receded close to their minimum extent.

Glacier A large mass of ice that creeps over Earth's surface.

Gneiss A high-grade *metamorphic rock* whose intergrown crystals resemble those of *igneous rock*, being granular rather than platy, but whose *minerals* tend to be segregated into wavy layers.

Gondwanaland A supercontinent, including South America, Africa, peninsular India, Australia, and Antarctica, that formed near the beginning of Cambrian time.

Graben A valley bounded by *normal faults* along which a central block of *crust* has slipped downward.

Grade, metamorphic A classification system based on the level of temperature and pressure responsible for *metamorphism*. *Metamorphic rocks* may be of high grade, medium grade, or low grade.

Graded bed A *sedimentary structure* in which grain size decreases from the bottom to the top.

Granite A *felsic igneous rock* that consists largely of quartz and feldspar and is the most abundant *intrusive* igneous rock in Earth's *crust*.

Granule A small piece of *gravel* (between 2 and 4 millimeters).

Grassland A *habitat* characterized by grass and very few trees.

Gravel *Siliciclastic sedimentary particles* that are larger than *sand* (larger than 2 millimeters).

Gravity spreading The lateral spreading of a mountain chain when it becomes so tall that the *rocks* within it deform under their own weight.

Graywacke A dark gray *siliciclastic rock* consisting of *sand-* and *silt-*sized grains that include dark rock fragments, as well as substantial amounts of *clay*.

Greenhouse effect The trapping of solar radiation by Earth's atmosphere that warms Earth's surface.

Greenhouse gases Gases such as CO_2, methane, and water vapor that contribute to the *greenhouse effect*.

Greenstone belt A podlike body of *rock* characteristic of Archean *terranes*. It consists of volcanic rocks and associated *sediments* that have commonly been metamorphosed so that they have a greenish color.

Group A *rock unit* of a rank higher than a *formation*.

Guide fossil *See* **Index fossil**.

Guyot A flat-topped volcanic seamount in the deep sea. Guyots form in shallow water when wave action truncates the upper part of a volcano, and they are transported to deeper water by lateral *plate* movements.

Gymnosperms Plants whose seeds are lodged in exposed positions on cones or on other reproductive organs.

Gypsum A *mineral* that consists of calcium sulfate with water molecules attached ($CaSO_4 \cdot H_2O$), or the *rock* that consists primarily of that mineral.

Gyre A large-scale circular flow of winds or ocean currents.

Habitat A setting on or close to Earth's surface that is inhabited by life.

Half-life The time required for a particular radioactive *isotope* to decay to half its original amount. That time is consistent for any isotope, regardless of the amount of that isotope present at the outset.

Halite A *mineral* that consists of sodium chloride (NaCl), popularly known as rock salt, or the *rock* that consists primarily of that mineral.

Herbivore An animal that feeds on plants or plantlike organisms; a *consumer* that feeds on *producers*.

Hexacorals A group of *corals* that includes both colonial reef builders and solitary species and that flourishes in present-day seas.

Homology The presence in two different groups of animals or plants of organs that have the same ancestral origin but serve different functions.

Hot spot A small area of heating and *igneous* activity in Earth's *crust* where a *thermal plume* rises from the *mantle*. The Hawaiian Islands represent a hot spot.

Humus Organic matter in *soils*, formed largely by the decay of leaves, woody tissues, and other plant materials.

Hydrothermal metamorphism Local *metamorphism* caused by the percolation of hot, watery fluids through *rocks*, as happens along a *mid-ocean ridge*, where seawater circulates through the hot, newly formed *lithosphere*.

Hypersaline Having a *salinity* that is higher than that of normal seawater (containing more than 40 parts salt per 1000 parts water).

Ichthyosaurs Fishlike *reptiles* that resembled modern dolphins (which are *mammals*) and bore live young.

Igneous arc A linear zone adjacent to a *subduction zone* along which volcanoes and *igneous intrusions* form.

Igneous rocks *Rocks* formed by the cooling of molten material.

Impression A *fossil* that consists of the flattened imprint of a soft or semihard organism, such as an *insect* or a leaf.

Inclination The angle of a compass needle below horizontal. This angle is low at low latitudes and high at high latitudes.

Index fossil (guide fossil) A *species* or genus whose *fossils* provide for especially precise *correlation*. An ideal index fossil is easily distinguished from other *taxa*, is geographically widespread, is common in many kinds of *sedimentary rocks*, and is restricted to a narrow stratigraphic interval.

Insects A group of *arthropods* that breathe air through a system of tubes, have bodies divided into head, thorax, and abdomen, and usually have two pairs of wings.

Interior drainage A drainage pattern in which lakes and rivers fail to carry runoff from a region beyond the borders of that region, commonly because the rainfall there is so light that streams and rivers are temporary.

Intertidal zone The belt that is alternately exposed and flooded as the *tide* ebbs and flows along a coast.

Intertropical convergence zone The tropical zone where the northern and southern *trade winds* converge. Because of the tilt of Earth's axis, this zone shifts seasonally with the location of maximum solar heating, from a few degrees north of the equator during the northern summer to a few degrees south of the equator during the southern summer.

Intrusion (pluton) A body of coarse-grained *igneous rock* that formed when *magma* displaced or melted its way into preexisting rock.

Intrusive igneous rock *Rock* formed by the cooling of *magma* within Earth.

Intrusive relationships, principle of The principle that an *intrusive igneous rock* is always younger than the rock that it invades.

Invertebrate An animal that lacks a backbone.

Ionic bond A chemical bond in which one atom loses an electron to another atom.

Iron meteorite A *meteorite* of which iron is the primary component.

Island arc A curved chain of islands produced by volcanism at a site where *magma* rises through the *lithosphere* from a *subducted plate*.

Isostasy The mechanism whereby areas of Earth's *crust* rise or subside to keep the crust in gravitational equilibrium as it floats on the *mantle*. Thus a mountain is balanced by a root of crustal material.

Isotope One of two or more varieties of an element that differ in the number of neutrons within the atomic nucleus.

Isotope stratigraphy The dating of rock *strata* by measurement of the isotopic composition of mineral grains and fossils.

Lagoon A ponded body of water along a marine coastline, usually landward of a *barrier island* or *organic reef*.

Laterite A *soil* rich in *oxides* of aluminum, iron, or both of these elements. Iron gives laterite a rusty red color.

Laurasia A supercontinent, including North America, Europe, and Asia, that formed in middle Paleozoic time.

Lava Molten *rock* (*magma*) that has reached Earth's surface.

Life habit The mode of life of an organism, or the way it functions within its *ecological niche*—how it obtains nutrients or food, reproduces, and stations itself or moves about within its environment.

Limestone A *sedimentary rock*, either *biogenic* or chemical in origin, consisting primarily of calcium carbonate.

Limiting factor An environmental condition, such as temperature, that restricts the distribution of a *species* in nature.

Lithification The consolidation of loose *sediment* by compaction, precipitation of mineral cement, or a combination of those processes to form *sedimentary rock*.

Lithologic correlation *Correlation* between *stratigraphic units* based on *rock* type.

Lithology The physical and chemical characteristics of *rock*.

Lithosphere Earth's outer rigid shell, situated above the *asthenosphere* and consisting of the *crust* and upper *mantle*. The lithosphere is divided into *plates*.

Lithostratigraphic unit *See* **Rock unit**.

Lobe-finned fishes A group of largely freshwater fishes with paired fins, the bones of which are attached to their bodies by a single shaft. They declined after the Devonian Period but are the ancestors of all terrestrial vertebrates.

Longshore current An ocean current that flows along a coast, often sweeping *sand* in a direction parallel to the coastline.

Lungfishes A group of *lobe-finned fishes* with lungs, which allow them to gulp air when they are trapped in stagnant pools during the dry season. Lungfishes were abundant in the Devonian Period, but only three genera survive today.

Lycopods Spore-bearing plants that flourished in late Paleozoic time as trees that occupied swamps and were primary contributors of organic matter that turned into coal.

Mafic rock Dark, dense *igneous rock* that is relatively poor in silicon and rich in iron and magnesium. *Basalt*, the characteristic igneous rock of oceanic *crust*, is an example.

Magma Naturally occurring molten *rock* found within Earth.

Magma ocean Molten *silicates* that rose to Earth's surface early in Earth's history when denser material sank to the center to form a predominantly iron *core* and a *mantle* of denser silicates.

Magnetic field The field of magnetism that results from motions of Earth's iron-rich outer *core*; those motions cause Earth to behave like a giant bar magnet, with a north and south pole. Reversals in polarity provide for accurate *correlation* throughout the world.

Magnetic stratigraphy The use of magnetic properties of *rocks* for *correlation*.

Mammals A class of *endothermic vertebrates* characterized by body hair, legs positioned fully under the body, and sweat glands, some of which are modified to secrete milk to nourish their young, which they bear live.

Mantle The zone of Earth's interior between the *core* and the *crust*, ranging from depths of approximately 40 to 2900 kilometers. It is composed of dense *ultramafic silicates* and divided into concentric layers.

Marble A homogeneous, granular *metamorphic rock* that consists of *calcite*, *dolomite*, or a mixture of these two minerals, and forms by the *metamorphism* of *sedimentary* carbonates.

Marker bed A distinctive bed that is useful for stratigraphic *correlation*.

Marsh, intertidal A *habitat* in the *intertidal zone* that is dominated by low-growing plants and is alternately flooded by the *tide* and exposed to the air. The remains of the plants usually accumulate to form *peat*.

Marsupial mammals *Mammals* that carry their immature offspring in a pouch.

Mass extinction A global episode of large-scale *extinction* in which large numbers of *species* disappear in a few million years or less.

Meandering river A river that winds back and forth like a ribbon, depositing *sediment* on the inside of each curve and eroding sediment on the outside.

Mediterranean climate A climate characterized by dry summers and wet winters, often found along coasts lying about 40° from the equator. Much of California and much of the Mediterranean region of Europe have this kind of climate.

Mélange A chaotic, deformed mixture of *rocks*, such as often forms where *subduction* occurs along a *deep-sea trench*.

Meltwater The water that issues from the front of a melting *glacier*.

Member A *rock unit* of a rank lower than a *formation*.

Metamorphic belt The *metamorphic* zone parallel to the long axis of a mountain chain and near the *igneous* core of the mountain chain. A metamorphic belt is a zone of *regional metamorphism*.

Metamorphic rock *Rock* formed by *metamorphism*.

Metamorphism The alteration of *rocks* within Earth under conditions of temperature and pressure high enough to change their mineralogy.

Meteorite An extraterrestrial object that has crashed to Earth's surface after being captured by Earth's gravitational field.

Methane hydrate Methane produced by archaebacteria that is frozen within a cage of water molecules.

Microplate A small *lithospheric plate*, usually of predominantly *felsic* composition.

Mid-ocean ridge A ridge on the ocean floor where oceanic *crust* forms and from which it moves laterally in each direction.

Mineral A naturally occurring inorganic solid element or compound with a particular chemical composition or range of compositions and a characteristic internal structure.

Mitochondrion (plural, **mitochondria**) A body within a *eukaryotic cell* in which complex compounds are broken down by oxidation to yield energy and, as a byproduct, carbon dioxide. The mitochondrion is apparently an evolutionary descendant of a small bacterium that became trapped within a larger one.

Moho *See* **Mohorovičić discontinuity**.

Mohorovičić discontinuity (Moho) The boundary between the *crust* and *mantle*, marked by a rapid increase in the velocity of seismic waves.

Molasse Nonmarine and shallow marine sediments—representing such depositional environments as *alluvial fans*, river systems, and *barrier island–lagoon complexes*—that accumulate in front of a mountain system after heavy sedimentation from the mountains has driven deep marine waters from the *foreland basin* there.

Mold A *fossil* that consists of a three-dimensional imprint of an organism or part of an organism.

Molecular clock A method of dating times of divergence of pairs of modern taxa based on their accumulation of *mutations* of neutral genes, which are unaffected by *natural selection.*

Mollusks The group of *invertebrates* that includes snails, clams, and octopuses. Most mollusks secrete skeletons of calcium carbonate.

Monoplacophorans The most primitive mollusks, with cap-shaped shells and a broad foot.

Monsoon Strong onshore or offshore winds near the margin of a continent, caused by the difference in temperature between land and water.

Moraine A ridge of *till* plowed up in front of a *glacier.*

Mosasaurs Marine lizards of the Cretaceous Period that sometimes reached 15 meters (45 to 50 feet) in length.

Mud An aggregate consisting of *silt-* or *clay-* sized *siliciclastic sedimentary* particles or a combination of the two.

Mudcracks *Sedimentary structures* that form, often in hexagonal patterns, as fine-grained, *clay-*rich sediments dry out and shrink.

Mudstone *Rock* formed largely of *mud.*

Mutation A chemical change in a genetic feature. Such changes provide much of the variability on which *natural selection* operates.

Natural levee A gentle ridge bordering a *meandering river* or a *distributary channel* of a *delta* and composed of *sand* and *mud* deposited by the river or distributary channel when it overflows its banks.

Natural selection The process recognized by Charles Darwin as the primary mechanism of *evolution.* The selection process, which operates on heritable variability, results from differences among individuals in longevity and in rate of production of offspring.

Negative feedback A result of a change that suppresses that change.

Nekton Fishes and other marine animals that move through the water primarily by swimming.

Niche, ecological *See* **ecological niche.**

Nonconformity An *unconformity* separating bedded *sedimentary rocks* above from *crystalline rocks* below.

Normal fault A *fault* whose *dip* is steeper than 45° and along which the *rocks* above have moved downward in relation to the rocks below.

Nothosaurs Marine *reptiles* of the Triassic Period that had paddlelike limbs resembling those of modern seals.

Oceanic realm The portion of the ocean that lies above the *deep-sea floor.*

Odd-toed ungulates A group of herbivorous animals that includes horses, tapirs, and rhinoceroses.

Onychophorans Animals intermediate in form between *segmented worms* and *arthropods.*

Oolite A *sediment* consisting of nearly spherical grains (ooliths) that grow by accumulating calcium carbonate particles while rolling about in shallow water where the seafloor is agitated by strong water movements.

Ophiolite A segment of seafloor that is elevated so as to rest on continental *crust.* An ophiolite usually includes *turbidites,* black *shales, cherts,* and *pillow basalts* along with *ultramafic rocks* from the *mantle.*

Opportunistic species A *species* that specializes in the rapid invasion of newly vacated *habitats,* where there is little *competition* from other species.

Organic reef A solid but porous *limestone* structure standing above the surrounding seafloor and constructed by living organisms, which contribute skeletal material to the reef framework.

Original horizontality, principle of The principle, enunciated in the seventeenth century, that all *strata* are at low angles when they form. (A more accurate statement would be that almost all strata are initially more nearly horizontal than vertical when they form.)

Original lateral continuity, principle of The principle that similar *strata* found on opposite sides of a valley or some other erosional feature were originally connected.

Orogen A body of *rocks* affected by mountain building.

Orogenesis The process of mountain building.

Orogenic stabilization The compression and *metamorphism* of *sediments* that have accumulated along a *continental shelf,* which thickens the *crust* and hardens sediments and *sedimentary rocks.*

Orogeny A mountain building event, which results from compression of continental *crust* and addition of *igneous* material to it.

Ostracoderms A group of small fishes of early and middle Paleozoic time with paired eyes, bony armor, and small mouths, but no jaws.

Outcrop A portion of a body of *rock* that is visible at Earth's surface. (Some geologists restrict this term to rocks laid bare by natural processes and apply the term *exposure* to artificially exposed areas of rock.)

Outwash Well-stratified glacial *sediment* deposited by a stream of *meltwater* issuing from a melting *glacier.*

Overturned fold A fold in which at least one limb has been rotated more than 90° from its original position.

Oxide A *mineral* consisting of a compound formed by the combination of oxygen with one or more positive ions.

Paleogeography Geography of Earth at some time in the geologic past.

Paleomagnetism The magnetism of a *rock,* developed from Earth's *magnetic field* when the rock formed.

Pangaea A supercontinent, formed near the end of the Paleozoic Era, that contained nearly all of Earth's continental *lithosphere.*

Parasite An organism that derives its nutrition from other organisms, usually without killing them.

Particulate inheritance The presence of hereditary factors, called *genes,* that retain their identities while being passed on from parent to offspring.

Passive margin A continental margin that is not affected by *rifting, subduction, transform faulting,* or other large-scale *tectonic* processes, but instead forms a shelf that accumulates *sediments.*

Patch reef A mound-shaped reef growing in the *lagoon* behind a *barrier reef.*

Peat Plant debris that is not buried deeply enough to have metamorphosed into *coal.* It accumulates in water than contains little oxygen, and therefore few bacteria that would cause decay.

Pebble A piece of *gravel* measuring between 4 and 8 millimeters.

Pelagic life Oceanic life that exists above the seafloor.

Pelagic sediment Fine-grained *sediment* that settles through the oceanic water column to the *deep-sea floor.* Some of this sediment is *biogenic.*

Pelycosaurs Fin-backed *reptiles* and their relatives, which were related to therapsids, the ancestors of *mammals.*

Period The most commonly used unit of geologic time, representing a subdivision of an *era.*

Permineralization A mode of fossilization in which spaces within part of an organism (such as bony or woody tissue) become filled with *mineral* material.

Photic zone The upper layer of the ocean, where enough light penetrates the water to permit *photosynthesis.*

Photosynthesis The process by which plants and plantlike organisms employ the compound chlorophyll to convert carbon dioxide and water from their environment into energy-rich sugar, which fuels chemical reactions essential to life.

Phylogeny A segment of the tree of life that includes two or more evolutionary branches.

Phytoplankton *Plankton* that is photosynthetic. Most phytoplankton *species* are single-celled algae.

Pillow basalt *Basalt* with a hummocky surface formed by rapid cooling of *lava* beneath water.

Placental mammals Mammals whose prenatal offspring are nourished by a placenta, an internal organ that unites the fetus to the mother's uterus.

Placoderms Large, heavily armored, jawed fishes of the Paleozoic.

Placodonts Blunt-toothed *reptiles* of early Mesozoic seas that had broad, armored bodies that gave them the appearance of enormous turtles.

Plankton Floating aquatic life; organisms that float in the ocean or in lake waters.

Plantae A kingdom of multicellular *eukaryotes* whose members produce their own food by means of *photosynthesis*.

Plate A segment of the *lithosphere* that moves independently over Earth's interior.

Plate tectonics The movements and interactions of *lithospheric plates*.

Playa lake A temporary lake in a region of *interior drainage*. When such a lake dries up, *evaporite* deposits form.

Plesiosaurs Aquatic *reptiles* that evolved from the *nothosaurs* in mid-Triassic time and in Cretaceous time attained the proportions of modern predatory whales, reaching some 12 meters (40 feet) in length.

Plunging fold A fold whose *axis* plunges (lies at an angle to the horizontal) so that the *beds* of the fold have a curved *outcrop* pattern if they are truncated by *erosion*.

Pluton *See* **Intrusion**.

Point bar An accretionary body of *sand* on the inside of a bend of a *meandering river*.

Polarity time-rock unit A *time-rock unit* in which the polarity of Earth's *magnetic field* was either the same as it is today (a so-called normal interval) or the opposite of what it is today (a reversed interval).

Population A group of individuals that live in the same area and interbreed.

Positive feedback A result of a change that accelerates that change.

Precambrian shield A Precambrian portion of a *craton* that is exposed at Earth's surface.

Prodelta The gently seaward-sloping bottomset area of a *delta front* where *clay* accumulates in deep water.

Producer A plant or plantlike organism that manufactures its own food.

Prograde To grow seaward by the accumulation of *sediment* or *sedimentary rocks*. *Deltas* often prograde, as do *organic reefs*. Progradation produces *regression*, or seaward migration, of the shoreline.

Prokaryote An organism whose cells do not contain a nucleus or certain other internal structures characteristic of the cells of higher organisms.

Protista A kingdom of *eukaryotes* whose members do not fit within any of the other three eukaryotic kingdoms. These simple, mostly unicellular *species* include the groups that were ancestral to plants, fungi, and animals.

Protocontinent A small body of continental *lithosphere*.

Protozoan A single-celled animal-like *eukaryote*.

Pseudoextinction The disappearance of a *species*, not by dying out, but by evolving to the point at which it is recognized as a different species.

Pterosaurs Flying and gliding reptiles of the Mesozoic Era.

Punctuation model The theory that most evolutionary change occurs rapidly, through *speciation*.

Quartzite A *metamorphic rock* formed by the *metamorphism* of quartz *sandstone* and consisting of almost pure quartz.

Radioactive decay The spontaneous breakdown of certain kinds of atomic nuclei into one or more nuclei of different elements, with a release of energy and subatomic particles.

Radiocarbon dating *Radiometric dating* by means of carbon 14, a radioactive *isotope* with a *half-life* so short that its decay can be used to date materials younger than about 70,000 years.

Radiolarians Marine protozoans that capture food with threadlike extensions of protoplasm that radiate from skeletons of opal.

Radiometric dating Measurement of the amounts of naturally occurring radioactive *isotopes* present in *rocks* in relation to the amounts of their daughter isotopes (products of *radioactive decay*) to ascertain the ages of the rocks.

Rain forest, tropical *See* **Tropical rain forest**.

Rain shadow A region that is on the leeward side of a mountain and receives little rain because the winds rise as they pass over the mountain, cooling and dropping most of their moisture before they reach the leeward side.

Ray-finned fishes A group of jawed fishes with thin bones that radiate from the body to support the fins. Ray-finned fishes dominated Mesozoic and Cenozoic seas and are widely represented today. They include, for example, trout, bass, herring, and tuna.

Red beds *Sediments* of any grain size that are reddish, usually because of the presence of iron oxide cement.

Redshift An increase in the length of light waves as they travel through space.

Reduced carbon Carbon atoms in organic compounds that contain little or no oxygen.

Reef flat The flat upper surface of an *organic reef*, usually standing close to sea level (often in the *intertidal zone*).

Reef, organic *See* **organic reef**.

Regional metamorphism The metamorphism of *rocks* over an area whose dimensions are measured in hundreds of kilometers; usually associated with mountain building.

Regional strike In deformed terrain, the prevailing orientation of fold *axes* or of the lines of *outcrop* of tilted *beds*.

Regression A seaward migration of a marine shoreline and of nearby environments.

Remobilization *Regional metamorphism* and deformation affecting a segment of *crust* previously altered by similar processes.

Reptiles A class of air-breathing, *ectothermic vertebrates* that evolved from *amphibians* through the development of *amniote eggs* with protective shells. Among them are turtles, lizards, snakes, and *crocodiles*.

Reservoir A body of one or more chemical entities that occupies a particular space.

Respiration Oxidation of organic compounds to obtain energy (the opposite of *photosynthesis*).

Rift A juncture between two *plates* where *lithosphere* forms and the plates diverge.

Ripples Small dunelike structures formed on the surface of *sediment* by moving water or wind.

Rock An aggregate of interlocking or bonded grains, most of which are composed of a single *mineral*.

Rock cycle The endless pathway along which *rocks* of various kinds change into rocks of other kinds.

Rock unit A body of *rock* that is formally recognized as a *formation, member, group*, or *supergroup*.

Rodents A group of mammals, typically of small body size, that make use of continually growing front teeth to gnaw hard food substances.

Rudists A group of bivalve mollusks that grew to large size and formed *organic reefs* during the Cretaceous Period. They died out with the dinosaurs.

Rugose corals A group of Paleozoic corals, some of which were colonial and formed *organic reefs*.

Salinity The saltiness of natural water. The salinity of normal seawater is 35 parts salt per 1000 parts water.

Sand *Siliciclastic sedimentary* particles ranging in diameter from $\frac{1}{16}$ to 2 millimeters.

Sandstone A *siliciclastic sedimentary rock* consisting primarily of *sand*—usually sand that is predominantly quartz.

Sauropods Lizard-hipped *herbivores* of the Jurassic Period that moved about on all fours; among them were the largest of the *dinosaurs*.

Savannah A broad grassland, which typically forms where there is enough rainfall to sustain grass and scattered trees but not so much as to sustain enough trees to form woodlands or forests.

Scavenger An organism that feeds on other organisms after they have died of causes unrelated to the scavenger's activities.

Schist A *metamorphic rock* of low to medium *grade* that consists largely of platy grains of *minerals*, often including mica; because of its strong *foliation*, schist tends to break along parallel surfaces.

Sea urchins A group of *echinoderms* that have a rigid skeleton to which numerous spines are attached by ball-and-socket joints.

Sediment Material deposited on Earth's surface by water, ice, or air.

Sedimentary rock A *rock* formed by the consolidation of loose *sediment* or by precipitation of *minerals* from a watery solution.

Sedimentary structure A distinctive arrangement of grains in a *sedimentary rock*.

Seed A reproductive structure of a plant—produced by the union of *gametes* and then released from the plant—that has the potential to grow into a new plant. The seed is actually a juvenile stage of the *spore*-bearing generation of the plant.

Seed ferns *Ferns* of the Carboniferous Period that reproduced by means of *seeds* rather than *spores*. They varied widely in size, from small bushy plants to large tree-like ones.

Segmented worm An advanced worm with a series of segments, each with its own fluid-filled coelom, which serves as a primitive skeleton under the pressure of muscular contractions. Many segmented worms are marine, but others, including earthworms, are terrestrial.

Seismic stratigraphy The study of *sedimentary rocks* by means of seismic reflections generated when artificially produced seismic waves bounce off physical discontinuities within buried *sediments*.

Sequence A large body of marine *sediment* deposited on a continent when the ocean rose in relation to the level of the continental surface and then receded again.

Series A *time-rock unit* consisting of all the *rocks* that represent a geologic *epoch*.

Sexual recombination The mixing of *chromosomes* from generation to generation, which continually creates new genetic combinations and hence new kinds of individuals on which *natural selection* can operate.

Shale A *fissile sedimentary rock* consisting primarily of *clay*.

Sharks A group of cartilaginous fishes that were especially well represented in early Mesozoic seas, but still exist today.

Shelf break The edge of a *continental shelf*, where it meets the *continental slope*.

Silicates The *mineral* group that includes the most abundant minerals in Earth's *crust* and *mantle*. The basic building block of silicates is a tetrahedral structure consisting of four oxygen atoms surrounding a silicon atom.

Siliceous ooze Fine-grained sediment of the deep sea formed largely of skeletons of single-celled planktonic organisms consisting of the soft form of silica known as opal.

Siliciclastic rock Sedimentary rock composed of clasts of *silicate* minerals.

Sill A sheetlike or tabular body of *intrusive igneous rock* that has been injected between sedimentary layers.

Silt *Siliciclastic sedimentary* particles with diameters between $\frac{1}{256}$ and $\frac{1}{16}$ millimeter.

Sink A chemical *reservoir* on Earth that grows rapidly enough to take up a chemical as rapidly as it is produced.

Slab An area of *lithosphere* that has been *subducted*.

Slate A fine-grained *metamorphic rock* of very low *metamorphic grade* that is *fissile*, like the *sedimentary rock* shale, but whose fissility results from alignment of platy materials by deformational pressures rather than by the depositional orientation of particles.

Soil Loose *sediment* that accumulates in contact with the *atmosphere*.

Solar nebula The kind of dense cloud of cosmic dust—the remains of a *supernova*—from which the sun formed.

Speciation The origin of a new *species* from two or more individuals of a preexisting species.

Species A group of individuals that interbreed or have the potential to interbreed in nature and that do not breed with other interbreeding groups.

Sphenopsids *Spore* plants characterized by branches that radiate from discrete nodes along the vertical stem and by horizontal underground stems that bear roots.

Spore A reproductive structure, not produced from *gametes*, that is released from a plant and has the potential to grow into a new plant.

Spreading zone A zone along which new *lithosphere* forms as *mafic magma* of relatively low density rises from the *ultramafic asthenosphere* and cools.

Stable isotope An *isotope* that does not decay to form another isotope.

Stage The *time-rock unit* that ranks below a *series* and consists of all the *rocks* that represent a geologic *age*.

Starfishes A group of flexible *echinoderms* that grasp their prey with their tube feet.

Stony-iron meteorite A *meteorite* consisting of a mixture of rocky and metallic material.

Stony meteorite A *meteorite* of rocky composition.

Stratification The arrangement of *sedimentary rocks* in discrete layers (or *strata*).

Stratigraphic section A local *outcrop* or series of adjacent outcrops that display a vertical sequence of *strata*.

Stratigraphic unit A *stratum* or group of adjacent strata distinguished by some physical, chemical, or paleontological property, or the unit of time that is based on the age of such strata.

Stratigraphy The study of stratified *rocks*, especially their geometric relations, compositions, origins, and age relations.

Stratum (plural, **strata**) A distinct layer of *sediment*.

Strike The compass direction that lies at right angles to the *dip* of a tilted *bed* or *fault*; that is, the compass direction of a horizontal line lying in the plane of a tilted bed or fault.

Strike-slip fault A high-angle *fault* along which the *rocks* on one side have moved horizontally in relation to rocks on the other side with a shearing motion.

Stromatolite An organically produced *sedimentary* structure that consists of alternating layers of organic-rich and organic-poor *sediment*. The organic-rich layers have usually been formed by sticky thread-like algae, which have trapped the sediment of the organic-poor layers.

Stromatoporoids A group of sponges that secreted massive skeletons and were important *organic reef* builders beginning in Ordovician time.

Subduction Descent of a *slab* of *lithosphere* into the *asthenosphere* along a *deep-sea trench*.

Subduction zone A region where *subduction* of the *lithosphere* occurs.

Substratum The surface—*sediment*, *rock*, or another organism—on which or within which a *benthic* aquatic organism lives.

Subtidal zone The belt positioned seaward of the *intertidal zone*.

Sulfate mineral A *mineral* in which the basic building block is a sulfur atom linked to four oxygen atoms. Most sulfate minerals are highly soluble in water and form by the evaporation of natural waters.

Supergroup A *rock* unit of a rank higher than *group*.

Supernova An exploding star that casts off matter of low density.

Superposition, principle of The principle that in an undisturbed sequence of *strata*, the oldest lie at the bottom and the progressively younger strata are successively higher.

Supratidal zone The belt along a coast just landward of the *intertidal zone* that is flooded only occasionally, during storms or unusually high *tides*.

Suspension feeder A member of an aquatic species that strains small particles of food from the water in which it lives.

Suture A juncture between two continents along a *subduction zone*.

Symbiotic relationship A relationship between species that is mutually beneficial.

Syncline A fold that is concave in an upward direction—that is, the vertex is the lowest point.

System *See* **Geologic system**.

Tabulate corals An extinct group of *corals* that secreted colonial skeletons of *calcite* and were important *organic reef* builders of Ordovician, Silurian, and Devonian time.

Talus The pile of rubble sloping seaward from the living surface of an *organic reef*.

Taxon (plural, **taxa**) A formally named group

of related organisms of any rank, such as phylum, class, family, or *species*.

Taxonomic group *See* **Taxon**.

Taxonomy The study of the composition and relationships of *taxa* of organisms.

Tectonics The study of *rock* deformation entailing large-scale features such as mountains.

Teleost fishes Marine and freshwater fishes characterized by symmetrical tails, round scales, specialized fins, and short jaws. Most living fish species belong to this group.

Temperate forest A forest dominated by deciduous trees (trees that lose their leaves in winter). This kind of forest typically grows under slightly warmer climatic conditions than an *evergreen coniferous forest*.

Tempestite A *bed* deposited on a *continental shelf* or on the floor of an *epicontinental sea* as a result of the settling of *sediment* stirred up by a hurricane or other major storm.

Temporal correlation Demonstration that *strata* in two different geographic areas are the same age.

Terrane A geologically distinct region of Earth's *crust* that has behaved as a coherent crustal block.

Therapsids An extinct group of animals that occupied an intermediate evolutionary position between *reptiles* and *mammals*.

Thermal plume A column of *magma* rising from the *mantle* through the *lithosphere*.

Thrust fault A fault along which the upper block of *crust* has moved uphill along the fault surface.

Thrust sheet A large, tabular block of *crust* that moves along a *thrust fault* during mountain building.

Tidal flat A surface where *mud* or *sand* accumulates in the *intertidal zone*.

Tide A major movement of the ocean that results primarily from the gravitational attraction of the moon. Tides ebb and flow in particular regions as Earth rotates beneath a bulge of water created by the pull of the moon.

Till A poorly sorted, gravelly deposit produced by a *glacier* from material it has plowed up.

Tillite Lithified *till*.

Time-rock unit A *stratigraphic unit* that includes all the *strata* in the world that were deposited during a particular interval of time.

Time unit The interval during which a *time-rock unit* formed.

Tissue A connected group of similar *cells* that perform a particular function or group of functions.

Topsoil The upper zone of many *soils*, consisting primarily of *sand* and *clay* mixed with *humus*.

Trace fossil A track, trail, or burrow left in the geologic record by a moving animal.

Trade winds Winds that blow diagonally westward toward the equator in each hemisphere between about 20° and 30° latitude. They result from a zone of high pressure that forms where air that has risen near the equator builds up.

Transform fault A *strike-slip fault* along which two segments of *lithosphere* move in relation to each other. Many transform faults offset *mid-ocean ridges*.

Transgression A landward migration of a marine shoreline and of nearby environments.

Transpiration The emission of water by plants into the *atmosphere*, primarily through their leaves.

Trilobite Any of an extinct group of marine *arthropods* with a body divided into three lobes.

Triple junction A point where three *lithospheric plates* meet.

Tropical climate A climate in which the average annual temperature is in the range of 18° to 20°C (64° to 68°F) or higher. Most tropical climates lie within about 30° of the equator.

Tropical rain forest A forest that develops in an equatorial region where heavy, regular rainfall results from the cooling of air that has ascended after being warmed and picking up moisture near Earth's surface.

Trough cross-stratification *Cross-bedding* of *sediments* in which one set of *beds* is truncated by *erosion* in such a way that the next set of beds laid down accumulates on a curved surface.

Tuff A *rock* deposited as a *sediment* but consisting of volcanic particles.

Tundra A terrestrial environment where air temperatures rise above freezing during the summer but a layer of soil beneath the surface remains frozen. Tundras are characterized by low-growing plants that require little moisture.

Turbidite A *graded bed*, with poorly sorted coarse *sediment* at the base and *mud* at the top, formed when a *turbidity current* slowed down and spread out.

Turbidity current A flow of dense, *sediment*-charged water that moves down a slope under the influence of gravity.

Type section A *stratigraphic section* at a particular locality that is designated to define a *rock unit*.

Ultramafic rock Very dense *rock* that is even poorer in silicon and richer in iron and magnesium than *mafic rock*. Ultramafic rocks characterize Earth's *mantle*.

Unconformity A surface between a group of *sedimentary strata* and the *rocks* beneath them, representing an interval of time during which *erosion*, rather than deposition, occurred.

Ungulates Hoofed herbivorous mammals, which may be either *odd-toed ungulates* or *even-toed ungulates*.

Uniformitarianism The principle that there are inviolable laws of nature that have not changed in the course of time.

Upwelling Ascent of cold water from the deep sea to the *photic zone*, usually providing nutrients for rich growth of *phytoplankton*. Upwelling is most common where ocean currents drag surface waters away from continental margins.

Varves Alternating layers of coarse and fine *sediments* that accumulate in a lake in front of a *glacier*. A coarse layer forms each summer, when streams of *meltwater* carry *sand* and *silt* to the lake. A fine layer forms each winter, when the surface of the lake is covered by ice, so that only *clay* and organic matter settle to the bottom.

Vascular plant A plant that has vessels in its stem to transport water and nutrients and to distribute the food it manufactures from those raw materials.

Vent An opening in Earth's *crust* through which *lava* emerges at the surface.

Vertebrates A group of animals that possess backbones.

Vestigial organ An organ that serves no apparent function but was functional in ancestors of the organism that now possesses it.

Volcanic igneous rock *See* **Extrusive igneous rock**.

Walther's law The principle that when depositional environments migrate laterally, *sediments* of one environment come to lie on top of sediments of an adjacent environment.

Water cycle The endless flow of H_2O between *reservoirs* at or near Earth's surface.

Weathering A collective term for the group of chemical and physical processes that break down rocks at Earth's surface.

Westerlies Winds that flow toward the northeast between about 30° and 60° from the equator in each hemisphere. These winds result from the same high-pressure zone that produces the *trade winds*, which flow in the opposite direction.

Zone A *rock unit* whose upper and lower boundaries are based on the ranges of one or more *taxa*—usually *species*—in the biostratigraphic record.

Zooplankton *Plankton*, or floating aquatic life, that is animal-like in its mode of nutrition (feeds on other organisms).

Index

Note: Page numbers followed by f indicate figures.